Introduction to Chemical Engineering Analysis Using Mathematica

For Karin

Introduction to Chemical Engineering Analysis Using Mathematica

For Chemists, Biotechnologists and Materials Scientists

Second Edition

Henry C. Foley

New York Institute of Technology
New York, NY, United States

ELSEVIER

ACADEMIC PRESS
An imprint of Elsevier

Academic Press is an imprint of Elsevier
125 London Wall, London EC2Y 5AS, United Kingdom
525 B Street, Suite 1650, San Diego, CA 92101, United States
50 Hampshire Street, 5th Floor, Cambridge, MA 02139, United States
The Boulevard, Langford Lane, Kidlington, Oxford OX5 1GB, United Kingdom

Notices

Knowledge and best practice in this field are constantly changing. As new research and experience broaden our understanding, changes in research methods, professional practices, or medical treatment may become necessary.

Practitioners and researchers must always rely on their own experience and knowledge in evaluating and using any information, methods, compounds, or experiments described herein. In using such information or methods they should be mindful of their own safety and the safety of others, including parties for whom they have a professional responsibility.

To the fullest extent of the law, neither the Publisher nor the authors, contributors, or editors, assume any liability for any injury and/or damage to persons or property as a matter of products liability, negligence or otherwise, or from any use or operation of any methods, products, instructions, or ideas contained in the material herein.

Library of Congress Cataloging-in-Publication Data
A catalog record for this book is available from the Library of Congress

British Library Cataloguing-in-Publication Data
A catalogue record for this book is available from the British Library

ISBN: 978-0-12-820051-3

For information on all Academic Press publications
visit our website at https://www.elsevier.com/books-and-journals

Publisher: Susan Dennis
Acquisitions Editor: Anita A. Koch
Editorial Project Manager: Liz Heijkoop
Production Project Manager: Bharatwaj Varatharajan
Designer: Mark Rogers

Typeset by VTeX

Contents

Introduction

A bit of an explanation of this book

Welcome to "An Introduction to Chemical Engineering Analysis Using *Mathematica*." This book is a guide for anyone who is interested in mathematical modeling, and especially of systems that involve chemistry. While aimed at chemical engineering students and professionals, it may be even more useful to chemists, biochemists, biotechnologists, materials scientists, biologists, and anyone else who wants to build models using basic concepts of analysis and Wolfram's Computing language, *Mathematica*.

Modeling is based on *computational thinking* the skill of analysis combined with the use of computational methods, both symbolic and numerical, and with graphical information displays. This combination makes building models and using them to convey the behavior of a system faster and easier to do than it ever has been. The process of improving a model is a cycle of comparison of the model's behavior to measurements made of the real system. Successive improvements bring the model closer to reality with each cycle of comparison and subsequent change. A model can begin at a relatively primitive state with an analysis that accounts for only the grossest features of the system, and then it can be refined as we learn more about the system and incorporate that learning into the analysis. That the first model developed is often a "learning" model is fine, because the learning model is the starting point for the construction of more sophisticated model that may eventually bring together a set of submodels that are integrated into one overall model. Ultimately, a model apprehends that which we know, and need to know, about the system.

The reason the learning model is important is that it lets us begin the thought processes of analysis. Modeling, is somewhat like writing - both have to start by putting ideas down on paper. Getting started, is the most important steps in the whole process. Being able to get started becomes easier, if we begin with a learning model. This book teaches you how to build simple learning models by doing straightforward analysis and computing. If you can get the spirit of doing models at the level of this book, then it is certain that you will have an important and useful skill and you will be readily able to transition to more complex and sophisticated analyses and models as your knowledge grows and as the problem demands.

I will not limit the readers of this book because most of what is needed for the modeling done here is provided here. I also know that a truly inspired learner can overcome any gaps in background that they may have. To do so all they need is a strong sense of curiosity, high ambition, the desire to know more and access to a library. That said, this book is mainly aimed at those readers who have had at least two (or three) semesters of calculus, one semester of differential equations would be helpful, and a good course in general chemistry. Having had two semesters of physical chemistry would be useful, but not necessary. Chemical engineering students have this background, typically, by the end of their second year of study for the B.S., but many other disciplines also have adequate background, including engineers in other fields, chemists, biologists, physicists, and materials scientists.

Analysis as defined here, is the combination of the conservation of mass (or energy) with calculus and differential equations used to write concise mathematical statements that describe the behavior of a chemical system in time and space. The system under analysis can be a process unit, or it may be a volume within the unit; it can be real or abstract. Put in other words, the conservation of mass can be applied at the macroscopic level or at the microscopic level of a differential control volume. The key to the analysis is the choice of a control volume. That control volume is separated from its surroundings by a control surface that also may be real or abstract. The appropriate level of analysis for the problem at hand drives the choice of the control volume. It is the combination of conservation principles i.e. for mass, energy, or momentum, within an appropriate control volume that leads to the great power of chemical engineering analysis in general. In this book, we will modestly use mostly the conservation of mass in anticipation of using other conservation principles later.

What about computing background? Since the first edition of this book computing has become ubiquitous. Computing is now everywhere and everything is becoming intelligent. Desktop and laptop computers have become ever more powerful and even easier to use, and more importantly, the idea of having computing devices in everyday life, like smartphones, is a commonplace. Hence, there is a much greater familiarity with technology today than there was twenty years ago. Some readers will be familiar with certain computer languages. They may even be expert in the use of HTML, Python, Java, or other languages, making it easy for them to begin programming in *Mathematica*. But, no computing background is necessary to read and use this book. If you have never written a computer program, you should still use this book and get started with one of the most high level and versatile programming languages available today. But *Mathematica*, is really much more than a programming language, as you will see. It is a powerful computing environment that is relatively easy to learn to use and that promotes thinking computationally, and I hope to demonstrate that to you.

That last point is what this book is really about - it is about learning how to think computationally. It seeks to help the reader start thinking differently, to refine what they see in their mind's eye and to translate that to math and computation. Hopefully the reader will become ambitious about doing analysis and computing, and about building models, whether they are for work, research, or just for the fun of it. Let's get started!

Chapter subjects in brief

Chapter 1 is a primer, it is an introduction to *Mathematica*. It could just as easily have been made the last chapter as the first. It is a wide ranging overview of the language, the computing environment and how to use it. The subject material is general and not based on chemical engineering or chemistry. It is a stand-alone chapter and can be considered a book within a book. The chapter begins by deploying *Mathematica* as a powerful calculator used interactively and then presses on to writing programs.

Among the most important ideas of this chapter are how to use the basic commands of *Mathematica*, why there are different kinds of assignment operators i.e. the various forms of equal signs, and lists because lists are a foundational element of *Mathematica*. The Table command is compared to the Do command and that is a very important comparison because of lists. The Table command automatically creates lists as output, the Do command does not.

How to render graphics in two and three dimensions, as well as contour and parametric plots is explained. Both analytical (symbolic) and numerical solutions of algebraic differential equations are explained using Solve, NSolve, DSolve, and NDSolve. DSolve and NDSolve are compared to doing Integration. Linear and non-linear fitting of data to a model is demonstrated. A fit to a Fourier series is illustrative of the power of the options that one has available to change methods of numerical computation within a command. Lastly, we use the very primitive function, Point, to "sketch," computationally, a solid region bounded by a parabolic cylinder and two planes; this is a classic calculus problem.

If the reader opens the book and works through this chapter by doing the examples shown, then modifying them and "playing" with them, they will be ready to do all the basic kinds of actions in *Mathematica*. Reading the example and then trying to duplicate it on one's own helps to dramatically increase one's facility with *Mathematica*, just as doing scales and etudes helps one to learn to play the piano. But unlike the piano, *Mathematica* has a guide to the language and system that is always available to the user in real-time. So, you can just jump in and get started without doing anything else. You can even skip Chapter 1 and go right to Chapter 2.

Chapter 2 grounds everything else that will be done in the chapters that follow. Put another way, this is where we explain the control volume, the control surface, the flow of mass across these boundaries, and the idea of net accumulation of mass within that control volume. This leads to the rate-based the concept of the mass balance. We compare and explain both the differential and the integral mass balance.

Next, we get down to specifics, by considering the filling of a reactor with a solid catalyst. This is the same as filling a tube with sand. The control volume is bounded by the walls of the vessel and the mass flow into the unit is considered along with the change in level of the solid particles in the unit with time. It is an "easy" problem that most students could solve without analysis. But, because it is an easy problem, it lets us connect its intuitive algebraic solution to the powerful process of analysis. Through analysis, we write a description of the time dependence of the level in terms of the differential change in mass (the rate of mass and level change) that is equal to the flow rate of mass into the unit. This naturally introduces the separation of considerations of the left-hand side from those of the right-hand side of the mass balance. We also see that the key characterizing variable is the level in the unit and that we can convert from the change in mass to the change in level. The analysis necessitates understanding mass flow rate, density and volumetric flow rates. So, a very easy problem covers a lot of ground conceptually and quickly.

We move from the filling of a tube with solid particles to the filling of a vessel with an ideal gas. Fundamentally, we see that is it the same problem, meaning the same form of the mass balance describes it. But, the key characterizing variable now is the pressure in the vessel. To go from mass to the measurable variable, pressure, we need a *constitutive equation*, i.e. a relationship between the pressure and the mass. Without this relationship the analysis cannot be completed. The ideal gas law comes in as the missing relationship and once again we look at density.

In each of the first two cases the mass flows were taken as constant, so in this instance we introduce a fluctional, i.e. time-dependent mass flow. This makes the right-hand side of the equation a function of time and we show how this is solved and how the time dependence of the input stream is reflected in the variation in the level of the liquid in the tank. We also notice that we need to be careful about how we use the results of the analysis. The model includes not just one way flow in, but two way flow, both in and out. If we don't want that, if we want only one way flow into the unit, then we need to exclude the possibility of two-way flow from the solution. The model does not know what we want to impose as constraints so we have to impose them explicitly as a part of the analysis and modeling process.

The last part of the chapter takes time to elucidate how the shape, or the geometry, of the tank is handled. The notion that the geometry is a part of the left-hand side of the equation helps to separate the problem as we leave the right-hand side the same. This also anticipates the transition from a series of stirred tanks that are each perfectly mixed to one continuous plug flow reactor. These problems are not hard problems either, but they do make one think because they bring together in the analysis, the consideration of key characterizing variable(s), and the use of geometrical reasoning to reduce the number of variables to one, and the calculus all in the same fundamental mass balance.

The last problem we handled looks very different from a filling tank and, yet, it is not that different from the perspective of the conservation of mass and, that, is the point. By doing a careful analysis of the very same kind, we can find the thickness of a layer that is deposited on a moving substrate. The control volume in this case is a bit more abstract. Again, not a hard problem, but it does require thinking about applying the same tools to a different situation.

In chapter 2 the right-hand side of the mass balance equation stayed mostly the same (except in one case) and much of the analytical effort was expended on the treatment of different geometries on the left-hand side. In chapter 3 the emphasis is on the right-hand side and the dependence of the flow rate out of the control volume, a flow rate that depends on physics, namely, gravity.

The control volume in this case is not abstract, it is defined by the walls of the unit itself, i.e. the tank. Unlike the case of sinusoidally time-dependent flow in chapter 2, here we do not know what the time dependence will be. From experience and even intuition, we do expect there to be time dependence in the flow from a tank under the force of gravity. It should be faster at first and then become slower. Hence, we choose to introduce experiments to get at this dependence. We could have considered the physics first, but we chose to do that later and as a rationalization of the experimental findings. Either order works.

The experiments show that the flow is quite time dependent. When we take the numerical derivative of the level as a function of time, a standard analytical operation based on the calculus and physics, we find that the rate of change in level versus time is linear. No matter at what level the experiment begins, this rate has a constant slope. More interestingly, when we reanalyze the rate of level change in terms of the level at any time, all the data falls on one curve. Looking at the rate of change of a variable in terms of that variable, rather than in terms of time explicitly, is in and of itself, a new idea and powerful. When we fit that data, we find that the rate of change of the level is dependent of the square root of the level at any time. This leads to Torricelli's Law for the exit flow rate. Going back to the start of the analysis, we see that with Torricelli's law as a constitutive equation, we can build a complete model for the draining tank system with the correct behavior. We also rationalize this finding with a consideration of the physics of a falling body, which in this case is a falling liquid. As simple as it is, this is a physics-based model; the square root dependence of the rate on level is not just the outcome of a statistical fit to the data, it is baked in to the physics of the situation. That point is significant and is emphasized.

Next, we bring input flow and output flows together to analyze the behavior of such a system. A fluxional input with gravity-driven exit flow is analyzed. This leads naturally to a consideration of control of the exit flow. By allowing the area of the orifice in Torricelli's Law to become a controllable variable, we can build a model for proportional control of the system and show that with the "gain" set properly, we can flatten the variation of the level in the system.

Thus, we show the interplay of experiment and analysis, the idea of a physics-based model, how the analysis changes with input and output taking place simultaneously, and, finally, the proportional control of the level in a tank with a fluxional input flow. This covers quite of bit of conceptual, analytical, computational, and engineering ground, but it is a gradual build-up of complexity.

To look at more interesting problems like those that arise in separations and chemical reactions, multiple different chemicals, or chemical species are involved simultaneously in mixtures or solutions. We refer to these different chemical as *components of the system,* or just as *components*, and we introduce these in Chapter 4. In the cases considered here, we take the control volume to be bounded by the unit. We do this to keep the left-hand side issues simple so that we can focus on the concept of a mixture, a solution and what it means to be well-mixed. We are seeking to conjure in the mind's eye pictures of these systems, so that the mathematics and the picture start to merge, intuitively.

We begin with a mixture of a liquid and an insoluble solid in particulate form. We are able to develop separately the mass balances for the solid and the liquid, and then combine these to develop the total mass balance equation for the system. The condition of being well-mixed because of the expenditure of energy at the agitator is familiar.

Next, we move to a solution of a liquid and a soluble solid, like salt and water. The analysis of this system is quite similar to that of the previous case, however, there is one critical issue that we must handle - how does the density of the salt solution vary with the concentration of the salt. To ask this question presupposes some knowledge of the physical chemistry of ionic solution of salts and water. We learn that depending on the salt, the density may increase with the concentration of salt, decrease, or stay about the same.

Then we look at the very real problem of washing a solution out of vessel as if we were doing so before its use for another process. This leads to the well-known result that we approach "zero" concentration of the salt in the washings asymptotically, meaning we have to know how close to zero is good enough for our purposes to know when we are "done." That brings us to a consideration of the "holding" time and mixing in a unit, and, again, what it means to be "perfectly" mixed. The degree of mixing is considered with the injection of a dye, or marker, into the system and analyzing the concentration of that marker in the exit flow as a function of time. These problems lets us use some of the power of *Mathematica* in new ways.

Just as real systems of importance to chemical engineers include multiple components, they also often include multiple phases, which we introduce in Chapter 5. Liquid and solid, liquid and immiscible liquid, gas and solid, liquid and gas, and gas-liquid-solid. In each case there is a boundary between the phases. Mass moves across these boundaries, if the systems are not at equilibrium. We have to sharpen our mind's eye to analyze these systems. We have to be able to picture the interface, quantify it, consider what causes mass to move from one phase to the other, what the rate function should be and to what extent thermodynamics plays a role.

We begin with a "simple" problem that we are already familiar with - a salt, NaCl dissolving in a solvent like water. The starting point is a pile of salt particles sitting at the bottom of a beaker filled with pure water; the end point is a colorless and clear solution (if we do not exceed the solubility of the salt). How does the system go from solid salt and liquid water to soluble salt in water? We have to think about salt as it is in the solid and salt as it is in the water solution. We must consider the interface between the solid and the solution and how that interface affects the dissolution process and its rate. What is that area? Is it constant? If it is not constant, how do we handle that? No salt is infinitely soluble, there is a limiting concentration. How does that limiting concentration or saturation concentration figure into the rate of dissolution? What about the mixing of the salt and the solvent? We build a model that includes all of these factors and then we seek to fit it to experimental data.

We also include another approach - the equilibrium stage approach. Some mass transfer processes are so fast that they can be considered to go to equilibrium instantaneously after the two phases are brought into contact. That means that the mixing must be excellent, the interfacial area is very high and that the linear driving force that is "pushing" a component from one phase to the other is very large. How fast is fast? It depends on the magnitude of the time constant for mass transfer relative to some other key time constant for the system. That other key time constant could be the holding time in the unit. If two phases are brought together in a well-mixed unit and the holding time is 1 second, then instantaneous mass transfer might be a good approximation if the transfer between the two phases takes place in a millisecond to 10 milliseconds. Why? Because the holding time would be 100 to 1000 times longer than the time to reach equilibrium between phases.

If we assume that the mass transfer processes are fast, then we can use equilibrium thermodynamics to assess the mass of the phases in the unit. The unit may also be considered to be made up of a series of equilibrium stages or sub-units. The plates of a distillation column may be treated as separate equilibrium stages connected in series. By contrast, a rate-based approach would look at the explicit rates of mass transfer at each plate.

Adsorption is a fascinating process at the fundamental level and it is also the basis of many different separation and purification processes in practice and we take it up in Chapter 6. By introducing adsorption, we can begin to think molecularly about surfaces, abstract sites on those surfaces, and adsorption of molecules onto those sites. We use Langmuir's classic analysis of these processes to get at their rates. We show how the net rate of adsorption at equilibrium leads to the Langmuir isotherm.

We also introduce the concept of a porous solid, because that is the adsorbent onto which adsorption takes place. We analyze a unit containing such an adsorbent, and the adsorption of a component from a stream entering a semi-batch adsorption unit. We push the analysis from a semi-continuous to a continuous adsorber unit. In so doing we use even more of the power of *Mathematica* to build a simple simulation of this unit with the rate constant for adsorption as a variable parameter.

Both of these systems were taken to be perfectly mixed throughout the gas and adsorbent phases. Though this let us focus on the right-hand side of the mass balances, such perfectly well-mixed systems would be difficult to achieve in practice, unless the systems were very large in cross section and shallow in depth. In fact that picture leads to the consideration of an adsorber unit with "plug" flow. In the case of plug flow, the mixing is perfect in the radial direction (cross-section), but there is no mixing in the axial direction, which is why it is oft said that plug flow is flow with zero back mixing. We can consider solving for concentration in the axial dimension and in time both analytically in the simplest case of first order, irreversible adsorption and numerically in the more complex case of Langmuirian kinetics. Now the control volume is both macroscopic radially, and microscopic, or differential, axially. Also, in this analysis we go from a set of coupled differential difference equations to a single partial differential equation for the whole of the system. This is a nice bit of intuitive mathematics that we then explore with *Mathematica*.

Permeation involves the transfer of mass from one side of a membrane to the other. Perm-selective membranes are those that transfer one component exclusively or preferentially. Palladium foils for hydrogen, certain metal oxides for oxygen, some polymers, and nanoporous materials for gases separations are all examples of perm-selective membranes with different mechanisms of transport and varying degrees of selectivity from nearly 100% to lesser levels. We treat the general case of permeation first. Then we end the chapter with a simple analysis in which a membrane forms a cell containing a salt solution. Pure water surrounding the cell is transferred into the cell by osmosis and that "flow" swells the cell.

Chapter 7 is all about chemical reactions and their rates. Firstly, we consider kinetic theory to get a quantitative sense of how often molecules collide in the gas phase. We want to understand this, because reactants react when they collide. While there is action at a distance, there is no reaction at a distance. But, collisions alone are not the whole story as we will see. In other words, a collision is necessary, but it is not a sufficient criterion for reaction.

Next we develop chemical rates of irreversible reaction with dependencies on concentration that range from first order to nth order. We also introduce the non-dimensionalization of these simple, batch reactor expressions. Reversibility, is important as it forms the basis for understanding that chemical equilibrium is anything but static, it is dynamic. We also see that the non-linearity of chemical kinetics, quickly forces us to move from looking for analytical solutions to using numerical methods. We can move from this base to more complex networks of series and parallel reactions. We finish with a consideration of the effect of high dilution on selectivity in a reaction network.

Having already explored Langmuir's rate of adsorption without reaction, we can introduce the Langmuir-Hinshelwood-Hougen-Watson rate of chemical reaction on a catalyst surface. (The background on this naming is that Langmuir and Hinshelwood derived the kinetics first. Then Hougen and Watson rederived the kinetics, but without knowing of Langmuir and Hinshelwood's prior work.) These rate expressions lead to highly non-linear equations with fascinating properties that illuminate the dynamics of surface reactions. This leads us to a biological example: the kinetics of substrate consumption by a growing population of microorganisms. It is fascinating that Minod and Michaelis-Menton developed these enzyme kinetics independently of Langmuir and Hinshelwood and Hougen and Watson, yet we can show that they are all mathematically equivalent. Hence, many also now bestow the names of Michaelis and Menton on these kinetics, as they are known as Langmuir-Hinshelwood-Hougen-Watson-Michaelis-Menton (or LLHWMM for short) kinetics because they all derived them independently, did so in different fields and at about the same time.

We then loop back to the effect of temperature on chemical reaction and the fact that the dependence of chemical rates on temperature is exponential because of the Boltzmann distribution of speeds of molecules at a given temperature. This leads us to consider how chemical reaction and heat transfer effects and their coupling together can lead to "thermal runaway" and an explosion. All this is rounded out with an analysis of Rice-Herzfeld kinetics for chain reactions involving free radicals. These are also referred to as free-radical chain kinetics.

Everything that is done in Chapter 7, is done as if the reactions are taking place in a simple, batch reactor with no flow in or out. This was to keep reaction flow effects distinct from intrinsic chemical kinetics. In Chapter 8, flow reactors are considered. First, we analyze the fed-batch or semi-batch reactor. This is a very useful reactor in practice because it gives a great deal of control over the course of the reaction. We give a simple example of the dilution of sulfuric acid and show how the reaction proceeds, produces considerable heat, but the heat transfer away from the vessel is easily managed by controlling the rate of addition of the acid. Then we examine the case wherein the concentration of the reactant flowing into the reactor is high enough that there is not a large change in the volume of solution in the reactor. But this is not general enough, so we reanalyze the system relaxing the small volume change constraint. Among the interesting features of the fed-batch reactor, is that the equations describing it, even with relatively simple kinetics lead to quite complex analytical solutions. In fact except for the simplest kinetics, numerical solutions must be used.

We then compare the batch and fed-batch reactor, the use of a fed batch reactor to produce alumina by precipitation, and the change in economic potential as less valuable reactants are converted to more valuable products. We introduce the idea of a maximum economic potential.

The continuous stirred tank reactor (CSTR) is a continuous flow reactor. We begin with a transient analysis and this leads to the steady state condition, which is the natural state for any flow reactor. It is essential here to show the simplicity of doing kinetics with the CSTR. We also show how to incorporate LHHWMM kinetics in such a system.

Linking of a series of CSTRs together leads to set of differential difference equations. These equations are ordinary differential equations in time and algebraically coupled difference equations in space. As we let the axial dimension of each unit shrink toward zero, we once again find that a single partial differential equation describes the set of coupled differential difference equations. This is once again the plug flow reactor (PFR).

The steady state equation for the PFR looks incredibly similar to the transient equation for the batch reactor. The denominator of the differential rate for the PFR is the holding time, τ, whereas for the batch reactor it is t, the actual time.

It is typical to compare the results of analyses of the PFR versus the CSTR with different kinds of kinetics and we do that here. We also give consideration to mixing effects and the results with a series of CSTRs with a total volume equal to one PFR. As the number of CSTRs in series increase, while holding total volume constant, they come to mimic one PFR.

The last section of the chapter examines the transient PFR and the solution of its governing equation. We solve these equations with different kinetics numerically and what we observe is interesting in that the PFR comes to steady state quickly. That is reassuring given that we usually use only the steady state version of the equation. We also see that we can solve the equation analytically with simple first order kinetics.

The last chapter (9) is a collection of assorted problems and models that the author has developed over the years. The first problem is a return to the tank with level control, but this time there is a jump (unit-step) in the inlet flow. We examine how to handle the piecewise solution of this problem. The proportional control of the tank is extended to proportional-integral-differential (PID) control.

The next two sections are classical general chemistry problems and should be quite familiar. Firstly, we analyze the titration of a strong acid with a strong base. Doing this with *Mathematica* is so easy and we do not need to make the usual simplify assumptions because we can solve higher order polynomial equations for the proton concentration directly. Similarly, in the next section, when we compute the pH of a weak acid, we do so directly. It is also amazingly convenient to have all the tools in one place to solve the equations, to do functional programming and to create outstanding graphical representations.

The relationship between salt solution density and salt concentration comes up early in our studies. Although a linear equation seems to work, it is not necessary that this is the best representation for this relationship. Examination of data in the CRC Handbook of Chemistry and Physics for the density and concentration of different salt solutions, led to the calculation of partial molal volumes of these solutions. Interestingly, and unexpectedly, the results show anomalies when we plot the partial molal volume versus molality. Namely there are oscillations in the partial molal volume at the lower molalities for sodium chloride solutions. This is not only the case for sodium chloride. Indeed the effect seems the most apparent for NaCl and is less apparent for LiCl and KCl, but it is quite apparent for CsCl. Are these "oscillations" simply due to experimental error? We cannot rule that out, but it seems unlikely. Is there a physical explanation for the behavior? Maybe, but what would that be? Spheres of hydration come to mind, but the data seem not to trend in a way that would support that as a hypothesis.

The design of chemical reactors is regularly studied in courses devoted to reaction and reactor engineering. In this case we have taken a slightly different approach by defining the economics of the process and the reactor and we seek to find the size that maximizes the profit in terms of the holding time in a CSTR.

Economics are also the focus of the next example, which is a linear optimization among starting materials and products with different values. A two dimensional optimization was chosen so that the graphical representations would be straightforward.

Often in analysis, a non-linear problem can be simplified mathematically with a linear approximation. A classic problem in mechanics is the undamped pendulum, or oscillator. Conservation of momentum leads to a non-linear equation with the Sine of the angular displacement on the right-hand side. Simple though the equation appears to be, it's not. By assuming the displacements are small, the Sine of the angle can be replaced by the angle itself. This reduced the non-linear equation to a simple linear one with a well-known solution. The question arises, do we need to make this assumption? In the past, it was expedient to do so, but now with *Mathematica,* it is not necessary. We go on to solve the original non-linear equation numerically and then compare the results from the linear equation and the non-linear equation to see what we lost in the approximate case.

Another question arises and that is whether, or not to use Laplace transforms. In this section we consider equations with impulse functions. We solve one equation using the Laplace transform method. But we also show that the equation can be solved analytically by DSolve without the necessity of the transform. We test DSolve on two more cases and it is robust enough to handle those cases as well. Thus it seems more expedient to use DSolve directly.

Like many others, the author was interested in expert systems and neural networks in the 1980s and 1990s. For many decades there was a lull in activity, but now artificial intelligence has come roaring back and is being used successfully in a plethora of applications. Out of curiosity, the author began to examine the roots of this phenomenon, to try to better understand why AI came back so suddenly and successfully. That led to a reading of the earliest work on AI and the idea of the "perceptron." The central idea is that of the transfer function and that is what we explore here. This leads to logic gates that are well known. From the history of the field at its beginning, one can better understand modern neural networks and how the incredible advancements in sheer computing power, exemplified by Moore's law and in algorithmic sophistication have advanced the field to where it is today. Frank Rosenblatt was the brilliant inventor of the perceptron. He made some claims about its applicability that would not be realized in machine learning for decades, but it is understandable. Minsky and Parpert, in their book entitled "Perceptrons" demonstrated that a single layer perceptron is limited in scope because it cannot do an XOR function. Apparently, that so dampened enthusiasm that the field languished for a decade or more. That was probably an over reaction because Rosenblatt and others were already working on multilayer perceptrons. The sudden accidental death of Rosenblatt in 1971 did not help matters. The story of the perceptron is as fascinating and this gives the reader a bit of grounding for deeper study.

Conclusion

All-in-all, this book is a collection of models derived by thinking both analytically and computationally. Computational thinking brings analytical thinking together with algorithmic thinking. *Mathematica* makes it so much easier to do both and to do them well. I think it is safe to predict that *Mathematica* will continue to evolve and as it does, it will become even more "Intelligent."

So, I hope you come to enjoy computing this way as much as I do and if you do, then this has been a successful enterprise.

CHAPTER 1

A primer of Mathematica

Getting started in Mathematica

We will use *Mathematica* throughout the text. The goal of chapter 1 is to give you a wide introduction to *Mathematica* that is enjoyable and above all interesting. We will cover a range of topics ranging from the basics to more advanced ideas. One of the great aspects of *Mathematica*, is just the sheer joy of computing that it provides. While, it does take some effort to get started, it will not be long before you can sit down at the computer and begin to do exploratory computing. Being able to do that is well worth your effort. Reading through some, or all of this chapter, and trying the things out on your own, will get you on the path to being a *Mathematica* expert!

Basics

In *Mathematica* when you type x = 1000 as an input you will see:

In[1] :=   x = 1000

When you press SHIFT RETURN, the output will look like this:

Out[1] := 1000

The In[1]:= is the input cell label and the Out[1]:= is the output label. As you continue to work inside this notebook and do successive calculations, the increment will increase by 1 each time. While this works beautifully synchronously on the computer, it does not work well when providing an asynchronous rendition of such work in a text book.

So, whereas, when seeking to evaluate Sin[2.33333 π] the input and output during a session would look as follows:

In[1] := Sin[2.33333 π]

Out[1] :=   0.86602

When we see this in the textbook the cell labels will not be displayed and the reader will see that input comes first and will be in bold and the output follows in plain face font as such:

Introduction to Chemical Engineering Analysis Using *Mathematica*
https://doi.org/10.1016/B978-0-12-820051-3.00006-1

Sin[2.33333π]

0.86602

Commands in *Mathematica* are given in natural language, so they have names such as "**Solve**" or "**Simplify**" etc. The format of a command is the word starting with a capital letter and enclosing the argument in square brackets:

Command[argument]

Parentheses are used arithmetically and algebraically in the usually way:

$$3a(x-2)^2$$

On the other hand curly brackets are distinct. They are used to designate lists, or vectors as in:

$$\{1, 2, 3, 4...\}$$
$$\{\{1, 1\}, \{2, 2\}, \{3, 3\}...\}$$

The three, curly brackets, square brackets, and parentheses must not be interchanged.

To clear the value of a given named variable there are three options. First, type the variable name followed by **= .** Second, type **Clear** with the variable name enclosed in square brackets, or, third, type **Remove** with the variable name inside the square brackets:

variable name = .

Clear[variable name]

Remove[variable name]

The first two simply clear the current value while the last removes the name entirely. You need to remember this because if you start a session and assign a value to a variable, then that value will be retained in that variable until you change it or clear it.

Equals (=), equal equals (==), and colon equals (:=)

The basic equation for a line is **y = m x + b**, where the intercept is **b**, the slope is **m**, and the equation is linear in **x**, meaning that it is raised to the first power. To be more explicit with our notation, this is **y [x] = m x + b**, meaning that **y** is a function of **x**. For each **x value** there is an associated **y value**. Hence the line that is formed is a set of paired values, or dyads, in the two dimensional space. When we say solve in this context, the process to be undertaken is to input **x** values and to output of **y** for set values of **m** and **b**. Thus **m** and **b** are distinguished from **x** and **y**, and are referred to as parameters, whereas **x** and **y** are described as variables. Because we defined **y [x]**, this makes x the independent variable and **y [x]** the dependent variable. In fact the distinction is formal and merely a convention that orders how we use the equation. We could just as easily let **y** be independent and find the **x** values, wherein we would rearrange the equation to put the dependent variable **y** on the right-hand side and **x [y]** on the left-hand side. We could also let **x** be constant and solve for **m** and **b** pairs, as in **b = - x m + y**. If we interpret this equation, it would tell something about lines that can go through the point **x** and **y**. There are of course an infinite number of these. Let's explore this.

We begin by learning how to use the different versions of equal signs that *Mathematica* provides, namely **=**, **==**, and **:=** by looking at something very simple the equation for a line where **y = m x + b**, where the intercept is **b**, the slope is **m** and the equation is linear in **x**.

The first thing we will do is to use the **?** to see how these equal signs are described in the documentation:

```
? =
```

lhs = *rhs* evaluates *rhs* and assigns the result to be the value of *lhs*. From then on, *lhs* is replaced by *rhs* whenever it appears.
$\{l_1, l_2, \ldots\} = \{r_1, r_2, \ldots\}$ evaluates the r_i, and assigns the results to be the values of the corresponding l_i. >>

```
? ==
```

lhs == *rhs* returns True if *lhs* and *rhs* are identical. >>

```
? :=
```

lhs := *rhs* assigns *rhs* to be the delayed value of *lhs*. *rhs* is maintained in an unevaluated form. When *lhs* appears, it is replaced by *rhs*, evaluated afresh each time. >>

These are the formal descriptions of each of the equal signs, and as such they are written in very careful language. Note that anything to the left of the equals sign is referred to as the *lhs* and to the right as the *rhs*. In the text we will use RHS and LHS in the same way.

When we use "=" as in LHS = RHS, it evaluates RHS and then assigns the result to be the value of LHS. From then on, LHS is replaced by RHS whenever it appears. The definition of equals is the one we expect from our experience. Simply put, it sets the value, numerically or functionally, of whatever we put to its left by that which we place on its right. In *Mathematica* this is referred to as Set. Once it is set equal, the LHS will always be given by the RHS, until such time as we undo this using for example Clear[LHS], LHS =. or by going to the menu bar Evaluation → Quit Kernal → Local. We will use this to set parameter values as we will see shortly.

Mathematica also provides the "==" as in LHS == RHS and then it will return True if the LHS and RHS are identical. This version, the "equal equals" sign, is a logical operator. When we see the LHS == RHS, *Mathematica* examines the two sides and, if they are equal, it returns either TRUE or FALSE. In *Mathematica* the "equal equals" sign is simply referred to as Equal. This is a very powerful operation because it can be extended so far in scope. We can use it to test our symbolic solutions to equations to be sure that they are correct. We also use this in the Solve, NSolve and the DSolve and NDSolve commands.

We will regularly employ the ":=" which is the "colon equals" sign and it is referred to as set delayed in *Mathematica*. In this instance when we write LHS := RHS, *Mathematica* delays the assigning the value of the LHS to the RHS. The RHS is maintained in an unevaluated form. When the LHS appears, it is replaced by the RHS, and the evaluation is done anew each time.

Put another way, when we use the colon equals, we create a LHS that is defined by the RHS, but the dependent variable is left unevaluated until it is specified later i.e. the definition of the variable is set, but setting the value of the variable is delayed. Because the LHS is defined in this way, it is a function name and the RHS is its functional definition. The function is not evaluated until its parameters are specified and a value, or values, of its variable is, or variables are, given. This is why the word "delayed" is used. The function's value is not immediately computed, that computation is delayed.

These definitions are accurate and precise, but they may not make the distinctions obvious. To really understand when to use the different connectors between the LHS and the RHS of an equation, you can read further, or we can learn by seeing what they do in order to understand operationally how they work and when they should be deployed. That is the purpose of this section. Doing is more fun than merely reading. So we will experiment with these core concepts.

= (equals or Set)

We begin with the most simple approach. Let's set y equal to m x + b and see what happens, and how it can be used. We begin by entering this equation and looking at the output.

Note that we start by clearing the values of all the parameters and variables that we will be using. This is done by putting the variable and parameter names that we will use inside of square brackets within the ***ClearAll*** *command. More will be said of this shortly. Also note that comment statements that are used to explain a line of input script can be inserted anywhere by placing the text inside of (*...*)*

ClearAll[y, m, x, b]

(* This clears all the values of these parameters and variables. *)

$\mathbf{y = mx + b}$

$\mathbf{y}$

$b + mx$

$b + mx$

We entered **y = m x + b** and the output for **y** matches that exactly. When we enter y^2, we see that *Mathematica* returns exactly the result $(mx + b)^2$:

$\mathbf{y^2}$

$\mathbf{y^2}$

$(b + mx)^2$

$(b + mx)^2$

Solve will do exactly what it says; it will symbolically solve equations. We could seek to find x expressed in terms of y, in other words, x = (y - b)/ m. We try it and it fails.

Solve[y, x]

Solve[$b + mx, x$]

```
··· Solve:
 b + m x is not a quantified system of equations and inequalities.
```

Solve[$b + mx, x$]

Maybe we needed to place the RHS in the brackets?

Solve[$b + mx, x$]

Solve[$b + mx, x$]

··· Solve:
$b + m x$ is not a quantified system of equations and inequalities.

Solve$[b + mx, x]$

Again, this does not work. Even if we place the full equation into the Solve function, it still fails.

Solve[$y = mx + b, x$]

Solve$[b + mx, x]$

··· Solve:
$b + m x$ is not a quantified system of equations and inequalities.

Solve$[b + mx, x]$

Along the way, *Mathematica* has told us why it did not work. It tells us that b + mx is not a quantified system of equations and inequalities. So, let's leave this aside for now and try a simple arithmetic evaluation of y. To do this, we set the values of b, m, each with the = sign and then we evaluate y, y = m x + b and y^2, we see the outputs are each based on the RHS:

$b = 3;$

$m = 5;$

y

$y = mx + b$

y^2

$3 + 5x$

$3 + 5x$

$(3 + 5x)^2$

We can also set the value of x and then evaluate again:

$x = 7;$

y

$y = mx + b$

y^2

38

38

1444

The evaluation finds that for these values, y = 38 and y^2= 1444. Now, we can go back to the first two inputs and repeat them to see what happens. At this point, the symbolic outputs that were returned before, are not returned. Instead, the numerical value of y is expressed, twice. This is because when we evaluated **y** with all its parameters specified that computes to 38.

$y = mx + b$ **(* our first two inputs *)**

y

38

38

How can we undo this? We have to use **Clear** y:

Clear[*y*]

y

y

When we **Clear** a symbol, it no longer has any value other than itself. Thus after clearing y, when we evaluate it, *Mathematica* simply echoes it back; input y gives output y. The value of y did not go back to m x + b because that assignment was "lost" when cleared **y**. So, when = is used, the LHS will take on the latest, or last value of the RHS. But we did not clear m, x, and b. Hence, if we evaluate m x + b we see:

m

x

b

$mx + b$

5

7

3

38

This happens because these symbols, have been set to specific numerical values using the single equals sign. Since we didn't clear them, when we call them again in this notebook, they will be given these values. To clear these, we can set each symbol equal to the period . as in **b = .** or we can use **ClearAll**:

ClearAll[*y*, *m*, *x*, *b*]

== (Equals Equals or Equal)

The next version of equals is equal equals. We can start with the equation for a line again, make it input and observe the output.

y==mx + b

$y == b + mx$

Here the output is the same as the input, and it is the full equation LHS == RHS; please notice that the output is not just the RHS as it was when we used = alone.

Note: If you are working interactively, and if you did not Clear y, m, x, and b from the previous section, then when you enter y == m x + b now, the output will be True. If you cleared y only, but not m, x, and b, then the output will be 38. If you did ClearAll [y, m, x, b], then the output will be m x + b.

Consistent with this, is that now, when we enter y, the output is just y:

y

y

Having used the ==, rather than =, now when we solve for x, the operation works as expected:

Solve[$y == mx + b, x$]

$$\left\{\left\{x \to \frac{-b+y}{m}\right\}\right\}$$

If we set the parameter values and the value of x, then the input equation y == m x + b gives us the proper value

$m = 5;$

$x = 7;$

$b = 3;$

$y == mx + b$

$y == 38$

When we used the single equals sign **Solve** did not work, but when we used the double equals sign, it did. That is significant and it shows the operational difference between the two symbols in *Mathematica* and the distinction is important.

:= (Set Delayed)

We can now explore the use of the Set Delayed further, the colon combined with the equals sign and the equation for a line once again. The key word here is delayed.

If we use Set Delay to let y be equal to m x + b, when y (LHS) is input, then m x + b (RHS) is returned as output. This appears to be operating like the single equal sign, but that is not the whole story. Firstly, we clear all the symbols of any values. Then we let y be equal to mx + b with the set delay equals sign. When we call for y, we get back m x + b:

ClearAll[y, m, x, b]

$y := mx + b$

y

$b + mx$

Next, we apply values to the parameters m, and b with the single equals sign. After doing this, the output from a call for y is 3 + 5x and the output of y == mx + b with the double equals sign, is now the logical result, True.

$m = 5;$

$b = 3;$

y

$y == mx + b$

$3 + 5x$

True

We used all three versions of the equals sign in this example.

Having set **m** and **b** to numerical values, y may be computed in a Table function, graphed with the Plot function, and dyads of x and y values may be plotted with ListPlot. So now we can compute values of y for input values of x from - 10 to 10 in increments of one, and plot y over the same range:

```
Table[y, {x, -10, 10}]
(* Table will calculate y given the values of x from - 10 to 10. *)
```

{−47, −42, −37, −32, −27, −22, −17, −12, −7, −2, 3, 8, 13, 18, 23, 28, 33, 38, 43, 48, 53}

```
Plot[y, {x, -10, 10}]
(* Plot will compute values of yfor each xand then connect them in a graph. *)
```

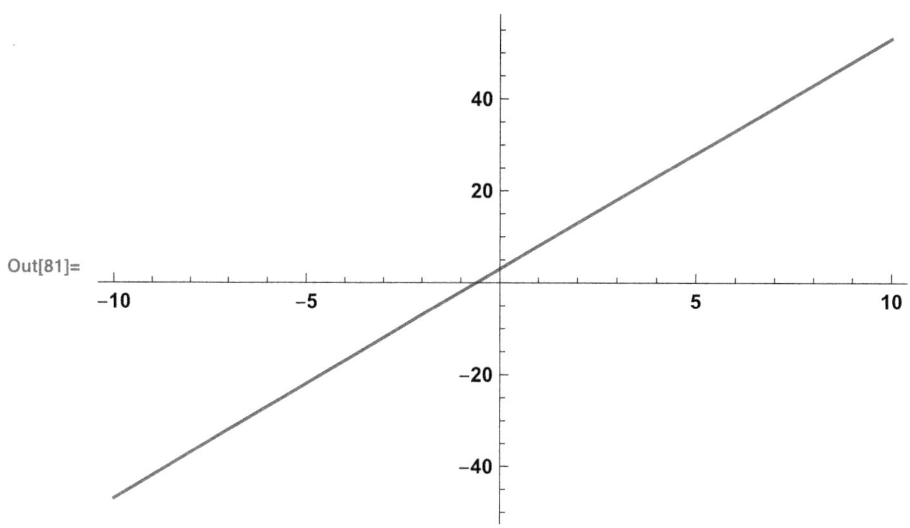

This shows us the line as a continuous function, but, it does not show us individual values of the function on the graph, i.e., we may want to see specific points. To do this we again use the Table function, but this time instead of reporting on just the y values, we will report the x value associated with each y value, that is to say Table will report out { x, y } as dyads that can then be examined and graphed with **ListPlot**:

Table[{*x*, *y*}, {*x*, −10, 10}]

(* In this statement Table is computing all the values of *y* at the given *x* values and it is associating each *y* with an *x* value. *)

{{−10, −47}, {−9, −42}, {−8, −37}, {−7, −32}, {−6, −27}, {−5, −22}, {−4, −17}, {−3, −12}, {−2, −7}, {−1, −2}, {0, 3}, {1, 8}, {2, 13}, {3, 18}, {4, 23}, {5, 28}, {6, 33}, {7, 38}, {8, 43}, {9, 48}, {10, 53}}

ListPlot[%]

(* % means the latest output. Here we take the latest or last output, the list of dyads of *x*and *y*and we plot just those points using ListPlot. *)

Out[83]=

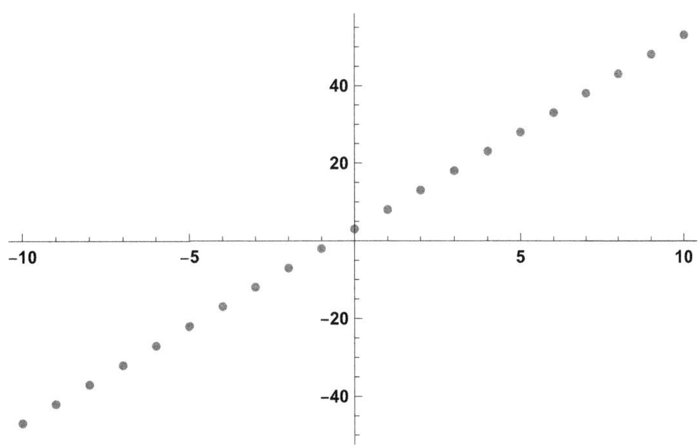

Alternatively, the Plot and the ListPlot can be shown together as one graph by combining them with Show:

Show[ListPlot[Table[{*x*, *y*}, {*x*, −10, 10}]],

Plot[*y*, {*x*, −10, 10}]]

(* The Show command literally shows both the output of the ListPlot and of the Plot commands in one graph. *)

Out[84]=

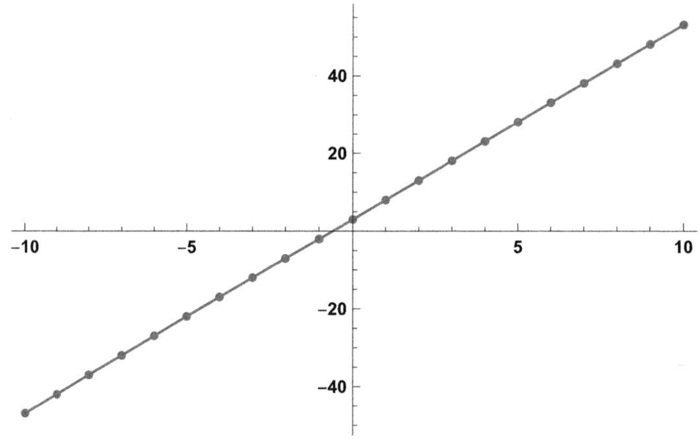

To summarize, the single equal sign is used to set or assign specific values. The double equals sign determines if the entities to its left and to its right are the same, or not, logically, and the colon equal sign, delays the assignment, or defers the assignments on the right-hand side until they are later specified. Through use and application, these distinctions become clear.

Simple commands

Mathematica is enhanced by **Palettes** (see the menu at the top of the notebook page) that are very handy tools. There are several Palettes. If one wants to use a trigonometric function, for example, we can either type its name or go to the Basic Math Assistant palette and then to the Basic or Advanced tab. Should we want to evaluate the Sine of 2.6 π, then we can do so as follows:

Sin[2.33333 π]

0.86602

Should we need to know the Sine of 120 degrees, then we include this unit name in the argument of the function:

Sin[120 Degree]

$\frac{\sqrt{3}}{2}$

To rationalize this fraction, we need to numerically evaluate it. We do so by surrounding the **Sin** function with **N**, where **N** is the command for numerical evaluation:

***N*[Sin[120Degree]]**

0.866025

When working interactively, the **N** can be appended with a post-fix double slash **//**:

Sin[120Degree]//*N*

0.866025

For logarithmic, and all other functions, we do the same as we have done with **Sin**. It is important to know, however, that the function **Log** in *Mathematica* is the natural logarithm and not that of log base ten. Hence we see:

***N*[Log[10]]**

***N*[Log[100]]**

***N*[Log[1000]]**

2.30259

4.60517

6.90776

To find the log base ten (or for any other base), we must specify it, base n, first as part of the argument:

Log[10, 10]

Log[10, 100]

Log[10, 1000]

1

2

3

Mathematica has a large number of built-in functions from the usual to the less usual, but we can work with them all more or less in the same way. As an example the **Riemann Zeta** function is readily accessible, as are many more special functions:

```
Zeta[10]//N
Zeta[20I]//N
```

1.00099

$-0.0538734 - 1.73066i$

```
Clear[y]
(* y had been previously assigned a value *)
```

```
Solve[Zeta[0.5 + I y] == 0, {y}]
```

$\{\{y \to 0. + 2.5i\}\}$

When this solution is used in the Zeta function, indeed the output is zero:

```
Zeta[0.5 + I (2.5I) ]
```

$0. + 0.i$

The Riemann hypothesis states that all non-trivial zeros of the Zeta function have their real part equal to 0.5, exactly.

Table

The **Table** function is the workhorse of *Mathematica* because it does so much with so little. In contrast to a procedural language in which we would have to write a looping structure to evaluate a function at several different values, or for a range of values, **Table** embeds all this and provides a vector of output values, which is a very useful format for subsequent operations.

We begin with the **Table** command to evaluate x^2 from zero to 20 at every integer:

```
Table[x^2, {x, 0, 20}]
```

{0, 1, 4, 9, 16, 25, 36, 49, 64, 81, 100, 121, 144, 169, 196, 225, 256, 289, 324, 361, 400}

Perhaps, we wanted the values of x^2 for every other whole number between 0 and 20? We can get these too:

Table$[x^2, \{x, 0, 20, 2\}]$

{0, 4, 16, 36, 64, 100, 144, 196, 256, 324, 400}

Should we need all the values for every integer value and the mid point between them, we would specify this:

Table$[x^2, \{x, 0, 20, .5\}]$

{0., 0.25, 1., 2.25, 4., 6.25, 9., 12.25, 16., 20.25, 25., 30.25, 36., 42.25, 49., 56.25, 64., 72.25, 81., 90.25, 100., 110.25, 121., 132.25, 144., 156.25, 169., 182.25, 196., 210.25, 225., 240.25, 256., 272.25, 289., 306.25, 324., 342.25, 361., 380.25, 400.}

Notice that the output in each case is a list that is to say a group of numbers within the left and right curly brackets, **{** and **}**, with comma separations. It is important to note that most outputs are in the form of lists, even if they have only one element. Let's redo the last computation, and call the output list "**ls1**."

ls1 = Table$[x^2, \{x, 0, 20, .5\}]$

{0., 0.25, 1., 2.25, 4., 6.25, 9., 12.25, 16., 20.25, 25., 30.25, 36., 42.25, 49., 56.25, 64., 72.25, 81., 90.25, 100., 110.25, 121., 132.25, 144., 156.25, 169., 182.25, 196., 210.25, 225., 240.25, 256., 272.25, 289., 306.25, 324., 342.25, 361., 380.25, 400.}

This variable name is now assigned to this list, until we clear it or remove it:

ls1

{0., 0.25, 1., 2.25, 4., 6.25, 9., 12.25, 16., 20.25, 25., 30.25, 36., 42.25, 49., 56.25, 64., 72.25, 81., 90.25, 100., 110.25, 121., 132.25, 144., 156.25, 169., 182.25, 196., 210.25, 225., 240.25, 256., 272.25, 289., 306.25, 324., 342.25, 361., 380.25, 400.}

In the next line, we clear **ls1** and then show that it is no longer assigned to the list of values.

ls1=.

ls1

ls1

We may also have occasion to want to generate the vector of values and to assign them to a list name, but we may not want to see all of them. For example suppose we wanted all the values for x^x from between 1 an 100? This can be done and the list can be named, but we may not want this sent to the screen. To suppress it, we place a semi-colon after the command:

ls2 = Table[x^x, {x, 1, 100}];

The number of elements in this list is, obviously, 100. We can see this by using the Length command:

Length[ls2]

100

To see what we have "missed," by not printing this out to the screen we can now do so by typing **ls2** as input:

ls2

{1, 4, 27, 256, 3125, 46656, 823543, 16777216, 387420489, 10000000000, 285311670611,
8916100448256, 302875106592253, 11112006825558016, 437893890380859375,
18446744073709551616, 827240261886336764177, 39346408075296537575424,
1978419655660313589123979,...
3697296376497267726571879056288054405956687642817411024302599724 2
35525704552775234214106500101282327279409788895483265401194299967694943594 51
621570193644014418071060667659301384999779999159200499899,100000000000000000
0000000000000000000000000000000000000000000000000000000000000000000000000000
0000000000000000000000000000000000000000000000000000000000000000000000000000
0000000000000000000000000000000}

We can also suppress the output to just the elements of the list we want to see by using the **Short** command and specifying the number of shown elements:

Short[ls2, 5]

{1, 4, ⟨⟨97⟩⟩,
1000000000000000000000000000000000000000000000000000000000000000000000000000
0000000000000000000000000000000000000000000000000000000000000000000000000000
0000000000000000000000000000000000000000000000000}

This leads to several other points. Firstly, we could also operate on all of the values in **ls2** by operating on **ls2** alone. This property of **ls2** is called "listability," and it is a very important attribute of such objects in *Mathematica.* So for instance we could divide each value we just found by dividing **ls2** by 10^{100}:

$N\left[\frac{\mathbf{ls2}}{10^{100}}\right]$

$\{1. \times 10^{-100}, 4. \times 10^{-100}, 2.7 \times 10^{-99},$
$2.56 \times 10^{-98}, 3.125 \times 10^{-97}, 4.6656 \times 10^{-96},$
$8.23543 \times 10^{-95}, 1.67772 \times 10^{-93}, 3.8742 \times 10^{-92},$
$1. \times 10^{-90}, 2.85312 \times 10^{-89}, 8.9161 \times 10^{-88},$
$3.02875 \times 10^{-86}, 1.1112 \times 10^{-84}, 4.37894 \times 10^{-83},$
$1.84467 \times 10^{-81}, 8.2724 \times 10^{-80}, 3.93464 \times 10^{-78},$
$1.97842 \times 10^{-76}, 1.04858 \times 10^{-74}, 5.84259 \times 10^{-73},$
$3.41428 \times 10^{-71}, 2.08805 \times 10^{-69}, 1.33374 \times 10^{-67},$
$8.88178 \times 10^{-66}, 6.15612 \times 10^{-64}, 4.43426 \times 10^{-62},$
$3.31455 \times 10^{-60}, 2.56769 \times 10^{-58}, 2.05891 \times 10^{-56},$
$1.70692 \times 10^{-54}, 1.4615 \times 10^{-52}, 1.2911 \times 10^{-50},$
$1.17566 \times 10^{-48}, 1.10251 \times 10^{-46}, 1.06387 \times 10^{-44},$
$1.05551 \times 10^{-42}, 1.07591 \times 10^{-40}, 1.12595 \times 10^{-38},$
$1.20893 \times 10^{-36}, 1.33088 \times 10^{-34}, 1.50131 \times 10^{-32},$
$1.73438 \times 10^{-30}, 2.05077 \times 10^{-28}, 2.48064 \times 10^{-26},$
$3.06803 \times 10^{-24}, 3.87792 \times 10^{-22}, 5.00702 \times 10^{-20},$
$6.60097 \times 10^{-18}, 8.88178 \times 10^{-16}, 1.21921 \times 10^{-13},$
$1.70677 \times 10^{-11}, 2.43568 \times 10^{-9}, 3.54212 \times 10^{-7},$
$0.0000524745, 0.00791643, 1.21581, 190.031,$
$30218.2, 4.88737 \times 10^{6}, 8.03748 \times 10^{8},$
$1.34365 \times 10^{11}, 2.28273 \times 10^{13}, 3.9402 \times 10^{15},$
$6.90825 \times 10^{17}, 1.22998 \times 10^{20}, 2.22337 \times 10^{22},$
$4.07949 \times 10^{24}, 7.59604 \times 10^{26}, 1.43504 \times 10^{29},$
$2.75006 \times 10^{31}, 5.3449 \times 10^{33}, 1.05334 \times 10^{36},$
$2.10449 \times 10^{38}, 4.26182 \times 10^{40}, 8.74647 \times 10^{42},$
$1.8188 \times 10^{45}, 3.83159 \times 10^{47}, 8.17599 \times 10^{49},$
$1.76685 \times 10^{52}, 3.86622 \times 10^{54}, 8.56517 \times 10^{56},$
$1.9208 \times 10^{59}, 4.35973 \times 10^{61}, 1.0014 \times 10^{64},$
$2.32738 \times 10^{66}, 5.47236 \times 10^{68}, 1.30159 \times 10^{71},$
$3.1312 \times 10^{73}, 7.61773 \times 10^{75}, 1.87399 \times 10^{78},$
$4.66101 \times 10^{80}, 1.17196 \times 10^{83}, 2.97864 \times 10^{85},$

$7.65143 \times 10^{87}, 1.98627 \times 10^{90}, 5.21025 \times 10^{92},$
$1.38088 \times 10^{95}, 3.6973 \times 10^{97}, 1. \times 10^{100}\}$

Secondly, we also note that the output from this last computation is in scientific notation, whereas **ls2** was not; it was written in standard form. It is worth noting that we could have had **ls2** in scientific notation, simply by changing the command as follows:

ls3 = Table [x^x, {x, 1., 100}]

(* By changing the first value of x
from the integer 1 to the rational 1.,
the output switches to scientific notation. *)

$\{1., 4., 27., 256., 3125., 46656., 823543.,$
$1.67772 \times 10^{7}, 3.8742 \times 10^{8}, 1. \times 10^{10},$
$2.85312 \times 10^{11}, 8.9161 \times 10^{12}, 3.02875 \times 10^{14},$
$1.1112 \times 10^{16}, 4.37894 \times 10^{17}, 1.84467 \times 10^{19},$
$8.2724 \times 10^{20}, 3.93464 \times 10^{22}, 1.97842 \times 10^{24},$
$1.04858 \times 10^{26}, 5.84259 \times 10^{27}, 3.41428 \times 10^{29},$
$2.08805 \times 10^{31}, 1.33374 \times 10^{33}, 8.88178 \times 10^{34},$
$6.15612 \times 10^{36}, 4.43426 \times 10^{38}, 3.31455 \times 10^{40},$
$2.56769 \times 10^{42}, 2.05891 \times 10^{44}, 1.70692 \times 10^{46},$
$1.4615 \times 10^{48}, 1.2911 \times 10^{50}, 1.17566 \times 10^{52},$
$1.10251 \times 10^{54}, 1.06387 \times 10^{56}, 1.05551 \times 10^{58},$
$1.07591 \times 10^{60}, 1.12595 \times 10^{62}, 1.20893 \times 10^{64},$
$1.33088 \times 10^{66}, 1.50131 \times 10^{68}, 1.73438 \times 10^{70},$
$2.05077 \times 10^{72}, 2.48064 \times 10^{74}, 3.06803 \times 10^{76},$
$3.87792 \times 10^{78}, 5.00702 \times 10^{80}, 6.60097 \times 10^{82},$
$8.88178 \times 10^{84}, 1.21921 \times 10^{87}, 1.70677 \times 10^{89},$
$2.43568 \times 10^{91}, 3.54212 \times 10^{93}, 5.24745 \times 10^{95},$
$7.91643 \times 10^{97}, 1.21581 \times 10^{100}, 1.90031 \times 10^{102},$
$3.02182 \times 10^{104}, 4.88737 \times 10^{106}, 8.03748 \times 10^{108},$
$1.34365 \times 10^{111}, 2.28273 \times 10^{113}, 3.9402 \times 10^{115},$
$6.90825 \times 10^{117}, 1.22998 \times 10^{120}, 2.22337 \times 10^{122},$
$4.07949 \times 10^{124}, 7.59604 \times 10^{126}, 1.43504 \times 10^{129},$
$2.75006 \times 10^{131}, 5.3449 \times 10^{133}, 1.05334 \times 10^{136},$
$2.10449 \times 10^{138}, 4.26182 \times 10^{140}, 8.74647 \times 10^{142},$
$1.8188 \times 10^{145}, 3.83159 \times 10^{147}, 8.17599 \times 10^{149},$
$1.76685 \times 10^{152}, 3.86622 \times 10^{154}, 8.56517 \times 10^{156},$
$1.9208 \times 10^{159}, 4.35973 \times 10^{161}, 1.0014 \times 10^{164},$

$2.32738 \times 10^{166}, 5.47236 \times 10^{168}, 1.30159 \times 10^{171},$
$3.1312 \times 10^{173}, 7.61773 \times 10^{175}, 1.87399 \times 10^{178},$
$4.66101 \times 10^{180}, 1.17196 \times 10^{183}, 2.97864 \times 10^{185},$
$7.65143 \times 10^{187}, 1.98627 \times 10^{190}, 5.21025 \times 10^{192},$
$1.38088 \times 10^{195}, 3.6973 \times 10^{197}, 1. \times 10^{200}\}$

In fact by changing any of the x-range values to rational, the first value of x, the last value of x, or the increment of x from 1 (exact integer value) to 1. (the exact rational value), the output will go over from integers to rational numbers, and *Mathematica* uses scientific notation because of their size.

We can finish this section with a small bit of matrix algebra and the concept of nesting a Table function inside of another Table Function. Firstly, we create a list of dyadic pairs, a list of lists:

a1 = Table[{*x*, *y*}, {*x*, 0, 2}, {*y*, 0, 2}]

(* This creates a list of lists *)

{{{0, 0}, {0, 1}, {0, 2}}, {{1, 0}, {1, 1}, {1, 2}}, {{2, 0}, {2, 1}, {2, 2}}}

In linear algebra, this would be displayed as a matrix, so we can output this in MatrixForm:

MatrixForm[a1] (* Lists of lists can be put into matrix form *)

$$\begin{pmatrix} \begin{pmatrix} 0 \\ 0 \end{pmatrix} & \begin{pmatrix} 0 \\ 1 \end{pmatrix} & \begin{pmatrix} 0 \\ 2 \end{pmatrix} \\ \begin{pmatrix} 1 \\ 0 \end{pmatrix} & \begin{pmatrix} 1 \\ 1 \end{pmatrix} & \begin{pmatrix} 1 \\ 2 \end{pmatrix} \\ \begin{pmatrix} 2 \\ 0 \end{pmatrix} & \begin{pmatrix} 2 \\ 1 \end{pmatrix} & \begin{pmatrix} 2 \\ 2 \end{pmatrix} \end{pmatrix}$$

We can transpose the elements, which means that the x and y values of each pair are exchanged. So in the case of { 0, 1 } its transpose is { 1, 0 }. Because this is a list, **Transpose** will do this exchange on all the elements:

b1 = Transpose[a1]

(* Then list of lists can be transposed *)

{{{0, 0}, {1, 0}, {2, 0}}, {{0, 1}, {1, 1}, {2, 1}}, {{0, 2}, {1, 2}, {2, 2}}}

Again we can put this into MatrixForm, where the exchange is a bit easier to notice:

b2 = MatrixForm[b1]

(* And the transpose may be put into matrix form *)

$$\begin{pmatrix} \begin{pmatrix} 0 \\ 0 \end{pmatrix} & \begin{pmatrix} 1 \\ 0 \end{pmatrix} & \begin{pmatrix} 2 \\ 0 \end{pmatrix} \\ \begin{pmatrix} 0 \\ 1 \end{pmatrix} & \begin{pmatrix} 1 \\ 1 \end{pmatrix} & \begin{pmatrix} 2 \\ 1 \end{pmatrix} \\ \begin{pmatrix} 0 \\ 2 \end{pmatrix} & \begin{pmatrix} 1 \\ 2 \end{pmatrix} & \begin{pmatrix} 2 \\ 2 \end{pmatrix} \end{pmatrix}$$

Now, let's consider the nesting of a Table inside of another table function. Because we have two variables this is easy to do in essence we will be doing the following two steps in one step:

Table[{*x*, *y*}, {*x*, 0, 2}]

(* Creates pairs with *x* specified and *y* left unspecified *)

{{0, *y*}, {1, *y*}, {2, *y*}}

Table[%, {*y*, 0, 2}]

(* Takes the last list, % and then specifies *y* for each *x* *)

{{{0, 0}, {1, 0}, {2, 0}}, {{0, 1}, {1, 1}, {2, 1}}, {{0, 2}, {1, 2}, {2, 2}}}

Instead of two steps, this is done in one step as follows:

Table[Table[{*x*, *y*}, {*x*, 0, 2}], {*y*, 0, 2}]

(* The computation with nested Tables *)

{{{0, 0}, {1, 0}, {2, 0}}, {{0, 1}, {1, 1}, {2, 1}}, {{0, 2}, {1, 2}, {2, 2}}}

We put this into MatrixForm to see it more clearly:

a2 = MatrixForm[%]

(* Again convert to matrix form *)

$$\begin{pmatrix} \begin{pmatrix}0\\0\end{pmatrix} & \begin{pmatrix}1\\0\end{pmatrix} & \begin{pmatrix}2\\0\end{pmatrix} \\ \begin{pmatrix}0\\1\end{pmatrix} & \begin{pmatrix}1\\1\end{pmatrix} & \begin{pmatrix}2\\1\end{pmatrix} \\ \begin{pmatrix}0\\2\end{pmatrix} & \begin{pmatrix}1\\2\end{pmatrix} & \begin{pmatrix}2\\2\end{pmatrix} \end{pmatrix}$$

What is interesting about this, is that the nested Tables gave us the transpose of the matrix that results from the single Table function with two iterators:

Table [{ x, y } , { x, 0, 2 } , { y, 0, 2}]

is the transpose of

Table[Table[{ x, y }, { x, 0, 2 }], { y, 0, 2}]

We created b2 as the MatrixForm of the transpose of a1, and we created a2 from the nested Tables also in MatrixForm. *Mathematica* will show us that they are in fact identical:

b2 == a2 (* Test if the two matrices are equal, this shows that the nested tables yield the transpose of the two dimensional table. *)

True

The script below repeats this process with a slightly more complex case - wherein the function x y is associated with an x and a y value. Again, we see that the nested tables give the transpose of the

Table[{*x*, *y*, *xy*}, {*x*, 0, 2}, {*y*, 0, 2}]

{{{0, 0, 0}, {0, 1, 0}, {0, 2, 0}}, {{1, 0, 0}, {1, 1, 1}, {1, 2, 2}}, {{2, 0, 0}, {2, 1, 2}, {2, 2, 4}}}

MatrixForm[%]

$$\begin{pmatrix} \begin{pmatrix} 0 \\ 0 \\ 0 \end{pmatrix} & \begin{pmatrix} 0 \\ 1 \\ 0 \end{pmatrix} & \begin{pmatrix} 0 \\ 2 \\ 0 \end{pmatrix} \\ \begin{pmatrix} 1 \\ 0 \\ 0 \end{pmatrix} & \begin{pmatrix} 1 \\ 1 \\ 1 \end{pmatrix} & \begin{pmatrix} 1 \\ 2 \\ 2 \end{pmatrix} \\ \begin{pmatrix} 2 \\ 0 \\ 0 \end{pmatrix} & \begin{pmatrix} 2 \\ 1 \\ 2 \end{pmatrix} & \begin{pmatrix} 2 \\ 2 \\ 4 \end{pmatrix} \end{pmatrix}$$

b = Transpose[%]

{{{0, 0, 0}, {1, 0, 0}, {2, 0, 0}}, {{0, 1, 0}, {1, 1, 1}, {2, 1, 2}}, {{0, 2, 0}, {1, 2, 2}, {2, 2, 4}}}

MatrixForm[%]

$$\begin{pmatrix} \begin{pmatrix} 0 \\ 0 \\ 0 \end{pmatrix} & \begin{pmatrix} 1 \\ 0 \\ 0 \end{pmatrix} & \begin{pmatrix} 2 \\ 0 \\ 0 \end{pmatrix} \\ \begin{pmatrix} 0 \\ 1 \\ 0 \end{pmatrix} & \begin{pmatrix} 1 \\ 1 \\ 1 \end{pmatrix} & \begin{pmatrix} 2 \\ 1 \\ 2 \end{pmatrix} \\ \begin{pmatrix} 0 \\ 2 \\ 0 \end{pmatrix} & \begin{pmatrix} 1 \\ 2 \\ 2 \end{pmatrix} & \begin{pmatrix} 2 \\ 2 \\ 4 \end{pmatrix} \end{pmatrix}$$

Table[Table[{*x*, *y*, *xy*}, {*x*, 0, 2}], {*y*, 0, 2}]

{{{0, 0, 0}, {1, 0, 0}, {2, 0, 0}}, {{0, 1, 0}, {1, 1, 1}, {2, 1, 2}}, {{0, 2, 0}, {1, 2, 2}, {2, 2, 4}}}

MatrixForm[%]

$$\begin{pmatrix} \begin{pmatrix}0\\0\\0\end{pmatrix} & \begin{pmatrix}1\\0\\0\end{pmatrix} & \begin{pmatrix}2\\0\\0\end{pmatrix} \\ \begin{pmatrix}0\\1\\0\end{pmatrix} & \begin{pmatrix}1\\1\\1\end{pmatrix} & \begin{pmatrix}2\\1\\2\end{pmatrix} \\ \begin{pmatrix}0\\2\\0\end{pmatrix} & \begin{pmatrix}1\\2\\2\end{pmatrix} & \begin{pmatrix}2\\2\\4\end{pmatrix} \end{pmatrix}$$

b == % (* Test of validity *)

True

The power and versatility of the Table function underlies the ease of use and applicability of *Mathematica* for many different kinds of calculations. We have only touched the surface of what lies beneath its utility.

Do

The **Table** function is one of the major differences that *Mathematica* provides versus a purely procedural language like Fortran, or C, wherein this kind of computation is done with a **Do** function or the Do loop. Let's linger here to see how **Do** works versus the **Table** function so that we understand their similarities and differences.

The **Do** function in its simplest form has this syntax, **Do**[expression, n], which evaluates the expression n times. When we do this here is what happens:

Do[*x*, 3]

There is no evidence that anything has happened because we did not specify an output. To observe what the Do function has done, we typically use the Print command, so that we can observe an output. In this case we specify and index the function g [i]:

Do[Print[*g*[*i*]], {*i*, 1, 4}]

g[1]

g[2]

g[3]

g[4]

Next, we can let g be specified and in this case we make it x^2:

$g = x^2$

Do[Print[$g[i]$], {i, 1, 4}]

g

x^2

$x^2[1]$

$x^2[2]$

$x^2[3]$

$x^2[4]$

x^2

Notice that because we specified g with the set equals sign, the output was x^2 of the index and not the square of the index. To obtain the square of the index, we write that explicitly into the Print statement:

g=. (* This clears g *)

Do[Print[i^2], {i, 1, 4}]

1

4

9

16

This is explicit, but it is not as powerful as having used the **Table** function to generate the squares. Let's see why:

b = Table[i^2, {i, 1, 4}]

{1, 4, 9, 16}

We see that the output from the **Table** function is not just the numbers, but it is an ordered list, enclosed in brackets. Furthermore, because we set it equal to b, we can immediately use that list for other purposes as it has been assigned to b. To do this, all we have to do is to call for b.

b

{1, 4, 9, 16}

We can accomplish this with the **Do** function also, but it takes more to do it. The command structure to be used is this one

b = { } ; Do [AppendTo[b, i^2], { i, 1, 4 }] ; b

This says define b as a set { } that has no elements, initially, then **AppendTo** that set the value of i^2, and then repeat this four times with 1, 2, 3, and 4, and finally show b. We put this into action:

***b*=. (* Clear b *)**

$b = \{\}$; Do[AppendTo[b, i^2], {i, 1, 4}]; b

{1, 4, 9, 16}

In many cases we want pairs of the number that was the index and the value of the function, dyads. Here we do this with the Table function and with the Do function:

Table[{i, i^2}, {i, 1, 4}]

(* Table of dyadic pairs *)

{{1, 1}, {2, 4}, {3, 9}, {4, 16}}

***b*=. (* Clear b *)**

***t*=. (* Clear t *)**

$t = \{\}$;

Do[AppendTo[t, {i, i^2}], {i, 1, 4}];

***t* (* Do version for dyadic pairs *)**

{{1, 1}, {2, 4}, {3, 9}, {4, 16}}

Without seeing the code that underlies the Table function, we can see that because it provides an ordered list output that it embodies the script that we just created for the Do function to also create such a list. Both the Do and the Table are very powerful. The Do function is more versatile and less specific, but it requires writing more code to make it useful. We need to know how to use both of them.

Plot, Plot3D, and ContourPlot

We turn now from the **Table** and **Do** commands to **Plot**. These have nearly identical syntax. One types the command name **Plot**, then the open square bracket **[**, then the function to be evaluated, in this case x^2, a comma separator and then the dependent variable and its range inside of curly brackets **{ ... }**. Let's return to the example of x^2 and x^x to see how this works:

Plot$\left[x^2, \{x, 0, 10\}\right]$

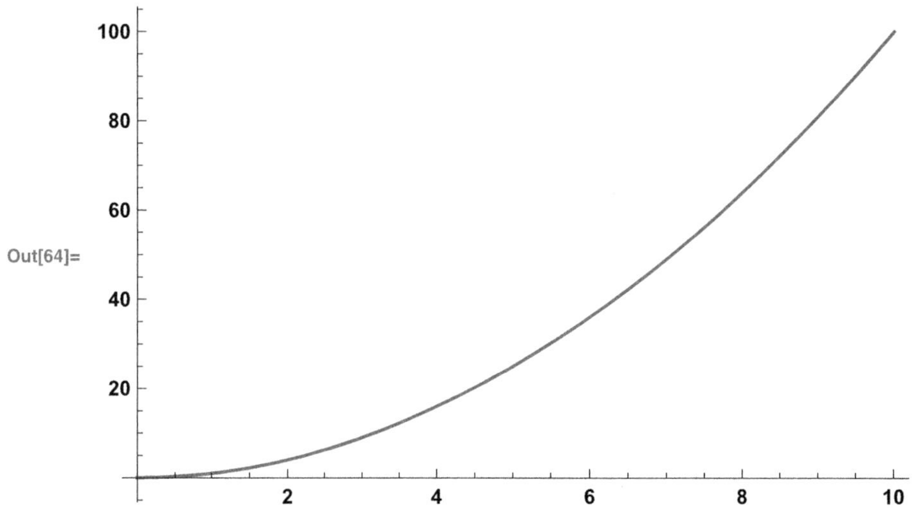

Plot$[x^x, \{x, 1, 10\}]$

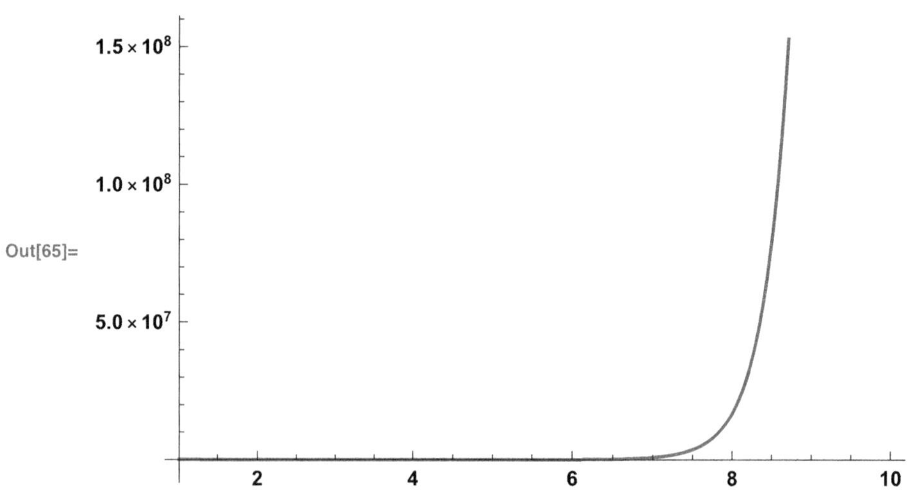

We can spruce these plots up with axes labels and other attributes, but to do so at this point would lead us off the track we are on. Notice that in both cases, we did not specify an increment value. In fact with plot we cannot. The reason is that *Mathematica* adjusts the increment as it moves through the function, making it smaller when the slope is large and larger when it is small. Hence, we do not have to set the increment; it is handled internally by the routine. We can be sure that in the vicinity of 7.5, the increment begins to become very small for x^x.

We can also plot in three dimensions. For example if we have a function of two variables, then it is simple matter to see how it looks in space. As an example, we can utilize the product of functions of **x** and **y** to see how they will appear in this space. The syntax is similar to the **Plot** command in that we type Plot3D, then the open square bracket, the function to be visualized and then two sets of ranges, one for each variable, separated by a comma:

Plot3D$\left[x^2y^2, \{x, -10, 10\}, \{y, -10, 10\}\right]$

(* Surface of x^2y^2 ***)**

Out[66]=

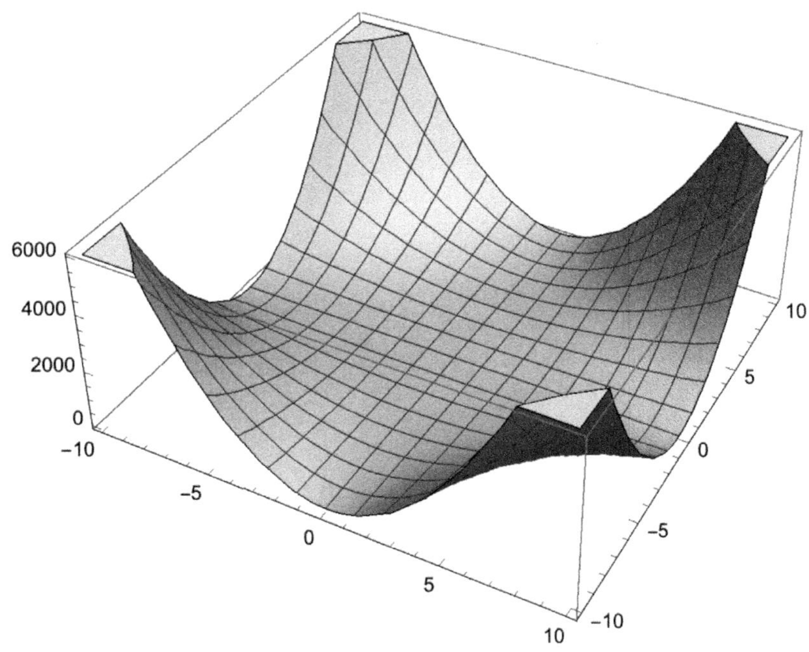

Plot3D[x^3y^2, {x, −10, 10}, {y, −10, 10}]

(* Surface of x^3y^2 *)

Out[67]=

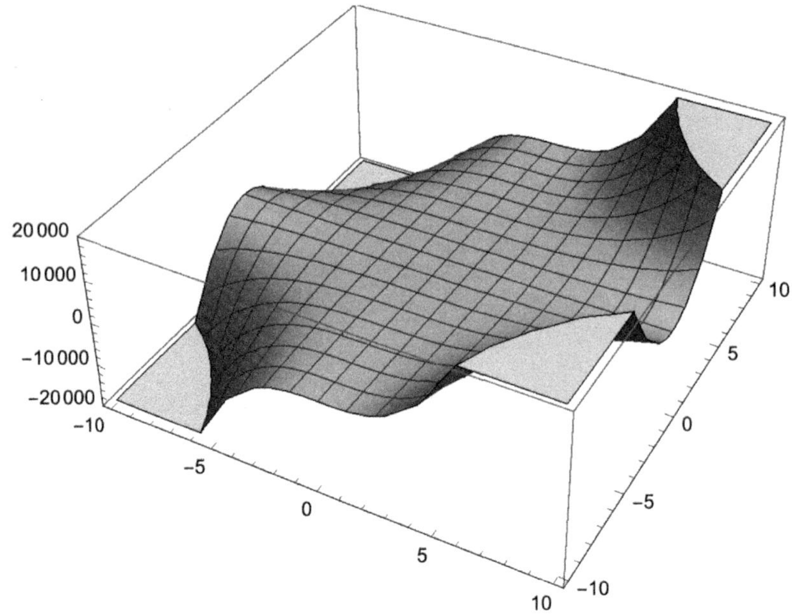

Plot3D[x^3y^3, {x, −10, 10}, {y, −10, 10}]

(* Surface of x^3y^3 *)

Out[68]=

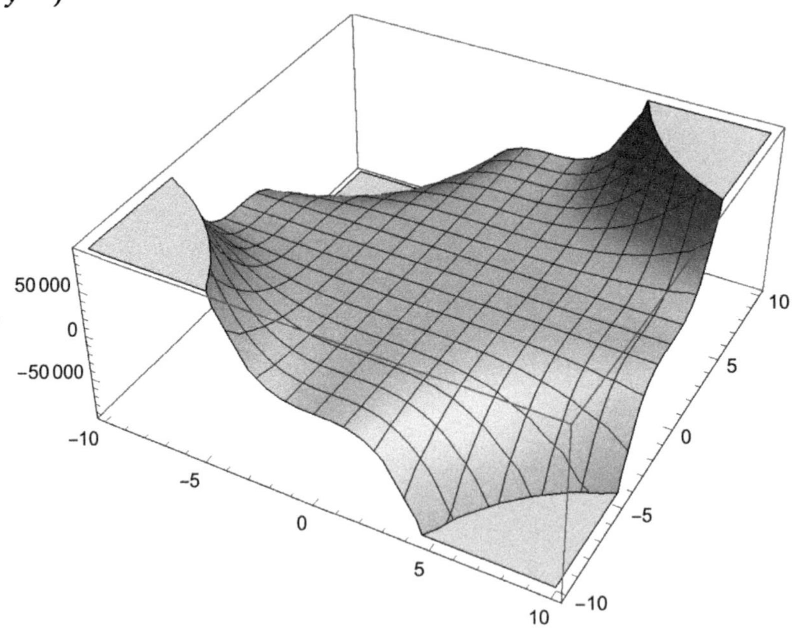

Plot3D[*x*Sin[*x*]*y*Cos[*y*], {*x*, −10, 10}, {*y*, −10, 10}]

(* Surface of *x* Sin[*x*] ∗ *y* Cos[*y*] *)

Out[69]=

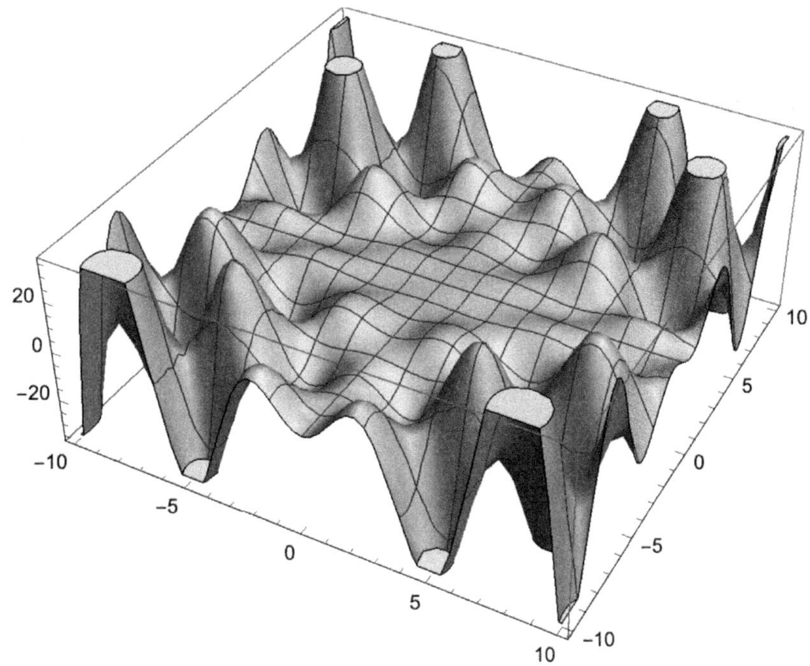

What we see is that *Mathematica* has plotted the functions, fitted them with surfaces, placed a grid on the fitted surfaces and enhanced them with coloration. All of this was done by the routine operating in its default mode, i.e. with a minimum of input from us. Here too we could spend time enhancing these graphics, but instead we shall postpone that and move on to the next subject.

The last plotting method that is of direct interest to us is the ContourPlot3D command. In this case the syntax is:

ContourPlot3D[function[x, y, z], {x, xmin, xmax}, {y, ymin, ymax}, {z, zmin, zmax}]

In differential geometry, solutions to certain partial differential equations are derived that look like functions of three variables the rendering of which leads to minimal surfaces. The structure and curvature of such surfaces are immensely interesting and beautiful. Several such surfaces were derived by Hermann Schwartz. These included the P ("primitive"), the D ("diamond"), the H ("hexagonal"), and the CIP (crossed layers of parallels.) We will look at the P and D minimal surfaces. The functional forms are given as follows:

P: Cos[x] + Cos[y] + Cos[z] = 0

D: Sin[x] Sin[y] Sin[z] + Sin [x] Cos[y] Cos [z] +
Cos[x] Sin[y]Cos[z] + Cos[x]Cos[y]Sin[z] = 0

They are equal to zero because the mean curvature of each surface is minimized. Let's begin with the P surface. We specify the function for P and then we seek to generate the surface from -2 Pi to 2 Pi. We also will want to know the timing for doing these renderings:

ContourPlot3D[Cos[*x*] + Cos[*y*] + Cos[*z*],

{*x*, −2Pi, 2Pi}, {*y*, −2Pi, 2Pi}, {*z*, −2Pi, 2Pi}]

Out[70]=

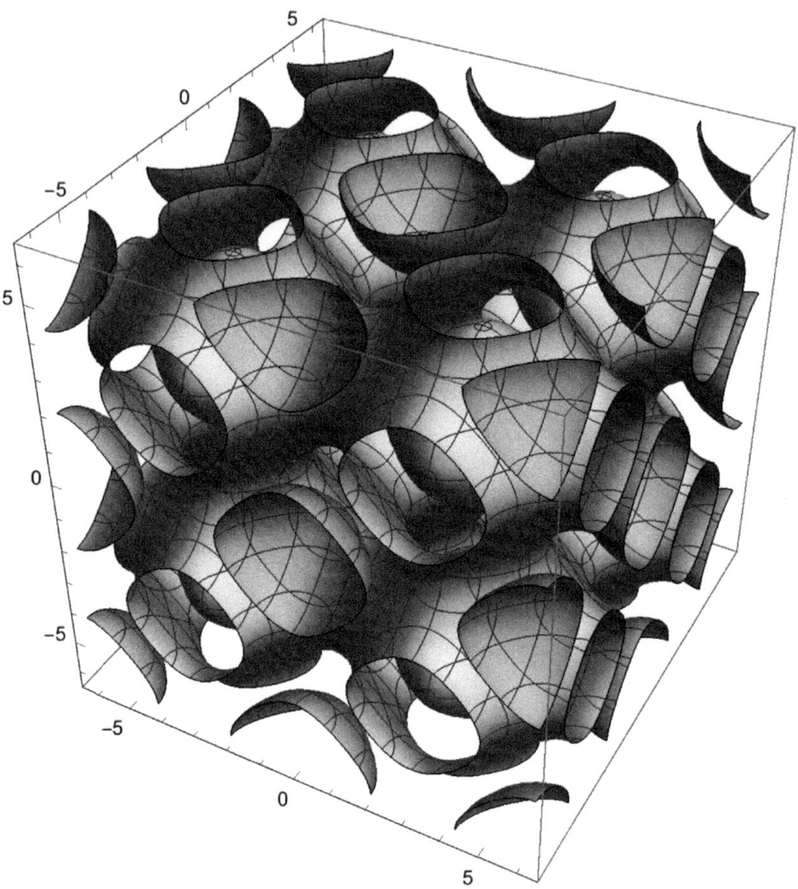

That gives us good sense of the surface, however, because it is well known, it is obvious that we need more points to render it more accurately. Also, we can suppress the axes and the box around the surface. Likewise we do not need the mesh on the surface, we just want to render the surface in and of itself:

```
ContourPlot3D[
Cos[x] + Cos[y] + Cos[z] == 0,
{x, −2Pi, 2Pi}, {y, −2Pi, 2Pi}, {z, 2Pi, −2Pi},
PlotPoints → 20,
Axes → False,
Boxed → False,
Mesh → None,
BoundaryStyle → None]
```

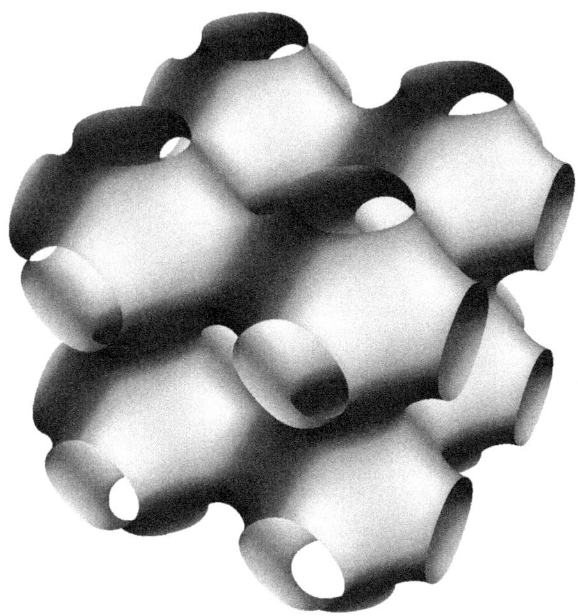

Now this looks like the renderings that are shown for this surface elsewhere. It is a beautiful object. On a computer you will notice that if you hover over the rendering with the cursor and then depress the mouse button, you can rotate it in space to see different viewpoints. As a matter of fact one can set the **ViewPoint** for the rendering. But we do not need to deal with that here. Instead let's use the very same script for the D minimal surface:

```
ContourPlot3D[
Sin[x]Sin[y]Sin[z] + Sin[x]Cos[y]Cos[z]+
Cos[x]Sin[y]Cos[z] + Cos[x]Cos[y]Sin[z]==0,
{x, −2Pi, 2Pi}, {y, −2Pi, 2Pi}, {z, 2Pi, −2Pi},
PlotPoints → 10,
Axes → False,
Boxed → False,
Mesh → None,
BoundaryStyle → None]
```

This is an even more complex surface and it appears to be made up of three dimensional saddles. It too is an object of beauty. So, as you can see, ContourPlot3D is a very powerful rendering tool.

Module

The Module function in its most general and simple form, contains a list of relevant variables { x, y = yo, z... } and some expression, or set of expressions, within it. The key important fact about Module is that variables and expressions are named and reside only within it. This is very important for scripts that act as stand alone programs that can be called to do operations on data. As we move from using *Mathematica* as an interactive assistant toward using it as programming language, the use of Module becomes central to that process.

We will use a simple example first to explore the use of Module in calculating, plotting and graphing. Let's say we are interested in exploring the function f [x, y] = Sin [a x] Cos [b y]. We have two functions Sin and Cos making up one function with two parameters and two variables { a, b, x , y }. We can say that the behavior of f [x, y] is parametric in a and b. We can specify the two parameters and then calculate values for f [a, b, x, y] and the repeat with a new pair of parameter values to learn how they affect the functional output, which is to say we can experiment with the function's behavior. One way to understand functional behavior is by visualization. To visualize f [a, b, x, y] it is logical to examine it over a couple of cycles in the x and y directions, say from - 2 π to +2 π, and then we can look at the surface that this generates for any set of { a, b }. We know intuitively that when the parameters are equal we will observe a symmetric "hill and valley" surface.

To start, let's plot this function interactively to be sure we know what we are seeking to do. To visualize f [x , y] we will use Plot3D in this way:

a = 1; b = 1; Plot3D[Sin[a x] Cos[b y], { x, -2 Pi, 2 Pi }, { y, -2 Pi, 2 Pi }]

```
a = 1; (* Set parameter *)

b = 1; (* Set parameter *)

f[ a_, b_,x_, y_]:= Sin[a x] Cos[b y]
(* Define function with parameters {a, b} and variables {x, y} *)

Plot3D[f[a, b, x, y], {x, −2Pi, 2Pi}, {y, −2Pi, 2Pi}]
(* Plot function *)
```

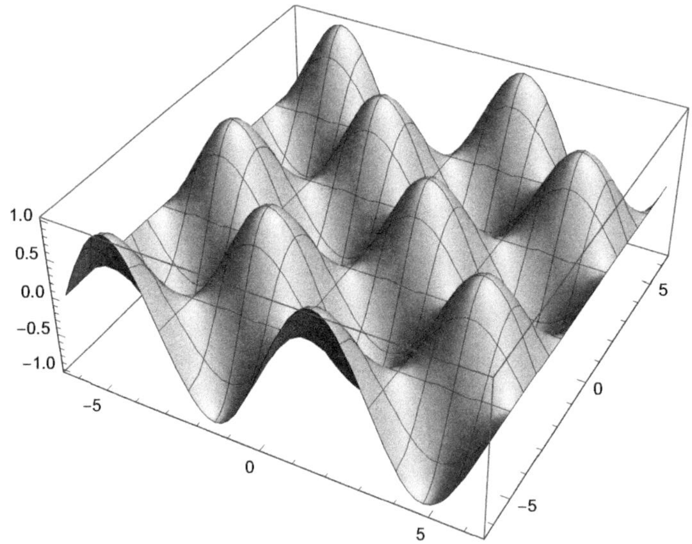

If we increase the parameter values of a and b, we see that the oscillations increase as expected and the function takes on a spikey structure:

```
a = 10; (* Set parameter *)

b = 10; (* Set parameter *)

f[a_, b_, x_, y_]:= Sin[a x] Cos[b y]
(* Define function with parameters {a, b} and variables {x, y} *)

Plot3D[f[a, b, x, y], {x, −2Pi, 2Pi}, {y, −2Pi, 2Pi}]
(* Plot function *)
```

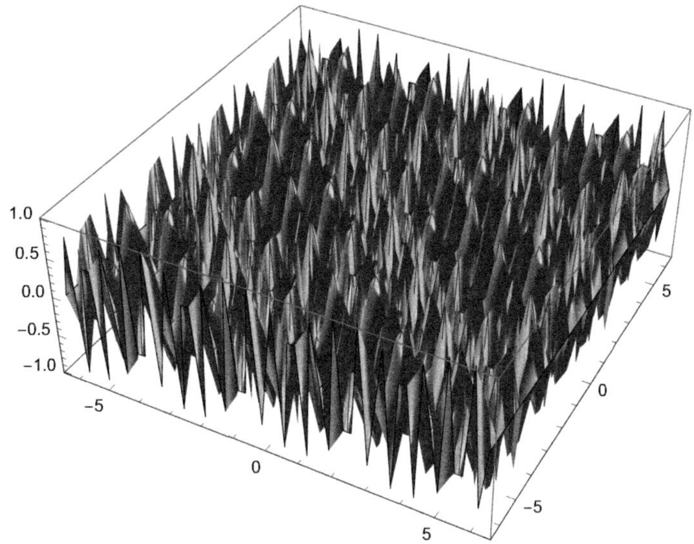

Now, while we may not mind doing this copy and paste approach interactively, sometimes, there are other times that we don't want to have to copy and paste more than two or three times. This is where the Module approach to writing a script comes in. We will write a new function using Module and have that function available to call and use as if it were an internal (preprogrammed) *Mathematica* function.

We will specify a function named **sincos**. To do this we will write the Module using **a**, **b**, **x** and **y** as inputs, and we will declare only one internal variable **f**, by which we mean **f** will be defined functionally and then evaluated only within the Module. What this means is that after we use the Module for **sincos**, if we then evaluate **f**, **f** will not have any residual value or definition. Inside of **sincos** on the left-hand side, we will use the set-delayed approach to the variables, e.g. **a_**, and we will wrap the two parameters and two variables in curly brackets **{ }** to indicate they are a set, and all of which must have numerical values for **sincos** to have a numerical value. Also included is **//N**, which is necessary to derive numerical output. After the new function is defined and scripted, it is evaluated and then to prove the point, the value of f is tested:

```
sincos[{a_, b_, x_, y_}]:=Module[{f}, f = Sin[ax] * Cos[by]//N](* Module function *)

sincos[{1, 1, 1, 1}](* Parameters and variables, specified *)

f(* This shows that f was defined only inside the Module *)
```

0.454649

f

For a = 1, b = 1, x = 1 and y = 1, **sincos** evaluates to 0.45469 and because **f** was inside **Module**, it evaluates only to itself and not to **Sin [a x] Cos [b y]** and not to the value 0. 45469.

This gives us one point for the set of two parameters and two variables. We can also **Map sincos** over a list of values as we can with any internal *Mathematica* function, because it is listable

sincos/@{{1, 1, 1, 1}, {1, 1, 2, 2}, {3, 2, 4, 5}, {10, 10, 10, 10}}

(* Map sincos over a list *)

{0.454649, −0.378401, 0.450223, −0.436649}

We can now use **sincos** as any other internal function and specify two values of a and b and set ranges for the variables and then plot it over the range of x and y values:

a **= 1; (* Set parameter *)**

b **= 1; (* Set parameter *)**

Plot3D[sincos[{*a*, *b*, *x*, *y*}], {*x*, −2 Pi, 2Pi}, {*y*, −2Pi, 2Pi}]

(* Plot function *)

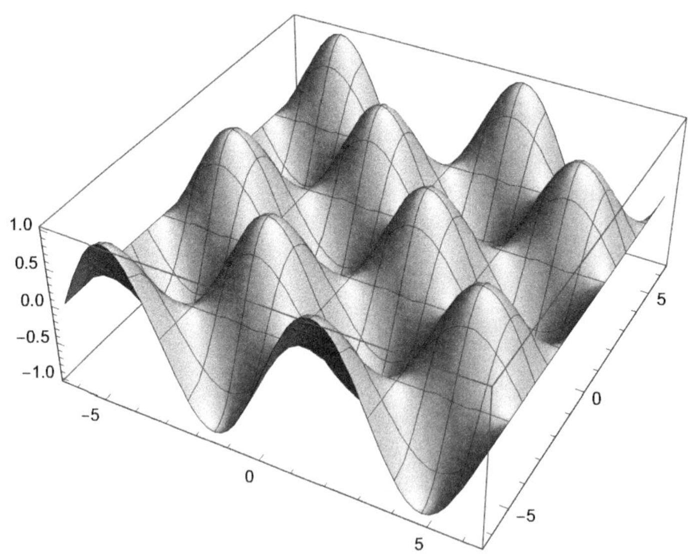

We can copy and paste this and change the a and b parameters to see what happens:

$a = 10$; (* Set parameter *)

$b = 10$; (* Set parameter *)

Plot3D[sincos[{a, b, x, y}], {x, −2Pi, 2Pi}, {y, −2Pi, 2Pi}]

(* Plot function *)

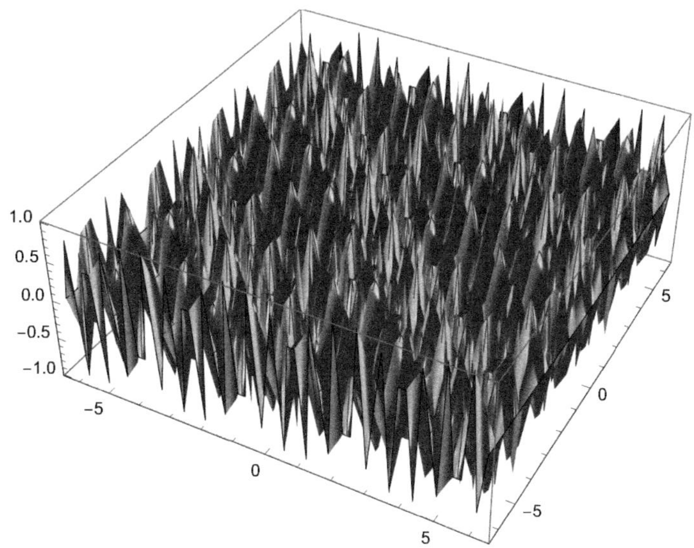

By increasing the values we get the same results that we have seen before, which reassures us that **sincos** functions as it should.

Just as we can use **sincos** to compute one set of points, we can use it inside the Table function to create a list of points. To keep this simple we can specify a and b up front, so they will not change. Then we can let x and vary from 0 to 2. To see what will happen inside the **Table** function, we compute three points:

sincos[{10, 10, 2, 0}]

sincos[{10, 10, 2, 1}]

sincos[{10, 10, 2, 2}]

0.912945

−0.766026

0.372557

Now we compute **sincos** within the **Table** function with a and b once again fixed, but letting **x** and **y** vary from 0 to 2. The default increment is 1. Hence, this will produce three values for each of the three **x** values giving us nine in total. These will be groups into three groups of three values inside one list of the form { { a1, a2, a3 }, { b1, b2, b3 }, { c1 , c2, c3 } }. Here is that result:

```
a = 10; (* Set parameter *)

b = 10; (* Set parameter *)

Table[sincos[{a, b, x, y}], {x, 0, 2}, {y, 0, 2}]
(* Table of values of sincos *)
```

{{0., 0., 0.}, {−0.544021, 0.456473, −0.222005}, {0.912945, −0.766026, 0.372557}}

The first output group within the overall list is { 0, 0, 0 }, which is the result when x is 0 and y is 0, 1, and 2. Then, when x is 1, and y is 0, 1, 2 the result is { -0.544021, 0.456473, -0.222005 } and finally with x equals 2, the result is as we computed before, namely { 0.912945, -0.766026, 0.372557 }. We could do the calculation over the same ranges by nesting one Table function varying over x within one varying over y.

```
a = 10; (* Set parameter *)

b = 10; (* Set parameter *)

Table[
Table[sincos[{a, b, x, y}], {x, 0, 2}],
 {y, 0, 2}]
(* Table embedded in a Table *)
```

{{0., −0.544021, 0.912945}, {0., 0.456473, −0.766026}, {0., −0.222005, 0.372557}}

Once again the structure of the output is the same; it is a primary list of three elements, each of which has a secondary list of three elements: { { { { a1, a2, a3 }, { b1, b2, b3 }, { c1 , c2, c3 } }, but we notice that the elements are not the same. Why? In this nested mode first the y value is fixed and the x values are varied. With a little more effort we find that Table [Table [{ x, y }, { x, a, b }] { y, a, b }]] gives the Transpose of Table [{ x, y }, { x, a, b }, { y, a, b }].

With built-in functions such as Sin and Cos, we can compute a number of points and then plot them with **ListPlot**. When we have two variables in a function, a list of values can be computed and these can be plotted with **ListPlot3D**. When we do this, each point for ListPlot3D must have 3 values: { x , y , z } where z can be the value of the function, f [x, y] = z at x and y. So, we can compute the value of **sincos** at different x and y values, but we should specify those values along with the function value. In this case this will look like this:

Table [{x, y, sincos [{ a, b, x, y }] }, { x, -2 Pi, 2 Pi }, { y, - 2 Pi, 2 Pi }]

where the semi-colon at the end suppresses the output. We also notice that the output of the table function has three sets of curly brackets viz. { { { a1, b1, c1 }, { d1, e1, f1 }, ... { x1, y1, z1 } }, { a2, b2, b2 }, { d2, e2, f2 }, ... { x2, y2, z2 } } ... {a13, b13, c13 }, { d13, e13, f13 }, ... { x13, y 13, z 13 } } } }. So the output is a list with a thirteen elements making up a sublist within which there are thirteen more elements, within which are three elements, for a total of 13 * 13 * 3 or 507 total elements. We can see this by pulling out **z [[1]]**, **z [[1, 1]]**, and **z [[1, 1, 1]]**, and then measuring the **Length** of each of these elements. We will do each step separately to explain each outcome:

First, we generate the table of values and let it be called z and we do not suppress the output:

```
a = 1; (* Set parameter *)

b = 1; (* Set parameter *)

z = Table[{x//N,y//N,sincos[{a,b,x,y}]}, { x,-2 Pi,2 Pi}, {y, - 2 Pi,2 Pi}]
(* This Table provides the values of x and y at each new value of sincos *)

{{{−6.28319, −6.28319, 0.}, {−6.28319, −5.28319, 0.}, {−6.28319, −4.28319, 0.},
{−6.28319, −3.28319, 0.}, {−6.28319, −2.28319, 0.}, {−6.28319, −1.28319, 0.},
{−6.28319, −0.283185, 0.}, {−6.28319, 0.716815, 0.}, {−6.28319, 1.71681, 0.},
{−6.28319, 2.71681, 0.}, {−6.28319, 3.71681, 0.}, {−6.28319, 4.71681, 0.},
{−6.28319, 5.71681, 0.}}, {{−5.28319, −6.28319, 0.841471},
```

{−5.28319, −5.28319, 0.454649}, {−5.28319, −4.28319, −0.350175},
{−5.28319, −3.28319, −0.83305}, {−5.28319, −2.28319, −0.550022},
{−5.28319, −1.28319, 0.238693}, {−5.28319, −0.283185, 0.807955},
{−5.28319, 0.716815, 0.634387}, {−5.28319, 1.71681, −0.122434},
{−5.28319, 2.71681, −0.76669}, {−5.28319, 3.71681, −0.706054},
{−5.28319, 4.71681, 0.0037241}, {−5.28319, 5.71681, 0.710079}},
{{−4.28319, −6.28319, 0.909297}, {−4.28319, −5.28319, 0.491295},
{−4.28319, −4.28319, −0.378401}, {−4.28319, −3.28319, −0.900198},
{−4.28319, −2.28319, −0.594356}, {−4.28319, −1.28319, 0.257933},
{−4.28319, −0.283185, 0.87308}, {−4.28319, 0.716815, 0.685521},
{−4.28319, 1.71681, −0.132303}, {−4.28319, 2.71681, −0.828488},
{−4.28319, 3.71681, −0.762966}, {−4.28319, 4.71681, 0.00402428},
{−4.28319, 5.71681, 0.767314}}, {{−3.28319, −6.28319, 0.14112},
{−3.28319, −5.28319, 0.0762475}, {−3.28319, −4.28319, −0.0587266},
{−3.28319, −3.28319, −0.139708}, {−3.28319, −2.28319, −0.0922422},
{−3.28319, −1.28319, 0.0400304}, {−3.28319, −0.283185, 0.135499},
{−3.28319, 0.716815, 0.106391}, {−3.28319, 1.71681, −0.020533},
{−3.28319, 2.71681, −0.128579}, {−3.28319, 3.71681, −0.11841},
{−3.28319, 4.71681, 0.000624555}, {−3.28319, 5.71681, 0.119085}},
{{−2.28319, −6.28319, −0.756802}, {−2.28319, −5.28319, −0.408902},
{−2.28319, −4.28319, 0.314941}, {−2.28319, −3.28319, 0.749229},
{−2.28319, −2.28319, 0.494679}, {−2.28319, −1.28319, −0.214676},
{−2.28319, −0.283185, −0.726659}, {−2.28319, 0.716815, −0.570555},
{−2.28319, 1.71681, 0.110115}, {−2.28319, 2.71681, 0.689546},
{−2.28319, 3.71681, 0.635011}, {−2.28319, 4.71681, −0.00334938},
{−2.28319, 5.71681, −0.638631}}, {{−1.28319, −6.28319, −0.958924},
{−1.28319, −5.28319, −0.518109}, {−1.28319, −4.28319, 0.399053},
{−1.28319, −3.28319, 0.949328}, {−1.28319, −2.28319, 0.626795},
{−1.28319, −1.28319, −0.272011}, {−1.28319, −0.283185, −0.920731},
{−1.28319, 0.716815, −0.722935}, {−1.28319, 1.71681, 0.139524},
{−1.28319, 2.71681, 0.873705}, {−1.28319, 3.71681, 0.804606},
{−1.28319, 4.71681, −0.00424391}, {−1.28319, 5.71681, −0.809192}},
{{−0.283185, −6.28319, −0.279415}, {−0.283185, −5.28319, −0.150969},
{−0.283185, −4.28319, 0.116278}, {−0.283185, −3.28319, 0.276619},
{−0.283185, −2.28319, 0.182638}, {−0.283185, −1.28319, −0.0792596},
{−0.283185, −0.283185, −0.268286}, {−0.283185, 0.716815, −0.210652},
{−0.283185, 1.71681, 0.040655}, {−0.283185, 2.71681, 0.254584},
{−0.283185, 3.71681, 0.23445}, {−0.283185, 4.71681, −0.00123661},
{−0.283185, 5.71681, −0.235786}}, {{0.716815, −6.28319, 0.656987},

{0.716815, −5.28319, 0.354971}, {0.716815, −4.28319, −0.273403},
{0.716815, −3.28319, −0.650412},
{0.716815, −2.28319, −0.429435},
{0.716815, −1.28319, 0.186362}, {0.716815, −0.283185, 0.630819},
{0.716815, 0.716815, 0.495304}, {0.716815, 1.71681, −0.0955916},
{0.716815, 2.71681, −0.5986}, {0.716815, 3.71681, −0.551259},
{0.716815, 4.71681, 0.00290762}, {0.716815, 5.71681, 0.554401}},
{{1.71681, −6.28319, 0.989358}, {1.71681, −5.28319, 0.534553},
{1.71681, −4.28319, −0.411718}, {1.71681, −3.28319, −0.979457},
{1.71681, −2.28319, −0.646688}, {1.71681, −1.28319, 0.280644},
{1.71681, −0.283185, 0.949952}, {1.71681, 0.716815, 0.745879},
{1.71681, 1.71681, −0.143952}, {1.71681, 2.71681, −0.901434},
{1.71681, 3.71681, −0.830142}, {1.71681, 4.71681, 0.0043786},
{1.71681, 5.71681, 0.834874}}, {{2.71681, −6.28319, 0.412118},
{2.71681, −5.28319, 0.222669}, {2.71681, −4.28319, −0.171502},
{2.71681, −3.28319, −0.407994}, {2.71681, −2.28319, −0.269379},
{2.71681, −1.28319, 0.116902}, {2.71681, −0.283185, 0.395704},
{2.71681, 0.716815, 0.310697}, {2.71681, 1.71681, −0.0599633},
{2.71681, 2.71681, −0.375494}, {2.71681, 3.71681, −0.345797},
{2.71681, 4.71681, 0.00182391}, {2.71681, 5.71681, 0.347768}},
{{3.71681, −6.28319, −0.544021}, {3.71681, −5.28319, −0.293936},
{3.71681, −4.28319, 0.226393}, {3.71681, −3.28319, 0.538577},
{3.71681, −2.28319, 0.355596}, {3.71681, −1.28319, −0.154318},
{3.71681, −0.283185, −0.522353}, {3.71681, 0.716815, −0.410139},
{3.71681, 1.71681, 0.0791551}, {3.71681, 2.71681, 0.495674},
{3.71681, 3.71681, 0.456473}, {3.71681, 4.71681, −0.00240767},
{3.71681, 5.71681, −0.459074}}, {{4.71681, −6.28319, −0.99999},
{4.71681, −5.28319, −0.540297}, {4.71681, −4.28319, 0.416143},
{4.71681, −3.28319, 0.989983}, {4.71681, −2.28319, 0.653637},
{4.71681, −1.28319, −0.283659}, {4.71681, −0.283185, −0.960161},
{4.71681, 0.716815, −0.753895}, {4.71681, 1.71681, 0.145499},
{4.71681, 2.71681, 0.911121}, {4.71681, 3.71681, 0.839063},
{4.71681, 4.71681, −0.00442565}, {4.71681, 5.71681, −0.843846}},
{{5.71681, −6.28319, −0.536573}, {5.71681, −5.28319, −0.289912},
{5.71681, −4.28319, 0.223293}, {5.71681, −3.28319, 0.531203},
{5.71681, −2.28319, 0.350727}, {5.71681, −1.28319, −0.152205},
{5.71681, −0.283185, −0.515201}, {5.71681, 0.716815, −0.404524},
{5.71681, 1.71681, 0.0780714}, {5.71681, 2.71681, 0.488888},

{5.71681, 3.71681, 0.450223}, {5.71681, 4.71681, −0.00237471},
{5.71681, 5.71681, −0.452789}}}

Normally, we would add a semicolon to end of the specification for z to suppress the output, but for learning, we need to look at the whole list. If you look carefully, and if you click on the first curly bracket of each pair (it will be highlighted in green) **{** {, then the closing curly bracket for that element will also be highlighted } **}**. If you go through the list in this way, you will find 13 groups of 13 elements. You could also say 13 sets with 13 sets within which are sets of three. But, counting by hand is not sensible in most cases, so we use the **Part** function **z [[l, m, n]]** to find each set of z at each level in the hierarchy using: **z [[1]]**, **z [[1, 1]]**, and **z [[1, 1, 1]]**.

z[[1]](* The first part of z *)

z[[1, 1]](* The first set in the first part of z *)

z[[1, 1, 1]](* The first element in the first set in the first part of z *)

{{−6.28319, −6.28319, 0.}, {−6.28319, −5.28319, 0.}, {−6.28319, −4.28319, 0.},
{−6.28319, −3.28319, 0.}, {−6.28319, −2.28319, 0.}, {−6.28319, −1.28319, 0.},
{−6.28319, −0.283185, 0.}, {−6.28319, 0.716815, 0.}, {−6.28319, 1.71681, 0.},
{−6.28319, 2.71681, 0.}, {−6.28319, 3.71681, 0.},
{−6.28319, 4.71681, 0.}, {−6.28319, 5.71681, 0.}}

{−6.28319, −6.28319, 0.}

−6.28319

Here we see that at the first level we have a set of 13 with three elements each, at the second level **z** is made up of sets of three and at the last level these sets are made up of single elements. *Mathematica* keeps everything very nicely organized for us. Next, we can query for the **Length** of each element in the hierarchy as follows by combining **Length** with **Part**:

Length[z](* number of overall elements in z *)

Length[z[[1]]](* the number of sets in the first part of z *)

Length[z[[1, 1]]](* the number of elements in the first set of the first part of *z* *)

13

13

3

As neat and tidy as the structure of **z** is, it is not in the form needed to do a **ListPlot3D** on its elements. The reason is that **z** has that extra set of external curly brackets **{** { { a1, b1, c1 }, ... { x13, y 13, z 13 } } **}** and we need to remove them for the ListPlot3 function. This is simple to do; we first remove all but the outside brackets from **z** using the function **Flatten** and then we repartition the result into groups of three using **Partition[% , 3]**. Again for learning how these functions work, the outputs will not be suppressed, but we can suppress most of the output using the **Short** function:

Flatten[*z*]//Short

{−6.28319, −6.28319, 0., ⟨⟨502⟩⟩, 5.71681, −0.452789}

Flatten[z] creates one list of 507 elements as one list. With **Short** appended to **Flatten[z]**, six of the elements are displayed, the first three and the last three, while 501 are hidden. Now we simply repartition this flattened list into sets of three with **Partition[% , 3]**, where % means the last result and the appended **Short** suppresses all but the first and last element from the output display (and only the display):

Partition[%, 3]//Short(* Repartitions z after flattening it *)

{{−6.28319, −6.28319, 0.}, ⟨⟨167⟩⟩, {5.71681, ⟨⟨18⟩⟩, −0.452789}}

Now the list is ready to do a ListPlot3D on this last result using **ListPlot3D[%]**:

ListPlot3D[%]

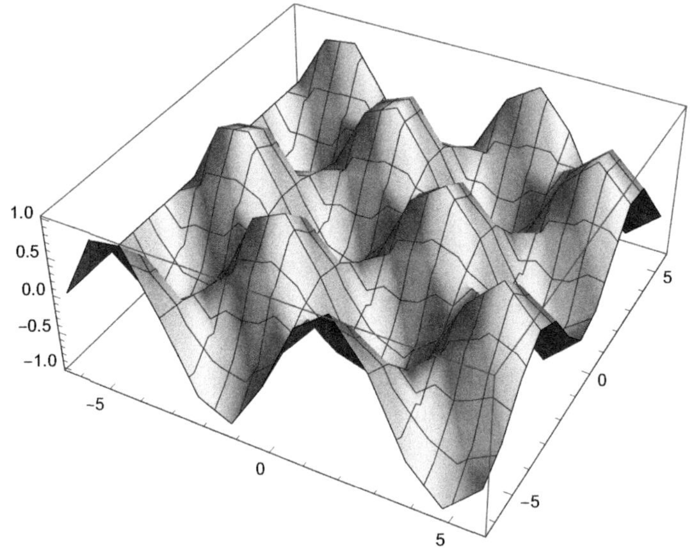

This looks somewhat similar to the result we produced with Plot3D, but it is far more jagged and irregular. This is because we have computed only 507 points and **ListPlot3D** has interpolated between them, whereas in the case of **Plot3D**, many more points are automatically generated making the graphical output look smoother.

We can see the effect that more points have on the features of the graphical output of **ListPlot3D** simply by computing more points. The simplest way to accomplish this is to specify the increments between points in the **x** and **y** directions. This is done by adding one more term to the ranges for these increments as in **{ x, -2 Pi, 2 Pi, 0.25 Pi }, { y, - 2 Pi, 2 Pi, 0.25 Pi }** add then we add the **Timing** function around **ListPlot3D** to learn how long it takes to do the computations and rendering. (Note that when appending **Timing** in this way the output graphic will be initially small. Pull the bottom right corner out with the cursor to expand it to a larger size for inspection):

```
a = 1; (* Set Parameter *)

b = 1; (* Set parameter *)

Table[{x//N, y//N, sincos[{a, b, x, y}]}, {x, −2Pi, 2Pi, 0.25Pi},
{y, −2Pi, 2Pi, 0.25Pi}]; (* Generate the table of 283 sincos values *)
```

Flatten[%]//Short(* Remove all curly brackets from the table that form subsets *)

Partition[%, 3]; (* Partition the last set into groups of three values, triads *)

ListPlot3D[%](* Plot the points *)

{−6.28319, −6.28319, ⟨⟨863⟩⟩, 6.28319, −2.4492935982947064`*^-16}

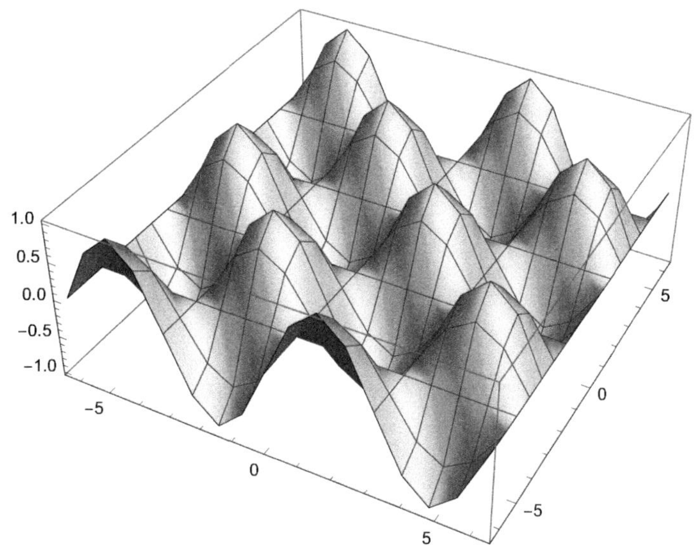

This version of the plot looks even better and smoother and that is because we have gone up from 169/3 ~ 56 **z** points to 865/3 ~ 283 This took about 0.01 seconds to compute (this will vary). What if we ramp this up even further, much further, using increments of 0.025 Pi? We know it will make the surface smoother but how much longer will it take?

a **= 1; (* Set parameter *)**

b **= 1; (* Set parameter *)**

Table[{*x*//*N*, *y*//*N*, sincos[{*a*, *b*, *x*, *y*}]}, {*x*, −2Pi, 2Pi, 0.025Pi},

{*y*, −2Pi, 2Pi, 0.025Pi}];

(* Generate the table of >25000 sincos values *)

Flatten[%]//Short(* Remove all curly brackets from the table that form subsets *)

Partition[%, 3]; (* Partition the last set into groups of three values, triads *)

Length[%]

ListPlot3D[%%]

(* Plot the points *)

{−6.28319, −6.28319, ⟨⟨77759⟩⟩, 6.28319, −2.4492935982947064`*^-16}

25921

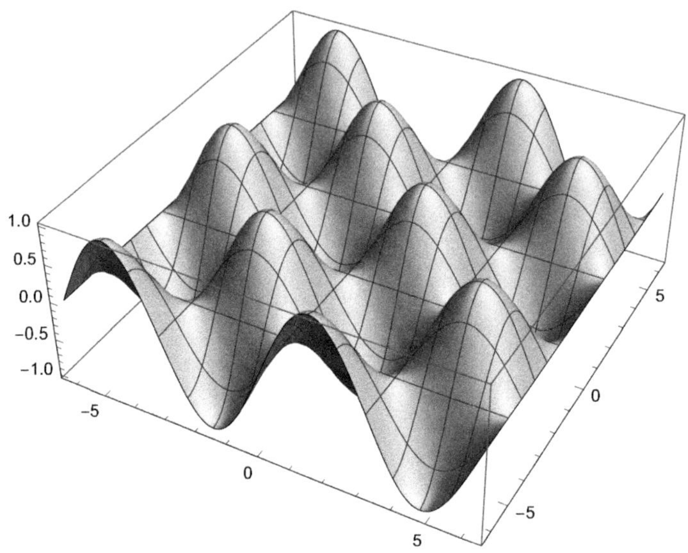

This looks very smooth and basically just the same as that which we obtained with Plot3D. Note that now have 77, 763/3 ~ 25,921 **z** points. It took 1.45 seconds of computing time to do this graphic which 1.2/0.01 ~ 120 times more compute time. When we repeat the computation using **Plot3D** it takes 0.1 seconds, indicating that **Plot3D** is an efficient algorithm that finds how many points are needed to render the graph "smooth".

$a = 1$; **(* Set parameter *)**

$b = 1$; **(* Set parameter *)**

```
Plot3D[sincos[{a, b, x, y}], {x, −2 Pi, 2Pi}, {y, −2Pi, 2Pi}]
(* Plot the function directly without first generating points *)
```

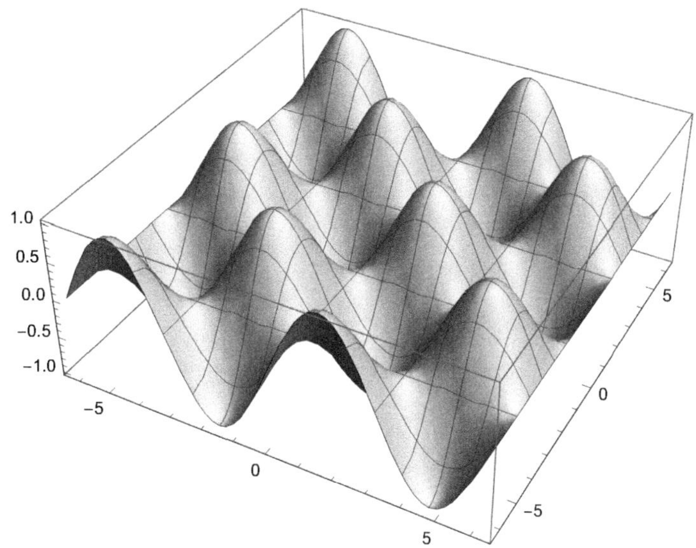

We can take this use of Module even further to visualize the effects of different parameters on the surface formed by Sin [x] Cos [y]. Now, we will write a Module script that will bundle Plot3D within it like so:

Module[{ f = Sin[a x]*Cos[b y]//N}, Plot3D[f, { x, -2 Pi, 2 Pi}, { y, -2 Pi, 2 Pi }].

Here is that implementation:

```
(* Using the Module function to create SinCos2 with the function calls and
other parts implemented locally within the command *)

sincos2[{a_, b_}]:=Module[
{f = Sin[a x]*Cos[b y]//N},
Plot3D[f, {x, −2 Pi, 2Pi}, {y, −2Pi, 2Pi},
PlotRange → {{x, −2 Pi, 2Pi}, {y, −2Pi, 2Pi}}]
        ]
```

Then we use it with a few different values of a and b interactively to be sure this works as intended:

sincos2[{1, 1}]

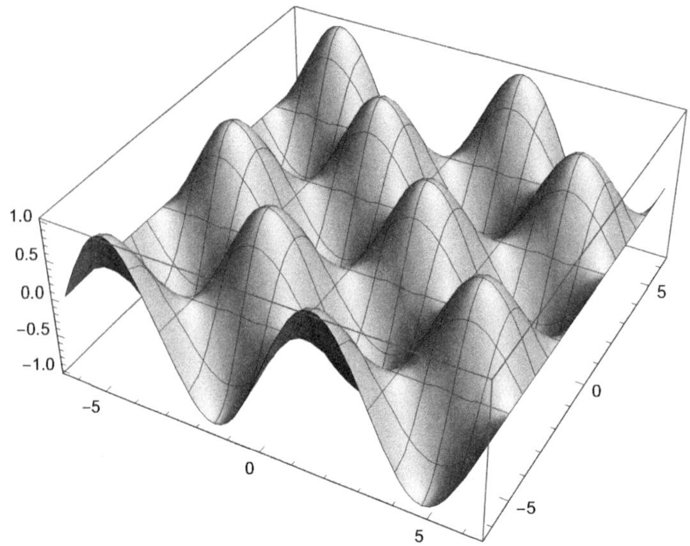

This looks fine, therefore let's try it with a = 10 and b = 10:

sincos2[{10, 10}]

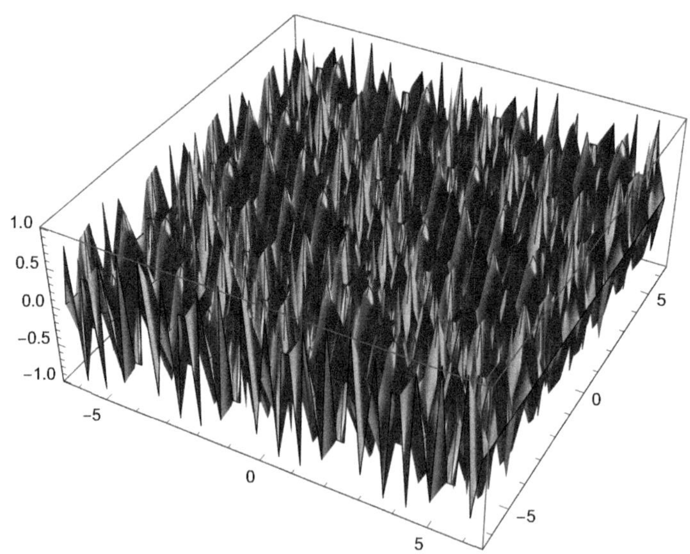

The two values of a and b do not have to be equal, nor do they need they be integers:

sincos2[{1.33, 5.576}]

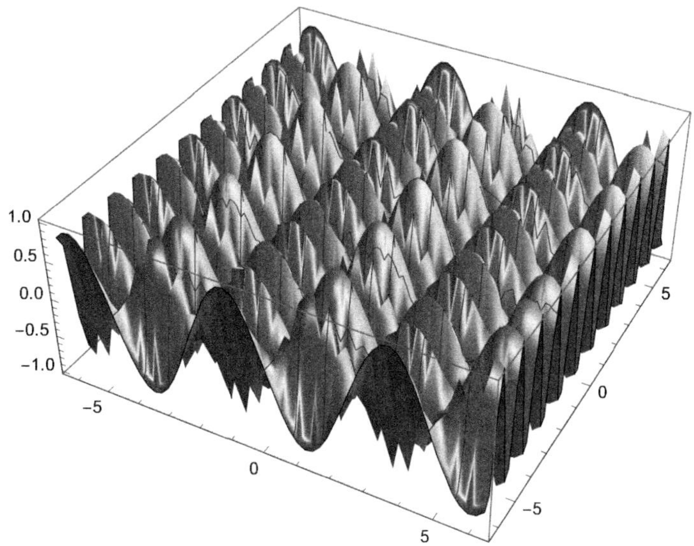

Clearly these Module scripts can do a lot of work for us and quickly. That makes them tremendously important to numerical analysis and visualizations, and so much more.

ClearAll[*a*, *b*, sincos, sincos2, *f*]

ListPlot, Fit, and Show

Often we will have data rather than a function and we wish to plot it, so that we can find a function by analysis that describes the data. In such cases, we can manipulate the data by bringing it into a matrix form and then plot it with **ListPlot**. We also can compare it to the behavior of functions that are meant to represent the data. Let's say that the following is a typical set of data obtained from an experiment. This could have been imported to *Mathematica* by any number of different means. The first column is the independent variable, say, time and the second column shows the measured value of what will be come the dependent variable in our analysis.

t	Measurement
0	10
2	8.2
4	6.7
6	5.5
8	4.5
10	3.7
12	3.
14	2.5
16	2.
20	1.4
24	0.9
28	0.6
32	0.4
36	0.3
40	0.2
44	0.1
50	0.1

Firstly, we write a vector of time values (**tim1**) at which measurements we made and we do the same with the dependent variable values (**dat1**) and we input these:

```
tim1 = {0, 2, 4, 6, 8, 10, 12, 14, 16, 20, 24, 28, 32, 36, 40, 44, 50}
(* Assign time values to tim1 *)
```

{0, 2, 4, 6, 8, 10, 12, 14, 16, 20, 24, 28, 32, 36, 40, 44, 50}

```
dat1 = {10, 8.2, 6.7, 5.5, 4.5, 3.7, 3.0, 2.5, 2.0, 1.4, 0.9, 0.6, 0.4, 0.3, 0.2, 0.1, 0.1}
(* Assign measured values to dat1 *)
```

{10, 8.2, 6.7, 5.5, 4.5, 3.7, 3., 2.5, 2., 1.4, 0.9, 0.6, 0.4, 0.3, 0.2, 0.1, 0.1}

To plot these we must join these into pairs of x,y values, so that they can be plotted by **ListPlot**. To do this we will use three commands interactively, they are **Join**, **Partition** and **Transpose**. Here is how it is done in stepwise fashion. First we join the two sets of values into one new list:

```
Join[tim1, dat1]
```

{0, 2, 4, 6, 8, 10, 12, 14, 16, 20, 24, 28, 32, 36, 40, 44, 50, 10, 8.2, 6.7, 5.5, 4.5, 3.7, 3., 2.5, 2., 1.4, 0.9, 0.6, 0.4, 0.3, 0.2, 0.1, 0.1}

The output from this operation is a single vector composed of time values followed by the dependent variable values. As such they have lost their identity. Hence we need to pair them in order to plot them. Thus we first break this vector into two vectors within one. The first is for the time values and the second for the dependent variable values. To get this right we need to partition time only with time values, therefore we need to state how many elements from the list should be in each partition. We can do this if we know the length of the time list. We get this information by asking for the number of elements in **tim1** with **Length**:

Length[tim1]

17

So, we have 17 time values and 17 measured values. We need to partition the joined set into two sets"

pdata = Partition[Join[tim1, dat1], Length[dat1]]

{{0, 2, 4, 6, 8, 10, 12, 14, 16, 20, 24, 28, 32, 36, 40, 44, 50}, {10, 8.2, 6.7, 5.5, 4.5, 3.7, 3., 2.5, 2., 1.4, 0.9, 0.6, 0.4, 0.3, 0.2, 0.1, 0.1}}

Now, look closely at the result. You notice at the left end and at the right end there are pairs of curly brackets **{ {** and **} }**. That is because we have two lists in one. Look at the middle of the set and you will see two more curly brackets that are not paired i.e. **} {** meaning that we really have two lists in one or a matrix. We can see this by adding "**//MatrixForm**" after the **Partition** command:

pdata//MatrixForm

$$\begin{pmatrix} 0 & 2 & 4 & 6 & 8 & 10 & 12 & 14 & 16 & 20 & 24 & 28 & 32 & 36 & 40 & 44 & 50 \\ 10 & 8.2 & 6.7 & 5.5 & 4.5 & 3.7 & 3. & 2.5 & 2. & 1.4 & 0.9 & 0.6 & 0.4 & 0.3 & 0.2 & 0.1 & 0.1 \end{pmatrix}$$

Since it is a matrix we can do a very simple and yet powerful operation on it – we can transpose it. When we transpose a matrix we exchange the rows for columns. Here is a simple example:

m1 = {{a, b, c, d}, {1, 2, 3, 4}}

{{a, b, c, d}, {1, 2, 3, 4}}

m1//MatrixForm

$$\begin{pmatrix} a & b & c & d \\ 1 & 2 & 3 & 4 \end{pmatrix}$$

Transpose[m1]//MatrixForm

$$\begin{pmatrix} a & 1 \\ b & 2 \\ c & 3 \\ d & 4 \end{pmatrix}$$

Returning to our example, we can see that by transposing the partitioned set "**pdata**" we will have the pairs of independent and dependent variables we seek to plot:

dataset = Transpose[pdata]

{{0, 10}, {2, 8.2}, {4, 6.7}, {6, 5.5}, {8, 4.5}, {10, 3.7}, {12, 3.}, {14, 2.5}, {16, 2.}, {20, 1.4}, {24, 0.9}, {28, 0.6}, {32, 0.4}, {36, 0.3}, {40, 0.2}, {44, 0.1}, {50, 0.1}}

Now we have dyads, or pairs, of values of independent and dependent variables. We also learned how lists in *Mathematica* are matrices and the usual operations on matrices can be performed on our lists. Much of the output from *Mathematica* that we will see is by default a list, unless, we force it not to be. We can always remind ourselves of this by putting lists into matrix or vector form:

dataset//MatrixForm

$$\begin{pmatrix} 0 & 10 \\ 2 & 8.2 \\ 4 & 6.7 \\ 6 & 5.5 \\ 8 & 4.5 \\ 10 & 3.7 \\ 12 & 3. \\ 14 & 2.5 \\ 16 & 2. \\ 20 & 1.4 \\ 24 & 0.9 \\ 28 & 0.6 \\ 32 & 0.4 \\ 36 & 0.3 \\ 40 & 0.2 \\ 44 & 0.1 \\ 50 & 0.1 \end{pmatrix}$$

Although we did each step interactively, we can do it all at once as follows:

dataset = Transpose[Partition[Join[tim1, dat1], Length[dat1]]]

{{0, 10}, {2, 8.2}, {4, 6.7}, {6, 5.5}, {8, 4.5}, {10, 3.7}, {12, 3.}, {14, 2.5}, {16, 2.}, {20, 1.4}, {24, 0.9}, {28, 0.6}, {32, 0.4}, {36, 0.3}, {40, 0.2}, {44, 0.1}, {50, 0.1}}

We can now use **ListPlot** to display this data:

ListPlot[dataset]

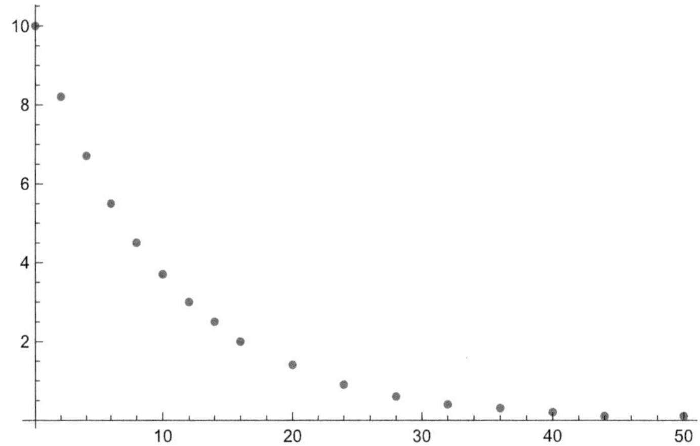

Instead of a graph of a function, we now have discreet points corresponding to the paired values of the independent and dependent variables. We can see that this data looks like an exponential decay of the **y** values with increasing **x**. A simple test of this would be to take the natural log of the **y-values** and plot them against **x**. Look once again at the paired values in data set. To do this, we want the **Log** of the second value in each of the pairs.

There are several ways in which can proceed. We could go back to the set of y-values, dat1, take the log of these and then redo all the steps we did above. That is an acceptable, but inelegant approach. It is acceptable because it works; it is inelegant because we already have the dataset in the form in which we need only take the log of every second value. Therefore a more elegant approach is to operate directly on dataset using the power of *Mathematica*'s rule and function based programming language. In the process of doing this we will use more of the language and we will see why *listability* is so important.

When we want to take an element from a set it is simply a matter of using the correct syntax. For example to take the fifth element from the dataset we simply type **dataset** with a five after it in double square brackets:

dataset[[5]]

{8, 4.5}

If we wanted to take out of **dataset** the **y-value** of the 5th point we would type 5 and 2 separately in double square brackets after dataset:

dataset[[5]][[2]]

4.5

Similarly if we wanted to take out the **x-value** from the fifth data point in the set:

dataset[[5]][[1]]

8

It is clear that we are close to what we need in this function. We could extract all the y-values from dataset by incorporating this syntax into a **Table** function. For example:

Table[dataset[[*n*]][[2]], {*n*, 1, Length[dataset]}]

{10, 8.2, 6.7, 5.5, 4.5, 3.7, 3., 2.5, 2., 1.4, 0.9, 0.6, 0.4, 0.3, 0.2, 0.1, 0.1}

Taking the **Log** of this and using **N** to evaluate numerically, we can have a vector of the **Log** of y-values from dataset because it is *listable*:

***N*[Log[Table[dataset[[*n*]][[2]], {*n*, 1, Length[dataset]}]]]**

{2.30259, 2.10413, 1.90211, 1.70475, 1.50408, 1.30833, 1.09861, 0.916291, 0.693147, 0.336472, −0.105361, −0.510826, −0.916291, −1.20397, −1.60944, −2.30259, −2.30259}

But, now we have violated our original goal, we have taken dataset apart. We can be even more savvy than this and avoid having to join, partition and transpose again. We do this by writing a function in *Mathematica* that will do what we want from the start. The syntax for a function in *Mathematica* or a rule is **f [x_]: = f [x]**. This function will take only single values for **x**. We have a set of paired values as the argument of our function, so we will follow the dummy variable on the left-hand side by a double underbar instead of a single under bar: **g [x__]: = g [x]**. The function or rule that we want is this in English:

"Take an element from dataset, keep the x-value as it is,
but take the Log of the y-value and numerically evaluate it, and
keep the two values xn and yn paired as they originally were."

Writing this in English first makes it fairly obvious what we need to do; this is our algorithm. In *Mathematica* we translate this algorithm directly into a rule or function. That rule will look like this for the nth element of any set:

lgf[x__]:={*x*[[*n*]][[1]], *N*[Log[*x*[[*n*]][[2]]]]}

Let's not move too fast on this because this little rule is a program, and it is little rules like this one that form the bricks that we can build larger structures from later. Note the left-hand side has function syntax with the dummy variable followed by a double-underbar and set off from the right-hand side by a colon and equal sign. This is called the *set delayed* structure in *Mathematica*; it means that until a specific argument is given within the brackets, this function in unevaluated that is its evaluation is delayed until we give it an argument. The form of the function is stored and will work with most any argument, provided we also give it a value for n. On the right-hand side, we find a set of curly brackets around two commands that now should look familiar. The first just takes the x-value of the nth element and the pairs it with the log of the y-value of the nth element. If we now put this inside the table function, we can operate on **dataset** from n equals one to the end which occurs at the value of **Length[dataset]**. Here it is:

lgdatset = Table[lgf[dataset], {*n*, 1, Length[dataset]}]

(* lgdatset is just a name for the data *)

{{0, 2.30259}, {2, 2.10413}, {4, 1.90211}, {6, 1.70475}, {8, 1.50408}, {10, 1.30833}, {12, 1.09861}, {14, 0.916291}, {16, 0.693147}, {20, 0.336472}, {24, −0.105361}, {28, −0.510826}, {32, −0.916291}, {36, −1.20397}, {40, −1.60944}, {44, −2.30259}, {50, −2.30259}}

pllgdatset = ListPlot[lgdatset]

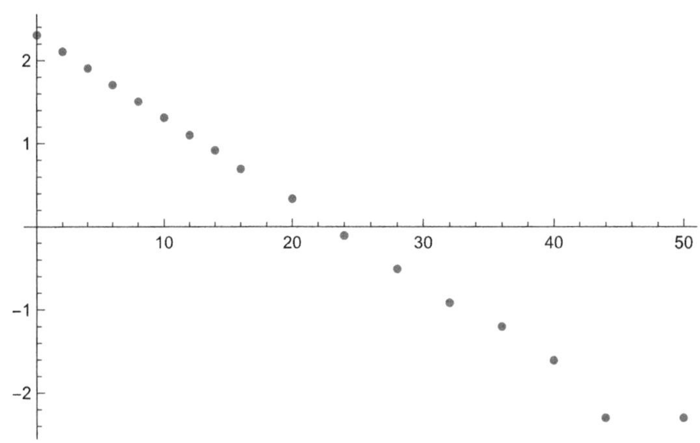

As we can see this looks quite linear indicating that the data follows an exponential decay. If we want to be more precise about this we can use *Mathematica* to find a fit to the log data. That is we can find the equation for the best fit line to **lgdatset**. Once we have this function we can then plot it and graph it with the data to once again visualize the goodness of fit. We introduce now the **Fit** command.The syntax for **Fit** is as follows–The argument consists of three elements, the first is the name of the matrix of data to be fit, the second is enclosed in curly brackets and it states that we want to fit to a linear equation (we can use any polynomial we like), and the last names the independent variable. The output is a line (or polynomial) in x. Since we will want to **Plot** this, we should give it a function name. We can call it **ftlg** for fit to the **Log** of the data and plot it from zero to 50 in x. (Firstly, we do it **as** such without the function name to show the output and then again with the function name. There is no reason to do the first, except to see the values of the slope an intercept). We can give the plot a name also **plftlg** plot of fit to log:

```
(* Here the Fit function takes the data and then fits it to a linear function of x *)
Fit[lgdatset, {1, x}, x];

ftlg = Fit[lgdatset, {1, x}, x];
(* Here the result from the Fit function is assigned the name ftlg *)

ftlg(* This lets us see ftlg *)

Table[ftlg, {x, 0, 50}](* this provides values of ftlg at different times *)

plftlg = Plot[ftlg, {x, 0, 50}](* This shows the line that has been fit to the data *)
```

2.27476 − 0.0975478*x*

{2.27476, 2.17722, 2.07967, 1.98212, 1.88457, 1.78703, 1.68948, 1.59193, 1.49438, 1.39683, 1.29929, 1.20174, 1.10419, 1.00664, 0.909095, 0.811547, 0.713999, 0.616451, 0.518904, 0.421356, 0.323808, 0.22626, 0.128712, 0.0311646, −0.0663832, −0.163931, −0.261479, −0.359027, −0.456574, −0.554122, −0.65167, −0.749218, −0.846766, −0.944314, −1.04186, −1.13941, −1.23696, −1.3345, −1.43205, −1.5296, −1.62715, −1.7247, −1.82224, −1.91979, −2.01734, −2.11489, −2.21244, −2.30998, −2.40753, −2.50508, −2.60263}

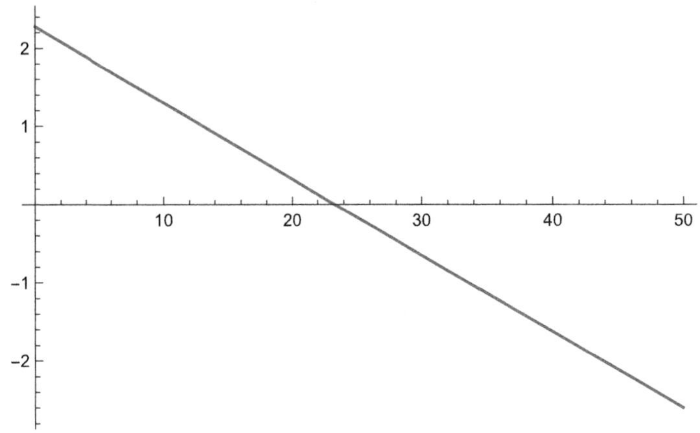

Finally, we can put the data points and this line on the same graph by calling for the **ListPlot**, **pllgdatset**, and the **Plot**, **plftlg**, within the **Show** command:

Show[plftlg, pllgdatset]

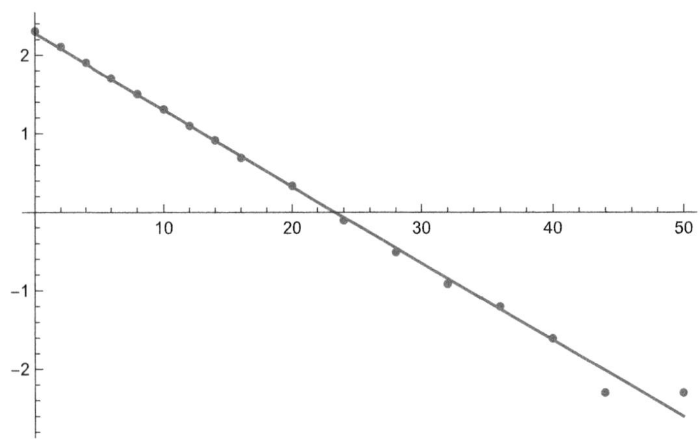

This looks good, but it is hard to see the data points when the line runs through them and we should label the axes. We can make the data points larger in radius so that they will show up more nicely and we could also change the color of the line through the points. To do these two tasks we introduce the **PlotStyle**, **PointSize** and **Hue** commands. One way to find out what these do and what they require for syntax and arguments is to use the "**??**" command. Specifically, whenever you are unsure of what a command does or requires type a double question mark followed by the command name. We can do this now for each of these three commands:

?PlotStyle

Symbol

PlotStyle is an option for plotting and related functions that specifies styles in which objects are to be drawn.

??PointSize

Symbol

PointSize[*d*] is a graphics directive which specifies that points which follow are to be shown if possible as circular regions with diameter *d*. The diameter *d* is given as a fraction of the total width of the plot.

Documentation Local » | Web »

Attributes {Protected, ReadProtected}

Full Name System`PointSize

??Hue

Symbol

Hue[*h*] is a graphics directive which specifies that objects which follow are to be displayed, in a color corresponding to hue *h*.

Hue[*h*, *s*, *b*] specifies colors in terms of hue, saturation, and brightness.

Hue[*h*, *s*, *b*, *a*] specifies opacity *a*.

Documentation Local » | Web »

Attributes {Protected, ReadProtected}

Full Name System`Hue

Returning to the **ListPlot** and **Plot** we can use **PlotStyle** with **PointSize** and **Hue** in each, respectively, to make the graph easier to interpret and understand:

pllgdatset = ListPlot[lgdatset, PlotStyle → PointSize[0.015]]

(* Replotting the data with larger points *)

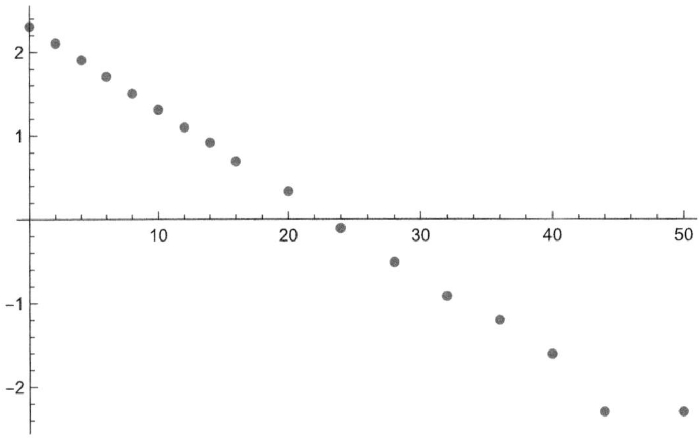

plftlg = Plot[ftlg, {x, 0, 50}, PlotStyle → Hue[0]]

(* Replotting the best fit line in red *)

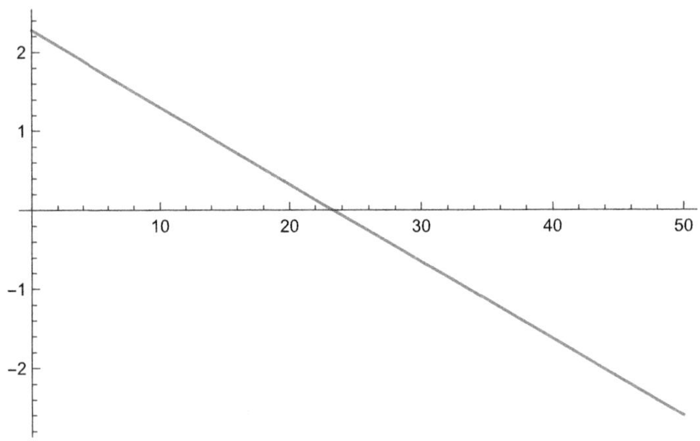

Show[pllgdatset, plftlg](* Showing the data and the line again together *)

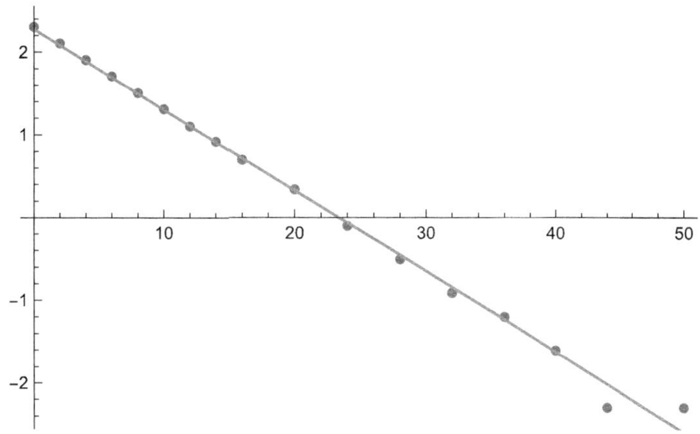

This looks much better than it did, but we still should give the **x** and **y** axes labels, why don't we call them **t** for time and **LogY(t)** for the log of position, **Y**, as a function of time. To do this we need the command **AxesLabel** which has the following attributes:

??AxesLabel

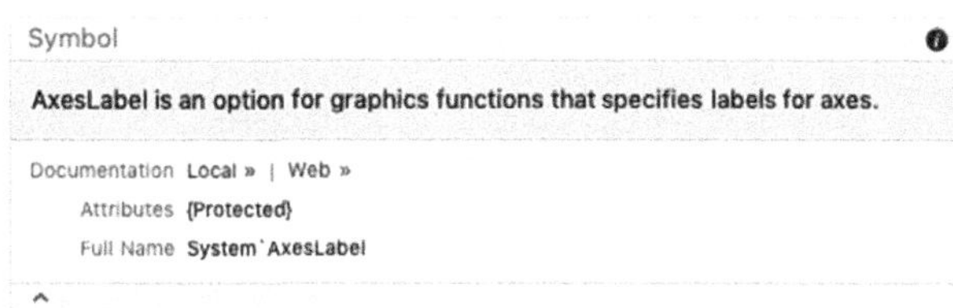

We will put the label for each axis within a set of curly brackets and then also within quotation marks so that they are not interpreted as a function to be evaluated but rather as simply strings.

Show[pllgdatset, plftlg, AxesLabel → {t, "Log[Y(t)]"}](* Add axes labels *)

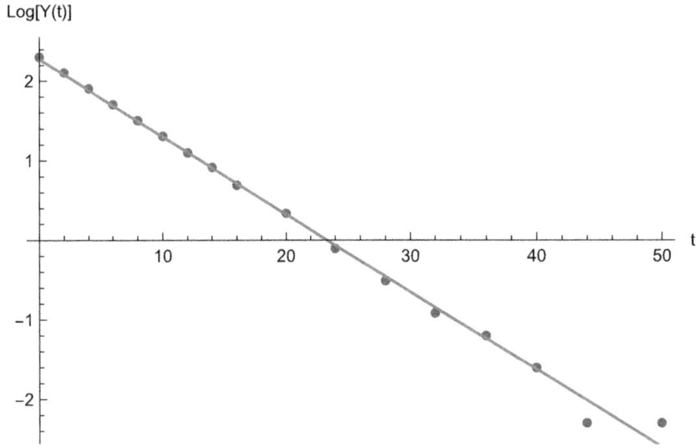

There is another difference between Plot and ListPlot that can be useful, especially when exploring the behavior of functions. For example, if we are interested in observing a function such as x Sin[x]^2, we can begin with the Plot command and the rendering is as follows:

Plot[*x*Sin[*x*]^2, {*x*, 0, 1000}]

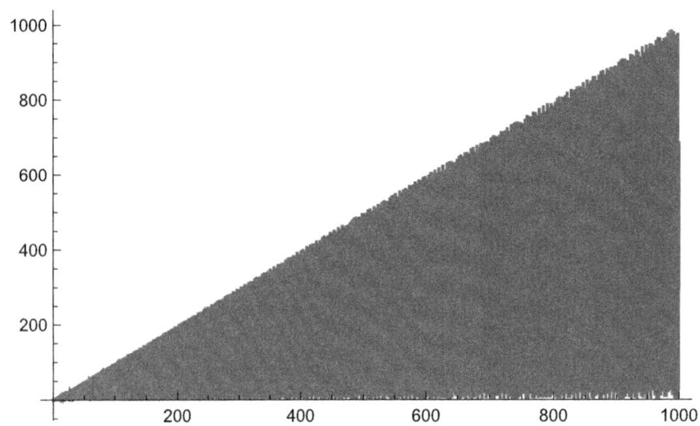

There are so many points that the graphic is not much more than a triangle with a rough-edged hypotenuse. Instead by using ListPlot on can render the same function and see more of its structure.

fnewdata = Table[{*x*, *x*Sin[*x*]^2}, {*x*, 0, 1000, .5}];

ListPlot[fnewdata]

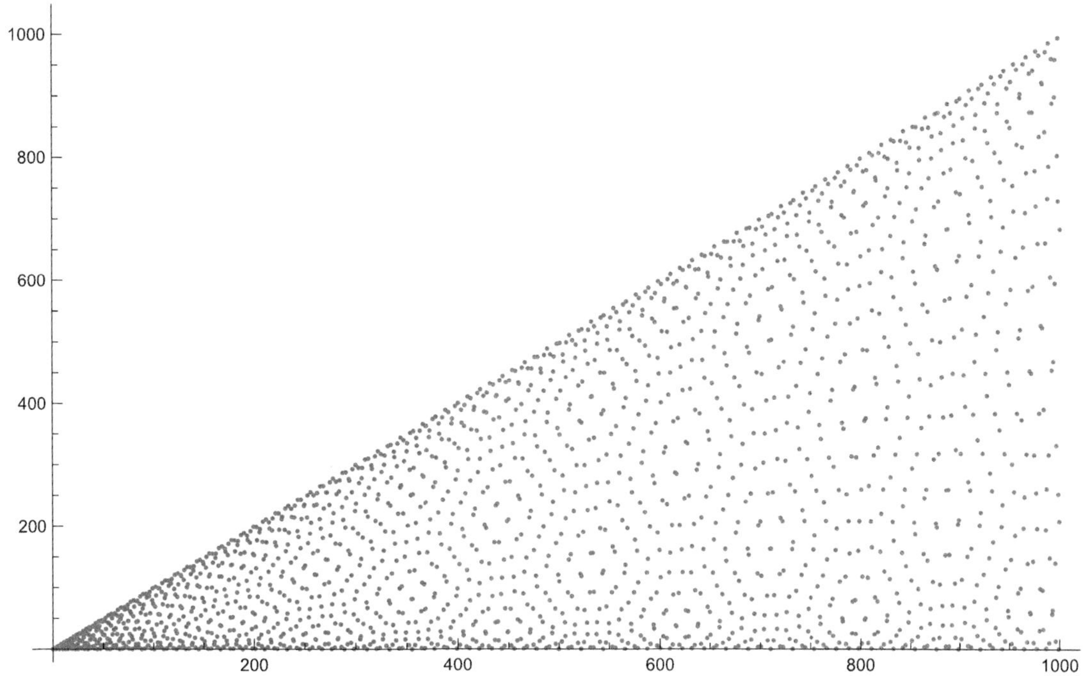

Then we can take the same function and ask how this behavior changes if we multiply by Cos[x] and then Cos $[x]^2$:

```
Table[{x, xSin[x]^2Cos[x]}, {x, 0, 1000, 0.25}];
ListPlot[%]
```

```
Table[{x, xSin[x]^2Cos[x]^2}, {x, 0, 1000, 0.25}];
ListPlot[%]
```

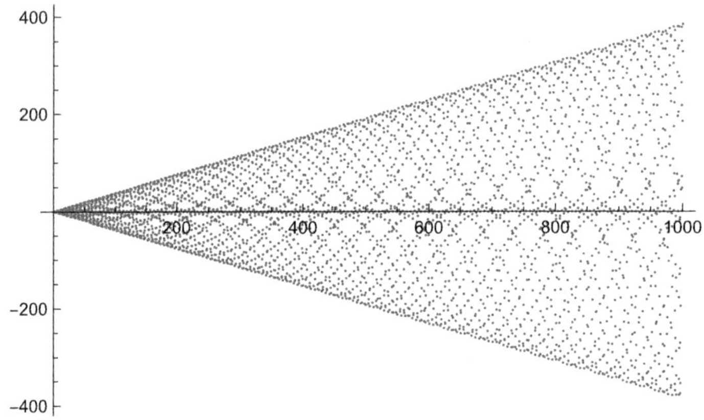

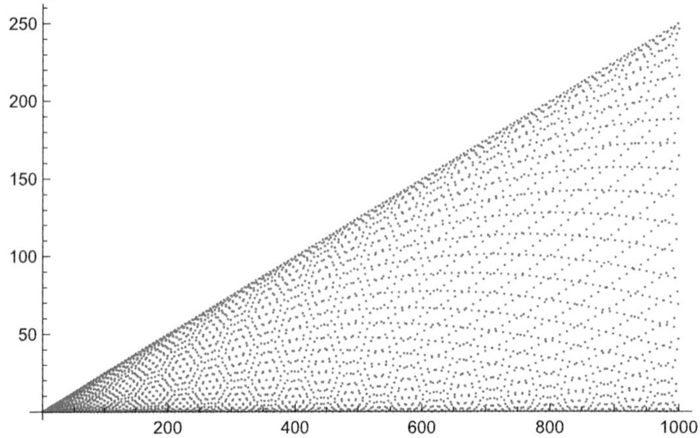

Or, perhaps, we are curious as to the effect that raising the power of x from 1 to 2 to 5 may have:

Table[{x, xSin[x]^2Cos[x]}, {x, 0, 1000, 0.25}];

ListPlot[%]

Table[{x, x^2Sin[x]^2Cos[x]}, {x, 0, 1000, 0.25}];

ListPlot[%]

Table[{x, x^5Sin[x]^2Cos[x]}, {x, 0, 1000, 0.25}];

ListPlot[%]

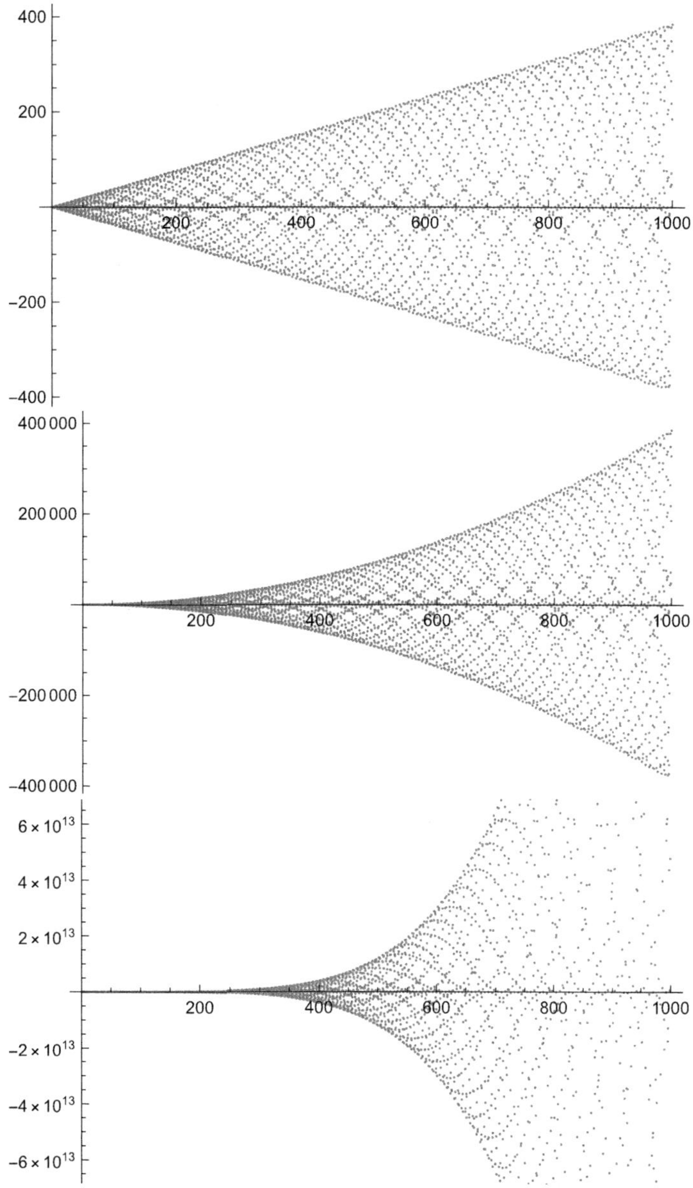

Here we look at variations on the general function $\text{Sin}[x]^l\text{Cos}[\text{ x }]^m\text{Tan}[x]^n$:

```
Table[{x, Sin[x]^2Cos[x]Tan[x]}, {x, 0, 1000, 0.25}];
```

```
ListPlot[%]

Table[{x, Sin[x]^2Cos[x]^2Tan[x]}, {x, 0, 1000, 0.25}];

ListPlot[%]

Table[{x, Sin[x]^2Cos[x]^2Tan[x]^2}, {x, 0, 1000, 0.25}];

ListPlot[%]
```

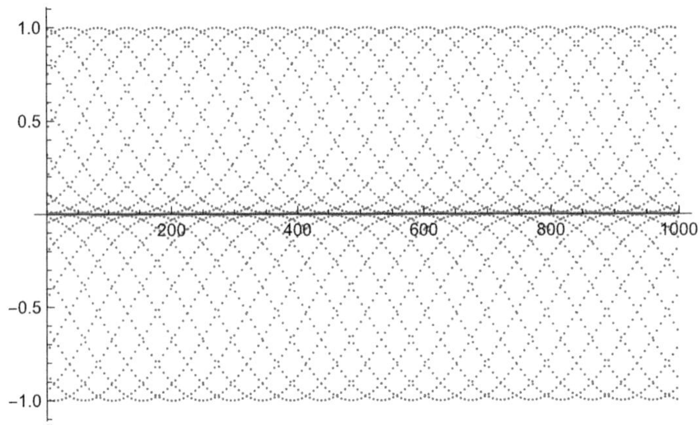

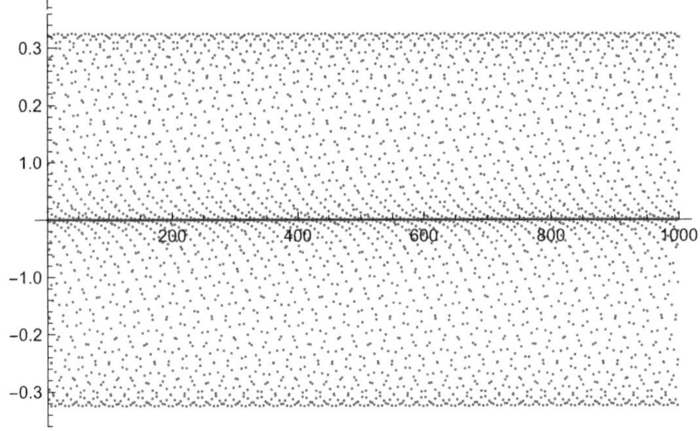

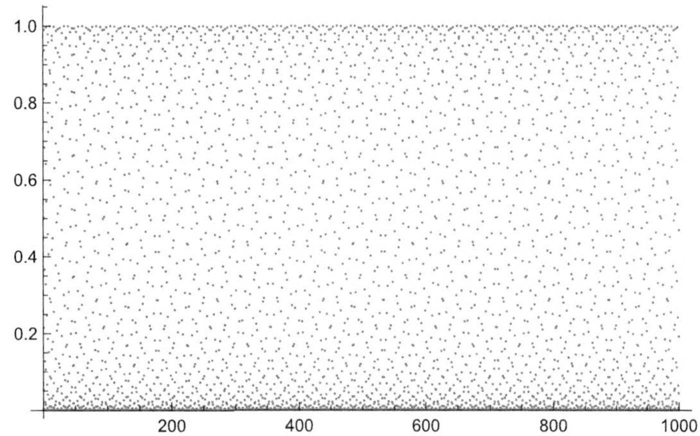

To round out this section, let's compare $\frac{\text{Sin}[x]}{x}$ with $\frac{\text{Sin}[\text{RandomReal}[x]]}{\text{RandomReal}[x]}$. First we will use Plot, then we will use ListPlot to see the difference and we will take the Fourier transform of each data set, and replot the Absolute Value squared to see what we can learn about regularity and behavior:

```
Plot[Sin[x]/x, {x, 0.1, 1000}]

Table[{x, Sin[x]/x}, {x, 0.1, 1000, .25}];

ListPlot[%]

ListPlot[%%, Joined → True]

Fourier[%%%, FourierParameters → {1, 1}];

Abs[%]^2;

ListPlot[Abs[%]^2, Joined → True]
```

```
Plot[Sin[RandomReal[x]]/RandomReal[x], {x, 0.1, 1000}]

Table[{x, Sin[RandomReal[x]]/RandomReal[x]}, {x, 0.1, 1000, 0.25}];

ListPlot[%]

ListPlot[%%, Joined → True]

Fourier[%%%, FourierParameters → {1, 1}];

Abs[%]^2;
```

ListPlot[Abs[%]^2, Joined → True]

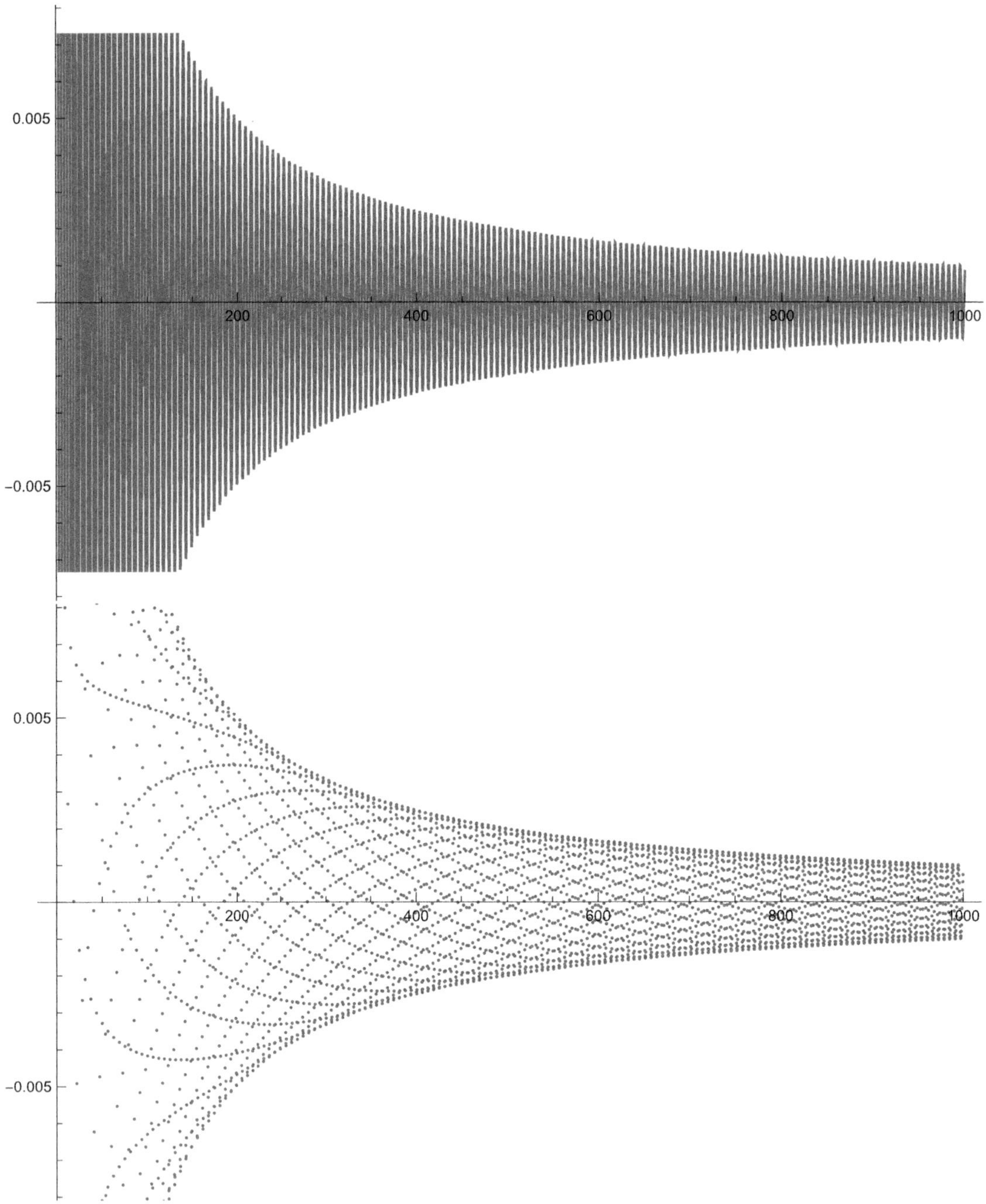

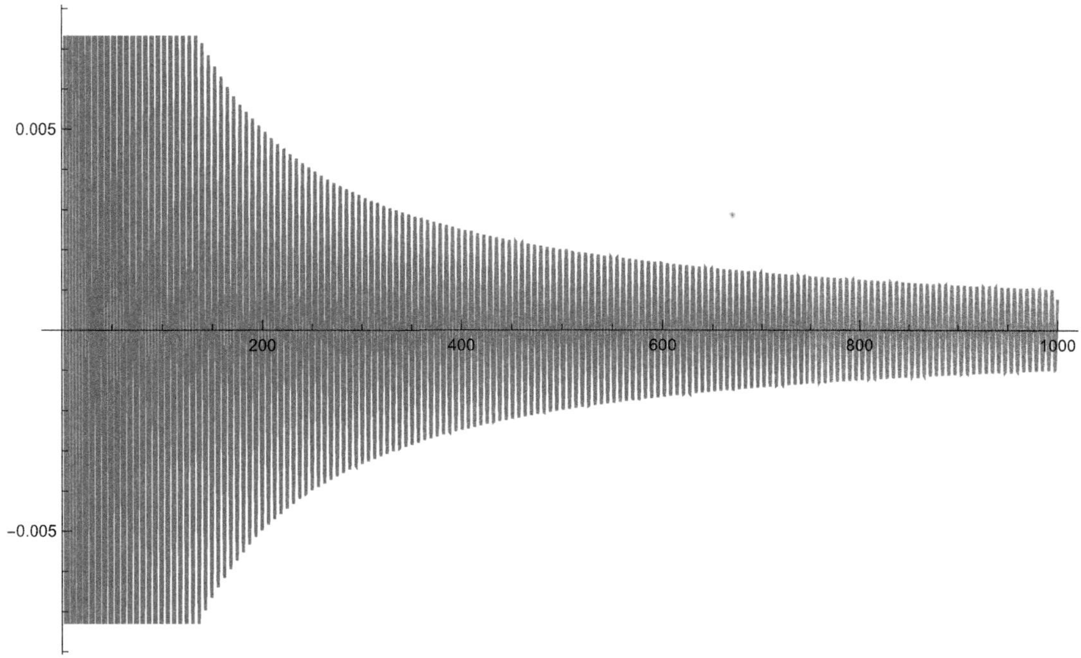
0.005
−0.005
200
400
600
800
1000

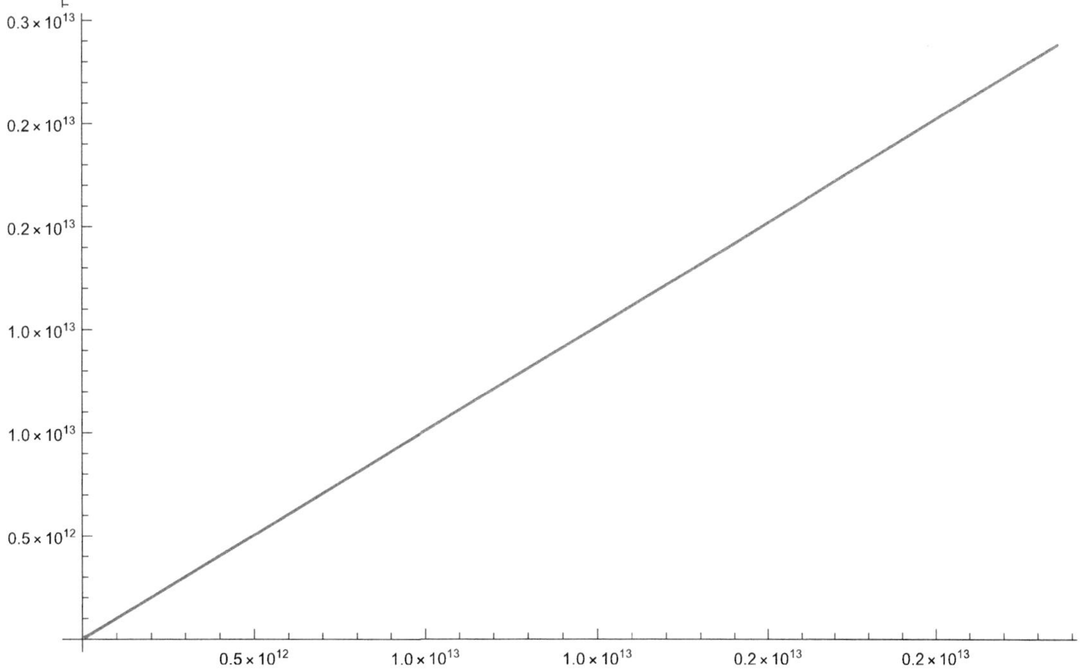
0.3×10^{13}
0.2×10^{13}
0.2×10^{13}
1.0×10^{13}
1.0×10^{13}
0.5×10^{12}
0.5×10^{12}
1.0×10^{13}
1.0×10^{13}
0.2×10^{13}
0.2×10^{13}

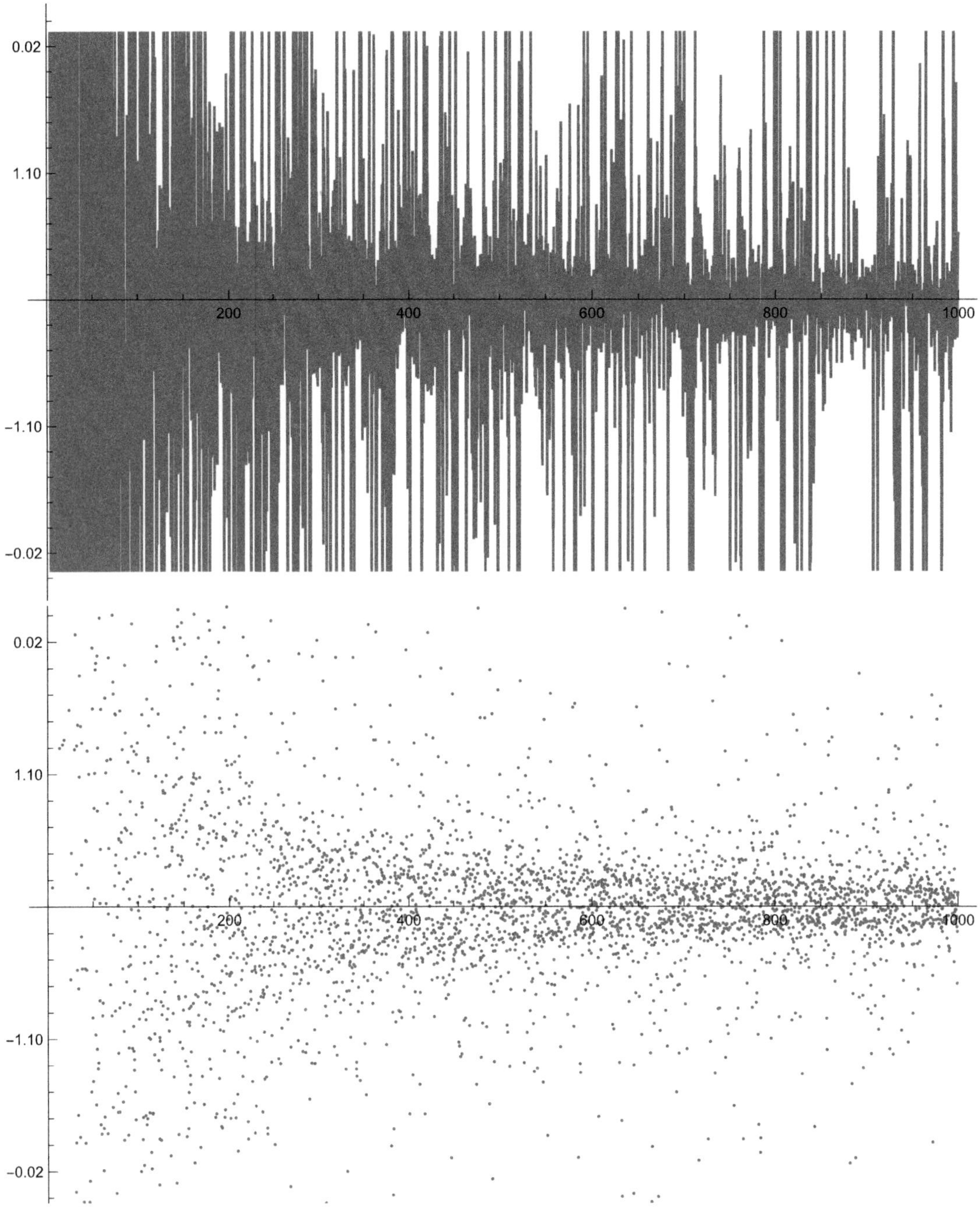
0.02
1.10
−1.10
−0.02
200
400
600
800
1000
0.02
1.10
−1.10
−0.02
200
400
600
800
1000

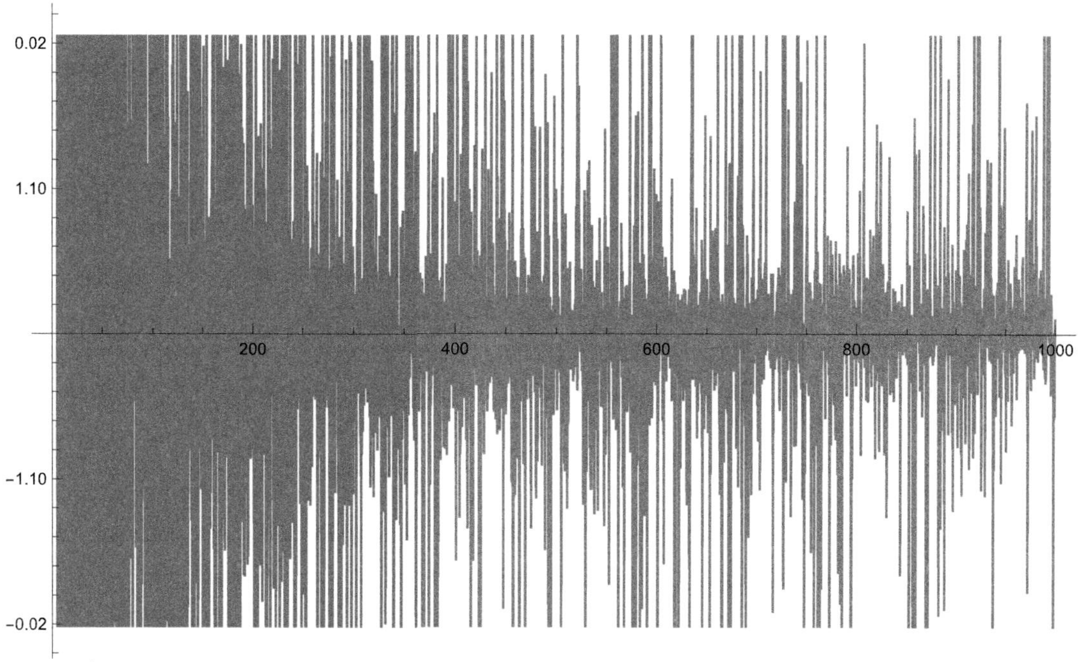

0.02
1.10
−1.10
−0.02
200
400
600
800
1000

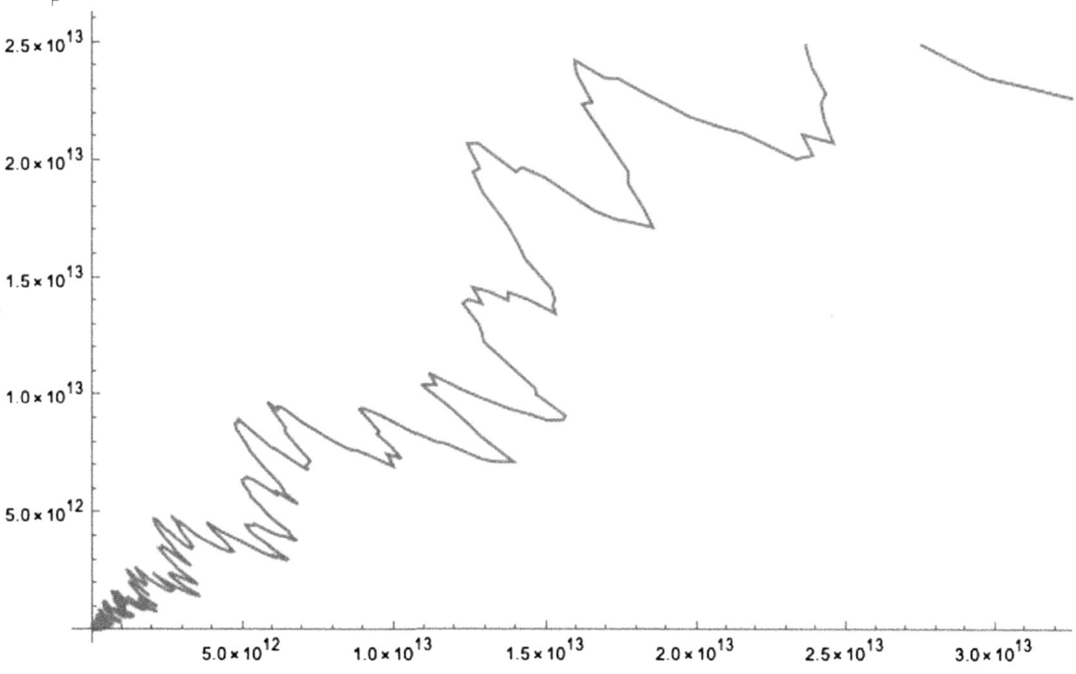

2.5 × 10^13
2.0 × 10^13
1.5 × 10^13
1.0 × 10^13
5.0 × 10^12
5.0 × 10^12
1.0 × 10^13
1.5 × 10^13
2.0 × 10^13
2.5 × 10^13
3.0 × 10^13

The function $\frac{\text{Sin}[x]}{x}$ is a continuously decreasing function (sinc function). The comparison between the rendering of Plot versus ListPlot is fascinating, with the latter being more beautiful to the eye. When we redo ListPlot and force the points to be joined, then the rendering looks exactly like that derived from Plot, as it should. In fact they are not different, they simply look different. The results of doing the Fourier transform are relatively unsurprising, the linearity of the results indicates that the function is quite regular as it is. The more interesting result is that involving RandomReal, which gives a random number between 0 and 1. The results are similar to that with increasing real numbers, but they are clearly not the same. As in the first case, when the points are joined the result appears to be the same as that derived from Plot. The Fourier transform looks like a random walk around the Fourier transform of the non-random case.

The reader might ask, is this important? The answer is that these particular results are probably not important, and to some who are better able to do mental visualizations, they may be expected. However, what is important is that with *Mathematica* and a minimum of programming one can begin to do mathematical experiments very easily. That is important because building small learning models, then analyzing them and elaborating them allows us to build an understanding that is visual and intuitive. While building intuitive understanding may or may not be useful for doing pure mathematics, it is undoubtedly important for doing physical sciences and engineering. And, finally, and unapologetically, the renderings of these functions with ListPlot are as beautiful as they are fascinating. There is an aesthetic appeal about them that must satisfy the brain's desire to find patters. Aesthetics and science have always gone together, so being able to indulge with immediacy might also be important, and plain fun. It is the joy of mathematics at its most primitive!

One of the many benefits of using *Mathematica* is that with a minimum of code a person can begin to do "experimental" mathematics with ease. For example, let's say a student wants to explore a family of functions based on Sin [x] and Cos [x }, this is easy to do. We begin with the Sin function with its well-known and predictable cyclic behavior:

Plot[Sin[*x*], {*x*, −10, 10}]

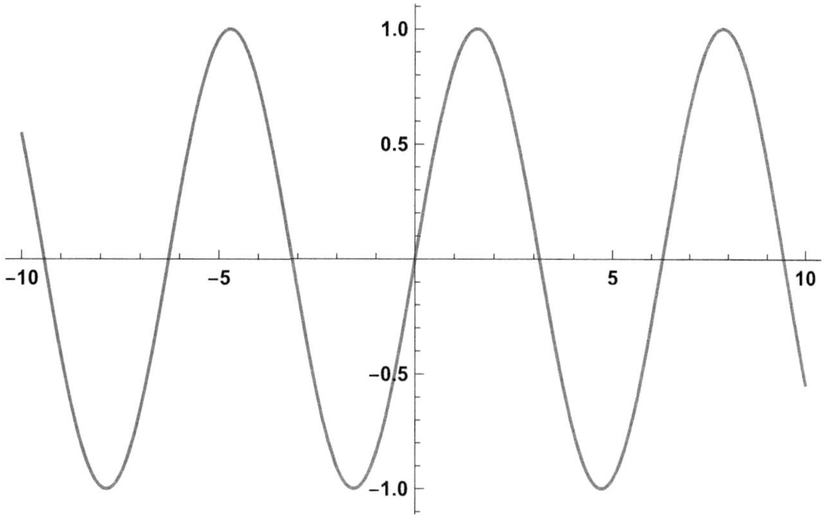

If we introduce a constant as a multiplier of the argument x, we increase the frequency of the wave function:

Plot[Sin[Pi *x*], {*x*, −10, 10}]

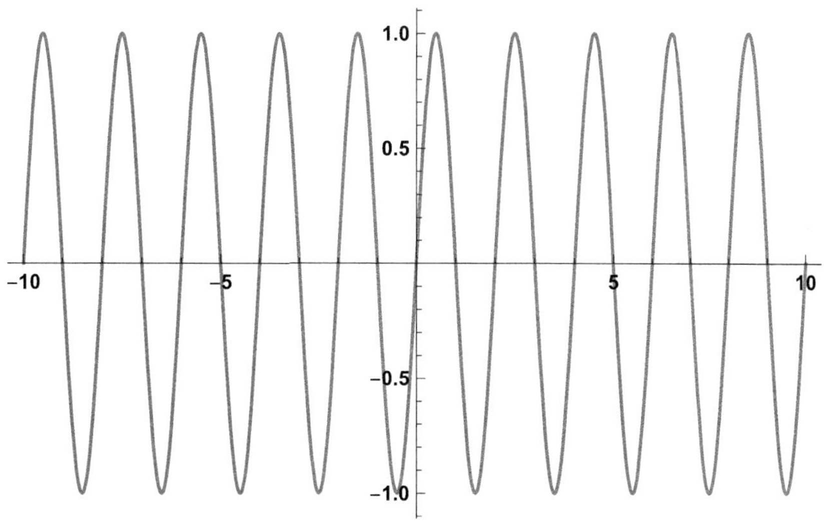

If we divide the Sin [x] by a constant, say for instance by π, then the behavior really does not change, only the amplitude of the wave is moderated:, which is not surprising:

Plot[Sin[Pi *x*]/Pi, {*x*, −10, 10}]

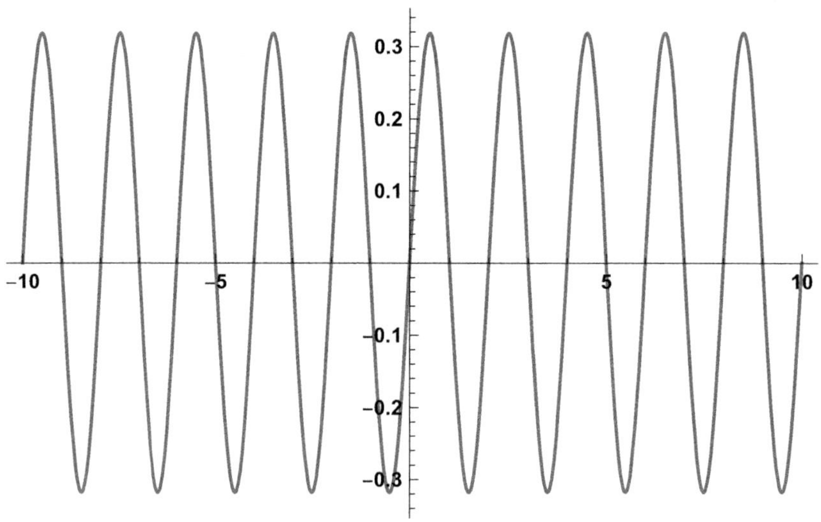

So, now what if we take x, the argument of the Sine function, and multiply that by the Sin[x], what will happen? Intuitively, we know that the function f [x] = x is a line with a slope of unity. So we will somehow combine that linear behavior with the sinusoidal behavior. Being able to visualize this is easy to do:

Plot[*x* Sin[Pi *x*]/Pi, {*x*, −10, 10}]

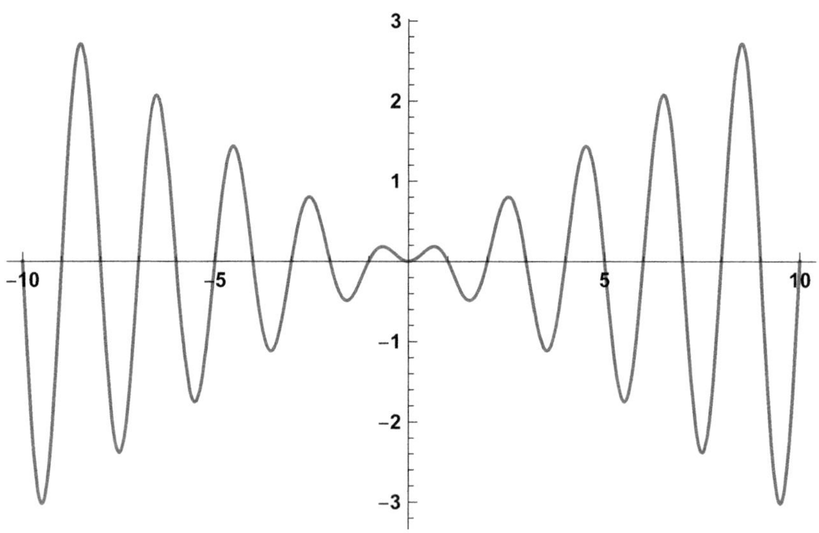

Notice that the function goes to zero at x = 0 and then maximizes symmetrically.

Plot[x Sin[Pi x]/Pi, {x, −1, 1}]

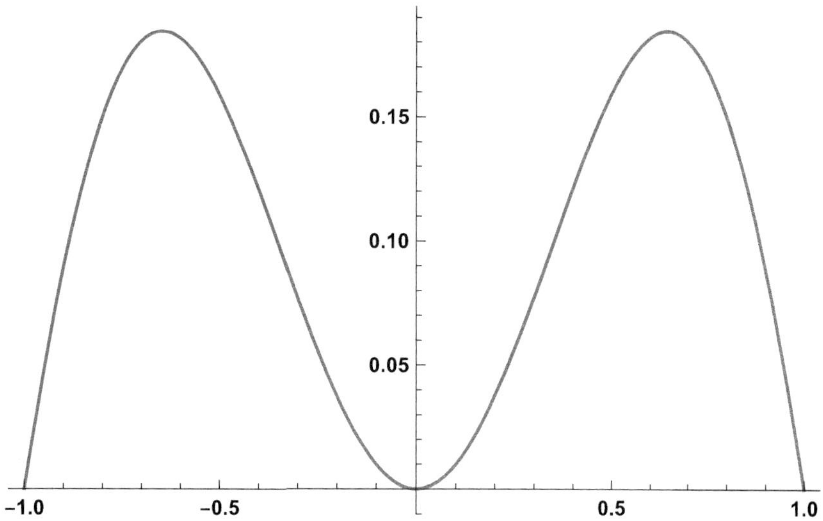

If we want to know where the region of the maximum is, we can find values of the function in a region around x = 0.7

Table[{x, x Sin[Pi x]/Pi}, {x, 0.6, 0.8, 0.025}]

{{0.6, 0.181638}, {0.625, 0.1838}, {0.65, 0.184351}, {0.675, 0.183198}, {0.7, 0.180263}, {0.725, 0.175482}, {0.75, 0.168809}, {0.775, 0.160212}, {0.8, 0.149678}}

We see that the positive maximum, the one closest to zero, is at x = 0.65 with f [0.65] = ∼ 0.1844. The question then becomes to what degree of accuracy is it necessary to specify the region of x where the function maximizes.

What we observe in this case is *amplitude modulation* of the sine wave, which is the basis for the transmittance of an AM radio signal. Looking at the behavior, it makes intuitive sense. Starting from zero in the positive, or the negative direction, the amplitude increases as we move to larger values of x. The picture is literally worth a thousand words.

If we now make the argument of the Sin [x], the product of the square of x and π, we see that the *frequency* increases in both the positive and negative direction, which is to say we see frequency modulation the basis of FM radio transmission:

Plot[Sin[Pi x^2]/Pi, {x, −10, 10}]

Plot[Sin[Pi x^2]/Pi, {x, −1, 1}]

Out[17]=

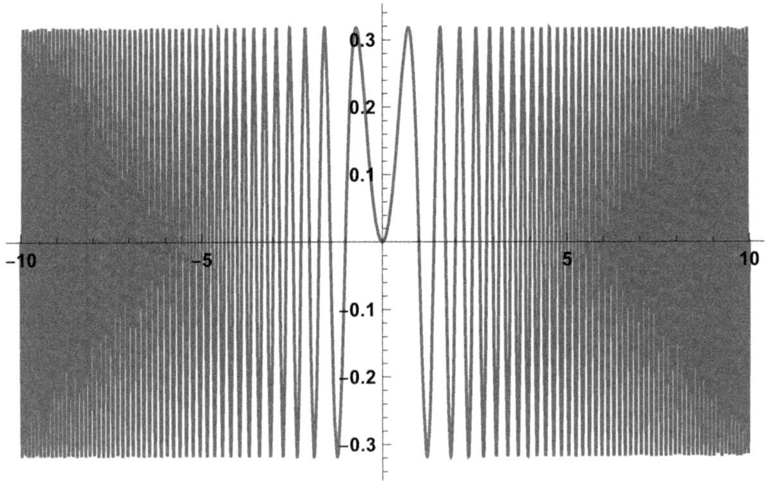

Out[18]=

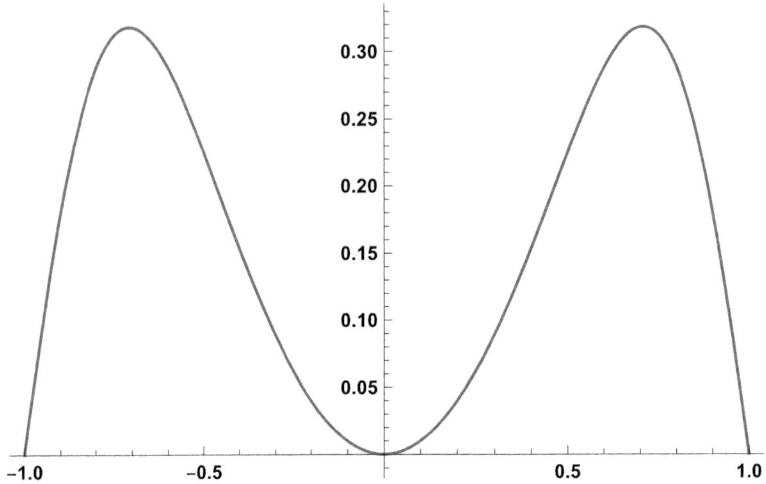

If we were to rotate this figure around the y-axis by 180 degrees (π radians), the result of the rotation would look exactly like the starting figure. We also notice the *all-positive* nature of the first two oscillations of this function. Once again at zero the function must be zero because Sin [0] = 0.

Table[{*x*, Sin[Pi x^2]/Pi}, {*x*, 0.6, 0.8, 0.015}]

{{0.6, 0.288015}, {0.615, 0.295299}, {0.63, 0.301758}, {0.645, 0.307297}, {0.66, 0.311817}, {0.675, 0.315222}, {0.69, 0.317413}, {0.705, 0.318296}, {0.72, 0.317778}, {0.735, 0.315772}, {0.75, 0.312194}, {0.765, 0.306969}, {0.78, 0.30003}, {0.795, 0.29132}}

We see that the maximum of the first oscillation on the positive side of zero is in the region of x = 0.705. Let's place x before the Sin [π x^2]. This will produce both *amplitude* and *frequency* modulation of the original Sine wave.

Plot[x Sin[Pi x^2]/Pi, {x, −10, 10}]

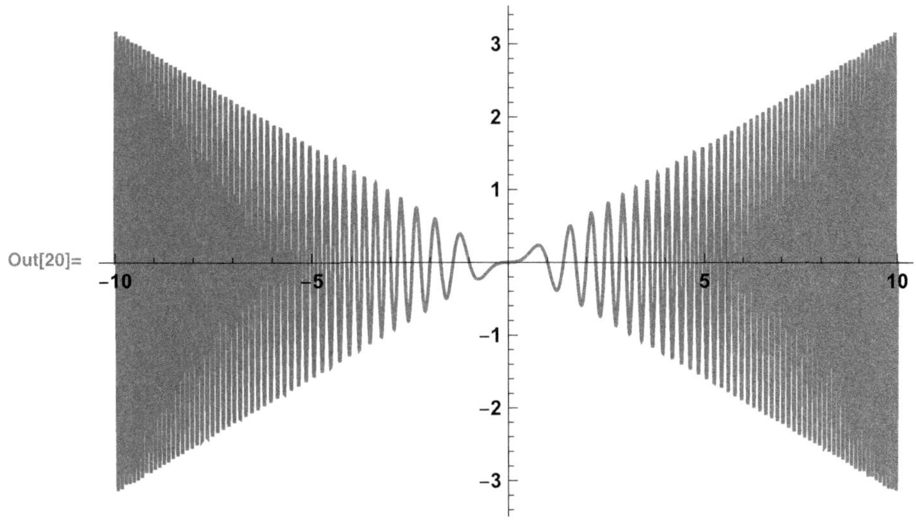

We can look at a narrow region around zero and we see that the first two oscillations are identical in shape, but they are inverted. A rotation of 360 degrees (2π radians) about the y-axis is necessary to bring the figure back to its original form. A rotation of 180 degrees about the y-axis merely inverts the figure. The function not only goes through zero, but it remains close to zero up to about x = 0.1 and then on the negative side of zero it goes to negative values.

Plot[x Sin[Pi x^2]/Pi, {x, −1, 1}]

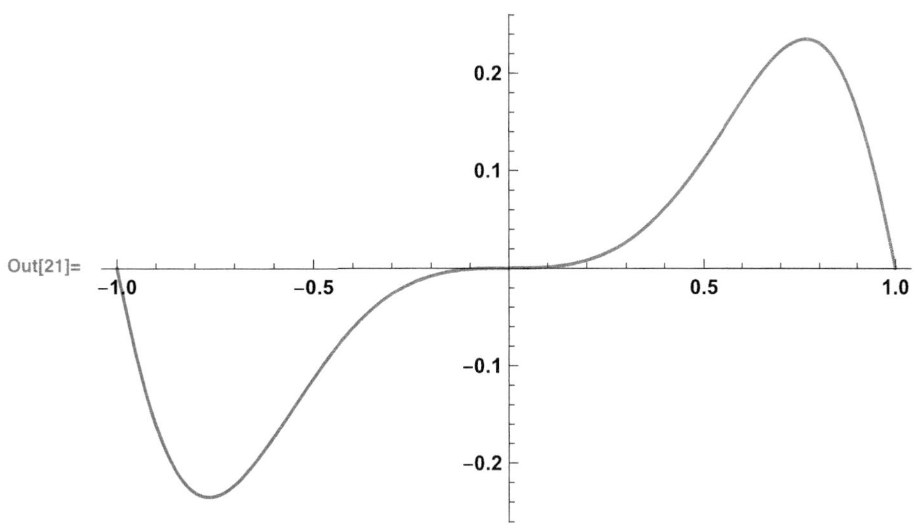

If we now place the argument Sin [x] over the argument x, then we see a new kind of amplitude modulation, namely that as x increases the amplitude decreases, which leads to Sinc-function behavior.

Plot[Sin[x]/x, {x, −100, 100}]

Out[23]=

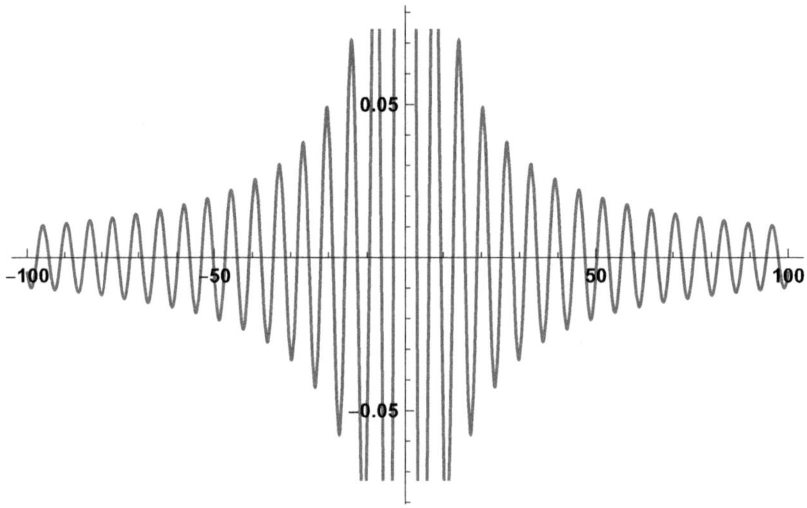

By placing x^2 in the denominator, the function *sincs* out faster.

Plot [Sin[x] /x^2 , {x, −100, 100}]

Out[24]=

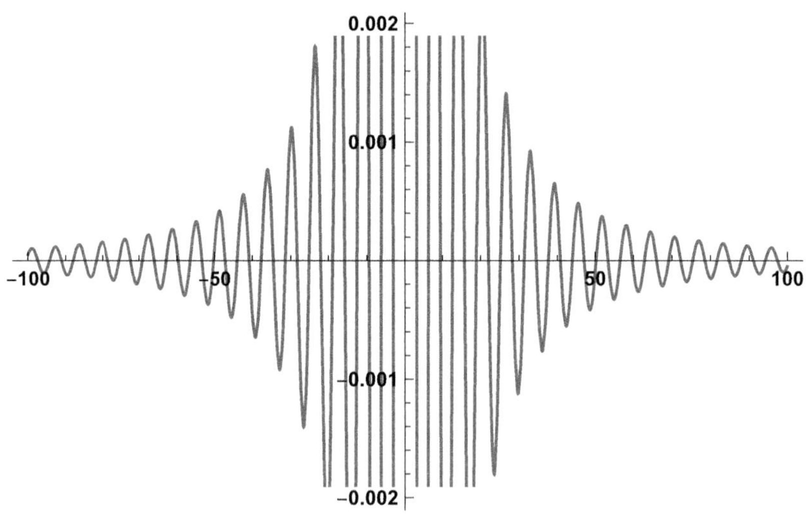

Squaring x inside the Sin[x] and in the denominator just decreases frequency and amplitude.

Plot[Sin[x^2]/x^2, {x, −100, 100}]

Out[25]=

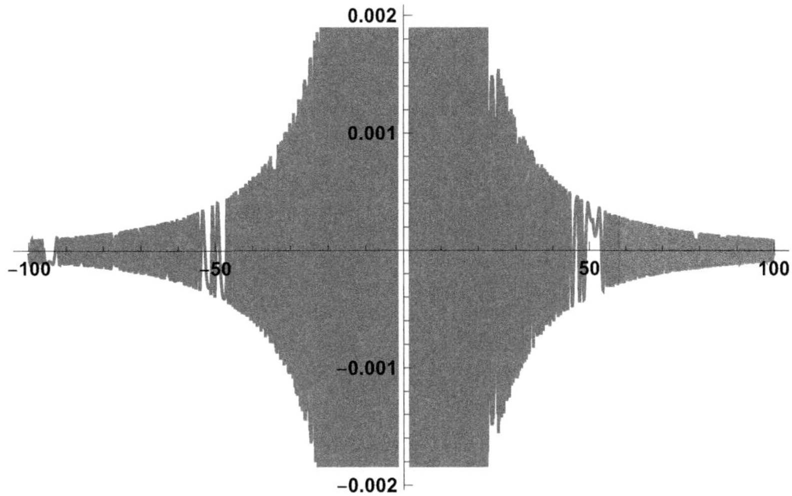

Naturally, it is of interest to ask what about the derivative of such functions. Let's review the derivative function given by the symbol D, with the argument made up of the function first followed by the variable over which the derivative is being taken. Here is how this is described in the documentation:

??Derivative

Symbol

f' represents the derivative of a function f of one argument. \nDerivative[n_1, n_2, ...][f] is the general form, representing a function obtained from f by differentiating n_1 times with respect to the first argument, n_2 times with respect to the second argument, and so on.

Documentation Local » | Web »

Attributes {NHoldAll}

Full Name System`Derivative

Taking the derivative of f [x] = $\frac{\text{Sin}[x]}{x}$ we obtain f ' [x] = $\frac{\text{Cos}[x]}{x} - \frac{\text{Sin}[x]}{x^2}$:

***D*[Sin[x]/x, x]**

$$\frac{\text{Cos}[x]}{x} - \frac{\text{Sin}[x]}{x^2}$$

Now, we can compare $\frac{\text{Sin}[x]}{x}$, $\frac{\text{Cos}[x]}{x} - \frac{\text{Sin}[x]}{x^2}$ as follows:

Plot[{Sin[x]/x}, {x, −10, 10}]

$$\text{Plot}\left[\left\{\frac{\text{Cos}[x]}{x}-\frac{\text{Sin}[x]}{x^2}\right\},\{x,-10,10\}\right]$$

$$\text{Plot}\left[\left\{\frac{\text{Cos}[x]}{x}-\frac{\text{Sin}[x]}{x^2},\text{Sin}[x]/x\right\},\{x,-10,10\}\right]$$

Out[28]=

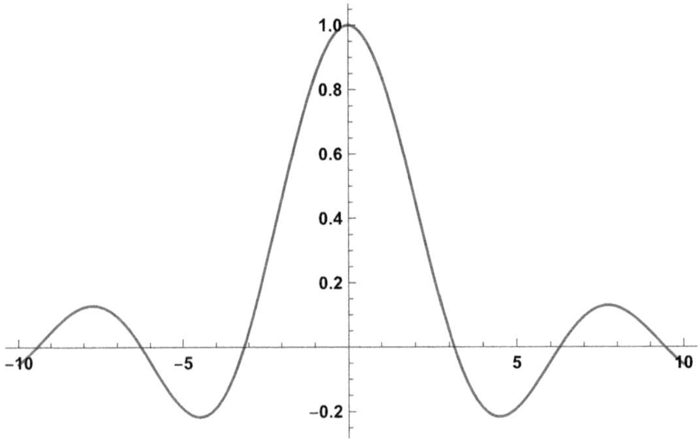

Out[29]=

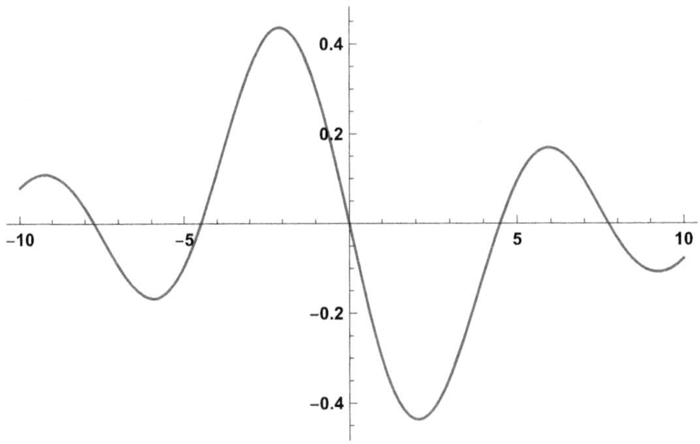

Out[30]=

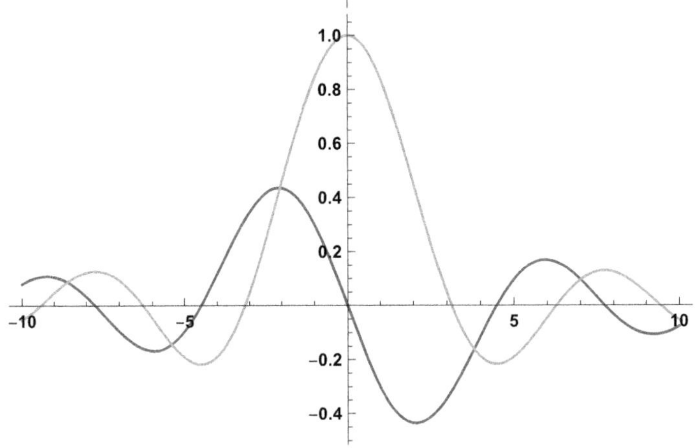

Finally, we can integrate $\frac{\text{Cos}[x]}{x} - \frac{\text{Sin}[x]}{x^2}$ and we obtain $\frac{\text{Sin}[x]}{x}$ back again.

***D*[*x*Sin[*x*], *x*]**

$x\text{Cos}[x] + \text{Sin}[x]$

$$\textbf{Integrate}\left[\frac{\textbf{Cos}[x]}{x} - \frac{\textbf{Sin}[x]}{x^2}, x\right]$$

$$\frac{\text{Sin}[x]}{x}$$

This was a short excursion that begins to illustrate how to use *Mathematica*. By experimenting, or literally, playing with these functions, a person can begin to build an intuition for the mathematics. Such intuition and the ability to visualize in the mind's eye are valuable skills for aspiring physical scientists and engineers to develop, and maybe for would-be mathematicians as well. While these are simple examples, they illustrate that *Mathematica* can be a computational laboratory, wherein the emphasis is on the mathematics rather than on the code.

DSolve

Most of the differential equations that we will be called upon to solve in this text are ordinary rather than partial. We will need to know the initial conditions in order to solve them for a function that describes the behavior of the system we are analyzing. Both **DSolve** and **NDSolve** can be used seamlessly to accomplish this. They can also be used for multiple coupled equations. Their syntax follows essentially that which we have seen for the commands that we have used to this point.

Early on we will find that many of the differential equations that we seek to solve belong to a general class that can be "separated." This means that all the independent variables can be placed on one side of the equation and the dependent ones on the other. An example of such an equation is:

$$\frac{df[x]}{dx} = -\text{C1} f[x]$$

This can be rewritten as:

$$\frac{df[x]}{f[x]} = -\text{C1}dx$$

The solution can be found by integrating both sides, on the left, over **f [x]** and on the right, over **x**. Hence the first equations that we will want to solve and be solved via separation and integration. We can solve this equation, even though **f [x]** is left unspecified, over some interval from f[x1] to f[x2]. When we undertake doing the integral of the LHS, we find that *Mathematica* takes a very long time (> 200 seconds) to give us the obvious answer in terms of the Log [f [x]], and when it does give the answer, it generates lines and lines of conditions that are meant to condition the expression. Why? Because the definite integral that we know so well is conditioned by being defined on the set of real numbers, and the function values must be greater than zero and so forth. If this is left unspecified, then *Mathematica* will be explicit in its answer.

$$\int_{f[x1]}^{f[x2]} \frac{1}{f[x]} df[x]$$

ConditionalExpression[
−Log[f[x1]] + Log[f[x2]], ((Re[f[x1]] < 0&&
((Im[f[x1]] < 0&&(Im[f[x2]] < Im[f[x1]]||
(Im[f[x2]] == Im[f[x1]]&&
(Re[f[x2]] < (Im[f[x1]]^2−
Im[f[x1]]Im[f[x2]] +
Re[f[x1]]^2)/Re[f[x1]]||
Re[f[x2]] > (Im[f[x1]]^2−
Im[f[x1]]Im[f[x2]] +
Re[f[x1]]^2)/Re[f[x1]]))||
Im[f[x1]] < Im[f[x2]] < 0||
(f[x2] ∈ Reals&&
(Re[f[x2]] < 0||Re[f[x2]] > 0))||
(Im[f[x2]] > 0&&Re[f[x2]] >
$\frac{\text{Im}[f[\text{x2}]]\text{Re}[f[\text{x1}]]}{\text{Im}[f[\text{x1}]]}$)))||
(f[x1] ∈ Reals&&(Im[f[x2]] < 0||
(f[x2] ∈ Reals&&
(Re[f[x2]] < Re[f[x1]]||
Re[f[x1]] < Re[f[x2]] < $\frac{1}{2}$Re[
f[x1]] + $\frac{1}{2}\sqrt{\text{Re}[f[\text{x1}]]^2}$))||
Im[f[x2]] > 0))||(Im[f[x1]] > 0&&
((Im[f[x2]] < 0&&Re[f[x2]] >
$\frac{\text{Im}[f[\text{x2}]]\text{Re}[f[\text{x1}]]}{\text{Im}[f[\text{x1}]]}$)||
(f[x2] ∈ Reals&&(Re[f[x2]] <
0||Re[f[x2]] > 0))||
0 < Im[f[x2]] < Im[f[x1]]||
(Im[f[x2]] == Im[f[x1]]&&

```
(Re[f[x2]] < (Im[f[x1]]^2−
Im[f[x1]]Im[f[x2]] +
 Re[f[x1]]^2)/Re[f[x1]]||
Re[f[x2]] > (Im[f[x1]]^2−
Im[f[x1]]Im[f[x2]] +
 Re[f[x1]]^2)/Re[f[x1]]))||
Im[f[x2]] > Im[f[x1]]))))||
(Re[f[x1]] == 0&&((Im[f[x1]] < 0&&
(Im[f[x2]] < Im[f[x1]]||
(Im[f[x2]] == Im[f[x1]]&&
(Re[f[x2]] < 0||Re[f[x2]] > 0))||
Im[f[x1]] < Im[f[x2]] < 0||
(f[x2] ∈ Reals&&
(Re[f[x2]] < 0||Re[f[x2]] > 0))||
(Im[f[x2]] > 0&&Re[f[x2]] > 0)))||
(Im[f[x1]] > 0&&
((Im[f[x2]] < 0&&Re[f[x2]] > 0)||

(f[x2] ∈ Reals&&
(Re[f[x2]] < 0||Re[f[x2]] > 0))||
0 < Im[f[x2]] < Im[f[x1]]||
(Im[f[x2]] == Im[f[x1]]&&
(Re[f[x2]] < 0||Re[f[x2]] > 0))||
Im[f[x2]] > Im[f[x1]]))))||
(Re[f[x1]] > 0&&((Im[f[x1]] < 0&&
(Im[f[x2]] < Im[f[x1]]||
(Im[f[x2]] == Im[f[x1]]&&
(Re[f[x2]] < (Im[f[x1]]^2−
Im[f[x1]]Im[f[x2]] +
 Re[f[x1]]^2)/Re[f[x1]]||
Re[f[x2]] > (Im[f[x1]]^2−
Im[f[x1]]Im[f[x2]] +
 Re[f[x1]]^2)/Re[f[x1]]))||
Im[f[x1]] < Im[f[x2]] < 0||
(f[x2] ∈ Reals&&
(Re[f[x2]] < 0||Re[f[x2]] > 0))||
(Im[f[x2]] > 0&&Re[f[x2]] >
 (Im[f[x2]]Re[f[x1]])/(Im[f[x1]])))||
(f[x1] ∈ Reals&&(Im[f[x2]] < 0||
```

$$
\begin{array}{l}
(f[\mathrm{x2}] \in \mathrm{Reals}\&\& \\
\left(\frac{1}{2}\mathrm{Re}[f[\mathrm{x1}]] - \frac{1}{2}\sqrt{\mathrm{Re}[f[\mathrm{x1}]]^2} < \right. \\
\mathrm{Re}[f[\mathrm{x2}]] < \mathrm{Re}[f[\mathrm{x1}]] \| \\
\mathrm{Re}[f[\mathrm{x2}]] > \mathrm{Re}[f[\mathrm{x1}]])) \| \\
\mathrm{Im}[f[\mathrm{x2}]] > 0)) \| (\mathrm{Im}[f[\mathrm{x1}]] > 0\&\& \\
((\mathrm{Im}[f[\mathrm{x2}]] < 0\&\&\mathrm{Re}[f[\mathrm{x2}]] > \\
\left.\frac{\mathrm{Im}[f[\mathrm{x2}]]\mathrm{Re}[f[\mathrm{x1}]]}{\mathrm{Im}[f[\mathrm{x1}]]}\right) \| \\
(f[\mathrm{x2}] \in \mathrm{Reals}\&\&(\mathrm{Re}[f[\mathrm{x2}]] < \\
0 \| \mathrm{Re}[f[\mathrm{x2}]] > 0)) \| \\
0 < \mathrm{Im}[f[\mathrm{x2}]] < \mathrm{Im}[f[\mathrm{x1}]] \| \\
(\mathrm{Im}[f[\mathrm{x2}]] == \mathrm{Im}[f[\mathrm{x1}]]\&\& \\
\left(\mathrm{Re}[f[\mathrm{x2}]] < \left(\mathrm{Im}[f[\mathrm{x1}]]^2 - \right.\right. \\
\mathrm{Im}[f[\mathrm{x1}]]\mathrm{Im}[f[\mathrm{x2}]] + \\
\left.\mathrm{Re}[f[\mathrm{x1}]]^2\right) / \mathrm{Re}[f[\mathrm{x1}]] \| \\
\mathrm{Re}[f[\mathrm{x2}]] > \left(\mathrm{Im}[f[\mathrm{x1}]]^2 - \right. \\
\mathrm{Im}[f[\mathrm{x1}]]\mathrm{Im}[f[\mathrm{x2}]] + \\
\left.\left.\mathrm{Re}[f[\mathrm{x1}]]^2\right) / \mathrm{Re}[f[\mathrm{x1}]]\right)) \| \\
\mathrm{Im}[f[\mathrm{x2}]] > \mathrm{Im}[f[\mathrm{x1}]])))))\&\& \\
\left(\left(\frac{f[\mathrm{x1}]}{f[\mathrm{x1}]-f[\mathrm{x2}]} \neq 0\&\& \right.\right. \\
\mathrm{Re}[\\
\left.\frac{f[\mathrm{x1}]}{-f[\mathrm{x1}]+f[\mathrm{x2}]}\right] \geq \\
0) \| \\
\mathrm{Im}\left[\frac{f[\mathrm{x1}]}{-f[\mathrm{x1}]+f[\mathrm{x2}]}\right] \neq \\
0 \| \\
\mathrm{Re}[\\
\left.\frac{f[\mathrm{x1}]}{f[\mathrm{x1}]-f[\mathrm{x2}]}\right] > \\
1)\&\& \\
((\mathrm{Im}[f[\mathrm{x1}]] \geq \\
0\&\& \\
\mathrm{Re}\left[\frac{f[\mathrm{x1}]}{f[\mathrm{x1}]-f[\mathrm{x2}]}\right] \geq \\
1\&\& \\
\mathrm{Im}[f[\mathrm{x1}]] \leq \\
\mathrm{Im}[\\
f[\mathrm{x2}]]) \| \\
\left(\mathrm{Im}[f[\mathrm{x1}]] \geq 0\&\&\mathrm{Re}\left[\frac{f[\mathrm{x1}]}{-f[\mathrm{x1}]+f[\mathrm{x2}]}\right] \geq 0\&\& \right. \\
\mathrm{Im}[f[\mathrm{x1}]] \leq \mathrm{Im}[f[\mathrm{x2}]]) \|
\end{array}
$$

$$
\begin{array}{l}
(\mathrm{Im}[f[\mathrm{x1}]] \geq 0\&\&\mathrm{Im}[f[\mathrm{x1}]] \leq \mathrm{Im}[f[\mathrm{x2}]]\&\& \\
\mathrm{Im}\left[\frac{f[\mathrm{x1}]}{-f[\mathrm{x1}]+f[\mathrm{x2}]}\right] \neq 0\Big) \| \\
(\mathrm{Im}[f[\mathrm{x1}]] \geq \mathrm{Im}[f[\mathrm{x2}]]\&\&\mathrm{Im}[f[\mathrm{x2}]] \geq 0\&\& \\
\mathrm{Re}\left[\frac{f[\mathrm{x1}]}{f[\mathrm{x1}]-f[\mathrm{x2}]}\right] \geq 1\Big) \| \\
(\mathrm{Im}[f[\mathrm{x1}]] \geq \mathrm{Im}[f[\mathrm{x2}]]\&\&\mathrm{Im}[f[\mathrm{x2}]] \geq 0\&\& \\
\mathrm{Re}\left[\frac{f[\mathrm{x1}]}{-f[\mathrm{x1}]+f[\mathrm{x2}]}\right] \geq 0\Big) \| \\
(\mathrm{Im}[f[\mathrm{x1}]] \geq \mathrm{Im}[f[\mathrm{x2}]]\&\&\mathrm{Im}[f[\mathrm{x2}]] \geq 0\&\& \\
\mathrm{Im}\left[\frac{f[\mathrm{x1}]}{-f[\mathrm{x1}]+f[\mathrm{x2}]}\right] \neq 0\Big) \| \\
(\mathrm{Im}[f[\mathrm{x1}]] \geq \mathrm{Im}[f[\mathrm{x2}]]\&\& \\
\mathrm{Re}\left[\frac{f[\mathrm{x1}]}{f[\mathrm{x1}]-f[\mathrm{x2}]}\right] \geq 1\&\&\mathrm{Im}[f[\mathrm{x1}]] \leq 0\Big) \| \\
\Big(\mathrm{Im}[f[\mathrm{x1}]] \geq \mathrm{Im}[f[\mathrm{x2}]]\&\&\mathrm{Re}\left[\frac{f[\mathrm{x1}]}{f[\mathrm{x1}]-f[\mathrm{x2}]}\right] \geq \\
1\&\&\mathrm{Im}[f[\mathrm{x2}]]\mathrm{Re}[f[\mathrm{x1}]] \leq \\
\mathrm{Im}[f[\mathrm{x1}]]\mathrm{Re}[f[\mathrm{x2}]]) \| \\
(\mathrm{Im}[f[\mathrm{x1}]] \geq \mathrm{Im}[f[\mathrm{x2}]]\&\& \\
\mathrm{Re}\left[\frac{f[\mathrm{x1}]}{-f[\mathrm{x1}]+f[\mathrm{x2}]}\right] \geq 0\&\&\mathrm{Im}[f[\mathrm{x1}]] \leq 0\Big) \| \\
\Big(\mathrm{Im}[f[\mathrm{x1}]] \geq \mathrm{Im}[f[\mathrm{x2}]]\&\&\mathrm{Re}\left[\frac{f[\mathrm{x1}]}{-f[\mathrm{x1}]+f[\mathrm{x2}]}\right] \geq \\
0\&\&\mathrm{Im}[f[\mathrm{x2}]]\mathrm{Re}[f[\mathrm{x1}]] \leq \\
\mathrm{Im}[f[\mathrm{x1}]]\mathrm{Re}[f[\mathrm{x2}]]) \| \\
(\mathrm{Im}[f[\mathrm{x1}]] \geq \mathrm{Im}[f[\mathrm{x2}]]\&\&\mathrm{Im}[f[\mathrm{x1}]] \leq 0\&\& \\
\mathrm{Im}\left[\frac{f[\mathrm{x1}]}{-f[\mathrm{x1}]+f[\mathrm{x2}]}\right] \neq 0\Big) \| \\
(\mathrm{Im}[f[\mathrm{x1}]] \geq \mathrm{Im}[f[\mathrm{x2}]]\&\& \\
\mathrm{Im}[f[\mathrm{x2}]]\mathrm{Re}[f[\mathrm{x1}]] \leq \mathrm{Im}[f[\mathrm{x1}]]\mathrm{Re}[f[\mathrm{x2}]]\&\& \\
\mathrm{Im}\left[\frac{f[\mathrm{x1}]}{-f[\mathrm{x1}]+f[\mathrm{x2}]}\right] \neq 0\Big) \| \\
(\mathrm{Im}[f[\mathrm{x2}]]\mathrm{Re}[f[\mathrm{x1}]] \geq \mathrm{Im}[f[\mathrm{x1}]]\mathrm{Re}[f[\mathrm{x2}]]\&\& \\
\mathrm{Re}\left[\frac{f[\mathrm{x1}]}{f[\mathrm{x1}]-f[\mathrm{x2}]}\right] \geq 1\&\& \\
\mathrm{Im}[f[\mathrm{x1}]] \leq \mathrm{Im}[f[\mathrm{x2}]]) \| \\
(\mathrm{Im}[f[\mathrm{x2}]]\mathrm{Re}[f[\mathrm{x1}]] \geq \mathrm{Im}[f[\mathrm{x1}]]\mathrm{Re}[f[\mathrm{x2}]]\&\& \\
\mathrm{Re}\left[\frac{f[\mathrm{x1}]}{-f[\mathrm{x1}]+f[\mathrm{x2}]}\right] \geq 0\&\& \\
\mathrm{Im}[f[\mathrm{x1}]] \leq \mathrm{Im}[f[\mathrm{x2}]]) \| \\
(\mathrm{Im}[f[\mathrm{x2}]]\mathrm{Re}[f[\mathrm{x1}]] \geq \mathrm{Im}[f[\mathrm{x1}]]\mathrm{Re}[f[\mathrm{x2}]]\&\& \\
\mathrm{Im}[f[\mathrm{x1}]] \leq \mathrm{Im}[f[\mathrm{x2}]]\&\& \\
\mathrm{Im}\left[\frac{f[\mathrm{x1}]}{-f[\mathrm{x1}]+f[\mathrm{x2}]}\right] \neq 0\Big) \| \\
\Big(\mathrm{Re}\left[\frac{f[\mathrm{x1}]}{f[\mathrm{x1}]-f[\mathrm{x2}]}\right] \geq 1\&\& \\
\mathrm{Im}[f[\mathrm{x1}]] \leq \mathrm{Im}[f[\mathrm{x2}]]\&\&\mathrm{Im}[f[\mathrm{x2}]] \leq 0) \|
\end{array}
$$

$\left(\text{Re}\left[\frac{f[\text{x1}]}{-f[\text{x1}]+f[\text{x2}]}\right] \geq 0\&\&$
$\text{Im}[f[\text{x1}]] \leq \text{Im}[f[\text{x2}]]\&\&\text{Im}[f[\text{x2}]] \leq 0) \|$
$(\text{Im}[f[\text{x1}]] \leq \text{Im}[f[\text{x2}]]\&\&\text{Im}[f[\text{x2}]] \leq 0\&\&$
$\left.\left.\text{Im}\left[\frac{f[\text{x1}]}{-f[\text{x1}]+f[\text{x2}]}\right] \neq 0\right)\right)\Big]$

As an aside, there will be times when a computation that you have launched seems to be taking too long to converge and to provide an answer. At such times, go to the Evaluation menu and click on Quit Kernel. A dialog box will appear and click Quit. This stops the kernel and allows you to regain control of the computation. You, then, can either restart the kernel, or reevaluate the script after making some changes to it.

In contrast, if we simply take the antiderivative, that is the indefinite form of the integral, the expected result comes back nearly instantly.

$\int \frac{1}{f[x]} df[x]$//Timing

{0.000414, Log[$f[x]$]}

One way around the problem with the definite integral, is to tell *Mathematica* to not provide all the conditions for the solution, as is done in what follows inside the **Integrate** command:

Integrate[$\frac{1}{f[x]}$, {$f[x]$, f[x1], f[x2]},
GenerateConditions → False]//Timing

(* By setting GenerateConditions to False, we save time,
but we lose information that in some cases may be essential. *)

{0.539682, −Log[f[x1]] + Log[f[x2]]}

The integral of the RHS is just as follows:

Clear[C1](* Clear parameters *)

$\int_{\text{x1}}^{\text{x2}} -\text{C1}dx$//Simplify

C1(x1 − x2)

This can be simplified as follows, if we seek to find f [x2] with the initial condition that f [x1] is fo at x1 equal to zero:

(* Set the LHS equal to the RHS, set f[x1] to fo, and x1 to zero, then Solve. *)

Solve[−Log[f[x1]] + Log[f[x2]]== − C1(−x1 + x2),

f[x2]]/.{f[x1] → fo, x1 → 0}

$\{\{f[\text{x2}] \to \text{ConditionalExpression}\left[e^{-\text{C1x2}}\text{fo}, -\pi < \text{Im}[-\text{C1x2} + \text{Log}[\text{fo}]] \leq \pi\right]\}\}$

So **f [x 2] = fo** $e^{-\mathbf{C1x2}}$. We can test this solution by placing it back in the differential equation on the left-hand side to see if the derivative will equal the right-hand side. To do this verification, we define the function for **f [x]** and then take its derivative and test if the derivative of the solution is the same as the original right-hand side of the equation. We do this last operation by placing the derivative and the right-hand side of the equation astride of the double equal sign and all of this is placed within the Simplify command. If the two elements on either side of "==" are in fact the same, then *Mathematica* returns a "**True**" statement.

f[x_]:=$e^{-\mathbf{C1}x}$fo (* Specifies the function *)

Simplify [$\partial_x f[x] == -\mathrm{C1} f[x]$]

(* Tests the validity of the solution *)

True

We have learned several important new concepts from this example.

- many differential equations are separable and require nothing more than the integration of the left-hand and right-hand sides

- we can use **Integrate** or the palette equivalent to carry out this operation on the separated form of the equation

- the solution we obtain can be made specific for the initial conditions by adding them at the end of the appropriate **Solve** statement

- solutions can typically be simplified

- the solution must be verified by testing its validity in the original differential equation.

The last point may not seem important, but it is, especially, when we derive analytical solutions that are far more complex. A slightly more complex form of a separable equation is one that involves a sum on the right-hand side, such as:

$$\frac{dg[x]}{dx} = \text{C1} + \text{C2}g[x]$$

This is also separable:

$$\frac{dg[x]}{C1 + \text{C2}g[x]} = dx$$

This is amenable to the techniques we have just used for the simpler equation, except that now that we know what we are doing we will combine the steps including the verification. First, we clear all variables and parameters. Then we set up the two integrals and let them be equal:

Clear[C1, C2, g, go, x1, x2]

Integrate$\left[\frac{1}{\text{C1}+\text{C2}g[x]}, \{g[x], g[\text{x1}], g[\text{x2}]\},\right.$
GenerateConditions → False$\left.\right]$==$\int_{\text{x1}}^{\text{x2}} dx$

$$\frac{-\text{Log}[\text{C1}+\text{C2}g[\text{x1}]]+\text{Log}[\text{C1}+\text{C2}g[\text{x2}]]}{\text{C2}} == -\text{x1}+\text{x2}$$

Next, we use Solve to rearrange the solution for g[x2], subject to the initial and final conditions as specified:

Solve$\left[\frac{-\text{Log}[\text{C1}+\text{C2}g[\text{x1}]]+\text{Log}[\text{C1}+\text{C2}g[\text{x2}]]}{\text{C2}} == -\text{x1}+\text{x2},\right.$

g[x2]]/.{g[x1] → go, x1 → 0, x2 → x}

$$\left\{\left\{g[x]\to \text{ConditionalExpression}\left[\frac{-\text{C1}+e^{\text{C2}x}(\text{C1}+\text{C2go})}{\text{C2}}, -\pi < \text{Im}[\text{C2}x+\text{Log}[\text{C1}+\text{C2go}]]\leq \pi\right]\right\}\right\}$$

We take the expression from this last result and assign it to the function g[x]:

g[x_]:=Evaluate[g[x]/.%]

(* Assign result from the last line (%)
to g[x] *)

(* See g[x] *)

g[x]

$$\left\{\text{ConditionalExpression}\left[\frac{-\text{C1}+e^{\text{C2}x}(\text{C1}+\text{C2go})}{\text{C2}}, -\pi < \text{Im}[\text{C2}x+\text{Log}[\text{C1}+\text{C2go}]]\leq \pi\right]\right\}$$

Lastly, we put g[x] back into the original equation to determine if the solution was correct:

Simplify $[\partial_x g[x] == \mathrm{C1} + \mathrm{C2}g[x]]$

(* This tests the solution *)

True

The result is that the LHS is equal to the RHS and we see that this evaluates to True.

In one set of statements, we have solved the separated equation, rearranged the function subject to the initial conditions, defined the function, verified it, and then restated the solution. Generally, we can use **DSolve** to find an analytical solution when one is possible. This is more general because **DSolve** can find solutions to much more complex cases than we have examined to this point, that is for those equations that are not separable. If no analytical solution exists, then we can solve the equation numerically with **NDSolve**. Let's see how these two powerful commands work.

We can redo the problem that we have just finished to see what is similar and different about using **DSolve**. The syntax is such that we place the equation and the initial condition in curly brackets, followed by the name of the function we seek and the name of the independent variable:

Clear[C1, C2, g**, go, x1, x2]**

(* We clear all the parameters *)

(* The syntax of DSolve is straightforward: differential equation, initial condition, function to be solved for and the independent variable *)

DSolve $[\{\partial_x g[x] == \mathrm{C1} + \mathrm{C2}g[x], g[0] == \mathrm{go}\}, g[x], x]$

$$\left\{\left\{g[x] \to \frac{-\mathrm{C1}+\mathrm{C1}e^{\mathrm{C2}x}+\mathrm{C2}e^{\mathrm{C2}x}\mathrm{go}}{\mathrm{C2}}\right\}\right\}$$

Verification can be done as we did before:

$$g[\mathrm{x\_}] := \frac{-\mathrm{C1} + \mathrm{C2}e^{\mathrm{C2}x}\left(\frac{\mathrm{C1}}{\mathrm{C2}} + \mathrm{go}\right)}{\mathrm{C2}}$$

Simplify $[\partial_x g[x] == \mathrm{C1} + \mathrm{C2}g[x]]$

True

This was much briefer than the previous method.

Another type of equation that we are likely to encounter is a linear first order differential equation (**LFODE**). An example is given below:

Clear[C1, y, yo, x]

DSolve[{$\partial_x y[x]$ + C1$y[x]$ == $g[x]$, y[0]==yo},

$y[x]$, x]

$$\{\{y[x] \to e^{-\mathrm{C1}x}\left(\mathrm{yo} + \int_0^x e^{\mathrm{C1}K[1]}g[K[1]]dK[1]\right)\}\}$$

This solution is implicit, meaning that it is not fully evaluated. We can see that this is so from the fact that on the right-hand side we have an integral that is over the function g and is left in terms of the dummy variable **K[1]**. Also we see that the exponential involves this variable as well. Until **g[x]** is specified, we cannot find the full solution to this problem. We can see what happens when **g [x]** = x^2 or **Sin [x]**, that is for specific functional forms:

Clear[C1, yo]

DSolve[{$\partial_x y[x]$ + C1$y[x]$ == x^2, y[0]==yo}, $y[x]$, x]

(* Let $g[x]$ = x^2 *)

$$\left\{\left\{y[x] \to \frac{e^{-\mathrm{C1}x}\left(-2+2e^{\mathrm{C1}x}-2\mathrm{C1}e^{\mathrm{C1}x}x+\mathrm{C1}^2e^{\mathrm{C1}x}x^2+\mathrm{C1}^3\mathrm{yo}\right)}{\mathrm{C1}^3}\right\}\right\}$$

Clear[C1, yo]

DSolve[{$\partial_x y[x]$ + C1$y[x]$ == Sin[x], y[0]==yo},

$y[x]$, x](* Let $g[x]$ = Sin[x] *)

$$\left\{\left\{y[x] \to \frac{e^{-\mathrm{C1}x}\left(1+\mathrm{yo}+\mathrm{C1}^2\mathrm{yo}-e^{\mathrm{C1}x}\mathrm{Cos}[x]+\mathrm{C1}e^{\mathrm{C1}x}\mathrm{Sin}[x]\right)}{1+\mathrm{C1}^2}\right\}\right\}$$

These solutions are involved and so it is critical that we verify them before applying them.

Clear[C1, yo]

y1[x_]:=$\dfrac{e^{-\mathrm{C1}x}\left(-2+2e^{\mathrm{C1}x}-2\mathrm{C1}e^{\mathrm{C1}x}x+\mathrm{C1}^2e^{\mathrm{C1}x}x^2+\mathrm{C1}^3\mathrm{yo}\right)}{\mathrm{C1}^3}$

(* Assign $y[x]$ *)

Simplify[∂_xy1[x] + C1y1[x] == x^2]

(* Test the validity *)

True

Clear[C1, yo]

$$\text{y2[x\_]:=}\frac{e^{-\text{C1}x}\left(1+\text{yo}+\text{C1}^2\text{yo}-e^{\text{C1}x}\text{Cos}[x]+\text{C1}e^{\text{C1}x}\text{Sin}[x]\right)}{1+\text{C1}^2}$$

Simplify $[\partial_x \text{y2}[x] + \text{C1y2}[x] == \text{Sin}[x]]$

True

Finally, given a set of parameter values for **C1** and **yo**, we can **Plot** the two solutions simultaneously, with red (mid gray in the print version) for **y1** and blue (dark gray in the print version) for **y2** to see how they behave with increasing x in a specific range:

C1 = 1;

yo = 10;

Plot[{y1[*x*], y2[*x*]}, {*x*, 0, 10},

PlotStyle → {Hue[0], Hue[0.6]},

PlotLabel → {"red = y1[x]", "blue = y2[x]"},

AxesLabel → {*x*, "yn[x]"}]

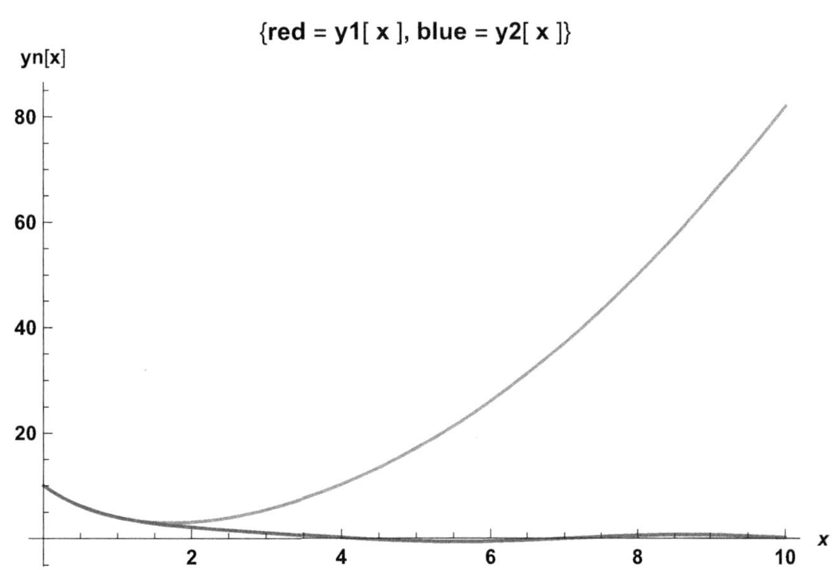

As expected, we can see that the two solutions overlap considerable at small x values up to about 2 and then their behaviors deviate markedly. If we want to get a sense of parametric sensitivity, we can drop the value of **C1** by 10^2 and then raise it by 10 and replot the graphs for these two cases:

```
C1 = .01;
yo = 10;
Plot[{y1[x], y2[x]}, {x, 0, 10},
PlotStyle → {Hue[0], Hue[0.6]},
PlotLabel → {red"y1[x]=", blue"y2[x]="},
AxesLabel → {x, "yn[x]"}]
```

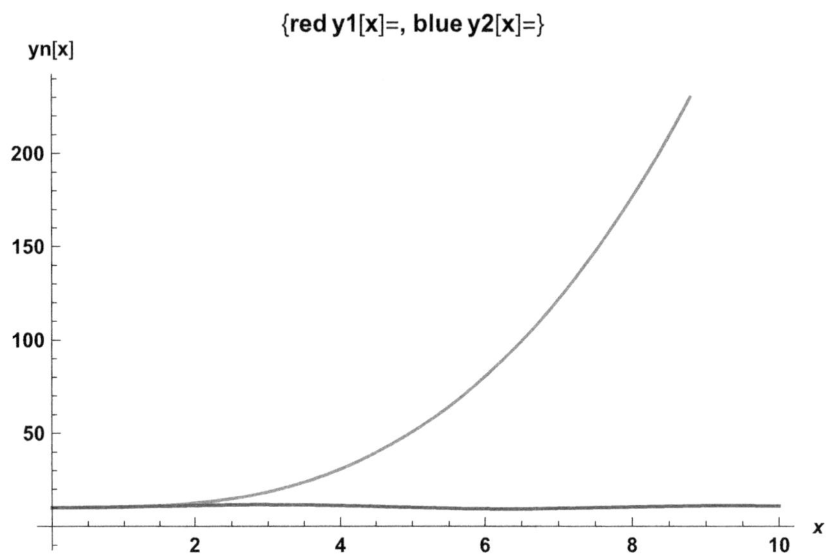


```
C1 = 10;
yo = 10;
Plot[{y1[x], y2[x]}, {x, 0, 10},
PlotStyle → {Hue[0], Hue[0.6]},
PlotLabel → {red"y1[x]=", blue"y2[x]="},
AxesLabel → {x, "yn[x]"}]
```

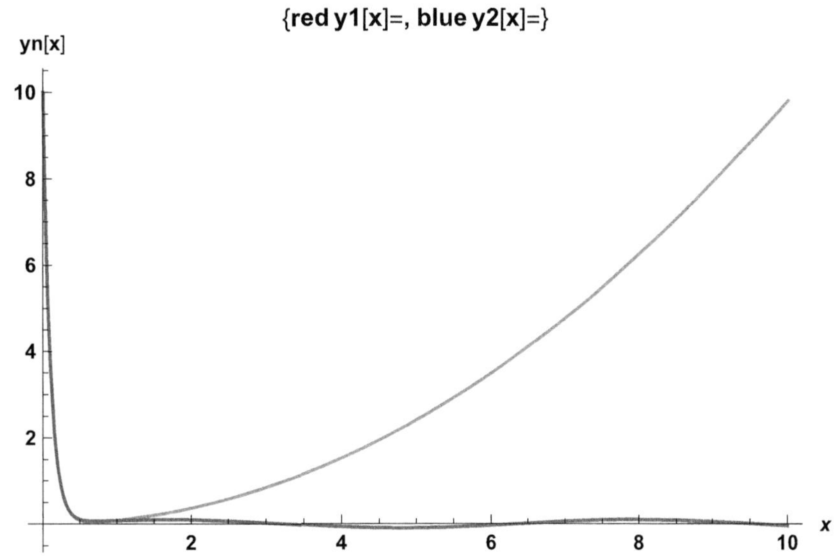

Remove[y1, y2, yo, C1, C2, *g*, go, x1, x2]

NDSolve

We turn now to **NDSolve** for the numerical solution of differential equations. A good starting point would be to begin to solve the equations that we just solved symbolically with **DSolve**. Instead of simply solving the equation, we are going to name the solution, we can call it **y2**:

Clear[C1, yo, soln, *y*, yo]

(* Clear all parameters *)

yo = 10;

(* Set parameters *)

C1 = 1;

(* Specify the equation, initial conditions and variable range *)

soln = NDSolve $[\{\partial_x y[x] + C1\, y[x] == \mathrm{Sin}[x], y[0] == \mathrm{yo}\},$

$y[x], \{x, 0, 10\}]$

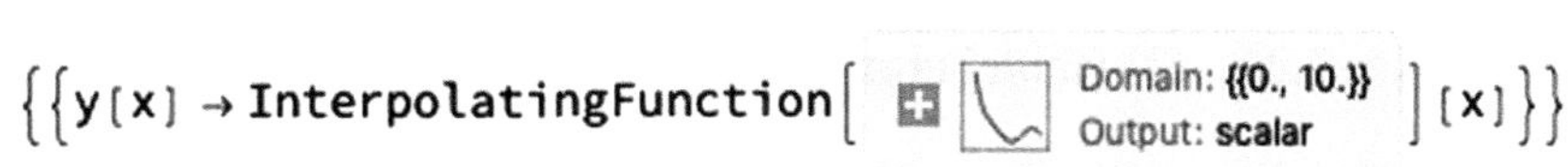

What we find is that the numerical solution is presented in the form of an **Interpolation** function, which is valid over the range of integration. This is much cleaner than having a table or list of values. To use the interpolated function we must assign it a function name and then we can apply it and explore the numerical solution's behavior. To do this we use a command structure that we have utilized before; it looks like this:

nyb[x_]:=Evaluate[*y*[*x*]/.soln]

(* This command assigns *a* new function to the interpolating function *)

What this says in simple terms is to assign to **ny2[x]** the interpolating function **y[x]** found in the solution called **y2**. This function can now be plotted:

Plot[nyb[*x*], {*x*, 0, 10}]

(* Plot the interpolating function *)

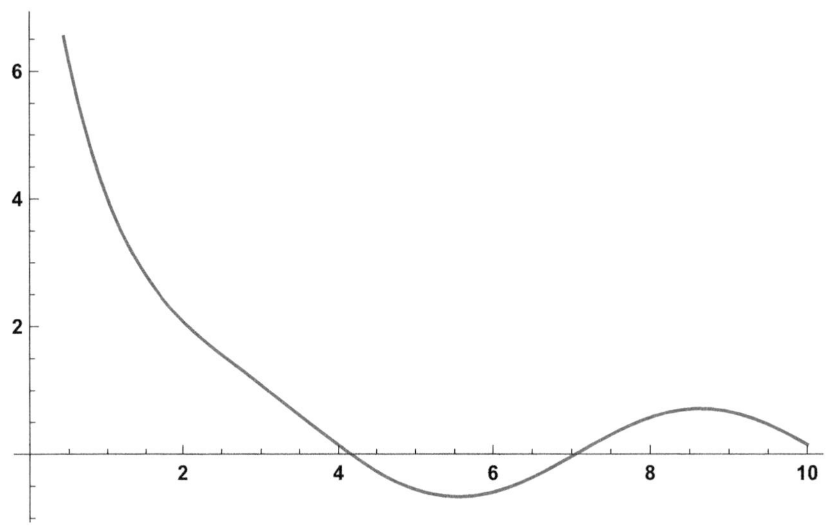

This looks identical to the plot we had before based upon the analytical solution. If we need to have a table of values for the function we can get this as follows:

Table[{*x*, nyb[*x*]}, {*x*, 0, 10, .5}]//TableForm

0.	10.
0.5	6.16949
1.	4.01332
1.5	2.80624
2.	2.08374
2.5	1.5617
3.	1.08832
3.5	0.609909
4.	0.140735
4.5	−0.266722
5.	−0.550544
5.5	−0.664194
6.	−0.593766
6.5	−0.364947
7.	−0.038883
7.5	0.301489
8.	0.570951
8.5	0.702386
9.	0.662919
9.5	0.461795
10.	0.148002

It is clear that for relatively simple linear equations such as these, that **DSolve** and **NDSolve** duplicate each other. When the equations become non-linear, however, it may not be possible to find an analytical solution. At that point **NDSolve** no longer merely duplicates, but, rather, it supplants **DSolve**. For example, if in the last differential equation **y[x]** appears quadratically rather than linearly, **DSolve** will not return a solution:

```
Clear[C1, yo]
DSolve[{∂x y[x] + C1 y[x]^2 == Sin[x], y[0] == yo}, y[x], x]
```

```
... Solve: Inverse functions are being used by Solve,
 so some solutions may not be found; use Reduce for complete solution information.
```

DSolve$\left[\left\{\text{C1}\,y[x]^2 + y'[x] == \text{Sin}[x], y[0] == \text{yo}\right\}, y[x], x\right]$

In contrast NDSolve will do so and it will do it well:

Clear[C1, yo]

yo = 10;

C1 = 0.01;

y3 = NDSolve$\left[\left\{\partial_x y[x] + \text{C1}\,y[x]^2 == \text{Sin}[x], y[0] == \text{yo}\right\}, y[x], \{x, 0, 10\}\right]$**;**

ny3[x_]:=Evaluate[$y[x]$/.y3]

plny3 = Plot[ny3[x], {x, 0, 10}, PlotStyle → Hue[0.6]]

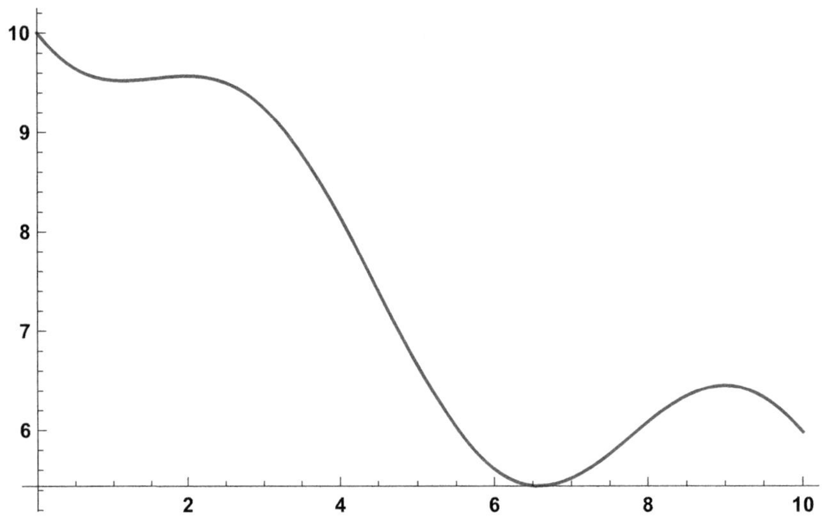

For the sake of learning we can now go back and compare the solution of this non-linear equation to the linear version. To do so numerically, we must solve the equation with the new value of **C1** set to 0.01:

Clear[C1, yo]

yo = 10;

```
C1 = 0.01;
(* Numerically solve the linear equation *)
y2 = NDSolve[{∂x y[x] + C1 y[x] == Sin[x], y[0] == yo}, y[x], {x, 0, 10}];
(* Assign the function *)
ny2[x_]:=Evaluate[y[x]/.y2]
plny2 = Plot[ny2[x], {x, 0, 10},
PlotRange → {{0, 10}, {0, 12}}, PlotStyle → Hue[0]];
(* Plot the solutions to the linear and the non – linear equations *)
Show[plny2, plny3, PlotLabel → {red"ny2=", blue"ny3"},
AxesLabel → {x, "nyi[x]"}]
```

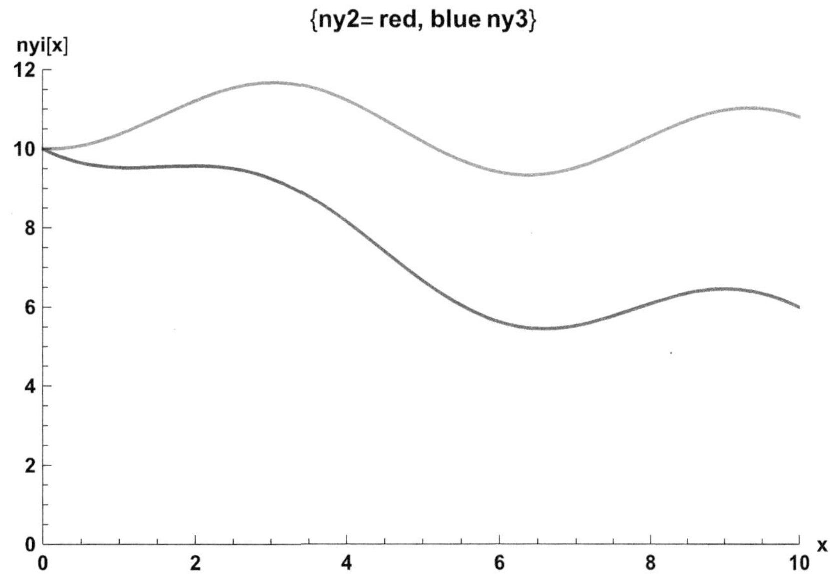

NDSolve is a very powerful method for solving one or a set of differential equations. It can be used for complex ordinary differential equations and to solve partial differential equations. We have used it to solve simple equations and with all of its options set to their default values. However, for more difficult problems, the ability to vary the options makes it even more likely to find solutions. Refer to NDSolve to read about these options and how they can be varied.

```
?? NDSolve
```

Symbol

NDSolve[*eqns*, *u*, {*x*, x_{min}, x_{max}}] finds a numerical solution to the ordinary differential equations *eqns* for the function *u* with the independent variable *x* in the range x_{min} to x_{max}.

NDSolve[*eqns*, *u*, {*x*, x_{min}, x_{max}}, {*y*, y_{min}, y_{max}}] solves the partial differential equations *eqns* over a rectangular region.

NDSolve[*eqns*, *u*, {*x*, *y*} ∈ Ω] solves the partial differential equations *eqns* over the region Ω.

NDSolve[*eqns*, *u*, {*t*, t_{min}, t_{max}}, {*x*, *y*} ∈ Ω] solves the time-dependent partial differential equations *eqns* over the region Ω.

NDSolve[*eqns*, {u_1, u_2, ...}, ...] solves for the functions u_i.

Documentation Local » | Web »

Options › AccuracyGoal → Automatic...(16 total)

Attributes {Protected}

Full Name System`NDSolve

NonlinearModelFit

Another powerful method for modeling is to use the function NonlinearModelFit. With this function we can fit non-linear data to functions of our own choice. The syntax is given as follows:

NonlinearModelFit [Data, Model, Parameters, Independent Variable]

As with other functions, there are many different options that can be varied to make the fitting process more robust for any given problem or set of data. To get an idea of how to use NonlinearModelFit, let's go back to the differential equation that we just solved:

$$\partial_x y[x] + C1\, y[x]^2 == \mathrm{Sin}[x],\ y[0] == \mathrm{yo}$$

We can use the solution to this equation to generate a data set, then we can fit to a function. In this case, we will fit the data to Fourier Series. By doing this example, we will have a fair grasp of how to do non-linear fitting with *Mathematica* and we can also learn some other new tricks along the way.

Firstly, we can begin by solving the equation again:

(* Clear all parameters and variables *)

Clear[ao, *t*, *b*, *c*, *e*, *f*, ny3]

Clear[z1, yo]

```
(* Set the initial conditions and parameter *)
yo = 10;
z1 = 0.01;
(* Solve the equation numerically *)
y3 = NDSolve[{∂x y[x] + z1 y[x]^2 == Sin[x], y[0] == yo}, y[x], {x, 0, 10}];
(* Create the needed function *)
ny3[x_]:=Evaluate[y[x]/.y3]
```

We can use ny3[x] to create the data set called **newdata**:

```
newdata = Table[{x, ny3[x][[1]]}, {x, 0, 10, 0.25}]
```

{{0., 10.}, {0.25, 9.78668}, {0.5, 9.64234}, {0.75, 9.55804}, {1., 9.52206}, {1.25, 9.52051}, {1.5, 9.53812}, {1.75, 9.55915}, {2., 9.56831}, {2.25, 9.55172}, {2.5, 9.49772}, {2.75, 9.39754}, {3., 9.24578}, {3.25, 9.04072}, {3.5, 8.78426}, {3.75, 8.48187}, {4., 8.1421}, {4.25, 7.77608}, {4.5, 7.39685}, {4.75, 7.01857}, {5., 6.65568}, {5.25, 6.32207}, {5.5, 6.03024}, {5.75, 5.79051}, {6., 5.61043}, {6.25, 5.49422}, {6.5, 5.44247}, {6.75, 5.452}, {7., 5.51604}, {7.25, 5.62453}, {7.5, 5.7648}, {7.75, 5.92229}, {8., 6.08151}, {8.25, 6.22701}, {8.5, 6.34441}, {8.75, 6.42126}, {9., 6.44789}, {9.25, 6.41791}, {9.5, 6.32862}, {9.75, 6.18106}, {10., 5.97999}}

Next, we want to create a model to fit to this data based on a Fourier series with the following form:

$$\text{fs}[\,t\,] = \frac{1}{2}a_0 + \sum_{i=1}^{\infty} a_i \text{Cos}[\,ikt] + \sum_{i=1}^{\infty} b_i \text{Sin}[\,ikt]$$

We can build this function with 16 total terms (8 Sin and 8 Cos) easily with this script:

```
ClearAll[a, b, k]
(* Sum is the function and i is the index in the summation terms *)
modelFS = 1/2 a0 + Sum[a[i] * Cos[i k t], {i, 1, 8}]+Sum[b[i] * Sin[i k t], {i, 1, 8}]
```

$\frac{a0}{2} + a[1]\text{Cos}[kt] + a[2]\text{Cos}[2kt] + a[3]\text{Cos}[3kt] + a[4]\text{Cos}[4kt] + a[5]\text{Cos}[5kt] + a[6]\text{Cos}[6kt] + a[7]\text{Cos}[7kt] + a[8]\text{Cos}[8kt] + b[1]\text{Sin}[kt] + b[2]\text{Sin}[2kt] + b[3]\text{Sin}[3kt] + b[4]\text{Sin}[4kt] + b[5]\text{Sin}[5kt] + b[6]\text{Sin}[6kt] + b[7]\text{Sin}[7kt] + b[8]\text{Sin}[8kt]$

To be clear, our approach is to fit this function to the data by finding the best values for all the parameters, that is for all the a and b parameters that are outside of Sin and Cos terms and k the parameter (frequency) inside these functions. We will need to specify the parameters separately in the function call, so this too can be done easily:

```
(* This line of code provides all the parameters in one term *)
paramsFS = Join[{a0}, Table[a[i], {i, 1, 8}],
Table[b[i], {i, 1, 8}], {k}]
```

{a0, a[1], a[2], a[3], a[4], a[5], a[6], a[7], a[8], b[1], b[2], b[3], b[4], b[5], b[6], b[7], b[8], k}

```
(* This line of code calls NonlinearModelFit with the data,  the model,
the parameters and the independent variable *)
fs[t] = NonlinearModelFit[newdata, modelFS,paramsFS, t]
```

FittedModel[7.76562 + 2.49021 Cos[0.415721 t] + «19» + 0.00655213 Sin[«19» t] − 0.0020775 Sin[3.32577 t]]

The model is fitted to the data and we have assigned it to fs[t]. By clicking to the right of the fitted function, a new suggestion bar appears. By clicking on parameter estimates and then on parameter table a list of all the fitted constants and their statistics appears.

To make fs[t] into a working function, we need to use the Normal command and assign it to a new function name.

```
ffs[t] = Normal[fs[t]]

(* This provides a plot of ffs[t],  but we suppress it here to show it later with the data *)
pffs = Plot[ffs[t], {t, 1, 10},PlotStyle → RGBColor[0, 1, 0]];
(* Plot of the data points *)
pl = ListPlot[newdata, PlotStyle → RGBColor[1, 0, 0]];

(* Show the data and the fit *)
Show[{pffs, pl}]
```

7.76562 + 2.49021Cos[0.415721t] + 0.521968Cos[0.831442t] − 0.272279Cos[1.24716t] − 0.401435Cos[1.66288t] − 0.135318Cos[2.0786t] + 0.0110735Cos[2.49432t] +

0.017556Cos[2.91005*t*] + 0.00253409Cos[3.32577*t*] + 0.490634Sin[0.415721*t*] − 0.145299Sin[0.831442*t*] − 1.01361Sin[1.24716*t*] − 0.131291Sin[1.66288*t*] + 0.113591Sin[2.0786*t*] + 0.0644729Sin[2.49432*t*] + 0.00655213Sin[2.91005*t*] − 0.0020775Sin[3.32577*t*]

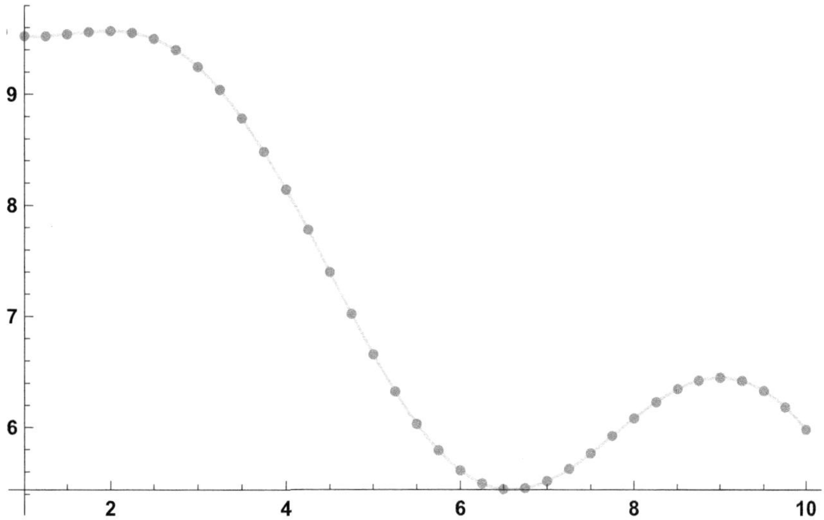

As expected, we have a very good fit to the data. This was all very routine in that we did not have to make any adjustments to the routine by specifying different options and we have enough points to get a good fit to the data. It is not always this easy.

Here is data for a different example:

```
yvals = {10.6534, 9.6646, 8.7137, 8.2863, 8.2863,
8.7137, 9.0000, 9.5726, 11.0000, 12.7137,
13.4274, 13.2863, 13.0000, 12.7137, 12.5726,
13.5726, 15.7137, 17.4274, 18.0000, 18.0000,
17.4274, 15.7137, 14.0297, 12.4345};

Length[yvals];
(* Next we create the x − values *)
xvals = Table[x, {x, 1, 24}];
(* Join the x and y values into dyads called "dat" *)
```

dat = Transpose[Partition[Join[xvals, yvals],Length[xvals]]];

(* ListPlot the "data" *)

ListPlot[dat]

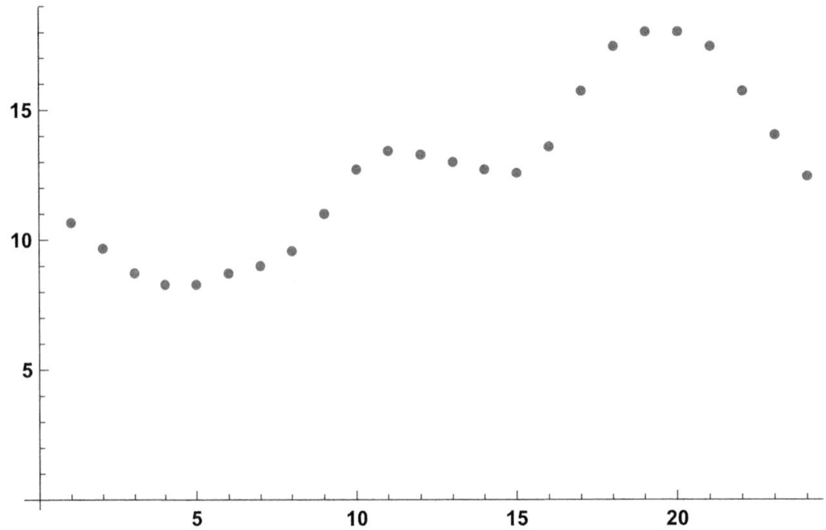

There is a bit more structure to this data and it is not as smooth as the example that we just completed. We will set up the same kind of calculation as we did previously. This time we place it all in one script. Let's see what we get:

(* Set the number of terms in the Fourier Series *)

trms = 8;

(* Clear the value of the first constant *)

a0=.

k=.

Clear[*a*, *b*, *e*, *i*]

(* Create the Fourier Series *)

model4 = a0 + Sum[*b*[*i*] ∗ Cos[*i* ∗ *t* ∗ *k*], {*i*, 1, trms}]+Sum[*e*[*i*] ∗ Sin[*i* ∗ *t* ∗ *k*], {*i*, 1, trms}];

(* Create the parameter table *)

params4 =

Flatten[Join[{a0}, Table[*b*[*i*], {*i*, 1, trms}],

```
Table[e[i], {i, 1, trms}], {k}]]
(* Call the NonLinearFit function and create m[t] *)
m[t_] = NonlinearModelFit[dat, model4, params4, t]
(Extract the Fitted Fourier Series *)
mm[t_] = Normal[m[t]]
(* Plot the Fitted Function *)
gg = Plot[mm[t], {t, 0, 25},PlotStyle->RGBColor[0, 1, 0]];
(* Show the "data" and the Fourier Series *)
Show[ListPlot[dat], gg]
```

{a0, b[1], b[2], b[3], b[4], b[5], b[6], b[7], b[8], e[1], e[2], e[3], e[4], e[5], e[6], e[7], e[8], k}

FittedModel[]

12.3101 − 0.868783Cos[1.02391t] − 0.401446Cos[2.04781t] + 0.0679157Cos[3.07172t] + 0.671671Cos[4.09562t] + 1.87606Cos[5.11953t] − 4.61127Cos[6.14344t] − 2.97884Cos[7.16734t] − 0.789547Cos[8.19125t] + 3.41542Sin[1.02391t] + 1.16644Sin[2.04781t] + 0.1662Sin[3.07172t] − 0.34132Sin[4.09562t] − 0.852824Sin[5.11953t] + 0.117358Sin[6.14344t] − 0.74963Sin[7.16734t] − 0.0954014Sin[8.19125t]

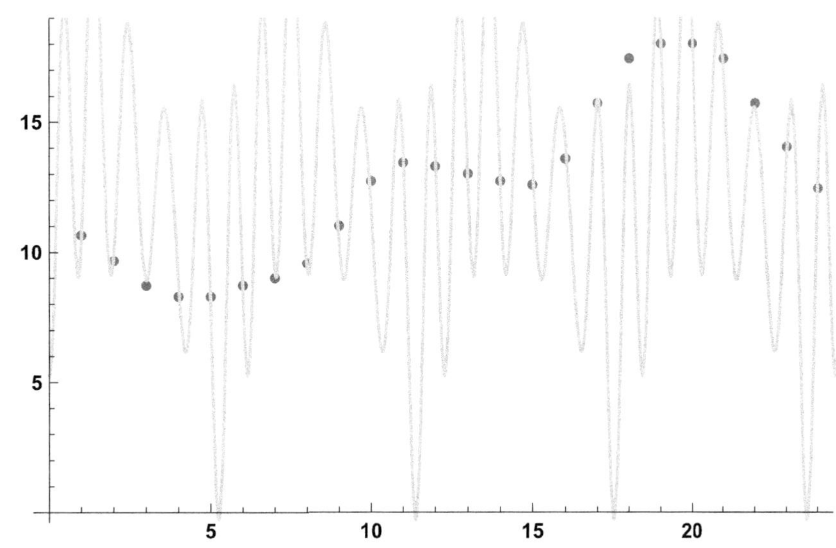

Clearly, this is not what we are looking for with such a fit. We could go back and add more terms, but the data does not look that different from the previous problem to necessarily warrant doing so. We might think that with more iterations, the fit might be better. We can copy and paste the script from above and call the option MaxIterations in the NonlinearModelFit command and set this to infinity.

```
(* Set the number of terms in the Fourier Series *)

trms = 8;

(* Clear the value of the first constant *)

a0=.k=.

Clear[a, b, e]

(* Create the Fourier Series *)

model4 = a0 + Sum[b[i] * Cos[i * t * k], {i, 1, trms}]+Sum[e[i] * Sin[i * t * k], {i, 1, trms}];

(* Create the parameter table *)

params4 =

Flatten[Join[{a0}, Table[b[i], {i, 1, trms}],Table[e[i], {i, 1, trms}], {k}]]

(* Call theNonLinearFit function and create m[t] *)

m[t_] = NonlinearModelFit[dat, model4, params4,t, MaxIterations → ∞]

(* Extract the Fitted Fourier Series *)

mm[t_] = Normal[m[t]]

(* Plot the Fitted Function *)

gg = Plot[mm[t], {t, 0, 25},PlotStyle → RGBColor[0, 1, 0]];

(* Show the "data" and the Fourier Series *)

Show[ListPlot[dat], gg]
```

{a0, b[1], b[2], b[3], b[4], b[5], b[6], b[7], b[8], e[1], e[2], e[3], e[4], e[5], e[6], e[7], e[8], k}

FittedModel[]

12.3101 − 0.868783Cos[1.02391t] − 0.401446Cos[2.04781t] + 0.0679157Cos[3.07172t] + 0.671671Cos[4.09562t] + 1.87606Cos[5.11953t] − 4.61127Cos[6.14344t] −

2.97884Cos[7.16734*t*] − 0.789547Cos[8.19125*t*] + 3.41542Sin[1.02391*t*] + 1.16644Sin[2.04781*t*] + 0.1662Sin[3.07172*t*] − 0.34132Sin[4.09562*t*] − 0.852824Sin[5.11953*t*] + 0.117358Sin[6.14344*t*] − 0.74963Sin[7.16734*t*] − 0.0954014Sin[8.19125*t*]

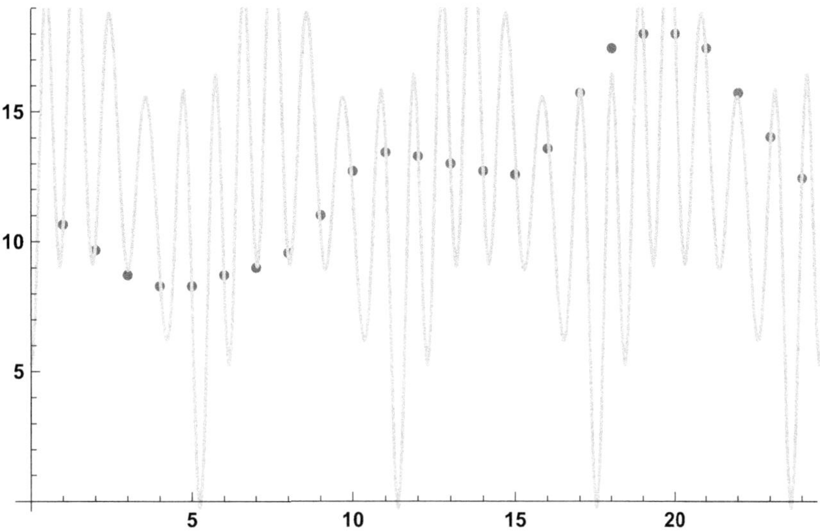

Again, we achieved the same results. One of the many other changes that we can make is to change the method used to find the minimum. In this case let's specify "QuasiNewton".

```
(* Set the number of terms in the Fourier Series *)
trms = 8;
(* Clear the value of the first constant *)
a0=.
k=.
Clear[a, b, e]
(* Create the Fourier Series *)
model4 = a0 + Sum[b[i] * Cos[i * t * k], {i, 1, trms}]+Sum[e[i] * Sin[i * t * k], {i, 1, trms}];
(* Create the parameter table *)
params4 =
Flatten[Join[{a0}, Table[b[i], {i, 1, trms}],Table[e[i], {i, 1, trms}], {k}]]
(* Call the NonLinearFit function and create m[t] *)
```

m[t_] = NonlinearModelFit[dat, model4, params4,t, MaxIterations → ∞,

Method → "QuasiNewton"]

(* Extract the Fitted Fourier Series *)

mm[t_] = Normal[m[t]]

(* Plot the Fitted Function *)

gg = Plot[mm[t], {t, 0, 25},PlotStyle → RGBColor[0, 1, 0]];

(* Show the "data" and the Fourier Series *)

Show[ListPlot[dat], gg]

{a0, b[1], b[2], b[3], b[4], b[5], b[6], b[7], b[8], e[1], e[2], e[3], e[4], e[5], e[6], e[7], e[8], k}

```
FittedModel[ «1» ]
```

12.647 − 0.113831Cos[0.258048t] − 0.165776Cos[0.516097t] − 0.617721Cos[0.774145t] + 0.247154Cos[1.03219t] − 0.0152026Cos[1.29024t] − 0.137775Cos[1.54829t] + 0.118902Cos[1.80634t] − 0.00223331Cos[2.06439t] − 3.95897Sin[0.258048t] − 1.61797Sin[0.516097t] + 0.590996Sin[0.774145t] − 0.0888776Sin[1.03219t] + 0.0648387Sin[1.29024t] + 0.0862469Sin[1.54829t] − 0.0280426Sin[1.80634t] − 0.0296926Sin[2.06439t]

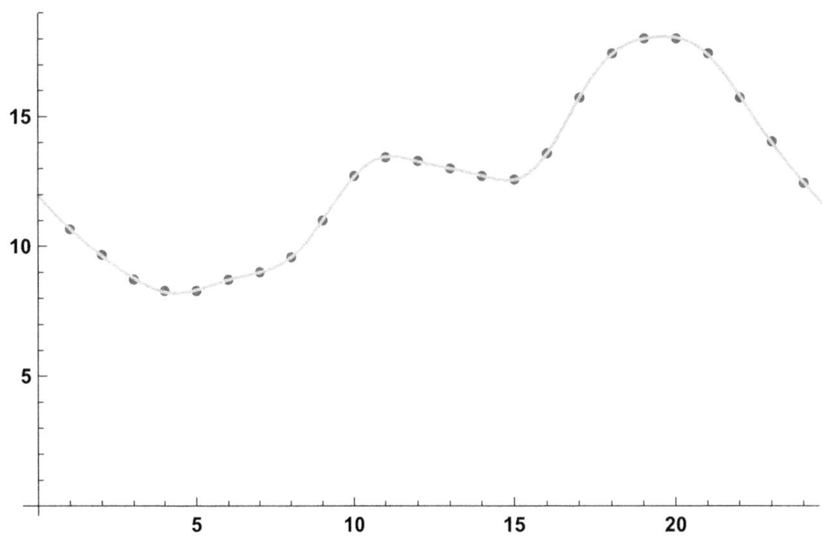

Now, the fit seems quite good overall. According to the documentation for NonlinearModelFit:

"Possible settings for Method, include ConjugateGradient, Gradient, LevenbergMarquardt, Newton, NMinimize, and QuasiNewton, with the default being Automatic."

In this case when we selected the method to be QuasiNewton the fitting proceeded nicely. Along with Method, one can change each of these parameters:

> AccuracyGoal, ConfidenceLevel, EvaluationMonitor, gradient, MaxIterations, PrecisionGoal, StepMonitor, VarianceEstimatorFunction, Weights, and WorkingPrecision.

The message here is that using NonlinearModelFit may not always be simple, and it may require a much deeper understanding of each of these parameters and their optimal setting in order to provide the requisite result.

Point: another versatile graphics method

There are times when we will want to graph and visualize three dimensional objects that are not readily available. For example, we may want to visualize a region of space that is bounded by different functions. Academically, this kind of problem arises in the Calculus when doing triple integrals. As well as doing the integrations, it is (was) typical to sketch the solid region. The quality of the results depended greatly on the ability of the person to translate what was in the mind's eye onto paper, and even if this were done well, each new view had to be rendered again. Instead, we want not to be dependent on such abilities, and we would like to be able to create such an object in-silico and to rotate it at will around all three axes. We may even want to produce a 3D print of the object if it is sufficient interest.

In the textbook "Calculus and Analytical Geometry," E. J. Purcell, Appleton-Century-Crofts (1965), we can read the following problem on page 646:

> Find the value of the triple integral of the function f, defined by f [x, y, z] = 2 x y z, over the solid region S bounded by the parabolic cylinder z = 2 - $\frac{1}{2}x^2$ and by the planes z = 0, y = x, and y = 0.

The triple integration is easy to do:

Clear[x, y, z]

$$\int_0^2 \int_0^x \int_0^{2-} 2xyz\,dz\,dy\,dx$$

$$\frac{4}{3}$$

It is useful to duplicate the text and to do each step individually:

Expand $\left[\int_0^{2-\frac{x^2}{2}} 2xyz\,dz\right]$

$\frac{x^5 y}{4} - 2x^3 y + 4xy$

Expand $\left[\int_0^x \%\,dy\right]$

$\frac{x^7}{8} - x^5 + 2x^3$

$\int_0^2 \%\,dx$

$\frac{4}{3}$

Along with this result, in figure 405, Purcell provides a very nice sketch of the region looking through the parabolic cylinder's surface toward the origin. How can we make this sketch in *Mathematica*? It may not be immediately obvious how to do so. One way that is useful is based on using the Graphics Primitive Point. We can work through this to see how this can be done.

Firstly, Point can be used in Graphics for two dimensional renderings, or in Graphics3D for three dimensional visualizations. Here are two examples:

(* The placement of the point in $2D$ **is given by the dyad inside the curly brackets. The Point primitive is called by Graphics *)**

Graphics[Point[{0, 0}]]

(* We can make changes to the size of the Point *)

Graphics[{PointSize[Large], Point[{0, 1}]}]

(* We can also change the color of the Point *)

Graphics[{PointSize[Large], Red, Point[{0, 1}]}]

We can Map Points onto an array of values and then plot those values either in two or three dimensions:

Graphics[{PointSize[Large], Red,

Point[Table[{x, 0.03x^3 + .01x^2 + .02x + 10},

{x, −10, 10}]]}]

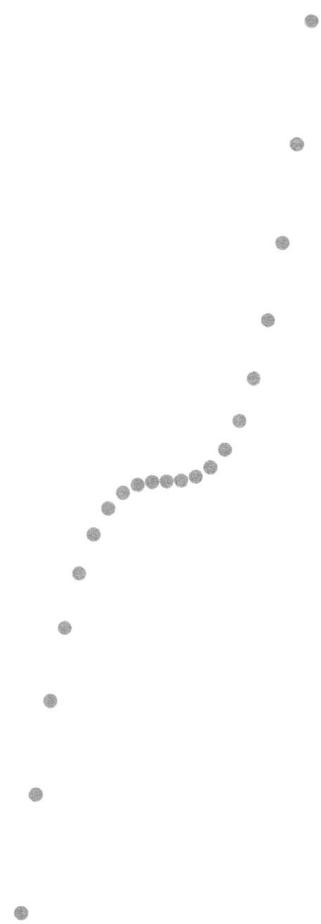

Graphics3D[{PointSize[Large], Red,

Point[Table[{x, x, 0.03x^3 + .01x^2 + .02x + 10},

{x, −10, 10}]]}]

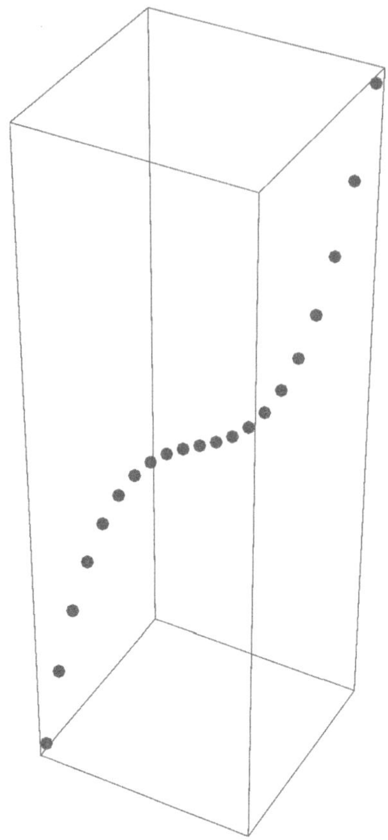

Now, we can go back to Purcell's problem, "the solid region S, is bounded by the parabolic cylinder z = 2 - $\frac{1}{2}x^2$ and by the planes z = 0, y = x, and y = 0." We can get a view of this region by plotting points on its edges. The shape of the parabolic cylinder is given by z = 2 - $\frac{1}{2}x^2$, so we know that all of the z values will fall between the edges of this shape that are made up of the plane zx when y = 0 and the plane below the parabolic edge when y = x. The bottom of the region will be a triangle in the xy plane when z = 0 and where y = x.

```
(* Here we create the triangle in the xy plane by creating a line on y = x,
from {0, 0, 0} out to the point {2, 2, 0} *)

Table[{x, x, 0}, {x, 0, 2, .05}];

(* Note that x = y and z = 0 here *)

Point/@%;
```

c1 = Graphics3D[%, Axes → True,

AxesLabel → {x, y, z}]

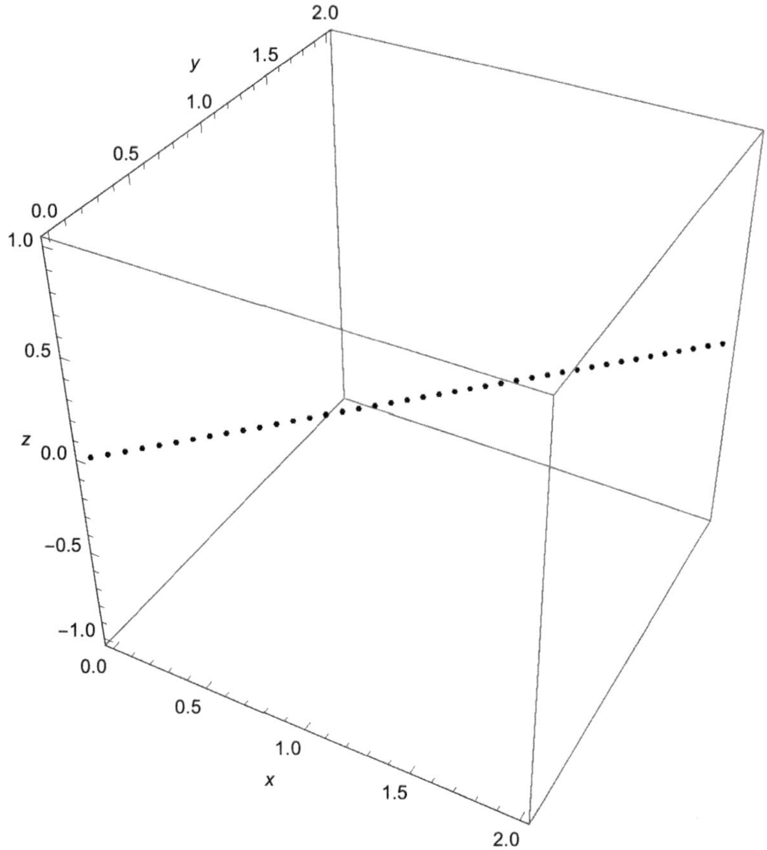

(* Next we can plot the edge of the region that exists in the xz plane *)

Table[{x, 0, 2 − 0.5x^2}, {x, 0, 2, .05}];

(* Note that y = 0 here *)

Point/@%;

b1 = Graphics3D[%, Axes → True,

AxesLabel → {x, y, z}]

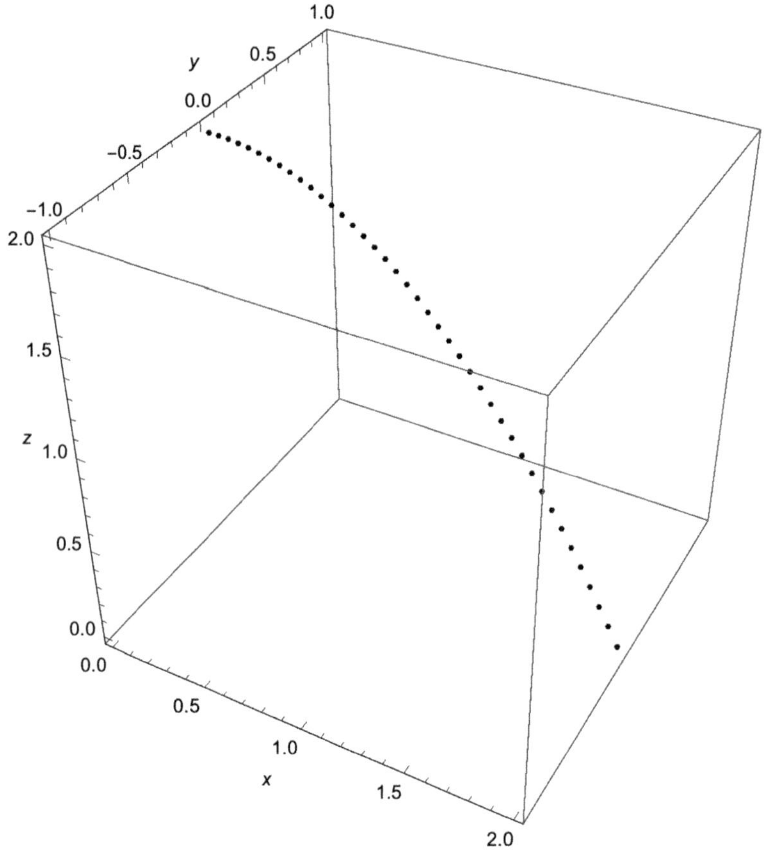

(* Finally, we can create the other edge that lies on the plane cuts the xy plane in half *)

Table[{x, x, 2 − 0.5x^2}, {x, 0, 2, .05}];

(* note that $x = y$ here *)

Point/@%;

a1 = Graphics3D[%, Axes → True, AxesLabel → {x, y, z}]

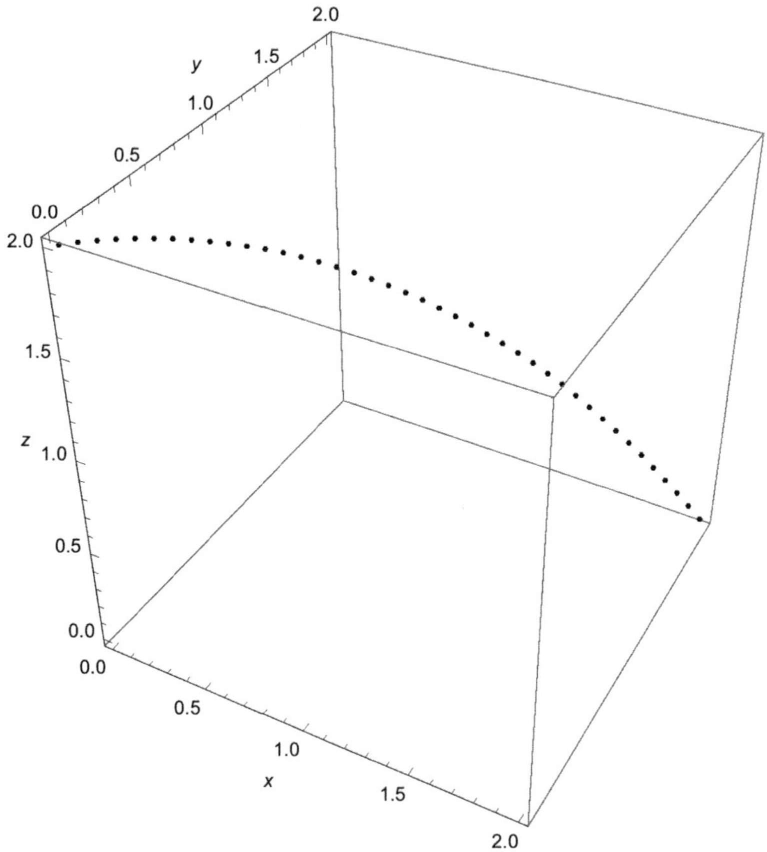

(* Now we can show all three together to see the bounded region *R* *)

edges = Show[a1, b1, c1]

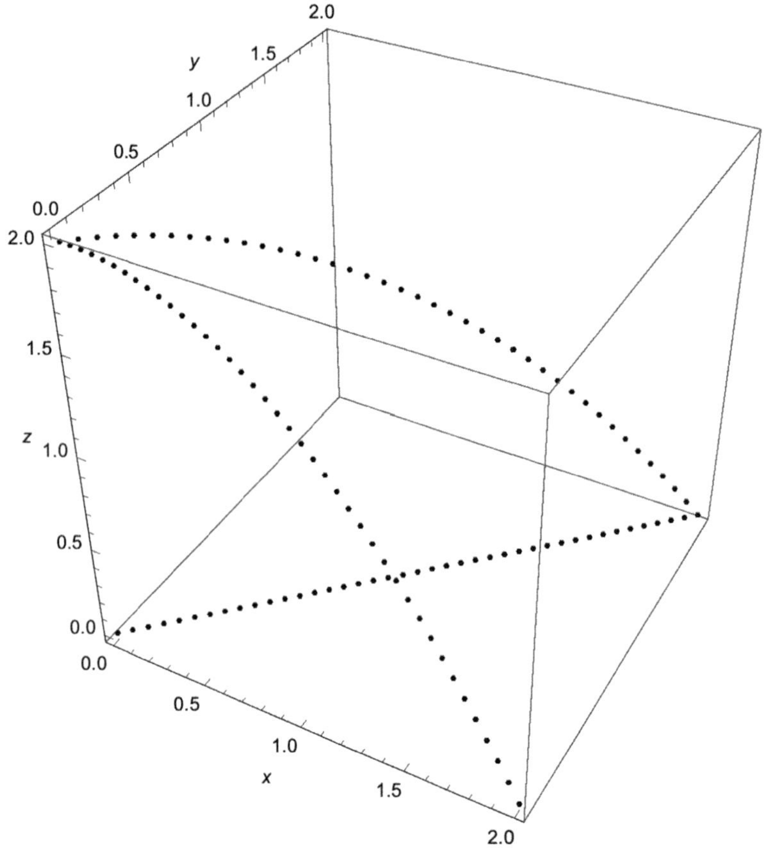

That is the region R that we seek to define. You can hover over the figure and rotate it by dragging it around its principal axes. We want more. We want to fill this region of space with points to better understand it. To do that, in the mind's eye, picture creating points on the surface of the parabolic cylinder first. Imagine starting at the point {0,0,2} and creating parabolas that lie between the two that we have created so far. To do that all we have to do is to adjust the y values incrementally to in essence rotate around the z axis. This script shows how to do this one at a time to solidify the idea. Notice the constants range between zero and unity:

Table[{x, .25x, 2 − 0.5x^2}, {x, 0, 2, .05}];

Point/@%;

e1 = Graphics3D[%, Axes → True, AxesLabel → {x, y, z}];

Table[{x, .375x, 2 − 0.5x^2}, {x, 0, 2, .05}];

```
Point/@%;
f1 = Graphics3D[%, Axes → True,
AxesLabel → {x, y, z}];

Table[{x, .5x, 2 − 0.5x^2}, {x, 0, 2, .05}];
Point/@%;
g1 = Graphics3D[%, Axes → True,
AxesLabel → {x, y, z}];

Table[{x, .625x, 2 − 0.5x^2}, {x, 0, 2, .05}];
Point/@%;
h1 = Graphics3D[%, Axes → True,
AxesLabel → {x, y, z}];

Table[{x, .75x, 2 − 0.5x^2}, {x, 0, 2, .05}];
Point/@%;
i1 = Graphics3D[%, Axes → True,
AxesLabel → {x, y, z}];

Table[{x, .875x, 2 − 0.5x^2}, {x, 0, 2, .05}];
Point/@%;
j1 = Graphics3D[%, Axes → True,
AxesLabel → {x, y, z}];

(* We already have a, b, c, and d. We reuse them. *)
Show[a1, b1, c1, e1, f1, g1, h1, i1, j1]
```

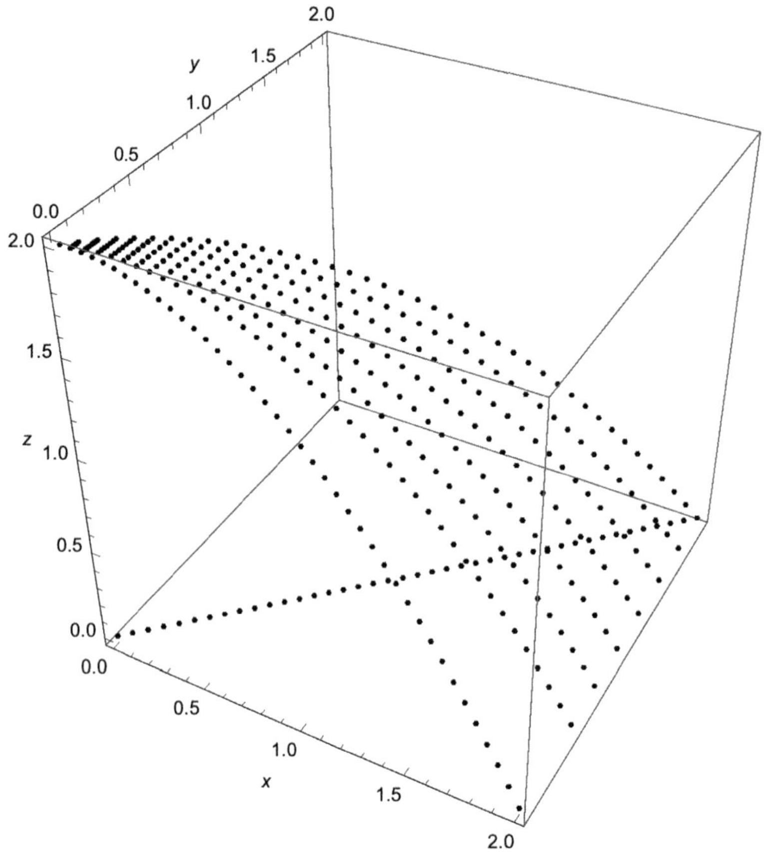

While doing this one-by-one gives a good understanding of what we are doing, it's much easier to do this automatically with Table and to add many more points. We do this by making the constant, k, a parameter that is incremented as finely as we wish it to be:

Clear[crv, surf1, surf2, surf3])

Table[Table[{x, kx, 2 − 0.5x^2}, {x, 0, 2, .05}],{k, 0, 1, 0.05}];

Point/@%;

crv = Graphics3D[%, Axes → True,AxesLabel → {x, y, z}]

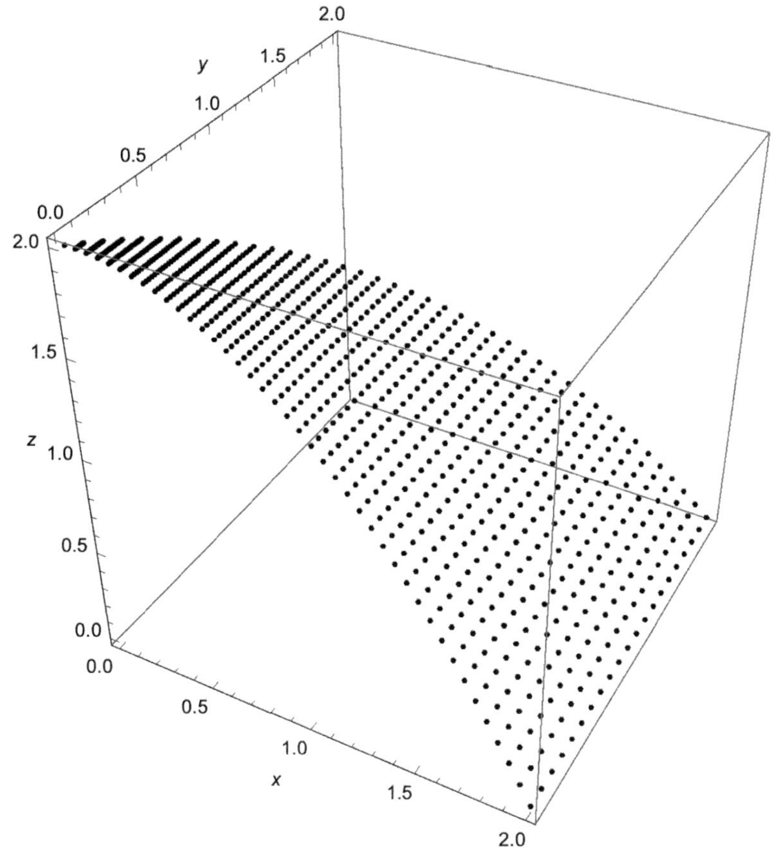

Show[edges, crv]

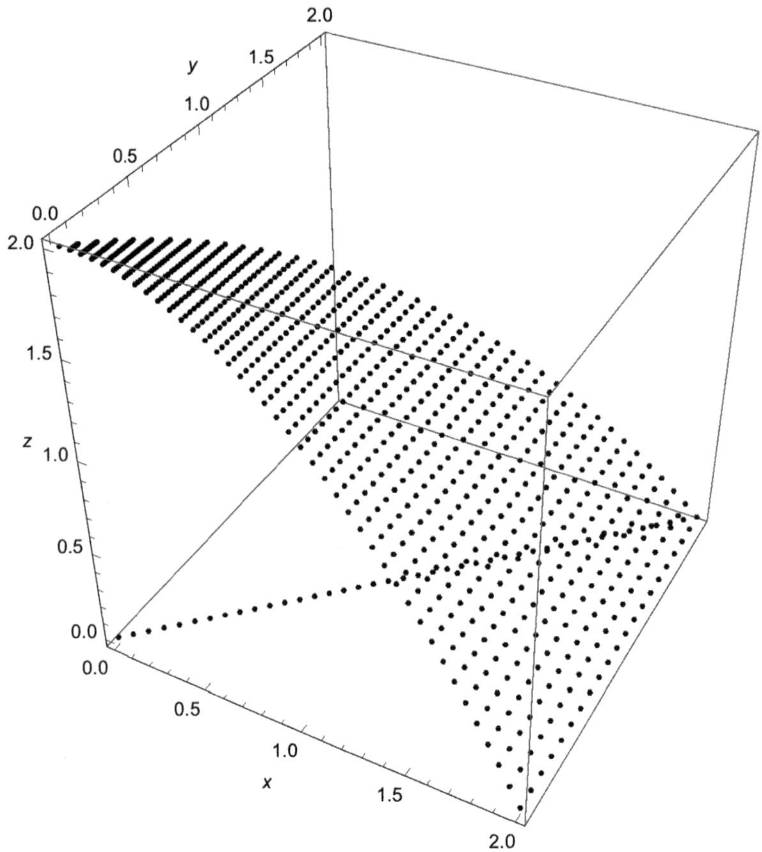

The parabolic surface is nicely filled, though we could fill it further by using even smaller increments of k. Now we turn to the two planes forming the two edge surfaces of the region R. Then we would put points on the surface in the xz plane below the parabola and below the parabola on the other plane that divides the xy plane. The script for doing this will take advantage of what we learned earlier about embedding a Table in a Table. In words, this script will compute at a given x and at y = 0 a column of points up to the parabola and not beyond it. Then the x value will be incremented to a slightly higher value and the column of z points will be created up to the parabola at the x value. The calculation is creating a column of points at each x up to the parabola and it steps over to the next x and does it again. As the steps the right increase x, the parabolic term lessens the z value for each successive column. It takes many more words to explain this than are required to do it. Here is the script to create these positions for the points:

```
Short[Table[
Table[{x, x, z}, {z, 0, 2 – 0.5x^2, 0.05}], {x, 0, 2, 0.05}], 20]
```

```
{{{0.,0.,0.},{0.,0.,0.05},{0.,0.,0.1},{0.,0.,0.15},{0.,0.,0.2},{0.,0.,0.25},{0.,0.,0.3},
{0.,0.,0.35},{0.,0.,0.4},{0.,0.,0.45},{0.,0.,0.5},{0.,0.,0.55},{0.,0.,0.6},{0.,0.,0.65},
{0.,0.,0.7},{0.,0.,0.75},{0.,0.,0.8},{0.,0.,0.85},{0.,0.,0.9},{0.,0.,0.95},{0.,0.,1.},
{0.,0.,1.05},{0.,0.,1.1},{0.,0.,1.15},{0.,0.,1.2},{0.,0.,1.25},{0.,0.,1.3},{0.,0.,1.35},
{0.,0.,1.4},{0.,0.,1.45},{0.,0.,1.5},{0.,0.,1.55},{0.,0.,1.6},{0.,0.,1.65},{0.,0.,1.7},
{0.,0.,1.75},{0.,0.,1.8},{0.,0.,1.85},{0.,0.,1.9},{0.,0.,1.95},{0.,0.,2.}},⟨⟨39⟩⟩,{{2.,2.,0.}}}
```

We can see that at the position x = 0 and y = 0, the script is increasing z up to the parabola with a maximum of 2. After that, we Map (/@) Point onto the list and then we scan view it with Show:

```
x=.
(* This does all that we need to do to get the first surface above the xy plane *)
Point/@
Table[
Table[{x, x, z}, {z, 0, 2 – 0.5x^2, 0.05}],{x, 0, 2, 0.05}];
sidexyz = Graphics3D[%, Axes → True,
AxesLabel → {x, y, z}];
surf1 = Show[%]
```

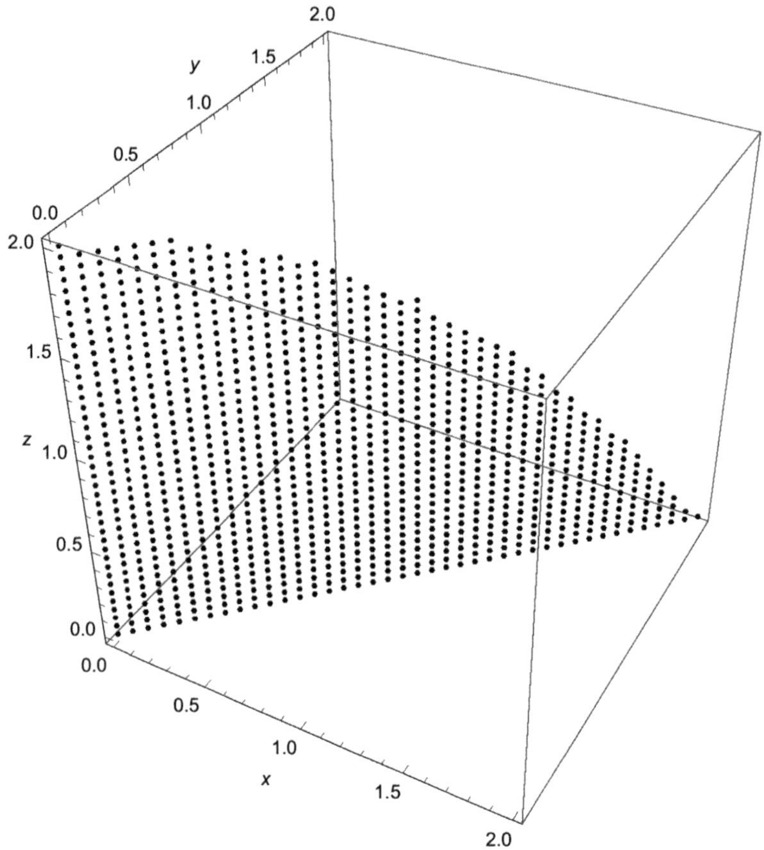


```
x=.

(* This does all that we need to do to get the surface in the xz plane *)

Point/@

Table[

Table[{x, 0, z}, {z, 0, 2 − 0.5x^2, 0.05}],

{x, 0, 2, 0.05}];

sidexyz = Graphics3D[%, Axes → True,

AxesLabel → {x, y, z}];

surf2 = Show[%]
```

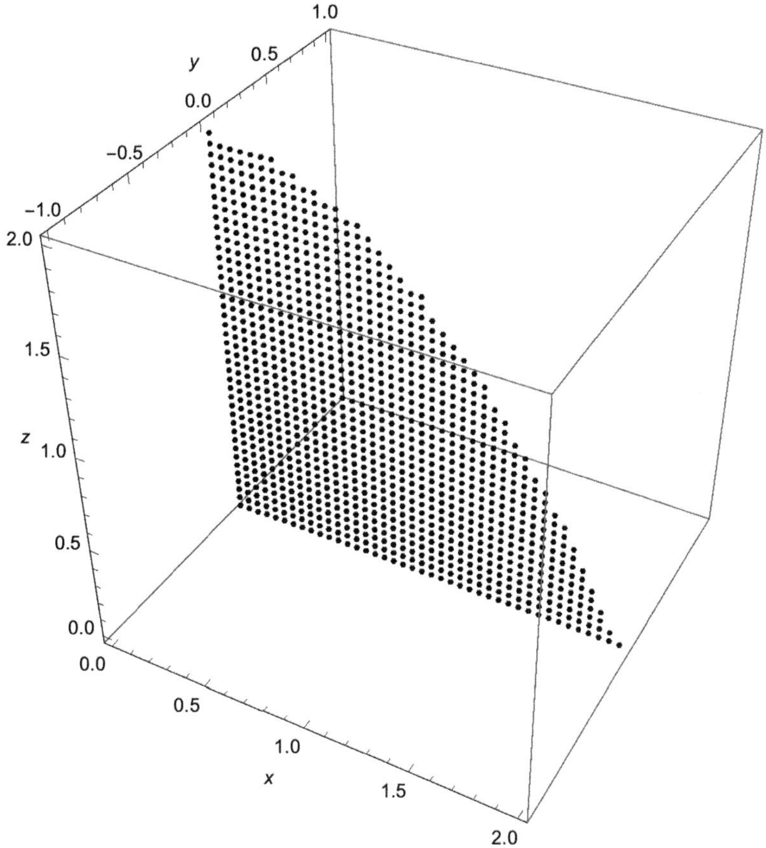

We can show the three surfaces together now. By rotation we can see that all the parts join nicely together. But we lack one surface and that is the triangular floor.

Show[edges, crv, surf1, surf2]

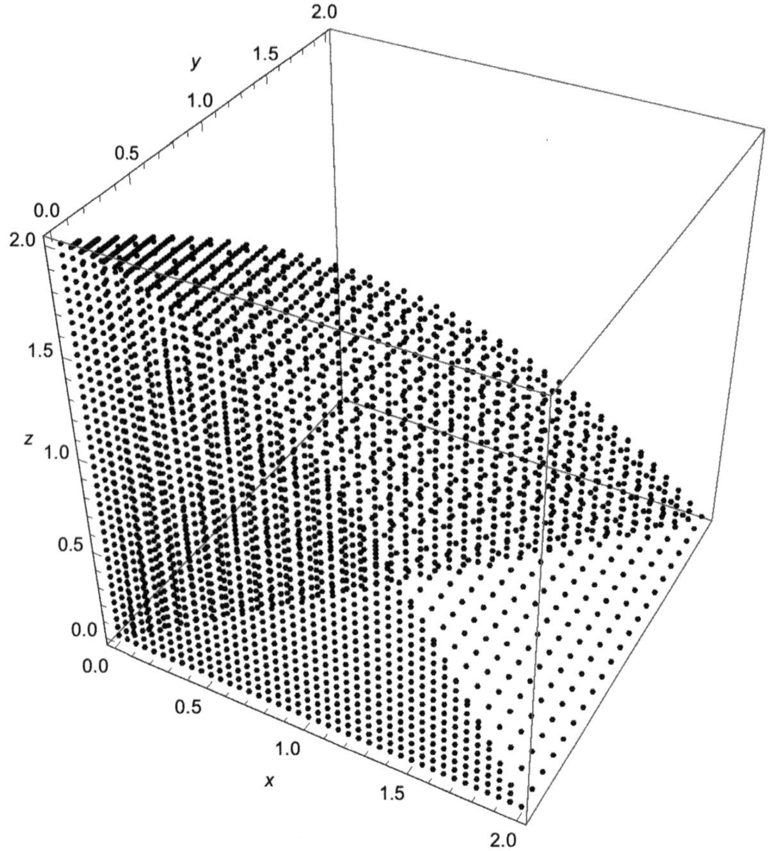

To create the triangular floor, we use the same approach as for the other three surfaces. We know that all the points will be in the xy plane, hence z= 0 throughout. We will want to select an x value and then create a column of points up to the x=y line. So at x = 0, y = 0. At x = 1, y=1 at the maximum. So we need points like { 1, 0 }, { 1, 0.25 }, { 1, 0.5 }, { 1, 0.75 }, and many points in between. Again by embedding a Table in a Table we get this:

```
Point/@
Table[
Table[{x, y, 0}, {y, 0, x, 0.05}],
{x, 0, 2, 0.05}];
```

```
sidexyz = Graphics3D[%, Axes → True,
AxesLabel → {x, y, z}];

surf3 = Show[%]
```

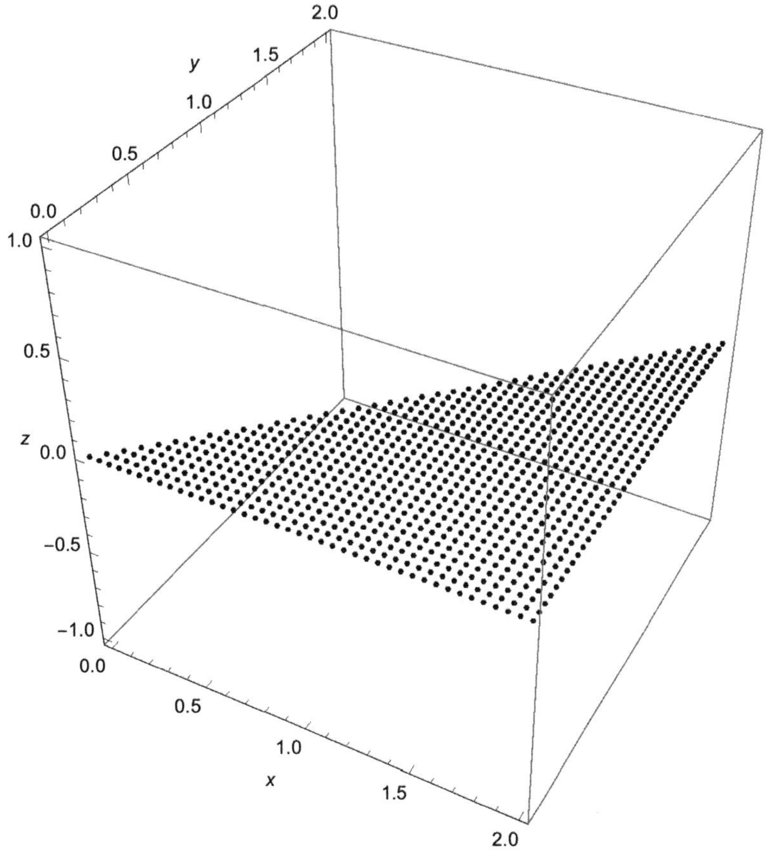


```
Show[edges, crv, surf1, surf2]
```

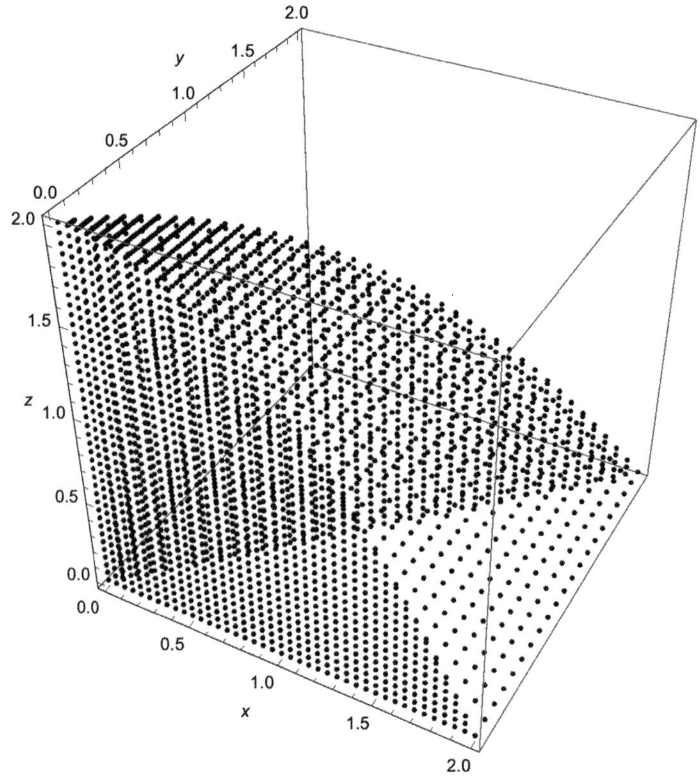
x
y
z
0.0
0.5
1.0
1.5
2.0

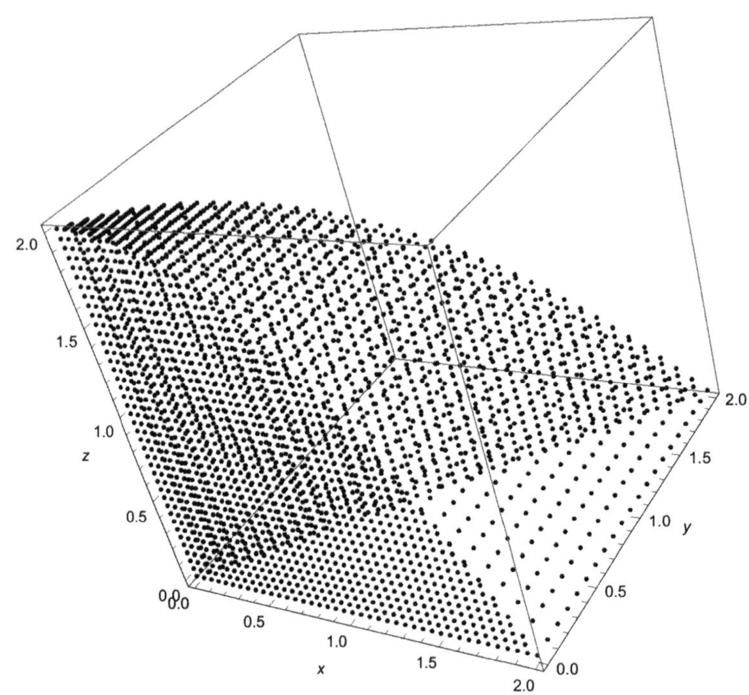
x
y
z
0.0
0.5
1.0
1.5
2.0

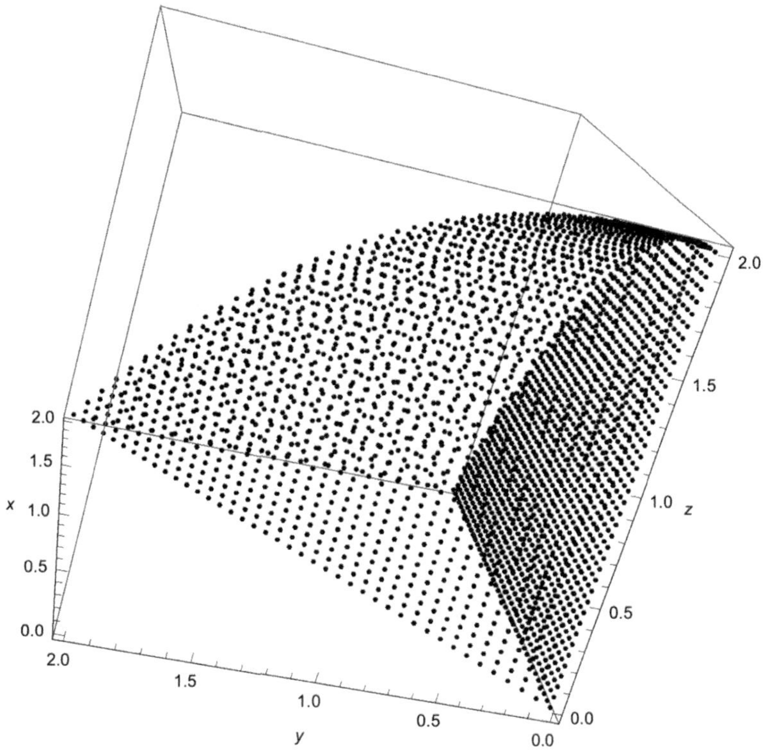
2.0
1.5
x 1.0
0.5
0.0
2.0
1.5
1.0
0.5
0.0
y
2.0
1.5
1.0 z
0.5
0.0

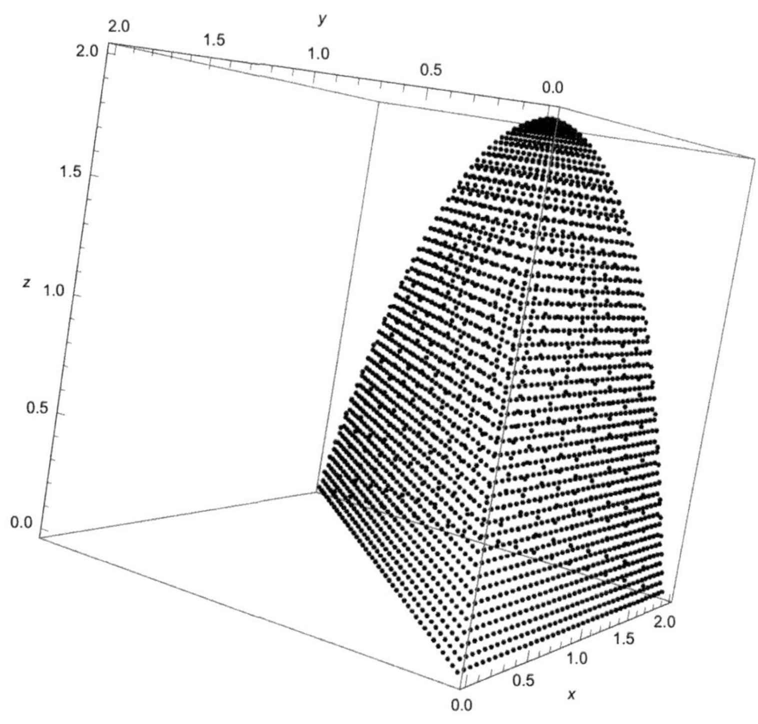
y
2.0
1.5
1.0
0.5
0.0
2.0
1.5
z 1.0
0.5
0.0
0.0
0.5
1.0
1.5
2.0
x

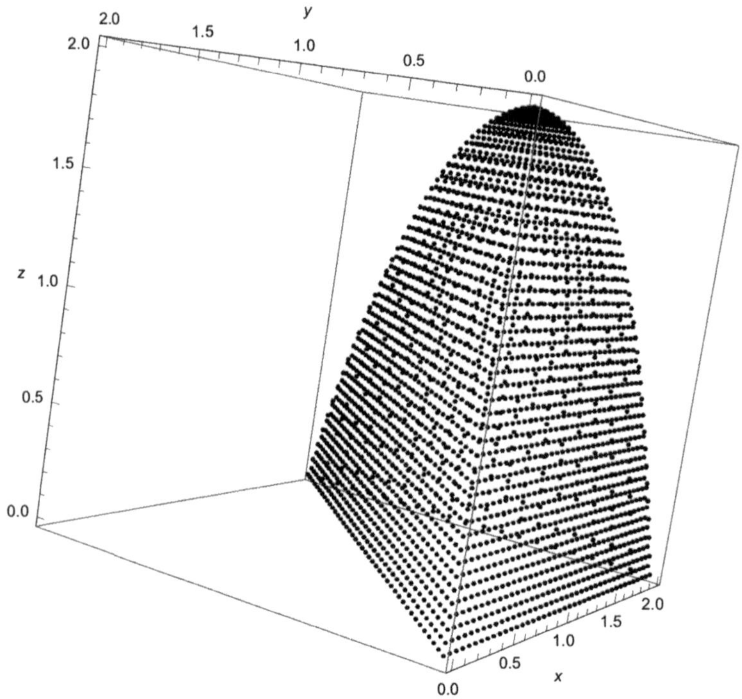
y
2.0
1.5
1.0
0.5
0.0
z
2.0
1.5
1.0
0.5
0.0
x
0.0
0.5
1.0
1.5
2.0

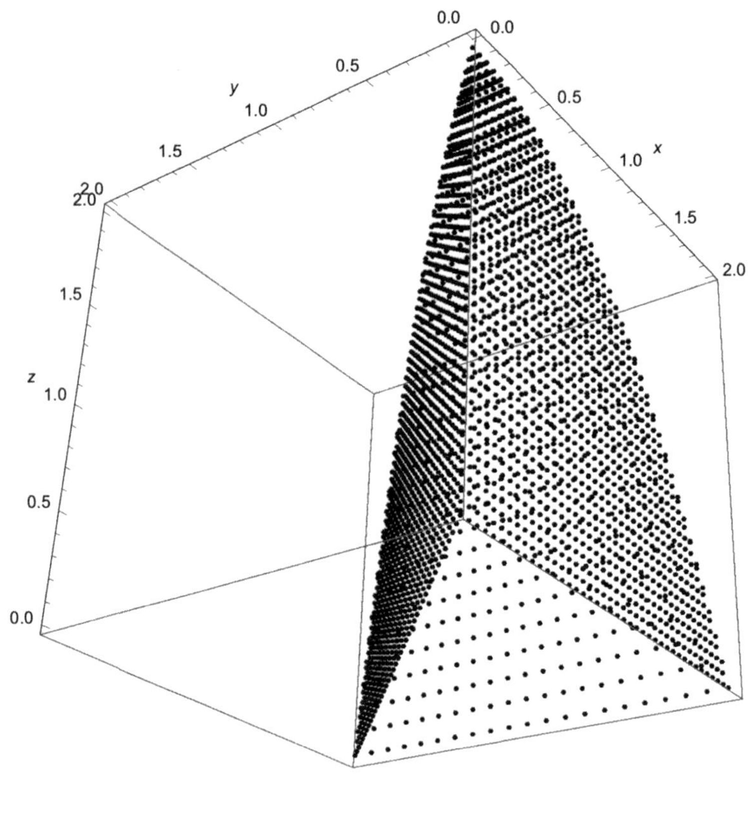
y
0.0
0.5
1.0
1.5
2.0
x
0.0
0.5
1.0
1.5
2.0
z
2.0
1.5
1.0
0.5
0.0

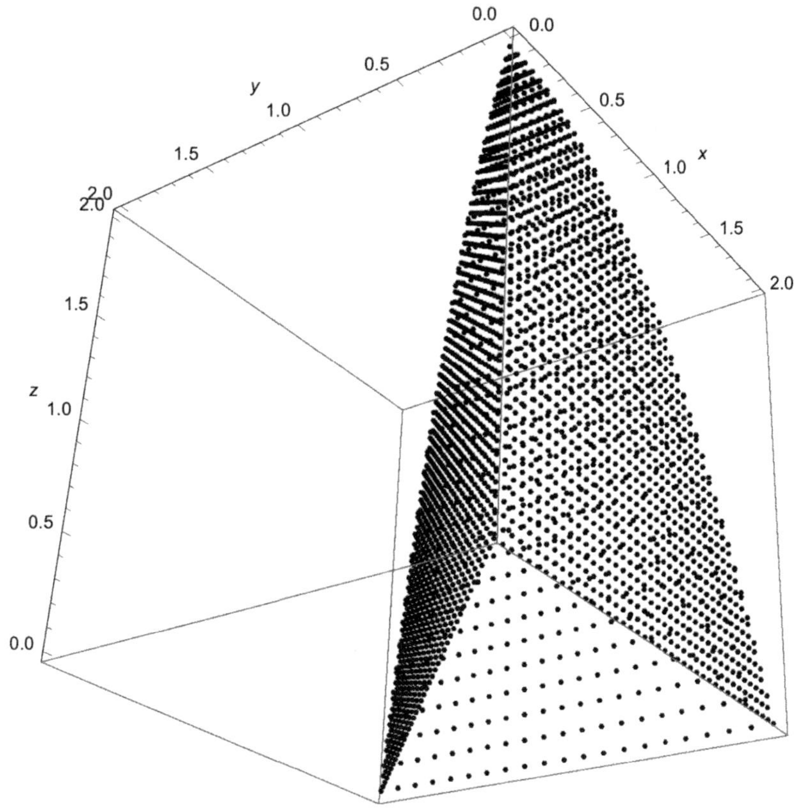

Now we could put all this together into one script to use, reuse, and to modify. Here we bring the incrementer in each Table calculation to the top of the script as a set parameter named inc. We also dropped the edges, since they are not needed, and we enlarged the final rendering.

```
inc = 0.02; (* Incrementer *)

(* Provides the Curved surface of the parabolic cylinder *)

Table[
Table[{x, kx, 2 - 0.5x^2},
{x, 0, 2, inc}],
{k, 0, 1, inc}];
```

```
Point/@%;

pcyl = Graphics3D[%, Axes → True,
AxesLabel → {x, y, z}];

(* First surface, above the xy plane *)
Point/@
Table[
Table[{x, x, z}, {z, 0, 2 − 0.5x^2, inc}],
{x, 0, 2, inc}];

s1 = Graphics3D[%, Axes → True, AxesLabel → {x, y, z}];


(* Second surface, in the xz plane *)
Point/@
Table[
Table[{x, 0, z}, {z, 0, 2 − 0.5x^2, inc}],
{x, 0, 2, inc}];

s2 = Graphics3D[%, Axes → True, AxesLabel → {x, y, z}];

(* Surface in the xy plane *)
Point/@
Table[
Table[{x, y, 0}, {y, 0, x, inc}],
{x, 0, 2, inc}];

s3 = Graphics3D[%, Axes → True, AxesLabel → {x, y, z}];

Show[pcyl, s1, s2, s3]
```

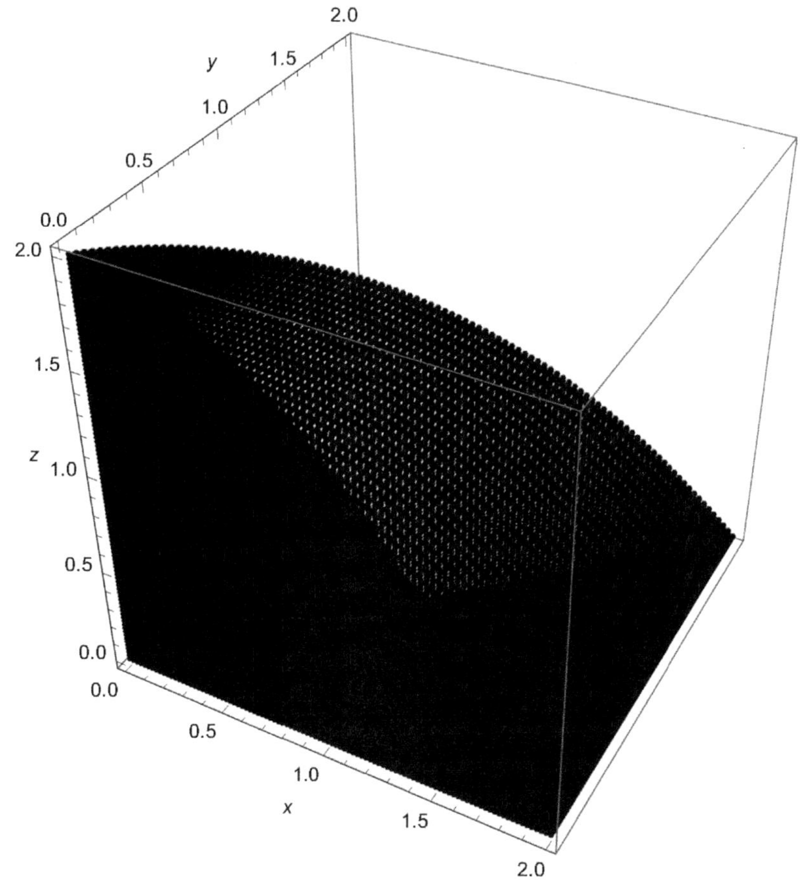

Conclusion

In this primer, we have moved from using the simple scientific calculator-like functions of *Mathematica* to more advanced topics including graphics, statistics and programming code; yet, we still have barely touched on the full power of this computational tool and environment. Rather than adding more to this chapter, we will close here with the hope and expectation that these examples have given you insight into how *Mathematica* works, both in terms of its syntax and functionality, and how to make it work. As we move forward, we will begin to examine problems that are more germane to chemical engineering and at the same time we will try to continue to advance the utilization of *Mathematica* as we do so.

CHAPTER 2

Elementary single component systems

Elementary single component systems are those that have just one chemical, or component, involved in the process. Filling of a vessel is an example of this kind. The component can be a solid, liquid, or gas. Regardless of the phase of the component, the time dependence of the process, how the system changes, is captured by the same statement of the conservation of mass within a well-defined region of space that we will refer to as the *control volume*.

In this chapter we will apply the conservation of mass principle to a number of different kinds of systems. While the systems are different, by the process of analysis they will each be reduced to their most common features and we will find that they are more the same than they are different in terms of how they change with time. When we have completed this chapter you will understand the concept of a control volume, the conservation of mass and you will be able to write and solve total material balances for single component systems.

The conservation of mass principle, the concept of a control volume

The conserved quantities that are of utmost importance are mass, energy, and momentum. It is the objective of this text to teach you how to utilize the conservation of mass in the analysis of units and processes that involve mass flows, mass transfer, and chemical reaction. Each conserved quantity is governed by the same principle – they are neither created nor destroyed. For mass this principle holds for all cases except those involving nuclear reactions. In all other situations, the principle is never violated. So we can use it to the utmost as you will see in developing time dependent and time independent descriptions of various processes.

The principle is so seemingly obvious that you may wonder how it can be so useful to us. How does knowing that mass is neither created nor destroyed relate to a process' behavior, or to anything else for that matter? The key is that in order to use this principle, we must translate it into mathematics so that we can work with it and derive the descriptions of change that we need.

If mass is neither created nor destroyed, then that means if we detect its depletion in one region of space, this can only be the case if in some other region of space, the same mass was accruing. Simply put, mass does not go missing. In other words, we always inspect some region of space and draw conclusions based on our measurements within that region. If mass is increasing within this space it must be coming from somewhere else. Similarly, if we detect

Introduction to Chemical Engineering Analysis Using *Mathematica*
https://doi.org/10.1016/B978-0-12-820051-3.00007-3

that mass is decreasing, that is because it is leaving the region of our measurement. We have everyday experiences that correspond to these statements. The level of water in a glass left on a table at room temperature will slowly decrease as the water leaves via evaporation. Pulling the drain plug on a bathtub causes the water to flow out due to the force of gravity. When a stalk of corn grows all the mass that is accumulated within the plant had to be delivered to by adsorption through the roots from the soil and through leaf structures in the air. Each of these, the glass of water, the tub, or the corn plant, can be considered a control volume and as such we can measure the rate of change that occurs within them, whether it is through evaporative losses, flow, or growth. Each involves the transport and transfer of mass from outside of the control volume to inside of it and vice versa.

Another example is the case of the living cell. Nutrients are transported across the cellular membrane and are utilized in metabolism. The by-products of metabolism are transported out of the cell and back across the membrane to the surroundings. The young cell grows and increases in size and mass because the rate of by-product flow out is less than the rate of nutrient flow in. We know this because of the conservation of mass principle, we need no other information than to know that the cell grows, in order to reach this conclusion. As the mass of the cell increases, the size of the cell also increases. If the cell is nearly spherical as is the case for many simple, single cell organisms, then we can expect that its diameter is also increasing. Hence the most simple measurement to make to detect cell grow in an experiment might be to measure the cellular diameter at different times. The diameter becomes a proxy measure of the mass of the cell. The membrane of the cell defines the control volume. If we know the density of the fluid inside the cell, and if its shape is a regular function of the diameter, then we can calculate the mass in the cell from images taken at different times. When the cell matures, we find that the rate of nutrient flow in is balanced by the rate of by-product out, that is why the cell no longer is growing. In this condition, when the input rate is balanced by the output rate, there is no net accumulation of mass in the cell. The cell biologist refers to this state as the *homeostatic state*, the physical chemist and chemical engineer call it the *steady state*.

In these word statements that describe both our ordinary experiences on the one hand and the dynamics of cell growth on the other, we find the essence of that which we need to formalize at this point. The conservation of mass is applied to a ***control volume*** which is defined by a control surface, which separates the control volume from its surroundings. By defining the control volume and its boundaries, we know where "inside" is. The inside of the cell is that space within the membrane, just as the inside of the glass lies within its regular walls. The same is true for the corn plant. Even though it has a more complex geometry, we can ignore that by imagining a box placed around the whole plant, including its roots. This box, though imaginary, becomes its control surface.

Now to begin to bring mathematical descriptions to bear on the problem, we can restate the conservation of mass in terms of the control volume and its boundary:

> ***The net rate of mass accumulation within a control volume is equal to the rate at which mass enters the control volume by any process minus the rate at which it leaves the control volume by any process.***

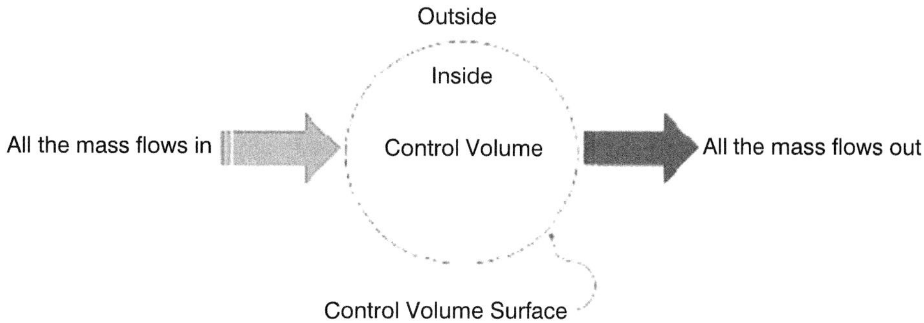

The mathematics that proceeds from this is at once simple and elegant. Since we are discussing rates we will write the mathematical statement in terms of rates also, the rate of change in mass:

$$\frac{dm}{dt} = \mathring{m}_{in} - \mathring{m}_{out}$$

The accumulation rate, on the left-hand side of the equation, is given as $\frac{dm}{dt}$, i.e. the net rate of change in mass within the control volume; the dimensions of this term are mass/time

We use the dot notation on the right-hand side of the equation to indicate a rate, $\mathring{m}_{in}$, this is the total rate of mass flow (the sum of all flows) into the control volume by any process; the dimensions of this term are also mass/time

The total rate of mass flow out (the sum of all flows) of the control volume by any process is $\mathring{m}_{out}$; the dimensions of this term are also mass/time

Notice that all three terms are rates. The rate of accumulation is the difference between the rate that mass flows into the control volume and the rate at which mass flows out of the control volume. This is the *differential* statement of the conservation of mass. Both sides of the equation are rates.

If the all the mass flows in are equal to all the mass flows out, then there will be not accumulation and the process is at a *steady state*. The equation and a little bit of integral calculus will lead to a very powerful result. We begin by integrating the left-hand and right-hand sides of the equation over a slice of time from t_o to t_f, and with corresponding masses at those times:

$$\frac{dm}{\mathrm{dt}} = \overset{\circ}{m}_{\mathrm{in}} - \overset{\circ}{m}_{\mathrm{out}}$$

$$\int_{t_o}^{t_f} \frac{dm}{\mathrm{dt}} dt = \int_{t_o}^{t_f} \overset{\circ}{m}_{\mathrm{in}} dt - \int_{t_o}^{t_f} \overset{\circ}{m}_{\mathrm{out}} dt$$

$$\int_{m_o}^{m_f} dm = \int_{t_o}^{t_f} \overset{\circ}{m}_{\mathrm{in}} dt - \int_{t_o}^{t_f} \overset{\circ}{m}_{\mathrm{out}} dt$$

Because the process is at steady state, the mass flows into and out of the system are constants:

$$\int_{m_o}^{m_f} dm = \overset{\circ}{m}_{\mathrm{in}} \int_{t_o}^{t_f} dt - \overset{\circ}{m}_{\mathrm{out}} \int_{t_o}^{t_f} dt$$

$$m_f - m_0 = \overset{\circ}{m}_{\mathrm{in}}(\mathrm{t} - t_o) - \overset{\circ}{m}_{\mathrm{out}}(\mathrm{t} - t_o)$$

$$\Delta \mathrm{m} = \overset{\circ}{m}_{\mathrm{in}}\ \Delta t - \overset{\circ}{m}_{\mathrm{out}}\ \Delta \mathrm{t}$$

If the mass flow into the control volume is both constant and equal to the mass flow out of the control volume, then the left-hand side of the equation is zero because the right-hand side is zero

$$\Delta \mathrm{m} = \overset{\circ}{m}_{\mathrm{in}} \Delta \mathrm{t} - \overset{\circ}{m}_{\mathrm{out}}\ \Delta \mathrm{t}$$

$$\Delta \mathrm{m} = (\overset{\circ}{m}_{\mathrm{in}} - \overset{\circ}{m}_{\mathrm{out}})\ \Delta \mathrm{t}$$

$$\text{if} \quad \overset{\circ}{m}_{\mathrm{in}} = \overset{\circ}{m}_{\mathrm{out}}$$

$$\text{then} \quad \Delta \mathrm{m} = (0)\ \Delta \mathrm{t}$$

$$\Delta \mathrm{m} = 0$$

At steady state there is no net change of mass inside the control volume. However, it does not mean that there were no flows of mass through the control volume, quite the opposite. The mass flows are equal and constant but they were not necessarily zero. Hence at steady state we can also see that this. First there is no net accumulation of mass in the control volume:

$$0 = \overset{\circ}{m}_{\mathrm{in}}\ \Delta \mathrm{t} - \overset{\circ}{m}_{\mathrm{out}}\ \Delta \mathrm{t} \quad \text{(steady state} \Rightarrow \text{no net accumulation)}$$

Second, the product of the rate of mass flow into the control volume and the time interval must be equal to the product of the rate of mass flow out of the control volume and the same time interval:

$$\overset{\circ}{m}_{\text{in}}\ \Delta t = \overset{\circ}{m}_{\text{out}}\ \Delta t$$

Third, when the rates of mass flow in and out are multiplied by time, the result is the total mass that flowed in and the total mass that flowed out and these are equal quantities.

$$m_{\text{in}} = m_{\text{out}}$$

The last equation is very simple and yet it is very powerful, and it is referred to as the steady state mass (or material) *balance*. It can also be thought of as the integral mass balance. The word balance appears because the mass in is *balanced* by the mass out at steady state.

Two more ideas need to be added at this point that are formalisms. First, on the left-hand side we wrote the derivative of mass with respect to time, the net mass accumulation term. Formally, for any process that is not at steady state, the left-hand side will be non-zero, and we will be solving a differential equation the solution of which will provide a time dependent function, either m[t] or some other function of time that is a measurable directly related to the time dependent mass function. Therefore, we will write m [t] instead of just m in the derivative. Second, recall that when we wrote the differential material balance, on the right-hand side we placed these two terms for the mass flows, $\overset{o}{m}_{\text{in}}$ and $\overset{o}{m}_{\text{out}}$, on the right-hand side. To be clear these represent the sum of all flows. Which is to say there may be many different flows in and many different flows out, and we can account for all of them formally with this equation. Putting these two together we rewrite the differential mass balance with mass as a function of time and with the sum of all flows in and out shown explicitly, and likewise for the integral mass balance:

$$\frac{dm[t]}{dt} = \sum_{i=1}^{x} \overset{\circ}{m}_{\text{in},i} - \sum_{j=1}^{y} \overset{\circ}{m}_{\text{out},j}$$

$$0 = \left(\sum_{i=1}^{x} \overset{\circ}{m}_{\text{in},i} - \sum_{j=1}^{y} \overset{\circ}{m}_{\text{out},j} \right) \Delta t$$

The differential and integral mass balance equations are the key unifying principles that we will use throughout this book. These will be all we need in order to analyze and model a wide array of elementary single and multicomponent systems, with and without reaction and across phases, and whether they change in time or stay the same. This is the foundation upon which everything else we do will be built. The best way to illustrate how to use this mathematical statement of conservation of mass is through examples.

Filling a vessel with a pelleted solid: conservation of mass

Heterogeneous catalysts are usually formed as pellets or as a variation of that shape. Catalysts promote the rates of chemical reactions and are used ubiquitously throughout the chemical and petroleum industries. Catalysts are not consumed in the course of the reaction that they promote. Never the less catalysts do eventually need to be replaced. This is because they become less and less active as contaminants react with the metals or their solid structures have become clogged. So at the end of its lifetime the catalyst must be replaced. The spent catalyst is removed from the reactor vessel, the reactor is cleaned and the space left open is ready for a charge of fresh solid catalyst. The solid will be delivered to the top of the reactor by a conveyor belt and dropped in. The process will proceed until the volume of the reactor vessel has been filled to the requisite level as is shown in the figure below. We can do an analysis of this simple process.

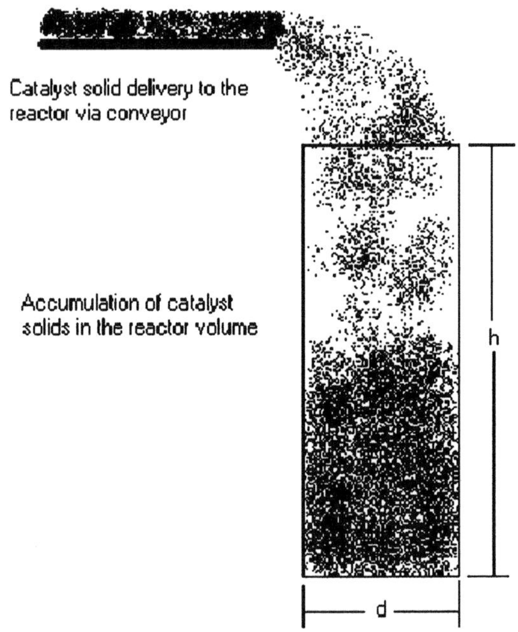

Flow reactors may be small or large in volume and are often cylindrical in shape. Let's represent the reactor then as a simple cylindrical volume. The height of the vessel is **h** and its diameter is **d**. The overall volume to be filled by the catalyst is given as:

$$V = \frac{\pi d^2 h}{4}$$

Catalyst is being delivered by conveyor belt at a constant mass flow rate. The question we would like to be able to answer is: How much catalyst mass is in the reactor vessel at any time. If we look into the reactor at any time we can measure the level to which the reactor is filled, and from that we could in principle compute the mass of catalyst if we had a density for the material. But remember this is a solid and it packs irregularly into the reactor, as we can see from the figure. So, the density is the density of the catalyst as it has dropped into the reactor. We can get an average value for this from packing experiments to determine the *compacted bulk density*.

We can apply the mathematical statement for the conservation of mass to this situation. In this case the reactor and its physical dimensions define the control volume. The rate of delivery is a constant which we will call $\overset{\circ}{m}_{\text{in}}$. The rate of mass flow out of the reactor is zero, that is $\overset{\circ}{m}_{\text{out}} = 0$. Therefore we have:

$$\frac{dm[t]}{\text{dt}} = \overset{\circ}{m}_{\text{in}}$$

The rate of accumulation of catalyst in the reactor is just equal to the rate of delivery, which is exactly what we would have said based on common sense. An equation of this kind is the simplest type of differential equation. It is separable and we integrate from **t** = 0 to **t** and from **m[0]** = 0 to **m[t]** here is a script that does it. All the input statements are in bold and the output statement in regular text face:

$\boldsymbol{m[0] = 0;}$

(* The mass in the reactor is zero at the start *)

$\boldsymbol{\overset{\circ}{m}_{\text{in}} = \overset{\circ}{m}_{\text{in,constant}};}$

(* The mass flow rate into the reactor is a constant $\overset{\circ}{m}_{\text{in}}$ = constant *)

$\boldsymbol{\int_{m[0]}^{m[t]} dm[t] == \int_0^t \overset{\circ}{m}_{\text{in}} dt}$

$m[t] == t\overset{\circ}{m}_{\text{in,constant}}$

The mass of catalyst in the reactor is a simple linear function of time because the mass flow rate of catalyst remains constant. The dimensions of $\overset{\circ}{m}_{\text{in}}$ are mass per unit time, $\frac{\text{mass}}{\text{time}}$, so we see that the resultant equation is dimensionally consistent and we can check it as follows, where the input statement is again in bold and the output statement (True) in regular face:

$$\text{mass} == \text{time} * \frac{\text{mass}}{\text{time}}$$

True

The mass of the catalyst in the reactor is the most critical factor in the design and operation of the reactor. When the catalyst particles are packed into the reactor, the whole mass is collectively called the catalyst bed. Eventually, once the process is started up, reactants will flow through the catalyst bed to be converted to products continuously. The mass of catalyst in the bed is computed to give the correct rate of reaction and, therefore, the mass of the catalyst bed controls the reactor's productivity. Clearly, we want to know the mass of catalyst in the reactor at any time during the filling process and we want to be sure that when full the mass of catalyst in the reactor will be equal to the mass of catalyst required by the design specifications of this process.

From experiment, we can find the amount of volume that a given mass of catalyst will occupy. When the particles are packed tightly, the volume that a given mass occupies is called the compacted bulk density. Because the percent conversion of the reactants to products at a given, temperature, pressure and flow rate is determined by the mass of the catalyst bed, the process of filling the reactor with catalyst is important to monitor and to control. If the mass flow of catalyst into the bed is too fast, then the particles may not pack as tightly as is necessary to achieve the full compacted bulk density. If the catalyst in the bed does not get to the compacted bulk density, then the reactor will appear to be full, but the actual mass of catalyst in the reactor will be below the mass that the reactor was designed to hold. With less catalyst in the bed there will be less conversion of reactants into product and the process will not operate at its design efficiency, which in turn will reduce profit, and time to pay back the capital invested in the reactor. The overall economics of the process will underperform expectations. So we would need to know the mass of catalyst that has flowed into the bed at any time and we need to know the volume that the mass is occupying throughout the filling process. If, at any time, the volume occupied by the catalyst is larger than it should be for the mass in the reactor, then the filling process must be stopped and the bed has to be packed by some other means to reach the expected compacted bulk density.

We can be sure that as catalyst is added by the conveyor belt, we will know the mass flow the catalyst. This is because the solid catalyst particles can be put in a hopper to supply the conveyor belt and the weight of the hopper can be measured continuously. If the flow of mass is constant, the change in mass of the hopper will also be a constant. So we will know the mass flow rate of catalyst into the reactor and the mass that has flowed into the reactor at any time. But that is not enough, we also need to know the volume occupied at each time. Since the reactor is a cylinder, we will measure the height to which the catalyst has filled the reactor at any time. Then we want to compare that level in the reactor to the level that the mass should attain if the particles are packing at their compacted bulk density. If the level of catalyst in the reactor is higher than it should be for that given mass, the process is stopped and the bed has to be packed more tightly. But the catalyst particles also must not be packed too tightly, that is to a compacted bulk density higher that the design calls for. The reasons for this are that a bed packed too tightly will produce a higher resistance to flow and hence a higher pressure will be needed to get the desired flow rate. When the catalyst is packed too tightly, we will notice that the catalyst bed provides too much back pressure. Another reason not to pack over the compacted bulk density is that the catalyst particle may break. Catalysts are typically very porous ceramic solids, because they be crushed and break rather easily, they are called mechanically friable. If the original catalyst particles begin to break into smaller particles, this process may continue during the reactor filling process and later with time on stream, leading to even tighter packing because the newly-produced, smaller particles take up spaces between the originally-sized particles. This can increase back pressure even more and fine (very small) catalyst particles may even be transported out of the bed. These are all bad outcomes that must be avoided to the greatest extent possible, or to within that which is tolerable by design. Hence, reactor filling is an important process.

Suppose that the reactor is 25 feet high and 5 feet in diameter. The catalyst delivery rate is 1000 pounds per hour. Let's say that experiments show that one cubic foot will hold 63.2 pounds of the catalyst when packed to the desired compacted bulk density. As the reactor is being filled, a person can literally stand at the top of the reactor and measure down to the top of the catalyst bed to determine how much of the reactor is still unfilled and, by difference, how high the catalyst bed stands. Ultimately we want a graph that has the mass on the x-axis, which we will know at any time and the level on the y-axis, which will be the level that any given mass of catalyst should attain if it is packing correctly. The degree of difference between actual level of catalyst in the bed and the calculated level for a given mass will be a measure of how well or poorly the catalyst is packing.

Firstly, we begin with the expression for the volume of the reactor:

$$V_{\text{reactor}} = \frac{\pi d^2 h}{4}$$

We can input values for the parameters in this expression to calculate the total volume:

$$\mathbf{V_{reactor}} = \frac{\pi(5\text{ft})^2 25\text{ft}}{4} // N$$

490.874 ft^3

The reactor has 490.9 cubic feet of volume to be filled.

The mass flow rate into the reactor, $\overset{\circ}{m}_{\text{in}}$, is 1000 pounds per hour. To know how long it will take to fill the reactor, we need to convert this to a volume flow rate, q_{in}. To get at this, we will divide the mass flow rate by the compacted bulk density ρ_b:

$$q_{\text{in}} = \frac{\overset{\circ}{m}_{\text{in}}}{\rho_b} = \frac{1000\text{lb}/h}{\frac{1}{63.2\text{lb}/\text{ft}^3}} = \frac{15.8\text{ft}^3}{h}$$

Here we use *Mathematica* as a calculator, and it will handle units correctly:

1000lb/*h* ∗ (1 /(63.2lb /ft³))

$\frac{15.8228\text{ft}^3}{h}$

The time required to fill the reactor is then:

$$t_{\text{fill}} = \frac{V_{\text{reactor}}}{\frac{\overset{\circ}{m}_{\text{in}}}{\rho_b}} = \frac{V_{\text{reactor}}}{q_{\text{in}}} = \frac{491\text{ft}^3}{\frac{15.8\text{ft}^3}{h}} = 31h$$

The mass of the catalyst needed to fill this volume is given by the product of the mass flow rate into the reactor and the overall time required to fill it:

$$m_{\text{catalyst}} = \overset{\circ}{m}_{\text{in}} t_{\text{fill}}$$

The input command is:

(1000 lb/*h*)(31.076 h)

31076. lb

The mass of catalyst that should be in the reactor after this period of time is about 31,076 lbs. The time required to fill the reactor will be about a day and a half, if the density of the bed always achieves its compacted bulk density. Now we should create a graph of the height that the bed should reach at each time during the 31 hours that it will take to fill the reactor, assuming that it packs properly. Again this is the height that the bed should be at, if the bed is at its compacted bulk density. The volume of the bed divided by the area of the reactor will give the height to which the bed should rise with time. The difference between the total height of the reactor and the height of the bed at any time should be the same as the measurement from the top of the reactor to the top of the bed. This, h_{diff}, is what will actually be measured.

$$h_{\text{bed}} = h_{\text{reactor}} - h_{\text{diff}}$$

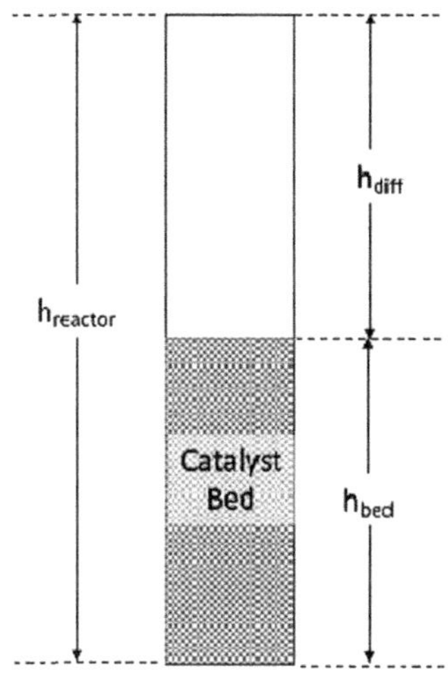

We now use the mass as a function of time to find a time dependent relationship for h_{diff}:

$$m_{\text{bed}}[t] = \overset{\circ}{m}_{\text{in}} t$$

$$\frac{m_{\text{bed}}[t]}{\rho_b} = \frac{\overset{\circ}{m}_{\text{in}}}{\rho_b} \text{ t}$$

$$V_{\text{bed}}[t] = \frac{\overset{\circ}{m}_{\text{in}}}{\rho_b} \text{ t}$$

$$h_{\text{bed}} = \frac{V_{\text{bed}}[t]}{\left(\frac{\pi d^2}{4}\right)} = \left(\frac{\overset{\circ}{m}_{\text{in}}}{\left(\frac{\pi d^2}{4}\right)\rho_b}\right) \text{ t}$$

$$h_{\text{diff}} = h_{\text{reactor}} - \left(\frac{\overset{\circ}{m}_{\text{in}}}{\left(\frac{\pi d^2}{4}\right)\rho_b}\right) \text{ t}$$

The relationship has one variable, time, and everything else is a parameter that will have a fixed value. The time dependence is time raised to the first power, t^1, which means that this is a line and we refer to the time dependence as *linear*. Next we implement a script to find h_{diff} for the given parameter values. Notice that we can give the units for all the parameters and *Mathematica* will handle them.

$h_{\text{reactor}} = 25\text{ft};$

$\overset{\circ}{m}_{\text{in}} = 1000\text{lb}/h;$

$d = 5\text{ft};$

$\rho_b = 63.2\text{lb}/\text{ft}^3\ ;$

$h_{\text{diff}}[\text{t\_}] := h_{\text{reactor}} - \frac{\overset{\circ}{m}_{\text{in}} t}{\frac{1}{4}(\pi d^2)\rho_b}$

(* The expression is function of time *)

We can test this script by computing h_{diff} at a few times to be sure it is functioning properly. At a time close to the beginning of the filling process the value should be close to 25 ft and at 30 h it should be small and close to zero, while at about 15 hr it should be near 12 ft.

$\{h_{\text{diff}}[0.1h], h_{\text{diff}}[15.h], h_{\text{diff}}[30.h]\}$

{24.9194ft, 12.9123ft, 0.824566ft}

These numbers look fine, but we may prefer to have these in feet and inches, rather than in decimal feet. This is trivial to calculate, but let's understand how to write a script to do it for us. To do that we extract the integer part of the number and then use the remaining decimal to compute the inches. Again, very easy to do by hand, but let's finding a way to program this is instructive.

First, let's see how **IntegerPart** a command in *Mathematica* works. Mapping it on the previous result should be illustrative:

IntegerPart/@ $\{h_{\text{diff}}[0.1h], h_{\text{diff}}[15.h], h_{\text{diff}}[30.h]\}$

{IntegerPart[24.9194ft], IntegerPart[12.9123ft], IntegerPart[0.824566ft]}

This did not work; the input was echoed back as output. The problem is that we used **IntegerPart** on a list of elements that each have two parts: the number and the unit feet. We wanted the units in our equation. Because "ft" appears after the number, this does not compute, 24.9 ft is not a pure number and **IntegerPart** takes pure numbers. We could redo the model without units, but that is a step backwards. We also could divide each element by "ft" to clear the dimension and to get back to a pure number, but that is inelegant. Still another approach is to map **IntegerPart** onto the first part of each element, which is to say the number. To implement this we place [[1]] after h_{diff} to extract its numerical result.

IntegerPart/@ {h_{diff}[0.1h][[1]], h_{diff}[15.h][[1]], h_{diff}[30.h][[1]]}

{24, 12, 0}

We can also extract just the decimal part with the command **FractionalPart** using the same approach:

FractionalPart/@ {h_{diff}[0.1h][[1]], h_{diff}[15.h][[1]], h_{diff}[30.h][[1]]}

{0.919415, 0.912283, 0.824566}

Then we can convert these decimal feet to inches by multiplying by 12 and the results can be rounded to the closest inch by mapping **Round** over the list:

Round[12 * FractionalPart/@{h_{diff}[0.1h][[1]], h_{diff}[15.h][[1]], h_{diff}[30.h][[1]]}]

{11, 11, 10}

Since these are inches we should label them as such by multiplying the symbol "in" over the list:

in * Round[12 * FractionalPart/@{h_{diff}[0.1h][[1]], h_{diff}[15.h][[1]], h_{diff}[30.h][[1]]}]

{11in, 11in, 10in}

Now, we can bring the two scripts together to see h_{diff} in terms of feet and inches. Notice that we have multiplied through by "ft" also:

ft * IntegerPart/@ {h_{diff}[0.1h][[1]], h_{diff}[15.h][[1]], h_{diff}[30.h][[1]]} +

in * Round[12 * FractionalPart/@{h_{diff}[0.1h][[1]], h_{diff}[15.h][[1]], h_{diff}[30.h][[1]]}]

{24ft + 11in, 12ft + 11in, 10in}

If we want to generate a table of feet and inches at a given set of times, then we can do this by implementing the same approach inside the table function:

Grid [Table [ft * IntegerPart/@ {h_{diff}[xh][[1]]} +

in * Round [12 * FractionalPart/@ {h_{diff}[xh][[1]]}] ,{x, 0.1, 31, 1}]]

24ft + 11in
24ft + in
23ft + 4in
22ft + 6in
21ft + 8in
20ft + 11in
20ft + in
19ft + 3in
18ft + 6in
17ft + 8in
16ft + 10in
16ft + in
15ft + 3in
14ft + 5in
13ft + 8in
12ft + 10in
12ft
11ft + 3in
10ft + 5in
9ft + 7in
8ft + 10in
7ft + 12in
7ft + 2in
6ft + 5in
5ft + 7in
4ft + 9in
3ft + 12in
3ft + 2in
2ft + 4in
ft + 7in
9in

The alternative to this table is a simple graph of the continuous function; we can obtain this simply by plotting h_{diff} as a function of time:

$$h_{\text{reactor}} = 25\text{ft};$$

$$\overset{\circ}{m}_{\text{in}} = 1000\text{lb}/h;$$

$$d = 5\text{ft};$$

$$\rho_b = 63.2\text{lb}/\text{ft}^3;$$

$$h_{\text{diff}}[\text{t\_}] := h_{\text{reactor}} - \frac{\overset{\circ}{m}_{\text{in}} t}{\frac{1}{4}(\pi d^2)\rho_b}$$

(* The expression is function *)

Plot $[h_{\text{diff}}[th], \{t, 0.1, 31\}]$

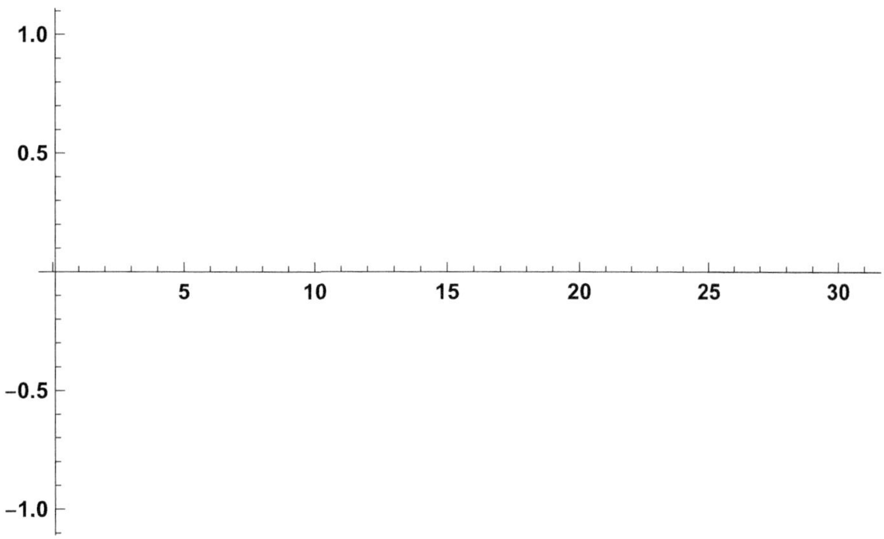

Interestingly, there is no output. We can go back and do a Table of this function to see if it is working:

Table $[h_{\text{diff}}[th], \{t, 0.1, 31\}]$

{24.9194ft, 24.1136ft, 23.3077ft, 22.5019ft, 21.696ft, 20.8902ft, 20.0843ft, 19.2785ft, 18.4726ft, 17.6668ft, 16.8609ft, 16.0551ft, 15.2492ft, 14.4434ft, 13.6375ft, 12.8317ft, 12.0259ft, 11.22ft, 10.4142ft, 9.60831ft, 8.80246ft, 7.99661ft, 7.19076ft, 6.38492ft, 5.57907ft, 4.77322ft, 3.96737ft, 3.16152ft, 2.35568ft, 1.54983ft, 0.743981ft}

Clearly, the function is working fine. The problem is similar to that observed above for IntegerPart; the function Plot expects pure numbers, but this function is providing a number and a unit "ft." We can fix the function in a way that it will provide only pure numbers and then use Plot. One way to do this is to simply formally divide each value by "ft" to convert back to a pure number:

$$\text{Plot}\left[\frac{h_{\text{diff}}[t\,h]}{\text{ft}}, \{t, 0.1, 31\}\right]$$

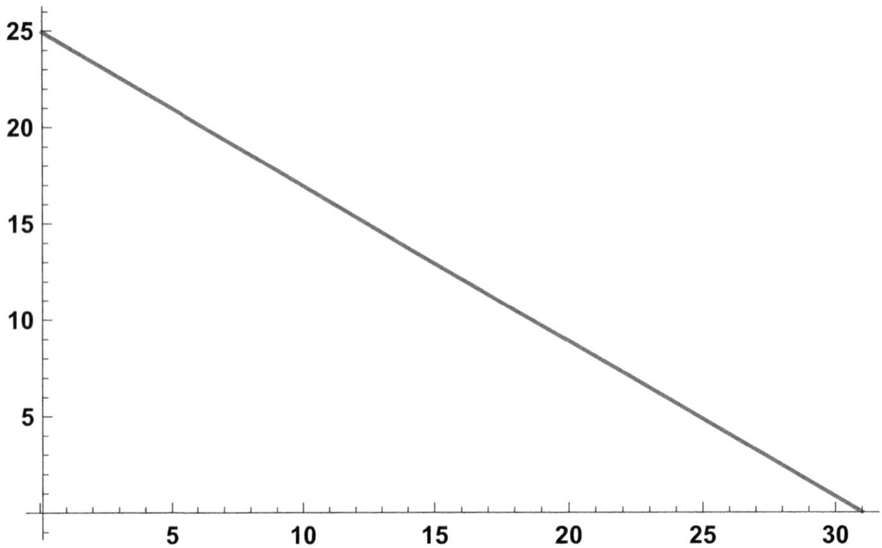

We can add labels to the two axes as follows:

$$\text{Plot}\left[\frac{h_{\text{diff}}[t\,h]}{\text{ft}}, \{t, 0.1, 31\},\right.$$

$$\left.\text{AxesLabel} \rightarrow \{\text{hrs}, \text{"}h_{\text{diff}}\text{,ft"}\}\right]$$

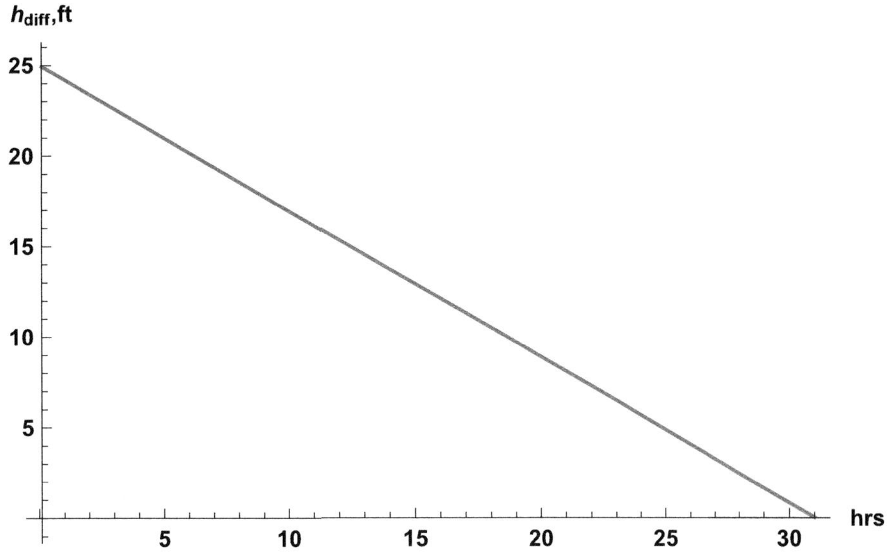

This is a model of how the system should behave and it is simply a linear function of time. We can go one step further and make this plot over with mass on the x-axis. This is handy, because if the filling process needs to be stopped at some point to pack the bed more tightly, it is easy to know exactly where the process left off.

$h_{reactor} = 25$ ft;

$\mathring{m}_{in} = 1000\ \text{lb}/h$;

$d = 5$ ft;

$\rho_b = 63.2\ \text{lb}/\text{ft}^3$;

$$h_{diff}[t\_] := h_{reactor} - \frac{\mathring{m}_{in} t}{\frac{1}{4}(\pi d^2)\rho_b}$$

(* The expression is a function *)

$m_{catalyst}[t\_] := \mathring{m}_{in} t$

Grid [Table [{$m_{catalyst}[th]$, $h_{diff}[th]$}, {t, 0.1, 31}]]

100.lb	24.9194ft
1100.lb	24.1136ft
2100.lb	23.3077ft
3100.lb	22.5019ft
4100.lb	21.696ft
5100.lb	20.8902ft
6100.lb	20.0843ft
7100.lb	19.2785ft
8100.lb	18.4726ft
9100.lb	17.6668ft
10100.lb	16.8609ft
11100.lb	16.0551ft
12100.lb	15.2492ft
13100.lb	14.4434ft
14100.lb	13.6375ft
15100.lb	12.8317ft
16100.lb	12.0259ft
17100.lb	11.22ft
18100.lb	10.4142ft
19100.lb	9.60831ft
20100.lb	8.80246ft
21100.lb	7.99661ft
22100.lb	7.19076ft
23100.lb	6.38492ft
24100.lb	5.57907ft
25100.lb	4.77322ft
26100.lb	3.96737ft
27100.lb	3.16152ft
28100.lb	2.35568ft
29100.lb	1.54983ft
30100.lb	0.743981ft

This is a decent result as it shows at each mass delivered what the measurement of h_{diff} should be. We would like to plot this, but to do so we have to drop the units. We can do that as we did earlier, by taking just the number and not the unit and then using ListPlot with the points joined to make the graph:

```
Table[{m_catalyst[th][[1]], h_diff[th][[1]]},
{t, 0.1, 31}];
ListPlot[%, Joined → True,
 AxesLabel → {"mass,lbs", "h_diff,ft"}]
```

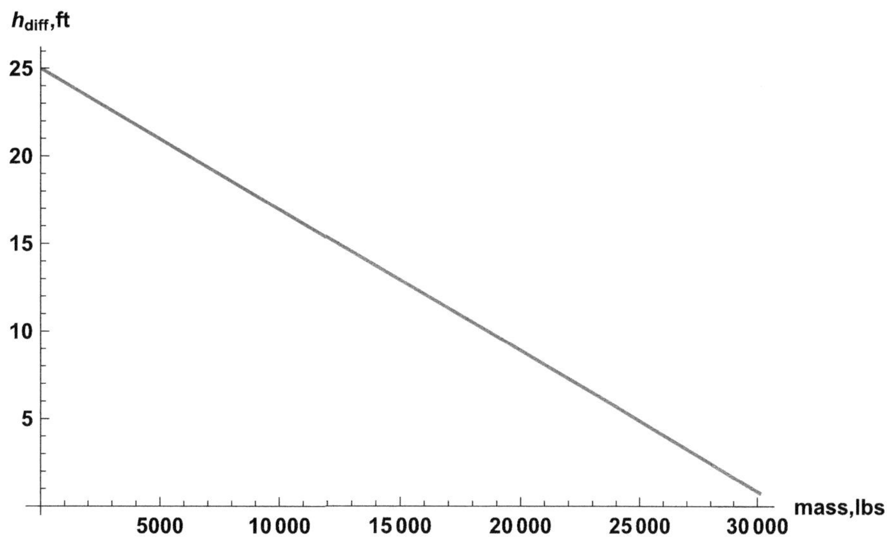

The plot is useful because for any given mass delivered, the expected level left to fill in the reactor, the measured variable, is specified.

Once the filling process has begun, we begin to accumulate data. After the first five hours a table generates the mass of catalyst in the reactor, the model prediction for h_{diff} and the measured value of h_{diff}:

mass (lb)	$h_{diff,mod}$ (ft)	$h_{diff,act}$ (ft)
100. lb	24.9	24.9
1100. lb	24.1	24.0
2100. lb	23.3	23.1
3100. lb	22.5	22.2
4100. lb	21.7	21.3
5100. lb	20.9	20.4
6100. lb	20.1	19.5
7100. lb	19.3	18.6
8100. lb	18.5	17.7
9100. lb	17.7	16.9

mass (lb)	$h_{diff,mod}$ (ft)	$h_{diff,act}$ (ft)
100. lb	24.9	24.9
1100. lb	24.1	24.0
2100. lb	23.3	23.1
3100. lb	22.5	22.2
4100. lb	21.7	21.3
5100. lb	20.9	20.4
6100. lb	20.1	19.5
7100. lb	19.3	18.6
8100. lb	18.5	17.7
9100. lb	17.7	16.9


```
Table[{m_catalyst[th][[1]], h_diff[th][[1]]}, {t, 0.1, 31}];
Show[ListPlot[%, Joined → True, AxesLabel → {"mass,lbs", "h_diff,ft"}],
Graphics[{Point[{100, 24.9}], Point[{4100, 21.3}],Point[{9100, 16.9}]}]]
```

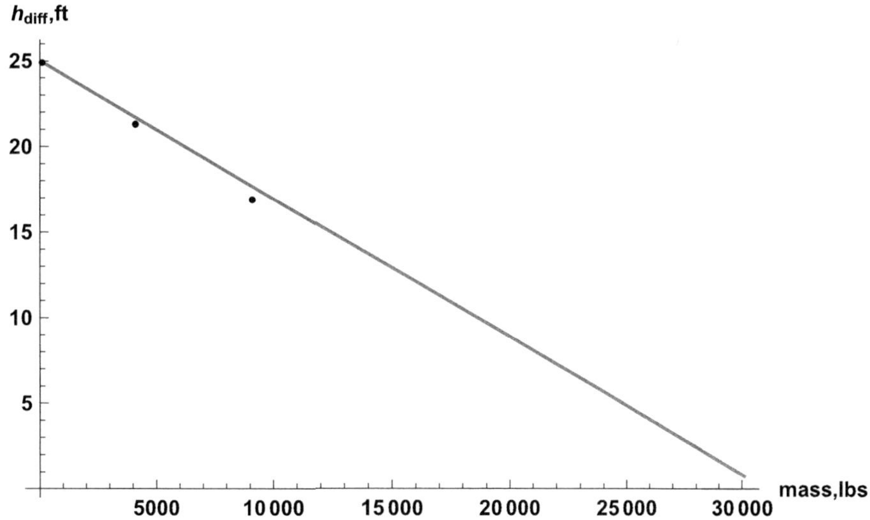

With about one third of the total mass of catalyst added, the difference between the model and the actual level in the reactor is between 9 and 10 inches. The question is should the process be continued, or should it be stopped, so that the bed can be compacted before adding more catalyst? How can this decision be made? One way to approach this would be let the process continue, but to fit the data and to make a prediction as to the likely deviation at the end of the process with an extrapolation. This makes the assumption that the packing density will not change. We can do this to see how far off the result would be.

First, we collect the data up to mass 9100 lb into a list:

```
data1 = {{100., 24.9}, {1100., 24.0}, {2100., 23.1},
{3100., 22.2}, {4100., 21.3}, {5100., 20.4},
{6100., 19.5}, {7100., 18.6}, {8100., 17.7},
{9100., 16.9}};
```

Then, this data is fitted to a line with the **Fit** command. The syntax is simple: the data set is named and the line is specified by { 1, x }.

```
line1 = Fit[data1, {1, x}, x]
```

$24.9749 - 0.000894545x$

The best fit line has the form - 0.0089 x + 24.97. We can extrapolate and plot this, then place it on the graph with the model and the experimental points. This equation predicts that at the present packing density, if it persists until the reactor is full, then the mass of catalyst will be at 28,056 pounds:

```
NSolve[0==24.97 – 0.00089x, x]
```

{{x → 28056.2}}

It is also easy to put this new best-fit line on the previous graph with the model values and a few of the data points:

```
Table[{m_catalyst[th][[1]], h_diff[th][[1]]},
{t, 0.1, 31}];

Show[
ListPlot[%, Joined → True,
 AxesLabel → {"mass,lbs", "h_diff,ft"}],

Graphics[{Point[{100, 24.9}], Point[{4100, 21.3}],
Point[{9100, 16.9}]}],

Plot[24.97 – 0.00089x, {x, 0, 30000},
PlotStyle → Red]
]
```

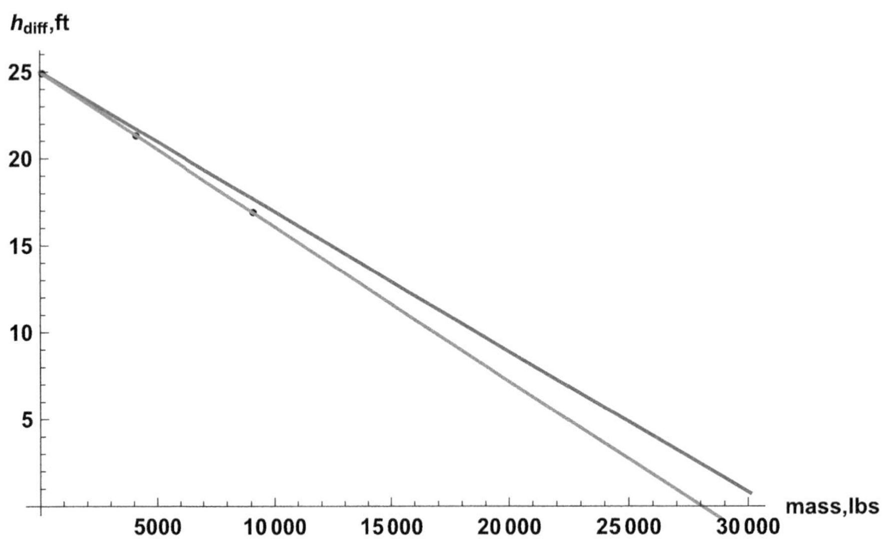

Clearly, from this result, the process would be stopped and the catalyst bed would be compacted by some means, such as vibration of the reactor, or by some other methodology. It is also likely that this will need to be done through out the process.

Reviewing this problem, we used the differential mass balance to find the mass of catalyst that would be expected in the reactor at any time. Then we found the functional time dependence of the mass accumulating in the reactor to get at the problem of packing so that we could make a measurement of the bed height during the process. A key part of the model was knowing the compacted bulk density of the catalyst particles, a number that was derived from experiment. Any deviation below this density would lead to more volume being occupied by a given mass of catalyst and that would be detrimental to the process and its economics. We learned and used a number of interesting *Mathematica* tricks. We saw that we can use dimensions and units in equations and they will be treated properly. But, we also learned that many of the functions that we use in *Mathematica* treat pure numbers. Here are a few of the tools that we used interactively in reverse order: Fit, NSolve, Grid, FractionalPart, and IntegerPart.

Pressurizing an initially evacuated tank with an ideal gas

Gas flows are also common and offer another opportunity for us to apply the total mass balance. Gases can be simple or complex. By simple, we mean that in some cases the atoms or molecules comprising the gas do not interact except at the point of collision. They behave as if they were nanoscopic billiard balls racing around colliding with one another and with the walls of the vessel in which they are contained. Furthermore, some gases continue to behave in the same way over a wide range of temperatures and pressures. We refer to gases of this kind as "ideal" and their characteristics are those of "hard-spheres." Most gases do not behave ideally, and they are referred to as real gases. If their properties are nearly ideal at low pressure, we find that they deviate from ideality at higher pressure. The reason for this is that these atoms or molecules do not behave merely as billiard balls bouncing off one another. Instead they are "sticky" or they are repulsive. They have size and they have shape. Some are polar, others are non-polar. It is these properties that give rise to the much more subtle and richer effects that are observed in "real" gases and which the hard sphere model could never predict. Nonetheless, the ideal gas is a good learning model and one from which we can gain a great deal and it does apply to some real gases under some ranges of conditions.

Recall that an ideal gas follows a very simple equation of state:

$$P\,V = n\,R\,T$$

Where **P** is the pressure, **V** the volume of the vessel, n is the number of moles of the gas in the vessel, **R** is the gas constant and **T** is the absolute temperature. With this we can calculate the pressure in vessel of volume **V** as a function of temperature, the volume of a gas at fixed pressure and temperature or the temperature at fixed pressure and volume by simple rearrangements:

$$\mathrm{P} = \frac{nRT}{V}$$

$$\mathrm{V} = \frac{nRT}{P}$$

$$\mathrm{T} = \frac{PV}{nR}$$

Even more, we can compute the concentration of the ideal gas in moles per unit volume or its density in mass per unit volume:

$$\mathrm{C} = \frac{n}{V} = \frac{P}{RT}$$

$$\rho = \frac{n\mathrm{MW}}{V} = \frac{P\mathrm{MW}}{RT}$$

Where **MW** refers to the molecular weight of the gas. This provides the link we need between the gas phase material and the overall mass balance. Let's see how this works.

Gases, or vapors, and liquids are both fluid phases, but they differ vastly in density. For example the density of liquid water at ambient conditions is about 1 gram per cm^3. The vapor pressure of water in equilibrium with the liquid is 0.43 psia at 75°F. (psia stands for pounds per square inch absolute meaning above vacuum.) We can compute the concentration of water in the liquid phase and compare this to that of the equilibrium vapor phase, then we can compute the density of the equilibrium vapor phase and compare that to the liquid density. In this way we can have a better sense of the magnitudes of these quantities and by how much they differ. We might call this having a "physical feel" for the numbers.

$$\textbf{C\_H2OLiq} = N\left[\frac{1\mathrm{g}}{\mathrm{cm}^3} * \frac{1\ \mathrm{g\text{-}mole}}{18g} * \frac{1000\ \mathrm{cm}^3}{L}\right]$$

(* This provides the concentration of liquid water in g-mole per liter *)

$$\frac{55.5556\mathrm{gmole}}{L}$$

UnitConvert[Quantity[75., "DegreesFahrenheit"], "Kelvins"]

(* This converts from Fahrenheit to Kelvin *)

$297.04K$

$$\textbf{C\_H2OEqVap} = \left(N\left[0.43\text{ psia} * \frac{1\text{ atmosphere}}{14.7\text{ psia}}\right] \Big/ \left(0.08205\frac{L\text{–atmosphere}}{\text{g-mole} * K} * 297.039K\right)\right)$$

$$\frac{0.00120022\text{g-mole}}{L}$$

Round[%[[1]], .0001] g-mole/L (* Here we round off the concentration *)

$$\frac{0.0012\text{g-mole}}{L}$$

We can compute the density of the equilibrium vapor of water:

$$\rho\textbf{\_H2OEqVap} == \left(\frac{0.0012\text{ g-mole}}{L}\right)\left(\frac{18g}{\text{g-mole}}\right)\left(\frac{1L}{1000\text{ cm}^3}\right) \text{// ScientificForm}$$

$$\rho\text{\_H2OEqVap} == \frac{(2.16\times 10^{-5})g}{\text{cm}^3}$$

The density of the water vapor is just over 2 x 10^{-5}grams in one centimeter. Thus the vapor is 50 thousand times less dense than the liquid:

$$\frac{1\frac{g}{\text{cm}^3}}{(2 * 10^{-5})\frac{g}{\text{cm}^3}} \text{//N}$$

50000.

Aside from getting the ratio of the density of a liquid to that of the vapor, we also have seen how to use UnitConvert and how to include a variety of units in interactive calculations with *Mathematica.*

A common operation in a pilot plant or a laboratory is the pressurization of a batch reaction vessel with a gas like hydrogen. A batch reactor is one that does not have flow into it, or out of it during reaction. It does have to be charged with reactants prior to operation. We can consider this process to be one that is amenable to the total mass balance equation. For simplicity we will sacrifice realism and assume that the gas remains ideal throughout the pressurization. We will see that in many ways this problem is similar to the filling of a tank with liquid phase fluid, or the filling of a reactor with catalyst solids, once we account for the gaseous nature of this fluid.

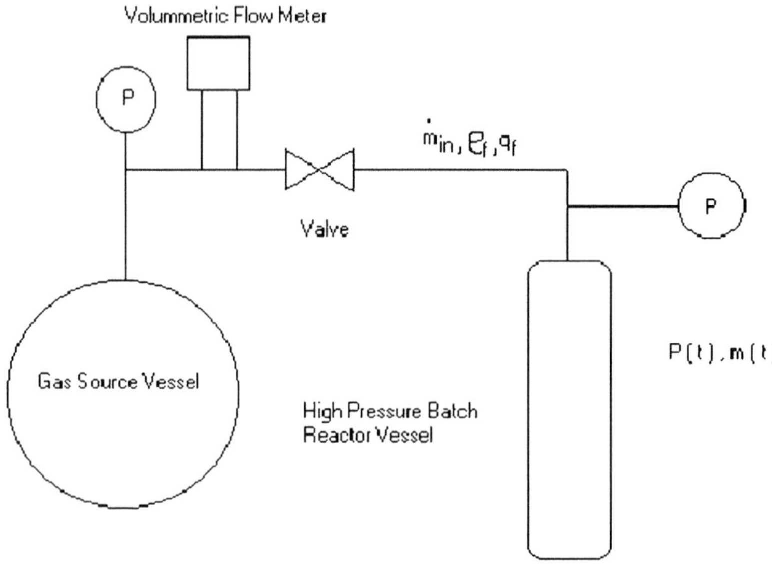

The source of the gas is a large volume vessel at high pressure. We assume that the whole system remains isothermal, that is at constant temperature, throughout the procedure. This source vessel must be at a pressure higher than the pressure we need to attain in the reactor and it must have sufficient gas within it that the pressurization of the smaller vessel will not substantially reduce its pressure. The pressure upstream of the valve is a constant, just after the valve and into the reactor, the pressure is time dependent. The reaction vessel is initially evacuated so that only the pure gas will be present in the gas phase. Opening the valve allows one to control the flow of gas into the vessel. By monitoring the pressure gauge, the flow can be stopped when the proper pressure has been attained. The rise in pressure at the reactor gauge as a function of time is also a measure of the mass flow of gas into the reactor vessel. Alternatively, if we know the mass flow, or the volumetric flow and the pressure upstream of the valve, then we can predict the time required to reach a set pressure in the vessel given its volume and temperature. We can consider the latter situation first.

The overall mass balance for the process of pressurizing the reactor vessel is:

$$\frac{dm[t]}{\mathrm{dt}} = \overset{\circ}{m}_{\mathrm{in}}$$

The mass accumulation of gas in the reactor is given by:

$$\frac{dm[t]}{\mathrm{dt}} = \frac{d\rho_g[t]V}{\mathrm{dt}}$$

This expresses the rate of change in the density of the gas as it accumulates in the fixed volume of the reactor. Assuming that the gas is ideal, we can re-express change in the gas density as a function of the change in pressure as a function of time because pressure is that which we can measure:

$$= \frac{d}{dt}\left[\frac{P[t]\text{MW}}{RT}V\right]$$

The volume of the vessel is constant, as are the temperature, molecular weight, and gas constant. These can be brought out of the derivative:

$$= \frac{V\text{MW}}{RT}\frac{dP[t]}{dt}$$

That takes care of the left-hand side of the equation, but what about the right-hand side? The right-hand side is the rate of mass flow into the reactor. The mass flow is a gaseous mass flow. Therefore it can be given as the product of the feed gas density and the volumetric flow rate into the reactor. The pressure is constant here because it is the pressure of the source vessel that extends to the valve. The pressure drops after the valve. We take the volume flow rate to also be a constant for all of the process until we shut the valve. The product of the density of the gas and the volume flow rate give us the mass flow rate. But, we can compute the density of the gas at the feed pressure from the equation of state (EOS). For this calculation, we use the ideal gas equation as the EOS, but that is not necessary. Thus the right-hand side of the equation becomes as follows:

$$\overset{\circ}{m}_{\text{in}} = \rho_f q_f = \frac{P_f \text{MW}}{RT} q_f$$

where P_f is the constant feed pressure, and q_f is the constant volume flow rate. The overall equation becomes:

$$\frac{V\text{MW}}{RT}\frac{dP[t]}{dt} = \frac{P_f\text{MW}}{RT}q_f$$

$$\frac{dP[t]}{dt} = \frac{P_f}{V}q_f$$

The final equation tells us that the rate of pressure rise in the reactor vessel will be a constant equal to the product of the ratio of the feed gas pressure to the vessel volume and the volumetric flow rate of the feed gas. (Assuming that the volume of the line leading from the valve to the vessel is negligible.) Because the right-hand side is constant, we can see immediately that the pressure rise will be linear.

$$P[t] = \left(\frac{P_f q_f}{V}\right)t$$

The linear rise in pressure is nearly the same as the linear rise in catalyst solids in the reactor. At one level both of these problems are similar because there is a constant rate of mass flow into a vessel and the conservation of mass principle holds no matter if the flow is that of a solid, a liquid or a gas.

Filling a cylindrical tank

Often the mass flow that we are concerned with will involve a liquid and liquid flows will be involved in the systems that we will consider in this book. When a liquid is flowing we typically measure its flow rate as a volumetric flow with dimensions of volume per unit time. The mass flow will be given as it was for a gas as the product of the volumetric flow and the density of the liquid in that stream. Hence we will consider next, the flow of a liquid into a tank, which is as simple a problem as we can find.

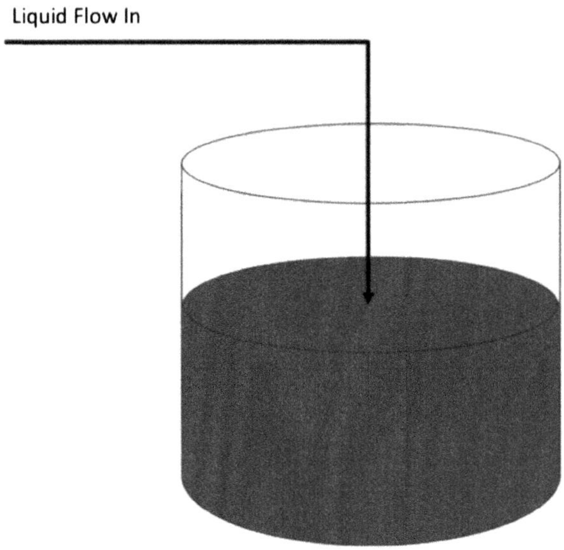

The control volume is the tank itself. This is because the liquid flowing into the tank is homogeneous, meaning that wherever we make a measurement of composition, density, temperature, it is always the same. The differential statement of the conservation of mass is as it was in the first case:

$$\frac{dm[t]}{\mathrm{dt}} = \overset{\circ}{m}_{\mathrm{in}}$$

The expression for the mass flow rate into the tank is the product of the density of the liquid, ρ_l, and the volumetric flow rate, q_{in}:

$$\frac{dm[t]}{\mathrm{dt}} = \rho_l q_{\mathrm{in}}$$

Let's do a dimensions check:

$$\frac{\text{mass}}{\text{time}} \equiv \left(\frac{\text{mass}}{\text{volume}}\right)\left(\frac{\text{volume}}{\text{time}}\right)$$
$$\frac{\text{mass}}{\text{time}} \equiv \left(\frac{\text{mass}}{l^3}\right)\left(\frac{l^3}{\text{time}}\right)$$

The dimensions are consistent across both sides of the equation.

The mass accumulated within the control volume is the product of the density of the liquid, the cross sectional area of the tank, A_C, and the height of the liquid in the tank at any time t, h [t]. Replacing this in the time derivative and rearranging we find:

$$\frac{d\rho_l A_C h[t]}{\mathrm{dt}} = \rho_l q_{\mathrm{in}}$$
$$\frac{dh[t]}{\mathrm{dt}} = \frac{q_{\mathrm{in}}}{A_C}$$

This equation states that the rate of change of liquid level in the tank is a constant, if the flow rate in is a constant. From this we know then that the change in level must be linear function of time. We show this explicitly by separation of the variables and integration:

$$\int_0^{h[t]} dh[t] == \int_0^t \frac{q_{\mathrm{in}}}{A_C} dt$$
$$h[t] == \frac{q_{\mathrm{in}}}{A_C} t$$
$$l \equiv \left(\frac{l^3}{t}\right)\left(\frac{1}{l^2}\right) t$$
$$l \equiv l$$

Here too we can do the integration trivially because the flow rate into the tank is a constant. Notice also that the units are consistent in the final expression. We can also see that *Mathematica* will do the operations for us as well.

$$\boldsymbol{l \equiv \left(\frac{l^3}{t}\right)\left(\frac{1}{l^2}\right) t}$$

$$l \equiv l$$

If we take the cross section area of the tank to be 10 meter^2 and the flow rate to be 2 meter^3 per minute, then a plot of the level of liquid versus time is as follows:

Plot$\left[\left(\frac{2}{10}\right)t, \{t, 0, 50\}\right.$,

AxesLabel → {"t/min", "h [t]meters "},

PlotStyle → Hue[0]]

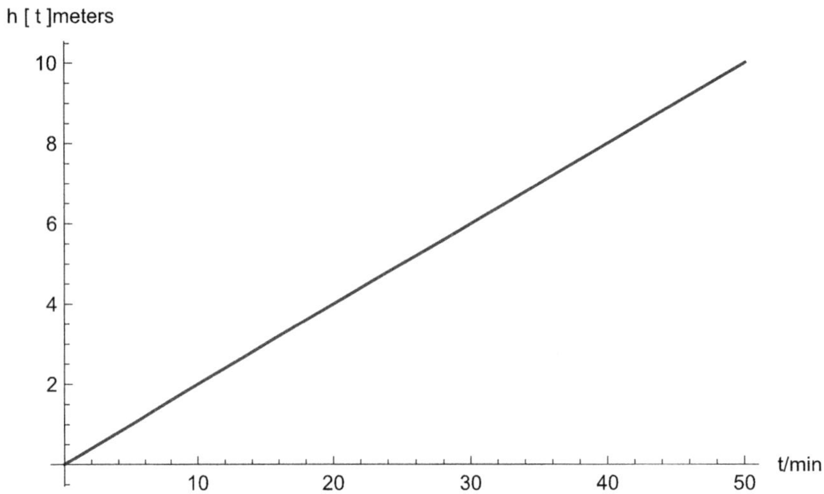

If the aspect ratio of the tank is 4, meaning the ratio of the height to the diameter is 4 to 1, then after how many minutes will it overflow under these conditions? We know the cross sectional area, so we can find the diameter of the tank which is one fourth of its height. Given this height we can solve for t_{critical}, the time at which overflow commences. The diameter is computed first:

Solve[10== $N[\pi \frac{d^2}{4}]$, d]

(* This solves for the diameter given the cross sectional area *)

$\{\{d \to -3.56825\}, \{d \to 3.56825\}\}$

The diameter of the tank is just over 3.5 meters and the height of the tank is four times the diameter, so the time to overflow is:

Solve[4 ∗ 3.57Meter == $\frac{2\text{Meter}^3\text{Minute}^{-1}}{10\text{Meter}^2}\left(\frac{60\text{Minute}}{\text{hour}}\right) t_{\text{critical}}, t_{\text{critical}}$]

$\{\{t_{\text{critical}} \to 1.19\text{hour}\}\}$

Thus the tank will begin to overflow after 1 hour and 11 minutes.

How would write the differential mass balance for that situation, when the liquid is overflowing continuously? We would do it just as we have before, except that now we would have the second term on the right-hand side:

$$\frac{dm[t]}{dt} = \overset{\circ}{m}_{in} - \overset{\circ}{m}_{out}$$

If we think about this situation,it is immediately obvious that the right-hand side is identically zero. This means that the net rate of change of level in the tank is also identically zero, meaning that it no longer can rise or fall, it stays at a steady state value.Thus the overall behavior of the unattended tank under these conditions is simply this:

$a = \text{Plot}\left[\left(\frac{2}{10}\right) t, \{t, 0, 71.5\},\right.$

AxesLabel → {"t / min", "h [t] /m"},

PlotStyle → Hue[0]];

$b = \text{Plot}\left[\left(\left(\frac{2}{10}\right) - \left(\frac{2}{10}\right)\right) t + 14.28, \{t, 71.5, 150\},\right.$

PlotStyle → Hue[0.7]];

Show[*a*, *b*, PlotLabel → "Onset of Steady State",

PlotRange → Automatic]

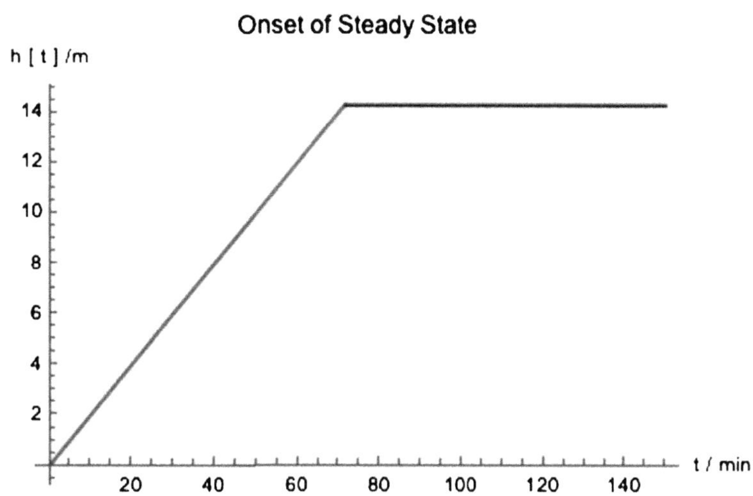

Once the liquid in the tank reaches the highest level, it begins to spill over and the rate into the tank is the same as the rate out of the tank and the level no longer changes. This system will be at steady state as long as nothing changes, and, perhaps, having negative consequences for the surroundings.

Time dependent flows

To this point we have considered all the inlet flows to be constants. We should now consider what happens when they are functions of time. When we specify the flow rate as a function of time we have said nothing about the mechanism that gives rise to the observed functionality. It is simply a statement based on observation. There are cases in which the consideration of the mechanism can make it plain that flows will be time dependent. For example the pumping of our hearts is periodic and gives rise to periodic, or pulsating flow of blood. As we open a valve or faucet, the flow grows in relation to the rate at which we open it and the opposite happens when we are closing the valve. There are other cases in which the flow may be periodic, pulsating, or otherwise time dependent.

An interesting case to begin examine is the periodic flow into a tank. We have already analyzed this for constant flow, but what would be different about for example a periodic flow? How would this affect the time dependence of the rate of mass accumulation in the tank? To begin let's consider a continuous, periodic flow with a sinusoidal dependence upon time. The flow rate could be described as:

$$\text{qf}[\text{t}] = \text{qfo}\,(1 + \alpha\,\text{Sin}[\beta\,\text{t}])$$

where qfo is a constant flow rate that is modulated by a sine wave, where α and β are constants and t has the usual meaning of time. The constant α is dimensionless, whereas the constant β has dimensions of inverse time. To see what this would look like we can plot it for some specific values of **qfo**, $\boldsymbol{\alpha}$, and $\boldsymbol{\beta}$.

```
Remove[qf, α, β, t]
qf[t_]:=qfo(1 + αSin[βt])
qfo = 10;
α = 0.5;
β = 0.25;
Plot[{qf[t], qfo}, {t, 0, 100}, PlotStyle → {Hue[0.1], Hue[0]},
AxesLabel → {"t, min", "qf[t], L/min"}, PlotRange → {{0, 16}}]
```

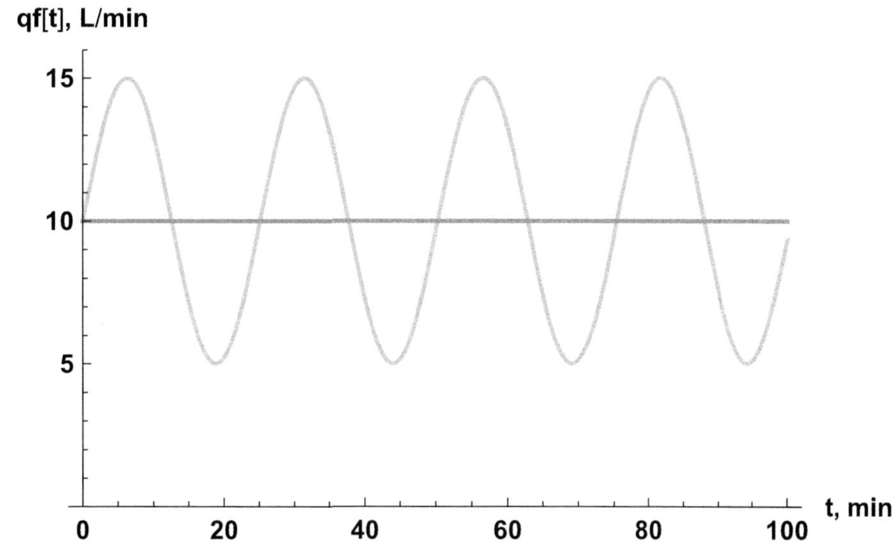

The dimensions of the constants $\boldsymbol{\alpha}$ and $\boldsymbol{\beta}$ are worth explaining further. The constant $\boldsymbol{\beta}$ is an element of the argument of the Sine, and Sine is a *transcendental* function. Because Sine is transcendental, its argument must be dimensionless; for the product [$\boldsymbol{\beta}$ t] to be dimensionless $\boldsymbol{\beta}$ must have the dimension of $\frac{1}{\text{time}}$ in the proper units. Thus $\boldsymbol{\beta}$ is an inverse time constant with some physical significance. What about $\boldsymbol{\alpha}$? We see that when we multiply through by **qfo**, the product of $\boldsymbol{\alpha}$ and **qfo** must have dimensions of volumetric flow rate. Because **qfo** already carries these dimensions, $\boldsymbol{\alpha}$ must be dimensionless. What are the physical significances of $\boldsymbol{\alpha}$ and $\boldsymbol{\beta}$? From the physics of waves, we know that $\boldsymbol{\alpha}$ is the amplitude of the wave and that $\frac{\mathbf{1}}{\boldsymbol{\beta}}$ is the peak-to-peak time, or the period of the wave.

If this is the input to our tank, how will the level as a function of time behave? The starting point is the overall material balance:

$$\frac{dm[t]}{\text{dt}} = \overset{\circ}{m}_{\text{in}}$$

$$\frac{d\rho Ah[t]}{\text{dt}} = \rho q_f[t]$$

The right-hand side is now a time dependent function:

$$\frac{dh[t]}{\text{dt}} = \frac{q_f[t]}{A_C}$$

$$\frac{dh[t]}{\text{dt}} = \frac{q_{\text{fo}}(1 + \alpha \text{Sin}[\beta t])}{A_C}$$

We can either separate and integrate, or use the command **DSolve** to solve the equation directly. Here is some annotated code to do that:

(* The first statement simply ensures that these variables and parameters are removed before use *)

Remove[h, qfo, α, β, t]

(* Next we place within DSolve the equation and the initial condition, $h[0]==0$, inside { } Then we specify the function that we seek, $h[t]$, and the independent variable, t *)

Simplify[

$$\text{DSolve}\left[\left\{\partial_t h[t] == \frac{\text{qfo}}{\text{Ac}}(1+\alpha\text{Sin}[\beta t]), h[0]==0\right\},\right.$$

$h[t], t$]]

$$\left\{\left\{h[t] \to \frac{\text{qfo}(\alpha + t\beta - \alpha\text{Cos}[t\beta])}{\text{Ac}\beta}\right\}\right\}$$

From this solution we can see that if the amplitude, α, is very small, in the limit as $\alpha \to 0$, then the result is the same as that which we had derived before.

$$\text{Limit}\left[\frac{\text{qfo}(\alpha + t\beta - \alpha\text{Cos}[t\beta])}{\text{Ac}\beta}, \alpha \to 0\right]$$

$$\frac{\text{qfo}t}{\text{Ac}}$$

But it is also clear that if the values of α and β are of certain ranges of magnitude, then this periodic flow will give rise to periodicity in the change in level of the tank. We can model this to see how this will look using the values that we had had for α and β earlier.

$$h[\text{t\_}] := \frac{\text{qfo}(\alpha + t\beta - \alpha\text{Cos}[t\beta])}{\text{Ac}\beta}$$

qf[t_]:=qfo(1 + αSin[βt])

qfo = 10;

$\alpha = .5;$

$\beta = 0.25;$

Ac = 10;

Plot[{*h*[*t*], qf[*t*]}, {*t*, 0, 70}, PlotStyle → {Hue[0], Hue[0.1]},

AxesLabel → {t, {"h [t]= red", "qf [t]= yellow"}}]

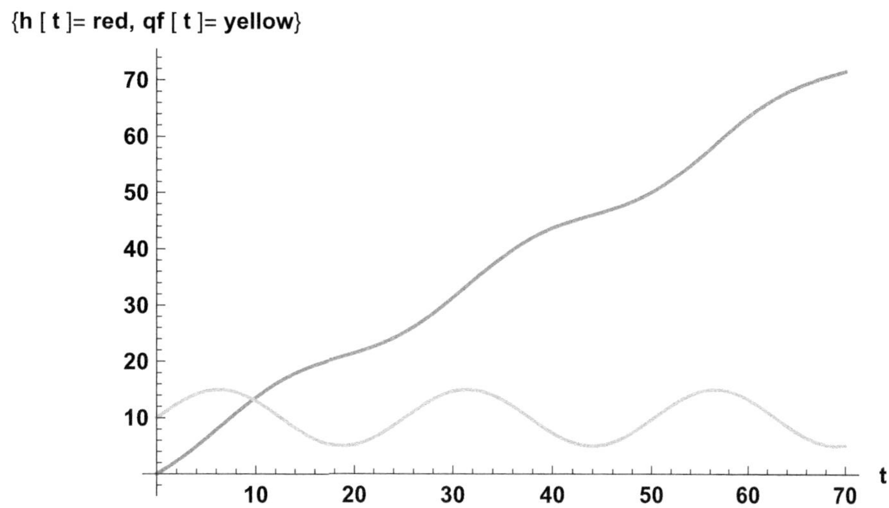

This shows us that the level will of course rise, but it will do so with varying rates depending upon the flow rate. This would actually be easier to see if we were to plot the dimensionless level and flow rates. We can get these by dividing **h [t]** by the maximum level in the tank and **q [t]** by **qfo**.

Plot[{*h*[*t*]/30, qf[*t*]/qfo}, {*t*, 0, 70}, PlotStyle → {Hue[0], Hue[0.1]},

$$\text{AxesLabel} \rightarrow \left\{\text{t}, \text{“}\frac{h[t]}{\text{hmax}}, \frac{\text{qf}[t]}{\text{qfo}}\text{”}\right\}\Bigg]$$

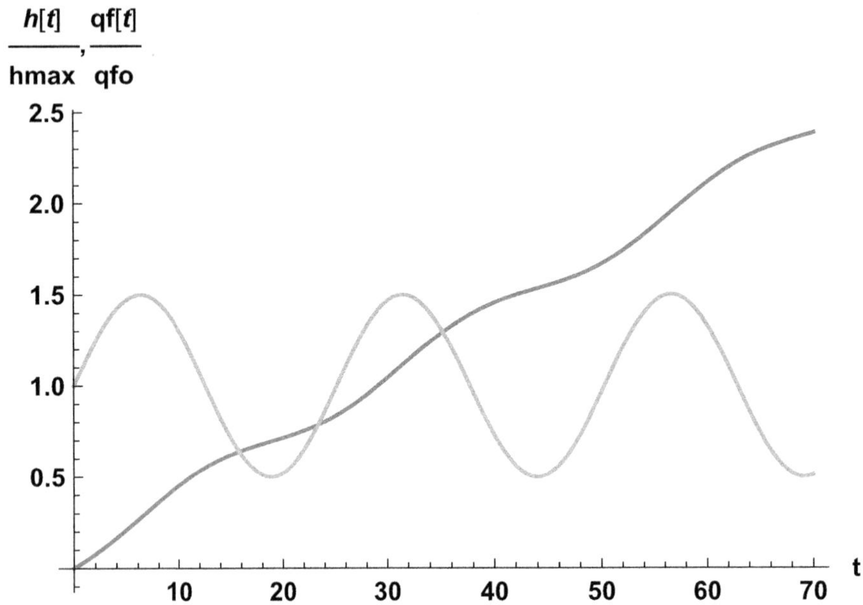

What would happen if we were to bring the amplitude up, say by a factor of five?

$\alpha = 2.5$;

Plot[{*h*[*t*]/30, qf[*t*]/qfo}, {*t*, 0, 70}, PlotStyle → {Hue[0], Hue[0.1]},

AxesLabel → $\left\{\text{t}, \text{“}\frac{h[t]}{\text{hmax}}, \frac{\text{qf}[t]}{\text{qfo}}\text{”}\right\}$]

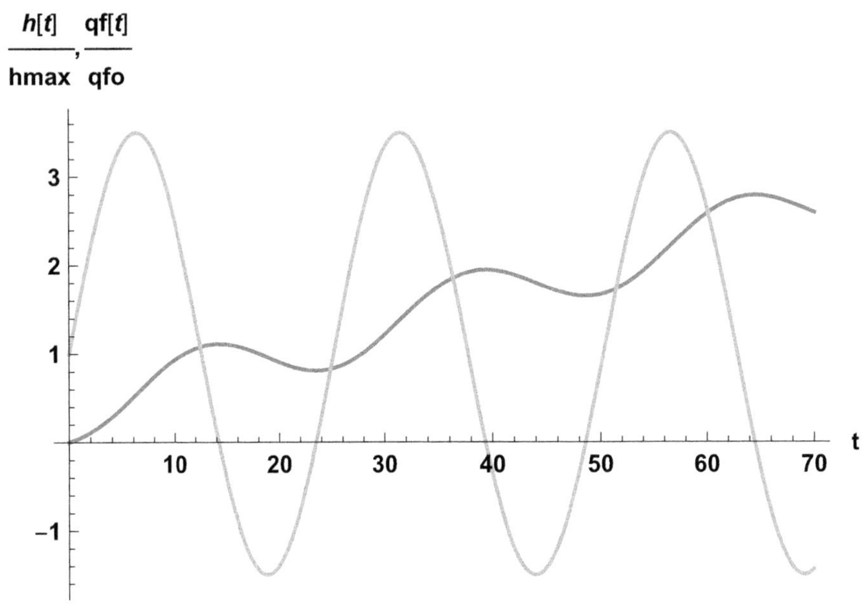

If we look closely, we see that in the time range between 15 and 20, the level is actually decreasing! But how can this happen? We have only accounted for flow into the tank in the total material balance. Once again, we need to be very careful with the model and the results that we derive from it. In this case when we increased the amplitude by a factor of five we went out of the region in which the solution gave physically meaningful results. If you look carefully, in the same region where the level is decreasing the flow rate is actually below zero, that is negative, or flow in the reverse direction of the feed, namely out of the tank. The change in sign of the input function has given rise to this negative rate of accumulation, i.e. negative slope, in the time domain. This is not the situation that we had in mind when we began the problem. It could correspond to some actual situation, but it does not correspond to the situation we were analyzing. One needs to be very mindful of the range of application of any model and should check the resulting behavior for its correspondence to the real system.

Geometry and the left-hand side of the mass balance equation

The triangular trough

To this point the situations that we have dealt with have all involved quite simple geometry – the right cylinder. We may ask the question how do we apply the new toolkit that we have to other geometrical shapes. The issues that we will encounter in these kinds of analyses are handled within the differential accumulation term through **V[t]**. For this reason we can think of these as "Left-hand Side" problems. The objective of this section is to demonstrate how to do that. We begin with an analysis of a tank that is shaped like a triangular trough shown below.

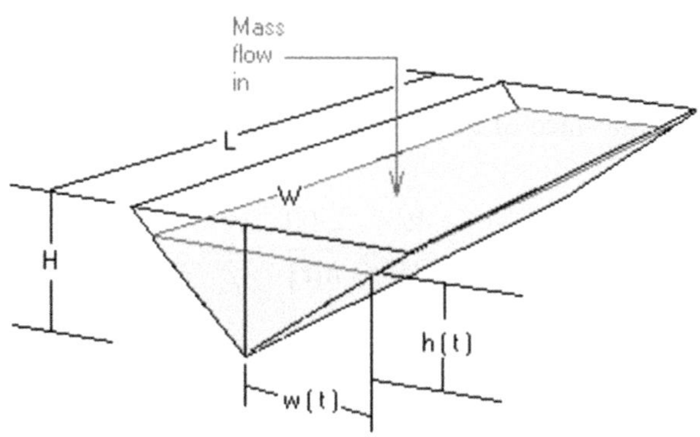

The mass flow into the tank is at a constant rate given by the density of the fluid and its volumetric flow rate. The mass in the tank at any time is the product of the density and the fluid volume. Notice that as the level of the fluid increases so does its width. Viewed from the top, the area of the liquid surface varies as a function of time. This is the main difference between this "tank" and that of a right cylinder standing on end. In that case the surface of the liquid viewed from above remained constant, so that the volume was only a function of the level. To summarize what we have so far:

$$\frac{dm[t]}{\mathrm{dt}} = \overset{\circ}{m}_{\mathrm{in}}$$

$$\frac{d\rho V[t]}{\mathrm{dt}} = \rho q_f[t]$$

$$\frac{dV[t]}{\mathrm{dt}} = q_f$$

The differential change in volume with time, **d V [t]**, can be viewed as taking place through differential changes in level, **d h [t]**. The key insight to the analysis is that the differential volume can be viewed as the product of an area of the fluid at the surface and a differential thickness given as **d h [t]**. This gives us:

$$\mathrm{d\,V[t] = A[t]\;d\,h[t]}$$

Of course that holds at some given time and some given level because the area of the surface of the fluid does change with level, and that change is given by the variation in the liquid's width multiplied by the length of the trough, which gives us:

$$\frac{dV[t]}{\mathrm{dt}} = 2\,\mathrm{L}\;\mathrm{w[t]}\,\frac{dh[t]}{\mathrm{dt}} = q_f$$

Now, we have two functions for the variable and we should have just one. To get to one variable function, we need a relationship between **w [t]** and **h [t]**. To find this, we can consider the geometry of the triangular face of the tank. From the law of similar triangles we find this:

$$\frac{W}{H} = \frac{w[t]}{h[t]}$$

The differential equation can now be rewritten in terms of only

h [t]:

$$\frac{dV[t]}{\mathrm{dt}} = \frac{2WL}{H} h[t] \frac{dh[t]}{\mathrm{dt}}$$

$$h[t]\frac{dh[t]}{\mathrm{dt}} = \frac{Hq_f}{2WL}$$

The equation is now readily soluble:

Remove[h, H, qf, W, L, t]

$$\mathbf{DSolve}\left[\left\{h[t]\partial_t h[t] == \frac{Hq_f}{2WL}, h[0] == 0\right\}, h[t], t\right]$$

$$\left\{\left\{h[t] \to -\frac{\sqrt{Ht10_f}}{\sqrt{L}\sqrt{W}}\right\}, \left\{h[t] \to \frac{\sqrt{Ht10_f}}{\sqrt{L}\sqrt{W}}\right\}\right\}$$

There are two solutions to this equation because it involves the square root of **h [t]**. Since the tank is filling, we select the positive solution. Also notice that all the terms on the right-hand side are under the radical. We can rewrite it as follows:

$$\text{h[t]} = \sqrt{\frac{Hq}{WL}t} = \text{constant } \sqrt{t}$$

The level change is not a linear function of time, but varies as the square root of time. This is a direct consequence of not having straight walls. We can also compute the maximum volume of the tank and the volume in the tank at any time from the equation now that we know how h[t] is expressed:

$$\frac{dV[t]}{\text{dt}} = \frac{2WL}{H}h[t]\frac{dh[t]}{\text{dt}}$$

$$\text{V [t]} = \frac{WL}{H}h[t]^2$$

Consider the level as function of time for a tank with a height of 10 ft, length of 10 ft, a width at the top of 10 ft, and an inlet flow rate of 5 $\text{ft}^3\ \text{min}^{-1}$. We compute the total volume in this tank first:

Clear[h, q, H, W, L]

$\mathbf{W = 10;}$

$\mathbf{L = 10;}$

$\mathbf{q = 10;}$

$\mathbf{H = 10;}$

$\mathbf{vmax = WLH\text{ft}^3}$

1000ft^3

The tank has a volume of 1000 ft^3. If the tank had straight sides, it would be rectangular and the volume would be 2 W L H or 2000 ft^3. The next step is to create a function for the level and another for the volume. We will call these "htt" and "vtt" where the tt stands for triangular trough. Then we can plot the level in the tank as a function of time with a linear rise for comparison and the maximum level marked:

$$\text{htt[t\_]:=}\sqrt{\frac{Hq}{WL}t}$$

$$\text{vtt[t\_]:=}\frac{WL}{H}\text{htt}[t]^2$$

```
(* This is the plot of the level in the triangular trough *)
plhtt = Plot[htt[t], {t, 0, 200},
PlotStyle → Hue[0.06],
AxesLabel → {"t / min", "h[ t ] / ft"}];

Show[plhtt, Graphics[Line[{{0, 10}, {200, 10}}]],
Graphics[{Dashed, Line[{{0, 0}, {200, 10}}]}]
]
```

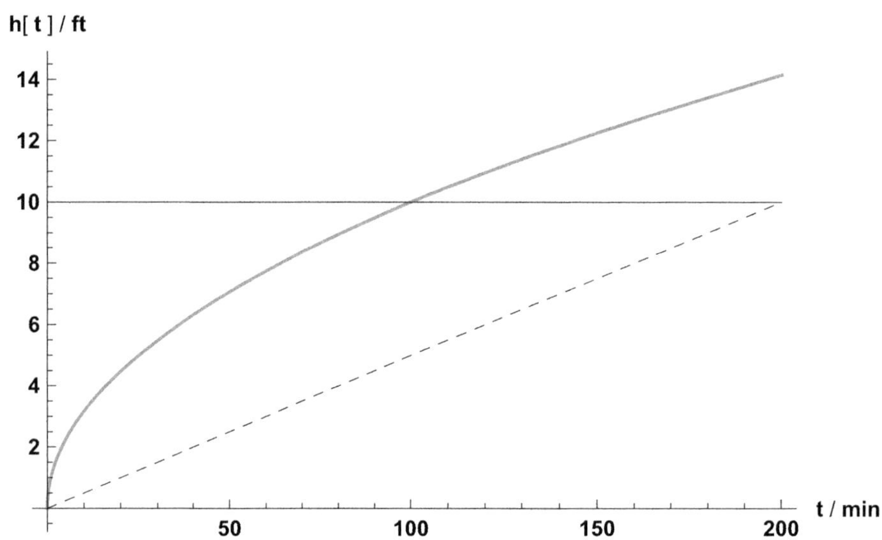

The Show command in the last script, combines the plot of htt[t] with a horizontal line indicating the highest level in the tank and with a line that corresponds to a linear filling of the tank. Early in the filling process, because of the square root dependence on time, the level rises very fast. Then after the tank is about half full and the process begins to look almost linear with time. The dashed line is a for a linear process and it is obvious that this non-linear, square root process is always "ahead" of the linear one.

Lastly, we plot the volume in the trough as a function of time:

plvtt = Plot[vtt[*t*], {*t*, 0, 200},

PlotStyle → Hue[0.6],

AxesLabel → {"t / min", "V[t] / ft^3"}];

(* This is a plot of the volume in the triangular trough *)

(* This allows us to put the top level line on the graph and a linear function *)

Show[plvtt, Graphics[Line[{{0, 1000}, {200, 1000}}]]]

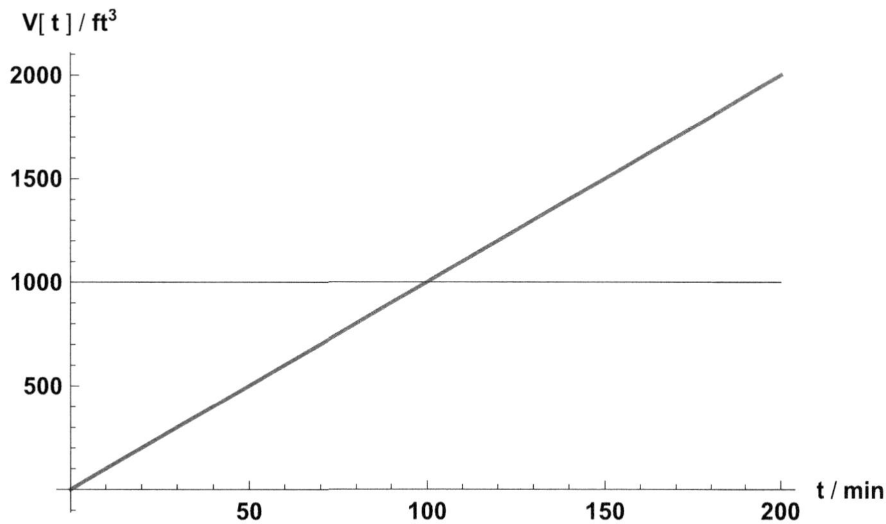

This is as we would expect - the volume changes with time in a linear fashion.

Now, we can put all the variables together to have a complete description of the triangular trough's behavior - the level, the width, the area of the fluid surface and the volume, each as functions of time.

Clear[*h*, *q*, *H*, *W*, *L*]

$$\text{htt[t\_]:=}\sqrt{\frac{Hq}{WL}t}$$

$$\text{wtt[t\_]:=}\frac{W}{H}\text{htt}[t]$$

$$\text{att[t\_]:=}L\text{wtt}[t]$$

$$\text{vtt[t\_]:=}\frac{WL}{H}\text{htt}[t]^2$$

$W = 10;$

$L = 10;$

$q = 10;$

$H = 10;$

Print["t/min", "—htt/ft", "—wtt/ft", ("–Att/ft²", "–Vtt/ft³"]

(* Quick way to put in column labels *)

Grid[

Table[{t, htt[t]//N, wtt[t]//N, att[t]//N, vtt[t]//N}, {t, 0, 200, 20}]]

t/min---htt/ft---wtt/ft--Att/ft^2--Vtt/ft^3

0	0.	0.	0.	0.
20	4.47214	4.47214	44.7214	200.
40	6.32456	6.32456	63.2456	400.
60	7.74597	7.74597	77.4597	600.
80	8.94427	8.94427	89.4427	800.
100	10.	10.	100.	1000.
120	10.9545	10.9545	109.545	1200.
140	11.8322	11.8322	118.322	1400.
160	12.6491	12.6491	126.491	1600.
180	13.4164	13.4164	134.164	1800.
200	14.1421	14.1421	141.421	2000.

Lastly, this trough could have another purpose - to be an analog computer for calculating the square root of a number. Rearranging the expression for the level in the tank so that the square root of time is isolated on the right-hand side, we can see how to do this. If W = L = H = q, then the fraction $\frac{WL}{Hq}$ is one and the level in the trough will be the square root of the time.

$$\mathrm{htt}[t] = \sqrt{\frac{Hq}{WL}t} \Rightarrow \mathrm{htt}[t]\sqrt{\frac{WL}{Hq}} = \sqrt{t}$$

when

$$W = L = H = q$$

then

$$\sqrt{\frac{WL}{Hq}} = 1$$

and

$$\mathrm{htt}[t] = \sqrt{t}$$

To do the "computation", we would start the input flow, stop at the time corresponding to the number we want the square root of and then measure the level. The level will be the square root of the time. The following script shows that. Assume all the values of the constants are 10, then when we look at the level, its value will be the square root of that time value. If we make the trough larger, then we make the accuracy of the value of the square root better because we can measure to more places beyond the decimal point. If we want the square root of a number up to 1000, just fill the tank for that period of time and measure the level. In principle, this would be relatively easy to do. For the example below, the values are computed for the whole numbers between 0 and 10 in jumps of 0.5. But one could get to any number in this range, in this way.

Clear[*h*, *q*, *H*, *W*, *L*]

(* This clears and expunges any values previously assigned to these variables *)

htt[t_]:= $\sqrt{\frac{Hq}{WL}t}$

(* This is the function for level *)

vtt[t_]:= $\frac{WL}{H}\mathrm{htt}[t]^2$

(* Keeping track of total volume *)

$W = 10;$

```
L = 10; (* Parameter values *)
q = 10;
H = 10;
vmax = W L H ft^3 (* Max Volume *)
Print["t/min", "–htt/ft""–vtt/ft"]
(* Quick Labels *)
Grid[Table[{t, htt[t], vtt[t]}//N,
{t, 0, 10, 0.5}]]
(* Computation of the level and the volume that we would measure *)
```

1000ft^3

```
t/min--htt/ft--vtt/ft
```

0.	0.	0.
0.5	0.707107	5.
1.	1.	10.
1.5	1.22474	15.
2.	1.41421	20.
2.5	1.58114	25.
3.	1.73205	30.
3.5	1.87083	35.
4.	2.	40.
4.5	2.12132	45.
5.	2.23607	50.
5.5	2.34521	55.
6.	2.44949	60.
6.5	2.54951	65.
7.	2.64575	70.
7.5	2.73861	75.
8.	2.82843	80.
8.5	2.91548	85.
9.	3.	90.
9.5	3.08221	95.
10.	3.16228	100.

All that is required for this to work well, is that we be able to control the flow rate into the trough and that we can stop the flow abruptly. We would need accurate measurements of time and level. Nothing else is necessary and that makes this very simple. If one wanted the square root of even larger numbers than 1000, then one would need to build a bigger tank, or do a little arithmetic.

The conical tank

Another geometry that can be useful to consider is that of the conical tank. The analysis is similar to that of the triangular trough, except that the cone is axially symmetric. Axially symmetric means that we can spin the cone around the central vertical axis and it will always look the same to us. This symmetry makes some difference in the outcome. The geometry for the tank is shown in the figure:

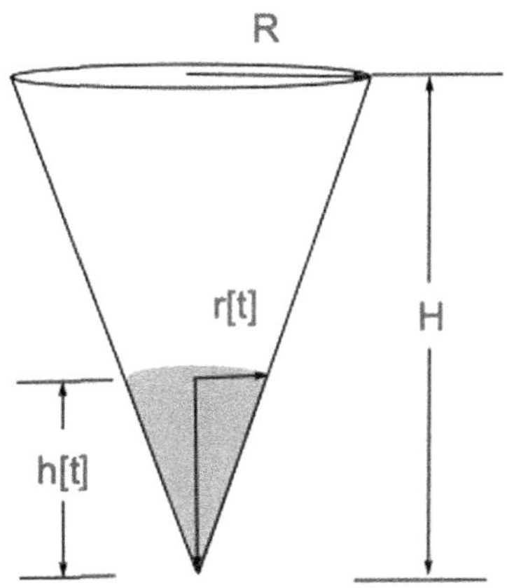

As in the case of the triangular trough, the area of the liquid surface changes with the level in the tank. We know that the mass balance will lead to the same equation for the differential change in volume with time. Therefore we must find the relationship that will render the differential volume change as a function of the level alone.

If we were to take a slice through the tank along the central axis, we would be left with a triangular face. From the similar triangles on that face we find:

$$\frac{R}{H} = \frac{r[t]}{h[t]}$$

If the volume were to change by some differential level change we would have one again:

$$dV[t] = A[t]\,dh[t]$$

However, in this case the area is given by the square of the radius:

$$A[t]\,dh[t] = \pi\, r[t]^2 dh[t]$$

But r [t] has to be expressed in terms of h [t] and that means the left-hand side of the mass balance equation becomes:

$$dV[t] = \frac{\pi R^2}{H^2} h[t]^2\, dh[t]$$

The solution to the level change as a function of time is the solution to this differential equation:

$$\frac{dm[t]}{dt} = \overset{\circ}{m}_{\text{in}}$$

$$\frac{d\rho V[t]}{dt} = \rho q_f[t]$$

$$\frac{dV[t]}{dt} = q_f$$

$$\frac{\pi R^2}{H^2} h[t]^2\, dh[t] = q_f$$

Remove[*h*, *R*, *H*, qf, *q*, *t*]

(* This will expunge all values and previous assignments *)

$$\textbf{DSolve}\left[\left\{\pi \frac{R^2}{H^2} h[t]^2 \partial_t h[t] == q_f, h[0] == 0\right\}, h[t], t\right]$$

(* Here we use DSolve in the usual manner *)

$$\left\{\left\{h[t] \to -\frac{\left(-\frac{3}{\pi}\right)^{1/3} (H^2 t q_f)^{1/3}}{R^{2/3}}\right\}, \left\{h[t] \to \frac{\left(\frac{3}{\pi}\right)^{1/3} (H^2 t q_f)^{1/3}}{R^{2/3}}\right\}, \left\{h[t] \to \frac{(-1)^{2/3} 3^{1/3} (H^2 t q_f)^{1/3}}{\pi^{1/3} R^{2/3}}\right\}\right\}$$

We note that there are three solutions because the level is raised to the third power. Of the three solutions, only the second is useful for this problem. It gives values that are positive and increasing.

$$h\,[t]^3 = \frac{3q_f H^2}{\pi R^2} t$$

$$h\,[t] = \sqrt[3]{\frac{3q_f H^2}{\pi R^2} t}$$

With the conical tank the change in level with respect to time is even faster than for the triangular trough.

Clear[h, q, H, R, L]

(* Clear previous settings *)

hcn[t_]:=$\sqrt[3]{\frac{3q_f H^2}{\pi R^2} t}$ (* Define the function *)

(* Set the parameter values *)

$R = 10;$

$q_f = 10;$

$H = 10;$

$\frac{\pi R^2}{3} H\text{ft}^3 // N$

(* Compute the total volume *)

plhcn = Plot[hcn[t], {t, 0, 104}, PlotStyle → Hue[0.6], AxesLabel → {t, "h [t] / ft"}] ;

(* Plot the level function in blue, but suppress it with ";" *)

Show[plhtt, plhcn,
Graphics[Line[{{0, 10}, {200, 10}}]],
Graphics[{Dashed, Line[{{0, 0}, {200, 10}}]}]]
(* Show allows us to put the graphs of the level in the conical tank (blue),
the triangular trough (orange), a linear process (dashed) and the maximum
level (horizontal) all at once. *)

1047.2ft^3

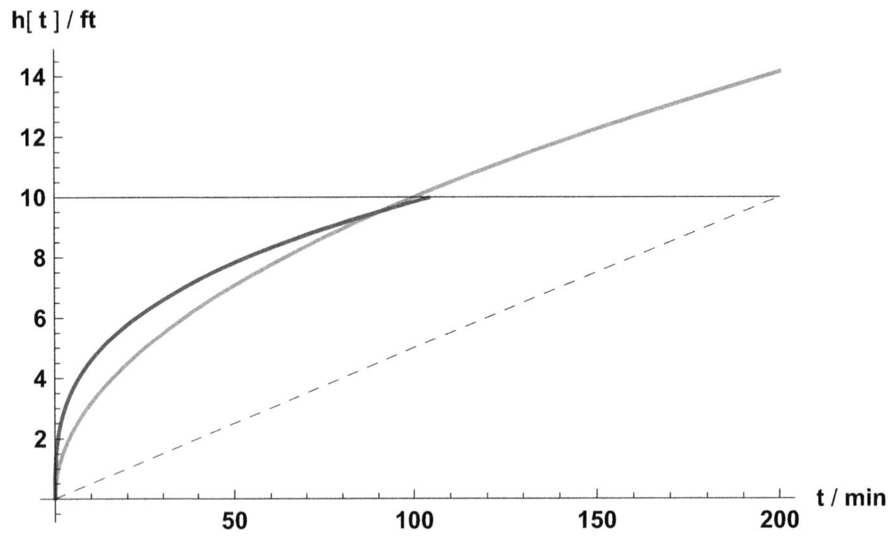

This is a very nice visual comparison. We see that the conical tank reaches the maximum level in about half the time with the same input flow rate. Next we can also plot and compare the volume as a function of time for the two systems. But we don't need the plot to know the answer. In both cases the input flow rate was constant and equal. Therefore the change in volume in both tanks is equal at each and every time.

Clear[h, q, H, R, L]

(* Clear all values *)

vcn[t_]:=$\frac{\pi R^2}{3H^2}$**hcn[t]^3**

(* Volume as a function of time for the conical tank*)

$R = 10;$

$q_f = 10;$

(* Parameter values *)

$H = 10;$

$\frac{\pi R^2}{3} H \text{ft}^3 // N$

(* The maximum volume *)

plvcn = Plot[vcn[t], {t, 0, 100}, PlotStyle → Hue[0.8], AxesLabel → {t, "V [t] / ft^3"}
Show[plvtt, plvcn, Graphics[Line[{{0, 1000}, {200, 1000}}]]]

$1047.2 ft^3$

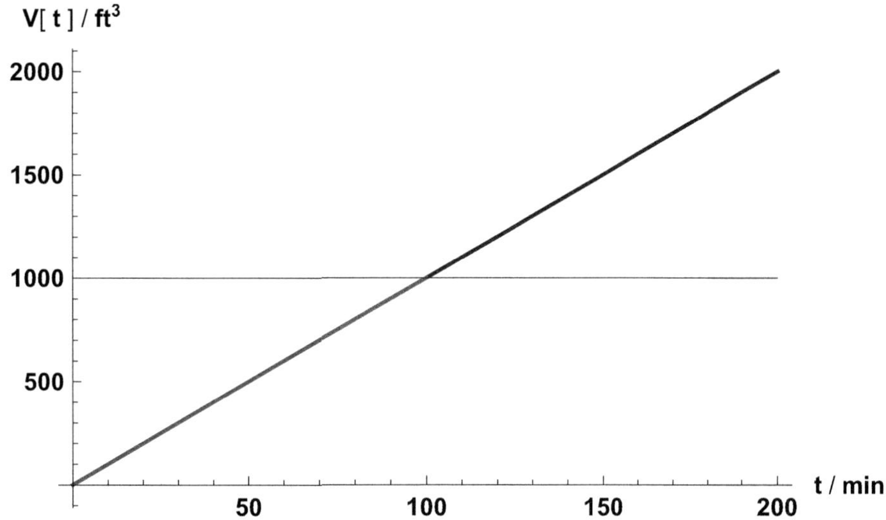

Both the conical tank and the triangular trough had volumes of 1000 ft^3 and the input flow rates for both were 10 ft^3/min. While the levels vary non-linearly with time, the volumes vary linearly.

The semi-cylindrical trough

Another variation of the left-hand side theme and the problem of geometry is that of the semi-cylindrical trough lying on its side. The physical situation is quite similar to that of the triangular trough, except that the walls follow a circular curve. The picture of the physical system is this:

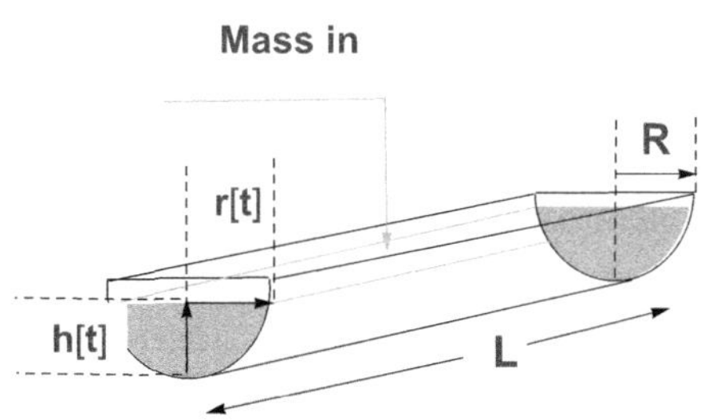

The new geometry is the new part of this problem. In view of that last two analyses that we have done, we know that we need the relationship between **r [t]** and **h [t]** once again. The diagram holds the key to this. The line that is the hypotenuse of the triangle which has **r [t]** for one leg and **R - h [t]** for the other. The Pythagorean theorem links the three:

$$\mathrm{R}^2 = (R - h[t])^2 + \mathrm{r}[\mathrm{t}]^2$$

Solving for **r [t]** in terms of **h [t]**:

Solve$\left[R^2 == \text{Expand}\left[(R - h[t])^2\right] + r[t]^2, r[t]\right]$

$$\left\{\left\{r[t] \to -\sqrt{20h[t] - h[t]^2}\right\}, \left\{r[t] \to \sqrt{20h[t] - h[t]^2}\right\}\right\}$$

The first solution is unphysical, so we substitute the second into the expression for **d V [t]**:

$$\mathrm{d\,V[t] = A[t]\,d\,h[t]}$$

$$= 2\,\mathrm{L\,r[t]\ d\,h[t]}$$

$$= 2\,\mathrm{L}\sqrt{2Rh[t] - h[t]^2}\ \mathrm{d\,h[t]}$$

Replacing this in the total mass balance (assuming all densities are constant and equal everywhere), we find:

$$\sqrt{2Rh[t] - h[t]^2}\,\frac{dh[t]}{\mathrm{dt}} = \frac{q_f}{2L}$$

To solve this equation we can turn to DSolve:

Clear[*t*, *h*, qf, *R*, *L*]

DSolve$\left[\left\{\sqrt{2Rh[t] - h[t]^2}==, h[0] == 0\right\},\ h[t], t\right]$

...DSolve: For some branches of the general solution, the given boundary conditions lead to an empty solution.

When this error message is generated, the *Mathematica* literature tells us this: “This message is generated when the equations are not complete to specify a solution, or are inconsistent.” One of the reasons that this may happen is that the solution starts with the initial condition at zero and that causes a division by zero problem. If we reset the initial condition so that it starts at some as yet unspecified value, ho, then we may find a solution, as shown below:

Clear[t, h, qf, R, L]

DSolve $\left[\left\{\sqrt{2Rh[t]-h[t]^2}==, h[0]==\text{ho}\right\}, h[t], t\right]$

$$\left\{\left\{h[t]\to\text{InverseFunction}\left[\frac{1}{2}\sqrt{-\#1(-2R+\#1)}\left(-R+\#1-\frac{2R^2\text{Log}\left[\sqrt{\#1}+\sqrt{-2R+\#1}\right]}{\sqrt{\#1}\sqrt{-2R+\#1}}\right)\&\right]\right.\right.$$
$$\left[\left(\sqrt{-\text{ho}(\text{ho}-2R)}\left(\text{ho}^{3/2}\sqrt{\text{ho}-2R}-\sqrt{\text{ho}}\sqrt{\text{ho}-2R}R-2R^2\text{Log}\left[\sqrt{\text{ho}}+\sqrt{\text{ho}-2R}\right]\right)\right)/\right.$$
$$\left.\left.\left.\left(2\sqrt{\text{ho}}\sqrt{\text{ho}-2R}\right)+\frac{tq_f}{2L}\right]\right\}\right\}$$

This is a result, but it is not one that we can readily implement since it involves an InverseFunction and PureFunctions, the alternative is to do the integration.

The antiderivative of right-hand side is obvious:

$$\int\frac{q_f}{2L}dt$$

$$\frac{q_f t}{2L}$$

This left-hand side is an integration of the form $\int\sqrt{ax^2+bx+c}\,dx$, which is readily integrable if the integral is indefinite, i.e. the antiderivative:

Clear[x, a, b, c]

$$\int\sqrt{ax^2+bx+c}\,dx$$

$$\frac{(b+2ax)\sqrt{c+x(b+ax)}}{4a}-\frac{\left(b^2-4ac\right)\text{Log}\left[b+2ax+2\sqrt{a}\sqrt{c+x(b+ax)}\right]}{8a^{3/2}}$$

We can find the antiderivative of the left-hand side as follows:

$$\int\sqrt{2h[t]R-h[t]^2}\,dh[t]$$

$$\frac{1}{2}\sqrt{-h[t](-2R+h[t])}\left(-R-\frac{R^2\text{ArcTan}\left[\frac{R-h[t]}{\sqrt{2R-h[t]}\sqrt{h[t]}}\right]}{\sqrt{2R-h[t]}\sqrt{h[t]}}+h[t]\right)$$

This looks a bit better and we can see that while this is not a simple function, it is also not that complex. Yet it does have some intricacies.

Remember that in this problem the face of the trough is semi-circular. This means that when the trough is full, the level will be at its maximum and equal to R the radius of the semi-circular face. The solution that we have derived has this fraction in it:

$$\frac{R^2 \text{ArcTan}\left[\frac{R - h[t]}{\sqrt{(2R - h[t])h[t]}}\right]}{\sqrt{(2R - h[t])h[t]}}$$

This fraction blows up when h[t] = R. We also see that the initial condition of h[0] = 0 is also a problem. Once again the fraction blows up. We also can now see that the InverseFunction was the ArcTangent function. So what do we have? We have a solution that will work between h[0] =0 and h [t] = R but not at either endpoint. There is no way to solve for h[t] explicitly in the region where the expression is well-behaved, we have an implicit solution for h [t].

One approach is to use the implicit solution to solve for **h [t]** numerically or graphically. The graphical method is not used much today, but it is worth illustrating because it reinforces a good "feel" for the functions and the numbers that result. We will look at the graphical solution first. The method, as you no doubt recall, is to find the solutions to this equation that make the left-hand and right-hand sides equal. We can view each side as a statement for two different functions that intersect at certain points; these intersections are the solutions. If we graph the two functions we can find these points. This is very easy to do in *Mathematica* as follows.

Let the radius of the tank be 5 feet and the length be 26 feet. We should find the total volume of the tank first so that we can choose a numerical value for the flow rate that does not require too long a time to make a real change in the tank level. To find the maximum volume we want to integrate the following equation from 0 to V_{max} on the left and from 0 to h_{max} = R on the right:

$$\text{d V [t]} = 2\,\text{L}\,\sqrt{2Rh[t] - h[t]^2}\,\text{d h [t]}$$

Clear[v, L, R, h, t]

$$\int_0^{\text{Vmax}} dv[t] == \int_0^R 2L\sqrt{2Rh[t] - h[t]^2}dh[t] //$$

Simplify

$$\text{Vmax} == L\sqrt{-R^2}\left(2\sqrt{R^2}\text{Log}\left[\sqrt{-R} + \sqrt{R}\right] - R\text{Log}[-2R]\right)$$

This result is not what we expected and it comes about as a more general solution to the integration. We want a solution for values of R that are positive, i.e. greater than zero. We can introduce **Assuming** at this point. We wrap this around the right-hand side and insert that R > 0:

$$\int_0^{\mathrm{Vmax}} dv[t] == \mathbf{Assuming}\left[R>0, \int_0^R 2L\sqrt{2Rh[t]-h[t]^2}dh[t]\right]$$

$$\mathrm{Vmax} == \frac{1}{2}L\pi R^2$$

The maximum volume comes out to be just what we knew it should be i.e. the product of half the area of a circular face and the length of the trough. This provides a very nice check on all our work to this point.

Vmax[R_, L_]:= $N\left[\frac{1}{2}L\pi R^2\right]$

Vmax[5, 26](* 5 feet radius and 26 feet in length *)

1021.02

This semi-circular trough has a volume of just over 1000 ft^3. We can choose a flow rate of 10 $ft^3\ min^{-1}$ as a reasonable value. The time to overflow to maximum capacity would be ~ 100 minutes. To solve the problem graphically, we write the left- and right-hand sides as two separate functions. Then we choose a value of **h** and graph this as a horizontal line versus the right-hand side as a function of time. The point of intersection provides the time at which that level would be reached subject to the chosen parameter values. We would choose another value of **h** and find a new value of **t** with a new graph. Before we begin it is worth noticing that the magnitude of **h [t]** can never exceed **R**, to do so would be unphysical. Unphysical, or not, it is very easy to begin plugging in values for **h** on the left-hand side that mistakenly range over the chosen value for **R**. If we were to do this, the mathematics will inform us of our error by providing an imaginary (complex) solution.

lhs[h_]:=

$$N\left[\frac{1}{2}\sqrt{(2R-h)h}\left(-R+\frac{2R^2\mathrm{ArcTan}\left[\frac{\sqrt{h}}{\sqrt{2R-h}}\right]}{\sqrt{2R-h}\sqrt{h}}+h\right)\right]$$

(* The left-hand side as function of h *)

```
rhs[t_]:= q/(2L) t
(* The right-hand side as a function of time *)

L = 26;
R = 5; (* Parameters *)
q = 10;
lhs[2.5]

Graphics[{RGBColor[1, 0, 0],
Line[{{0, lhs[2.5]}, {50, lhs[2.5]}}]}];

Plot[{rhs[z]}, {z, 0, 50}, PlotStyle → Hue[0.6]]

Show[%, %%, AxesLabel → {"t/min", "LHS==RHS"}]
```

7.67731

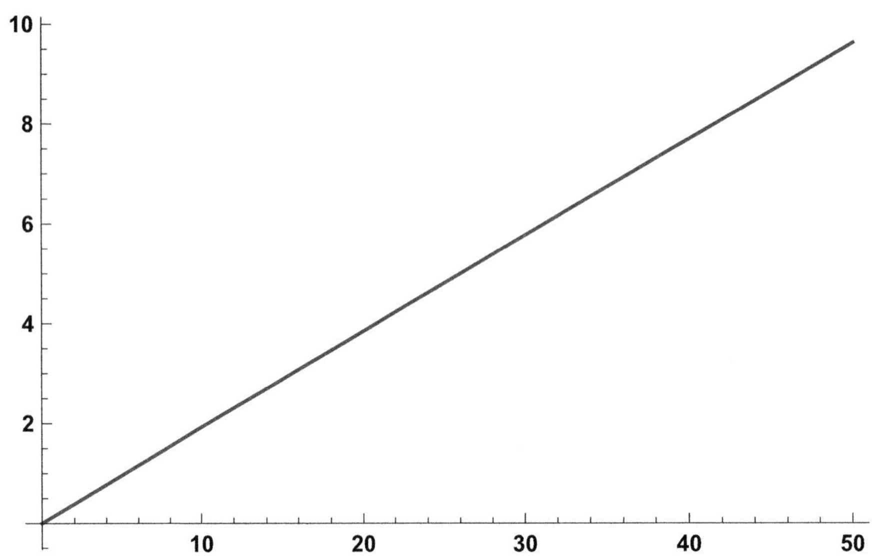

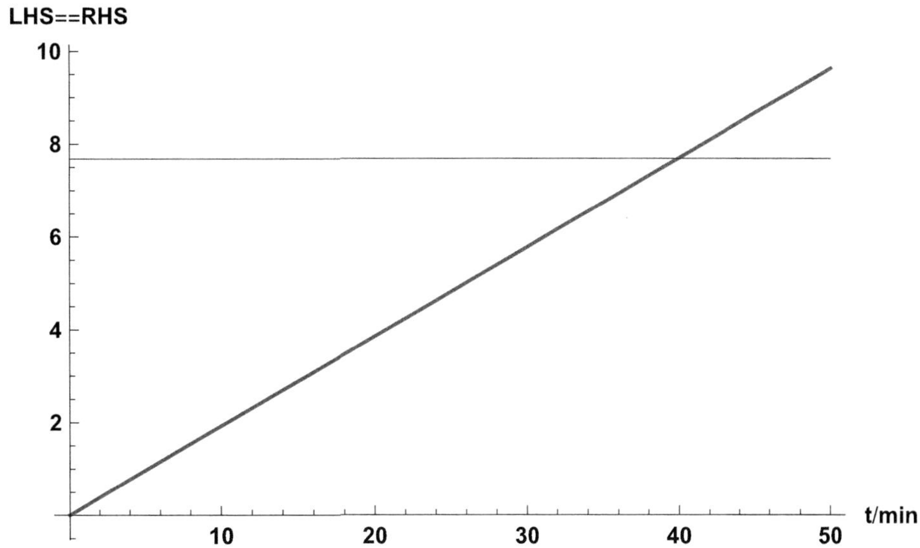

From this we can see that halfway point in terms of level will occur at just over 40 minutes. We could replot this in the vicinity of 40 minutes to get this more accurately, but there are better ways to do that as we shall see. To find a set of solutions beginning at a level of 1 foot in steps of 1 foot, we could use the following code to create the graphic we need:

```
Table[lhs[h], {h, 1, 5, 1}]

(* These are the values of the left-hand side (LHS) for values of h from 1 to
5 in increments of 1 *)

Graphics[

Table[{RGBColor[h/5, .35, 0],

Line[{{0, lhs[h]}, {80, lhs[h]}}]}, {h, 1, 5, 1}]];

(* This gives us a series of horizontal lines of different colors at each of
the five values of the LHS *)


Plot[{rhs[z]}, {z, 0, 80}, PlotStyle → Hue[0.6]];

(* Here we plot the right-hand side (RHS) versus time,
and then we suppress the output with ";" *)

Show[%, %%, AxesLabel → {"t/min", "LHS==RHS"}]
```

(* Here we show all the lines together. Wherever the line representing the RHS cross the horizontal lines representing the LHS, we have a solution *)

{2.04376, 5.59119, 9.90842, 14.6685, 19.635}

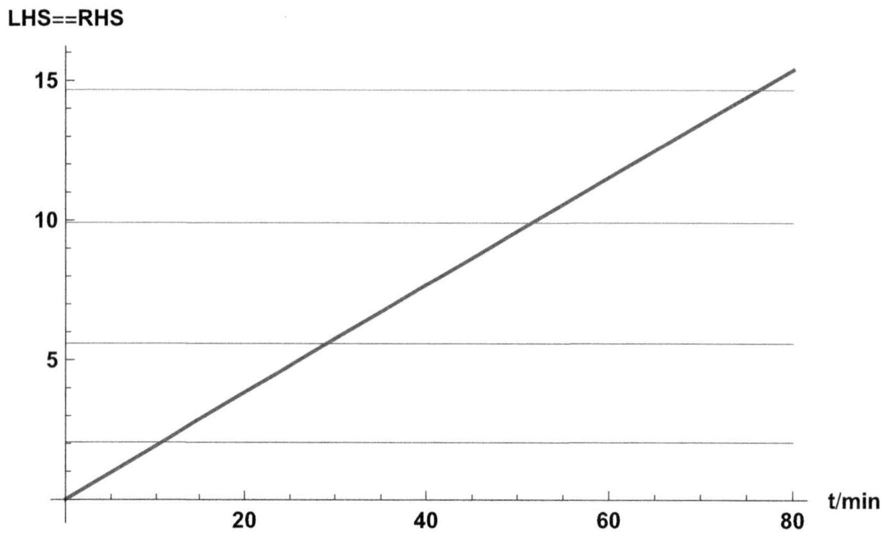

As interesting as this approach is, it is a bit archaic in comparison to that which we can obtain by solving the problem numerically.

To solve numerically, however, we really do the very same calculation: we choose a value of the level, evaluate the left-hand side and then back solve the resultant equation for the time. To do this we use the command **NSolve**. An example of this procedure is given for the halfway point at **h** = 2.5 feet:

Clear[*h*, *L*, *R*, qf]

h **= 2.5**

L **= 20; (* Parameter values *)**

R **= 5;**

qf = 10;

$$\text{NSolve}\left[\frac{1}{2}\sqrt{(2R-h)h}\left(-R+\frac{2R^2\text{ArcTan}\left[\frac{\sqrt{h}}{\sqrt{2R-h}}\right]}{\sqrt{2R-h}\sqrt{h}}+h\right)==\frac{\text{qf}t}{2L},t\right]$$

(* Here NSolve will solve for the time at which the two sides of the equation are equal *)

2.5

$\{\{t \to 30.7092\}\}$

We solve this to find that the exact time is 30.7 minutes. (We could have *Mathematica* limit the figures by enclosing the command in **NumberForm**.) Of course what we really want to see is a plot of the level versus time for this tank. To get that we need to repeat this procedure for many different levels and then plot the resultant time and level pairs.

$L = 20;$

$R = 5;$ **(* Parameters *)**

$q = 10;$

Print[" Here are the time values"]

ntimes = Table[NSolve[

$$\frac{1}{2}\sqrt{(2R-h)h}\left(-R+\frac{2R^2\text{ArcTan}\left[\frac{\sqrt{h}}{\sqrt{2R-h}}\right]}{\sqrt{2R-h}\sqrt{h}}+h\right)==\frac{qt}{2L},t\right],$$

$\{h, .005, 5, .25\}$]

(* ↑ We are repeating the previous calculation from a level of 0.005 to 5 ft with 0.25 ft increments, it gives a list of time values for each h value *)

Print[

"Here are the levels corresponding to the time values from above"]

levels = Table[h, $\{h, .005, 5, .25\}$]

```
(* Creates a corresponding list of the h values *)

Print["These are the dyads of time and level"]
timeleveldat =
Transpose[
Partition[Join[Flatten[t/.ntimes], levels],
Length[levels]]]
(* ↑ The times and their corresponding h values must be joined, then flattened,
partitioned and transposed to be in theproper form for ListPlot ↓ *)

pllevdat = ListPlot[timeleveldat,
PlotStyle → {PointSize[0.01], Hue[0.08]},
AxesLabel → {"t/min", "h[t]/Meter"}];
(* ↑ Here we plot the time – level pairs, but we suppress the output with ";" so
that we can Show it with a horizontal line that at the maximum level in the
tank and the cutoff for our solution. ↓ *)

Print[
"This is the plot of the dyads we computed and the maximum level in the tank"]
Show[pllevdat,
Epilog → {RGBColor[0, 0, 1], Line[{{0, 5}, {80, 5}}]}]
(* Epilog does as its name suggests, in this case it adds the line for the
maximum level of the tank, RGBColor is set to generate a blue line *)
```

Here are the time values

{{{t → 0.00298098}}, {{t → 1.07753}}, {{t → 2.97999}}, {{t → 5.40505}},
{{t → 8.23512}}, {{t → 11.399}}, {{t → 14.8464}}, {{t → 18.5388}},
{{t → 22.4448}}, {{t → 26.5381}}, {{t → 30.7959}}, {{t → 35.1977}},
{{t → 39.7253}}, {{t → 44.3618}}, {{t → 49.0915}}, {{t → 53.8996}},
{{t → 58.772}}, {{t → 63.6951}}, {{t → 68.656}}, {{t → 73.6418}}}

Here are the levels corresponding to the time values from above

{0.005, 0.255, 0.505, 0.755, 1.005, 1.255, 1.505, 1.755, 2.005, 2.255, 2.505, 2.755, 3.005, 3.255, 3.505, 3.755, 4.005, 4.255, 4.505, 4.755}

These are the dyads of time and level

{{0.00298098, 0.005}, {1.07753, 0.255}, {2.97999, 0.505}, {5.40505, 0.755}, {8.23512, 1.005}, {11.399, 1.255}, {14.8464, 1.505}, {18.5388, 1.755}, {22.4448, 2.005}, {26.5381, 2.255}, {30.7959, 2.505}, {35.1977, 2.755}, {39.7253, 3.005}, {44.3618, 3.255}, {49.0915, 3.505}, {53.8996, 3.755}, {58.772, 4.005}, {63.6951, 4.255}, {68.656, 4.505}, {73.6418, 4.755}}

This is the plot of the dyads we computed and the maximum level in the tank

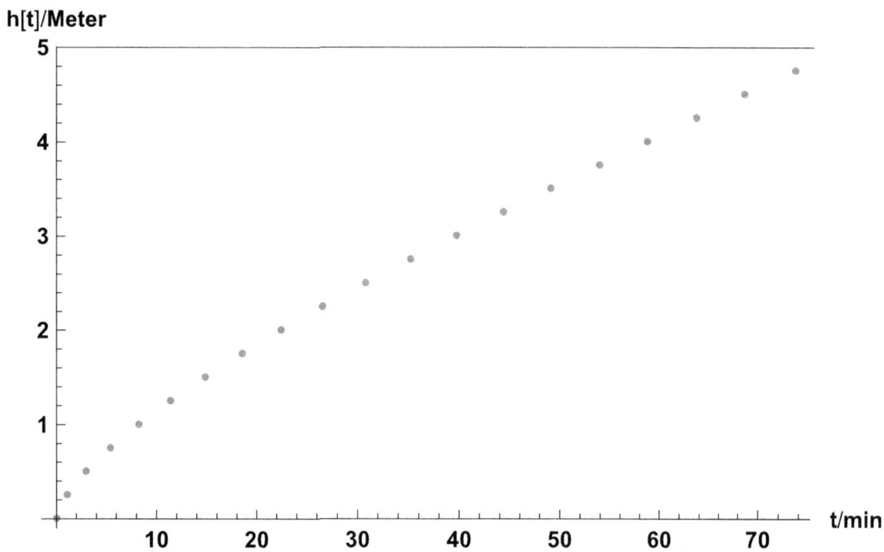

It was troublesome to create these points. We might want a graph of the level versus time that could be used at any time to find the level or vice versa. The logical thing to at this point is to fit these points to a function of time. We can do so using two parameters in a simple power law **h [t] = a** t^n:

timeleveldat

{{0.00298098, 0.005}, {1.07753, 0.255}, {2.97999, 0.505}, {5.40505, 0.755}, {8.23512, 1.005}, {11.399, 1.255}, {14.8464, 1.505}, {18.5388, 1.755}, {22.4448, 2.005}, {26.5381, 2.255}, {30.7959, 2.505}, {35.1977, 2.755}, {39.7253, 3.005}, {44.3618, 3.255}, {49.0915, 3.505}, {53.8996, 3.755}, {58.772, 4.005}, {63.6951, 4.255}, {68.656, 4.505}, {73.6418, 4.755}}

```
Clear[a, t, n]
FindFit [timeleveldat, at^n, {a, n}, t]
(* FindFit has the syntax of [data, model, parameters, independent variables ] *)
```

{a → 0.21636, n → 0.716687}

We have now fit a model to a model, and the new fitted function is simple and explicit, but specific. It says that for this semi-circular tank (sct) the level as a function of time closely follows this expression:

```
hsct[t_]:=0.216 t^0.717
(* The fitted function of time *)
Plot[hsct[t], {t, 0, 80},
AxesLabel → {"t/min", "h [ t ] / ft"}];
(* The plot of the fitted function, suppressed *)
Show[%, pllevdat,
Epilog → {RGBColor[0, 1, 0], Line[{{0, 5}, {80, 5}}]}]
```

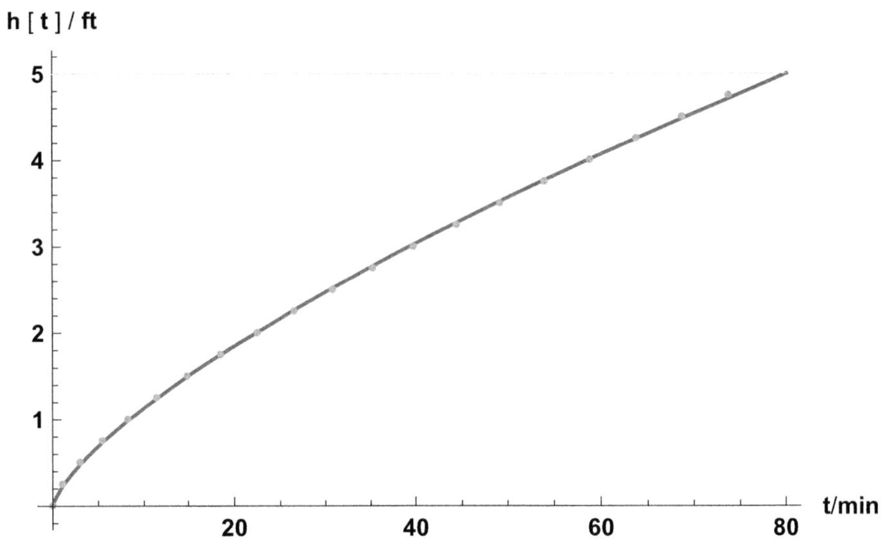

With the value of **a** at 0.216 and that of **n** at 0.716 we find a very nice fit to the data. It is interesting to note that this statistically derived function fits so well and, yet, offers us no insight into what is really taking place.

The spherical tank

The following is a simple case of filling a spherical tank. The liquid flow into the tank is again a constant and we wish to be able to predict the level as function of time.

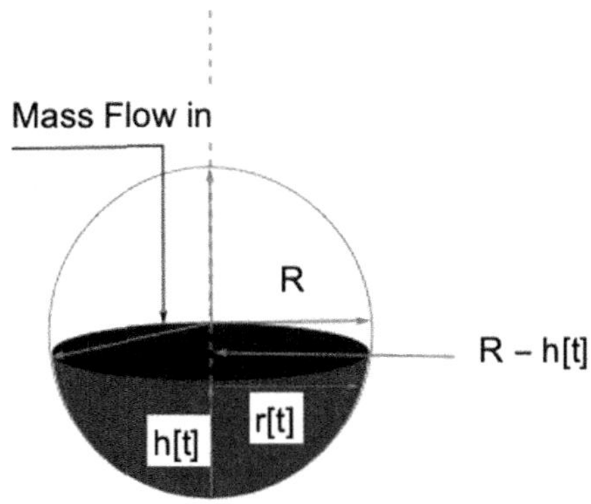

The total mass balance statement is as usual:

$$\frac{dV[t]}{\mathrm{dt}} = q$$

The differential change in volume differential level change:

$$dV[t] = Adh[t] = \pi r[t]^2 dh[t]$$

and therefore

$$r[t]^2 \frac{dh[t]}{\mathrm{dt}} = \frac{q}{\pi}$$

Next, a relationship between r [t] and h [t] is required. The triangle in the left lower quadrant of the circular face of the sphere has an hypotenuse of **R**, and two legs one of which is **r [t]** and the other is **R - h [t]**. Just as in the case of the semi-cylindrical trough, the Pythagorean theorem can be applied to this right triangle to give:

$$R^2 = r[t]^2 + (R - h[t])^2$$

And once again we can solve for r [t] in terms of h [t]:

Remove[R, r, h]

Solve$\left[R^2 == r[t]^2 + \text{Expand}\left[(R - h[t])^2\right], r[t]\right]$

$$\left\{\left\{r[t] \to -\sqrt{2R - h[t]}\sqrt{h[t]}\right\}, \left\{r[t] \to \sqrt{2R - h[t]}\sqrt{h[t]}\right\}\right\}$$

Returning to the differential equation we substitute this result in for r [t]. Then we need to find a solution for **h [t]**:

$$\sqrt{2Rh[t]-h[t]^2}\,\frac{dh[\,t\,]}{\mathrm{dt}}=\frac{q}{\pi}$$

This is just the same equation that we encountered in the last problem for the semi-cylindrical trough:

$$\sqrt{2Rh[t]-h[t]^2}\,\frac{dh[\,t\,]}{\mathrm{dt}}=\frac{q}{2L}$$

The solution method would be the same. The main difference is the tank is now axially symmetric as was the conical tank and that gives rise to a π on the right-hand side rather than **2L**. Recall that the sphere is made by rotating the semi-circular face around the central axis by $\boldsymbol{\pi}$ radians. The semi-cylindrical tank is constructed by translating the same semi-circular face by a distance **L** along a straight line. Symmetries of this kind are very interesting and will be useful to take advantage of in the solution of more sophisticated problems, but at this point we merely wish to point them out in passing.

To sum up this section, while the levels as a function of time are all different for the different geometries that we examined, if the input flow rate is constant and the same, then the volume in the tank will always be the same despite the geometry. The different geometries that we examined give rise to issues that we can think of as being on the left-hand side of the equation. Noting this, we recall that the control volume in each of these cases is bounded by the vessel itself, and the accumulation in the vessel is the left-hand side of the equation. So, it makes sense that the geometry of the vessel is embodied there. It all fits together nicely. Lastly, the geometry of the tank can be manipulated by design so that the tank becomes an analog computer. We saw how this can be done to find square roots and cube roots. Other geometries could be found that would provide other time dependencies that could be interesting to consider.

Depositing a polymer coating on a disk

In many industrial processes, one material is deposited onto another. From manufacture of non-stick frying pans to computer chips, this kind of process is common. Here we will consider the deposition of a fluid from a nozzle and onto a moving strip below it. In this particular process a substrate is translated under a spray nozzle at some velocity $\boldsymbol{v_z}$. At the same time a spray of small droplets containing a polymer impinges on the substrate. The droplets are created by sonicating the liquid mixture in the spray gun. The action of sonicating not only provides very small droplets (~0.5 microns diameter), the activator combines with oxygen in the air to initiate the conversion of monomer into polymer. As the substrate moves under the spray gun, the polymer droplets impinge and gel on the solid substrate creating a solid film on the surface which hardens quickly by the action of an activator added to the polymer that further crosslinks the polymer. The volume and mass of the activator compared to the volume and mass of polymer is small enough to be ignored at this level of analysis. Also the rate of the hardening process will be taken as instantaneous.

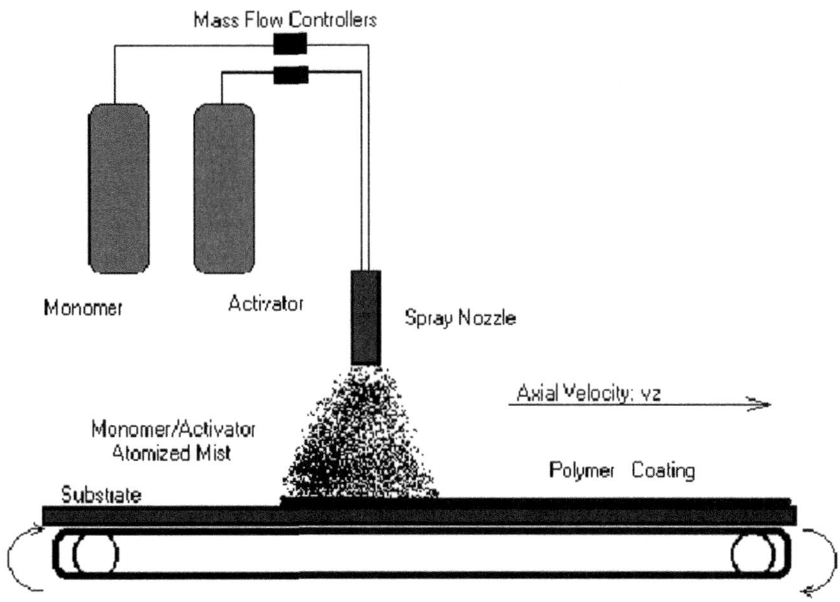

At the end of the spray nozzle there is an orifice with a circular area, $\boldsymbol{A_o}$. All of the polymer and the activator must pass through this orifice area. The mass flowing per unit time per unit area through this area is referred to as a *flux*. After leaving the orifice the spray area expands to form a cone-shaped region above the substrate. The spray area on a motionless substrate would be approximately circular. The radius of the circular impingement area would be related to the distance, **L**, between the nozzle and the substrate. Let's say that experiments have shown that the mass impinging on this area is homogeneous, that is the same everywhere. The substrate has a width equal to the diameter of the impingement area, therefore if it is moved under the nozzle at constant velocity, it will be evenly coated, excluding end effects.

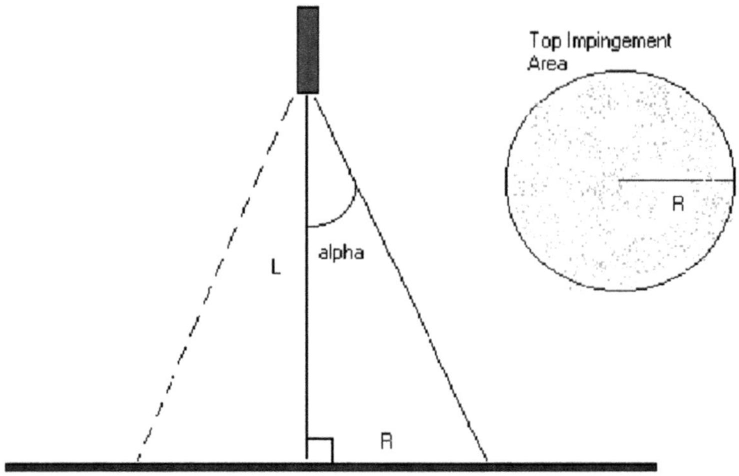

From this description we should be able to predict the thickness of the polymer coating as a function of the delivery rate and substrate velocity. The monomer and activator are taken to have equal densities. The total mass balance around the spray nozzle is:

$$\frac{dm[t]}{\mathrm{dt}} = \overset{\circ}{m}_{\mathrm{in}} - \overset{\circ}{m}_{\mathrm{out}}$$

Clearly, there should be no net accumulation in the spray nozzle, meaning that it will nearly always operate at a steady state. In fact a design criterion for a sprayer like this is that it have nearly instantaneous rise and shut-off times. So from this we can say that the mass flow rate in must equal the mass flow rate out.

$$\overset{\circ}{m}_{\mathrm{in}} = \overset{\circ}{m}_{\mathrm{out}}$$

The rate of mass flow into the spray gun is given as the sum of the mass flows of the monomer and activator. The densities of the two liquids are nearly identical, and the volumetric flow rate of the monomer is five times the flow rate of the activator. Therefore the mass out of the orifice is a constant:

$$\overset{\circ}{m}_{\mathrm{out}} = \rho_{\mathrm{monomer}} * q_{\mathrm{monomer}}$$

when

$$\rho_{\mathrm{activator}} * q_{\mathrm{activator}} \rightarrow small$$

then

$$\overset{\circ}{m}_{\mathrm{out}} \sim \rho_{\mathrm{monomer}} q_{\mathrm{monomer}} = \overset{\circ}{m}_{o}$$

The mass flux from the orifice, $\boldsymbol{J_o}$, is the mass flow rate out divided by the orifice area, $\frac{\overset{\circ}{m}_{\mathrm{out}}}{A_o}$. Since the spray spreads in the form of right cone, the mass flux at the substrate, J_s, is smaller than the flux from the orifice and it impinges on a larger area, A_s. To solve our problem we must know this. By definition, the product of the flux and the area is the mass flow rate. Due to the conservation of mass, the mass flow to the substrate must be the same as the mass flow from the orifice. Therefore we can proceed as follows:

$$\overset{\circ}{m}_s = \overset{\circ}{m}_o$$

$$J_o A_o = J_s A_s$$

$$J_s \frac{A_s}{A_o} = J_o$$

Let's review what we know. We know the mass flow out of the spray nozzle and we can calculate the area of impingement knowing the area of the orifice, A_o, the distance to the substrate, L, and the angle, alpha, made by the spray leaving the nozzle. So, the next step must be to find the area of impingement in terms of alpha, L and A_o.

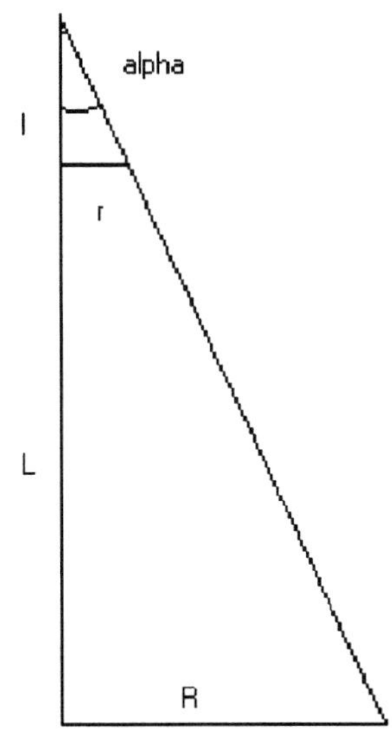

From this diagram of the geometry we see that the radius of the orifice is r which is related to l through the tangent of alpha. So too, the radius R of the impingement area is related to (l + L) through the tangent of alpha ($\tan\alpha = \frac{\text{opp}}{\text{adj}}$). The distance l is not known. It corresponds to the vertex of the triangle that is within the sprayer. We can solve for it in terms of alpha and **r**, both of which are known. Then we can set the two ratios of equal to one another and solve for R in terms of L.

Clear[R, L, l, r, l, α]

Solve$\left[\left\{\frac{r}{l} == \text{Tan}[\alpha], \frac{r}{l} == \frac{R}{(L+l)}\right\}, \{l, R\}\right]$

$\{\{l \to r\mathrm{Cot}[\alpha], R \to r + L\mathrm{Tan}[\alpha]\}\}$

The area of impingement at the substrate, A_s, is $\pi\ R^2$, so the impingement flux at the substrate is:

$$\mathbf{A_o = \pi (r + L\mathrm{Tan}[\alpha])^2;}$$

$$\mathbf{J_o == \frac{\mathring{m}_{out}}{A_o}}$$

$$J_o == \frac{\mathring{m}_{out}}{\pi (r + L\mathrm{Tan}[\alpha])^2}$$

The next part of the analysis is to find the mass accumulation of polymer on the substrate given J_o as the substrate moves under the nozzle at a velocity equal to v_z. If the substrate did not move, then the mass would accumulate on the circular impingement area. Given a rigid solid polymer we would begin to grow a deposit of circular cross section and increasing thickness. As the deposit grew away from the substrate, it would decrease L, and so the circular deposition area would decrease. We could imagine that we would grow a deposit that would roughly look like the frustum of a cone and, eventually, a cone. If we allowed the deposition to take place for some time, τ, such that the thickness grew to δ, and then moved the substrate by a distance 2R and deposited for a period τ again, and then continued this stepwise process we would have a series of circular deposits of nearly uniform thickness. We would also have areas between the circular deposits that were nearly uncovered. This would be a semi-continuous process. If τ were to become very short, then the process would begin to approach a continuous one, especially if we only moved the substrate by some fraction of 2R every time.

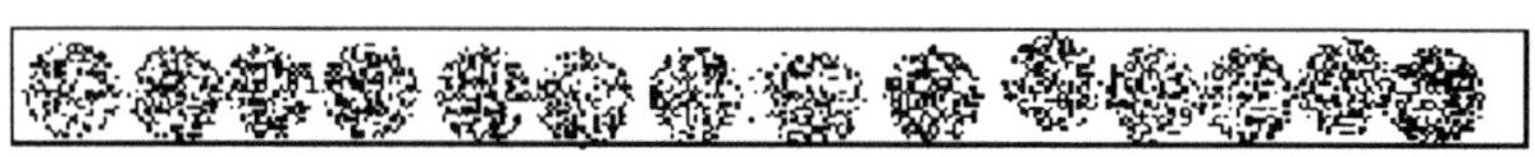

Coating period = 2R   Coating time = tau

Coating Period ~R   Coating time = tau

Continuous Coating

In the continuous process we let τ shrink to small values by moving at a continuous velocity. Now the time spent under the nozzle is just $\tau = \frac{2R}{v_z}$, that is the "residence time" under the sprayer. By translating in this way the circular spray area becomes nearly rectangular, save for the edges. We can write a mass balance for the problem in the usual way choosing the spray area and the growing polymer film as the control volume:

$$\frac{dm[t]}{\mathrm{dt}} = \overset{\circ}{m}_{\mathrm{in}}$$

$$\frac{d\left[\rho_{\mathrm{polymer}} V[t]_{\mathrm{deposit}}\right]}{\mathrm{dt}} = \overset{\circ}{m}_{\mathrm{in}}$$

$$\frac{d\left[\rho_{\mathrm{polymer}} A[t]_{\mathrm{deposit}} \delta[t]\right]}{\mathrm{dt}} = \overset{\circ}{m}_{\mathrm{in}}$$

$$\frac{d[wz[t]\delta[t]]}{\mathrm{dt}} = \frac{\overset{\circ}{m}_{\mathrm{in}}}{\rho_{\mathrm{polymer}}}$$

At this point we can recognize that w is related to R and substitute it in:

$$\mathrm{w} = 2\mathrm{R}$$

$$\frac{d[z[t]\delta[t]]}{\mathrm{dt}} = \frac{\overset{\circ}{m}_{\mathrm{in}}}{2R\rho_{\mathrm{polymer}}}$$

But R is related to the distance L, the angle α and the dimension of the orifice r.

$$2\,\mathrm{R} = 2\,(\mathrm{r} + \mathrm{L}\ \mathrm{Tan}[\ \alpha\ ])$$

This brings us to this equation

$$\frac{d[z[t]\delta[t]]}{\mathrm{dt}} = \frac{\overset{\circ}{m}_{\mathrm{in}}}{2(r + L\mathrm{Tan}[\alpha])\rho_{\mathrm{polymer}}}$$

Applying the chain rule to the left-hand side, we find:

$$\delta\,[\,\mathrm{t}\,]\,\frac{dz[t]}{\mathrm{dt}} + \mathrm{z}\,[\,\mathrm{t}\,]\,\frac{d\delta[t]}{\mathrm{dt}} = \frac{\overset{\circ}{m}_{\mathrm{in}}}{2(r + L\mathrm{Tan}[\alpha])\rho_{\mathrm{polymer}}}$$

If the substrate is translated at a fixed velocity then we can say that **z [t]**, the position of the substrate at any time, is just a linear function of time:

$$\mathrm{z}\,[\,\mathrm{t}\,] = v_z \mathrm{t} + \mathrm{c}$$

$$z[0] = c = z_o$$

This leaves us with this equation to solve for the thickness as a function of time. By rearranging it we can see that it is a non-linear differential equation and we must treat it accordingly:

$$\delta[t]\,v_z + (v_z t + z_o)\frac{d\delta[t]}{dt} = \frac{\overset{\circ}{m}_{\text{in}}}{2(r + L\text{Tan}[\alpha])\rho_{\text{polymer}}}$$

$$(v_z t + z_o)\frac{d\delta[t]}{dt} = \frac{\overset{\circ}{m}_{\text{in}}}{2(r + L\text{Tan}[\alpha])\rho_{\text{polymer}}} - \delta[t]\,v_z$$

$$\frac{d\delta[t]}{\left(\frac{\overset{\circ}{m}_{\text{in}}}{2(r + L\text{Tan}[\alpha])\rho_{\text{polymer}}} - \delta[t]v_z\right)} = \frac{dt}{(v_z t + z_o)}$$

$$\frac{dx}{(c1 - bx)} = \frac{dt}{(at + z_o)}$$

This shows us that the equation is separable and the form of both sides indicates that the antiderivatives will involve the log function. These two sides can be integrated indefinitely to give:

DSolve$[\{(v_z t + z_o)\,\partial_t\delta[t] == K - \delta[t]v_z, \delta[0] == 0\},$

$\delta[t], t]$

$$\left\{\left\{\delta[t] \to \frac{Kt}{t + z_o}\right\}\right\}$$

Thus the thickness of the coating will be proportional to the mass flow out of the nozzle and inversely proportional to the speed at which the substrate moves under the nozzle. With the initial position of the substrate being $z_o = 0$, we have:

$$\delta = \frac{K}{v_z} = \frac{\overset{\circ}{m}_{\text{in}}}{2v_z(r + L\text{Tan}[\alpha])\rho_{\text{polymer}}}$$

This problem boils down to just two terms. We should check the dimensions.

$$\text{length} = \frac{\left(\frac{\text{mass}}{\text{time}}\right)}{\left(\frac{\text{length}}{\text{time}}\right)(\text{length})\left(\frac{\text{mass}}{\text{length}^3}\right)}$$

$$= \frac{\left(\frac{\text{mass}}{\text{time}}\right)}{\left(\frac{\text{length}}{\text{time}}\right)\left(\frac{\text{mass}}{\text{length}^3}\right)}$$

$$= \frac{\left(\frac{\text{mass}}{\text{time}}\right)}{\left(\frac{\text{mass}}{\text{length}^2\text{time}}\right)}$$

$$= \left(\frac{\text{mass}}{\text{time}}\right)\left(\frac{\text{length}^2\ \text{time}}{\text{mass}}\right) = \text{length}$$

The dimensions check out nicely. If the density is given as g/cm^3, the mass flow as g/sec, and the velocity in cm/sec, all the units will be self-consistent and the thickness will be given in cm.

Let's assume that the nozzle is held 10 cm away from the substrate (**L** = 10), the nozzle radius, r, will be set at 0.03 cm, and the cone angle made by the spray is 20 degrees. The polymer may have a density of 2 g percm^3. The axial velocity should be small, let's choose 0.15 cm per sec (v_z = 10). The flow of polymer mass to the substrate may also be $\overset{\circ}{m}_{\text{in}}$ = 0.01 g per sec. The nozzle radius, r, will be set at 0.03 cm.

$\rho_{\text{polymer}} = 2$; (* g/cm^3 *)

$L = 10$; (* cm *)

$\alpha = \frac{20}{360} 2\,\pi$; (* dimensionless *)

$\overset{\circ}{m}_{\text{in}} = 0.01$; (* g/sec *)

$v_z = 1$; (* cm/sec *)

$r = 0.03$; (* cm *)

Print["The thickness of the layer will be..."]

$$\delta = \text{NumberForm}\left[\frac{\overset{\circ}{m}_{\text{in}}}{2 v_z (r + L\text{Tan}[\alpha])\rho_{\text{polymer}}}\text{cm}\ \left(\frac{10^4\text{micron}}{\text{cm}}\right), 3\right]$$

The thickness of the layer will be...

6.81micron

At this delivery rate of polymer, substrate dimensions, and velocity, the thickness of the layer will be 6.8 microns thick.

Conclusion

In this chapter we have covered quite a number of apparently different problems that all are readily attacked by use of the concepts of a control volume and the conservation of mass. For each case the same equation was used as the point of departure for the analysis – namely the differential statement of the total mass balance. It is surprising at first to find that so many different situations can be analyzed by the correct application of this equation along with some calculus and geometry along the way.

During the course of this chapter we also have learned how to use *Mathematica* interactively to do analysis. The methods we have used are general and we will use them throughout the text. The objective is to think and analyze the problem with the assistance of the computer in real time with you, the human, in the loop.

CHAPTER 3

The draining tank and related systems

The right-hand side of the mass balance equation

In chapter 2 we developed models based on analyses of systems that had simple inputs. The right-hand side was either a constant, or it was simple function of time. In those systems, we did not consider the cause of the mass flow. The cause of the flow was left implicit. The pump, or driving device, was upstream from the control volume, and all we needed to know was the magnitude of the flow and its time dependence. Given that information we could replace the right-hand side of the balance equation and integrate to derive the functional description of the system.

This level of simplicity is not the usual case in the systems that are of interest to chemical engineers. The complexity we will encounter will be much higher and will involve more detailed issues on the right-hand side of the equations with which we work. Instead of a constant, or some explicit function of time, the function will be an explicit function of one or more key characterizing variables of the system and implicit in time. The reason for this is that of cause. Time in and of itself is never a cause – it is simply the independent variable. When we need to deal with the analysis of more complex systems, the mechanism that causes the change we are modeling becomes all important. Therefore we look for descriptions that will be dependent on the mechanism of change.

In fact we can learn about the mechanism of change by testing our ideas about the physics or chemistry in the context of a model derived from an analysis based on some specific assumptions. Comparison of the model predictions with the actual behavior of the system provides a check on the analysis and its assumptions. Chemical engineers do this with almost every problem they encounter. Depending on the level of analysis and the nature of the problem, the results can be anything from a useful engineering description of the systems to new science. Regardless of the goals and objectives of the project, the chemical engineer uses the same powerful analysis paradigm to make progress and to solve the problem.

With this in mind we turn now to another problem that is seemingly naive at first glance, but which offers considerable insight into the next level of this process called analysis. The problem that we shall consider now is that of flow out of vessel due to the force of gravity. We will apply the same principles as in Chapter 2, but the cause of the flow will be an essential part of

Introduction to Chemical Engineering Analysis Using *Mathematica*
https://doi.org/10.1016/B978-0-12-820051-3.00008-5

the analysis. We also have the chance to see how this problem fits into the history and foundations of our physics. This is a three hundred year old problem, which is still full of fascination for us and from which we can learn much.

Mechanism of water flow from a tank: Torricelli's law - a constitutive relationship

In 1640 a young man educated at the Collegio di Sapienza in Rome, published a treatise entitled "Trate del Moto," that is the Treatise on Motion. This brought the author to the attention of the preeminent natural philosopher of the day - Galileo Galilei. Impressed by the work, Galileo invited the author, Evangelista Torricelli, to join him at the Florentine Academy in 1641. Torricelli worked as Galileo's personal secretary for just one year before Galileo died in 1642. The faculty at Florence immediately appointed Torricelli to succeed Galileo as professor of mathematics that same year. Torricelli is ensconced in physical science through the unit of pressure that bears his name, in honor of his experiments that led to the creation of the first partial vacuum and the barometer. But Torricelli did more than this and it is to some of his other work that we will turn to now.

Just as Galileo had studied gravitational effects by dropping small and large solid masses from towers, Torricelli followed in his footsteps by analyzing liquids "falling" out of tanks. What he did was to measure flow rate of liquids flowing out of tanks with holes in the bottom at different initial fluid levels. A physical system that corresponds to this is shown in the figure below. The vessel is a right cylinder with a cross sectional area **A**. At the bottom of the vessel is a hole with an area **Ao** that has plug in it. The initial liquid level has been set to **ho**.

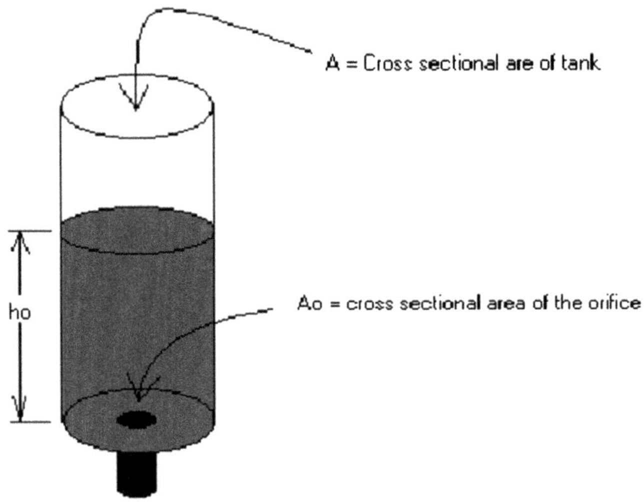

When the plug is pulled, the liquid flows out. One can measure the flow rate, or the level as a function of time to learn how the system behaves. If we did this, then we would notice from these experiments that the flow rate out is not constant, but diminishes with time. Logically, as the level drops, there is less driving force to push the liquid through the hole and so the rate at which mass leaves the tank decreases. Before we consider the data that comes from an experiment like this one, we should apply the principle of the conservation of mass to the system to see what it is that we know and do not know about it. We begin with the overall equation as always:

$$\frac{dm[t]}{\mathrm{dt}} = -\overset{\circ}{m}_{\mathrm{out}}$$

The mass flow in is zero, so we are left with just one term on the right-hand side. The liquid density is a constant inside and outside of the vessel and the cross-section of the liquid is just that of the tank, **A**. This leads us to the following equation for the change in level as a function of time:

$$\frac{dh[t]}{dt} = -\frac{q[t]}{A}$$

On the right-hand side, we have noted that the flow rate changes with time by writing **q** as a function of time, but that is all that we can do at this point. We have no way of knowing how **q** varies with time, or with the level in the tank. Therefore we cannot go any further with this analysis until we have some way to express this dependence. What we are seeking for **q**, is a *constitutive relationship* that will express it in terms of the level change, **h [t]**, so that the equation has only one dependent variable in time and not two. In other words the analysis we have just done is incomplete, but it has shown us what our experiments should be designed to do. We must find a functional relationship between the level in the tank and the flow rate out of it, or in mathematical terms we seek:

$$\mathrm{q}[\mathrm{t}] \longrightarrow \mathrm{f}[\mathrm{h}[\mathrm{t}]]$$

$$\frac{dh[t]}{dt} = -\frac{f[h[t]]}{A}$$

The first statement is the constitutive relationship that we seek and the second is just a restatement of the mass balance.

Experiment and the constitutive equation

What causes the fluid to flow or "fall" out of the tank? Gravity of course. That is easy for us to answer because we are educated in fundamental physics concepts from an early age. But let's turn the clock back to Torricelli's day and try to answer the question again in the year in which Galileo died and Isaac Newton was born. That means that Torricelli was working on this problem prior to the Calculus and Newtonian physics. So let's attack this problem from the "experimental" side and try to piece together the findings that may have led to Torricelli's Law. How Torricelli measured time in his experiments is an interesting question; perhaps, he may have used his pulse as did Galileo for certain observations.

In the Table are collected data from several experiments with different starting levels, but all with the same orifice area, and all done in the same cylindrical tank. Let's look at the data for some clues as to the behavior of this system. Note that it takes 60 seconds to drain the tank from an initial level of 10 cm. When the initial level is set to 50 cm, it takes 130 seconds for the tank to drain.

t/sec			Level/cm		
0	10	20	30	40	50
10	6.8	15.4	24.2	33.3	42.5
20	4.2	11.3	19.1	27.2	35.6
30	2.3	7.9	14.6	21.8	29.3
40	0.9	5.1	10.7	16.9	23.6
50	0.2	2.9	7.3	12.7	18.5
60	0	1.3	4.7	9.	14.1
70	0	0.4	2.6	6.	10.2
80	0	0	1.1	3.6	7.
90	0	0	0.3	1.8	4.4
100	0	0	0	0.7	2.4
110	0	0	0	0	1
120	0	0	0	0	0.2
130	0	0	0	0	0

We raised the initial level by a factor of five, and so too the volume that must be drained, and, yet, it took only twice the total time to complete the process. Looking even more closely at the data, we note that in the same experiment for which the initial level was 50 cm, when the level in that experiment had fallen to 10 cm, it took another 60 seconds to completely drain the tank. In other words once the level gets to 10 cm it takes the same time to completion as if the experiment had begun at 10 cm. Therefore the first 40 cm flowed (fell) out of the tank in 70 seconds, but the final 10 cm took another 60 seconds to flow out. Looking across the data sets we see the same thing in each. The time required to go to completion beginning at 20 cm is the same as the time required to empty the tank after having reached 20 cm beginning at 40 cm. But the time required to go from an initial level of 40 cm to 20 cm is just 30 seconds, or less than half the time required to drain the second 20 cm worth of fluid! This behavior had to have been very intriguing to Torricelli and the analysis must not have been obvious, at least at the outset of the work.

The relationships that we have just described in words are much easier to see if we plot the data as level versus time and do so all on one graph. We can use a few *Mathematica* functions to pull this all together. The data in the table has been entered as a matrix of levels at each time called "**totdat**":

```
totdat =
{{0, 10, 20, 30, 40, 50},
{10, 6.8, 15.4, 24.2, 33.3, 42.5},
{20, 4.2, 11.3, 19.1, 27.2, 35.6},
{30, 2.3, 7.9, 14.6, 21.8, 29.3},
{40, 0.9, 5.1, 10.7, 16.9, 23.6},
{50, 0.2, 2.9, 7.3, 12.7, 18.5},
{60, 0, 1.3, 4.7, 9.0, 14.1},
{70, 0, 0.4, 2.6, 6.0, 10.2},
{80, 0, 0, 1.1, 3.6, 7.0},
{90, 0, 0, 0.3, 1.8, 4.4},
{100, 0, 0, 0, 0.7, 2.4},
{110, 0, 0, 0, 0, 1},
{120, 0, 0, 0, 0, .2},
{130, 0, 0, 0, 0, 0}};
```

The matrix form of the data needs to be transformed into data sets having each time and each level in pairs. We can do this in one small program as follows and the new data set will be called "**sepdat**" for the separated data:

```
sepdat = Table[Table[{totdat[[n]][[1]], totdat[[n]][[m]]}, {n, 1, Length[totdat]}], {m, 2, 5}]
```

```
{{{0, 10}, {10, 6.8}, {20, 4.2}, {30, 2.3}, {40, 0.9}, {50, 0.2}, {60, 0}, {70, 0}, {80, 0},
{90, 0}, {100, 0}, {110, 0}, {120, 0}, {130, 0}}, {{0, 20}, {10, 15.4}, {20, 11.3}, {30, 7.9},
{40, 5.1}, {50, 2.9}, {60, 1.3}, {70, 0.4}, {80, 0}, {90, 0}, {100, 0}, {110, 0}, {120, 0},
{130, 0}}, {{0, 30}, {10, 24.2}, {20, 19.1}, {30, 14.6}, {40, 10.7}, {50, 7.3}, {60, 4.7},
{70, 2.6}, {80, 1.1}, {90, 0.3}, {100, 0}, {110, 0}, {120, 0}, {130, 0}}, {{0, 40}, {10, 33.3},
```

{20, 27.2}, {30, 21.8}, {40, 16.9}, {50, 12.7}, {60, 9.}, {70, 6.}, {80, 3.6}, {90, 1.8}, {100, 0.7}, {110, 0}, {120, 0}, {130, 0}}}

And then we can graph this data using ListPlot as follows:

ListPlot[Flatten[sepdat, 1],

PlotStyle → PointSize[0.02],

Epilog → {RGBColor[1, 0, 0],

Line[{{0, 10}, {120, 10}}]},

AxesLabel → {"t/s", "h[t]/cm"}]

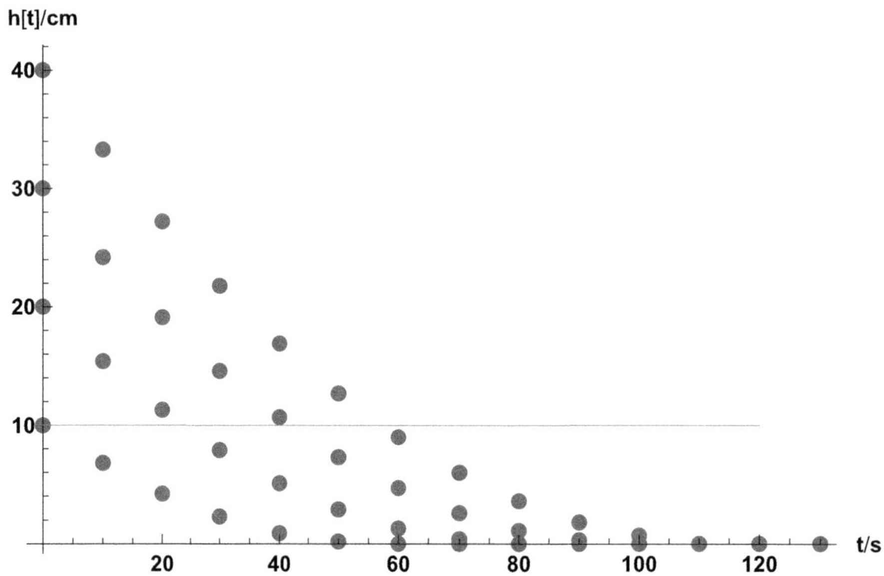

For reference the red (mid gray in the print version) line across the data is set at the 10 cm level. With this, we can see that once this level is reached, then independent of the starting point, it takes 50 seconds to finish the process. So the fluid moving out of the vessel has no "memory" of the level at which the process was initiated. What we seek now is a relationship between the rate of change in level and the level in the tank, as suggested by the material balance and shown by the experimental data. We can get to this relationship by computing the rate of change in level as a function of time for each experiment and then plotting this for comparison.

The approximate rate of change can be computed from the data by taking the slope between successive data points and plotting this versus the time at that second point. We can write a function to do this and then plot the data. The algorithm for implementing this procedure on the set "**sepdat**" is:

Take the nth set of data, from this extract the (m+1) data pair and from this take the second number, subtract from this the second number from the mth data point of the first data set; divide this by the difference between the first number from the (m+1) pair from the nth set and the first number in the mth data pair of the nth set; do this for all n datasets and all m pairs in each set;

This is a good illustration of why natural language is cumbersome compared to the succinctness of mathematical expression. The function to do what is stated above is this:

$$\text{rttot[x\_\_,m\_,n\_]:=}\frac{x[[n]][[m+1]][[2]]-x[[n]][[m]][[2]]}{x[[n]][[m+1]][[1]]-x[[n]][[m]][[2]]}$$

This is a numerical approximation to the derivative of the data level versus time. It is worth understanding because it is such a useful tool when analyzing data. The relationship is written in a general way that can be implemented often in our analyses. To implement this we simply place it inside a pair of nested **Table** commands. The inside command creates a loop around the **m** data pairs in each set and the outside command loops over the **n** data sets. Since we want the slopes associated with times for **sepdat** we also include a command that takes each of the times and pairs it with a slope, that looks like this:

{ sepdat[[1]][[m+1]][[1]], rttot[sepdat,m,n] }

{ time, numerical rate}

The following cell puts all of this together and makes a plot of the slopes versus time for each set of data. Each data set corresponds to a different initial level:

rttot[x__, m_, n_]:=

$$\frac{x[[n]][[m+1]][[2]]-x[[n]][[m]][[2]]}{x[[n]][[m+1]][[1]]-x[[n]][[m]][[1]]}$$

Table[

Table[{sepdat[[1]][[*m*]][[1]], rttot[sepdat, *m*, *n*]},

{*m*, 1, Length[sepdat[[1]]] – 1}], {*n*, 1, 4}];

ListPlot[Flatten[Abs[%], 1],

AxesLabel → $\left\{\text{"t/s"}, \text{"}\left|\frac{\Delta h[t]}{\Delta t}\right| \text{(cm sec}^{-1}\text{)"}\right\}$,

PlotStyle → PointSize[0.02],

PlotRange → {{0, 130}, {0, 0.7}}]

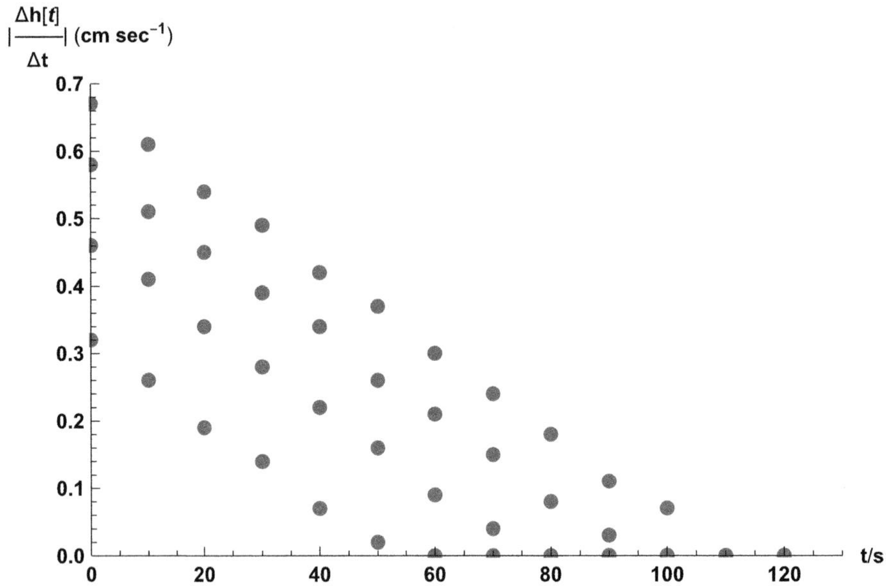

Now, each data set falls on a line and the slopes of each of these lines are identical. Because this is a plot of numerical rates versus time, the slopes of these plots are the approximate second derivative of the change in level with time.

From the graph we can see that the second derivatives are all constant and are each equal to the same value, or stated more precisely:

$$\frac{\Delta}{\Delta t}\left(\frac{\Delta h[t]}{\Delta t}\right) = \text{constant}$$

(or the second derivative is constant).

The constant we now realize involves the gravitational acceleration, **g**. Even without knowing this, it would be logical to replot the rate of change in level against the level rather than against the time. Our initial analysis of the system suggested this very approach, since level as a function of time is the key characterizing variable in this system. To do this, we reuse the last set cell, but with a change in the table function to make the plot one of rate versus level rather than rate versus time. Hence, the line "**sepdat[[1]][[m]] [[1]]**" becomes "**sepdat[[n]] [[m]] [[2]].**"

rttot[x__, m_, n_]:=

$$\frac{x[[n]][[m+1]][[2]] - x[[n]][[m]][[2]]}{x[[n]][[m+1]][[1]] - x[[n]][[m]][[1]]}$$

```
Table[
Table[{sepdat[[n]][[m]][[2]], rttot[sepdat, m, n]},
{m, 1, Length[sepdat[[1]]] – 1}], {n, 1, 4}];
ListPlot[Flatten[Abs[%], 1],
AxesLabel → {"h/m", "|Δh[t]/Δt| (cm sec^-1)"},
PlotStyle → PointSize[0.015],
PlotRange → {{0, 45}, {0, 0.7}}]
```

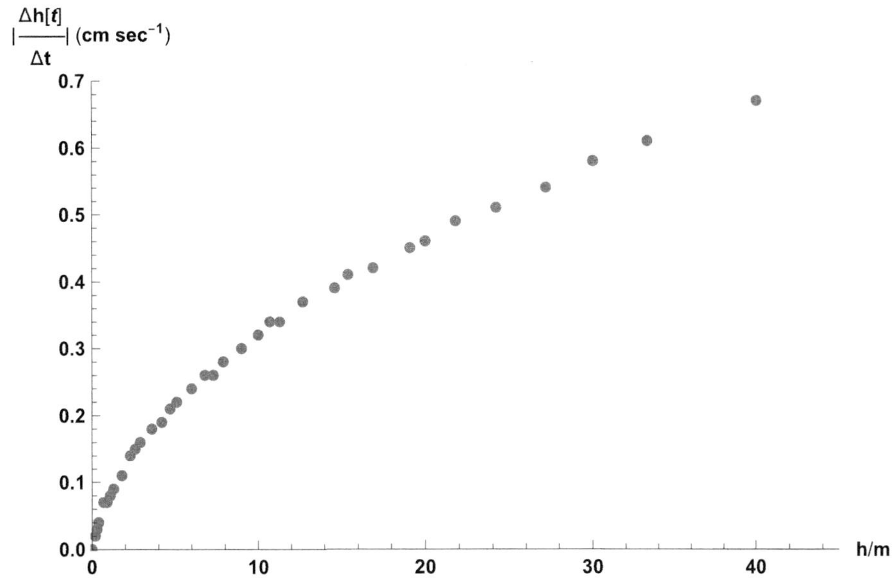

This is quite interesting–all the data from the different experiments now fall on one curve. We notice that the data have distinct functional dependence, which is neither linear nor logarithmic, but it is square root. If Torricelli pursued this approach, and no doubt he did in some fashion, then we can imagine that he felt quite a thrill when he made this graph (or some equivalent representation of the data) and realized that he had uncovered a fundamental physical law. To be sure that the rate of change data do follow a square root dependence on level, we can fit the data to this form and evaluate the constants. We will use **NonlinearModelFit** for doing this. The data set will be constructed from the nested "**Table**" code in the last cell. The function we seek to fit to will be $\mathbf{k}\sqrt{h[t]}$. The procedure to do this and the comparison to the data are done in the following routine:

```
lsdat = Flatten[
Table[
Table[{sepdat[[n]][[m]][[2]],
rttot[sepdat, m, n]},
{m, 1, Length[sepdat[[n]]] − 1}], {n, 1, 4}]
, 1];
nlm = NonlinearModelFit[Abs[lsdat], k√h, k, h]
Plot[nlm[h], {h, 0, 50}, PlotStyle → Hue[0.8]];
ListPlot[Abs[lsdat],
AxesLabel → {"h/m", "|Δh[t]/Δt| (cm sec^-1)"},
PlotStyle → PointSize[0.015],
PlotRange → {{0, 45}, {0, 0.7}}];
Show[%, %%]
```

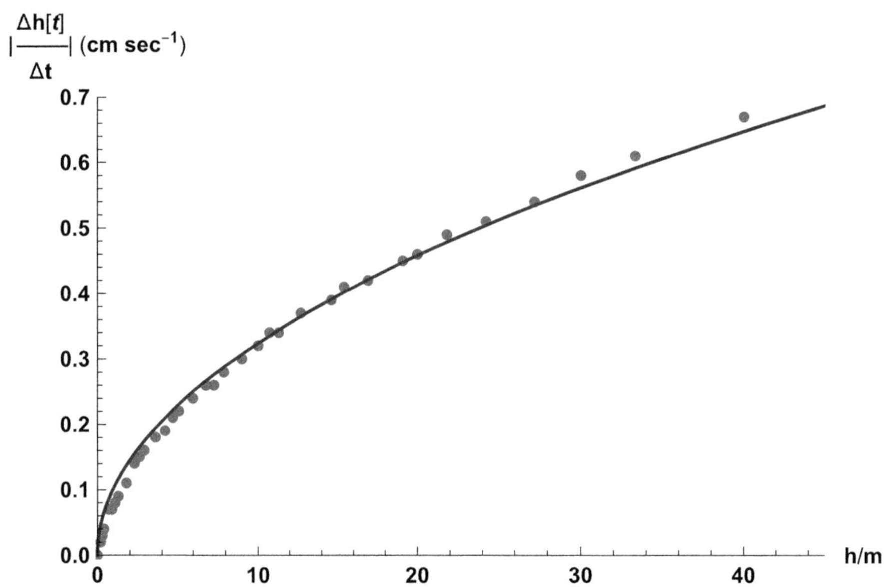

The fit to the experimental data is quite good with the function $0.10\sqrt{h}$. (We can get a report on the "goodness of fit" if we need to, but we do not need it here.) At this point we have found a function that relates the flow rate out of the tank to the level in the tank. The parameter value that fitted the data includes within it the cross section area of the tank, as you may recall from the original statement:

$$\frac{dh[t]}{\mathrm{dt}} = -\frac{f[h[t]]}{A} = -\frac{q[h[t]]}{A} =$$
$$-\mathrm{k}\sqrt{h[t]} = -\frac{m}{A}\sqrt{h[t]}$$
$$\Rightarrow \mathrm{k} = \frac{m}{A} = 0.10$$

At this point we should analyze the dimensions of the parameter on the right-hand side. The left-hand side has dimensions of length time^{-1}, so too must the right-hand side. Therefore we can solve for the dimensions on the parameter **m**:

$$\mathbf{Solve}\left[\frac{\mathbf{Length}}{\mathbf{time}} == \frac{m}{\mathbf{Length}^2}\sqrt{\mathbf{Length}}, m\right]$$

$$\left\{\left\{m \rightarrow \frac{\mathrm{Length}^{5/2}}{\mathrm{time}}\right\}\right\}$$

Now if we rerun the experiment we just described, but vary the orifice area, then we will find that the rate of flow increases in proportion to the increase in the orifice area.

From experiments like these, Torricelli finally deduced that the flow rate looked like this:

$$\mathrm{q}\,[\,\mathrm{h}\,] = \mathrm{b1Ao}\,\sqrt{h[t]}$$

It turns out that **b1** is another constant **b** times the square root of 2 **g** where, **g** is the acceleration due to gravity.

$$\mathrm{q}\,[\,\mathrm{h}\,] = \mathrm{b\ Ao}\,\sqrt{2g}\sqrt{h[t]}$$

Thus, in full form, Torricelli's law is:

$$\mathrm{q}\,[\,\mathrm{h}\,] = \mathrm{b\ Ao}\,\sqrt{2gh[t]}$$

The parameter **b** is found empirically. It too has physical meaning; it is related to the resistance to flow through the orifice. If the orifice were perfectly smooth, then **b** would have a value of unity, alternatively, if the orifice has roughness that disturbs the flow then, b is less than unity. It is a measure of the friction at the wall of the opening. We can check for dimensional consistency one more time and we find:

Simplify[

$$\frac{\text{Length}^3}{\text{time}} == (1)\ \text{Length}^2\text{PowerExpand}\left[\sqrt{\frac{\text{Length}}{\text{time}^2}\text{Length}}\right]]$$

True

Today it is clear that the dependence should be half order in **t** since we can do a simple energy balance to find this to be the case. If the fluid is falling out of the vessel when it flows, due to the force of gravity, then it truly is a falling body. The kinetic energy during the fall can never exceed the potential energy that the body had prior to the fall. So we can state that:

$$\text{m g h}[\text{t}] = \frac{1}{2}\text{m v}[\text{t}]^2$$

from which we can see that:

$$\text{v}[\text{t}] = \sqrt{2gh[t]}$$

In the case of the fluid, this is its velocity through the orifice. The product of this velocity and the area of the orifice is by definition the volumetric flow rate:

$$\text{q}[\text{h}[\text{t}]] = \text{Ao v}[\text{t}] = \text{Ao}\sqrt{2gh[t]}$$

Torricelli's Law is the constitutive relationship that we needed to complete our model. We return to that endeavor now.

Solving for level as a function of time

The equation that will yield the level as function of time is this one:

$$\frac{dh[t]}{\text{dt}} = -\frac{b\text{Ao}\sqrt{2gh[t]}}{A}$$

We can solve this for **h [t]**:

Clear[Ao, *g*, *A*, ho, *h*, *b*]

Simplify[

DSolve[{*h*′[*t*]==–, *h*[0] == ho},

***h*[*t*], *t*]]**

$$\left\{\left\{h[t]\to \text{ho}-\frac{\sqrt{2}\text{Ao}b\sqrt{g}\sqrt{\text{ho}}t}{A}+\frac{\text{Ao}^2b^2gt^2}{2A^2}\right\},\right.$$

$$\left.\left\{h[t]\to \text{ho}+\frac{\sqrt{2}\text{Ao}b\sqrt{g}\sqrt{\text{ho}}t}{A}+\frac{\text{Ao}^2b^2gt^2}{2A^2}\right\}\right\}$$

The functional form is what we should expect for the position of a body in motion; we find it to be a quadratic in time. Since we know the level is falling and not rising, the first solution is the appropriate one for this situation. We can simplify it further as follows:

$$\textbf{hoExpand}\left[\left(\textbf{ho}-\frac{\sqrt{2}\textbf{Ao}b\sqrt{g}\sqrt{\textbf{ho}}t}{A}+\frac{\textbf{Ao}^2b^2gt^2}{2A^2}\right)\Big/\textbf{ho}\right]$$

$$\text{ho}\left(1-\frac{\sqrt{2}\text{Ao}b\sqrt{g}t}{A\sqrt{\text{ho}}}+\frac{\text{Ao}^2b^2gt^2}{2A^2\text{ho}}\right)$$

This can be factored since the coefficient of **t** is the two times the square root of the coefficient of t^2. Therefore we find:

$$\text{h [t] = ho [1 - }\frac{\text{Ao}\,b\sqrt{gt}}{A\sqrt{2\text{ho}}}\text{]}^2$$

This version of the expression is interesting because it shows **ho** multiplied by $[\,1 - \mathbf{b}\left(\frac{\text{Ao}}{A}\right)\sqrt{\frac{g}{2\text{ho}}}\,\mathbf{t}\,]^2$. The quantity in the brackets must be dimensionless, because one is dimensionless and the left-hand side must have the same dimensions as the right-hand side, namely length or height. Further to be dimensionless, means that the group $\mathbf{b}\left(\frac{\text{Ao}}{A}\right)\sqrt{\frac{g}{2\text{ho}}}$ must have dimensions of one over time, $\frac{1}{t}$, or inverse time, so that when it is multiplied by time the product is dimensionless. If we look closely we see that the ratio of Ao to A is dimensionless, and that the dimensions of the square root of the gravitational acceleration over a length are the square root of $\frac{1}{\sqrt{t^2}}$ or $\frac{1}{t}$, or inverse time. Therefore the constant b is a pure number as we indicated before that empirically expresses the degree of friction between the orifice wall and the liquid passing through it. Lastly, a comparison of the function for h [t] given above to the data points gives an excellent fit:

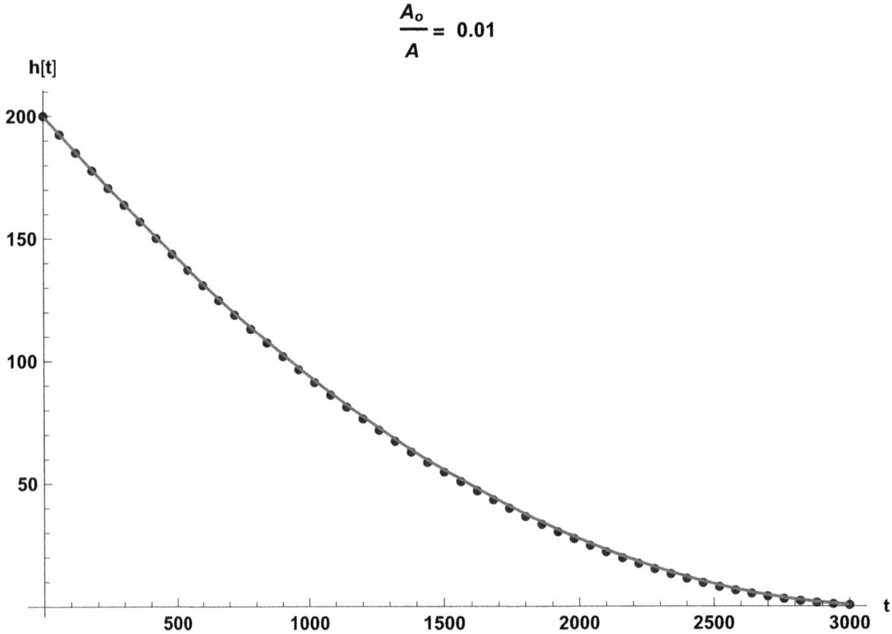

Mass input, output, and control

Constant input

A good next problem to consider is that of a vessel with a specified mass flow in and gravity mass flow out. The accumulation term is the response to the interaction of these two flows. The physical picture is shown below.

The mass balance equation has two terms on the right-hand side.

$$\frac{dm[t]}{\mathrm{dt}} = \mathring{m}_{\mathrm{in}} - \mathring{m}_{\mathrm{out}}$$

The mass flow term in can be taken as the product of the density of the fluid and its volume flow rate. The mass flow out can be specified by Torricelli's Law multiplied by the fluid density and the mass in the control volume is the product of the fluid density, the tank's cross sectional area and the level at any time **t**. This gives us the following equation:

$$\frac{d\rho A h[t]}{\mathrm{dt}} = \rho\ \mathrm{q} - \rho\ \mathrm{b}\ \mathrm{Ao}\ \sqrt{2gh[t]}$$

Simplification provides us with:

$$\frac{d\ h[t]}{\mathrm{dt}} = \frac{q - b\mathrm{Ao}\sqrt{2gh[t]}}{A}$$

This can be solved and we can examine the behavior of the time dependent solution. We will begin the analysis assuming that the tank is initially empty, that is that **h [0] = 0**, the initial condition. We will assume for the now that **q** is a constant:

```
Simplify[
DSolve[{d h[t]/dt ==, h[0] == 0},
h[t], t]]
```

$$\left\{\left\{h[t] \to \frac{Aqt + \mathrm{Ao}^2 b^2 g t^2 - \sqrt{\mathrm{Ao}^2 b^2 g t^3 \left(2Aq + \mathrm{Ao}^2 b^2 g t\right)}}{A^2}\right\},\right.$$

$$\left.\left\{h[t] \to \frac{Aqt + \mathrm{Ao}^2 b^2 g t^2 + \sqrt{\mathrm{Ao}^2 b^2 g t^3 \left(2Aq + \mathrm{Ao}^2 b^2 g t\right)}}{A^2}\right\}\right\}$$

Which of the two solutions should we use? The second solution is a net increasing function, because all the terms are additive. It is logical to use the first solution, which has a difference of terms. The best way to examine its behavior is graphically. Let's take the input flow rate to be 2 m^3 sec^{-1}, the radius of the tank is 2 m and the orifice area is 1% of the tank cross sectional area:

$$h[\mathrm{t\_}]\text{:=}\frac{Aqt + \mathrm{Ao}^2 b^2 g t^2 - \sqrt{\mathrm{Ao}^2 b^2 g t^3 \left(2Aq + \mathrm{Ao}^2 b^2 g t\right)}}{A^2}$$

```
qex[t_]:= bAo
r = 2;
A = N[πr^2];
Ao = 0.01A;
b = 1;
g = 9.80;
q = 2;
tmax = 10000;
h[tmax]
```

```
plh = Plot[h[t], {t, 0, tmax}, AxesLabel → {"t/sec", "h[t]/m"},
PlotStyle → Hue[0.02], PlotRange → All]
plq = Plot[qex[t], {t, 0, tmax}, AxesLabel → {"t/sec", "qex[t]/m sec^-1"},
PlotStyle → Hue[0.6], PlotRange → All]
```

12.7179

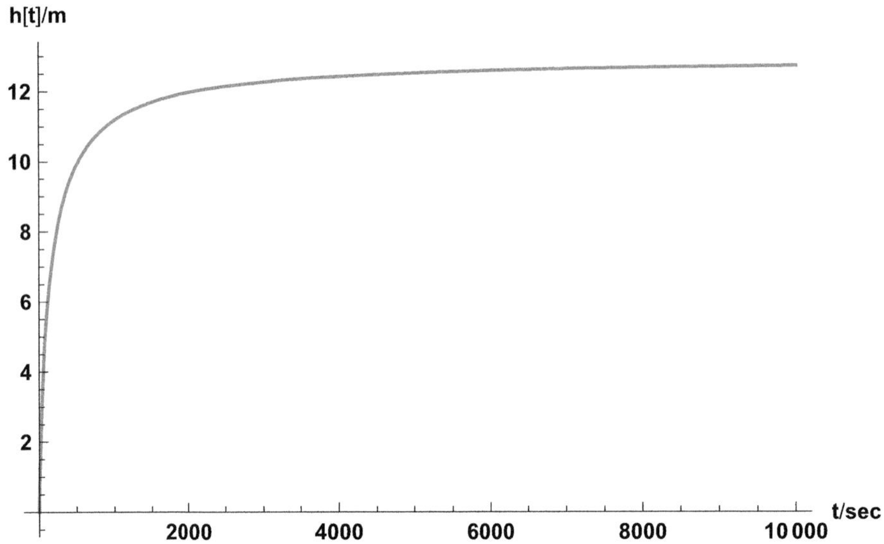

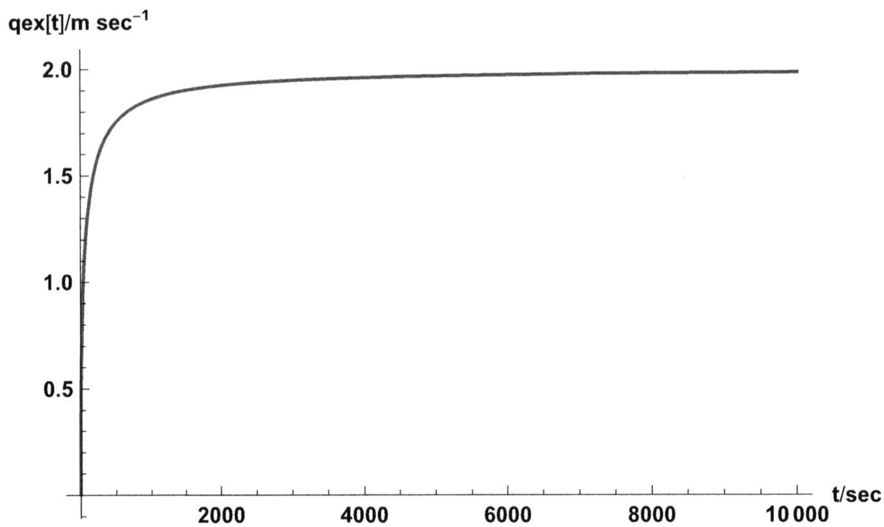

This solution indicates that the level in the tank rises to about 12 meters and then stops changing. Graphing the exit flow rate, $q_{ex}[t]$, as a function of time shows that the flow first grows and then stabilizes. This all makes sense based on the mathematics and it also makes good sense physically. The tank is initially empty, so at early times the level is low and the flow rate out is strongly level dependent, so it too is small. This leads to a filling of the tank, but, as the level rises, so too does the exit flow rate. After about 400 seconds the flow rate comes to a near constant value as does the level. This is called the *steady state*. At steady state, the net accumulation of mass is zero. The fluid mass, level, and volume remain constant, just so long as the parameters remain unchanged.

If the net accumulation is zero at the steady state, then we could have solved for the steady state level without solving the differential equation per se. In other words we could have solved for h_{ss} at the steady state from this equation:

$$0 = \frac{q - b\text{Ao}\sqrt{2gh_{ss}}}{A}$$

Clear[g, q, b, Ao, A]

Solve$\left[\frac{q-b\text{Ao}}{A} == 0, h_{ss}\right]$

$$\left\{\left\{h_{ss} \to \frac{q^2}{2\text{Ao}^2 b^2 g}\right\}\right\}$$

Applying the parameter values, that we used before, we find that the level at steady state will be 12.9 m which agrees well with the graph of the full time-dependent solution, showing that it approaches this value *asymptotically*. Putting the two on the same graphic we can see this:

$r = 2;$

$A = N\left[\pi r^2\right];$

$\text{Ao} = 0.01 A;$

$b = 1;$

$g = 9.80;$

$q = 2;$

$\text{hss} = \frac{q^2}{2\text{Ao}^2 b^2 g}$

Show[plh,

Graphics[{RGBColor[0, 0, 1],

Line[{{0, hss}, {10000, hss}}]}]]

12.9236

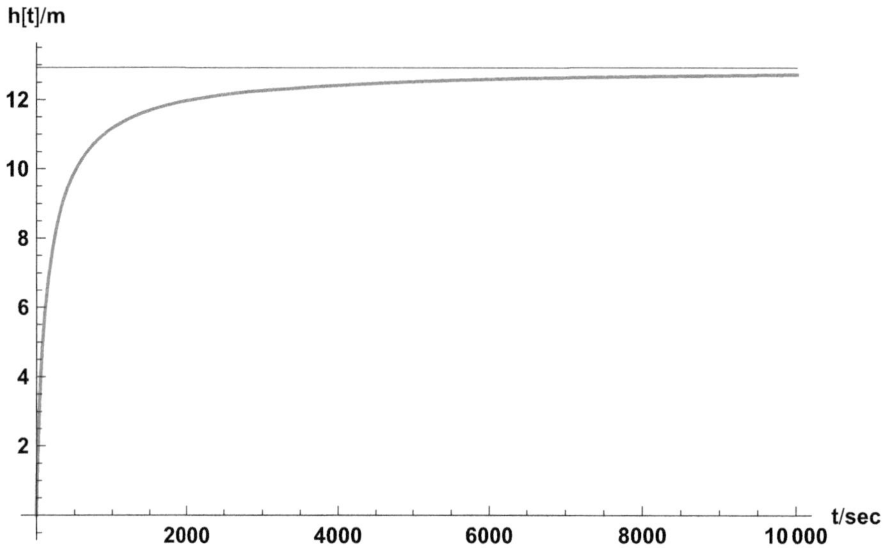

If we leave all else the same, but begin the process with either a higher, or a lower, volume flow rate what will be the result? Will the position of the steady state change? We can see the answer immediately by using the steady state solution. Choosing a higher or lower flow rate will markedly affect the position of the steady state, since this level depends of the square of the flow rate in.

Fluxional input

What would be the affect of fluxional input? Suppose for example that the input were sinusoidal as we saw in Chapter 2, what would the output look like given a gravity-driven exit flow? The virtue of *Mathematica* is that we can solve this problem with very little effort beyond what we have already done and we can compare the results with those from constant input. Here is how we do it.

Firstly, taking **q [t]** the input flow rate as:

$$\text{qo}\,(1 + \alpha\ \text{Sin}\,[\ \beta\ \text{t}\ ])$$

we will solve the new differential equation:

$$\frac{d\,h[t]}{\text{dt}} = \frac{\text{qo}(1 + \alpha\ \text{Sin}[\beta t]) - b\text{Ao}\sqrt{2gh[t]}}{A}$$

This can be solved numerically for specific value of the parameters. We then evaluate this the new definition of **h [t]** and call it **hsin [t]** to distinguish it from the earlier work we have done. We make similar changes to the plot names. The value for **qo** is taken as 20 to be that same and the constant flow case. The magnitudes of α and β are taken to be the same as they were in Chapter 2.

```
Clear[α, β, g, q, qo, b, Ao, A, h]
r = 2;
A = N[π r^2];
Ao = 0.01A;
b = 1;
g = 9.80;
qo = 2;
α = .25;
β = .01;
tmax = 10000;
soln0 =
NDSolve[
{∂_t h[t] == (qo(1 + αSin[βt]) - bAo√(2gh[t]))/A,
h[0] == 0}, h[t], {t, 0, tmax}];
hsin[t_]:=Evaluate[h[t]/.soln0]
qin[t]:=qo(1 + αSin[βt])
qexsin[t_]:= bAo
plqsinin = Plot[qin[t], {t, 0, tmax},
AxesLabel → {"t/sec", "qin[t]/m sec^-1"},
PlotStyle → Hue[0.6], PlotRange → {{0, tmax}, {0, 3}},
PlotRange → All]
plhsin = Plot[hsin[t], {t, 0, tmax},
```

AxesLabel → {"t/sec", "h[t]/m"}, PlotRange → All]

plqexsin = Plot[qexsin[t], {t, 0, tmax},

AxesLabel → {"t/sec", "qex[t]/m sec^{-1}"},

PlotStyle → Hue[0.3], PlotRange → All]

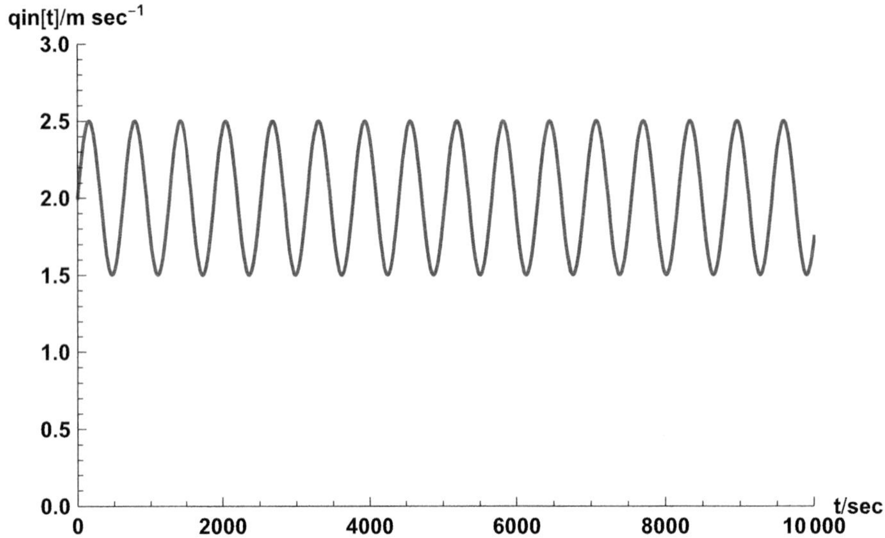

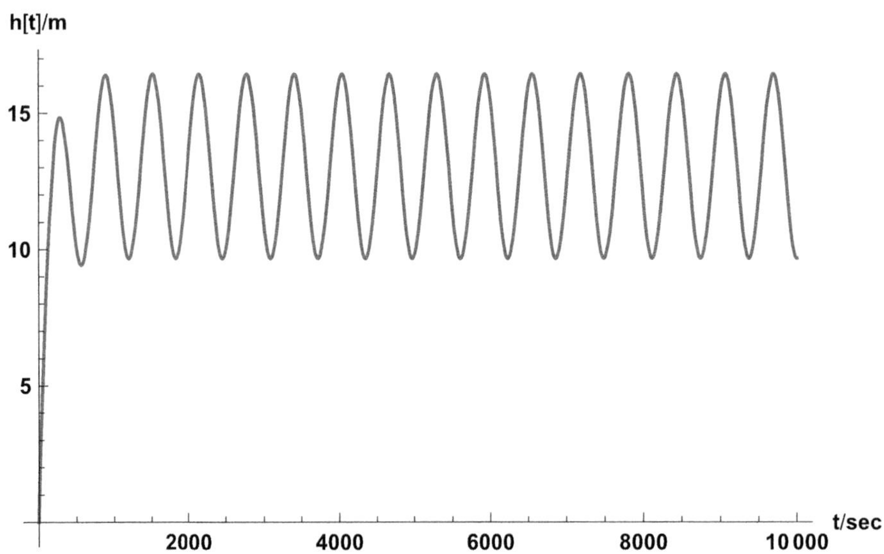

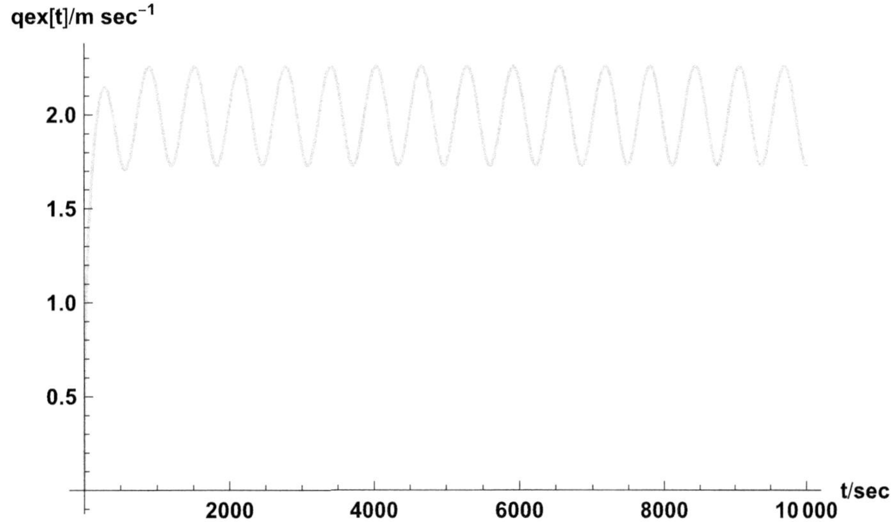

The results show that the output flow is coupled to the input flow quite tightly after an initial transient period. We can compare these responses to the sinusoidal input with those from the constant input case by plotting **qex [t]** and **h [t]** for both cases:

Show[plqexsin, plq]

Show[plhsin, plh]

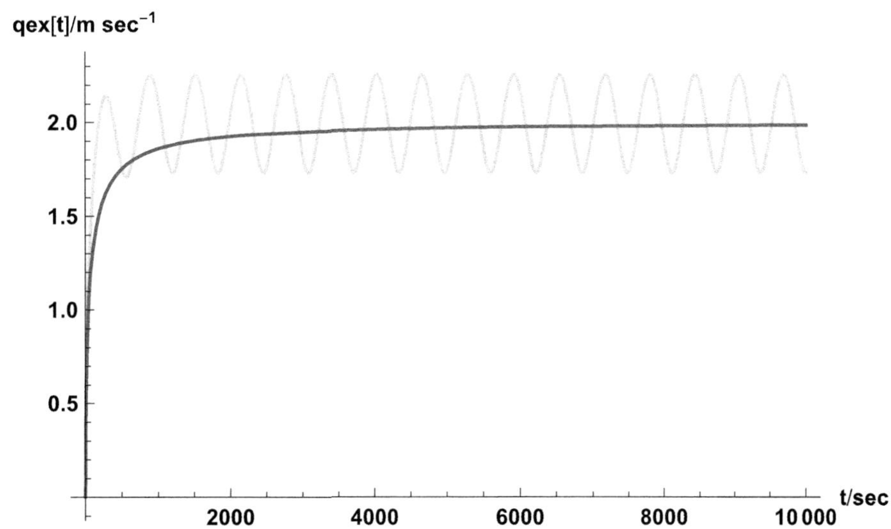

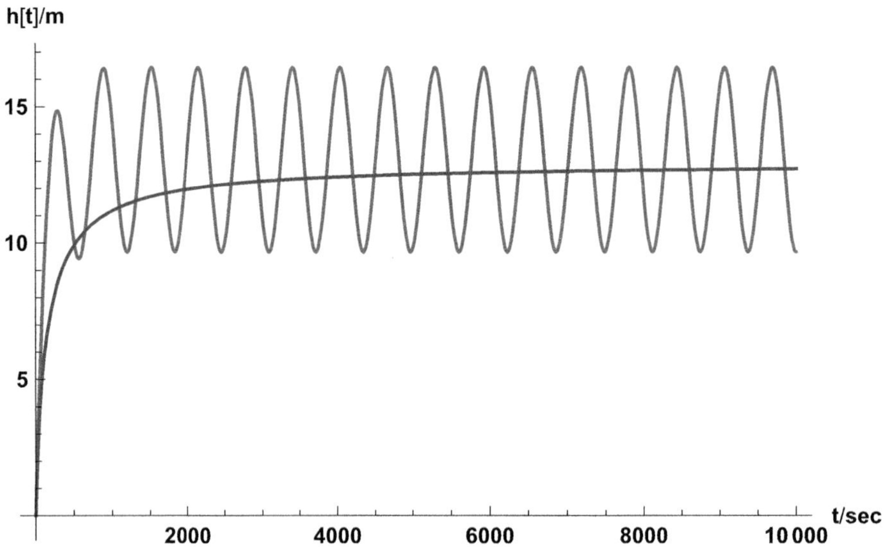

Since we chose all the parameters to be the same, we note the fluxional values of **h** and **qex** for the sinusoidal input case are larger earlier than they are for the fixed input. The reason is that the input flow in the sinusoidal case rises to a value well over its average value.

Show[plqsinin, Graphics[{RGBColor[1, 0, 0], Line[{{0, 2}, {10000, 2}}]}]]

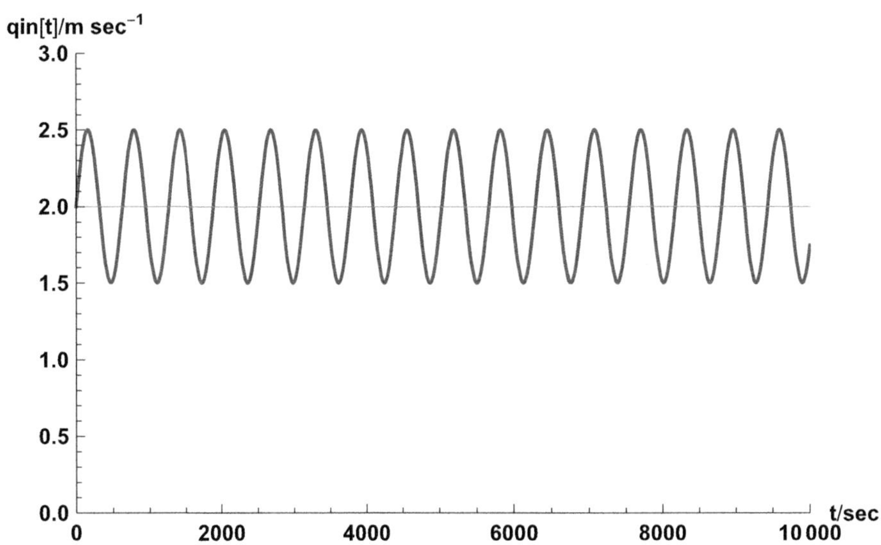

Control

It would be very nice to be able to dampen the input fluctuations and to smooth the output from this vessel. To do this requires a control function. One form of control would be to increase the flow rate out of the tank whenever the level in the tank rises or falls above or below a designed set point level, h_d. For example the set point level could be the steady state level that we found from the earlier example with constant input flow, which is also the red (mid gray in the print version) line in level the graph above. To increase the flow rate in the case of gravity driven flow, we must increase the size of the orifice. We can increase it in proportion to the difference between the actual level in the tank at any time and the design level. The actual implementation would involve having a level sensor tied to an actuator which would open the valve more or less depending on the level. The mathematical description of this control function can be given as:

Ao [h [t]] = Aoo (1 - K (hd - h [t]))

In this expression **Aoo** is the nominal aperture size to deliver at the design flow rate based on the constant set input flow rate. The second term in the parenthetical expression is the product of a proportionality constant, **K**, and the difference between the set point level and the actual level as a function of time. We substitute this for **Ao** in Torricelli's law and also in the equation describing system with sinusoidally fluctuating input flow:

$$\frac{d\,h[t]}{\mathrm{dt}} = \frac{\mathrm{qo}(1+\alpha\,\mathrm{Sin}[\beta t]) - b\mathrm{Aoo}(1-K(\mathrm{hd}-h[t]))\sqrt{2gh[t]}}{A}$$

Notice that if **K** were set to zero, the equation would revert back to that which we have already solved for the uncontrolled system. We can operate on the right-hand side to put it into a form that is more readily understood:

Clear[α, β, g, q, qo, b, Ao, A, h, K, hd, Aoo]

Collect[Simplify[PowerExpand[$\frac{\mathrm{qo}(1+\alpha\,\mathrm{Sin}[\beta t]) - b\mathrm{Aoo}(1-K(\mathrm{hd}-h[t]))\sqrt{2gh[t]}}{A}$]], qo]

$$-\frac{\sqrt{2}\mathrm{Aoo}\,b\sqrt{g}\sqrt{h[t]}(1-\mathrm{hd}K+Kh[t])}{A}+\frac{\mathrm{qo}(1+\alpha\mathrm{Sin}[t\beta])}{A}$$

$$-\frac{\sqrt{2}\mathrm{Aoo}b\sqrt{g}\sqrt{h[t]}(1+K(-\mathrm{hd}+h[t]))}{A}+\frac{\mathrm{qo}(1+\alpha\mathrm{Sin}[t\beta])}{A}$$

From this form of the equation, we can see that we have one term that is a Sine function of time alone, and we will put that one on the right-hand side; the other term is a function of h [t], and so that will go to the left-hand side:

$$\frac{d\,h[t]}{\mathrm{dt}} = -\frac{\sqrt{2}\mathrm{Aoo}\,b\sqrt{g}\sqrt{h[t]}(1+K(h[t]-\mathrm{hd}))}{A} + \frac{\mathrm{qo}(1+\alpha\mathrm{Sin}[t\beta])}{A}$$

$$\frac{d\,h[t]}{dt} + \frac{\sqrt{2}\mathrm{Aoo}\,b\sqrt{g}\sqrt{h[t]}(1+K(h[t]\;-\;\mathrm{hd}))}{A} = \frac{\mathrm{qo}(1+\alpha\mathrm{Sin}[t\beta])}{A}$$

We will use all the same parameter values that we have used in the previous problem, but we will have to pick a magnitude for **K**. This is best done by solving the problem and then resetting the value to get a sense of the solution's sensitivity to the magnitude of **K**. We start with a value of zero to be sure that the solution to this new equation reduces to that of the one we have solved already:

Clear[α, β, g, q, qo, b, Ao, A, h, K, hd, Aoo, t]

$r = 2;$

hd = 12;

$A = N\left[\pi r^2\right];$

Aoo = 0.01A;

$b = 1;$

$g = 9.80;$

qo = 2;

$\alpha = .25;$

$\beta = .01;$

$K = 0;$

tmax = 10000;

soln1 =

NDSolve[

$$\left\{\mathrm{hx}'[t] == -\frac{\sqrt{2}\mathrm{Aoo}b\sqrt{g}\sqrt{\mathrm{hx}[t]}(1+K(\mathrm{hx}[t]-\mathrm{hd}))}{A} + \frac{\mathrm{qo}(1+\alpha\mathrm{Sin}[t\beta])}{A}, \mathrm{hx}[0] == 0\right\},$$

```
hx[t],{t, 0, tmax}];

hc[t_]:=Evaluate[hx[t]/.soln1]

qin[t_]:=qo(1 + αSin[βt])

Ao[t_]:=Aoo(1 + K(hd − hc[t]))

qexc[t_]:= bAo[t]

plhc = Plot[{hc[t]}, {t, 0, tmax},PlotRange → {{0, tmax}, {0, 18}},

AxesLabel → {"t/sec", "h[t]/m"},PlotStyle → {Hue[0.], Hue[0.1]}]

plqcin = Plot[qin[t], {t, 0, tmax},AxesLabel → {"t/sec", "qin[t]/m sec^-1"},

PlotStyle → Hue[0.6],PlotRange → {{0.0, tmax}, {0.0, 2.5}}]

plqexc = Plot[qexc[t], {t, 0, tmax},AxesLabel → {"t/sec", "qex[t]/m sec^-1"},

PlotStyle → Hue[0.6]]
```

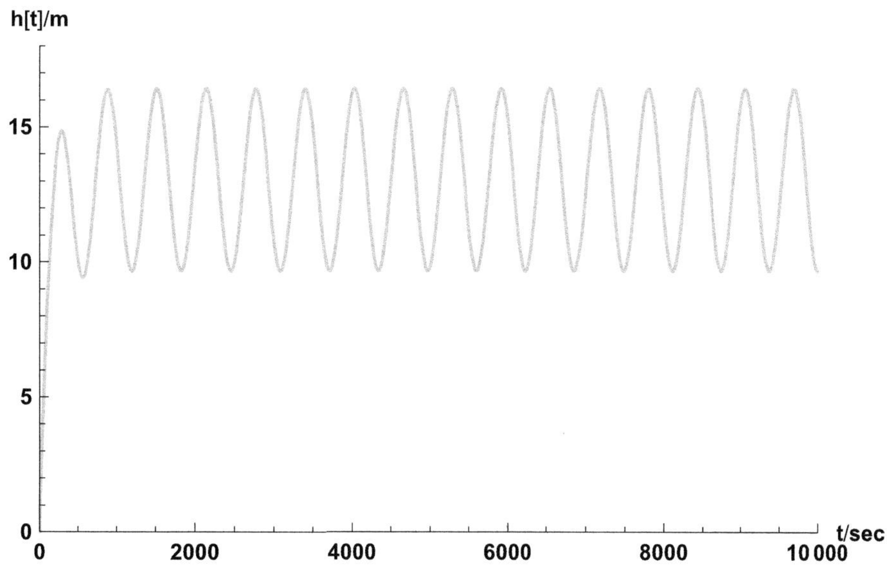

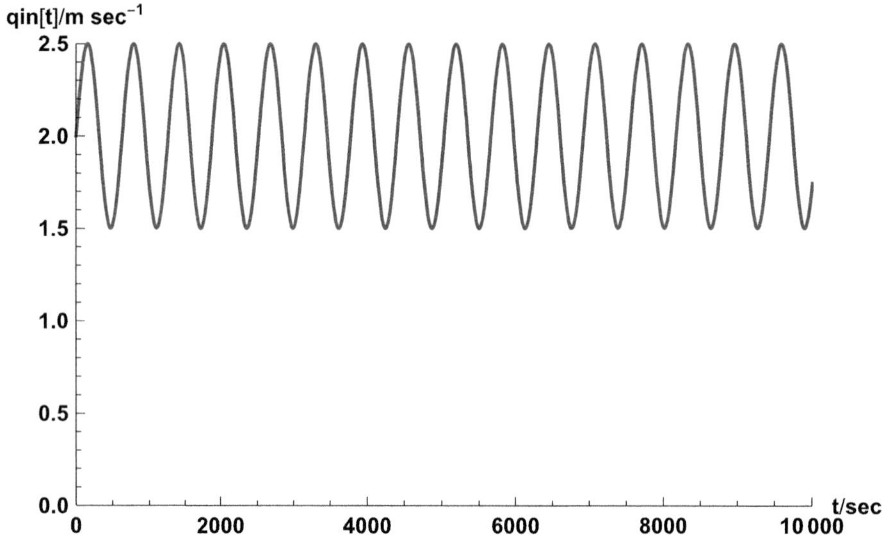

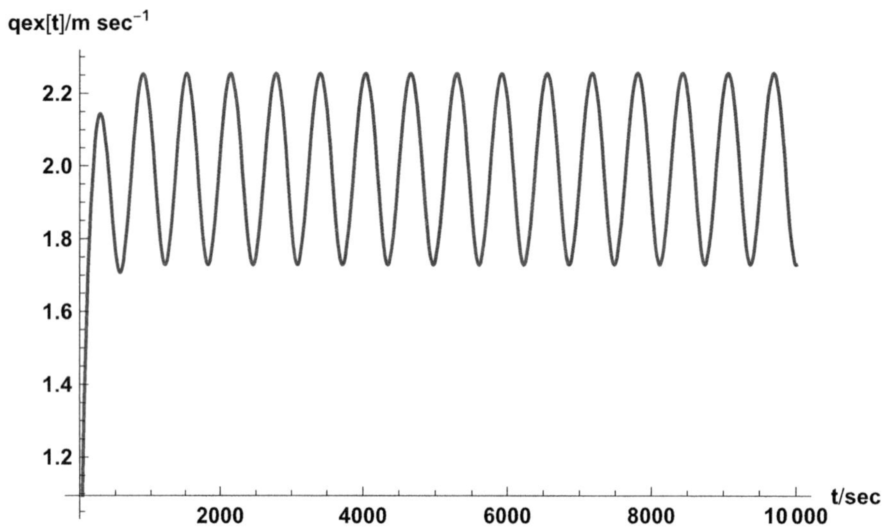


```
Show[plhc, plhsin, PlotLabel → "Red=Controlled"]
```

```
Show[plqexc, plqexsin, PlotLabel → "Blue=Controlled"]
```

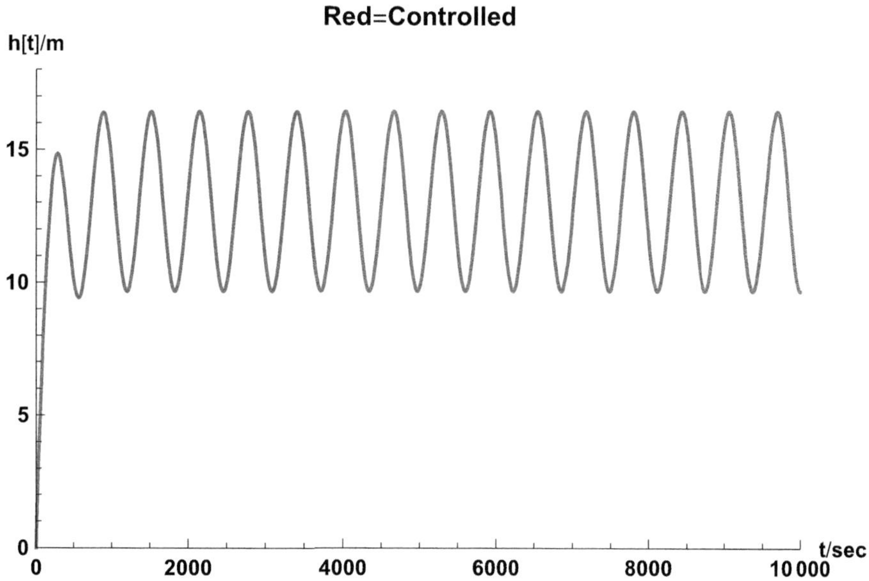

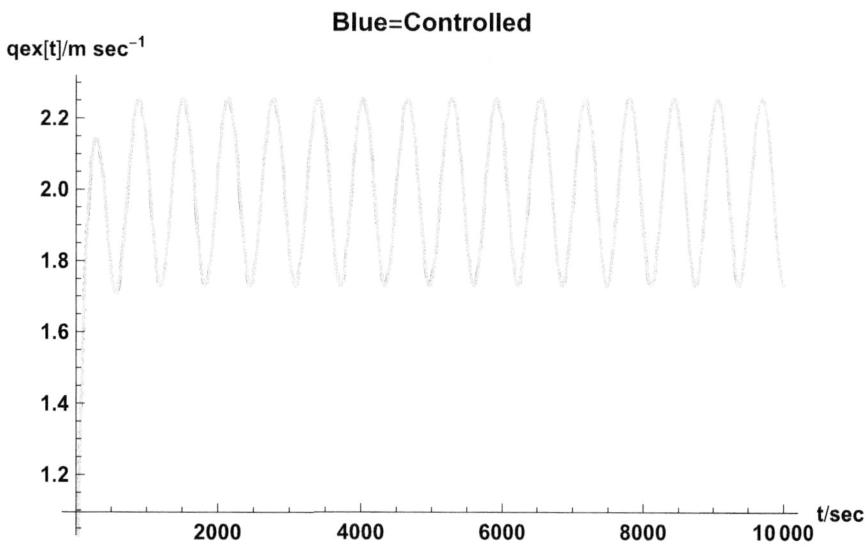

The graphs overlap perfectly, as they should if the model is working. The means that when K = 0 in this new model the results are the same as those we obtained in the earlier model, and this is what we expect.

Next, we set K = 0.1 to test its effect on the results:

Clear[α, β, g, q, qo, b, Ao, A, h, K, hd, Aoo, t]

Ao=.

```
r = 2;
hd = 12;
A = N[πr^2];
Aoo = 0.01A;
b = 1;
g = 9.80;
qo = 2;
α = .25;
β = .01;
K = .1;
tmax = 10000;
soln1 =
NDSolve[
{hx'[t]== -(√2 Aoo b √g √hx[t] (1+K(hx[t]-hd)))/A + (qo(1+αSin[tβ]))/A, hx[0] == 0}, hx[t],{t, 0, tmax}];
hc[t_]:=Evaluate[hx[t]/.soln1]

qin[t]:=qo(1 + αSin[βt])
Ao[t_]:=Aoo(1 + K(hd − hc[t]))
qexc[t_]:= bAo[t]
plhc = Plot[{hc[t]}, {t, 0, tmax},PlotRange → {{0, tmax}, {0, 18}},
AxesLabel → {"t/sec", "h[t]/m"},PlotStyle → {Hue[0.], Hue[0.1]},
];
plqcin = Plot[qin[t], {t, 0, tmax},AxesLabel → {"t/sec", "qin[t]/m sec^-1"},
PlotStyle → Hue[0.6],PlotRange → {{0, tmax}, {0, 25}},
];
plqexc = Plot[qexc[t], {t, 0, tmax}, AxesLabel → {"t/sec", "qex[t]/m sec^-1"},
```

PlotStyle → Hue[0.6]];

Show[plhc, plhsin,

PlotLabel → "Red=Controlled, Blue=Not"]

Show[plqexc, plqexsin, PlotRange → {{0, tmax}, {0, 2.5}},

PlotLabel → "Blue=Controlled, Green=Not"]

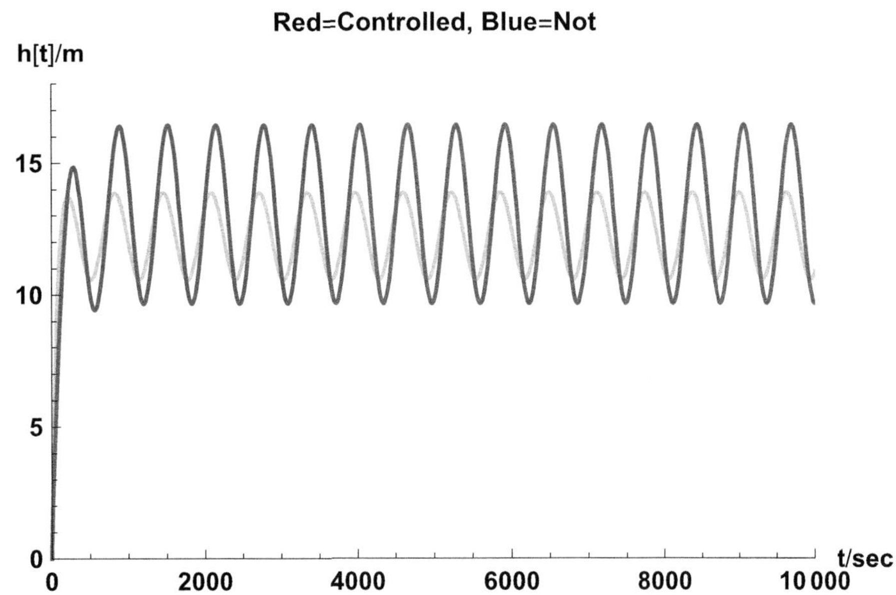

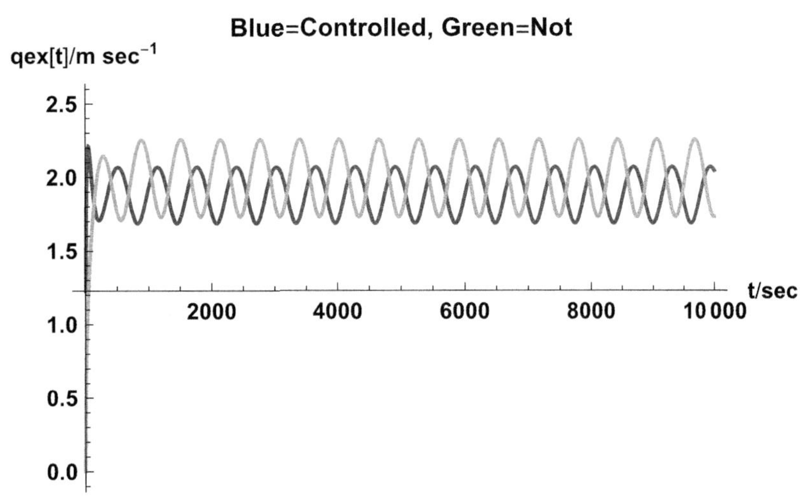

What we find is that in the case of the K = 0.1 versus K = 0, there is less deviation in the level and the exit flow rates. We can say that the there is a dampening effect on both.

Using a bit of code and ingenuity, we can have *Mathematica* compute the responses to incremental changes in **K** ranging from zero to two and one quarter in increments of a quarter unit. The code to do this is show below along with the outputs in two graphical forms– a stack plot:

Clear[$\alpha, \beta, g, q,$ qo, b, Ao, A, h, K, hd, Aoo, t]

Ao=.

$r = 2;$

hd = 12;

$A = N\left[\pi r^2\right];$

Aoo = 0.01A;

$b = 1;$

$g = 9.80;$

qo = 2;

$\alpha = .25;$

$\beta = .01;$

$K = .1;$

tmax = 10000;

solns =

Table[

NDSolve[

{hx′[t] ==

$$-\frac{\sqrt{2}\text{Aoo}b\sqrt{g}\sqrt{\text{hx}[t]}(1 + K(\text{hx}[t] - \text{hd}))}{A} +$$

$$\frac{\text{qo}(1 + \alpha\text{Sin}[t\beta])}{A}, \text{hx}[0] == 0\Big\}, \text{hx}[t],$$

{t, 0, tmax}], {K, 0, 2.25, .05}];

```
fns = Table[Evaluate[hx[t]/.solns[[n]]][[1]],
{n, 1, Length[solns]}];
plots = Table[Plot[{fns[[n]]}, {t, 0, tmax},
PlotRange → {{0, tmax}, {0, 20}},PlotStyle → Hue[n/10]
], {n, 1, Length[solns]}];

Show[plots]
```

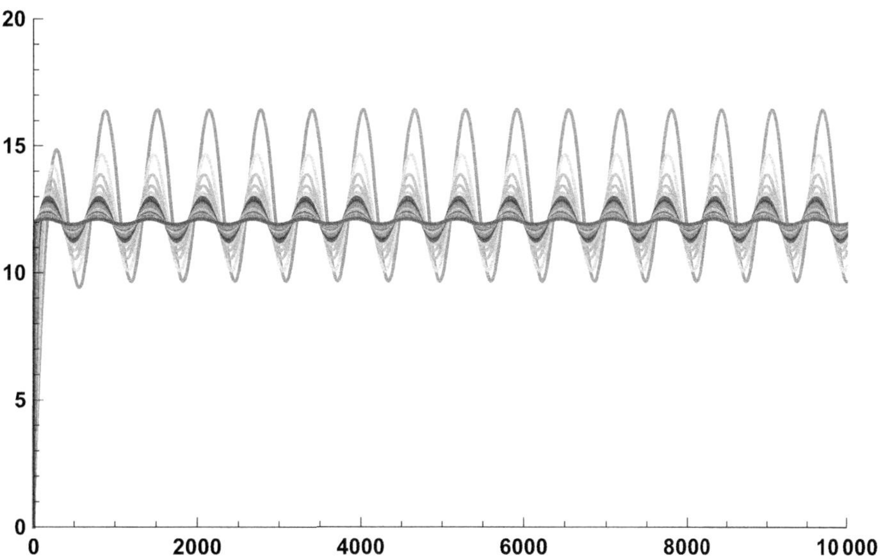

This set of plots shows that as the magnitude of K increases the deviation of level around the mean value decreases. As beautiful as this visualization is, it is easier to see the effect by showing the plots for K equal to zero and K equal to 2.5:

```
Show[Plot[fns[[1]], {t, 0, tmax},
PlotRange → {{0, tmax}, {0, 20}},
PlotStyle → Hue[1/10]],
Plot[fns[[Length[solns]]], {t, 0, tmax},
PlotRange → {{0, tmax}, {0, 20}},
PlotStyle → Hue[Length[solns]/10]],
PlotLabel → "Blue=Controlled"]
```

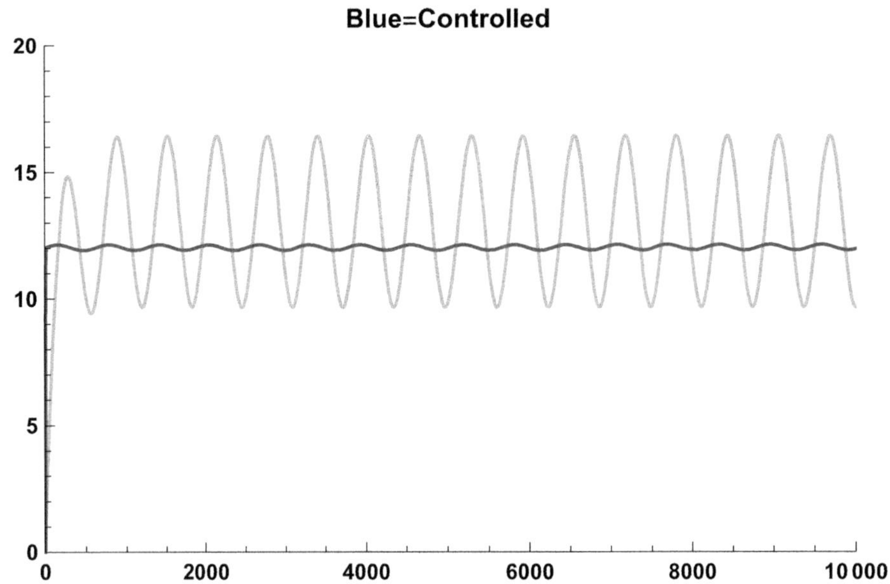

By increasing the magnitude of **K** we meet the original goal of dampening out the excursions that would take place had we not included the control function. There are other ways to express the control function besides this one in which the change is proportional to the deviation from the set value. An example of proportional-integral-differential control is given in chapter 9.

Conclusion

In this chapter we have extended the analyses that we can do well beyond the elementary systems of Chapter 2. We began with a fairly simple problem, the gravity driven flow of fluid from a tank, that led to Torricelli's Law. With this in the tool box we were able to analyze a series of systems with input and output fluid flows that were increasingly more complex, culminating in the proportional control of the level of a tank with sinusoidally driven input flow. As we moved through these examples, we have begun to use *Mathematica* in ever more sophisticated ways, providing us with new techniques to add to our arsenal of problem-solving methods.

CHAPTER 4

Multiple component systems

Real systems are comprised of many components and in multiple phases. The objective of this chapter is to build a strong foundation for problems of multicomponent systems of all kinds.

The concept of the component balance

A component is a chemical constituent of the system at hand. A beaker of pure water is a single component system. If the water is impure, as is seawater then it contains salt, NaCl. If that were all that was present, then this would be a two component system. We know it also will have iron, calcium, magnesium, and other elements in it. These are each components. Sea water is a multicomponent system. Phases are states of the system - solid, liquid, or gas/vapor. A sealed vessel of pure water with space above the liquid that is filled with water vapor and only water vapor is a single component system with two phases. If that system is cooled sufficiently to form ice, it becomes a single component, three phase system, of solid, liquid, and vapor. It is each chemically distinct species in a system that corresponds to a different component.

When we have a single component system, we have one mass balance and it is the total mass balance. When we have a multicomponent system, we will write a mass balance for each individual component and the total mass balance is the sum of each of the component balances.

Imagine we are playing a game tossing ping pong balls into a box on a scale. Each ball has the same mass, but they have different colors - red or green. The two differently colored balls are in essence two components. The scale tells us how fast the **total mass of balls**, both red and green, is changing. If we want to know how fast the mass of just red balls in the box is changing, then we need to know how many are being thrown per unit time over the period of the measurement. Similarly, for the green balls. The sum of the arrival rates of the red and green balls together is the rate of mass change in total within the box.

The total material balance for this system is:

$$\frac{dm_{\text{tot}}[t]}{dt} = \overset{\circ}{m}_{\text{red}} + \overset{\circ}{m}_{\text{green}}$$

Introduction to Chemical Engineering Analysis Using *Mathematica*
https://doi.org/10.1016/B978-0-12-820051-3.00009-7

The two component balances are:

$$\frac{dm_{\text{red}}[t]}{dt} = \overset{\circ}{m}_{\text{red}}$$

$$\frac{dm_{\text{green}}[t]}{dt} = \overset{\circ}{m}_{green}$$

Restating the total material balance, we have:

$$\frac{dm_{\text{tot}}[t]}{dt} = \frac{dm_{\text{red}}[t]}{dt} + \frac{dm_{\text{green}}[t]}{dt}$$

The sum of the component balances is the total material balance and the net rate of change of any component's mass within the control volume is the sum of the rate of mass input of that component minus the rate of mass output and these can occur by any process, including chemical reaction, as we will see in later chapters. This last part of the dictum is important, because, as we will see in chapter 6, chemical reactions within a control volume do not create or destroy mass, they merely redistribute it among the components. In a real sense, chemical reactions reassign where the mass shall appear.

Concentration versus density

To this point, we have had to deal only with the mass per unit volume in form of density, since we were concerned only with single component systems. Multiple components share the volume and because of this we must use concentration as well as density. For a multicomponent system the total density is the sum of the masses of the components per unit volume:

$$\rho_{\text{tot}} = \frac{\sum_{i=1}^{n} \text{mass}_i}{\text{Vol}}$$

The concentration of any component i can be either a mass concentration of a molar concentration:

$$C_i = \frac{m_i}{\text{Vol}} \equiv \frac{\text{mass}}{\text{Vol}}$$

$$M_i = \frac{N_i}{\text{Vol}} = \frac{1}{\text{Mw}_i}\frac{m_i}{\text{Vol}} = \frac{C_i}{\text{Mw}_i} \equiv \frac{\text{mole}}{\text{Vol}}$$

The component material balance for a system with input and output, but no chemical reaction, is written as follows:

$$\frac{dm_i}{\text{dt}} = \overset{\circ}{m}_{i,\text{in}} - \overset{\circ}{m}_{i,\text{out}}$$

If the mass flows are those of liquids, then in terms of mass concentrations, this becomes:

$$\frac{V}{V}\frac{dm_i}{\mathrm{dt}} = \overset{\circ}{m}_{i,\mathrm{in}} - \overset{\circ}{m}_{i,\mathrm{out}}$$

$$\frac{d}{\mathrm{dt}}\left[\frac{m_i}{V}V\right] = \overset{\circ}{m}_{i,\mathrm{in}} - \overset{\circ}{m}_{i,\mathrm{out}}$$

$$\frac{d}{\mathrm{dt}}[C_i V] = (C_{i,\mathrm{in}} q_{\mathrm{in}} - C_{i,\mathrm{out}}\, q_{\mathrm{out}})$$

The last statement is the typical form of a liquid phase component mass balance. When this is divided through by the molecular weight of species i, this becomes a differential mole balance since the concentrations are expressed in molarity units:

$$\frac{1}{\mathrm{Mw}_i}\frac{d}{\mathrm{dt}}[C_i V] = (C_{i,\mathrm{in}} q_{\mathrm{in}} - C_{i,\mathrm{out}}\, q_{\mathrm{out}})\frac{1}{\mathrm{MW}_i}$$

$$\frac{d}{\mathrm{dt}}[M_i V] = (M_{i,\mathrm{in}} q_{\mathrm{in}} - M_{i,\mathrm{out}}\, q_{\mathrm{out}})\frac{1}{\mathrm{MW}_i}$$

Typically, this last statement is written with the symbol C for molar concentration just as it is for mass concentration. Given that this is the case, the particular meaning of C must be understood from context. Fortunately, this is usually easy to do.

The well-mixed system

Once we move away from single component systems, there is the possibility that the components will partition themselves in different parts of the vessel due to different densities, solubilities, or miscibilities. Partitioned systems are also referred to as "distributed." That means that the properties are not everywhere the same over macroscopic length scales. To handle distributed systems, we typically have to choose a differential control volume that is an infinitesimal within the macroscopic system. We will see this when we consider plug flow down a tube.

Although partitioning is often encountered, and even though it may be advantageous in many cases, it is also true that many systems are either naturally homogeneous, or are forced to be by the action of vigorous mixing. When a system is homogeneous, it means that the density and concentration are everywhere the same throughout the control volume. This is referred to as the condition of being "well-mixed." For the math, it means that a well-mixed system can be described solely in terms of time dependence without spatial dependence. Put another way derivatives will be time derivatives and not position (x, y, z) derivatives. We turn now to problems of systems with multiple components and which are well-mixed.

Multicomponent systems

Liquid and an insoluble solid

Mixtures are combinations of two or more components that share the same volume, but retain there identity. A liquid plus an insoluble solid for example is a type of mixture that brings to mind cement or plaster. Two immiscible fluids may be mixed in way that disperses one fluid in the other, oil and water for instance. Preparing such a mixture may be done in a mixing tank. Often the mixing is done in batch mode, but it may also be done continuously in a system such as this one:

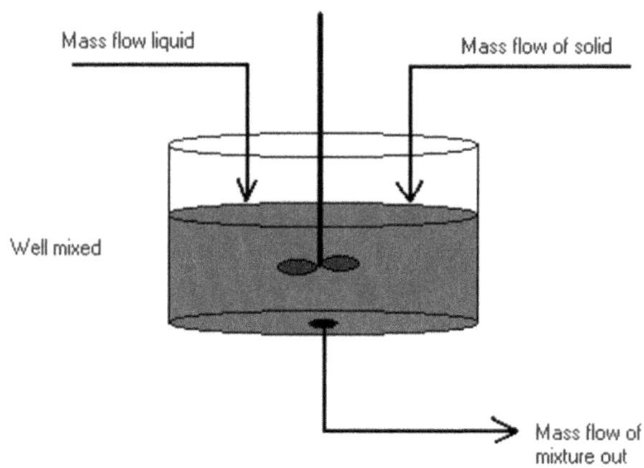

There is a mass flow of liquid into the tank and one of a solid; the two are mixed well and then flow out of the tank to their application. The total and component mass balances for the system are:

$$\text{Total:} \quad \frac{d\rho_{\text{mix}} V}{dt} = \rho_l\, q_l + \overset{\circ}{m}_s - \rho_{\text{mix}}\, q_{\text{mix}}$$

$$\text{Liquid:} \quad \frac{dC_l,\,\text{mix} V}{dt} = \rho_l\, q_l - C_{l,\text{mix}} q_{\text{mix}}$$

$$\text{Solid:} \quad \frac{dC_{s,\text{mix}} V}{dt} = \overset{\circ}{m}_s - C_{s,\text{mix}}\, q_{\text{mix}}$$

We have said that the total material balance is the sum of the component balances. Is that the case here? If so, what can it teach us? Let's check:

$$\frac{dC_l,\,\text{mix} V}{dt} + \frac{dC_{s,\text{mix}} V}{dt} = \rho_l\, q_l - C_{l,\text{mix}} q_{\text{mix}} + \overset{\circ}{m}_s - C_{s,mix}\, q_{\text{mix}}$$

$$\frac{d\left[C_l,\,\text{mix} + C_{s,\text{mix}}\right] V}{\text{dt}} = \rho_l\, q_l + \overset{\circ}{m}_s - (\, C_{l,\text{mix}} + C_{s,\text{mix}}\,)\, q_{\text{mix}}$$

For this to be equal to the total material balance, it must be true that the sum of the mass concentrations of solid and liquid are equal to the density of the mixture:

$$\frac{d\rho_{\text{mix}} V}{dt} = \frac{d\left[C_l, \text{mix} + C_{s,\text{mix}}\right] V}{\text{dt}};$$

iff

$$\rho_{\text{mix}} = \left[C_l, \text{mix} + C_{s,\text{mix}}\right]$$

This is the case. Remember the density is the sum of the masses occupying the same unit of volume. So we can see that:

$$\rho_{\text{mix}} = \frac{m_l + m_s}{V_{\text{mix}}} = \frac{m_l}{V_{\text{mix}}} + \frac{m_s}{V_{\text{mix}}} = C_{l,\text{mix}} + C_{s,\text{mix}}$$

There are really only two independent equations describing this system. We can solve these. But, before we do, it is important to inspect the component balances and the total balance for their other details - namely the consequence of the system being "well-mixed."

In the total material balance, the argument of the derivative involves the mixture density. This as we have seen is the sum of the two concentrations of the two components, it is not the density of the solid alone, nor is it that of the liquid alone. On the right-hand side of the same equation, we note that the two input terms do involve the densities of the solid and the liquid in their pure states. This is because they are being delivered to the system as pure "feeds." The outflow term, however, includes the mixture density, the same density that appears in the argument of the differential. This is critical to understand. It says that

Everywhere in the control volume, the density is the same at any time and that the material exiting the control volume also has the same density as that material in the tank.

This is the consequence of assuming the system is well-mixed. The same analysis can be made for the two component balances. They also are subject to the well-mixed assumption because they include the corresponding mixture concentrations in the differential and the outflow terms.

To solve these equations, we need to have a set of initial conditions for the system. We must decide, or know, whether the tank is initially empty, if both the solid and the liquid are added simultaneously, if the tank is initially loaded with pure liquid (or pure solid), or if the tank contains product mix from some previous production run. For the sake of this example, let's assume that we must start up the tank and the process from scratch, that is from an initially empty condition to production of a target mix. To do this we will follow three time intervals:

1. Fill the tank with liquid to a predetermined level, $h_{\max}$.

2. Feed only solid with good mixing until a target density for the mixture, ρ^*_{mix}, is reached.

3. Feed liquid with solid, and mix vigorously to maintain this density while product mix flows from the tank continuously.

We will want to know how long it will take to reach each stage for a given set of inputs such as the feed rates and the target mixture density. To get that information we need to solve the balance equations in each interval. We do that now.

Interval 1: No solid flow only liquid flow $\Rightarrow$ single component balance problem. The total material balance becomes:

$$\frac{d\rho_{\text{mix}} V}{dt} = \rho_l\, q_l + \overset{\circ}{m}_s - \rho_{\text{mix}}\, q_{\text{mix}}$$

$$\frac{d\rho_l V}{dt} = \rho_l\, q_l$$

$$\text{V(t)} = q_l \text{ t and h(t)} = \frac{q_l}{A}\text{t} \implies t_o = \frac{h_{\text{liq},o} A}{q_l}$$

Interval 2: Solid flow, no liquid flow to reach the critical density, ρ^*_{mix}, corresponding to $C^*_{l,\text{mix}}$ and $C^*_{s,\text{mix}}$. The density of the mixture will be changing with time as solid is fed to the system. We need to know how long it will take to reach the target density.

$$\frac{d\rho_{\text{mix}} V}{dt} = \overset{\circ}{m}_s$$

$$\text{V}\frac{d\rho_{\text{mix}}}{dt} + \rho_{\text{mix}}\frac{dV}{dt} = \overset{\circ}{m}_s$$

$$\text{V}\frac{d\left[C_{l,\text{mix}} + C_{s,\text{mix}}\right]}{dt} + \left[C_{l,\text{mix}} + C_{s,\text{mix}}\right]\frac{dV}{dt} = \overset{\circ}{m}_s$$

$$[\text{ V}\frac{dC_{l,\text{mix}}}{dt} + C_{l,\text{mix}}\frac{dV}{dt}\text{] + [V}\frac{dC_{s,\text{mix}}}{dt} + C_{s,\text{mix}}\frac{dV}{dt}\text{]} = \overset{\circ}{m}_s$$

$$\frac{dC_{l,\text{mix}} V}{dt} + \frac{dC_{s,\text{mix}} V}{\text{dt}} = \overset{\circ}{m}_s$$

$$\frac{dC_{l,\text{mix}} V}{dt} = 0$$

$$\frac{dC_{s,\text{mix}} V}{\text{dt}} = \overset{\circ}{m}_s$$

If we expand the total material balance equation, we get the sum of these two back. So we are unable to solve the problem. This is because we have three unknowns and only two independent equations. The unknowns are the concentrations of each component and the change in volume. Each are functions of time. We need another independent equation. To get this we need to think about what is happening more deeply.

When we add an insoluble solid to a liquid, the volume of the mixture must increase. In fact the volume must grow by the volume of solid that has been added according to Archimedes' Law.

$$V_{\text{mix}} = V_{\text{solid}} + V_{\text{liquid}} = \frac{\text{mass}_{\text{solid}}}{\rho_{\text{solid}}} + \frac{\text{mass}_{\text{liquid}}}{\rho_{\text{liquid}}}$$

During **Interval 2**, the volume of the liquid is a constant, but the volume of the solid in the mixture changes with time:

$$V_{\text{mix}}[\,t\,] = \frac{\overset{\circ}{m}_s t}{\rho_{\text{solid}}} + V_{\text{liquid}}$$

$$\frac{dV_{\text{mix}}[t]}{dt} = \frac{\overset{\circ}{m}_s}{\rho_{\text{solid}}}$$

Returning to the overall material balance equation, we find:

$$V_{\text{mix}}[\,t\,]\,\frac{d\rho_{\text{mix}}[t]}{dt} + \rho_{\text{mix}}[t]\,\frac{dV_{\text{mix}}[t]}{dt} = \overset{\circ}{m}_s$$

$$\left(\frac{\overset{\circ}{m}_s t}{\rho_{\text{solid}}} + V_{\text{liquid}}\right)\frac{d\rho_{\text{mix}}[t]}{dt} + \rho_{\text{mix}}[t]\,\frac{\overset{\circ}{m}_s}{\rho_{\text{solid}}} = \overset{\circ}{m}_s$$

This equation can be solved since it involves just one dependent variable (function) that of $\rho_{\text{mix}}[t]$. At the start of **Interval 2** the density of the "mixture" in the control volume is just that of the liquid. This provides the essential initial condition that we need for solving the problem:

Clear[ms, V]

Simplify[

DSolve$\left[\left\{\left(\frac{\dot{m}s}{\rho_s}t + V_l\right)\rho'_{\text{mix}}[t] == \dot{m}s\left(1 - \frac{\rho_{\text{mix}}[t]}{\rho_s}\right),\right.\right.$

$\left.\left.\rho_{\text{mix}}[0] == \rho_l\right\}, \rho_{\text{mix}}[t], t\right]\Big]$

$$\left\{\left\{\rho_{\text{mix}}[t] \to \frac{(t\dot{\text{m}}\text{s} + V_l\rho_l)\,\rho_s}{t\dot{\text{m}}\text{s} + V_l\rho_s}\right\}\right\}$$

We can also solve for the change in the concentration of the solid in the mixture as a function of time:

$$\frac{dC_{s,\text{mix}}[t]V_{\text{mix}}[t]}{\text{dt}} = \overset{\circ}{m}_s$$

$$V_{\text{mix}}[t]\,\frac{dC_{s,\text{mix}}[t]}{dt} + C_{s,\text{mix}}[t]\frac{dV_{\text{mix}}[t]}{\text{dt}} = \overset{\circ}{m}_s$$

$$\left(\frac{\overset{\circ}{m}_s t}{\rho_{\text{solid}}} + V_{\text{liquid}}\right)\frac{dC_{s,\text{mix}}[t]}{dt} + C_{s,\text{mix}}[t]\,\frac{\overset{\circ}{m}_s}{\rho_{\text{solid}}} = \overset{\circ}{m}_s$$

Simplify[

DSolve $\left[\left\{\left(\frac{\dot{\text{m}}\text{s}}{\rho_s}t + V_l\right)C'_{s,\text{mix}}[t] == \dot{\text{m}}\text{s}\left(1 - \frac{C_{s,\text{mix}}[t]}{\rho_s}\right),\right.\right.$

$\left.\left. C_{s,\text{mix}}[0] == 0\right\}, C_{s,\text{mix}}[t], t\right]\Big]$

$$\left\{\left\{C_{s,\text{mix}}[t] \to \frac{t\;\dot{\text{m}}\text{s}\;\rho_s}{t\;\dot{\text{m}}\text{s} + V_l\rho_s}\right\}\right\}$$

We can also solve for the change in concentration of the liquid in the mixture.

DSolve $\left[\left\{\left(\frac{\dot{\text{m}}\text{s}}{\rho_s}t + V_l\right)C'_{l,\text{mix}}[t] + \dot{\text{m}}\text{s}\;C_{l,\text{mix}}[t] == 0,\right.\right.$

$\left.\left. C_{l,\text{mix}}[0] == \rho_l\right\}, C_{l,\text{mix}}[t], t\right]$

$$\{\{C_{l,\text{mix}}[t] \to \rho_l\,(V_l\rho_s)^{\rho_s}\,(t\dot{\text{m}}\text{s} + V_l\rho_s)^{-\rho_s}\}\}$$

Our objective at this stage in the analysis was to solve for the time required to reach the target product mixture density of ρ^*_{mix}:

Simplify $\left[\textbf{Solve}\left[\rho_{\text{mix},p} == \frac{(t\;\dot{\text{m}}\text{s} + V_l\rho_l)\,\rho_s}{t\;\dot{\text{m}}\text{s} + V_l\rho_s}, t\right]\right]$

$$\left\{\left\{t \to \frac{V_l\rho_s\left(-\rho_l + \rho_{\text{mix},p}\right)}{\dot{\text{m}}\text{s}\left(\rho_s - \rho_{\text{mix},p}\right)}\right\}\right\}$$

At a given delivery rate of solid, fixed initial volume of liquid and solid density, we find that the time is:

$$\overset{*}{t} = \frac{V_l \rho_s \left(\overset{*}{\rho}_{\text{mix}} - \rho_l\right)}{\overset{\circ}{\text{ms}} \left(\rho_s - \overset{*}{\rho}_{\text{mix}}\right)}$$

We should also like to know the level in the tank of given volume and cross section, A, when this product density is reached. It is critical that we check this. In other words, the initial level of water in the tank must also be fixed by the maximum level of the target density of the mixture that can be held in the tank and mixed very well. We have an expression for the volume of the mixture in the tank as a function of time during this interval. We should divide it by A and then substitute in the time we just solved for to find the level that will be called for to achieve target density:

$$V_{\text{mix}} [\, t \,] = \frac{\overset{\circ}{m}_s t}{\rho_{\text{solid}}} + V_{\text{liquid}}$$

$$h_{\text{mix}} [\, t \,] = \frac{\overset{\circ}{m}_s t}{A \rho_{\text{solid}}} + h_{\text{liquid},o}$$

Clear[*A*]

$$\textbf{Solve}\left[h_{\text{mix}} == \frac{\overset{\circ}{m}_s t}{A \rho_{\text{solid}}} + h_{\text{liquid},o}, h_{\text{mix}}\right]$$

$$\left\{\left\{h_{\text{mix}} \to \frac{t\overset{\circ}{\text{ms}} + A \rho_{\text{solid}} h_{\text{liquid},o}}{A \rho_{\text{solid}}}\right\}\right\}$$

The new level in the tank at the end of this second interval will be:

$$\overset{*}{h}_{\text{mix}} = \frac{V_l \left(-\rho_l + \overset{*}{\rho}_{\text{mix}}\right) \rho_s}{A \left(-\overset{*}{\rho}_{\text{mix}} + \rho_s\right) \rho_s} + h_{\text{liquid},o}$$

but we know that V_l is just A $h_{\text{liquid},o}$ so the overall expression becomes:

$$\overset{*}{h}_{\text{mix}} = h_{\text{liquid},o} \left[\frac{(-\rho_l + \overset{*}{\rho}_{\text{mix}})\rho_s}{(-\overset{*}{\rho}_{\text{mix}} + \rho_s)\rho_s} + 1\right]$$

Interval 3: This interval begins when the product mixture reaches its target density, then the flow out of the tank begins. At the same time, the flow of the liquid feed must also be turned on in order to maintain the system in a steady state. We can solve for this:

$$\frac{d\rho_{\text{mix}} V}{dt} = \rho_l\, q_l + \overset{\circ}{m}_s - \rho_{\text{mix}}\, q_{\text{mix}}$$

$$0 = \rho_l\, q_l + \overset{\circ}{m}_s - \rho_{\text{mix}}\, q_{\text{mix}}$$

$$\rho_l\, q_l + \overset{\circ}{m}_s = \rho_{\text{mix}}\, q_{\text{mix}}$$

$$q_l = \frac{\rho_{\text{mix}} q_{\text{mix}} - \overset{\circ}{m}_s}{\rho_l}$$

Since the solid is insoluble in the liquid, we know that the volume flow rate of the mixture must be the sum of the volume flow rates of the two components. The volume flow rate of the solid is just the mass flow of the solid divided by its density, assuming that the latter is non-porous. Therefore we can solve for **ql** as follows:

Clear[ql, qs]

Simplify[

Solve$\left[\left\{\text{ql}==\frac{\rho_{\text{mix}}(\text{ql}+\text{qs})-\dot{\text{m}}\text{s}}{\rho_l}, \text{qs}==\frac{\dot{\text{m}}\text{s}}{\rho_s}\right\}\right.$,

{ql, qs}]]

$$\left\{\left\{\text{ql} \to \frac{\dot{\text{m}}\text{s}\,(\rho_{\text{mix}} - \rho_s)}{(\rho_l - \rho_{\text{mix}})\,\rho_s}, \text{qs} \to \frac{\dot{\text{m}}\text{s}}{\rho_s}\right\}\right\}$$

We see that the flow rate of the pure liquid should be the flow of solid into the tank multiplied by the ratio of the difference between the density of the mixture and the solid over the difference between the density of the pure liquid and the mixture:

$$\text{ql} = \frac{\dot{\text{m}}\text{s}\,(\rho_{\text{mix}} - \rho_s)}{\rho_s\,(\rho_l - \rho_{\text{mix}})} = \text{qs}\,\frac{(\rho_{\text{mix}} - \rho_s)}{(\rho_l - \rho_{\text{mix}})} \quad \text{where } \rho_l < \rho_{\text{mix}} < \rho_s$$

The density of the mixture will be greater than that of the pure liquid and less than that of the pure solid.

Let's use these solutions. The mixing will be done in a pilot-scale unit. The tank that is available is 3 meters high and has an aspect ratio of 3 : 1 : height : diameter. Its diameter (1 meter), area, and total volume can be immediately computed:

hmax = 3;

$d = N$**[hmax/3](* meter *)**

1.

$A = N\left[\pi \left(\frac{d}{2}\right)^2\right]$**(* meter^2 *)**

0.785398

Vtank = Ad **(* meter^3 *)**

0.785398

The tank volume is $\sim$ 0.8 meter^3, the cross sectional area is $\sim$ 0.8 meter^2 and the diameter is 1 meter. The density of the solid is 3 kg L^{-1}, the liquid is 1 kg L^{-1} and the target density, ρ^*_{mix}, is 2.5 kg L^{-1}. Experience indicates that the level of the mixture when settled should never rise to more than half the maximum level of the tank, to ensure that no mass leaves the tank during vigorous mixing. We can base our calculations on a mixture level of $\frac{\text{hmax}}{2}$. From this information and the parameters we can solve for the initial level of water the tank at the end of **Interval 1**:

ρmix = 2.5; (* kg/L *)

ρsolid = 3; (* kg/L *)

ρliq = 1; (* kg/L *)

hmixt = 1.5; (*meter*)

hliqmax =

$$\text{Solve}\left[\text{hmixt==hliqo}\left(\frac{(-\rho\text{liq} + \rho\text{mix})\rho\text{solid}}{(-\rho\text{mix} + \rho\text{solid})\rho\text{solid}} + 1\right), \text{hliqo}\right]\text{(*meter*)}$$

Vl = hliqmax[[1, 1, 2]] ∗ A **(* meter^3 *)**

{{hliqo → 0.375}}

0.294524

The volume of water that must be added during **Interval 1** is $\sim$ 0.3 m^3, which is the product of the area of the tank and the liquid level. The water can be fed at .25 m^3 min^{-1} so the time required to add the water is 1.2 minutes (Volume required/Volumetric flow rate):

$$\frac{\mathbf{V}}{\mathbf{ql}}$$

$$\frac{\mathbf{0.3}m^3}{\mathbf{.25}}$$

1.2 min

In **Interval 2**, solid is added to the liquid to create an initial mixture with the required density of 2.5 kg/L. The mixing of the solid with the water is best done slowly to ensure homogeneity. Therefore 10 minutes are to be allowed for **Interval 2**. Given this we can compute the mass flow of solid required from either of the following equations. We use both to verify the result. We note that the densities are in units of kg/L, whereas the volumes and levels are in units of meters, thus the need to make the units consistent arises. There are (as shown below) 1000 L per m^3. Therefore each of the densities must be multiplied by this factor to convert them to kg/m^3:

$$\frac{\mathbf{1}\ L}{\mathbf{1000\ cm^3}}\left(\frac{\mathbf{100\ cm}}{\mathbf{1}\ m}\right)^3 \ //N$$

$$\frac{1000.L}{m^3}$$

We now need to compute a value for the mass flow rate of solid $\dot{m}$s into the mixer during **Interval 2**. We know the time frame will be 10 minutes and we also know the target density of the mixture, so we can solver either of the following equations that we have derived:

$$\rho\text{mix} = \frac{(t\ \dot{m}\text{s} + \text{Vl}\ \rho\text{liq}1000)\ \rho\text{solid}}{t\dot{m}\text{s} + \text{Vl}\ \rho\text{solid}\ 1000}$$

or

$$t = \frac{1000\ \text{Vl}\ \rho\text{solid}\ 1000(\rho\text{mix} - \rho\text{liq})}{\dot{m}\text{s}\ 1000(\rho\text{solid} - \rho\text{mix})}$$

to find $\dot{m}$s as shown here:

$\mathbf{t = 10;}$

$\mathbf{Vl = 0.3;}$

$$\textbf{NSolve}\left[\mathbf{1000\rho mix{=}{=}\frac{(t\dot{m}s + Vl\rho liq\ 1000)\ \rho solid1000}{t\dot{m}s + Vl\rho solid1000}},\right.$$

$\mathbf{\dot{m}s]}$

(* kg/min *)

$$\text{NSolve}\left[==\frac{1000\text{Vl}\rho\text{solid}1000(\rho\text{mix}-\rho\text{liq})}{\dot{\text{m}}\text{s}1000(\rho\text{solid}-\rho\text{mix})},\dot{\text{m}}\text{s}\right]$$

(* kg/min *)

$\{\{\dot{\text{m}}\text{s}\to 270.\}\}$

$\{\{\dot{\text{m}}\text{s}\to 270.\}\}$

The mass flow of solid must be 270 kg min^{-1} during **Interval 2**. We can check our work to this point to be sure that at the end of the second interval we have a slurry with the target density of 2.5 kg/L.

Firstly, we compute the volume of the solid added to the mixer in **Interval 2**:

$$\frac{270\frac{\text{kg}}{\text{min}}10\,\text{min}//N}{3\frac{\text{kg}}{L}1000\frac{L}{m^3}}$$

(* Volume of the Solid Added During Interval 2 *)

$0.9m^3$

We know the volume of the liquid that was added during **Interval 1**, so the sum of the two volumes is the volume of the slurry (the mixture if solid and liquid) that has formed during the second interval:

$\mathbf{0.3m^3}$

(* Volume of the Liquid Added During Interval 1 *)

$0.3m^3$

$\mathbf{0.3m^3+0.9m^3}$

(* Total Volume of the Solid and Liquid at End of Interval 2 *)

$1.2m^3$

(* Level of the slurry in the mixer = slurry volume/tank cross sectional area *)

$$\frac{\mathbf{Vmixture}}{\boldsymbol{A}};$$

$$\frac{\mathbf{1.2}\boldsymbol{m}^3}{\mathbf{0.8}\boldsymbol{m}^2}$$

$1.5m$

The slurry volume will be 1.2 m^3 and it will be at a level of 1.5 m. Now we can check our work by computing the density of the slurry at the end of **Interval 2** of the process to be sure we have done everything correctly. Because this is a mixture, the solid is not soluble, but instead is suspended in the liquid. That means that not only are the mass of the liquid and the solid additive, but so too are their volumes as we saw above. Stated in terms of the variables the density of the slurry is:

$$\frac{\text{mass of solid } + \text{mass of liquid}}{\text{volume of solid} + \text{volume of liquid}}$$

$$\frac{\left(270\frac{\mathbf{kg}}{\mathbf{min}}10\,\mathbf{min}+0.3\boldsymbol{m}^3\frac{1000\mathbf{kg}}{\boldsymbol{m}^3}\right)}{\left(0.3\boldsymbol{m}^3+\frac{270\frac{\mathbf{kg}}{\mathbf{min}}10\,\mathbf{min}}{3\frac{\mathbf{kg}}{\boldsymbol{L}}1000\frac{\boldsymbol{L}}{\boldsymbol{m}^3}}\right)}\left(\frac{1\boldsymbol{m}^3}{1000\boldsymbol{L}}\right)$$

(* Mass Solid + Liquid /Total Volume *)

$$\frac{2.5\text{kg}}{L}$$

This number checks. It is the target density of the mixture and that confirms that we have the right quantities, flows and times to this point.

Our attention now turns to **Interval 3**, which will be the steady state production of slurry. At steady state we would like to be producing 1000 kg min^{-1} = $\rho^*_{\text{mix}}\, q^*_{\text{mix}}$ of slurry at the density of 2.5 kg/L. This would correspond to a volume flow rate equal to 400 L/min or 0.4 m^3/min slurry. The steady state material balance can be used to find the required solids mass flow rate need to achieve this production rate:

$$\rho_l\, q_l + \overset{\circ}{m}_s = \rho_{\text{mix}}\, q_{\text{mix}}$$

$$q_l = \frac{\rho_{\text{mix}} q_{\text{mix}} - \overset{*}{m}_s}{\rho_l}$$

Since the maximum water flow rate is 0.25 m^3 min^{-1}, we can compute the solids flow rate:

ρliq = 1;

ql = 0.25;

Solve[1000ρliqql + ṁs == 1000, ṁs]

{{ṁs → 750.}}

To make the production rate we seek, the solids must be flowed into the unit at a rate of 750 kg min^{-1}.

We can check the steady state quantities by recomputing the volume flow a rate of slurry from the mass flow rate solids, the volume flow rate of water and the target density of the product:

$$\frac{750\frac{\text{kg}}{\text{min}} + .25\frac{m^3}{\text{min}}\frac{1000\text{kg}}{m^3}}{2500\frac{\text{kg}}{m^3}}$$

(* Volume Flow of Slurry at St .St . *)

$$\frac{0.4m^3}{\text{min}}$$

This checks perfectly with the previously computed value.

A good question to ask at this point might be how would we transition from the end of **Interval 2** into the steady state of **Interval 3**? How would we program the increase in mass flows that would have to take place in order to maintain the product density and to maintain the steady level in that tank?

Liquid and soluble solid

A more complex problem is that of a soluble solid, a solute, and a liquid solvent. The physical situation is the same as in the previous problem, but we cannot assume the volumes of the solid and the liquid are additive. The initial equations are also the same except that now we are dealing with a solution rather than with a mixture:

$$\text{Total:} \quad \frac{d\rho_{\text{soln}} V}{dt} = \rho_l\, q_l + \overset{\circ}{m}_s - \rho_{\text{soln}}\, q_{\text{soln}}$$

$$\text{Liquid = Solvent:} \quad \frac{dC_l V}{dt} = \rho_l\, q_l - C_{l,} q_{\text{soln}}$$

$$\text{Solid = Solute:} \quad \frac{dC_{s,\text{soln}}V}{dt} = \overset{\circ}{m}_s - C_{s,\text{soln}}\, q_{\text{soln}}$$

Before we go on to solve these equations, let's review them term by term to be sure we understand them and what they are stating. The total material balance is:

$$\text{Total:} \quad \frac{d\rho_{\text{soln}}V}{dt} = \rho_l\, q_l + \overset{\circ}{m}_s - \rho_{\text{soln}}\, q_{\text{soln}}$$

The first term involves the mass of the solution in the control volume. It is given as the product of the density of the solution and the volume of the solution, $\rho_{\text{soln}}V$. The time derivative of this product is a rate and it is the rate of change of solution mass in the system. There are no derivatives with respect to spatial variables (x, y, z) because the system is well-mixed, all the dependent variables are the same at any time at every position. On the right-hand side the first term is the product of the liquid (solvent) density and its volumetric flow rate into the unit, $\rho_l\, q_l$. The dimensions of this product are mass per unit time. It is the rate at which liquid flows into the unit. Solid also flows into the unit. Its rate of entry is stated simply as $\overset{\circ}{m}_s$, which is meant to indicate mass of solid entering per unit time. Lastly, there is the term $\rho_{\text{soln}}\, q_{\text{soln}}$ and this is the term for product leaving the unit. The product is that of the solution density and flow rate. In this case the solid is fully dissolved. This is also a rate, it is the rate of solution leaving the unit in mass per unit time dimensions. It is important to note that the density of the solution, ρ_{soln}, appears both in the derivative on the left and the exit rate on the right. This is also a consequence of the well-mixed assumption. We are saying that for the purposes of building a model to simulate the real system, that the system is so well-mixed that the solid being added to the liquid dissolves instantaneously so that the density of the solution is the density in the unit and the density in the exit stream. This is an approximation. If it leads to a model that is inaccurate when compared to experimental data, we can go back to the model, add a term for the rate of dissolution and improve the model iteratively as needed. For now however, we will make the assumption.

The component balances look as follows:

$$\text{Solvent:} \quad \frac{dC_lV}{dt} = \rho_l\, q_l - C_{l,}q_{\text{soln}}$$

$$\text{Solute:} \quad \frac{dC_{s,\text{soln}}V}{dt} = \overset{\circ}{m}_s - C_{s,\text{soln}}\, q_{\text{soln}}$$

Taking the liquid (l) solvent balance first - we see that on the left-hand side is the time derivative of the product of concentration of the liquid in the solution and the volume of that solution, C_lV. Due to the system being well-mixed there are no spatial variables. On the right-hand side the first term is the rate of liquid solvent flowing into the system, given as $\rho_l\, q_l$, and the rate of solvent in solution leaving the system, given as $C_{l,}q_{\mathrm{soln}}$. Note again the concentration of liquid leaving the system in solution is the same as the concentration of liquid in solution within the unit itself. This is how the well-mixed assumption manifests itself in the component balance.

The component balance for the solid that will become the solute in the system gives the time rate of change in the solute mass on the left-hand side, that is the time derivative of the mass of solute given as the product of the concentration of solute in solution and the volume of the solution, $C_{s,\mathrm{soln}}V$. The right-hand side gives the rate of solute mass flow into the unit minus the mass flow rate of the solute out of the unit in the solution, i.e. $C_{s,\mathrm{soln}}\, q_{\mathrm{soln}}$. Here too the consequence of the well-mixed assumption is that the concentration of solute in solution is the same within the unit as it is in the product stream exiting the unit.

What is the difference between this case and the case in the previous section? The difference is solubility of the solid. In the case of the insoluble solid, a mixture is formed, but not a solution. When liquid and an insoluble solid are mixed, the volume of the solid and the liquid may be added, in other words, mixing does not change their volumes. When a solid such as a salt is added to a solvent, such as water, the volumes are not conserved. The volume of the liquid may increase with the addition of the solute, decrease, or stay the same. Understanding solutions, and their behavior so that predictions can be made is a topic in physical chemistry.

Case 1. Constant densities

If the concentration of the soluble solid does not reach too high a level in the solution, then experiment shows that it is reasonable to assume that the densities of the pure solvent and the solution are similar enough to treat as equal and constant. Doing this transforms the total material balance into:

$$\text{if } \rho_{\mathrm{soln}} = \rho_l, \text{ then } \quad \frac{dV}{dt} = q_l + \frac{\overset{\circ}{m}_s}{\rho_{\mathrm{soln}}} - q_{\mathrm{soln}}$$

This can be integrated immediately if the exit flow rate is a constant:

$$\mathrm{V}[\,\mathrm{t}\,] = \mathrm{Vo} + \left[\, q_l + \frac{\overset{\circ}{m}_s}{\rho_{\mathrm{soln}}} - q_{\mathrm{soln}} \,\right] \mathrm{t}$$

Looking at this equation, we see that it is linear with time; the volume in the tank will increase as time increased. If we left the system this way, it would overflow, which would be undesirable. If the system is to come to a steady state before it overflows, we would have to set the maximum volume. If we do that then when we do the integration we would integrate from Vo to Vmax, and the time to reach Vmax would be computed.

Alternatively, if we look at the equation and in particular at the part that is in the brackets, we see that it is the sum of the inlet liquid flow rate, the inlet mass flow rate minus the outlet flow rate of the solution. For the system to reach a steady state in volume, at some point this difference has to be zero or else the system will continue to increase in volume until it overflows.

Let's assume then that the difference between $(q_l + \frac{\overset{\circ}{m}_s}{\rho_{\text{soln}}})$ and q_{soln} is close to zero by design and that V[t] = Vo. This is a reasonable set of assumptions. Then the only transient, i.e. time dependence is in the evolution to steady state in the solution concentration, because at time = zero the liquid in the tank is pure liquid with a volume Vo. It takes time for this liquid to be replaced by the solution at its full concentration.

The component balances can be integrated. The initial condition is that the volume in the tank is Vo of pure solvent and the concentration of the solid is zero at t = 0. To find the analytical solutions to these equations, we specify V [t] = Vo, then we use DSolve to simultaneously solve for the concentrations, calling the set of solutions to the equations "a". Two functions are named, and then extracted from the solution set and assigned to these names. Finally, the two new functions are placed back into the original differential equations and tested for validity.

```
(* Remove all the variable names that will be used to be sure they are not
assigned from previous work. *)

Remove[a, V, cs, cl, ρ, ql, qsoln, t, Vo, cs1, cl1, ms];

(* This specifies V[t] *)

V[t_]:=Vo

(* These are the two component balances to be solved. *)


a = FullSimplify[DSolve[{

∂t(cs[t]V[t]) == ṁs − cs[t]qsoln, cs[0] == 0,

∂t(cl[t]V[t]) == ρql − cl[t]qsoln, cl[0] == ρ

},{cs[t], cl[t]}, t]]

(* a is the set of two solutions. We extract them and assign them to
 functions as follows. *)

cs1[t_]:=Evaluate[cs[t]/.a]

cl1[t_]:=Evaluate[cl[t]/.a]
```

$$\left\{\left\{\text{cs}[t]\to\frac{\left(1-e^{-\frac{\text{qsoln}t}{\text{Vo}}}\right)\dot{\text{m}}\text{s}}{\text{qsoln}},\ \text{cl}[t]\to\frac{\left(\text{ql}+e^{-\frac{\text{qsoln}t}{\text{Vo}}}(-\text{ql}+\text{qsoln})\right)\rho}{\text{qsoln}}\right\}\right\}$$

The two solutions for the concentration of liquid and the concentration of solute from the script above are as follows:

$$\text{cl}[t]=\frac{\left(1-e^{-\frac{\text{qsoln}t}{\text{Vo}}}\right)\dot{\text{m}}\text{s}}{\text{qsoln}}$$

$$\text{cs}[t]=\frac{\left(\text{ql}+e^{-\frac{\text{qsoln}t}{\text{Vo}}}(-\text{ql}+\text{qsoln})\right)\rho}{\text{qsoln}}$$

How do we know if they are correct? One way to determine that is to put them back into the equations and to let *Mathematica* determine if they are consistent with the equations from which they emerged.

Simplify $\left[\partial_t(\text{cl1}[t]V[t])==\rho\text{ql}-\text{cl1}[t]\text{qsoln}\right]$

Simplify $\left[\partial_t(\text{cs1}[t]V[t])==\dot{\text{m}}\text{s}-\text{cs1}[t]\text{qsoln}\right]$

True

True

These two results show that the solutions are valid, *Mathematica* returns true in both cases meaning the left and right sides are equal in both equations.

Parameter values are applied and the functions are plotted in time:

Clear[ρ, ql, qsoln, t, Vo]

ρ = 1;

Vo = 100;

ṁs = 2;

ql = 10;

tmax = 100;

qsoln = 12.;

V[t] = Vo

Plot[{cs1[t], cl1[t]}, {t, 0, tmax},

PlotStyle → {Hue[0], Hue[0.5]},

AxesLabel → {"t/min", "cl[t],cs[t]"},

PlotRange → {{0, tmax}, {0, 1.0}}, Frame → True]

Plot[$V[t]$, {t, 0, tmax}, PlotStyle → Hue[0.6],

AxesLabel → {"t/min", "V[t]"}, Frame → True]

100

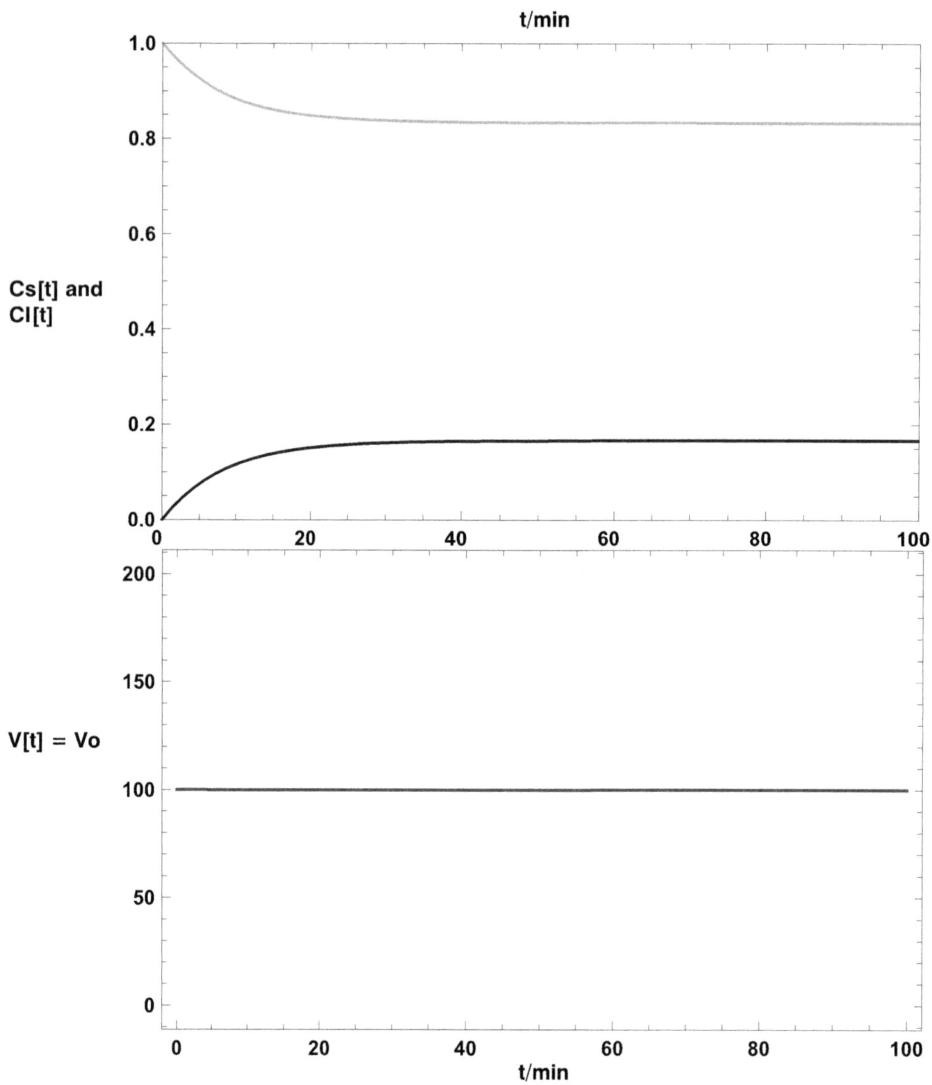

The solutions show two important aspects of this model as written: the concentrations come to a constant value and the volume remains constant indefinitely, a result that we forced.

If we choose to let the volume increase with time, we can recompute the solutions to the component balance equations by just adding in the equation for V [t]:

```
Remove[a1, V, Vo, cs, cs1, cl, cl1, ρ, ql, ms, qsoln];
```

$$V[\text{t\_}]:=\text{Vo}+t\left(q_l+\frac{\dot{\text{ms}}}{\rho_{\text{soln}}}-q_{\text{soln}}\right)$$

a1 = FullSimplify[DSolve[{

$$\partial_t(\text{cs}[t]V[t]) == \dot{\text{ms}}-\text{cs}[t]\text{qsoln}, \text{cs}[0]==0,$$

$$\partial_t(\text{cl}[t]V[t]) == \rho\text{ql}-\text{cl}[t]\text{qsoln}, \text{cl}[0]==\rho\}, \{\text{cs}[t], \text{cl}[t]\}, t]];$$

cs1[t_]:=Evaluate[cs[*t*]/.a1]

cl1[t_]:=Evaluate[cl[*t*]/.a1]

The two solutions obtained when we let the volume in the unit vary with time are complicated looking:

$$\text{cs1}[t] = \{(\dot{\text{ms}}\rho_{\text{soln}}(t\dot{\text{ms}}+(\text{Vo}+tq_l-tq_{\text{soln}})\ \rho_{\text{soln}})^{-\frac{\dot{\text{ms}}+(\text{qsoln}+q_l-q_{\text{soln}})\rho_{\text{soln}}}{\dot{\text{ms}}+(q_l-q_{\text{soln}})\rho_{\text{soln}}}}$$
$$\left(-(\text{Vo}\rho_{\text{soln}})^{1+\frac{\text{qsoln}\rho_{\text{soln}}}{\dot{\text{ms}}+(q_l-q_{\text{soln}})\rho_{\text{soln}}}}+(t\dot{\text{ms}}+(\text{Vo}+tq_l-tq_{\text{soln}})\ \rho_{\text{soln}})^{1+\frac{\text{qsoln}\rho_{\text{soln}}}{\dot{\text{ms}}+(q_l-q_{\text{soln}})\rho_{\text{soln}}}}\right)\Big)\Big/$$
$$(\dot{\text{ms}}+(\text{qsoln}+q_l-q_{\text{soln}})\ \rho_{\text{soln}})\}$$

$$\text{cl1}[t] = \{((t\dot{\text{ms}}+(\text{Vo}+tq_l-tq_{\text{soln}})\ \rho_{\text{soln}})^{-\frac{\dot{\text{ms}}+(\text{qsoln}+q_l-q_{\text{soln}})\rho_{\text{soln}}}{\dot{\text{ms}}+(q_l-q_{\text{soln}})\rho_{\text{soln}}}}$$
$$\Big(\rho\ (\text{Vo}\rho_{\text{soln}})^{\frac{\dot{\text{ms}}+(\text{qsoln}+q_l-q_{\text{soln}})\rho_{\text{soln}}}{\dot{\text{ms}}+(q_l-q_{\text{soln}})\rho_{\text{soln}}}}\ (\dot{\text{ms}}+(-\text{ql}+\text{qsoln}+q_l-q_{\text{soln}})\ \rho_{\text{soln}})\ +$$
$$\text{ql}\rho\rho_{\text{soln}}(t\dot{\text{ms}}+(\text{Vo}+tq_l-tq_{\text{soln}})\ \rho_{\text{soln}})^{1+\frac{\text{qsoln}\rho_{\text{soln}}}{\dot{\text{ms}}+(q_l-q_{\text{soln}})\rho_{\text{soln}}}}\Big)\Big)\Big/$$
$$(\dot{\text{ms}}+(\text{qsoln}+q_l-q_{\text{soln}})\ \rho_{\text{soln}})\}$$

Checking the solutions to learn if they are correct is best done computationally using the original differential equations and the new function forms of the solutions. If the solutions are valid, the derivative of the product on the left will be equal to the difference on the right once the new functions are substituted in for cs1[t] and for cl1[t]. If this is the case then result will be the logical result "True" for both cases. In fact that is what we see.

```
Simplify[∂t(cs1[t]V[t]) == ṁs − cs1[t]qsoln]
Simplify[∂t(cl1[t]V[t]) == ρql − cl1[t]qsoln]
```

True

True

Another case is one in which the outlet flow rate is given by Torricelli's Law, i.e. qsoln = b Ao$\sqrt{2gh[t]}$:

$$\frac{dV}{dt} = q_l + \frac{\overset{\circ}{m}_s}{\rho} - \text{b Ao}\sqrt{2gh[t]}$$

$$\frac{dh[t]}{dt} = \frac{q_l}{A} + \frac{\overset{\circ}{m}_s}{\rho \text{A}} - \text{b}\,\frac{\text{Ao}}{A}\sqrt{2gh[t]}$$

For this case, because Torricelli's Law is non-linear, we solve the equations using NDSolve and the results are given as interpolating functions:

```
Remove[a2, ρ, ql, A, ms, ρ, b, Ao, g, ho, tmax, Vo, V, t];
ql = 10;
A = 10;
ṁs = 5;
ρ = 1;
b = 1;
Ao = 0.1A; (* These are parameter values *)
g = 9.8;
ho = 10;
tmax = 100;
Vo = 100;

(* The differential equations to be solved *)

a2 = NDSolve[{
```

$$\partial_t h[t]==\frac{1}{A}\left(\text{ql}+\frac{\overset{\circ}{m}_s}{\rho}-bAo\sqrt{2gh[t]}\right),$$

$$h[0]==\text{ho},$$

$$\partial_t(\text{cs}[t]Ah[t])==\frac{\overset{\circ}{m}_s}{\rho}-\text{cs}[t]b\text{Ao}\sqrt{2gh[t]},$$

$$\text{cs}[0]==0,$$

$$\partial_t(\text{cl}[t]Ah[t])==\rho\text{ql}-\text{cl}[t]b\text{Ao}\sqrt{2gh[t]},$$

$$\text{cl}[0]==\rho$$

},{*h*[*t*], cs[*t*], cl[*t*]},{*t*, 0, tmax}]

```
{{h[t] →
  InterpolatingFunction[ Domain: {{0., 100.}} Output: scalar ][t],
 cs[t] → InterpolatingFunction[ Domain: {{0., 100.}} Output: scalar ][
  t], cl[t] →
  InterpolatingFunction[ Domain: {{0., 100.}} Output: scalar ][t]}}
```

The following equations assign function names to the interpolating functions from the solution, a2.

hn[t_]:=Evaluate[*h*[*t*]/.a2[[1]]]

Vn[t_]:=*A*hn[*t*]

csn[t_]:=Evaluate[cs[*t*]/.a2[[1]]]

cln[t_]:=Evaluate[cl[*t*]/.a2[[1]]]

Now, we can look at the variables for level, volume and concentrations of the solute and the solvent. First, if we plot the level in the tank, we note that it rises by about 15% as system goes to steady state.

Plot[{hn[*t*]/ho}, {*t*, 0, tmax}]

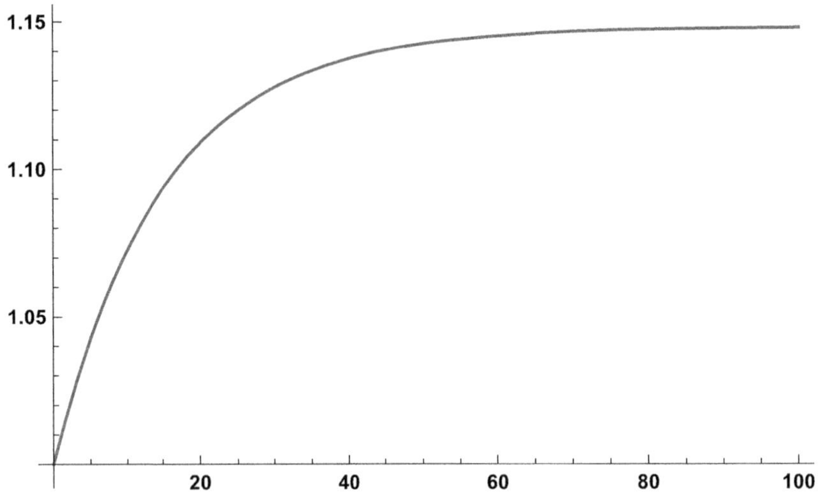

Another way to look at this is to divide the level and the volume in the unit by the initial values. This makes the variables dimensionless:

pla2V = Plot[{hn[*t*]/ho, Vn[*t*]/Vo}, {*t*, 0, tmax},

AxesLabel → {"t/min", "hn[t]/ho,Vn[t]/Vo"},

PlotStyle → {Hue[0.2], Hue[.6]},

PlotRange → {{0, tmax}, {0, 1.5}}]

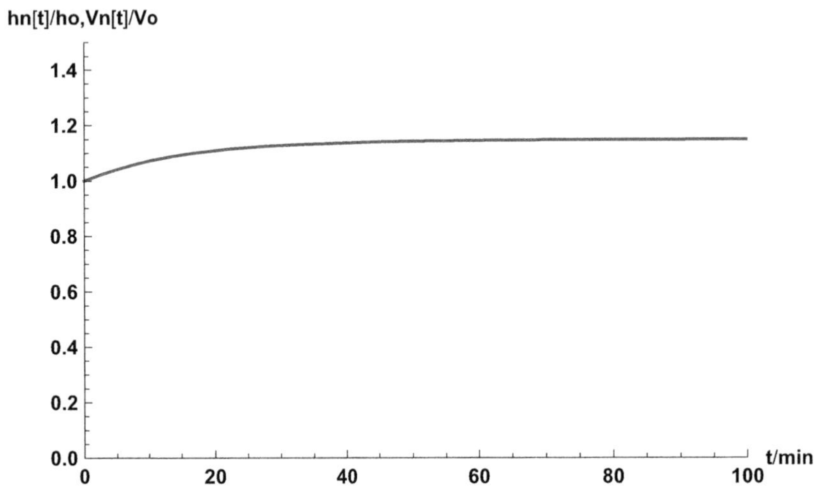

We observe that the change overall is small and that there is only one graph shown. This is because in non-dimensionalized form the change in level and volume with time is the same.

Next, we plot the change in solvent and solute concentration. These are shown below. The solute is in blue (dark gray in the print version) and the solvent is in red (mid gray in the print version). We notice that as the solute is added to the unit its concentration rises and then attains a steady state. At the same time, the concentration of the solvent in solution decreases in proportion to the solute increase. Both are smooth monotonic changes on the course to a steady state. Production of the solution begins at the steady or just after the attainment of the steady state at about 40 minutes.

```
pla2C = Plot[{cln[t], csn[t]}, {t, 0, tmax},
PlotStyle → {Hue[0], Hue[0.6]},
AxesLabel → {"t/min", "csn[ t ], cln[ t ]"},
PlotRange → {{0, tmax}, {0, 1.0}}]
```

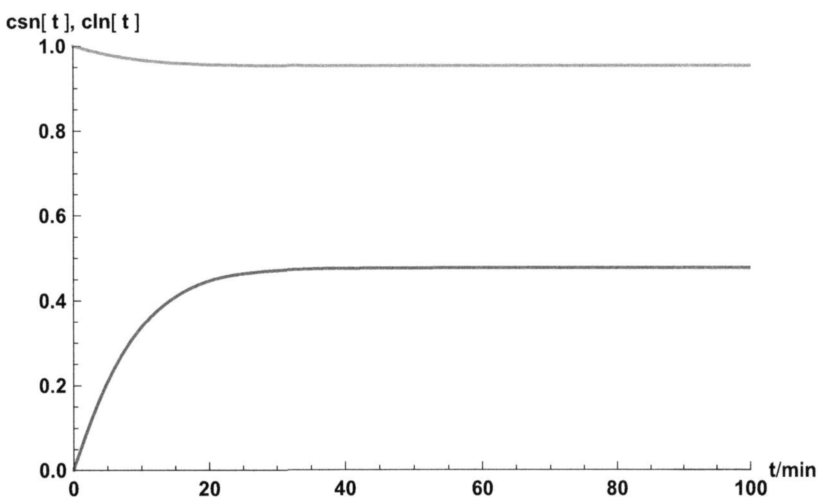

So, we have solved the full problem with gravity-driven flow. We see that the concentrations transition smoothly once again to steady state values, but now, so too, do the level and volume of liquid in the tank. The assumption that the density of solution does not change very much with concentration is quite restrictive. Therefore in the next section we deal with this problem explicitly.

Case 2. Variable densities ***The 'how and why' of variable density***

Assuming that the densities were all similar in magnitude was a restrictive simplification on the solution we derived. We can derive the solution again without this assumption. But, we need a relationship to functionally couple the density of the solution to the solute concentration. A suitable expression can be found by consulting either the CRC Handbook of Chemistry and Physics, or Perry's Handbook for data relating the concentration of various solutions of salts to their densities. From an analysis of these data we would find that the density of a solution is linearly related to the concentration of that salt over a wide range of concentrations. In their text, An Introduction to Chemical Engineering Analysis, Russell and Denn recognized this and expressed that function as:

$$\rho = \rho_o + \mathrm{b}\,\mathrm{C}$$

With some salts the volume of the solution expands as their concentration increases, this leads to a value of the constant b, which is less than unity. This tells us that the salt dissolves into the solvent and causes the volume expansion. In other cases the opposite occurs and the density increases, thus the value of b is larger than unity. If b were unity, then that would indicate that the salt dissolved in a given volume of solvent leaving the solvent volume unchanged.

These observations and the linear relationship they lead to can be rationalized by considering the definition of the density of a solution. Consider the density of a salt solution; it is the sum of the mass of the salt and the mass of the solvent divided by the total volume of the solution. This means that the density can be seen as a function of the mass of the salt in solution:

$$\rho[\,m_{\text{salt}}] = \frac{m_{\text{solvent}} + m_{\text{salt}}}{V_{\text{solution}}}$$

In the limit, when the mass of salt in solution is zero, the density of the solution is the density of the solvent. We used this concept in the approximation we made earlier:

$$\rho\,[\,0\,] = \frac{m_{\text{solvent}}}{V_{\text{solvent}}}\,(\,m_{\text{salt}} \rightarrow 0\,;\,V_{\text{solution}} \rightarrow V_{\text{solvent}}) = \rho_{\text{solvent}}$$

If the mass of dissolved salt causes no change in volume beyond that which would occur by adding the same mass of solvent, then the expression becomes:

$$\rho = \rho_{\text{solvent}} + \frac{m_{\text{salt}}}{V_{\text{solution}}} = \rho_{\text{solvent}} + C_{\text{salt}}\;;\;\text{iff}\;\;V_{\text{solution}} = V_{\text{solvent}}$$

The implication is that the constant "b" has value of unity. This can only be true if the salt simply adds to the solution and occupies no more or less space than the solvent molecules, as we can see from the equation.

We can rearrange this expression taking δV = Vsolution - Vsolvent and we find:

$$b = 1 - \left[\frac{\text{msolvent}}{\text{msalt}}\right]\left[\frac{\delta\text{V}}{\text{Vsolvent}}\right]$$

So in the case of a value of b equal to unity, there is no excess change, i.e. $\delta V = 0$. This expression for b is a dimensionless grouping which offers some insight into what this constant really means physically.

If the salt causes restructuring of the solvent by attracting solvent to its ions in ensembles that have higher (or lower) numbers of molecules per unit volume, that implies non-ideality, $\delta V \neq 0$ and b is no longer unity. Such non-idealities are handled by re-expressing in terms of activity rather than concentration:

$$\rho = \rho_{\text{solvent}} + a_{\text{salt}}$$

Activity is the "effective" concentration. It takes into account the fact that not all salts or solutes will behave as if their colligative properties were simple and based solely on the number of molecules present per unit volume. Instead, at a given concentration, the non-ideal solution acts as though there were more, or fewer molecules, per unit volume than really are there based upon the actual mass dissolved. The degree to which the activity and concentration deviate is captured in their ratio, which is termed the activity coefficient:

$$\gamma = \frac{a}{C}$$

We can substitute this into the statement for density and we find:

$$\rho = \rho_{\text{solvent}} + \gamma_{\text{salt}} C_{\text{salt}}$$

This expression has the same form as the empirical formula and it gives meaning to the terms for the intercept and the slope.

We can see the effect of the non-idealities with three cases of γ values (1.1, 0.9, and 1) in the graphs that follows:

```
Clear[ρ, ρ1, ρsol, γ1, γ2, c]

ρ1[c_]:=ρsol + γ1c

ρ2[c_]:=ρsol + γ2c

ρ[c_]:=ρsol + c
```

```
γ1 = 1.1;
γ2 = .9;
ρsol = 1;
Plot[{ρ1[c], ρ[c], ρ2[c]}, {c, 0, 1},
PlotStyle → {Hue[0.4], Hue[.6], Hue[0]},
PlotLabel → {"green:ρ1; blue:ρ; red:ρ2"},
Frame → True]
```

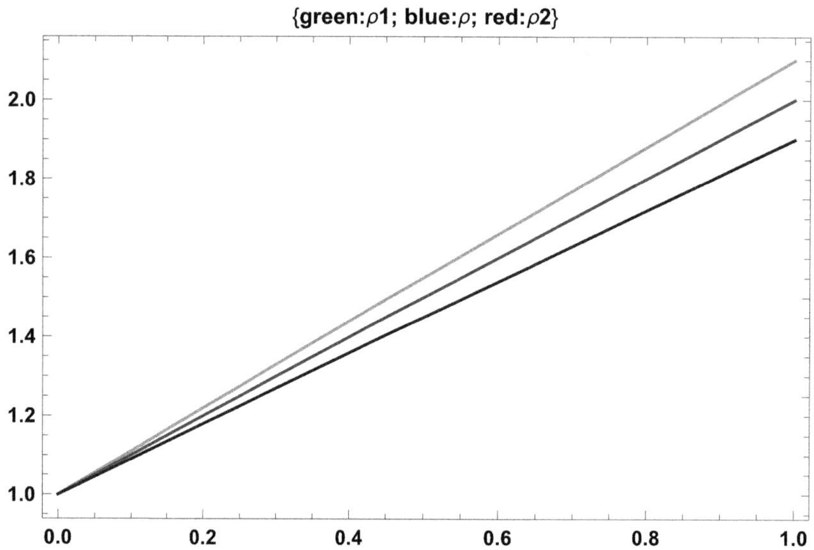

It is of interest to diverge at this point to examine the properties of "simple" electrolyte solutions for two reasons: firstly, doing so will give us some deeper insights into the thermodynamics and physical chemistry of these systems and the kinds of approximations that we use in engineering to describe them and, secondly, so that we may introduce more of the data analysis functions that are available in *Mathematica*.

That there is a linear relationship between the concentration of solute and the density of the solution, is intuitive. However, in the absence of a deeper understanding and model for the intermolecular forces, we could consider other functional forms. With that in mind we recognize that an empirical fit of data might include higher order dependence. Hence, a quadratic relationship of concentration of solute to solution density may be viable. We will examine this in the next section.

Density as a function of concentration

The data given here is for sodium chloride in water. These were taken from the CRC Handbook of Chemistry and Physics. An array of mass weight percentages of NaCl (gm NaCl per 100 gm of water) from 0.10 to 26 wt% and the corresponding solution densities are given as the set "**densconcdata**":

densconcdata = {{0.10, 1.0007}, {0.20, 1.0014},

{0.30, 1.0021}, {0.40, 1.0028}, {0.50, 1.0035},

{0.60, 1.0042}, {0.70, 1.0049}, {0.80, 1.0056},

{0.90, 1.0063}, {1.00, 1.0071}, {1.10, 1.0078},

{1.20, 1.0085}, {1.30, 1.0093}, {1.40, 1.010},

{1.50, 1.0107}, {1.60, 1.0114}, {1.70, 1.0121},

{1.80, 1.0128}, {1.90, 1.0135}, {2.00, 1.0143},

{2.10, 1.0150}, {2.20, 1.0157}, {2.30, 1.0164},

{2.40, 1.0171}, {2.50, 1.0178}, {2.60, 1.0185},

{2.70, 1.0193}, {2.80, 1.0200}, {2.90, 1.0207},

{3.00, 1.0214}, {3.10, 1.0221}, {3.20, 1.0229},

{3.30, 1.0236}, {3.40, 1.0243}, {3.50, 1.0250},

{3.60, 1.0257}, {3.70, 1.0265}, {3.80, 1.0272},

{3.90, 1.0279}, {4.00, 1.0286}, {4.10, 1.0293},

{4.20, 1.0301}, {4.30, 1.0308}, {4.40, 1.0315},

{4.50, 1.0322}, {4.60, 1.0330}, {4.70, 1.0337},

{4.80, 1.0344}, {4.90, 1.0351}, {5.00, 1.0358},

{5.20, 1.0373}, {5.40, 1.0387}, {5.60, 1.0402},

{5.80, 1.0417}, {6.00, 1.0431}, {6.20, 1.0446},

{6.40, 1.0460}, {6.60, 1.0475}, {6.80, 1.0490},

{7.00, 1.0504}, {7.20, 1.0519}, {7.40, 1.0533},

{7.60, 1.0548}, {7.80, 1.0563}, {8.00, 1.0578},

{8.20, 1.0592}, {8.40, 1.0607}, {8.60, 1.0622},

{8.80, 1.0637}, {9.00, 1.0651}, {9.20, 1.0666},

{9.40, 1.0681}, {9.60, 1.0696}, {9.80, 1.0711},

{10.00, 1.0726}, {10.50, 1.0763}, {11.00, 1.0801},

{11.50, 1.0838}, {12.00, 1.0876}, {12.50, 1.0914},

{13.00, 1.0952}, {13.50, 1.0990}, {14.00, 1.1028},

{14.50, 1.1066}, {15.00, 1.1105}, {16.00, 1.1182},

{17.00, 1.1260}, {18.00, 1.1339}, {19.00, 1.1418},

{20.00, 1.1498}, {21.00, 1.1579}, {22.00, 1.1660},

{23.00, 1.1742}, {24.00, 1.1825}, {25.00, 1.1909},

{26.00, 1.1993}};

Length[densconcdata]

96

We note several things about this data. Firstly, it is very complete with small variations between points at the low end of the weight percentages of salt. Secondly, the data are reported to two decimal places in the weight percent of salt and to four places after the decimal in density.

There are several ways in which we can convert this data set to one that includes the weight fraction of salt and the density instead of the weight percent and the density, one method is illustrated below. We rename the converted data set to "***newdata***."

newdat =

$$\textbf{Table}\left[\left\{\frac{\textbf{densconcdata[[}n\textbf{, 1]]}}{\textbf{100}}\textbf{, densconcdata[[}n\textbf{, 2]]}\right\}\textbf{,}\right.$$

{*n*, 1, Length[densconcdata]}]

{{0.001, 1.0007}, {0.002, 1.0014}, {0.003, 1.0021}, {0.004, 1.0028}, {0.005, 1.0035},
{0.006, 1.0042}, {0.007, 1.0049}, {0.008, 1.0056}, {0.009, 1.0063}, {0.01, 1.0071},
{0.011, 1.0078}, {0.012, 1.0085}, {0.013, 1.0093}, {0.014, 1.01}, {0.015, 1.0107},
{0.016, 1.0114}, {0.017, 1.0121}, {0.018, 1.0128}, {0.019, 1.0135}, {0.02, 1.0143},

{0.021, 1.015}, {0.022, 1.0157}, {0.023, 1.0164}, {0.024, 1.0171}, {0.025, 1.0178},
{0.026, 1.0185}, {0.027, 1.0193}, {0.028, 1.02}, {0.029, 1.0207}, {0.03, 1.0214},
{0.031, 1.0221}, {0.032, 1.0229}, {0.033, 1.0236}, {0.034, 1.0243}, {0.035, 1.025},
{0.036, 1.0257}, {0.037, 1.0265}, {0.038, 1.0272}, {0.039, 1.0279}, {0.04, 1.0286},
{0.041, 1.0293}, {0.042, 1.0301}, {0.043, 1.0308}, {0.044, 1.0315}, {0.045, 1.0322},
{0.046, 1.033}, {0.047, 1.0337}, {0.048, 1.0344}, {0.049, 1.0351}, {0.05, 1.0358},
{0.052, 1.0373}, {0.054, 1.0387}, {0.056, 1.0402}, {0.058, 1.0417}, {0.06, 1.0431},
{0.062, 1.0446}, {0.064, 1.046}, {0.066, 1.0475}, {0.068, 1.049}, {0.07, 1.0504},
{0.072, 1.0519}, {0.074, 1.0533}, {0.076, 1.0548}, {0.078, 1.0563}, {0.08, 1.0578},
{0.082, 1.0592}, {0.084, 1.0607}, {0.086, 1.0622}, {0.088, 1.0637}, {0.09, 1.0651},
{0.092, 1.0666}, {0.094, 1.0681}, {0.096, 1.0696}, {0.098, 1.0711}, {0.1, 1.0726},
{0.105, 1.0763}, {0.11, 1.0801}, {0.115, 1.0838}, {0.12, 1.0876}, {0.125, 1.0914},
{0.13, 1.0952}, {0.135, 1.099}, {0.14, 1.1028}, {0.145, 1.1066}, {0.15, 1.1105},
{0.16, 1.1182}, {0.17, 1.126}, {0.18, 1.1339}, {0.19, 1.1418}, {0.2, 1.1498},
{0.21, 1.1579}, {0.22, 1.166}, {0.23, 1.1742}, {0.24, 1.1825}, {0.25, 1.1909}, {0.26, 1.1993}}

We can compare a linear relationship between density and mass fraction with a quadratic fit to the density, i.e.,

$$\rho_{\text{linear}} = \text{b (wt fr salt)} + \text{c}$$

$$\rho_{\text{quadratic}} = a_2(\text{wt fr salt})^2 + b_2(\text{wt fr salt}) + c_2$$

The command Fit allows us to do this in one step. Then, we assign functions to the results and Plot them versus the actual data for NaCl. To do so we use the following three commands:

(* linear model *)

Clear[*l*, line]

***l* = LinearModelFit[newdat, *x*, *x*]**

Normal[*l*]

line[x_]:=Evaluate[Normal[*l*]]

FittedModel[]

$0.998469 + 0.75486x$

(* quadratic model *)

Clear[*a*, *b*, *c*, *q*, quad]

```
q = NonlinearModelFit[newdat, {c + bx + ax^2},
{a, b, c}, x]
Normal[q]
quad[x_]:=Evaluate[Normal[q]]
```

FittedModel []

$1.00017 + 0.699006x + 0.250775x^2$

```
fitplot1 =
Plot[line[x], {x, 0, newdat[[Length[newdat]]][[1]]},
PlotStyle → {Hue[0.2], Hue[0.6]},
AxesLabel → {"ω", "ρ"}]
```

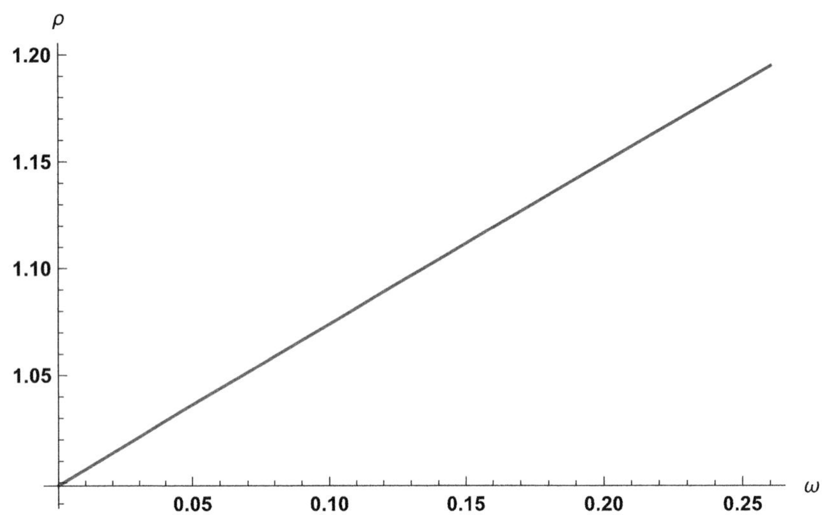


```
fitplot2 = Plot[{quad[x]},
{x, 0, newdat[[Length[newdat]]][[1]]},
PlotStyle → {Hue[0.2], Hue[0.6]},
AxesLabel → {"ω", "ρ"}]
```

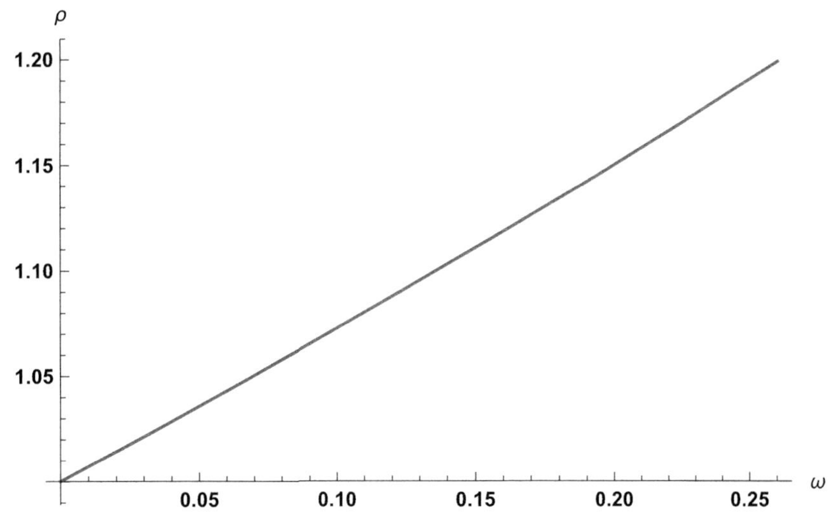


```
dataplot = ListPlot[newdat,
DisplayFunction → Identity,
PlotStyle → {Hue[0], PointSize[0.02]}
]
```

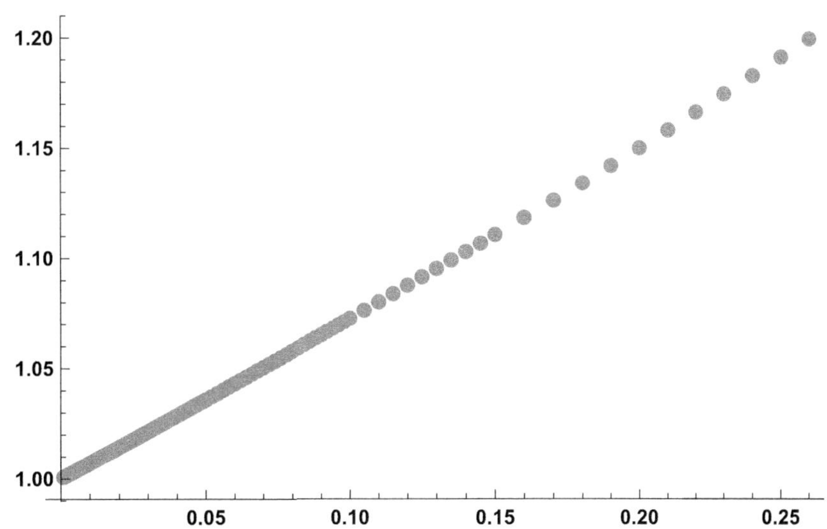

Show[fitplot1, dataplot]

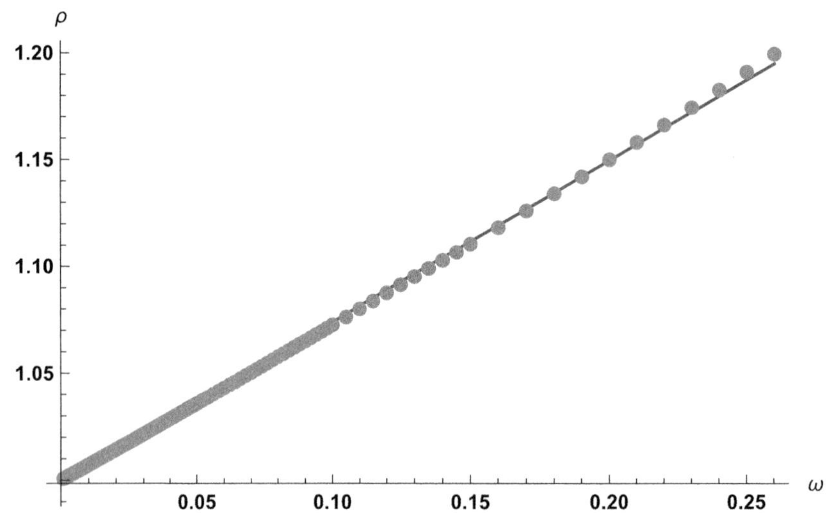

Show[fitplot2, dataplot]

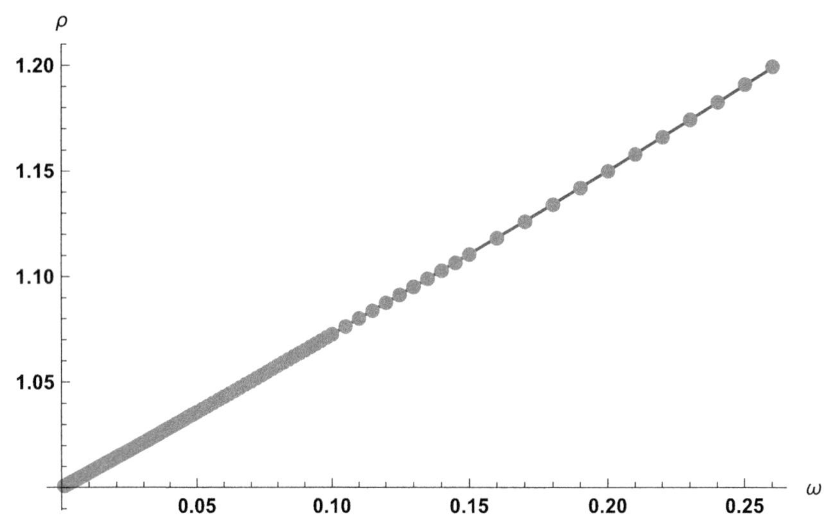

For this data both the linear and quadratic functions appear to fit the data well. For engineering purposes, we can feel comfortable with the first order dependence in this range of concentrations. On the other hand, if we were asked for a more quantitative statement of how well each function fit, we could provide that by examining ParameterTable and RSquared. This is done as is shown here:

l[{"ParameterTable", "RSquared"}]

	Estimate	Standard Error	t-Statistic	P-Value
1	0.998469	0.00017639	5660.57	7.626265092243176`*^-262
x	0.75486	0.00187099	403.455	4.9530918721416005`*^-154

, 0.999423

q[{"ParameterTable", "RSquared"}]

	Estimate	Standard Error	t-Statistic	P-Value
a	0.250775	0.00270755	92.6206	2.1463925902800284`*^-93
b	0.699006	0.000633721	1103.02	3.0849431561809707`*^-193
c	1.00017	0.0000259962	38473.8	0.

, 1.

The first command loads the parameter data and the second provides the r^2 values. Both models fit well with the quadratic fit doing slightly better.

We can now return the problem of the feeding a salt and water to a mixing vessel that initially contains water, and from which the flow is governed by gravity without the assumption of ideality:

$$\text{Total:} \quad \frac{d\rho_{\text{soln}}V}{dt} = \rho_l\, q_l + \overset{\circ}{m}_s - \rho_{\text{soln}}\, q_{\text{soln}}$$

$$\text{Solvent:} \quad \frac{dC_l, V}{dt} = \rho_l\, q_l - C_{l,}q_{\text{soln}}$$

$$\text{Solute:} \quad \frac{dC_{s,\text{soln}}V}{dt} = \overset{\circ}{m}_s - C_{s,\text{soln}}\, q_{\text{soln}}$$

The total material balance now becomes:

$$\text{Total:} \quad \frac{d\left[\rho_l + \gamma C_{s,\text{soln}}\right]}{dt} = \rho_l\, q_l + \overset{\circ}{m}_s - \left[\rho_l + \gamma C_{s,\text{soln}}\right] b\text{Ao}\sqrt{\frac{2g}{A}V}$$

Recall that ρ_l is just the density of the pure liquid solvent, and that h [t] = $\frac{V[t]}{A}$ can be replaced into Torricelli's Law. These equations have become complicated enough that we shall define the density up-front and let NDSolve handle the work of solving the simultaneous equations.

```
Remove[a3, ql, A, ms, ρ, b, Ao, g, ho, tmax, Vo,
V, t, cs, cl, γ];
(* These are the parameters *)
ql = 10;
A = 10;
ṁs = 5;
ρo = 1;
b = 1;
Ao = 0.1A;
g = 9.8;
ho = 10;
tmax = 100;
Vo = Aho;
γ = .9;

(* This is the density of solution. *)

ρ[t_]:=ρo + γcs[t]
```

Here we solve the mass balance equations numerically.

```
a3 = NDSolve[{cs[0] == 0.001,
∂t(V[t]ρ[t]) == ρoql + −ρ[t]bAo,
V[0] == Vo,
∂t(cs[t]V[t]) == −cs[t]bAo,
```

```
∂t(cl[t]V[t]) == ρoql − cl[t]bAo,
cl[0] == ρo },
{V[t], cs[t], cl[t]},
{t, 0, tmax}];
```

These equations define the functions for volume and concentration.

```
Vn[t_]:=Evaluate[V[t]/.a3]
csn[t_]:=Evaluate[cs[t]/.a3]
cln[t_]:=Evaluate[cl[t]/.a3]
```

We plot the time-dependent volume and concentration functions.

```
pla3V = Plot[{Vn[t]/Vo}, {t, 0, tmax},
PlotLabel → {"Vn[t]/Vo versus t"},
PlotStyle → {{Dashing[{0.15, 0.05}], Hue[.6]}},
PlotRange → {{0, tmax}, {0, 1.2}}, Frame → True]

pla3C = Plot[{cln[t], csn[t]}, {t, 0, tmax},
PlotStyle → {{Dashing[{0.15, 0.05}], Hue[0.3]},
{Dashing[{0.15, 0.05}], Hue[0.5]}},
PlotLabel → { γ"=γ", "green=cl[t],aqua=cs[t]"},
Frame → True]
```

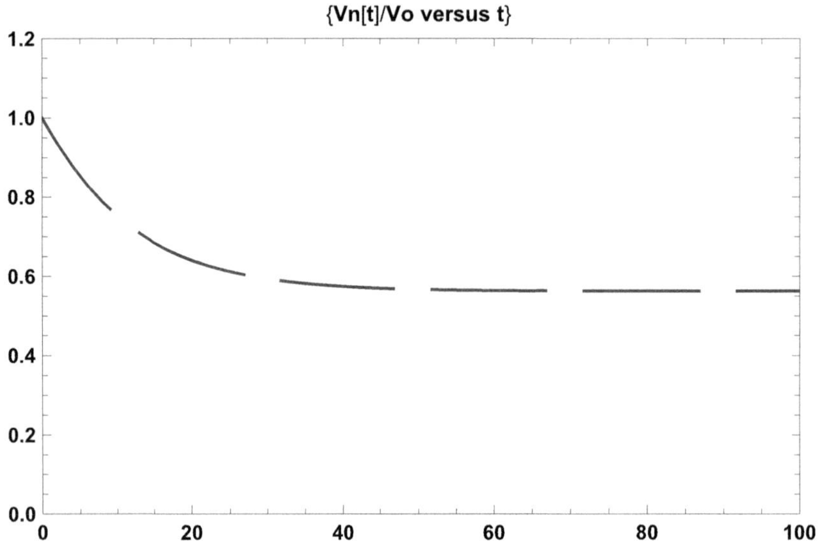

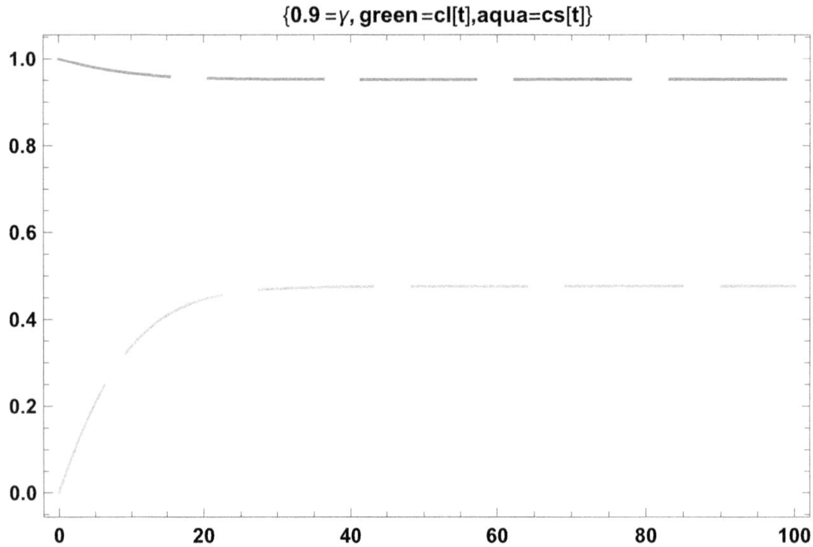

Lastly, we can compare the earlier solutions for concentration in which we assumed the density of the solution was constant to these new solutions wherein the density of solution was allowed to vary with salt concentration. Notice the graphs below appear to be dashed lines but they are not. They appear to be dashed because the earlier and current results overlap. Thus, we see is that with $\gamma = 0.9$, the constant density assumption gave results identical to the variable density case.

(* The Show command allows us to compare the constant and variable density results directly *)

```
Show[pla2C, pla3C,
PlotLabel → {γ"=γ(dash);rd=cl[t],aq=cs[t]"}]
```

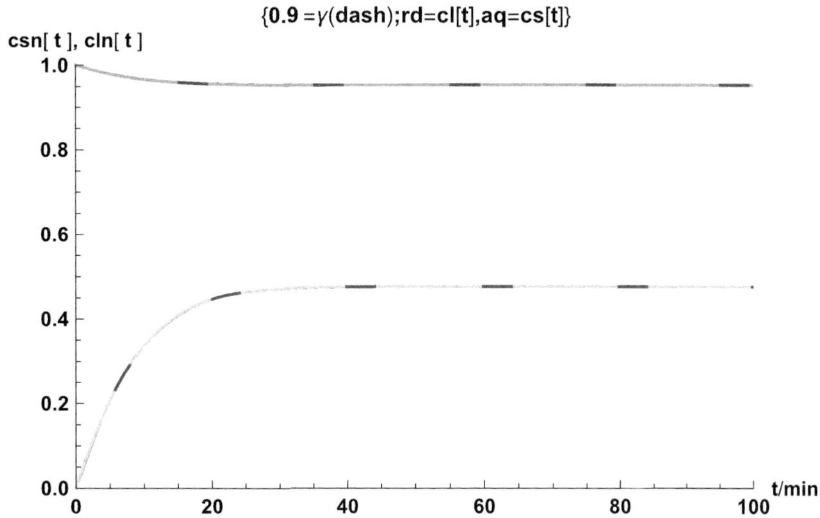

Washing a salt solution from a vessel

A simpler problem, than the one we have been solving, is that of washing out a vessel. Consider that at the end of the process of preparing, feeding and using the salt solution we now have to wash the unit in order to decontaminate it, so that it can be used for a different purpose? How long will it take? How much water should we use? Will the flow rate matter? These are all very relevant questions that we can answer and do so fairly easily. The figure shows the new situation; we have a feed of fresh water into a tank containing the salt solution as the initial condition.

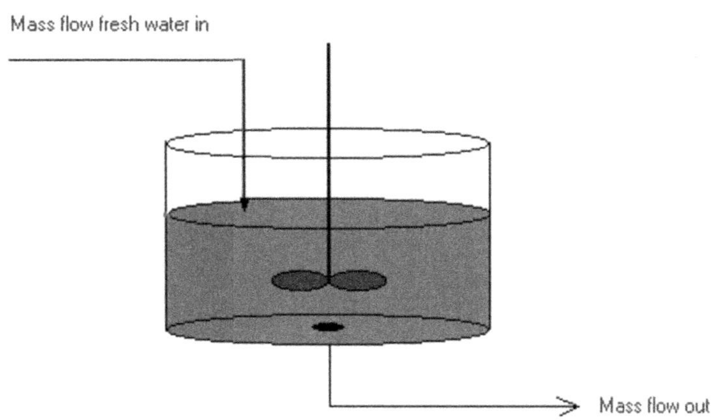

The total and component material balances for this systems are as follows:

$$\text{Total:}\quad \frac{d\rho_{\text{soln}}V}{dt} = \rho_l\, q_l - \rho_{\text{soln}}\, q_{\text{soln}}$$

$$\text{Solvent:}\quad \frac{dC_l,V}{dt} = \rho_l\, q_l - C_{l,}q_{\text{soln}}$$

$$\text{Solute:}\quad \frac{dC_{s,\text{soln}}V}{dt} = -\, C_{s,\text{soln}}\, q_{\text{soln}}$$

It is easy to see that if the wash out is done in such a way that the volume in the tank does not change and the concentration of salt is low enough to ignore the density effect, then the total differential material balance is zero because the volume flow rate in is equal to that out.

$$\text{Total:}\quad \frac{d\rho_{\text{soln}}V}{dt} = 0 \text{ because } \rho_l \sim \rho_{\text{soln}} \text{ and } q_l = q_{\text{soln}}$$

Therefore

$$\rho_l\, q_l = \rho_{\text{soln}}\, q_{\text{soln}} \text{ and } \rho_l\, q_l - \rho_{\text{soln}}\, q_{\text{soln}} = 0$$

thus

$$\frac{d\rho_{\text{soln}}V}{dt} = 0$$

The only equation that we need to solve is the last one for the salt concentration. Once this concentration reaches a specified minimum, the wash-out process is finished. We will, however, solve both the salt and the solvent balances.

$$\text{Solute:}\quad \frac{dC_{s,\text{soln}}V}{dt} = -\, C_{s,\text{soln}}\, q_{\text{soln}}$$

$$\text{Solvent:}\quad \frac{dC_l,V}{dt} = \rho_l\, q_l - C_{l,}q_{\text{soln}}$$

Remove[a4, qsoln, Vo, *t*, cs, cso, cl];

a4 = DSolve[

$$\left\{\mathbf{cl'[t]==\frac{(\rho l - cl[t])qsoln}{Vo}},\right.$$

cl[0] == clo,

$$\text{cs}'[t]==-\frac{\text{cs}[t]\ \text{qsoln}}{\text{Vo}},$$

$\text{cs}[0]==\text{cso}\},$

$\{\text{cl}[t],\text{cs}[t]\},t];$

cl4[t_]:=Evaluate[cl[t]/.a4[[1]]];

cs4[t_]:=Evaluate[cs[t]/.a4[[1]]];

We check these results by putting them both back into the differential equations to be sure that they are correct. If they are correct, then once again the response will be the logical True.

$$\text{Simplify}\left[\partial_t \text{cl4}[t]==\frac{(\rho\text{l}-\text{cl4}[t])\text{qsoln}}{\text{Vo}}\right]$$

$$\text{Simplify}\left[\partial_t \text{cs4}[t]==-\frac{\text{cs4}[t]\text{qsoln}}{\text{Vo}}\right]$$

True

True

Reasonable values of the parameters are applied, and we can then graph the results to see the behavior over time.

$\text{ql}=10;$

$A=10;$

$\text{qsoln}=\text{ql};$

$\text{ho}=10;$

$\text{tmax}=100;$

$\text{Vo}=A\text{ho};$

$\text{cso}=0.5;$

$\rho\text{l}=1;$

$\text{clo}=1+0.9\text{cso}-\text{cso};$

$\text{pla4C}=\text{Plot}[\{\text{cl4}[t],\text{cs4}[t]\},\{t,0,\text{tmax}\},$

```
PlotStyle → {{Dashing[{0.15, 0.05}], Hue[0]},
{Dashing[{0.15, 0.05}], Hue[0.6]}},
PlotLabel → " Cs[ t ] Washout vs Time/min",
Frame → True, PlotRange → {{0, tmax}, {0, cso + clo}}]
```

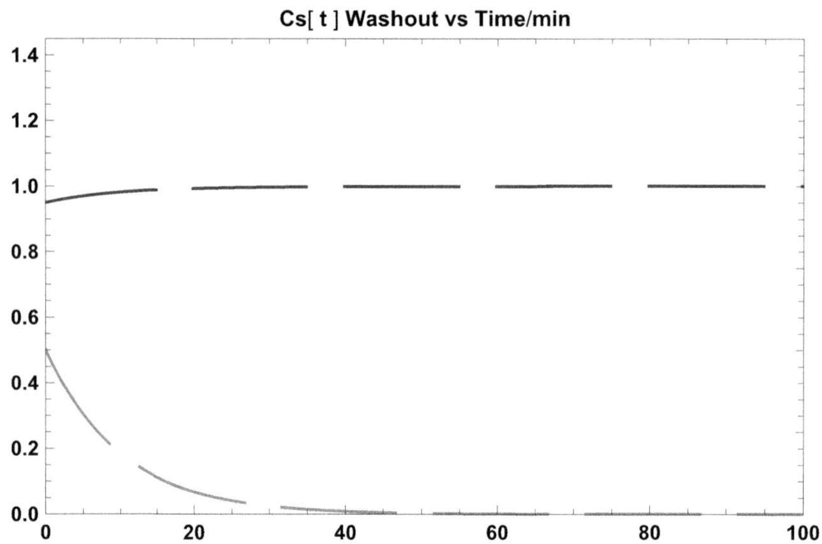

The blue-colored (mid gray in the print version) graph is that of the salt concentration and we can see that by 45-50 minutes the concentration of salt in the unit is close to zero. We also note that the concentration of the solvent increases, albeit not by much, as the salt is washed out of solution.

Let's not make any of these simplifying assumptions. Instead we will take the flow out as given by Torricelli's Law, and the density of the solution to be a linear function of the salt concentration. The all we need do is to modify the numerical routine we had for the previous problem by eliminating the term for the salt feed:

```
Remove[a5, ql, A, ρ, b, Ao, g, ho, Vo, V, t, cs, cl, γ];

ql = 10;
A = 10;
ṁs = 5;
ρo = 1;
b = 1;
```

```
Ao = 0.1A;
g = 9.8;
ho = 10;
tmax = 100;
Vo = Aho;
γ = .9;
cso = 0.5;
ρ[t_]:=ρo + γcs[t]
a5 = NDSolve[{
∂t(V[t]ρ[t]) == ρoql − ρ[t]bAo,
V[0] == Vo,
∂t(cs[t]V[t]) == −cs[t]bAo,
cs[0] == cso,
∂t(cl[t}V[t]) == ρoql − cl[t]bAo,
cl[0] == ρ[0] − cs[0]
{V[t], cs[t], cl[t]},
{t, 0, tmax}];
Vn[t_]:=Evaluate[V[t]/.a5]
pla5V = Plot[{Vn[t]/Vo}, {t, 0, tmax},
PlotLabel → "Vn[t]/Vo vs t",
PlotStyle → {{Dashing[{0.15, 0.05}], Hue[.6]}},
PlotRange → {{0, tmax}, {0, 1.2}}, Frame → True];
csn[t_]:=Evaluate[cs[t]/.a5]
cln[t_]:=Evaluate[cl[t]/.a5]
pla5C = Plot[{cln[t], csn[t]}, {t, 0, tmax},
PlotStyle → {{Dashing[{0.15, 0.05}], Hue[0.3]},
```

```
{Dashing[{0.15, 0.05}], Hue[0.8]}},
AxesLabel → {"t/min", "c[t]"},
PlotLabel →{ γ"=γ", "green=cl[ t ],magenta=cs[ t ]"},
Frame → True]
Show[pla5C, pla4C,
PlotLabel → "With and without variable densities"]
```

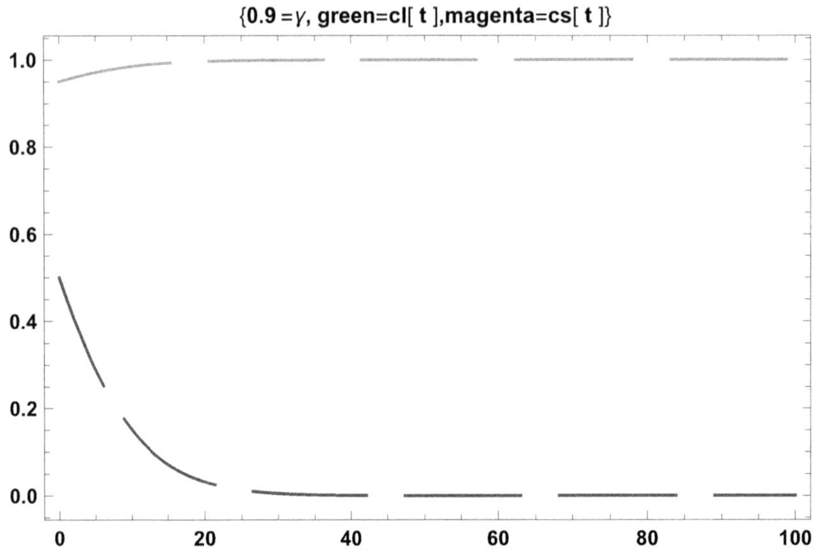

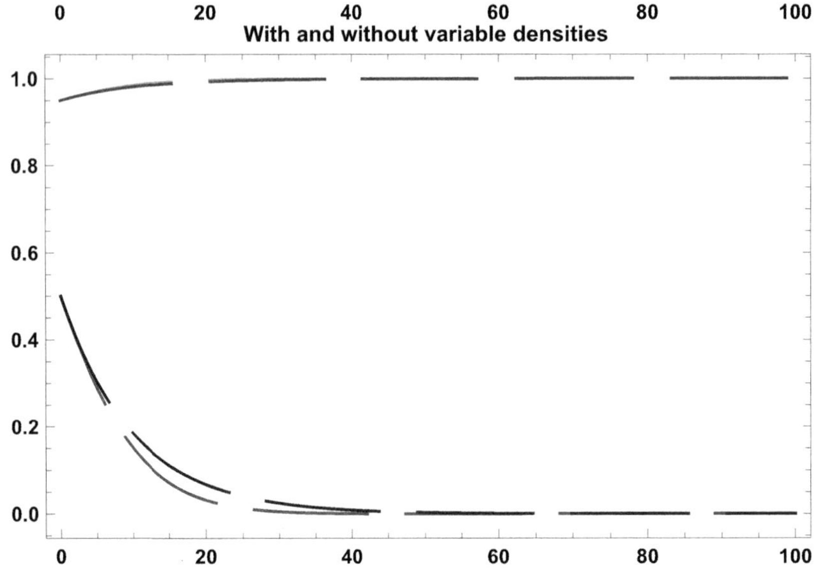

We can see from the last graph comparing the solution accounting for the change of density with concentration versus the simple solution without this taken into account, that for this value of $\gamma = 0.9$, the error made in the approximation is small, although the salt concentration is predicted to fall a bit faster when the variation in density is taken into account.

The pulse input tracer experiment and analysis

The key assumption we have made through out this chapter is that the solutions within the control volume are indeed homogeneous or well-mixed. Questions of the degree to which mixing occurs in system arise in sciences as seemingly diverse as medicine and environmental engineering. For the system that is well-mixed, if we inject a pulse of tracer, we should see a characteristic decay for the concentration in the system as a function of time. Cardiologists use this method to measuring the pumping speed of a heart, by inserting a catheter and injecting a tracer of known volume into the heart. The rate of decay of the concentration within the chambers of the heart, provides the flow rate away from this organ. Similarly an environmental engineer may need to know the flow rate and mixing dynamics in river or stream. By injecting a water soluble and harmless dye into the flowing water the diminution of the dye concentration at the point of "injection" can be used to compute the dynamics of the rivers flow.

This kind of experiment can be used in chemical reactors to find the "residence time distribution" of the reactor, a crucial piece of information that links microscopic flow behavior, i.e. fluid dynamics, to measurables of the system, such as chemical conversion and selectivity. For vessels that are not used for reaction processes, but are used for other operations that are also critically dependent upon mixing, the tracer experiment provides a great deal of insight into how the system behaves. We can analyze how a pulse of injected tracer would behave in the well-stirred vessel which we have been analyzing.

Imagine that an injection is made as a pulse of tracer the concentration of which can be measured in the tank and in the exit stream as a function of time. For a laboratory vessel, the injection might be done with a syringe full of tracer such as a dye or a radioactively tagged molecule. For larger vessels at pilot and production scale ingenious methods have been invented for putting a "pulse" of tracer into the unit. Ideally, the pulse should be added instantaneously, which means in as short a time period as is possible. In other words the time to add the tracer must be much shorter than the time required to "wash" it out of the unit.

For the case of the unit we have been considering, water would be flowing to the system continuously with stirring and the whole system would be at a steady state with respect to level and volume. The injection would be made at the top of the vessel with a very small volume of highly concentrated dye, nothing else would be or should be done. The high concentration is critical to making the measurements accurate and precise. It also makes it possible to use only a small volume of the dye, which is important so that the steady state is maintained with respect to volume. Finally, small volume, high concentration injections can be done fast.

How can we model such a problem? To do the analysis we need to introduce and become comfortable with two new functions: the Dirac Delta function and the UnitStep (or Heaviside) function. The Dirac Delta function is infinitely intense and infinitesimally narrow – like a pulse of laser light. We can imagine it arising in the following way. We begin by considering a pulse that is quite broad, such as the function that is potted below:

$$\mathrm{f}[\mathrm{t}] = \frac{1}{\theta\pi}\mathrm{Exp}\left[-\left(\frac{t}{\theta}\right)^2\right]$$

$\theta = 2;$

Plot$\left[\textbf{Sqrt[}\mathit{fraction}\mathbf{1/(\theta Pi)]Exp}\left[-\left(\frac{t}{\theta}\right)^2\right], \{t, -5, 5\},\right.$

PlotRange → All]

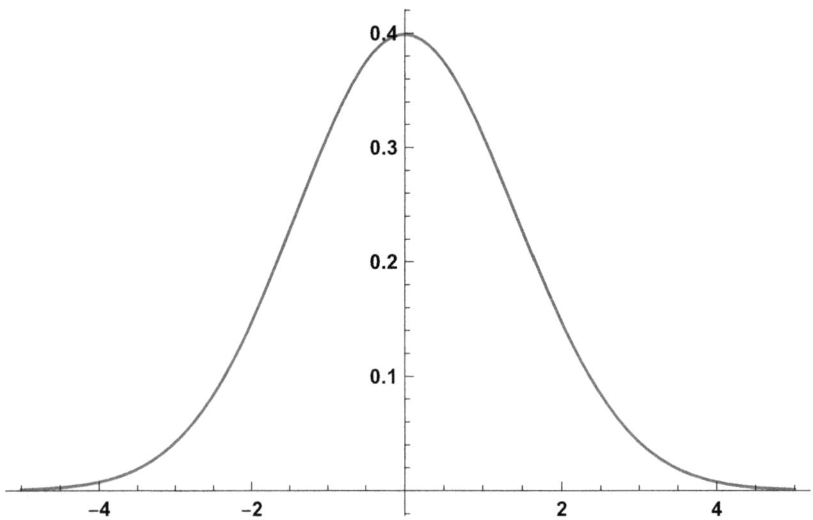

We can sharpen this function in time by decreasing the value of the "time constant" θ, as shown below in the three graphs:

Clear[θ, f, a]

$$f[\theta\_, \mathrm{t\_}] := N\left[\sqrt{\frac{1}{\pi\theta}}\mathrm{Exp}\left[-\left(\frac{t}{\theta}\right)^2\right]\right]$$

```
θ = {1, .5, .1};
a = Table[
Plot[f[θ, t][[n]], {t, −2, 2},
PlotRange → All,
PlotStyle → Hue[.2n + .1],
AxesLabel → {t, "I[t]"}
], {n, 1, Length[θ]}];
Grid[{a}]

Show[a]
```

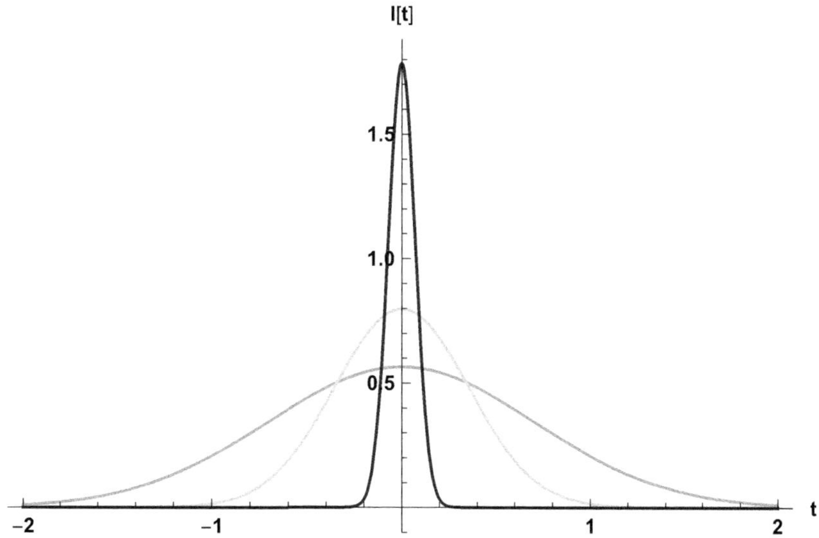

As θ decreases and as the function sharpens, more of the values of this function become zero in the range from - ∞ to $+\infty$.

```
Clear[θ, f, a]
f[θ_, t_]:=N[Sqrt[1/(θPi)]Exp[−(t/θ)^2]]
θ = {.1, .05, .01};
a = Table[
```

```
Plot[f[θ, t][[n]], {t, −2, 2},
PlotRange → All,
PlotStyle → Hue[.2n + .1],
AxesLabel → {t, "I[t]"}
],
{n, 1, Length[θ]}];
Grid[{a}]

Show[a]
```

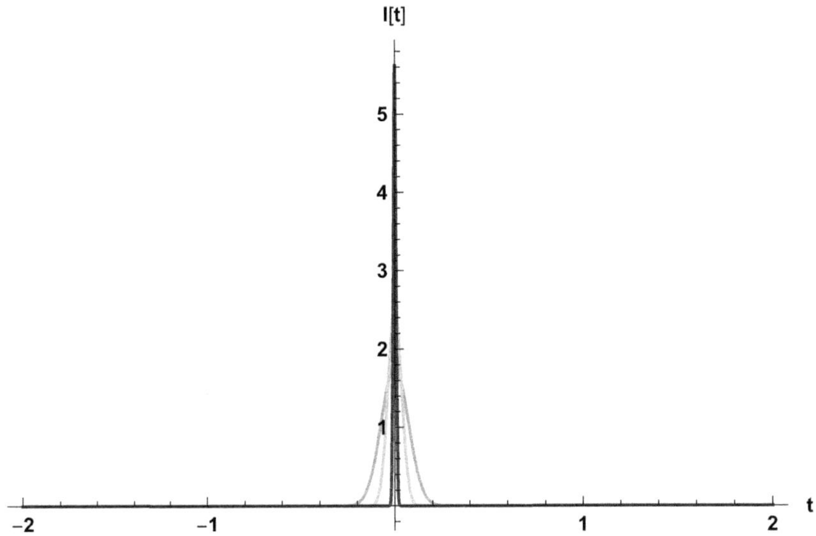

As we decrease the time constant, the function becomes more intense in and around the t = 0. Doing this in the limit of $\theta \rightarrow 0$ transforms this into the infinitely intense pulse of infinitely short time duration. We can use this Dirac Delta function, once we know more about its properties and how it is implemented in *Mathematica*.

If we begin with a simple Table function, we see that if we ask for "t" in the interval from -5 to 5, we get a simple vector of those integers back:

```
Clear[t, x]

Table[t, {t, −2, 2}]
```

{−2, −1, 0, 1, 2}

Taking f [t] over the same interval leads to the vector of elements each one of which is f [t] evaluated at the integer:

Clear[*t*, *x*, *f*]

Table[*f*[*t*], {*t*, −2, 2}]

{−2*f*[−2], −*f*[−1], 0, *f*[1], 2*f*[2]}

When we let f [t] be the distribution function and use a small value of $\theta \sim 0.1$, then

f[t_]:=Sqrt[1/(0.01Pi)]Exp$\left[-\left(\frac{t}{0.1}\right)^2\right]$

Table[*f*[*t*], {*t*, −2, 2}]

$\{1.0805187371933992 \text{ x } 10^{-173}$,

$2.0988281156772085 \text{ x } 10^{-43}$,

5.6419,

$2.0988281156772085 \text{ x } 10^{-43}$,

$1.0805187371933992 \text{ x } 10^{-173}\}$

We see that the values of f [t] are close to zero for each value of t tested, except at t = 0. Next, look at what happens when we use the DiracDelta [t] over the same range of t values. The values of Dirac[t] are identically zero, except at t = 0, where the result DiracDelta[0.] is formally left unevaluated:

Table[*N*[DiracDelta[*t*]], {*t*, −2, 2, 1}]

{0., 0., DiracDelta[0.], 0., 0.}

The only element that is non-zero is that which falls at the point t = 0, this is because at every other point the DiracDelta function is identically zero by definition. This is the consequence of being infinitely intense and infinitely short in duration. We can display the position of this pulse by placing another integer in the argument of the DiracDelta as follows:

Table[*N*[DiracDelta[*t* − 5]], {*t*, −5, 5}]

{0., 0., 0., 0., 0., 0., 0., 0., 0., 0., DiracDelta[0.]}

In this case the non-zero value of the function has been pushed to the positive extremum of this interval on t. If we integrate the product of the dye mass, m_{dye} and the DiracDelta, we obtain this:

Clear $[t, m_{\text{dye}}]$

Integrate $[\text{DiracDelta}[t] m_{\text{dye}}, \{t, -5, 5\}]$

m_{dye}

The integration returns just m_{dye} when integrated over time. Now, we can use this technique to find how the concentration of dye changes as a function of time in well-stirred vessel. We need to write and integrate the component balance on the dye to have the answer, since the volume of the tank is assumed not to change with time. We will use the following equation for the rate of change of the dye mass:

$$\frac{dc_{\text{dye}}[t]}{dt} = \frac{m_{\text{dye}}\text{DiracDelta}[t] - c_{\text{dye}}[t]\text{qex}}{V}$$

Notice how we have used the DiracDelta in the balance, it has the dimensions of inverse time. [This is because the integral of the DiracDelta function over time from negative infinity to positive infinity is 1 and unity is dimensionless. Therefore, the DiracDelta function has dimensions of inverse time in this case.] At time t = 0, theoretically, all the mass of the dye is injected and it is done "instantaneously." At all other times, the term for dye mass input is identically zero. We can integrate this analytically:

Clear[qex, t, to, V, cdye]

Simplify[

DSolve[

$$\left\{\text{cdye}'[t] == \frac{m_{\text{dye}}\text{DiracDelta}[t] - \text{cdye}[t]\text{qex}}{V},\right.$$

cdye[to] == 0}, cdye[t], t]

]

$$\left\{\left\{\text{cdye}[t] \to \frac{e^{-\frac{\text{qex}t}{V}}(\text{HeavisideTheta}[t] - \text{HeavisideTheta}[\text{to}])m_{\text{dye}}}{V}\right\}\right\}$$

We find that the integration looks like an exponential decay except that we now have a new function that has appeared – the HeavisideTheta function. The indefinite integral of the DiracDelta function is the HeavisideTheta function and it is dimensionless. To see how the HeavisideTheta function behaves in time, we can plot it:

Plot[HeavisideTheta[*t*], {*t*, −5, 5},

AxesLabel → {t, "Heaviside[t]"}, PlotStyle → Hue[0]]

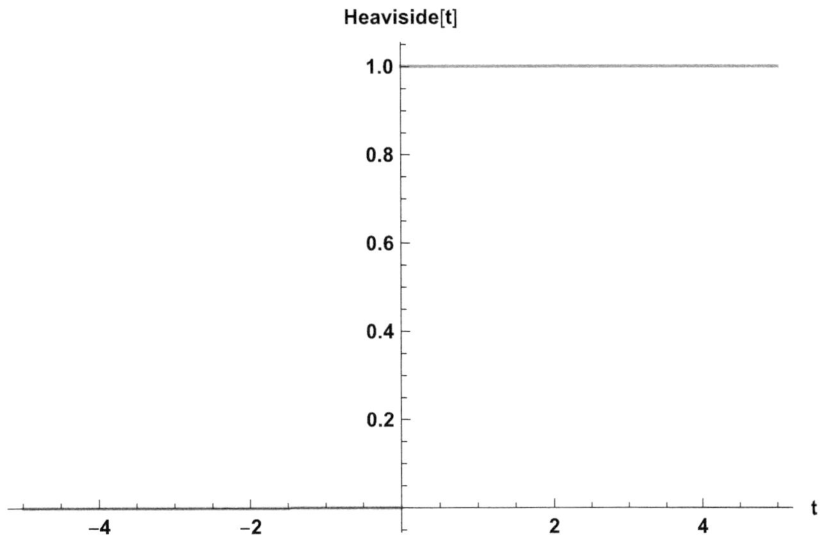

The Heaviside step function is everywhere zero until it comes to t = 0 and then it goes to a value of unity that it maintains *ad infinitum*. We can rewrite the solution as a function of time:

Clear[qex, *t*, to, *V*, cdye, *m*]

cdye[t_]:=

$$\frac{e^{-\frac{qex\,t}{V}}(\text{HeavisideTheta}[t] - \text{HeavisideTheta}[to])m_{\text{dye}}}{V}$$

$m_{\text{dye}} = 10;$

qex = 1;

$V = 100;$

to = −5;

pl1 = Plot[cdye[*t*], {*t*, −100, 500},

AxesLabel → {t, "cdye[t]"}, PlotStyle → Hue[0.8]]

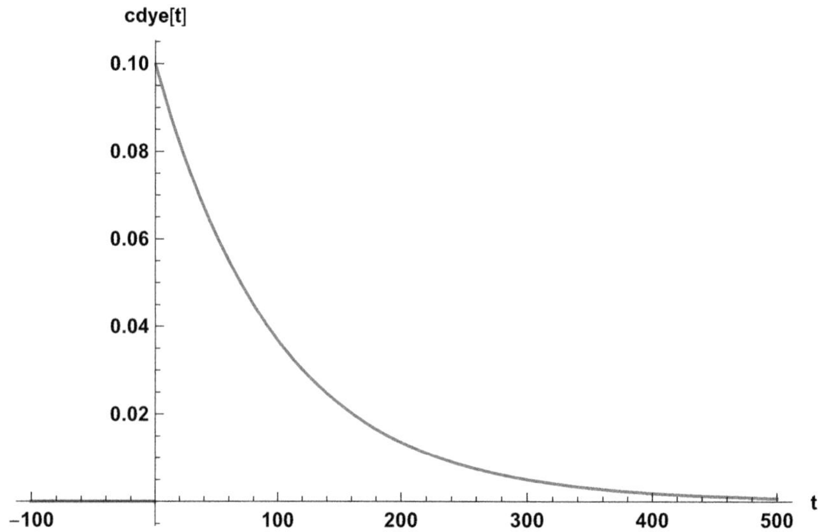

Here the pulse is injected at time "zero", the concentration instantaneously jumps discontinuously to a magnitude of 0.1 and then begins to descend exponentially as a function of time. This is the characteristic curve we should see if the unit is well-mixed.

To make this result more general we can non-dimensionalize both of the axes i.e. concentration and time. The concentration of dye can be referenced to the maximum concentration at time zero, $c_{\text{dye}}[0]$. But what of the time axis? How shall we non-dimensionalize this? We will use the "holding time" as the reference time. The holding time is the time required for a volume of liquid equal to the volume of the unit to pass entirely through the unit. The holding time is the ratio of the volume to the volume flow rate i.e. $\tau = \frac{V}{\text{qex}}$. We can remake the graph in non-dimensional form:

$\dot{m}_{\text{dye}} = 10;$

$\text{qex} = 1;$

$V = 100;$

$\text{tro} = -1;$

$\tau = \frac{V}{\text{qex}};$

```
cdye[0] = 1;

ndcdye[tr_]:= e^(-tr) ṁdye (UnitStep[tr] - UnitStep[tro]) / (cdye[0] V)

Plot[
ndcdye[t],
{t, -1, 5},
AxesLabel → {"tr", "n.d.cdye[tr]"},
PlotStyle → Hue[0.4]
]
```

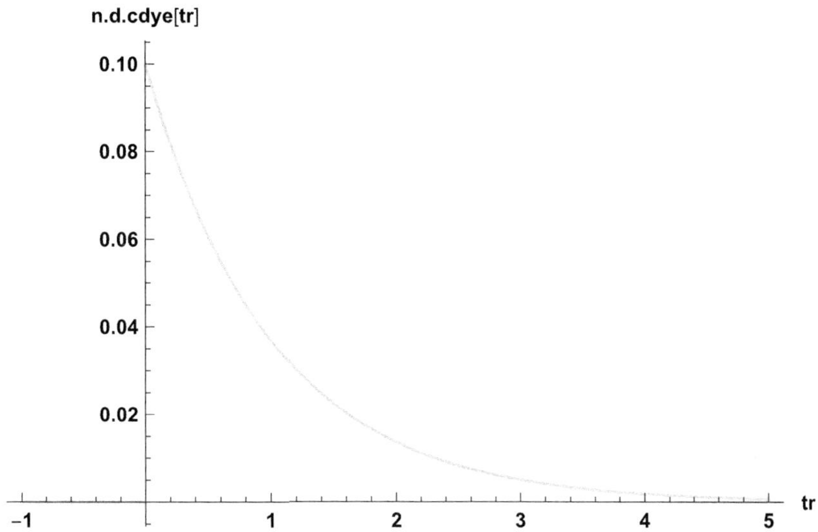

The time axis is the "reduced" time, which is the ratio of real time to the holding time, $t_r = \frac{t}{\tau}$, and the y-axis is the non-dimensionalized dye concentration. If we were to plot the experimental change in non-dimensionalized concentration versus reduced time, it should fall very near to this curve. The extent to which the real system deviates, is a measure of the degree to which the system veers from the ideally "well-mixed" limit.

Mixing

Consider the following case. In this experiment everything is the same as in the last one, except that there are two well-mixed tanks rather than one. The same overall flow, qex, is diverted through the two units with a 50:50 split and the same mass of tracer or dye is added instantaneously to the inlet flow. The volumes of the lines to and from the tanks are considered to be negligible. Each tank has exactly 50% of the volume of the previous tank, therefore their residence times are half that of the one large tank with the same overall volume. The lines from the two tanks come together prior to the tank and are assumed to be perfectly mixed when they do. The question is – will the time distribution of the tracer concentration look the same or different from that of one well-mixed tank?

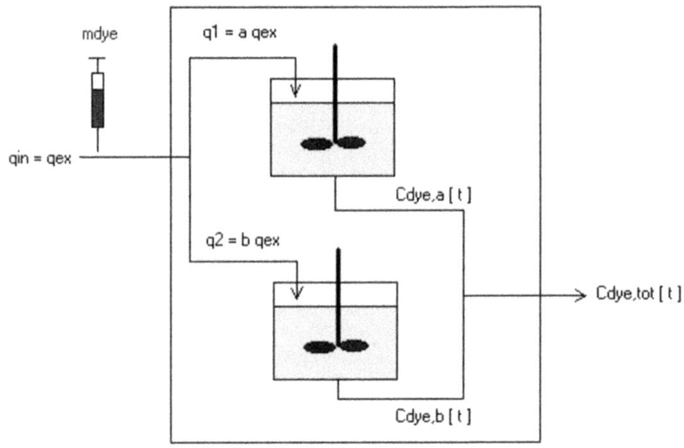

We do not need to redo the analysis. Instead we will take the solution for one well-mixed tank and apply it to the two tanks. We have to be careful about handling the splits at the input side where the stream divides into two and again when the two separate tank streams come back as one. The overall is divided into the two flows, q1 and q2, by the fraction a and b. The mass of the dye trace also will be split in the same way. The reason is that the mass of dye into the first splitter must be the same as the total mass of dye out, which is just the sums of the masses in the two streams:

$$
\begin{aligned}
m_{\text{dye,in}} &= C_{\text{dye,in}}\ \text{qin}\ \delta\text{t} \\
&= C_{\text{dye,1}}\text{q1}\ \delta\text{t} + C_{\text{dye,2}}\text{q2}\ \delta\text{t} \\
&= \text{a}\ C_{\text{dye,1}}\ \text{qex}\ \delta\text{t} + \text{b}\ C_{\text{dye,2}}\text{qex}\ \delta\text{t} \\
&= (\ \text{a}\ C_{\text{dye,1}} + \text{b}\ C_{\text{dye,2}}\ )\ \text{qex}\ \delta\text{t}
\end{aligned}
$$

Where δt is the instantaneous or infinitesimally short duration of the DiracDelta function pulse. This is also the time that for the mass of the dye to hit each tank, i.e. zero time.

The concentration of dye in each tank has the same form as that which we have derived for the single vessel.

$$\text{cdye2a[t\_]} := \frac{a e^{-\frac{a q e x t}{V1}} m_{\text{dye}}(\text{Heaviside}[t] - \text{Heaviside}[\text{to}])}{\text{V1}}$$

$$\text{cdye2b[t\_]} := \frac{b e^{-\frac{b q e x t}{V2}} m_{\text{dye}}(\text{Heaviside}[t] - \text{Heaviside}[\text{to}])}{\text{V2}}$$

Notice that we have rewritten q1 and q2 in the arguments of the exponentials as "a qex" and "b qex," where "a " and "b" are the splits fractions that set the stream flows. We can compute these two concentrations and plot them as functions of time, but what we really want is the time dependence of the dye concentration after the two stream are recombined. The reason is that we wish to compare the overall effect of two tanks on the tracer with that of one tank given that the sum of the volumes of the two are the same as the one. We also want to be able to split the flows between these two tanks in different ratios and with different tank volumes to see how this affects the overall time dependence. In other words, what if we did not know there were two tanks? What if all we knew were the inputs and the outputs at the dotted box around the two tanks? Would we be able to detect a difference for this system versus the one tank system on the basis of the tracer experiment? Let's continue the analysis to find out.

To find the concentration at the point where the streams come back together we again apply the conservation of mass. The mass of dye in the two lines coming into this mixing point per unit time must be equal to the total mass going out of it per unit time. The rate of mass flow in is just the sum of the products of the concentrations and flow rates of the two streams exiting the tanks, while the rate of mass flow out is the concentration of the dye times the total flow rate. The mathematical statement is much more succinct:

$$(\text{a Cdye2a [t]} + \text{b Cdye2b [t]})\ \text{qex} = \text{Cdye2tot [t]}\ (\text{a} + \text{b})\ \text{qex}$$

$$\text{Cdye2tot [t]} = \frac{(a\text{Cdye2a}[t] + b\text{Cdye2b}[t])}{(a+b)}$$

This last concentration is what we can measure if the tanks inside the blue (dark gray in the print version) box are hidden from view, so this is the computation that we want to make and to compare to the first case of one unit in plain view. We should also do one more calculation to be sure we are not making any errors. The mass of dye into the units must eventually come back out. So we should integrate the product of the exit concentrations and flows from each tank and sum these to be sure it is equal to the dye input mass. Yet, another application of the conservation of mass.

The code for doing all of this is shorter than the description of it. Once it is written we can use it over and over again. We could also put it into non-dimensional form, if we chose to, but instead we will make our comparisons in real time and concentration. Recall that "a" and "b" are the splits – these are entered as fractions, but they must sum to unity.

To check ourselves in the first case, we set the splits to one half each and the volumes are equal. We compare the concentration versus time curve for this case versus that for the one tank. We do all the calculations for the two tank case and then end off with a graph comparing it to the one tank case.

Clear[a, b]

Remove[cdye2b]

$a = 1/2;$

$b = 1/2;$

cdye2a[t_]:= $\frac{a e^{-\frac{a \text{qex} t}{\text{V1}}} m_{\text{dye}} (\text{UnitStep}[t] - \text{UnitStep}[\text{to}])}{\text{V1}}$

cdye2b[t_]:= $\frac{b e^{-\frac{b \text{qex} t}{\text{V2}}} m_{\text{dye}} (\text{UnitStep}[t] - \text{UnitStep}[\text{to}])}{\text{V2}}$

$m_{\text{dye}} = 10;$

qex = 1;

V1 = 50;

V2 = 50;

to = −5;

pl2 = Plot[{cdye2a[t], cdye2b[t]}, {t, −100, 500},

AxesLabel → {t, "cdye[t]"},

PlotStyle → {Hue[0.4], Hue[0.6]}, PlotRange → All];

pl2b = Plot$\left[\left\{\frac{a\,\text{cdye2a}[t] + b\,\text{cdye2b}[t]}{a+b}\right\}\right.$, {$t$, −100, 500},

AxesLabel → {t, "cdye[t]"}, PlotStyle → {Hue[0.5]},

PlotRange → All];

```
NIntegrate[cdye2a[t]aqex, {t, -100, 400}];

NIntegrate[cdye2b[t]bqex, {t, -100, 400}];

% + %%

NIntegrate[cdye[t]qex, {t, -100, 400}]

Show[pl2b, pl1]
```

9.81684

9.81399

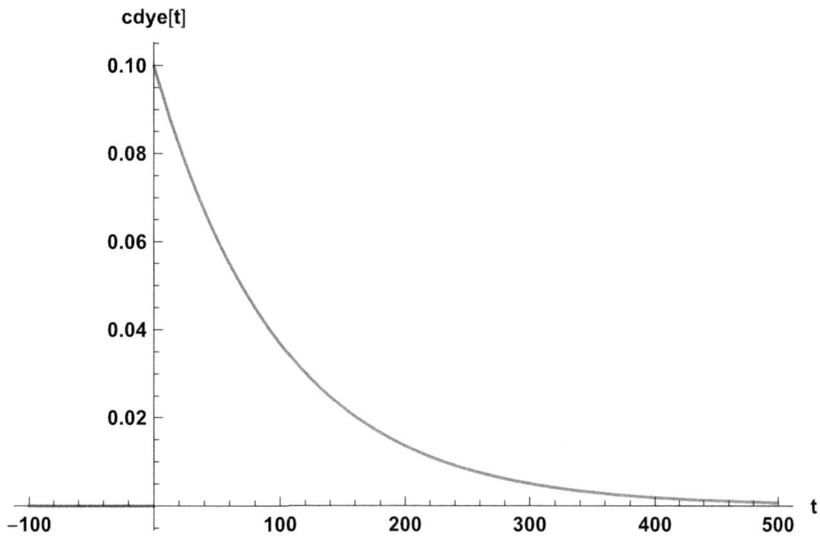

The tracer curves all overlap perfectly and the integrals are all approaching 10 after 400 time units. So the code is working and our derivations are verified. Now we can turn to a different case. Let's assume that the flows are not evenly split but that a is 2/3 and b is 1/3. Let us also take the volumes to be different, we will set V1 to 20 and V2 will be 80. (Recall that V1 and V2 have to sum to the same value as that of V in the one vessel case if we are to make valid comparisons.)

```
Clear[a, b]

Remove[cdye2b]

a = 2/3;

b = 1/3;

cdye2a[t_]:=
```

$$\frac{a e^{-\frac{a q e x t}{V1}} m_{dye}(\text{UnitStep}[t] - \text{UnitStep}[to])}{V1}$$

$$\text{cdye2b[t\_]:=}\frac{b e^{-\frac{b q e x t}{V2}} m_{dye}(\text{UnitStep}[t] - \text{UnitStep}[to])}{V2}$$

$m_{dye} = 10;$

```
qex = 1;

V1 = 20;

V2 = 80;

to = -5;

pl2 = Plot[{cdye2a[t], cdye2b[t]}, {t, -100, 500},

AxesLabel → {t, "cdye[t]"},

PlotStyle → {Hue[0.4], Hue[0.6]}, PlotRange → All];
```

$$\text{pl2b} = \text{Plot}\left[\left\{\frac{a\ \text{cdye2a}[t] + b\ \text{cdye2b}[t]}{a+b}\right\}, \{t, -100, 500\},\right.$$

```
AxesLabel → {t, "cdye[t]"}, PlotStyle → {Hue[0.5]},

PlotRange → All];

NIntegrate[cdye2a[t]aqex, {t, -100, 400}];

NIntegrate[cdye2b[t]bqex, {t, -100, 400}];

% + %%

NIntegrate[cdye[t]qex, {t, -100, 400}]

Show[pl2b, pl1]
```

9.3704

9.81399

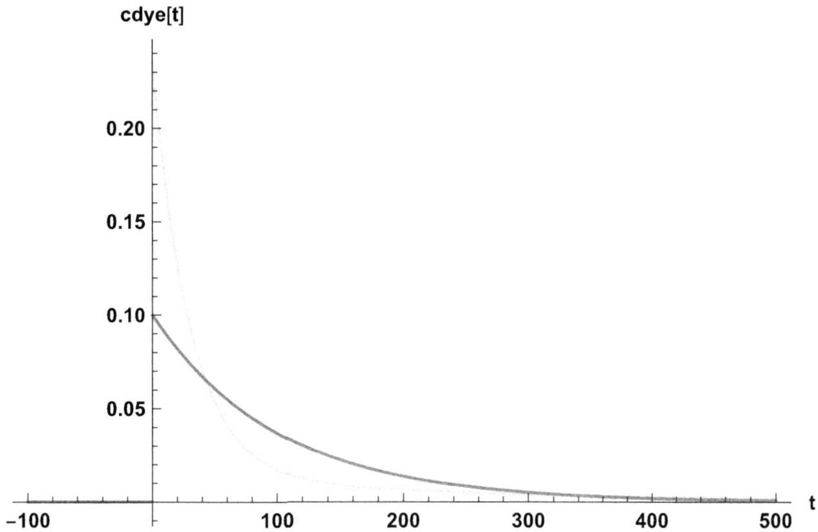

The time dependence of the tracer in this case is markedly different, sharper and narrower overall. Why? Because 2/3 of the flow is shunted through the small vessel with a much shorter holding time. So, even though the total flows through the two systems with equal volumes are the same, the behavior is quite different. Therefore even if we could not see the two tanks, we would have to know that there was a very different flow pattern in this second case. Try other values of a and b as well as V1 and V2 to see what happens.

To extend this model to three tanks would be straightforward, but so too would it be to extend it to n tanks where n was large. One of the points to note about the equations is that the argument of the exponential term is a ratio of the actual time to the holding time in each unit, since the holding time in the nth unit, θ_n, is $\frac{V_n}{q_n}$, that is the ratio of the flow volume of the unit to the flow rate through it;

$$\text{cdye}n[t] = \frac{\text{a n } e^{-\frac{t\text{qn}}{\text{Vn}}} m_{\text{dye}}(\text{UnitStep}[t] - \text{UnitStep}[\text{to}])\text{qtot}}{\text{Vn}}$$

$$= \frac{\text{a n } e^{-\frac{t}{\theta\text{n}}} m_{\text{dye}}(\text{UnitStep}[t] - \text{UnitStep}[\text{to}])}{\theta\text{n}}$$

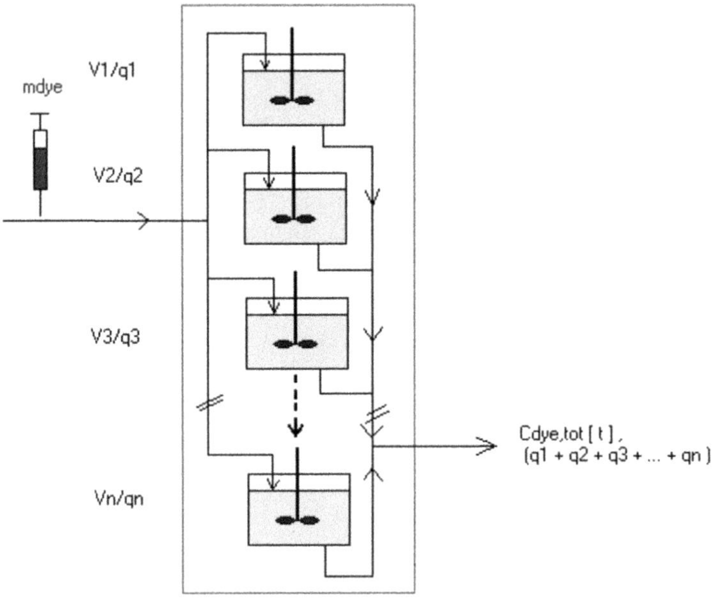

Now if we kept the total volume and total flow rate through these n different units the same as the one unit case, then we would have a very different tracer concentration-time curve, depending on the distribution of the flows and the volumes within the green (mid gray in the print version) box. There would be n different holding times in this overall unit, but the average would be the same as the single vessel. If n were a discrete number of units, then there would be a discrete distribution of holding times. But as n grew larger, say toward infinity, the volume in any one unit would be infinitely small and as would be the holding time. At this point the discrete distribution could be described nicely by one that was continuous in the holding time. The key would be to know how those discrete volumes, and hence residence times, making up the total volume were distributed.

Conclusion

We have now fully integrated the concept of a component and the rate of change of a component's mass into our analysis toolkit. Along the way, we have taken some time to understand the concepts and meaning of density and how it relates to the concentration of the solute or salt and the solvent. This included the notion of non-ideality when we realized that for most solutes the volume either expands or contracts with their dissolution from that which it would have had if the solute added simply as more solvent but of different mass per molecule. In going from a set of simplifying assumptions to a fuller analysis including these density changes with solute concentration, we had to introduce more computing methods, but we were able to go seamlessly from analytical solutions to numerical ones and compare the results. The last section of the chapter was devoted to some new ways of looking at the idea of mixing. In this analysis we learned to use the DiracDelta function. But we also defined the holding time and used this to construct a general non-dimensionalized solution to the tracer injection problem. These are all tools that we will see again!

CHAPTER 5

Multiple phases - mass transfer

A topic of utmost importance is that of mass transfer. Processes often require moving molecules between different phases in order for the desired outcome to take place. For example a "simple" catalytic hydrogenation of a liquid phase unsaturated molecule, such as benzene, can require many mass transfer steps to occur prior to reaction.

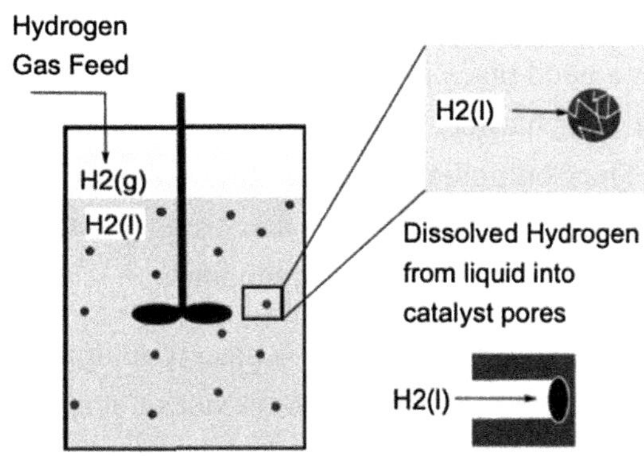

Here is a picture of mass transfer for liquid phase catalytic hydrogenation. Hydrogen gas is fed to the system in the gas phase. With vigorous agitation, the hydrogen in the gas phase is transferred into the liquid phase. This is because the chemical potential of hydrogen in the gas phase is higher than in the liquid. This induces the transport of gaseous hydrogen to the liquid phase. Put differently, the concentration of hydrogen is higher in the gas phase initially, so it partitions to the liquid phase until it reaches saturation. However, the hydrogen is also being consumed in the hydrogenation of a substrate, like an arene. As long as there is arene present, the hydrogen in solution will be consumed and its concentration and chemical potential will remain low. This creates the driving force in chemical potential that moves hydrogen from the gas phase to the liquid phase and into the catalyst, where hydrogenation takes place. The steps are hydrogen transfers to the liquid phase, diffuses through the liquid to a catalyst particle, then transfers through the pores of the particle to a metal surface and there it is chemisorbed and reacts with an unsaturated hydrocarbon. The slowest step of all these steps will be the step whose rate controls the observed rate, the rate that we measure. It may be that under some circumstances, it is the rate of hydrogen transfer from the

Introduction to Chemical Engineering Analysis Using *Mathematica*
https://doi.org/10.1016/B978-0-12-820051-3.00010-3

gas phase to the liquid phase that is the slowest rate in all the stepwise processes. However, if the mass transfer across the gas-liquid interface is large, due to vigorous agitation, then the rate measured will be the rate of one of the subsequent steps. If all the transport processes are fast, then the measured rate will be the rate of the reaction. This chapter sets out to provide a means to handling these types of interphase mass transfer problems in terms of their fundamental characterizing variables, the conservation of mass and appropriate constitutive relationships. To begin we will consider quite simple problems, such as the dissolution of a salt in water.

Salt dissolution

The dissolution of a salt is a good place to begin, since we already know quite a bit about this process. The solid, say sodium chloride, consists of cations and anions in a fixed 1:1 ratio that make up the solid lattice. The Coulombic forces of attraction - the Madelung energy - keep the lattice together in the solid state. These forces are strong enough to make the crystalline lattice an energetically favorable configuration for the ions.

When the lattice of ions held together in this way is placed in liquid hydrocarbon, such as hexane, nothing happens. It remains intact; the hexane does not affect it because hexane is not a solvent for the salt. Why? Hexane is a non-polar hydrocarbon, whereas the salt is made up of charged ions that are at the limits of polarity – one a cation and the other an anion. If the lattice were to fall apart into ions in hexane, it would do so only if the ions were more stable in solution than they were in the lattice. This is not the case with hexane because it lacks the polarity to interact with the ions to stabilize them.

On the other hand water, which is quite polar, will dissolve the salt very well. The reason that water is polar is that the oxygen is electronegative and carries a more negative partial charge than the hydrogens, which are partially positively charged. These charges are why water has high solvating potential for sodium chloride. The oxygen of the water molecule will associate with the cation thereby reducing its overall energy. The hydrogens will associate with the anion of the salt to partially dissipate its charge. Nature finds bare charges to be unfavorable, so this coordination by water is highly favorable. The interaction is called solvation.

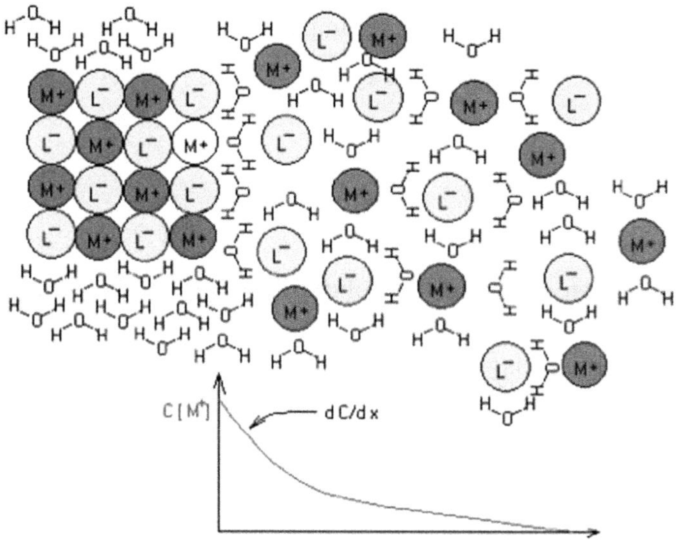

Quantitatively, we also know that the concentration of the ions in solution is given by their solubility product or K_{sp}. This is the equilibrium constant for the salt in water, rearranged to take up the activities of the pure water and the pure solid salt:

$$K_{solvation} = \frac{a_{cation}a_{anion}}{a_{salt}a_{H2O}}$$

$$K_{sp} = K_{solvation}\ a_{salt}a_{H2O} = a_{cation}a_{anion}$$

$$K_{sp} = \gamma_{cation}C_{cation}\ \gamma_{anion}\ C_{anion}$$

At low concentrations the activity coefficients are close to unity and we have:

$$K_{sp} \rightarrow C_{cation}\ C_{anion}$$

This information is purely thermodynamic, however, and does nothing to tell us how long a dissolution process may take. Even if a salt is soluble, we do not have a means to get at its rate of dissolution. Furthermore how do salts with smaller Ksp values compare with those with larger Ksp's? Will they dissolve faster, slower, or is the rate independent of this factor? What role does the physical form of the salt play in the rate of dissolution? Do smaller particles dissolve faster? If so, why? What about mixing? Does more vigorous mixing lead to fast rates of dissolution of the salt in the water? What about the temperature of the solvent? Does this also affect the dissolution rate of the salt? Intuitively, if all else is the same, we would say yes - smaller particles dissolve faster, more mixing also hastens the dissolution process and higher temperature increase the rate as well. But, how do we express this mathematically so that we can model such fundamentally simple and yet important systems? We seek more than a qualitative understanding, we want to be able to handle these processes quantitatively.

Batch Dissolution

Background

The dissolution of a salt into a surrounding solution is easiest to think of as taking place in a closed vessel, that is in a "batch" mode with no flows in, or out of the vessel. Let's consider a large particle of the solid salt hanging in the water by a thread. We can take the solid salt particle as one control volume and the volume of solvent as the other. If we have many particles in solution we treat them all as one phase and one component. The process is taken to occur at constant temperature. The physical situation looks like this:

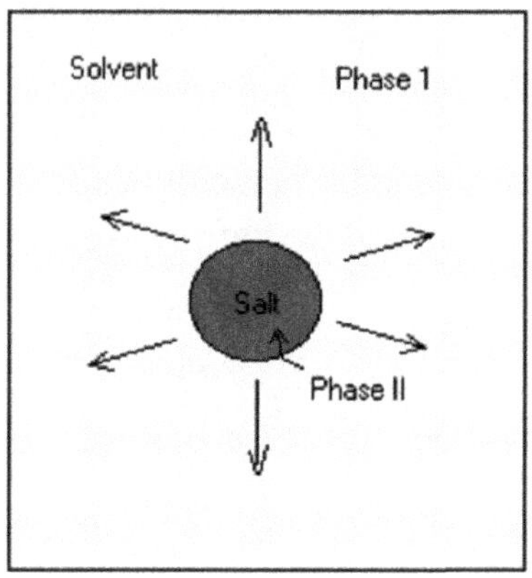

The process will continue until either all the salt has dissolved, or the saturation limit of the solvent has been reached. The ratio of the volume of the solvent to the mass of salt will be critical. If we were to do this experiment several times with the same volume of solvent and with the same mass of salt, but with different numbers of particles of the salt, we would find that the experiments with more, smaller particles would require less time to fully dissolve the salt, than would those with fewer, larger particles. Our common experience of dissolving salt in water is that stirring makes the dissolution process faster. So, if we were to do a series of salt dissolution experiments, keeping all else the same, but varying the rate of mixing, we would find that faster and better mixing would lead to more rapid dissolution. This may seem obvious, but let's be sure to write it mathematically before we go on:

$$\text{rate}_{\text{dissolution}} \alpha \; A_{\text{interface}}$$

This says rate is proportional to the interfacial area, which is the surface area of the salt particle or particles. We can also state that we observe this rate to increase if we mix the mixture:

$$\text{rate}_{\text{dissolution}} \uparrow \text{ as mixing } \uparrow$$

Let's cast this in a rate equation.

Rate of dissolution

Experiments also show that the rate of dissolution must stop when the solution reaches the saturation limit, and the rate will be at its highest early, in the process, when the concentration of salt in the solvent is low. We also find that if the temperature of the solvent is raised, not only can we dissolve more salt at saturation, the rate at which it dissolves increases. All of these experimentally observed behaviors can be accounted for in a simple rate for the dissolution process:

$$\text{rate}_{\text{dissolution}} = K_m A_{\text{interface}}[\, C_{\text{salt}}^{\text{sat}'d} - C_{\text{salt}}^{I}[\,\text{t}\,]\,]$$

Let's break this expression down. First, on the right-hand side, we see the difference between the concentration of salt at saturation $C_{\text{salt}}^{\text{sat}'d}$, which is a constant in time, and the changing concentration of salt in the process, $C_{\text{salt}}^{I}[\,\text{t}\,]$, which increases in time. The rate will be proportional to this difference. Early in the dissolution process the concentration of salt is close to zero and small compared to the saturation concentration. At the beginning of the dissolution process, using pure water, the difference is at its largest magnitude. With time, the dissolved salt concentration rises in the water phase. As it rises, the difference between the concentration in solution and the saturated concentration decreases. If there is enough salt present, eventually, the concentration of salt in solution rises to the saturation concentration. At that point the rate of dissolution goes to zero and the process is at equilibrium. If the mass of salt present is inadequate to reach saturation, the process will also stop because all the solid salt has been transferred to the solution phase. However, if more salt were added to the solution, the dissolution process would begin again and would continue, either until saturation was reached, or all the salt was dissolved.

This expression also helps us to understand why increasing the temperature of the water increases the rate of dissolution. At higher temperatures, the aqueous phase has more capacity for dissolved salt, which is to say that $C_{\text{salt}}^{\text{sat}'d}$ increases. As this increases the difference between the saturation concentration and the actual salt concentration increases. Since the rate is proportional to this difference, the rate of dissolution increases. The term in brackets is called a "linear driving force." Linear because the concentration dependence is first order, i.e. power unity, and driving force because the rate is proportional to the difference between the maximum and the actual concentrations. That difference "drives" the process to completion.

The rate expression also includes two constants: **Km** is called the mass transfer coefficient, with dimension of $\frac{\text{Length}}{\text{time}}$ i.e. velocity, and $\boldsymbol{A}_{\textbf{interface}}$, which is the area $\left(\text{Length}^2\right)$ of the solid in contact with the liquid (either for one particle or for n-particles). In most cases the mass transfer coefficient will be larger if the mixing or agitation is greater. If the system has very little mixing it will be at a minimum value. The interfacial area is linearly related to the rate – the more area, the higher the rate. For the same mass one large particle will dissolve more slowly than many smaller particles. The difference is that the surface area to volume ratio is lower for the one large particle, whereas many small particles have high surface area to volume ratios and higher interfacial area (the area shared by both water and the salt particles.)

Conservation of mass across phases

The next step is to apply the conservation of mass principle to this problem. We need to write a component material balance on salt for both phases - that is for the solid and for the solution. Any mass that leaves one phase, (the solid) must end up in the other phase, (the solution). So we can say the following regarding the rate of salt mass accumulation in the two phases:

$$\frac{dm^{I}_{\text{salt}}}{dt} = \frac{dC^{I}_{\text{salt}} V^{I}}{dt} = + A_i \, r_d$$

$$\frac{dm^{\text{II}}_{\text{salt}}}{dt} = \frac{dC^{\text{II}}_{\text{salt}} V^{\text{II}}}{dt} = - A_i \, r_d$$

Where $\boldsymbol{A_i} = \boldsymbol{A}_{\textbf{interface}}$ and $\boldsymbol{r_d} = \textbf{rate}_{\textbf{dissolution}}$ which is the particular mass transfer process we are analyzing. We can substitute in the constitutive expression for the rate of mass transfer between the two phases to obtain:

$$\frac{dm^{I}_{\text{salt}}}{dt} = \frac{dC^{I}_{\text{salt}} V^{I}}{dt} = + K_m A_i \left[C^{\text{sat}'d}_{\text{salt}} - C^{I}_{\text{salt}}[t]\right]$$

$$\frac{dm^{\text{II}}_{\text{salt}}}{dt} = \frac{dC^{\text{II}}_{\text{salt}} V^{\text{II}}}{dt} = - K_m A_i \left[C^{\text{sat}'d}_{\text{salt}} - C^{I}_{\text{salt}}[t]\right]$$

The material balance for the solid salt phase includes in the differential term, $\boldsymbol{C}^{\textbf{II}}_{\textbf{salt}}$. Since the salt is a pure solid, this is the same as the density of the solid, which remains constant throughout the process. What this means is the water is assumed not to disrupt the lattice by penetrating into it and expanding it to result in dissolution. Instead we envision that it is only the first few layers of the salt that are involved in the process and the interior of the particle is left unchanged until it becomes surface. In essence the salt particles are considered to be shrinking over time, that is they have a shrinking core of solid that finally disappears when the salt is fully dissolved. The process is like one of layer by layer lift-off and dissolution. If the mass transferred between the phases is in total small, then we can ignore the change in solution volume that comes with the density change as the salt concentration rises. If this is too restrictive, then we can relax it later, but for now it makes good sense to ignore it and to concentrate on the mass transfer problem. The equations can be rewritten as:

$$\frac{dm^I_{\text{salt}}}{dt} = V^I \frac{dC^I_{\text{salt}}[t]}{dt} = + K_m A_i \left[C^{\text{sat}'d}_{\text{salt}} - C^I_{\text{salt}}[t]\right]$$

$$\frac{dm^{\text{II}}_{\text{salt}}}{dt} = \rho_{\text{salt}} \frac{\mathrm{d}V^{\text{II}}[t]}{dt} = - K_m A_i \left[C^{\text{sat}'d}_{\text{salt}} - C^I_{\text{salt}}[t]\right]$$

$$\therefore \quad V^I \frac{dC^I_{\text{salt}}}{dt} = - \rho_{\text{salt}} \frac{\mathrm{d}V^{\text{II}}}{dt}$$

The first equation, describes the increase in salt mass going into phase I, the solution phase. The second equation shows that all the salt mass going into solution comes from the solid salts, phase II, that are dissolving. This is a consequence of the conservation of mass. We see that the rate at which salt increases in solution is equal to the rate at which the solid salt is diminished. We also see that the volume of the solution is V^I and that of the solid salt is V^{II}. The volume of the solid salt is shrinking with time, while the concentration of the salt in the solution is increasing and the solution volume is constant.

The last equation can be integrated assuming that at $\mathbf{t = 0}$, $\boldsymbol{C^I_{\text{salt}}[t] = 0}$ and that $\boldsymbol{V^{\text{II}}[t] = V^{\text{II}}_o}$ and rearranged to give:

$$\rho_{\text{salt}} \, (V^{\text{II}}_o - V^{\text{II}}[t]\,) = V^I \; C^I_{\text{salt}}[t]$$

The left side of the equation expresses the mass of the solid salt at any time that is being transferred into solution. The right side of the equation gives the magnitude of the salt that has been transferred into solution at any time. At time zero, to, the mass of salt transferred is zero because $V^{\text{II}}_o = V^{\text{II}}[\text{to}]$ and $C^I_{\text{salt}}[t] = 0$. As time evolves the $V^{\text{II}}[t]$ becomes smaller than V^{II}_o and $C^I_{\text{salt}}[t]$ increases. Rearranging, the equation for $V^{\text{II}}[t]$ eventually goes to zero if all the solid salt is transferred into solution.

$$V^{\text{II}}[t] = V^{\text{II}}_o - \frac{V^I C^I_{\text{salt}}[t]}{\rho_{\text{salt}}}$$

$$C^I_{\text{salt}}[t] = \frac{\rho_{\text{salt}} \left(V^{\text{II}}_o - V^{\text{II}}[t]\right)}{V^I}$$

The last equation relates the concentration of the salt in the liquid phase at any time to the volume of the salt remaining in the solid at the same time. This solution and the one for the concentration of salt in solution are implicit. To use these equations we would need to measure the actual volume of the solid salt initially and then as a function of time. The initial volume would be easily obtained by using Archimedes principle. Put the solid salt particles in a non-solvent like hexane and measure the increase in apparent volume. The displaced volume would be the volume of the solid. Density of the solid can be found this way. The measurement of the solid volume during dissolution would be hard to do. So what we really need is an explicit solution in time. To obtain this we must return to the statements of the material balance between the phases.

The salt leaving the solid follows this equation:

$$\rho_{\text{salt}} \frac{\mathrm{dV}^{\mathrm{II}}[t]}{dt} = - K_m A_i \left[C_{\text{salt}}^{\text{sat'd}} - C_{\text{salt}}^{I}[t] \right]$$

If we are concerned only with the case in which the total mass of salt transferred is small relative to the volume of solvent and the saturation limit, then the equation becomes:

$$\rho_{\text{salt}} \frac{\mathrm{dV}^{\mathrm{II}}[t]}{dt} = - K_m A_i C_{\text{salt}}^{\text{sat'd}}$$

This looks as though we should be able to integrate it immediately, but look again! The area between the two phases is the area of the salt particles; it must be changing with time, and quite considerably at that. Therefore we cannot integrate this as of yet. We need a relationship between the volume of the solid at any time and the surface area it projects. But, we can find this. If each particle of solid is the same in size then their initial interfacial areas are the same. We will assume that as the area changes, the particles areas all change the same way. Then we can write for the surface are of each particle:

$$A_i = \mathrm{N}\, A_{i,N} = \mathrm{N}\, \gamma \left(\frac{V^{\mathrm{II}}}{N} \right)^{\frac{2}{3}} = N^{\frac{1}{3}}\, \gamma\, V^{\mathrm{II}\frac{2}{3}}$$

Here **N** is the number of identical particles of solid, and $\boldsymbol{\gamma}$ is the surface area to volume ratio, or the shape factor which accounts for the geometry of the solid, assuming that it is a regular polytope. Notice that the interfacial area goes as the 2/3 power of the solid volume. That makes sense since area has dimensions of L^2 and volume has dimensions of L^3. This allows us to rewrite the rate of change in solid volume (dropping the notation for t-dependence):

$$\frac{\mathrm{dV}^{\mathrm{II}}}{dt} = - \frac{K_m N^{\frac{1}{3}} \gamma V^{\mathrm{II}\frac{2}{3}} C_{\text{salt}}^{\text{sat'd}}}{\rho_{\text{salt}}}$$

$$\frac{\mathrm{dV}^{\mathrm{II}}}{V^{\mathrm{II}\frac{2}{3}}} = - \frac{K_m N^{\frac{1}{3}} \gamma C_{\text{salt}}^{\text{sat'd}}}{\rho_{\text{salt}}} \mathrm{d\,t}$$

Separate, integrate, and rearrange to obtain:

$$V^{II}[\,t\,] = \left(\sqrt[3]{V_o^{II}} - \frac{K_m N^{\frac{1}{3}} \gamma C_{salt}^{sat'd}}{\rho_{salt}} t\right)^3$$

Interestingly, the change in solid salt volume goes as time cubed. We have already stated that the measurement of change in the solid volume with time would be a difficult experiment to conduct. The measurement of salt concentration as a function of time is easy to do, so we want an explicit equation for the concentration. To get this we use this equation for volume change with time to get the concentration change with time.

$$C_{salt}^{I}[t] = \frac{\rho_{salt}\left(V_o^{II} - \left(\sqrt[3]{V_o^{II}} - \frac{K_m N^{\frac{1}{3}} \gamma C_{salt}^{sat'd}}{\rho_{salt}} t\right)^3\right)}{V^I}$$

If we made a series of experiments in which we sought the mass transfer coefficient, then we would rearrange this so that we could plot a function of the salt concentration against the time:

$$\left(1 - \frac{C_{salt}^{I}[t] V^I}{\rho_{salt} V_o^{II}}\right)^{\frac{1}{3}} = \left(1 - \frac{K_m N^{\frac{1}{3}} \gamma C_{salt}^{sat'd}}{\rho_{salt} \sqrt[3]{V_o^{II}}} t\right)$$

$$\frac{1}{\theta} = \frac{K_m N^{\frac{1}{3}} \gamma C_{salt}^{sat'd}}{\rho_{salt} \sqrt[3]{V_o^{II}}} \equiv \frac{1}{\text{time}}$$

A plot of this left-hand side versus the time gives a graph whose slope is the coefficient of **t**. Everything in this group with inverse time as its dimensions should be known before the experiments are even conducted. If we knew the mass transfer coefficient, then the inverse of this group would provide the time required to dissolve all **N** particles of the salt. We can see that when the time **t** is equal in magnitude to the reciprocal of this group the right-hand side goes to zero identically.

$$\left(1 - \frac{C_{salt}^{I}[t] V^I}{\rho_{salt} V_o^{II}}\right)^{\frac{1}{3}} = \left(1 - \frac{t}{\theta}\right)$$

$$\left(\frac{1}{\theta}\right) = \frac{K_m N^{\frac{1}{3}} \gamma C_{salt}^{sat'd}}{\rho_{salt} \sqrt[3]{V_o^{II}}} \equiv \left(\frac{1}{\text{time}}\right)$$

The left-hand side must also be zero, therefore the concentration of the salt at that time will be the reciprocal of the product of the density of the salt and its initial volume divided by the volume of the liquid.

Fit to the batch data

The table that follows is data for an experiment in which 100 salt cubes ($\gamma = 6$) were dissolved in 100 cm^3 of water. The total volume of the salt was 1cm^3, its density was 2.2 g cm^3, the saturation limit of the salt was 0.38 g cm^{-3}. Data points were logged every 50 seconds for a total of 2500 seconds.

t/sec	Csalt[t]	t/sec	Csalt[t]
50	0.00171084	1300	0.0196429
100	0.00279533	1350	0.0192367
150	0.00424841	1400	0.0193228
200	0.00494699	1450	0.0193325
250	0.00635887	1500	0.0192889
300	0.00763428	1550	0.0201392
350	0.00835931	1600	0.0197141
400	0.010246	1650	0.0197589
450	0.0105354	1700	0.0203605
500	0.0116536	1750	0.0202748
550	0.0120036	1800	0.0200916
600	0.0135537	1850	0.0201311
650	0.0137191	1900	0.02012
700	0.0142037	1950	0.0200507
750	0.0150545	2000	0.0195248
800	0.0153611	2050	0.0202914
850	0.0165474	2100	0.0199335
900	0.0169347	2150	0.0204394
950	0.0167272	2200	0.0204604
1000	0.0177943	2250	0.0203341
1050	0.018098	2300	0.0200852
1100	0.0184004	2350	0.0196325
1150	0.0188067	2400	0.0205709
1200	0.0190497	2450	0.0198959
1250	0.019132	2500	0.020606

Putting the data into vector notation for manipulation we have:

csaltdata =

{{“0”, “0.00004044”}, {“50”, “0.001815”},

{“100”, “0.002589”}, {“150”, “0.004589”},

```
{"200", "0.005424"}, {"250", "0.006528"},
{"300", "0.00783"}, {"350", "0.008987"},
{"400", "0.009514"}, {"450", "0.01022"},
{"500", "0.01143"}, {"550", "0.01287"},
{"600", "0.01328"}, {"650", "0.01399"},
{"700", "0.01488"}, {"750", "0.01535"},
{"800", "0.01536"}, {"850", "0.0161"},
{"900", "0.01694"}, {"950", "0.01755"},
{"1000", "0.01704"}, {"1050", "0.0183"},
{"1100", "0.01851"}, {"1150", "0.01894"},
{"1200", "0.01894"}, {"1250", "0.01861"},
{"1300", "0.01951"}, {"1350", "0.01946"},
{"1400", "0.01949"}, {"1450", "0.01958"},
{"1500", "0.01997"}, {"1550", "0.01968"},
{"1600", "0.02014"}, {"1650", "0.01973"},
{"1700", "0.02041"}, {"1750", "0.02017"},
{"1800", "0.02026"}, {"1850", "0.02036"},
{"1900", "0.02042"}, {"1950", "0.02023"},
{"2000", "0.0204"}, {"2050", "0.01979"},
{"2100", "0.02013"}, {"2150", "0.02013"},
{"2200", "0.02035"}, {"2250", "0.0196"},
{"2300", "0.02043"}, {"2350", "0.02035"},
{"2400", "0.02025"}, {"2450", "0.02061"},
{"2500", "0.02048"}};
datpl = ListPlot[csaltdata,
AxesLabel → {"t/sec", "Csalt [ t ]"},
PlotStyle → {PointSize[.01], Hue[0]}]
```

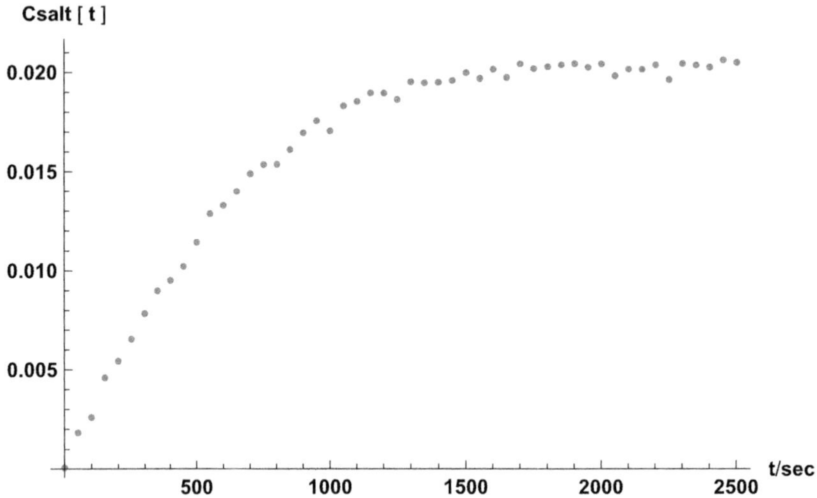

We shall want to fit this data "**csaltdata**" to the expression that we have derived since, as we can see from the data, the final concentration is less than 50% of that at saturation. We will fit the time data out to 1500 seconds. One way to do this is to do a one parameter, non-linear fit to the expression after we have simplified it by evaluating all the parameters. The first step is to get the fitted expression, evaluate it and then compare to the data:

Clear[*g*]

ρsalt = 2.2;

VoII = 1;

γ = 6;

Km=.

Csaltsatd = .38;

VI = 100;

***n* = 100;**

tmax = 2500;

$$\text{csalt[t\_]:=}\frac{\rho\text{salt}\left(\text{VoII}-\left(\sqrt[3]{\text{VoII}}-\frac{\text{Km}\,n^{\frac{1}{3}}\gamma\,\text{Csaltsatd}}{\rho\text{salt}}t\right)^{3}\right)}{\text{VI}}$$

csalt[*t*]

$0.022\left(1-(1-4.81037\text{Km}t)^{3}\right)$

g = NonlinearModelFit[csaltdata, csalt[*t*], Km, *t*]

FittedModel []

Normal[*g*]

$0.022\left(1-(1-0.00040787t)^{3}\right)$

plfit = Plot[*g*[*t*],

{*t*, 0, 3500}, PlotRange → {{0, 2500}, {0.0, .025}},

PlotStyle → Hue[0.6]];

Show[plfit, datpl]

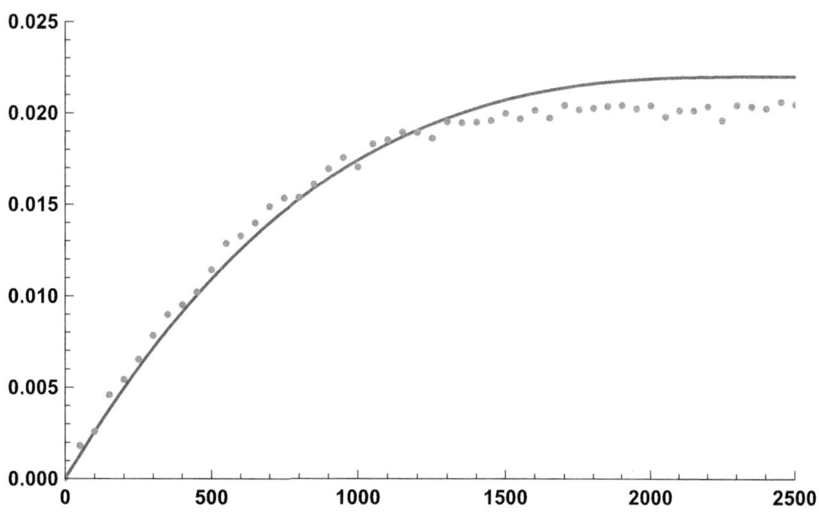

The coefficient of **t** is **Km** and that is:

Km = 0.00041 = 4.1 x 10^{-4}cm sec^{-1}

So that is how we can use the linear driving force model of mass transfer to analyze and model the dissolution of salt in water.

Semi-continuous: pseudo-steady state

We can imagine a situation where our goal is to remove a salt deposit that is sparingly soluble out of a unit. An example of such salts is the alkaline salts that deposit in boilers and heat exchangers as "scale." Nothing more than the accumulation of precipitate on the inner walls of the vessel over time, these can present a real hazard in that they change the heat conduction through the wall, since they are good insulators. As a result of this, boilers can develop hot spots and explode, heat exchangers can become much less efficient over time with similarly deleterious results. At the same time these salts may be sparingly soluble except in acidic solution, which, if the pH is too low, will etch away the vessel wall along with the salts over time. Therefore one may be forced to accept the low solubility in the less acidic pH range, and be willing to pump large volumes of solvent through for longer periods of time.

The essence of this problem, is the transfer of mass from the solid deposit to the liquid that is continuously flowing and slowly diminishing the mass of the solid. The flow of solvent does two things – it provides a large volume of solvent when the flow is integrated over time and, if it is a relatively high rates, it provides much better mass transfer rates than if the same large volume were merely standing in contact with the solid without flow. Flow gives mixing and mixing gives higher mass transfer coefficients, which means it will take less time to dissolve than it would with less or zero mixing. The physical situation is as shown in this figure:

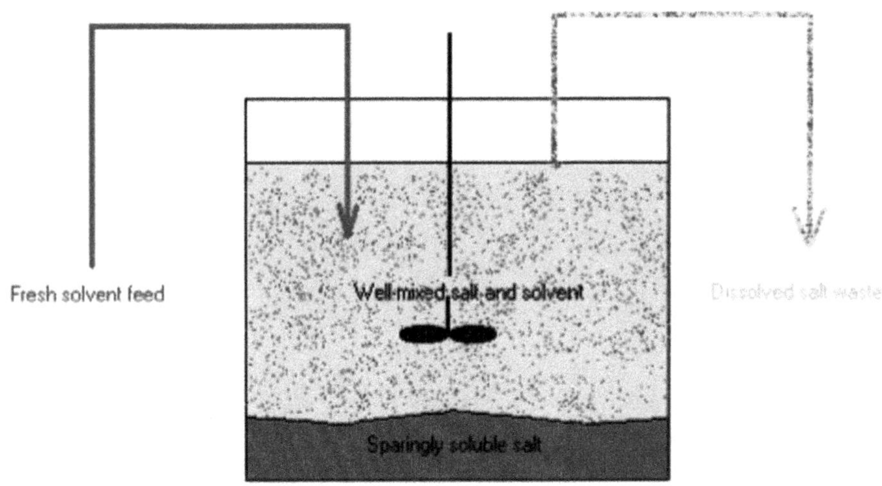

In this situation the salt is not in particulate form. Instead it is more like a layer that is solidly attached to the floor of the vessel. It is more like a slab of salt.

The equations look largely the same, except that the solution phase balance on the salt has a flow term for the mass of salt leaving the unit by this process:

Net change in salt mass = rate of dissolution - flow out

$$\frac{dm^{I}_{salt}}{dt} = V^{I}\frac{dC^{I}_{salt}[t]}{dt} = + K_m A_i \left[C^{sat'd}_{salt} - C^{I}_{salt}[t]\right] - C^{I}_{salt}[t]\ qex$$

$$\frac{dm^{II}_{salt}}{dt} = \rho_{salt}\frac{dV^{II}[t]}{dt} = -\ K_m A_i \left[C^{sat'd}_{salt} - C^{I}_{salt}[t]\right]$$

The interfacial area in this case will be taken to be a constant and well-approximated by the cross sectional area of the unit, πr^2. This means we will assume that the salt is removed by a process that removes layers at the same rate of dissolution at every point in the salt slabs surface. The change in salt volume then becomes a one dimensional problem of computing the salt thickness at any time. We can and will come back to this.

If the salt is sparingly soluble, then $C^{sat'd}_{salt}$ is small in magnitude, and if the product $K_m A_i$ is relatively large due to good mixing, then the salt concentration is likely to be a constant and close to, but not as large as $C^{sat'd}_{salt}$, depending on the magnitudes of the parameters. If the salt concentration is taken as a constant, then its rate of change is zero, and the salt in solution is at a steady state. The rate of change of the salt concentration is zero, but the concentration itself is not zero. Even so, the salt mass is changing steadily and constantly as a function of time. Because the rate of dissolution is the same at any point on the upper surface, it is the thickness of the salt slab that is changing. So the concentration of salt is the same, and the thickness of the salt is changing and both are simultaneously true. This is a mixed condition in this two phase system, as a whole it is said to be in a pseudo-steady state. The reason is that if we could only measure the concentration of salt exiting the reactor, then we would find it to be a constant at constant conditions. But we know the salt is coming from inside the control volume since we are not feeding it. That means, by the principle of conservation of mass, that the salt is emerging from a dissolving source within the control volume, and this mass must be decreasing with time.

The equations work out as follows for the pseudo-steady state. The rate of change in salt concentration with time is zero;

$$0 = + K_m A_i \left[C^{sat'd}_{salt} - C^{I}_{salt,stst}\right] - C^{I}_{salt,stst}\ qex$$

therefore:

$$K_m A_i \left[C^{sat'd}_{salt} - C^{I}_{salt,stst}\right] = C^{I}_{salt,stst}\ qex$$

and the rate of change in the salt mass and volume are changing with time

$$\rho_{salt}\frac{dV^{II}[t]}{dt} = -\ C^{I}_{salt,stst}\ qex$$

and therefore:

$$V^{\mathrm{II}}[t] = V_o^{\mathrm{II}} - C^{I}_{\mathrm{salt,stst}}\,\mathrm{qex}\ \ \mathrm{t}$$

Since the area of the salt deposit, A_{salt}, is constant, we can see that this gives the time dependence of the change in the salt deposit's thickness th[t]:

$$th_{\mathrm{salt}}[t] = \frac{V_o^{\mathrm{II}}}{A_{\mathrm{salt}}} - \frac{C^{I}_{\mathrm{salt,stst}}\mathrm{qex}t}{A_{\mathrm{salt}}} = th_o^{\mathrm{II}} - \frac{+C^{I}_{\mathrm{salt,stst}}\mathrm{qex}t}{A_{\mathrm{salt}}}$$

This very simple solution comes about as a result of the fact that at the steady state the concentration of salt is a constant and the exit mass flow has to be equal to the rate of salt mass transfer into the solvent.

Full solution

By solving the equation in the way we have just described, we make the mathematics much simpler, but we also place constraints on the solution. Instead of doing that, we now solve the equations without these assumptions, this way they are appropriate for the most general case–from short time to long, sparingly soluble to very soluble, etc.

$$\frac{d\rho^{I}[t]V^{I}[t]}{dt} = (\ \rho_{\mathrm{solvent}} - \rho^{I}[\ \mathrm{t}\ ]\ )\ \mathrm{qex}$$

$$\rho^{I}[\ \mathrm{t}\ ] = \rho_{\mathrm{solvent}}\ +\ \mathrm{a}\ C^{I}_{salt}[t]$$

$$\frac{dm^{I}_{\mathrm{salt}}}{dt} = \frac{dC^{I}_{salt}[t]V^{I}[t]}{dt} = +\,K_m A_i\left[C^{\mathrm{sat'd}}_{\mathrm{salt}} - C^{I}_{\mathrm{salt}}[t]\right] - C^{I}_{\mathrm{salt}}[t]\ \mathrm{qex}$$

$$\frac{dm^{\mathrm{II}}_{\mathrm{salt}}}{dt} = \rho_{\mathrm{salt}}\frac{\mathrm{d}V^{\mathrm{II}}[t]}{dt} = -\,K_m A_i\left[C^{\mathrm{sat'd}}_{\mathrm{salt}} - C^{I}_{\mathrm{salt}}[t]\right]$$

Here we have added one equation – the total mass balance – which includes the density of phase one as it flows out of the system. Recall also the for a double salt, $\boldsymbol{M_a\ L_b}$ we have the following:

$$K_{\mathrm{sp}} = C_{M^+}{}^{a} C_{L^-}{}^{b} = (\ \mathrm{a}\ C^{\mathrm{sat'd}}_{\mathrm{salt}}\ )^{a}\ \ (\ \mathrm{b}\ C^{\mathrm{sat'd}}_{\mathrm{salt}}\ )^{b}$$

We can solve for the saturation concentration of the salt in terms of its **Ksp** and the stoichiometric numbers:

```
Solve[Ksp == PowerExpand[(aCsatd)^a (bCsatd)^b],
Csatd]
```

$$\left\{\left\{\text{Csatd} \to \left(a^{-a} b^{-b} \text{Ksp}\right)^{\frac{1}{a+b}}\right\}\right\}$$

Turning once more to the equations, we write code that will solve these numerically and simultaneously by using this expression for the saturation concentration of the salt and the linear dependence of density upon concentration. The code that follows does just this. The tank parameters are specified along with the volumes of the solution and salt phases at time zero, (**VIo** and **VIIo**); the salt parameters, the mass transfer and flow rates, the maximum time for the integration to be done, the function calls for the exit flow rate in terms of the inlet flow rate, density of the solution and the saturation concentration of the salt, the material balance equations, the implementation of the numerical solution of the equations and the assignment of the interpolation functions to function names, and finally the graphical output routines.

The tank is 3.1 million cm^3 or 3100 liters in volume. The salt is an adherent slab at the bottom of the tank that is $\sim$ 3.2 cm thick.

```
"The tank parameters are:";
r = 100; (* 100 cm *)
Ai = N[πr^2];
VIo = 100Ai; (* the tank is 100 cm high *)
VIIo = 100000;
(* this is the volume of the salt *)
th = VIIo/Ai; (* this gives the thickness of the salt layer at the bottom of the tank *)

"These are the salt and solvent parameters";
ρsolvent = 1;
ρsalt = 2;
a = 1;
b = 1;
Ksp = 1 10^-4;
```

$\gamma = 1;$

$\text{clo} = 10^{-10}$; (* initial concentration of salt in the solvent phase *)
(* The mass transfer coefficient and flow rates *)

$\text{Km} = 7.3\ 10^{-3}$; (* cm/ sec *)

$\text{qo} = 60$; (* cm^3/sec *)

$f = .05$; (* flow factor sec^{-1} *)

(* This is the maximum time for the integration *)

$\text{tmax} = N\left[2.5\ 10^5\right];$

(* This specifies the exit flow in terms of
the inlet flow rate i.e. it is a constant *)

qex[t_]:=qo

(* The density of the solution exiting the
unit and the saturation concentration or
solubility of the salt *)

$\rho\text{I[t\_]:=}N[\rho\text{solvent} + \gamma\text{cI}[t]]$

(* This is a function for the saturation solubility in g/cm^3 *)

$\text{csatd[a\_, b\_, Ksp\_]:=}N\left[(\text{Ksp})^{\frac{1}{a+b}}\right]$

(* The set of mass balance equations to be solved *)

eqns = {

$\partial_t(\rho\text{I}[t]\text{VI}[t]) == \rho\text{solventqo} - \rho\text{I}[t]\text{qex}[t],$

$\text{VI}[0] == \text{VIo},$

$\partial_t(\text{cI}[t]\text{VI}[t]) ==$

$\text{KmAi}(\text{csatd}[a, b, \text{Ksp}] - \text{cI}[t]) - \text{cI}[t]\text{qex}[t],$

```
cI[0] == cIo,

∂t(VII[t]) == -KmAi(csatd[a,b,Ksp]-cI[t]) / ρsalt,

VII[0] == VIIo};

(* Numerical solutions of the equations and their assignments to functions *)

soln = NDSolve[

eqns,

{VI[t], VII[t], cI[t]},

{t, 0, tmax}];

(* Here we assign the solutions to functions *)

cOne[t_]:=Evaluate[cI[t]/.soln[[1]]]

vOne[t_]:=Evaluate[VI[t]/.soln[[1]]]

vTwo[t_]:=Evaluate[VII[t]/.soln[[1]]]

Plot[N[cOne[t]], {t, 0, tmax},

PlotRange → {{0, tmax},

{0, cOne[tmax]}},

AxesLabel → {t, "C^I_salt[ t ]"},

PlotStyle → Hue[0.4]]

Plot[vTwo[t]/Ai, {t, 0, tmax},

PlotRange → {{0, tmax},

{0, vTwo[0]/Ai}},

AxesLabel → {t, "TH^II[ t ]"},

PlotStyle → Hue[0.9],

PlotLabel → "Change in salt slab thickness, TH/cm"

  ]
```

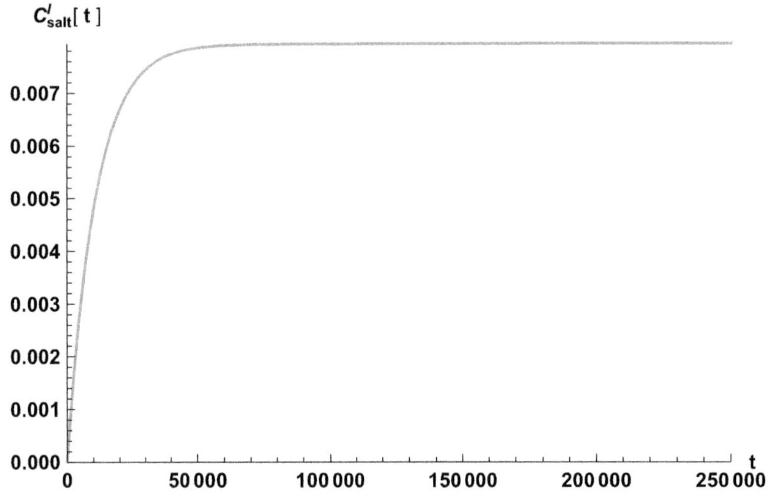

Change in salt slab thickness, TH/cm

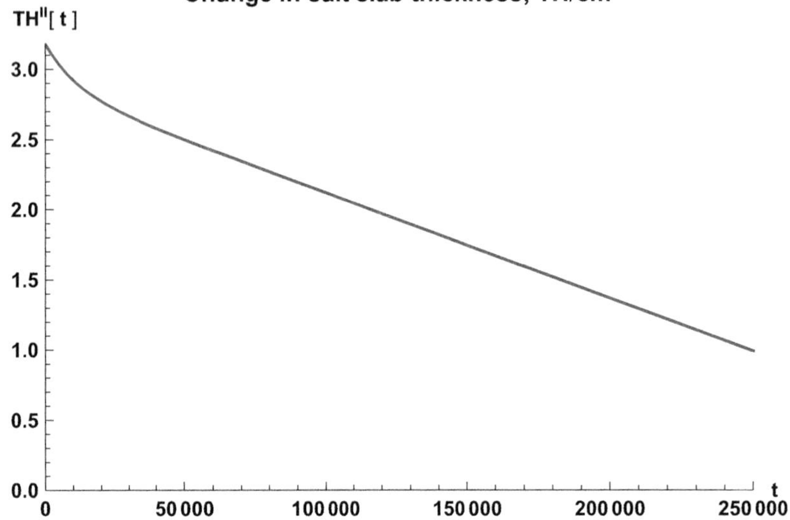

This can now be used interactively to experiment with parameter values in order to learn how they affect the observed behavior of this system. There are several points that must be noted about this code that bear explanation.

Firstly, the initial concentration of the salt in the solution, **cIo**, is not taken as zero; it is set to a very low value to simulate zero at time zero. If, we set this identically to zero, the numerical routine will come back with a complex infinity error, because it will have divided by zero at the start of the calculation.

Secondly, the exit flow rate has been made equal to the inlet concentration. This is one way to handle the problem of the exit flow rate. By doing it this way, what we see is that there is a short transient at the beginning of the process as the salt concentration comes to a steady state value. The thickness of the salt layer drops sharply during this transient and then drops linearly with time. For simplicity we have left everything in cm and seconds, hours would be more appropriate time units.

In the following code we have changed the parameters involving seconds to hours and time to hours.

"The tank parameters are:";

$r = 100$; (* 100 cm *)

$\text{Ai} = N[\pi r^2]$;

VIo = 100Ai; (* VIo is the volume of the tanks and its wall is 100 cm high *)

VIIo = 100000; (* VIIo is the volume of the salt *)

th = VIIo/Ai; (* This gives the thickness of the layer at the bottom of the tank *)

"These are the salt and solvent parameters";

ρsolvent = 1;

ρsalt = 2;

$a = 1$;

$b = 1$;

$\text{Ksp} = 1\ 10^{-3}$;

$\gamma = 1$;

$\text{cIo} = 10^{-10}$; (* The initial concentration of salt in the solvent phase *)

(* The mass transfer coefficient and flow rates *)

$\text{Km} = 7.3\ 10^{-3} * 60 * 60$;

(* cm/ sec *60 sec / min *60 min /hr *)

qo = 60 * 60; (* cm^3/ sec * 60 sec /min * 60 min /hr *)

(* This is the maximum time for the integration *)

$\text{tmax} = N[2.5\ 10^5/60/60]$;

(* This specifies the exit flow in terms of the inlet flow rate, which is a constant *)

```
qex[t_]:=qo
ρI[t_]:=N[ρsolvent + γcI[t]]

(* This is a function for the saturation solubility in g/cm^3 *)
csatd = N[(Ksp)^(1/(a+b))]

(* The set of mass balances to be solved *)
eqns = {
∂t( ρI[t]VI[t]) == ρsolventqo − ρI[t]qex[t],
VI[0] == VIo,
∂t(cI[t]VI[t]) == KmAi(csatd − cI[t])−
cI[t]qex[t], cI[0] == cIo,

∂t(VII[t])== −KmAi(csatd−cI[t])/ρsalt,
VII[0] == VIIo};
soln = NDSolve[
eqns,
{VI[t], VII[t], cI[t]},
{t, 0, tmax}];
(* Here we assign the solutions to functions *)
cOne[t_]:=Evaluate[cI[t]/.soln[[1]]]
vOne[t_]:=Evaluate[VI[t]/.soln[[1]]]
vTwo[t_]:=Evaluate[VII[t]/.soln[[1]]]
Plot[N[cOne[t]], {t, 0, tmax},
PlotRange → {{0, tmax},
{0, cOne[tmax]}},
AxesLabel → {"t/hr", "C^I_salt[ t ]"},
PlotStyle → Hue[0.4]]
Plot[vTwo[t]/Ai, {t, 0, tmax},
```

```
PlotRange → {{0, tmax},
{0, vTwo[0]/Ai}},
AxesLabel → {"t/hr", "TH^II[ t ]"},
PlotStyle → Hue[0.9],
PlotLabel → "Change in salt slab thickness, TH/cm"
]
```

0.0316228

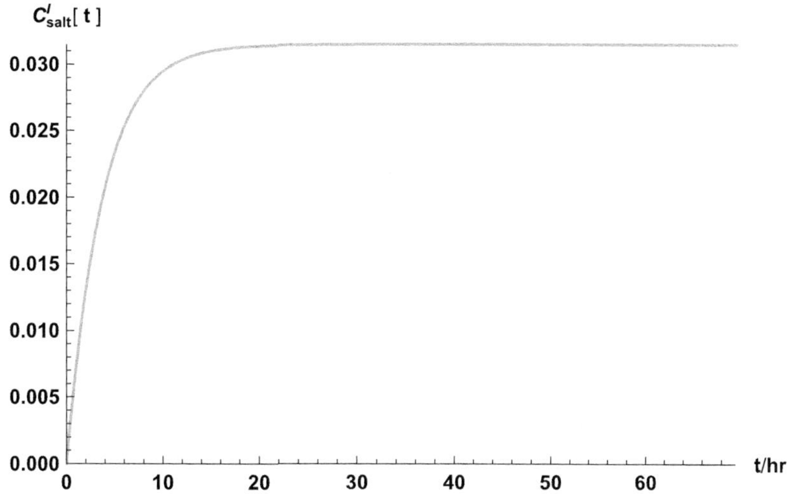

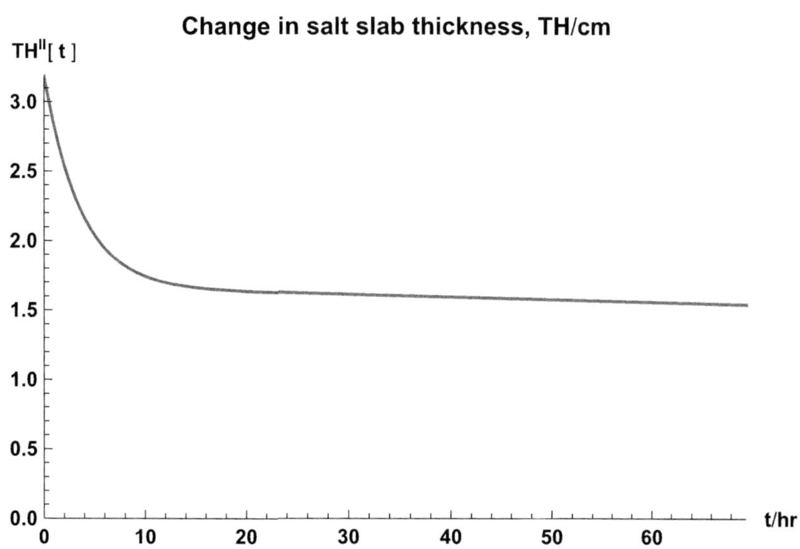

There is a significant decrease in the salt slab thickness in the first 10 or so hours. After that the rate diminishes to a slow linear decline. Hence the model shows that either the parameters controlling solubility have to be adjusted chemically or by temperature, or the process should be stopped and started every ten hours. Otherwise the rate of removal will be inordinately long, inefficient, and probably economically unacceptable. Since it may be difficult to measure the actual progress of the process if it is in a unit that cannot be opened for observation, the model becomes an essential tool for assessing the likely progress of the process.

Liquid-liquid system

Fully continuous

Steady state: equilibrium stage

Liquid-liquid extractions are used in may different applications from chemicals production to environmental clean-up. It is possible to extract organics from water by contacting the water with a better solvent for the impurities, which is also immiscible with the water. When there are two liquid phases involved we have new equilibrium considerations to take into account, whereas in the case of the salt we had only one, the solubility, since the second phase was the pure salt. The phases will not change in volume within the unit that is used for contacting them. There are now two "solubilities" of the transferred component– one for each phase. These are better termed the equilibrium concentrations of the component dissolved separately in each phase.

This kind of liquid-liquid extraction is done routinely in chemical synthesis work at the bench level. The solution to be extracted is typically aqueous and contains the desired compound. This is added to a separatory funnel first. Then a less dense, immiscible organic solvent, that has a higher affinity for the target compound is added as a layer on top. This solvent's "higher affinity" for the target, means that the target is more soluble in it than in water. After capping the funnel and inverting it, the mixture is shaken vigorously for some time. Then the funnel is returned to a stand and the two immiscible liquids are allowed to separate. The lighter organic solvent, now containing the target molecule, is decanted or siphoned off the top. The process is typically repeated three times. Then the organic solvent is evaporated or reduced in volume to recover the product and the aqueous phase is discarded.

Interestingly, at the scale of a process all the same things are done, but typically continuously for large scale production. Batch processes, however, can be scaled up to larger volumes, and this is done in processes that yield specialty chemicals, such as flavors, fragrances, or pharmaceuticals with high added value. We will consider the continuous process run at steady state. The physical situation looks as follows:

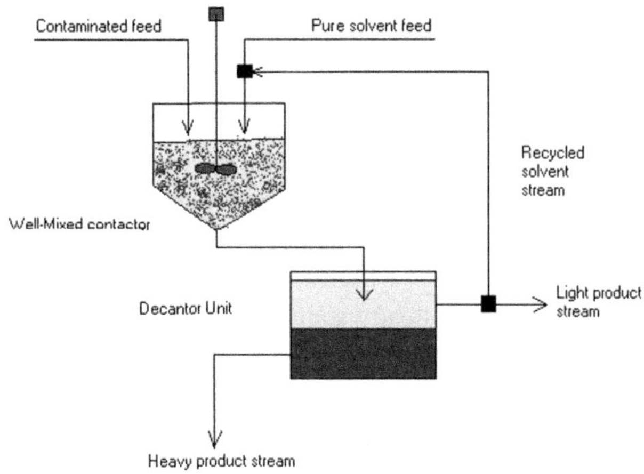

The denser aqueous feed containing the desired product (or a contaminant) is mixed with the less dense pure solvent in a contactor. The well-mixed stream emerges from this unit and flows into the decanter unit where the two phases are given enough time to fully separate. If the two phases separate rapidly and completely, then this can be done continuously, if not, then the process is semi-continuous, or not continuous (batch, mode). Once in the unit the two liquids are well-mixed, but in the decantor unit they are well-separated, as shown in the schematic. The top layer consists of the less dense organic solvent with the product in it, and the heavier aqueous phase can be extracted again if there is more product to be recovered economically, if pure it can be discarded, or reused elsewhere. The organic solvent can be reduced in volume by vaporization in another unit. The product can be recovered as a solid (or liquid) and the organic solvent can be condensed and sent back to the separation unit for reuse. The analysis of this unit calls for analyses of the sub-units that make it up.

We will call the contactor where the two immiscible liquids, one aqueous and the other organic are forced to mix well, unit 1. The mix of the two liquids will be fed to a decantor unit, which is a large holding tank in which the two immiscible liquids can separate. This is unit 2. The light liquid in the decantor is enriched in product i because the product is more soluble in the organic solvent than in the aqueous liquid. Because the two liquids split and the organic layer is lighter, and less dense, it is suspended on top of the aqueous liquid. As a result it can be siphoned away from the heavier aqueous phase.

After leaving the decantor, the organic liquid is transported to another unit, unit 3, that strips the solvent leaving the product as a microcrystalline solid, and ready for further downstream purification and processing. The stipped solvent is sent back to the feed of solvent into the contactor.

The aqueous liquid is transported out of the bottom of the decantor, unit 2. Depending on the whether there is any residual product i in this liquid, it may be extracted again if it is economical to do so. If it is not, then this water may be used elsewhere or discarded in compliance with all environmental protection laws.

We begin the analysis at the top of unit 1 with the inlet stream pure organic solvent. The stream of solvent (s) has a density, ρ_s, and a flow rate, q_s. This is mixed with the solvent from the recycle stream (rs) that has a density, ρ_{rs}, and a flow rate q_{rs}, and it may have small residual concentration of the product, i, hence the concentration $C_{i,rs}$. The two streams are mixed combined into one with a flow rate, $q_{s,f}$, where sf stands for solvent feed, a density of $\rho_{s,f}$, and a concentration of i of $C_{i,s,f}$. The steady state mass balances at the mixing tee (where the pure and recycled streams are merged) are:

(mass flow of pure solvent into the tee) + (mass flow of recycled solvent into the tee) = (total mass flow of solvent out of the tee)

$$\rho_s\, q_s + \rho_{rs}\, q_{rs} = \rho_{s,f}\, q_{s,f}$$

(mass flow of i into the tee) = (mass flow of i out of the tee)

$$C_{i,rs}\, q_{rs} = C_{i,s,f}\, q_{s,f}$$

Next we note that the heavy aqueous stream containing the product and the solvent stream are being fed to the top of the well-mixed contactor, unit 1. Mathematically, because we assume the two streams are "well-mixed", they are considered to be perfectly mixed instantaneously. So the perfect mixing in this case is the perfect dispersion of the aqueous liquid in the organic. We can imagine the aqueous liquid as being dispersed as infinitely small spheres within the organic liquid. The transfer of product i from the aqueous to the organic liquid takes place across the spherical surface and it is driven by the fact that the product is much more soluble in the organic liquid than in water. At the outlet of unit 1, the two immiscible liquids have been mixed. We can also assume that the transfer of the product i is instantaneous, which is to say that we can set aside the kinetics of the mass transfer process for the initial analysis by assuming that equilibrium is established between the two liquids.

At steady state, the material balances for this well-mixed unit are:

(mass flow of the heavy phase (h,f: heavy feed))
+ (mass flow of the light phase i.e. solvent (s, f: solvent,feed))
= (mass flow of the mixture (m,c: mixture from contactor) out of the contactor)

$$\rho_{h,f}\, q_{h,f} + \rho_{s,f}\, q_{s,f} = \rho_{m,c} q_{m,c}$$

(mass flow of the product i in the heavy phase (i,h,f: product i, in the heavy phase, feed)
+(mass flow of product i in the light phase i.e. solvent (sf))
= (mass flow of the product i in the mixture (mc) out of the contactor)

$$C_{i,h,f}\, q_{h,f} + C_{i,s,f}\, q_{s,f} = C_{i,m,c}\, q_{m,c}$$

Notice that the mass flows out of unit 1 are written as if the mixture is truly perfect and as if there is only one phase present and not two - the organic and the aqueous. This is an approximation that simplifies the equations and remains consistent internally with the perfectly mixed criterion with unit 1.

The decantor, unit 2, has these equations associated with it. Remember we are not concerned with its internals, only with the mass flows into and out of it. The mass flow in is that of the mixture feed from the contactor. The flows out are those of the impure solvent and the purified heavy stream:

(mass flow of mixture into the decantor) = (mass flow of the light liquid out of the decantor) + (mass flow of the heavy liquid out of the decantor)

$$\rho_{m,c}\, q_{m,c} = \rho_{d,s}\, q_{d,s} + \rho_{d,h}\, q_{d,h}$$

$$C_{i,m,c}\, q_{m,c} = C_{i,s,d}\, q_{d,s}\, C_{i,h,d}\, q_{d,h}$$

The light, organic solvent is now enriched with the product i, and, ideally, the heavy aqueous liquid has little or no residual product i.

Finally, the solvent stream is separated at another unit (3) shown as a simple tee in the diagram. Entering this unit is the solvent stream enriched with the product i. The unit strips the solvent either by raising the temperature, or by reducing pressure, or both, so that the product is mostly separated from the solvent, and is left in a highly concentrated solution for further processing i.e. crystallization. The nearly pure or fully pure solvent is sent back to the unit 1 through the inlet solvent tee. The equations that describe this are:

(mass flow out of unit 2 of light solvent phase with product i)
= (mass flow of pure solvent out of separator)
+ (mass flow of the concentrated product out of separator)

$$\rho_{d,s}\, q_{d,s} = \rho_{r,s}\, q_{r,s} + \rho_{s,p} q_{s,p}$$

(mass flow of product i out of unit 2 and into the separator)
= (mass flow of product i in the recycle solvent stream) +
(mass flow product i in concentrated product stream)

$$C_{i,s,d}\, q_{d,s} = C_{i,r,s}\, q_{r,s} + C_{i,s,p}\, q_{s,p}$$

The schematic of the flow sheet is shown once again with all the streams labeled and with a red (mid gray in the print version) imaginary box around the unit which cuts all the streams that either enter or leave this unit:

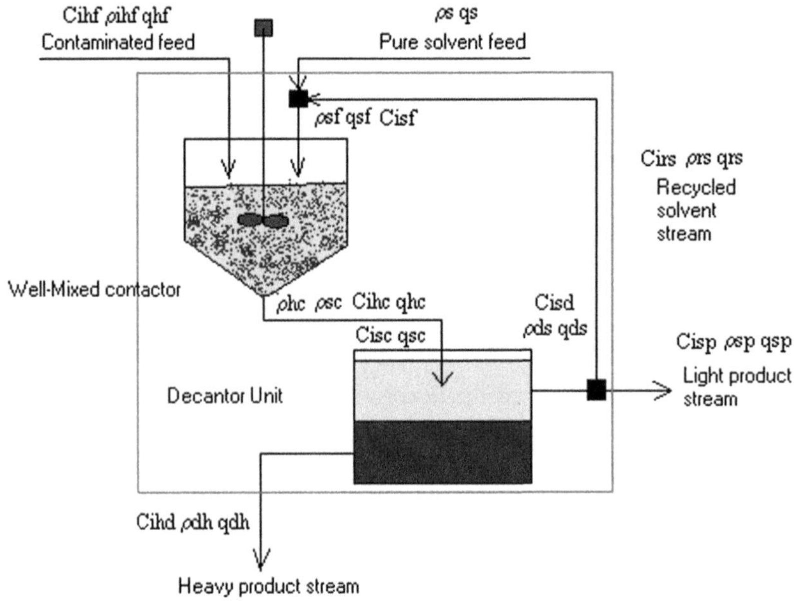

The red (mid gray in the print version) box is an imaginary *control surface* for the unit as a whole. Despite all the details that we have just considered, there is one overall set of mass balances for the unit as a whole. This treats the unit as a so-called "black-box", which means that even if the internal workings were hidden from view, we would be able to do an overall balance on the system as shown below:

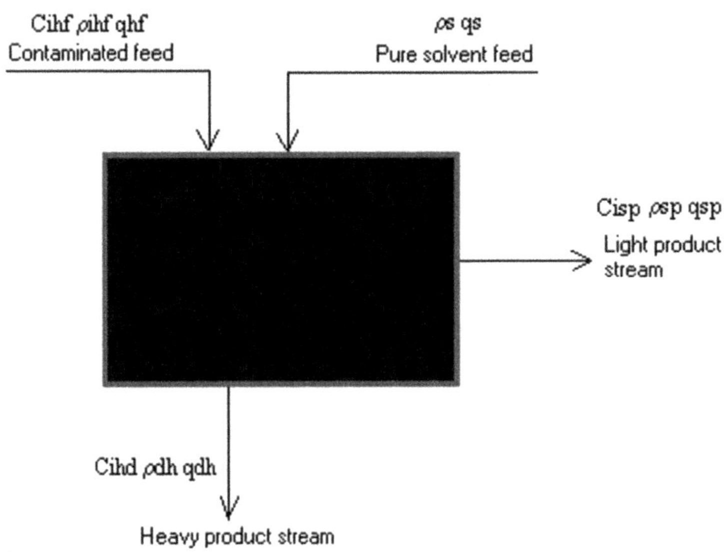

This greatly simplifies the initial stages of this problem, so it is a logical place to begin. From this diagram and the conservation of mass, we can write that:

(mass flow of the heavy phase in) + (mass flow of light phase)

= (mass flow of heavy phase out) + (mass flow of light phase out)

$$\rho_{i,h,f}\, q_{h,f} + \rho_s q_s = \rho_{d,h}\, q_{d,h} + \rho_{s,p}\, q_{s,p}$$

$$C_{i,h,f}\, q_{h,f} = C_{i,h,d}\, q_{d,h} + C_{i,s,p}\, q_{s,p}$$

In these two equations we have a total of 14 parameters and variables. We need to reduce this number. How can we do this? The most crucial assumption we can make is that the unit runs at equilibrium. This means that the concentrations of the product in both liquid phases emerging from the unit are at equilibrium. To understand this let's pretend that the separation was done stepwise rather than continuously. The volume of light solvent added would be $q_s \Delta t$. The volume of the product-containing stream would be $q_{i,h,f} \Delta t$. The initial concentration of the product in the heavy phase is $C_{i,h,f}$. If the two phases are in contact, then the product will transfer spontaneously to the light phase where its solubility is higher. This transfer will occur until the concentrations of the product in the two phases is no longer changing, in other words, until the product concentration comes to equilibrium between the two phases. For this reason, a unit assumed to operate at the limit of equilibrium is referred to as an *equilibrium stage*, and this level of analysis is the *equilibrium stage analysis*. If ratio of these two concentrations at equilibrium is a constant over a range of different concentrations, then the constant is referred to as the partition coefficient, K_d:

$$K_d = \frac{C_{i,h,e}}{C_{i,s,e}} = \frac{\text{Concentration of i in the aqueous, heavy phase at equilibrium}}{\text{Concentration of i in the organic, light phase at equilibrium}}$$

If the system we are examining also comes to an equilibrium condition, then the concentrations of impurity in the two outlet streams are coupled.

$$K_d = \frac{C_{i,h,d}}{C_{i,s,p}}$$

It is also reasonable to expect that the density of the contaminated streams is not too different from their pure densities, since the contaminant is usually at low concentrations.

$$\rho_h\, q_{h,f} + \rho_s\, q_s = \rho_h\, q_{d,h} + \rho_s\, q_{s,p}$$

$$C_{i,h,f}\, q_{h,f} = C_{i,h,d}\,\left(q_{d,h} + \frac{1}{\text{Kd}}\, q_{s,p}\right)$$

Simplify[

Solve$[\{\rho_h q_{h,f} + \rho_s q_s == \rho_h q_{d,h} + \rho_s q_{s,p},$

$C_{i,h,f} q_{h,f} == C_{i,h,d}(q_{h,d} + q_{s,p})\},$

$\{q_{d,h}, C_{i,h,d}\}]]$

$$\left\{\left\{q_{d,h} \to \frac{q_s\rho_s + \rho_h q_{h,f} - \rho_s q_{s,p}}{\rho_h}, C_{i,h,d} \to \frac{K_d q_{h,f} C_{i,h,f}}{K_d q_{h,d} + q_{s,p}}\right\}\right\}$$

Thus we can show that the flow rate of the heavy phase from the unit is equal to its flow rate in plus a factor related to the ratios of the densities of the light and heavy liquids:

$$q_{d,h} = \frac{q_{h,f}\rho_h + (q_s - q_{s,p})\rho_s}{\rho_h} = q_{h,f} + \frac{\rho_s}{\rho_h}(q_s - q_{s,p})$$

If the flow rate of the solvent is the same going in as coming out, then the flow rate of the heavy is the same in and out. Therefore we have that concentration of the impurity in exit flow is:

$$C_{i,h,d} = \frac{C_{i,h,f}K_d q_{h,f}\rho_h}{(K_d q_{h,f} + q_{s,p})\rho_h + K_d\rho_s(q_s - q_{s,p})}$$
$$= \frac{C_{i,h,f}K_d q_{h,f}\rho_h}{(K_d q_{h,f} + q_{s,p})\rho_h}$$
$$= C_{i,h,f}\left(\frac{1}{1 + \frac{q_{s,p}}{K_d q_{h,f}}}\right)$$

One way to use this result would be to compute the flow rate of the solvent that we would need in order to achieve a certain exit product concentration, $C_{i,h,d}$, in the heavy stream, given K_d, the flow rate of the impure heavy feed and its impurity level, $C_{i,h,d}$.

$C_{i,h,d} = 10^{-8}$; **(* g/L *)**

$C_{i,h,f} = 10^{-3}$; **(* g/L *)**

$K_d = 10^{-5}$; **(* dimensionless *)**

$q_{h,f} = 100$; **(* L/min *)**

NSolve$[C_{i,h,d} == C_{i,h,f}\,(), q_{s,p}]$

$\{\{q_{s,p} \to 99.999\}\}$

We find that at these conditions, given that the impurity is 10^5 times more soluble in the light solvent than in the heavy liquid, to reduce the product concentration in the heavy phase from 10^{-3}g/l to 10^{-8} g/L would require a solvent flow of about $100L/$ min for a product feed stream flow of 100 L/min. Again all this assumed instantaneous transfer and equilibrium between phases.

Mass transfer analysis: non-equilibrium

The previous calculation was helpful because using the equilibrium assumption is a good place to start an analysis. But it assumes that the two streams really do come to equilibrium with respect to their impurity concentrations instantaneously. Will they? How can we know this? What does it depend upon? The equilibrium stage analysis does not involve time, it is based on thermodynamics. Yet, we know that thermodynamics can in some cases, be misleading because we can compute the equilibrium position correctly, but for a real process attaining that equilibrium may be not be instantaneous and may take time. In other words for the next level of design we need to take account of time and the rate processed involved. The equilibrium case is very important and should not be minimized as such because it always provides "best case" numbers. When rates are needed, it means that the processes do not attain equilibrium but fall short of that maximum limit.

To handle rate processes we must return to the analysis of unit 1, the contactor. This is the unit where the mixing and interphase mass transfer take place and these may not get to equilibrium in the time allowed for the two phases to be in contact. To do even a first order design of this unit, we need to analyze the mass transfer rate explicitly.

If we recall the material balances that we wrote around the contactor, then you may well be wondering where the rates of mass transfer come in:

$$\rho_{h,f}\, q_{h,f} + \rho_{s,f}\, q_{s,f} = \rho_{h,c} q_{h,c} + \rho_{s,c} q_{s,c}$$

$$C_{i,h,f}\, q_{h,f} + C_{i,s,f}\, q_{s,f} = C_{i,h,c}\, q_{h,c} + C_{i,s,c}\, q_{s,c}$$

The way we can answer this is to go back to our usual approach to this kind of problem and to write the time dependent mass balances for component **i** in each of the phases:

$$\text{Heavy phase:} \quad \frac{dC_{i,h,c} V_{h,c}}{dt} = C_{i,h,f} q_{h,f} - C_{i,h,c} q_{h,c} - K_m\, A_i\, [C_{i,h,c} - K_d\, C_{i,s,c}]$$

$$\text{Solvent phase:} \quad \frac{dC_{i,s,c} V_{s,c}}{dt} = C_{i,s,f} q_{s,f} - C_{i,s,c} q_{s,c} + K_m\, A_i\, [\, C_{i,h,c} - K_d\, C_{i,s,c}]$$

Subscripts:

i ⇒ product

h ⇒ heavy phase

s ⇒ lighter solvent phase

f ⇒ feed

c ⇒ contactor

If the contactor is at steady state, the left-hand side of each equation is identically zero. Adding the two equations and placing the terms for the heavy phase and the light solvent phase on opposite sides of the equation leads to the "steady state" material balance we had before. Now we can see where the rates of mass transfer come in.

The rate of mass transfer that we introduced in this analysis requires some explanation. The constant K_d is the distribution coefficient for product i between the two phases. It is K_m and A_i have their usual meanings, of mass transfer coefficient and the interfacial area. But what about the driving force term? Why is it written as the difference between the actual concentration of i in the first phase minus the actual concentration of i in the second phase multiplied by K_d?

$$\text{Driving force term} = [\ C_{i,h,c} - K_d\ C_{i,s,c}\ ]$$

The reason is this. The driving force to transfer product i from the heavy to the light phase is dependent upon the concentration of i in the light phase. Remember the reason the product i transfers is that it is more soluble in the light organic phase. So, the concentration of product i in the aqueous heavy phase may be well below its solubility limit, but it will still transfer to the light phase because it is even further below its solubility limit in that phase.

Let's put this in words:

The rate of mass transfer of i from phase h to s is proportional to the difference between the actual concentration of i in phase h and the concentration of i in phase s that would be in equilibrium with the actual concentration of i in phase h.

If we put this into a mathematical sentence it would look like this:

$$r_{i,h\rightarrow s} = K_m\ A_i\ [\ C_{i,h,c} - C_{i,h,c,\mathrm{eq}}[\ C_{i,s,c}\ ]\ ]$$

where $C_{i,h,c,\mathrm{eq}}[\ C_{i,s,c}\ ]$ means the concentration of i that *would be in equilibrium* with the actual concentration of i in phase s, that is the theoretical concentration of i is a function of the concentration of i in the light solvent phase. But that concentration is calculable from the partition constant:

$$C_{i,h,c,\mathrm{eq}}[\ C_{i,s,c}\ ] = K_d\ C_{i,s,c}$$

$$\therefore\quad r_{i,s+} = K_m\ A_i\ [\ C_{i,h,c} - K_d\ C_{i,s,c}\ ]$$

The rate of departure of the product i from the heavy phase to the light phase is just negative of the rate of appearance of i in the light phase:

$$r_{i,h-} = - K_m \, A_i \, [\, C_{i,h,c} - K_d \, C_{i,s,c} \,]$$

This is because any mass that appears in the second phase had to leave the first phase and it must appear in the second phase at the same absolute rate that it disappears from the first phase.

We can simplify these two equations by recognizing that the mass transferred between the two phases does not significantly affect the density of either phase, nor of its volume flow rate:

h, Heavy phase:

$$\frac{dC_{i,h,c}}{dt} = (\, C_{i,h,f} - C_{i,h,c} \,) \frac{q_{h,c}}{V_{h,c}} - \frac{K_m A_i}{V_{h,c}} [\, C_{i,h,c} - K_d \, C_{i,s,c} \,]$$

s, Solvent phase:

$$\frac{dC_{i,s,c}}{dt} = (\, C_{i,s,f} - C_{i,s,c} \,) \frac{q_{s,c}}{V_{s,c}} + \frac{K_m A_i}{V_{s,c}} [\, C_{i,h,c} - K_d \, C_{i,s,c} \,]$$

These two equations are soluble, but before we do that we should discuss them further. Notice that the flow rates are divided by the volumes of each phase i.e. $\frac{q_{h,c}}{V_{h,c}}$ and $\frac{q_{s,c}}{V_{s,c}}$. These two terms are the reciprocal holding times for the two phases in the contactor, $\theta_{\mathrm{hc}}{}^{-1}$ and $\theta_{\mathrm{sc}}{}^{-1}$. The coefficients of the two driving force terms are the ratios of the product of the mass transfer coefficient and the interfacial area to the volume of the phase, i.e. $\frac{K_m A_i}{V_{h,c}}$ and $\frac{K_m A_i}{V_{s,c}}$. Recalling K_m has dimensions of $\frac{\text{Length}}{\text{time}}$, we can see that this group is also an inverse time constant, but now this is a reciprocal characteristic time for mass transfer, τ^{-1}. If we multiply through on both sides by the holding time we obtain:

h, Heavy phase:

$$\theta_{\mathrm{hc}} \frac{d\mathrm{Cihc}}{dt} = (\, C_{i,h,f} - \mathrm{Cihc} \,) - \frac{\theta_{\mathrm{hc}}}{\tau \mathrm{hc}} [\, \mathrm{Cihc} - \mathrm{Kd\ Cisc} \,]$$

s, Solvent phase:

$$\theta_{\mathrm{sc}} \frac{d\mathrm{Cisc}}{dt} = (\, \mathrm{Cisf} - \mathrm{Cisc} \,) + \frac{\theta_{\mathrm{sc}}}{\tau \mathrm{sc}} [\, \mathrm{Cihc} - \mathrm{Kd\ Cisc} \,]$$

We could go one more step and refer all the concentrations to the inlet concentration of the impurity in the heavy feed, $C_{i,h,f}$, which is a constant. If we do this, we would be dividing both sides of both equations by this quantity to give the non-dimensionalized concentrations, X:

h, Heavy phase:

$$\theta_{hc} \frac{dX_{i,h,c}}{dt} = (1 - X_{i,h,c}) - \frac{\theta_{hc}}{\tau hc} [X_{i,h,c} - K_d X_{i,s,c}]$$

s, Solvent phase:

$$\theta_{sc} \frac{dX_{i,s,c}}{dt} = (X_{i,s,f} - X_{i,s,c}) + \frac{\theta_{sc}}{\tau sc} [X_{i,h,c} - K_d X_{i,s,c}]$$

Finally, we can see that the time constants can also be used in the same way, we can multiply the second equation on both sides by $\frac{\theta_{hc}}{\theta_{hc}}$ and then re-express both time derivatives in terms of the "reduced time," that is the ratio of real time to holding time:

h, Heavy phase:

$$\theta_{hc} \frac{dX_{i,h,c}}{dt} = (1 - X_{i,h,c}) - \frac{\theta_{hc}}{\tau hc} [X_{i,h,c} - K_d X_{i,s,c}]$$

s, Solvent phase:

$$\left(\frac{\theta_{hc}}{\theta_{hc}}\right)\theta_{sc} \frac{dX_{i,s,c}}{dt} = (X_{i,s,f} - X_{i,s,c}) + \frac{\theta_{sc}}{\tau sc} [X_{i,h,c} - K_d X_{i,s,c}]$$

h, Heavy phase:

$$\frac{dX_{i,h,c}}{d\theta} = (1 - X_{i,h,c}) - \frac{\theta_{hc}}{\tau hc} [X_{i,h,c} - K_d X_{i,s,c}$$

s, Solvent phase:

$$\frac{\theta_{sc}}{\theta_{hc}} \frac{dX_{i,s,c}}{d\theta} = (X_{i,s,f} - X_{i,s,c}) + \frac{\theta_{sc}}{\tau sc} [X_{i,h,c} - K_d X_{i,s,c}]$$

$$\frac{dXisc}{d\theta} = (X_{i,s,f} - X_{i,s,c}) + \frac{\theta_{hc}}{\tau sc} [X_{i,h,c} - K_d X_{i,s,c}]$$

At first, it may seem odd to reduce time to a non-dimensional unit. However, this is actually quite useful because the reduced time is really the number of holding times that have ensued. When we graph the results the x-axis (abscissa) will be literally the number of holding times.

Now, we can get quite a general non-dimensionalized solution for this case, which can be used for specific cases simply by computing the time constants from the parameters. Let's take this step by step. Firstly, we need to make the equations appear simpler and without all the subscripts, all of which are essential to doing the analysis, but are cumbersome when we write the code to solve the equations. Let the following new function names be defined:

$$X_{i,h,c} = \mathrm{y}[\ \theta\ ]$$

$$X_{i,s,c} = \mathrm{z}[\ \theta\ ]$$

$$\frac{dX_{i,h,c}}{d\theta} = (1 - X_{i,h,c} - \frac{\theta_{\mathrm{hc}}}{\tau \mathrm{hc}}[\ X_{i,h,c} - K_d\ X_{i,s,c}] ==$$

$$y'[\theta] == (1 - y[\theta]) - \frac{\theta \mathrm{hc}}{\tau \mathrm{hc}}(y[\theta] - \mathrm{Kd} * z[\theta])$$

$$\frac{d\mathrm{Xisc}}{d\theta} = (\ X_{i,s,f} - X_{i,s,c}\ ) + \frac{\theta_{\mathrm{hc}}}{\tau \mathrm{sc}}[\ X_{i,h,c} - K_d X_{i,s,c}] ==$$

$$z'[\theta] == (\mathrm{zf} - z[\theta]) + \frac{\theta \mathrm{hc}}{\tau \mathrm{sc}}(y[\theta] - \mathrm{Kd} * z[\theta])$$

The equations for y'[θ] and z'[θ] become the equations we need to solve in θ for a given set of initial conditions. The initial concentrations of product i in the heavy and the light phases within unit 1 (not the feed streams) are what we specify as the initial conditions. These are the initial conditions within the control volume, which in this case is the whole of the liquid volume in Unit 1. Rather than using specific numerical values for these two conditions, we will let them be parameters that can be set to different values i.e. y[0]==yo, z[0]==zo.

The code for solving these analytically is shown below. We begin by clearing all the variable and parameters. Then we use DSolve to find the theta-dependent solutions. Remember theta, θ, is the reduced time, the ratio of real time to the holding time of the heavy phase in Unit 1. We should have solutions for both y [θ] and z [θ] and we name the solution set "sol". Let's see what we get:

```
Clear[Kd, zf, yo, θhc, τhc, zo, y, z]

sol =

DSolve[

{y'[θ]==(1 − y[θ]) − (y[θ] − Kd ∗ z[θ]),

z'[θ]==(zf − z[θ]) + (y[θ] − Kd ∗ z[θ]),

y[0] == yo, z[0] == zo},
```

{y[θ], z[θ]},

θ]

$$\{\{y[\theta]\rightarrow$$

$$\left(e^{-\theta-\frac{\theta(\theta\text{hc}+\tau\text{hc}+\text{Kd}\theta\text{hc}\tau\text{hc})}{\tau\text{hc}}}\left(e^{\theta}\text{yo}\theta\text{hc}-e^{\frac{\theta(\theta\text{hc}+\tau\text{hc}+\text{Kd}\theta\text{hc}\tau\text{hc})}{\tau\text{hc}}}\right.\right.$$

$$\text{Kdzf}\theta\text{hc}+e^{\theta+\frac{\theta(\theta\text{hc}+\tau\text{hc}+\text{Kd}\theta\text{hc}\tau\text{hc})}{\tau\text{hc}}}\text{Kdzf}\theta\text{hc}-$$

$$e^{\theta}\text{Kdzo}\theta\text{hc}+e^{\frac{\theta(\theta\text{hc}+\tau\text{hc}+\text{Kd}\theta\text{hc}\tau\text{hc})}{\tau\text{hc}}}\text{Kdzo}\theta\text{hc}-$$

$$e^{\theta}\tau\text{hc}+e^{\theta+\frac{\theta(\theta\text{hc}+\tau\text{hc}+\text{Kd}\theta\text{hc}\tau\text{hc})}{\tau\text{hc}}}\tau\text{hc}+e^{\theta}\text{yo}\tau\text{hc}+$$

$$e^{\theta}\text{Kdzf}\tau\text{hc}-e^{\frac{\theta(\theta\text{hc}+\tau\text{hc}+\text{Kd}\theta\text{hc}\tau\text{hc})}{\tau\text{hc}}}\text{Kdzf}\tau\text{hc}-$$

$$e^{\theta}\text{Kdzo}\tau\text{hc}+e^{\frac{\theta(\theta\text{hc}+\tau\text{hc}+\text{Kd}\theta\text{hc}\tau\text{hc})}{\tau\text{hc}}}\text{Kdzo}\tau\text{hc}-$$

$$e^{\frac{\theta(\theta\text{hc}+\tau\text{hc}+\text{Kd}\theta\text{hc}\tau\text{hc})}{\tau\text{hc}}}\text{Kd}\theta\text{hc}\tau\text{hc}+$$

$$e^{\theta+\frac{\theta(\theta\text{hc}+\tau\text{hc}+\text{Kd}\theta\text{hc}\tau\text{hc})}{\tau\text{hc}}}\text{Kd}\theta\text{hc}\tau\text{hc}+e^{\theta}\text{Kdyo}$$

$$\theta\text{hc}\tau\text{hc}+e^{\frac{\theta(\theta\text{hc}+\tau\text{hc}+\text{Kd}\theta\text{hc}\tau\text{hc})}{\tau\text{hc}}}\text{Kdyo}\theta\text{hc}\tau\text{hc}-$$

$$e^{\frac{\theta(\theta\text{hc}+\tau\text{hc}+\text{Kd}\theta\text{hc}\tau\text{hc})}{\tau\text{hc}}}\text{Kd}^2\text{zf}\theta\text{hc}\tau\text{hc}+$$

$$e^{\theta+\frac{\theta(\theta\text{hc}+\tau\text{hc}+\text{Kd}\theta\text{hc}\tau\text{hc})}{\tau\text{hc}}}\text{Kd}^2\text{zf}\theta\text{hc}\tau\text{hc}-$$

$$e^{\theta}\text{Kd}^2\text{zo}\theta\text{hc}\tau\text{hc}+e^{\frac{\theta(\theta\text{hc}+\tau\text{hc}+\text{Kd}\theta\text{hc}\tau\text{hc})}{\tau\text{hc}}}\text{Kd}^2$$

$$\text{zo}\theta\text{hc}\tau\text{hc}-e^{\frac{\theta(\theta\text{hc}+\tau\text{hc}+\text{Kd}\theta\text{hc}\tau\text{hc})}{\tau\text{hc}}}\text{Kd}\tau\text{hc}^2+$$

$$e^{\theta+\frac{\theta(\theta\text{hc}+\tau\text{hc}+\text{Kd}\theta\text{hc}\tau\text{hc})}{\tau\text{hc}}}\text{Kd}\tau\text{hc}^2+$$

$$e^{\frac{\theta(\theta\text{hc}+\tau\text{hc}+\text{Kd}\theta\text{hc}\tau\text{hc})}{\tau\text{hc}}}\text{Kdyo}\tau\text{hc}^2-$$

$$e^{\frac{\theta(\theta\text{hc}+\tau\text{hc}+\text{Kd}\theta\text{hc}\tau\text{hc})}{\tau\text{hc}}}\text{Kd}^2\theta\text{hc}\tau\text{hc}^2+$$

$$e^{\theta+\frac{\theta(\theta\text{hc}+\tau\text{hc}+\text{Kd}\theta\text{hc}\tau\text{hc})}{\tau\text{hc}}}\text{Kd}^2\theta\text{hc}\tau\text{hc}^2+$$

$$e^{\frac{\theta(\theta\text{hc}+\tau\text{hc}+\text{Kd}\theta\text{hc}\tau\text{hc})}{\tau\text{hc}}}\text{Kd}^2\text{yo}\theta\text{hc}\tau\text{hc}^2\Bigg)\Bigg)\Bigg/$$

$$((1+\text{Kd}\tau\text{hc})(\theta\text{hc}+\tau\text{hc}+\text{Kd}\theta\text{hc}\tau\text{hc})),$$

$$z[\theta]\rightarrow$$

$$\left(e^{-\theta-\frac{\theta(\theta\text{hc}+\tau\text{hc}+\text{Kd}\theta\text{hc}\tau\text{hc})}{\tau\text{hc}}}\left(-e^{\frac{\theta(\theta\text{hc}+\tau\text{hc}+\text{Kd}\theta\text{hc}\tau\text{hc})}{\tau\text{hc}}}\text{zf}\theta\text{hc}+\right.\right.$$

$$e^{\theta+\frac{\theta(\theta\text{hc}+\tau\text{hc}+\text{Kd}\theta\text{hc}\tau\text{hc})}{\tau\text{hc}}}\text{zf}\theta\text{hc}+e^{\frac{\theta(\theta\text{hc}+\tau\text{hc}+\text{Kd}\theta\text{hc}\tau\text{hc})}{\tau\text{hc}}}$$

$$\text{zo}\theta\text{hc}-e^{\frac{\theta(\theta\text{hc}+\tau\text{hc}+\text{Kd}\theta\text{hc}\tau\text{hc})}{\tau\text{hc}}}\text{zf}\tau\text{hc}+$$

$$e^{\theta+\frac{\theta(\theta\text{hc}+\tau\text{hc}+\text{Kd}\theta\text{hc}\tau\text{hc})}{\tau\text{hc}}}\text{zf}\tau\text{hc}+e^{\frac{\theta(\theta\text{hc}+\tau\text{hc}+\text{Kd}\theta\text{hc}\tau\text{hc})}{\tau\text{hc}}}$$

$$\text{zo}\tau\text{hc}-e^{\frac{\theta(\theta\text{hc}+\tau\text{hc}+\text{Kd}\theta\text{hc}\tau\text{hc})}{\tau\text{hc}}}\theta\text{hc}\tau\text{hc}+$$

$$e^{\theta+\frac{\theta(\theta\text{hc}+\tau\text{hc}+\text{Kd}\theta\text{hc}\tau\text{hc})}{\tau\text{hc}}}\theta\text{hc}\tau\text{hc}-e^{\theta}\text{yo}\theta\text{hc}\tau\text{hc}+$$

$$e^{\frac{\theta(\theta\text{hc}+\tau\text{hc}+\text{Kd}\theta\text{hc}\tau\text{hc})}{\tau\text{hc}}}\text{yo}\theta\text{hc}\tau\text{hc}-$$

$$e^{\frac{\theta(\theta\text{hc}+\tau\text{hc}+\text{Kd}\theta\text{hc}\tau\text{hc})}{\tau\text{hc}}}\text{Kdzf}\theta\text{hc}\tau\text{hc}+$$

$$e^{\theta+\frac{\theta(\theta\text{hc}+\tau\text{hc}+\text{Kd}\theta\text{hc}\tau\text{hc})}{\tau\text{hc}}}\text{Kdzf}\theta\text{hc}\tau\text{hc}+e^{\theta}\text{Kdzo}$$

$$\theta\text{hc}\tau\text{hc}+e^{\frac{\theta(\theta\text{hc}+\tau\text{hc}+\text{Kd}\theta\text{hc}\tau\text{hc})}{\tau\text{hc}}}\text{Kdzo}\theta\text{hc}\tau\text{hc}+$$

$$e^{\theta}\tau\mathrm{hc}^2 - e^{\frac{\theta(\theta\mathrm{hc}+\tau\mathrm{hc}+\mathrm{Kd}\theta\mathrm{hc}\tau\mathrm{hc})}{\tau\mathrm{hc}}}\tau\mathrm{hc}^2 - e^{\theta}\mathrm{yo}\tau\mathrm{hc}^2 +$$

$$e^{\frac{\theta(\theta\mathrm{hc}+\tau\mathrm{hc}+\mathrm{Kd}\theta\mathrm{hc}\tau\mathrm{hc})}{\tau\mathrm{hc}}}\mathrm{yo}\tau\mathrm{hc}^2 - e^{\theta}\mathrm{Kdzf}\tau\mathrm{hc}^2 +$$

$$e^{\theta+\frac{\theta(\theta\mathrm{hc}+\tau\mathrm{hc}+\mathrm{Kd}\theta\mathrm{hc}\tau\mathrm{hc})}{\tau\mathrm{hc}}}\mathrm{Kdzf}\tau\mathrm{hc}^2 +$$

$$e^{\theta}\mathrm{Kdzo}\tau\mathrm{hc}^2 - e^{\frac{\theta(\theta\mathrm{hc}+\tau\mathrm{hc}+\mathrm{Kd}\theta\mathrm{hc}\tau\mathrm{hc})}{\tau\mathrm{hc}}}\mathrm{Kd}\theta\mathrm{hc}\tau\mathrm{hc}^2 +$$

$$e^{\theta+\frac{\theta(\theta\mathrm{hc}+\tau\mathrm{hc}+\mathrm{Kd}\theta\mathrm{hc}\tau\mathrm{hc})}{\tau\mathrm{hc}}}\mathrm{Kd}\theta\mathrm{hc}\tau\mathrm{hc}^2 -$$

$$e^{\theta}\mathrm{Kdyo}\theta\mathrm{hc}\tau\mathrm{hc}^2 + e^{\frac{\theta(\theta\mathrm{hc}+\tau\mathrm{hc}+\mathrm{Kd}\theta\mathrm{hc}\tau\mathrm{hc})}{\tau\mathrm{hc}}}\mathrm{Kd}$$

$$\mathrm{yo}\theta\mathrm{hc}\tau\mathrm{hc}^2 + e^{\theta}\mathrm{Kd}^2\mathrm{zo}\theta\mathrm{hc}\tau\mathrm{hc}^2))/$$

$$((1+\mathrm{Kd}\tau\mathrm{hc})(\theta\mathrm{hc}+\tau\mathrm{hc}+\mathrm{Kd}\theta\mathrm{hc}\tau\mathrm{hc}))\}\}$$

The solutions for y [θ] and z [θ] are lengthy and complex looking. They are set inside of two sets of curly brackets {{ ... }}. To try and simplify these expressions, we will wrap the equations in FullSimplify and to remove the brackets we will wrap them in Flatten.

```
Clear[Kd, zf, yo, zo, θhc, τhc, τsc, y, z, θ]

sol = Flatten[
FullSimplify[
DSolve[
{y'[θ]==(1 − y[θ]) − (y[θ] − Kd * z[θ]),
z'[θ]==(zf − z[θ]) + (y[θ] − Kd * z[θ]),
y[0] == yo, z[0] == zo},
{y[θ], z[θ]},
θ]]]
```

$$\Big\{y[\theta] \to \Big(e^{-\theta\left(2+\theta\mathrm{hc}\left(\frac{1}{\tau\mathrm{hc}}+\frac{\mathrm{Kd}}{\tau\mathrm{sc}}\right)\right)}\Big)\Big(e^{\theta+\theta\mathrm{hc}\left(\frac{1}{\tau\mathrm{hc}}+\frac{\mathrm{Kd}}{\tau\mathrm{sc}}\right)}\mathrm{Kd}((-1+\mathrm{yo})\tau\mathrm{hc}$$

$$+(-\mathrm{zf}+\mathrm{zo})\tau\mathrm{sc})(\mathrm{Kd}\theta\mathrm{hc}\tau\mathrm{hc}+(\theta\mathrm{hc}+\tau\mathrm{hc})\tau\mathrm{sc}) + e^{\theta}\tau\mathrm{sc}(\mathrm{Kd}(\mathrm{yo}-\mathrm{Kdzo})\theta\mathrm{hc}$$

$\tau\text{hc} + (-\tau\text{hc} + \text{Kdzf}\tau\text{hc} + \text{yo}(\theta\text{hc} + \tau\text{hc}) - \text{Kdzo}(\theta\text{hc} + \tau\text{hc}))\tau\text{sc}) +$
$e^{\theta\left(2+\theta\text{hc}\left(\frac{1}{\tau\text{hc}}+\frac{\text{Kd}}{\tau\text{sc}}\right)\right)}(\text{Kd}\tau\text{hc} + \tau\text{sc})(\tau\text{hc}\tau\text{sc} + \text{Kd}\theta\text{hc}(\tau\text{hc} + \text{zf}\tau\text{sc}))\Big)\Big)/$
$((\text{Kd}\tau\text{hc} + \tau\text{sc})(\text{Kd}\theta\text{hc}\ \tau\text{hc}\ + (\theta\text{hc} + \tau\text{hc})\tau\text{sc})),$
$z[\theta] \to$
$\left(e^{-\theta\left(2+\theta\text{hc}\left(\frac{1}{\tau\text{hc}}+\frac{\text{Kd}}{\tau\text{sc}}\right)\right)}\right)\left(e^{\theta+\theta\text{hc}\left(\frac{1}{\tau\text{hc}}+\frac{\text{Kd}}{\tau\text{sc}}\right)}((-1 + \text{yo})\tau\text{hc} + (-\text{zf} + \text{zo})\tau\text{sc}\right)$
$(\text{Kd}\theta\text{hc}\tau\text{hc} + (\theta\text{hc} + \tau\text{hc})\tau\text{sc}) + e^{\theta\left(2+\theta\text{hc}\left(\frac{1}{\tau\text{hc}}+\frac{\text{Kd}}{\tau\text{sc}}\right)\right)}(\text{Kd}\tau\text{hc} + \tau\text{sc})(\theta\text{hc}\tau\text{hc} + \text{zf}(\theta\text{hc} +$
$\tau\text{hc})\tau\text{sc}) + e^{\theta}\tau\text{hc}(\text{Kd}(-\text{yo} + \text{Kdzo})\theta\text{hc}\tau\text{hc} + (\tau\text{hc} - \text{Kdzf}\tau\text{hc} - \text{yo}(\theta\text{hc} + \tau\text{hc}) +$
$\text{Kdzo}\quad(\theta\text{hc} + \tau\text{hc}))\tau\text{sc})))/((\text{Kd}\tau\text{hc} + \tau\text{sc})(\text{Kd}\theta\text{hc}\tau\text{hc} + (\theta\text{hc} + \tau\text{hc})\tau\text{sc}))\}$
$\Big\{y[\theta] \to \left(e^{-\theta\left(2+\theta\text{hc}\left(\frac{1}{\tau\text{hc}}+\frac{\text{Kd}}{\tau\text{sc}}\right)\right)}\right)$
$\left(e^{\theta+\theta\text{hc}\left(\frac{1}{\tau\text{hc}}+\frac{\text{Kd}}{\tau\text{sc}}\right)}\text{Kd}((-1 + \text{yo})\tau\text{hc}\right.$
$\tau\text{sc})(\text{Kd}\theta\text{hc}\tau\text{hc} + (\theta\text{hc} + \tau\text{hc})\tau\text{sc}) +$
$e^{\theta}\tau\text{sc}(\text{Kd}(\text{yo} - \text{Kdzo})\theta\text{hc}\tau\text{hc} +$
$(-\tau\text{hc} + \text{Kdzf}\tau\text{hc} + \text{yo}(\theta\text{hc} + \tau\text{hc}) -$
$\text{Kdzo}(\theta\text{hc} + \tau\text{hc}))\tau\text{sc}) +$
$e^{\theta\left(2+\theta\text{hc}\left(\frac{1}{\tau\text{hc}}+\frac{\text{Kd}}{\tau\text{sc}}\right)\right)}(\text{Kd}\tau\text{hc} + \tau\text{sc})$
$(\tau\text{hc}\tau\text{sc} + \text{Kd}\theta\text{hc}(\tau\text{hc} + \text{zf}\tau\text{sc}))))/$
$((\text{Kd}\tau\text{hc} + \tau\text{sc})(\text{Kd}\theta\text{hc}\tau\text{hc} + (\theta\text{hc} + \tau\text{hc})\tau\text{sc})),$
$z[\theta] \to \left(e^{-\theta\left(2+\theta\text{hc}\left(\frac{1}{\tau\text{hc}}+\frac{\text{Kd}}{\tau\text{sc}}\right)\right)}\right)$
$\left(e^{\theta+\theta\theta\text{hc}\left(\frac{1}{\tau\text{hc}}+\frac{\text{Kd}}{\tau\text{sc}}\right)}\right.$
$((-1 + \text{yo})\tau\text{hc} + (-\text{zf} + \text{zo})\tau\text{sc})$
$(\text{Kd}\theta\text{hc}\tau\text{hc} + (\theta\text{hc} + \tau\text{hc})\tau\text{sc}) +$
$e^{\theta\left(2+\theta\text{hc}\left(\frac{1}{\tau\text{hc}}+\frac{\text{Kd}}{\tau\text{sc}}\right)\right)}(\text{Kd}\tau\text{hc} + \tau\text{sc})$
$(\theta\text{hc}\tau\text{hc} + \text{zf}(\theta\text{hc} + \tau\text{hc})\tau\text{sc}) +$
$e^{\theta}\tau\text{hc}(\text{Kd}(-\text{yo} + \text{Kdzo})\theta\text{hc}\tau\text{hc} +$
$(\tau\text{hc} - \text{Kdzf}\tau\text{hc} - \text{yo}(\theta\text{hc} + \tau\text{hc}) +$
$\text{Kdzo}(\theta\text{hc} + \tau\text{hc}))\tau\text{sc})))/$
$((\text{Kd}\tau\text{hc} + \tau\text{sc})(\text{Kd}\theta\text{hc}\tau\text{hc} + (\theta\text{hc} + \tau\text{hc})\tau\text{sc}))\}$

$$\Big\{y[\theta] \to \Big(e^{-\theta\left(2+\theta\text{hc}\left(\frac{1}{\tau\text{hc}}+\frac{\text{Kd}}{\tau\text{sc}}\right)\right)}$$

$$\Big(e^{\theta+\theta\theta\text{hc}\left(\frac{1}{\tau\text{hc}}+\frac{\text{Kd}}{\tau\text{sc}}\right)}\text{Kd}((-1+\text{yo})\tau\text{hc}+(-\text{zf}+\text{zo})$$

$$\tau\text{sc})(\text{Kd}\theta\text{hc}\tau\text{hc}+(\theta\text{hc}+\tau\text{hc})\tau\text{sc})+$$

$$e^{\theta}\tau\text{sc}(\text{Kd}(\text{yo}-\text{Kdzo})\theta\text{hc}\tau\text{hc}+$$

$$(-\tau\text{hc}+\text{Kdzf}\tau\text{hc}+\text{yo}(\theta\text{hc}+\tau\text{hc})-$$

$$\text{Kdzo}(\theta\text{hc}+\tau\text{hc}))\tau\text{sc})+$$

$$e^{\theta\left(2+\theta\text{hc}\left(\frac{1}{\tau\text{hc}}+\frac{\text{Kd}}{\tau\text{sc}}\right)\right)}(\text{Kd}\tau\text{hc}+\tau\text{sc})$$

$$(\tau\text{hc}\tau\text{sc}+\text{Kd}\theta\text{hc}(\tau\text{hc}+\text{zf}\tau\text{sc}))))$$

$$((\text{Kd}\tau\text{hc}+\tau\text{sc})(\text{Kd}\theta\text{hc}\tau\text{hc}+(\theta\text{hc}+\tau\text{hc})\tau\text{sc})),$$

$$z[\theta] \to \Big(\Big(e^{-\theta\left(2+\theta\text{hc}\left(\frac{1}{\tau\text{hc}}+\frac{\text{Kd}}{\tau\text{sc}}\right)\right)}$$

$$\Big(e^{\theta+\theta\theta\text{hc}\left(\frac{1}{\tau\text{hc}}+\frac{\text{Kd}}{\tau\text{sc}}\right)}((-1+\text{yo})\tau\text{hc}+(-\text{zf}+\text{zo})$$

$$\tau\text{sc})(\text{Kd}\theta\text{hc}\tau\text{hc}+(\theta\text{hc}+\tau\text{hc})\tau\text{sc})+$$

$$e^{\theta\left(2+\theta\text{hc}\left(\frac{1}{\tau\text{hc}}+\frac{\text{Kd}}{\tau\text{sc}}\right)\right)}(\text{Kd}\tau\text{hc}+\tau\text{sc})$$

$$(\theta\text{hc}\tau\text{hc}+\text{zf}(\theta\text{hc}+\tau\text{hc})\tau\text{sc})+$$

$$e^{\theta}\tau\text{hc}(\text{Kd}(-\text{yo}+\text{Kdzo})\theta\text{hc}\tau\text{hc}+$$

$$(\tau\text{hc}-\text{Kdzf}\tau\text{hc}-\text{yo}(\theta\text{hc}+\tau\text{hc})+$$

$$\text{Kdzo}(\theta\text{hc}+\tau\text{hc}))\tau\text{sc})))/((\text{Kd}\tau\text{hc}+\tau\text{sc})(\text{Kd}\theta\text{hc}\tau\text{hc}+(\theta\text{hc}+\tau\text{hc})\tau\text{sc}))\Big\}$$

The expressions for y[θ] and z[θ] are now simpler, not simple and they are inside of one set of curly brackets {{ ... }}. Next, we use the two expressions from "sol" to create two new function names yy[x] and zz [x] so that we can test the solutions for their veracity. To do so, we will place the two functions back in the differential equations, and wrap them in Simplify. If they are consistent the left- and right-hand sides of both equations will be equal and the logical result will be True:

yy[x_]:= sol[[1,2]]

zz[x_]:= sol[[2,2]]

$$\textbf{Simplify}\left[\frac{\partial \text{yy}(\theta)}{\partial\theta} = (1-\text{yy}(\theta)) - \frac{\theta\text{hc}(\text{yy}(\theta)-\text{Kdzz}(\theta))}{\tau\text{hc}}\right]$$

$$\textbf{Simplify}\left[\frac{\partial \text{zz}(\theta)}{\partial\theta} = \frac{\theta\text{hc}(\text{yy}(\theta)-\text{Kdzz}(\theta))}{\tau\text{sc}} + (\text{zf}-\text{zz}(\theta))\right]$$

True

True

In both instances the logical result was True. So these are verified as solutions to the two equations. We may now wonder what these two functions evolve to as $\theta \longrightarrow \infty$. We can determine that by taking the limits of both expressions:

Limit[yy[θ], θ → ∞]

Limit[zz[θ], θ → ∞]

$$\text{ConditionalExpression}\left[\frac{\tau\text{hc}\tau\text{sc} + \text{Kd}\theta\text{hc}(\tau\text{hc} + \text{zf}\tau\text{sc})}{\text{Kd}\theta\text{hc}\tau\text{hc} + (\theta\text{hc} + \tau\text{hc})\tau\text{sc}},\right.$$
$$((-1+\text{yo})\tau\text{hc}+(-\text{zf}+\text{zo})\tau\text{sc}|\text{Kd}(\text{yo}-\text{Kdzo})\theta\text{hc}\tau\text{hc}+(-\tau\text{hc}+\text{Kdzf}\tau\text{hc}+\text{yo}(\theta\text{hc}+\tau\text{hc})-\text{Kdzo}(\theta\text{hc}+\tau\text{hc}))\tau\text{sc}|\text{Kd}\theta\text{hc}(\tau\text{hc}+\text{zf}\tau\text{sc}))\in\mathbb{R}\&\&\theta\text{hc}\neq 0\&\&\tau\text{hc}\neq 0\&\&\tau\text{sc}\neq 0\&\&\theta\text{hc}(\text{Kd}\tau\text{hc}+\tau\text{sc})(\text{Kd}\theta\text{hc}\tau\text{hc}+\theta\text{hc}\tau\text{sc}+2\tau\text{hc}\tau\text{sc})>0\&\&\tau\text{hc}\tau\text{sc}(\text{Kd}\theta\text{hc}\tau\text{hc}+\theta\text{hc}\tau\text{sc}+2\tau\text{hc}\tau\text{sc})>0\&\&\text{Kd}^2\theta\text{hc}^2\tau\text{hc}^2+2\text{Kd}\theta\text{hc}^2\tau\text{hc}\tau\text{sc}+3\text{Kd}\theta\text{hc}\tau\text{hc}^2\tau\text{sc}+\theta\text{hc}^2\tau\text{sc}^2+3\theta\text{hc}\tau\text{hc}\tau\text{sc}^2+2\tau\text{hc}^2\tau\text{sc}^2>0\Big]$$

$$\text{ConditionalExpression}\left[\frac{\theta\text{hc}\tau\text{hc} + \text{zf}(\theta\text{hc} + \tau\text{hc})\tau\text{sc}}{\text{Kd}\theta\text{hc}\tau\text{hc} + (\theta\text{hc} + \tau\text{hc})\tau\text{sc}},\right.$$
$$((-1+\text{yo})\tau\text{hc}+(-\text{zf}+\text{zo})\tau\text{sc}|\text{zf}(\theta\text{hc}+\tau\text{hc})\tau\text{sc}|\text{Kd}(-\text{yo}+\text{Kdzo})\theta\text{hc}\tau\text{hc}+(\tau\text{hc}-\text{Kdzf}\tau\text{hc}-\text{yo}(\theta\text{hc}+\tau\text{hc})+\text{Kdzo}(\theta\text{hc}+\tau\text{hc}))\tau\text{sc})\in\mathbb{R}\&\&\theta\text{hc}\neq 0\&\&\tau\text{hc}\neq 0\&\&\tau\text{sc}\neq 0\&\&\theta\text{hc}(\text{Kd}\tau\text{hc}+\tau\text{sc})(\text{Kd}\theta\text{hc}\tau\text{hc}+\theta\text{hc}\tau\text{sc}+2\tau\text{hc}\tau\text{sc})>0\&\&\tau\text{hc}\tau\text{sc}(\text{Kd}\theta\text{hc}\tau\text{hc}+\theta\text{hc}\tau\text{sc}+2\tau\text{hc}\tau\text{sc})>0\&\&\text{Kd}^2\theta\text{hc}^2\tau\text{hc}^2+2\text{Kd}\theta\text{hc}^2\tau\text{hc}\tau\text{sc}+3\text{Kd}\theta\text{hc}\tau\text{hc}^2\tau\text{sc}+\theta\text{hc}^2\tau\text{sc}^2+3\theta\text{hc}\tau\text{hc}\tau\text{sc}^2+2\tau\text{hc}^2\tau\text{sc}^2>0\Big]$$

What we learn is this: at long reduced times, the expressions simplify considerably and are dependent on the holding time in Unit 1, θhc and the characteristic time for transfer between the phases, τhc.

$$\text{yy}[\infty] == \frac{\tau\text{hc}\tau\text{sc} + \text{Kd}\theta\text{hc}(\tau\text{hc} + \text{zf}\tau\text{sc})}{\text{Kd}\theta\text{hc}\tau\text{hc} + (\theta\text{hc} + \tau\text{hc})\tau\text{sc}} \equiv \text{dimensionless}$$

$$\text{zz}[\infty] == \frac{\theta\text{hc}\tau\text{hc} + \text{zf}(\theta\text{hc} + \tau\text{hc})\tau\text{sc}}{\text{Kd}\theta\text{hc}\tau\text{hc} + (\theta\text{hc} + \tau\text{hc})\tau\text{sc}} \equiv \text{dimensionless}$$

And we can go back and check these long time solutions at what would be a steady state, but solving the two equations when both derivatives are equal to zero. Remember, when the derivatives are zero, the condition of the system is that it is no longer changing with time. Here yy_{ss} stands for the steady state concentration of product i in the aqueous, heavy phase and zz_{ss} is the concentration in the organic, light phase.

Clear[Kd, zf, yo, θhc, τhc, zo, y, z]

FullSimplify[

Solve$\left[0 == (1 - yy_{ss}) - (yy_{ss} - Kd * zz_{ss}) \&\&\right.$

$0 == (zf - zz_{ss}) + (yy_{ss} - Kd * zz_{ss}), \{yy_{ss}, zz_{ss}\}]]$

$$\left\{\left\{yy_{ss} \to \frac{\tau hc\tau sc + Kd\theta hc(\tau hc + zf\tau sc)}{Kd\theta hc\tau hc + (\theta hc + \tau hc)\tau sc}, zz_{ss} \to \frac{\theta hc\tau hc + zf(\theta hc + \tau hc)\tau sc}{Kd\theta hc\tau hc + (\theta hc + \tau hc)\tau sc}\right\}\right\}$$

Once again everything checks with the previous result. So, we now know that we have both reduced time dependent and steady state solutions for the concentrations of product i in the light and heavy phases inside Unit 1.

We can apply parameter values to get to the steady state values of product i concentration in the aqueous, heavy phase, which is the phase in which it started and in the organic, light phase. For the sake of the example, we will let the concentration of product i in the light organic feed (zf) to be 0, the time constants for transfer between phases (τhc, τsc) will be 0.1, the holding time in Unit 1 (θhc) will be 100 times the time constant for the mass transfer process and the distribution constant (Kd) will be 0.001.

(* Concentration of *i* in the organic solvent feed (*g*/*L*) *)

zf = 0;

(* Mass transfer time constant $(\mathbf{min^{-1}})$ ***)**

τhc = $\mathbf{10^{-1}}$**;**

(* Mass transfer time constant $(\mathbf{min^{-1}})$ ***)**

τsc = $\mathbf{10^{-1}}$**;**

(* Holding time of the heavy, aqueous phase *i*, Unit 1,
$(\mathbf{min^{-1}})$ ***)**

θhc = 10;

(* Distribution coefficient (dimensionless) *)

Kd = $\mathbf{10^{-2}}$**;**

(* yss : steady state conc. of *i* in the aqueous phase (g/L) *)

$$\frac{\tau\text{hc}\tau\text{sc}+\text{Kd}\theta\text{hc}(\tau\text{hc}+\text{zf}\tau\text{sc})}{\text{Kd}\theta\text{hc}\tau\text{hc}+(\theta\text{hc}+\tau\text{hc})\tau\text{sc}}//N;$$

(* Print command *)

Print$\left[\text{yy}_{\text{ss}}, \text{“=”}, \%\right]$

(* zss : steady state conc. of *i* in the organic phase (g/L) *)

$$\frac{\theta\text{hc}\tau\text{hc}+\text{zf}(\theta\text{hc}+\tau\text{hc})\tau\text{sc}}{\text{Kd}\theta\text{hc}\tau\text{hc}+(\theta\text{hc}+\tau\text{hc})\tau\text{sc}}//N;$$

(* Print command *)

Print$[z_{\text{ss}}, \text{“=”}, \%]$

$\text{yy}_{\text{ss}} = 0.0196078$

$z_{\text{ss}} = 0.980392$

We might want to repeat these calculations for different values of the parameters. To do so we build a function, and the values of the parameters will be set delayed within the code. It will look like this for yy_{ss}:

Clear[zf, τhc, θhc, Kd]

(* yss: steady state conc. of *i* in the aqueous phase,
(g/L) *)

yss[zf_, τhc_, τsc_, θhc_, Kd_]:=

$$N\left[\frac{\tau\text{hc}\tau\text{sc}+\text{Kd}\theta\text{hc}(\tau\text{hc}+\text{zf}\tau\text{sc})}{\text{Kd}\theta\text{hc}\tau\text{hc}+(\theta\text{hc}+\tau\text{hc})\tau\text{sc}}\right]$$

yss[0, 0.1, 0.1, 10, .001]

0.0108803

Recall because we divided through our equations by the inlet concentration of the impurity in the heavy feed, $C_{i,h,f}$, that this is now a dimensionless, fractional concentration. It tells us that the steady state concentration of product i exiting Unit 1 is about 1% of the inlet concentration in this phase.

One more way to handle the use of these results for repetitive calculations is to build Module functions in which the values of the variables will be help privately or locally within the Module. Looking at the Module for yss, we see that we use the set delayed feature to write the key parameters that we can vary. Within the module we represent the formula for yss with the internal variable name yy. We declare the new variable name before we use it in the first line of the Module and we wrap it in curly bracket { }. After writing the function, we have included a Print statement that writes the variable names and its computed magnitude. We do the same for zss:

yss[zf_, τhc_, τsc_, θhc_, Kd_]:=Module[{yy},

$$\mathbf{yy} = N\left[\frac{\mathbf{\tau hc\tau sc + Kd\theta hc(\tau hc + zf\tau sc)}}{\mathbf{Kd\theta hc\tau hc + (\theta hc + \tau hc)\tau sc}}\right];$$

Print["yy = ", yy]]

zss[zf_, τhc_, τsc_, θhc_, Kd_]:=Module[{zz},

$$\mathbf{zz} = N\left[\frac{\mathbf{\theta hc\tau hc + zf(\theta hc + \tau hc)\tau sc}}{\mathbf{Kd\theta hc\tau hc + (\theta hc + \tau hc)\tau sc}}\right];$$

Print["zz = ", zz]]

To use these Modules, we call the function name and populate the parameter values:

yss[0, 0.1, 0.1, 10, 0.001]

zss[0, 0.1, 0.1, 10, 0.001]

yy = 0.0108803

zz = 0.98912

Not only is the exit concentration of product i in the aqueous phase reduced by about 99%, the fractional concentration in the organic solvent phase leaving the unit is about 99% of what was inlet concentration in the aqueous phase.

Perhaps, we would want to see how the magnitudes of yss and zss change as we vary the value of the partition coefficient over a range of values:

Table[yss[0, 0.1, 0.1, 10, *x*],

$\{x, \{10^{-5}, 10^{-4}, 10^{-3}, 10^{-2}, 10^{-1}, 1\}\}]$

Table[zss[0, 0.1, 0.1, 10, *x*],

$\{x, \{10^{-5}, 10^{-4}, 10^{-3}, 10^{-2}, 10^{-1}, 1\}\}]$

yy = 0.00991079

yy = 0.00999901

yy = 0.0108803

yy = 0.0196078

yy = 0.0990991

yy = 0.502488

zz = 0.990089

zz = 0.990001

zz = 0.98912

zz = 0.980392

zz = 0.900901

zz = 0.497512

These numbers inform us that for values of Kd between 10^{-5} and 10^{-3} the residual concentration of product i in the aqueous, heavy phase is about the same at 0.01. We also notice that even as the partition coefficient changes by several orders of magnitude, the concentration in the organic, light phase of product i is about the same, implying that there is plenty of holding time and capacity.

Now, we can go back to the full time dependent solutions to see how they behave in reduced time. First we assign these solutions to functions for y [θ] and z [θ].

y[θ_]:=

$$\left(e^{-\theta\left(2+\theta\mathrm{hc}\left(\frac{1}{\tau\mathrm{hc}}+\frac{\mathrm{Kd}}{\tau\mathrm{sc}}\right)\right)}\right.$$

$$\left(e^{\theta+\theta\theta\mathrm{hc}\left(\frac{1}{\tau\mathrm{hc}}+\frac{\mathrm{Kd}}{\tau\mathrm{sc}}\right)}\mathrm{Kd}((-1+\mathrm{yo})\tau\mathrm{hc}+(-\mathrm{zf}+\mathrm{zo})\tau\mathrm{sc})\right.$$

$$(\mathrm{Kd}\theta\mathrm{hc}\tau\mathrm{hc}+(\theta\mathrm{hc}+\tau\mathrm{hc})\tau\mathrm{sc})+$$

$$e^{\theta}\tau\mathrm{sc}(\mathrm{Kd}(\mathrm{yo}-\mathrm{Kdzo})\theta\mathrm{hc}\tau\mathrm{hc}+$$

$$(-\tau hc + Kdzf\tau hc + yo(\theta hc + \tau hc)-$$
$$Kdzo(\theta hc + \tau hc))\tau sc)+$$
$$e^{\theta\left(2+\theta hc\left(\frac{1}{\tau hc}+\frac{Kd}{\tau sc}\right)\right)}(Kd\tau hc + \tau sc)$$
$$(\tau hc\tau sc + Kd\theta hc(\tau hc + zf\tau sc))))/$$
$$((Kd\tau hc + \tau sc)(Kd\theta hc\tau hc + (\theta hc + \tau hc)\tau sc))$$

$$z[\theta\_]:=$$
$$\left(e^{-\theta\left(2+\theta hc\left(\frac{1}{\tau hc}+\frac{Kd}{\tau sc}\right)\right)}\right.$$
$$\left(e^{\theta+\theta\theta hc\left(\frac{1}{\tau hc}+\frac{Kd}{\tau sc}\right)}((-1+yo)\tau hc + (-zf+zo)\tau sc)\right.$$
$$(Kd\theta hc\tau hc + (\theta hc + \tau hc)\tau sc)+$$
$$e^{\theta\left(2+\theta hc\left(\frac{1}{\tau hc}+\frac{Kd}{\tau sc}\right)\right)}(Kd\tau hc + \tau sc)$$
$$(\theta hc\tau hc + zf(\theta hc + \tau hc)\tau sc)+$$
$$e^{\theta}\tau hc(Kd(-yo + Kdzo)\theta hc\tau hc+$$
$$(\tau hc - Kdzf\tau hc - yo(\theta hc + \tau hc)+$$
$$Kdzo(\theta hc + \tau hc))\tau sc)))/$$
$$((Kd\tau hc + \tau sc)(Kd\theta hc\tau hc + (\theta hc + \tau hc)\tau sc))$$

Next, we can examine the time dependent behavior of these solutions. One reason to do so, is to examine how long the model predicts for the process to reach a steady state. We choose parameter values, and in this case we let the initial concentration of product i in the aqueous phase within Unit 1 be zero, yo = 0:

$zf = 0;$

$zo = 0;$

$yo = 0;$

$\tau hc = 10^{-1};$ (* Parameters *)

$\theta hc = 10;$

```
Kd = 10^-2;
Plot[y[t], {t, 0.00001, 7},
PlotRange → {{0, 8}, {0, 0.03}},
AxesLabel → {"θ", "xihc[θ]"}]
Plot[z[t], {t, 0.00001, 7},
PlotRange → {{0, 8}, {0, 2}}, PlotStyle → Hue[0],
AxesLabel → {"θ", "xisc[θ]"}]
```

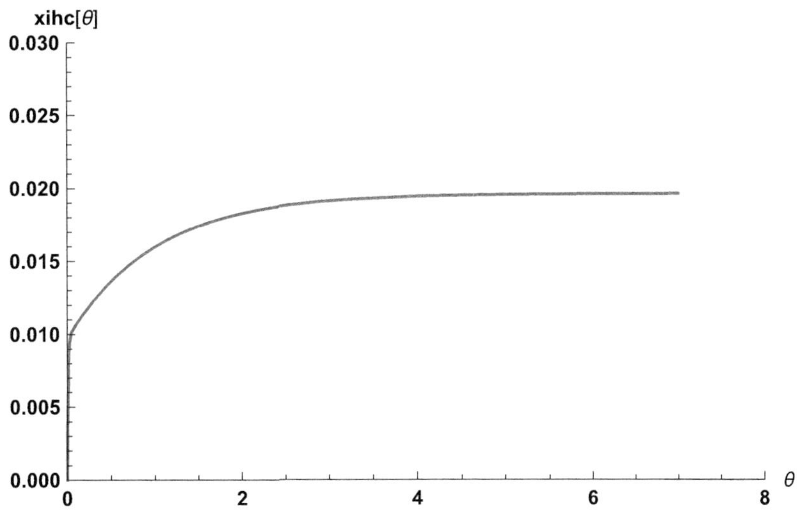

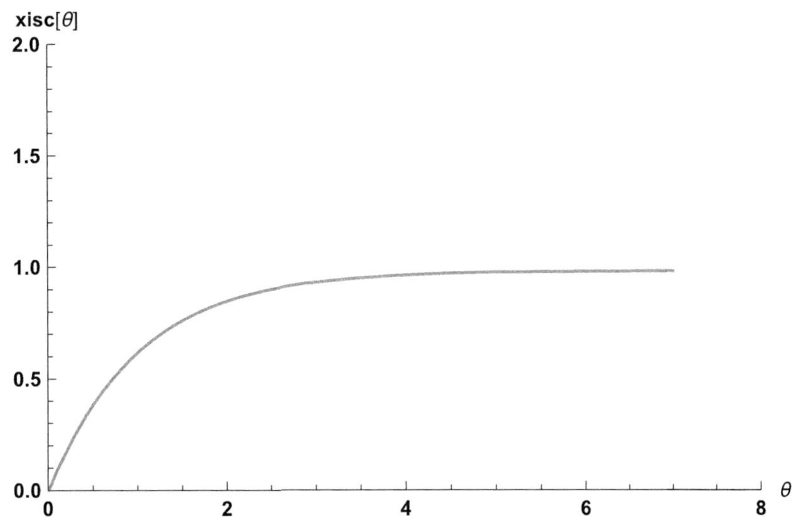

Remember that the abscissa in these plots is the dimensionless time, θ. The two concentrations rise, the concentration of product i in the aqueous phase rises very sharply from zero, while that of product i in the organic phases takes more time. The very sharp rise in the concentration in the aqueous phase is due to new aqueous feed with product i being added to aqueous phase in the unit that is void of product i. The two concentrations come to steady values in about 5-6 holding times, (recall that the abscissa magnitudes are numbers of holding times). We may have expected the concentration in the heavy phase to drop before rising to a steady value, instead of simply rising. The rise we see here is simply a consequence of the initial condition, yo = 0 that we chose for the initial concentration of product i in the heavy phase in the tank. It may be more realistic to assume that the initial concentration of product i in the heavy, aqueous phase in Unit 1, is not zero, but is the same as it is in the feed, yo = 1. Let's see what happens:

```
zf = 0;

zo = 0;

yo = 1.0;

τhc = 10^-1; (* Parameters *)

θhc = 10;

Kd = 10^-2;

Plot[y[t], {t, 0.00001, 7},

PlotRange → {{0, 8}, {0, 0.03}},

AxesLabel → {"θ", "xihc[θ]"}]

Plot[z[t], {t, 0.00001, 7},

PlotRange → {{0, 8}, {0, 2}}, PlotStyle → Hue[0],

AxesLabel → {"θ", "xisc[θ]"}]
```

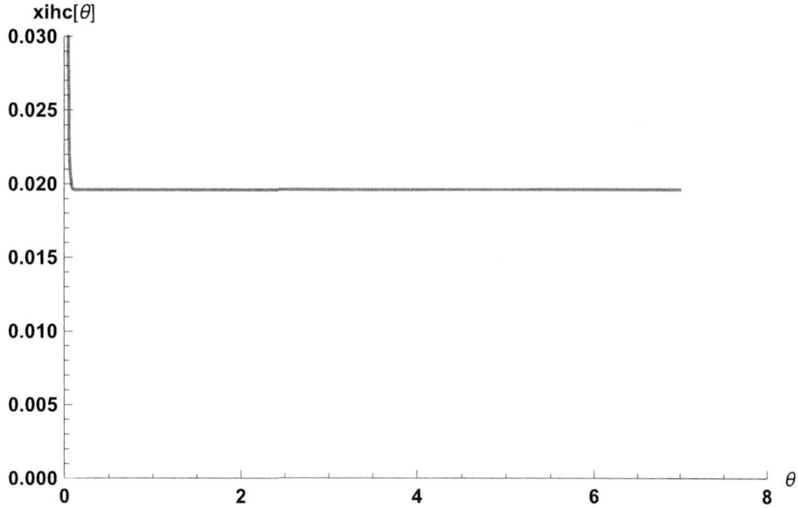

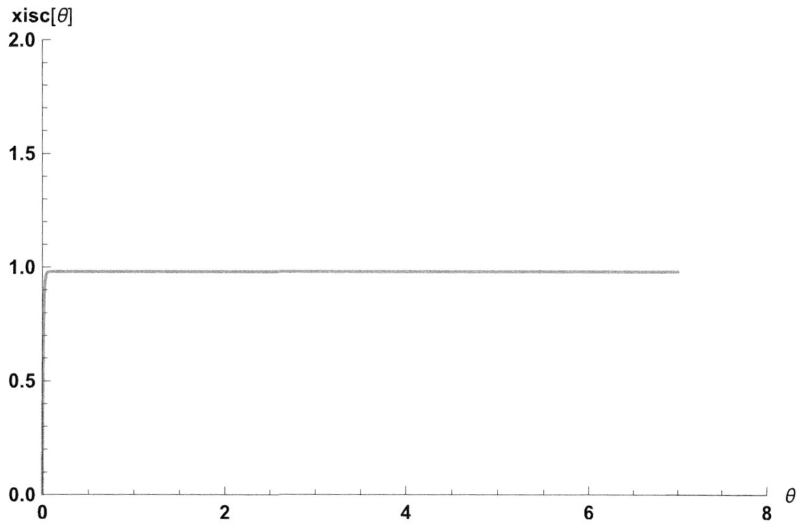

When we let the initial concentration of product i in the unit be the same as it is in the feed stream, everything changes. The onset of the steady state is almost instantaneous. Why? Because the mass transfer process is treated now as beginning immediately and is not dependent on the flow of material into the system.

If we do this directly, this gets kind of sloppy after a few cases and we get annoying error messages about the choice of variable names that are somewhat too similar for *Mathematica*'s checker to be silent. So a better way to do this sort of calculation repetitively, is to write a function call using "Module."

```
iinh[zf_, zo_, yo_, τhc_, τsc_, θhc_, Kd_, tmax_]:=
```

Module[

{yy, pl1},

yy[θ_]:=

$$\left(e^{-\theta\left(2+\theta hc\left(\frac{1}{\tau hc}+\frac{Kd}{\tau sc}\right)\right)}\right.$$

$$\left(e^{\theta+\theta\theta hc\left(\frac{1}{\tau hc}+\frac{Kd}{\tau sc}\right)}Kd((-1+yo)\tau hc+(-zf+zo)\tau sc)\right.$$

$(Kd\theta hc\tau hc+(\theta hc+\tau hc)\tau sc)+$

$e^{\theta}\tau sc(Kd(yo-Kdzo)\theta hc\tau hc+$

$(-\tau hc+Kdzf\tau hc+yo(\theta hc+\tau hc)-$

$Kdzo(\theta hc+\tau hc))\tau sc)+$

$$e^{\theta\left(2+\theta hc\left(\frac{1}{\tau hc}+\frac{Kd}{\tau sc}\right)\right)}(Kd\tau hc+\tau sc)$$

$(\tau hc\tau sc+Kd\theta hc(\tau hc+zf\tau sc))))/$

$((Kd\tau hc+\tau sc)(Kd\theta hc\tau hc+(\theta hc+\tau hc)\tau sc));$

pl1 = Plot[{yy[*t*]}, {*t*, 0, tmax},

PlotStyle → Hue[0],

AxesLabel → {"θ", "xihc[θ]"},

PlotLabel → θhc"= θhc",

PlotRange → All]

]

iins[zf_, zo_, yo_, τhc_, τsc_, θhc_, Kd_, tmax_]:=Module[{yy, pl2},

zz[θ_]:=

$$\left(e^{-\theta\left(2+\theta hc\left(\frac{1}{\tau hc}+\frac{Kd}{\tau sc}\right)\right)}\right.$$

$$\left(e^{\theta+\theta hc\left(\frac{1}{\tau hc}+\frac{Kd}{\tau sc}\right)}((-1+yo)\tau hc+(-zf+zo)\tau sc)\right.$$

```
(KdθhcτHc + (θhc + τhc)τsc)+
e^(θ(2+θhc(1/τhc + Kd/τsc))) (Kdτhc + τsc)
(θhcτhc + zf(θhc + τhc)τsc)+
e^θ τhc(Kd(−yo + Kdzo)θhcτhc+
(τhc − Kdzfτhc − yo(θhc + τhc)+
Kdzo(θhc + τhc))τsc)))/
((Kdτhc + τsc)(Kdθhcτhc + (θhc + τhc)τsc));
pl2 = Plot[{zz[t]}, {t, 0, tmax},
PlotStyle → Hue[0.6],
AxesLabel → {"θ", "xihc[θ]"},
PlotLabel → θhc"= θhc",
PlotRange → All]
]

iinh[0, 0, 0, .1, .1, 10, .01, 7]
iins[0, 0, 0, .1, .1, 10, .01, 7]
```

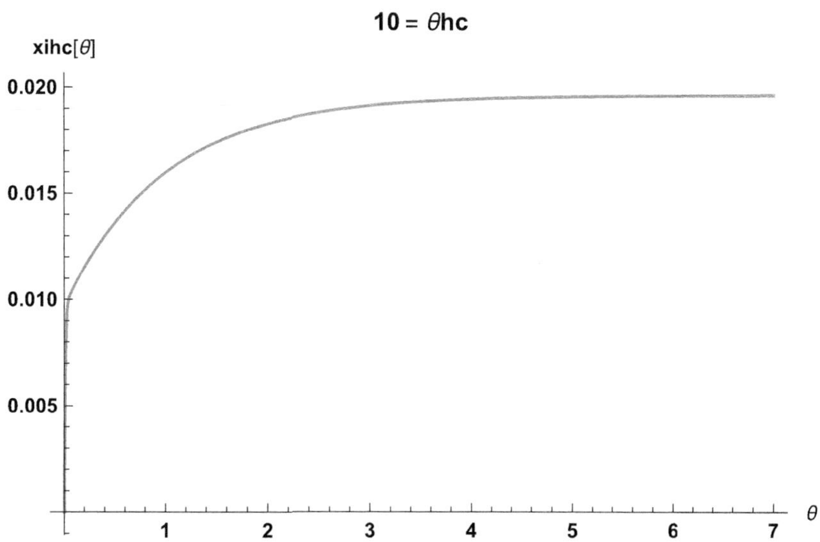

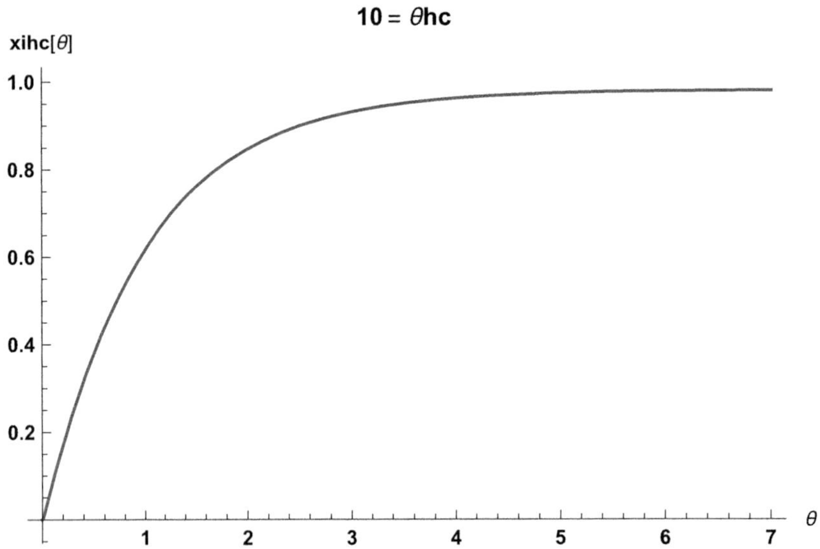

These solutions look reasonable and go to the steady state values by six holding times. However, there is mathematical instability lurking in these solutions. Look at what happens if we push these out to ten holding times:

iinh[0, 0, 0, .1, .1, 10, .01, 10]

iins[0, 0, 0, .1, .1, 10, .01, 10]

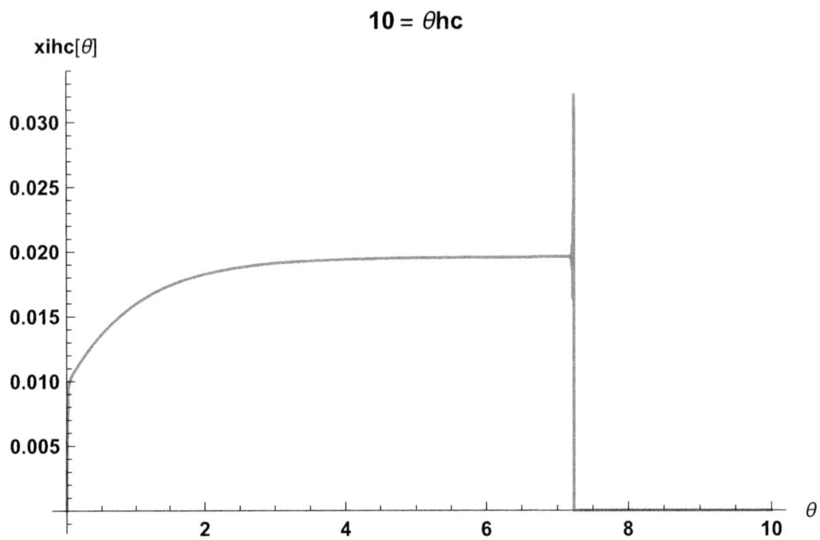

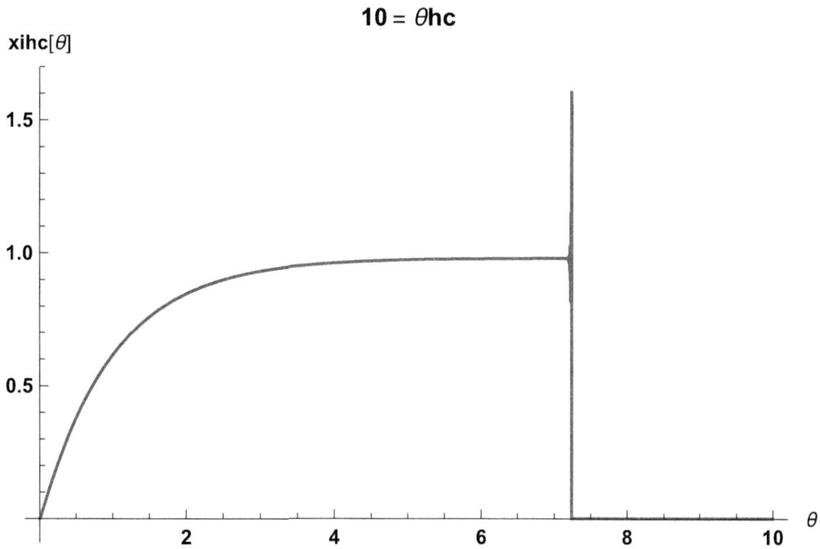

We see that both of the time dependent solutions "blow up" just beyond seven holding times. To get a closer look at this unstable region of the plot, we can change the plot range in the Module function to go from (tmax - 0.01) to tmax:

iinh2[zf_, zo_, yo_, τhc_, τsc_, θhc_, Kd_, tmax_]:=

Module[

{yy, pl1},

yy[θ_]:=

$$\left(e^{-\theta\left(2+\theta \mathrm{hc}\left(\frac{1}{\tau \mathrm{hc}}+\frac{\mathrm{Kd}}{\tau \mathrm{sc}}\right)\right)}\right.$$

$$\left(e^{\theta+\theta\theta \mathrm{hc}\left(\frac{1}{\tau \mathrm{hc}}+\frac{\mathrm{Kd}}{\tau \mathrm{sc}}\right)}\mathrm{Kd}((-1+\mathrm{yo})\tau \mathrm{hc}+(-\mathrm{zf}+\mathrm{zo})\tau \mathrm{sc})\right.$$

$$(\mathrm{Kd}\theta \mathrm{hc}\tau \mathrm{hc}+(\theta \mathrm{hc}+\tau \mathrm{hc})\tau \mathrm{sc})+$$

$$e^{\theta}\tau \mathrm{sc}(\mathrm{Kd}(\mathrm{yo}-\mathrm{Kdzo})\theta \mathrm{hc}\tau \mathrm{hc}+$$

$$(-\tau \mathrm{hc}+\mathrm{Kdzf}\tau \mathrm{hc}+\mathrm{yo}(\theta \mathrm{hc}+\tau \mathrm{hc})-$$

$$\mathrm{Kdzo}(\theta \mathrm{hc}+\tau \mathrm{hc}))\tau \mathrm{sc})+$$

$$e^{\theta\left(2+\theta \mathrm{hc}\left(\frac{1}{\tau \mathrm{hc}}+\frac{\mathrm{Kd}}{\tau \mathrm{sc}}\right)\right)}(\mathrm{Kd}\tau \mathrm{hc}+\tau \mathrm{sc})$$

```
(τhcτsc + Kdθhc(τhc + zfτsc))))/
((Kdτhc + τsc)(Kdθhcτhc + (θhc + τhc)τsc));
pl1 = Plot[{yy[t]}, {t, tmax − .01, tmax},
PlotStyle → Hue[0],
AxesLabel → {"θ", "xihc[θ]"},
PlotLabel → θhc"= θhc"]
]

iins2[zf_, zo_, yo_, τhc_, τsc_, θhc_, Kd_, tmax_]:=
Module[
  {yy, pl2},
zz[θ_]:=
```

$$\left(e^{-\theta\left(2+\theta\mathrm{hc}\left(\frac{1}{\tau\mathrm{hc}}+\frac{\mathrm{Kd}}{\tau\mathrm{sc}}\right)\right)}\right.$$

$$\left(e^{\theta+\theta\theta\mathrm{hc}\left(\frac{1}{\tau\mathrm{hc}}+\frac{\mathrm{Kd}}{\tau\mathrm{sc}}\right)}((-1+\mathrm{yo})\tau\mathrm{hc}+(-\mathrm{zf}+\mathrm{zo})\tau\mathrm{sc})\right.$$

$$(\mathrm{Kd}\theta\mathrm{hc}\tau\mathrm{hc}+(\theta\mathrm{hc}+\tau\mathrm{hc})\tau\mathrm{sc})+$$

$$e^{\theta\left(2+\theta\mathrm{hc}\left(\frac{1}{\tau\mathrm{hc}}+\frac{\mathrm{Kd}}{\tau\mathrm{sc}}\right)\right)}(\mathrm{Kd}\tau\mathrm{hc}+\tau\mathrm{sc})$$

```
(θhcτhc + zf(θhc + τhc)τsc)+
e^θ τhc(Kd(−yo + Kdzo)θhcτhc+
(τhc − Kdzfτhc − yo(θhc + τhc)+
Kdzo(θhc + τhc))τsc)))/
((Kdτhc + τsc)(Kdθhcτhc + (θhc + τhc)τsc));
  pl2 = Plot[{zz[t]}, {t, tmax − .01, tmax},
PlotStyle → Hue[0.6],
AxesLabel → {"θ", "xihc[θ]"},
PlotLabel → θhc"= θhc"]
```

```
]

iinh2[0, 0, 0, .1, .1, 10, .01, 7.24]

iins2[0, 0, 0, .1, .1, 10, .01, 7.24]
```

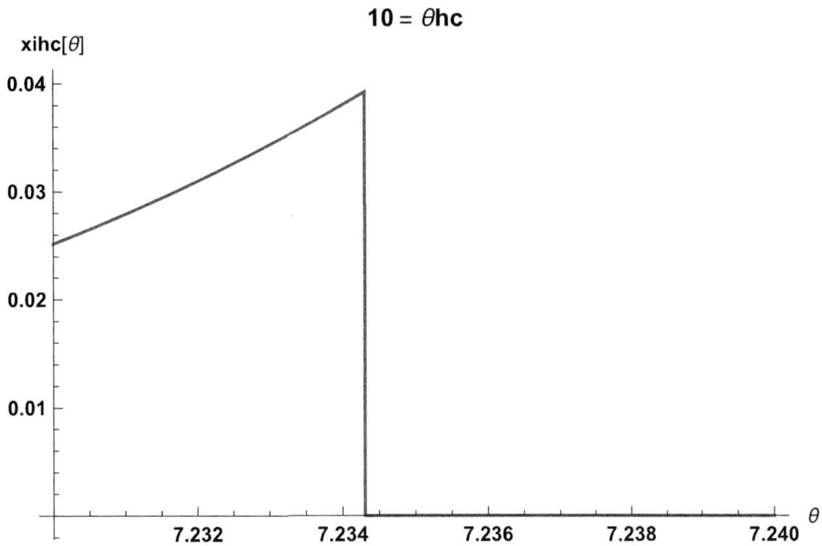

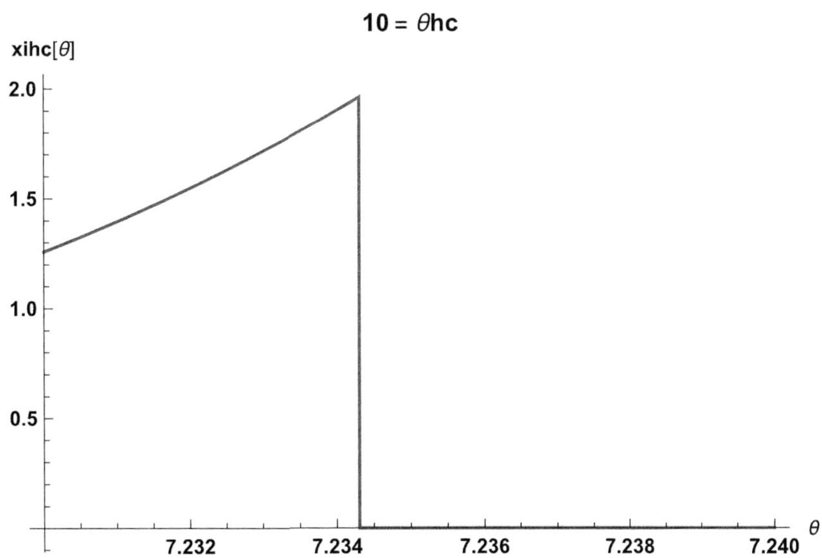

If we decrease the holding time in Unit 1, say we make it equal to the characteristic time for mass transfer, then the instability is pushed out to over 245 holding times, but it is still there in the functions.

iinh[0, 0, 0, .1, .1, .1, .01, 300]

iins[0, 0, 0, .1, .1, .1, .01, 300]

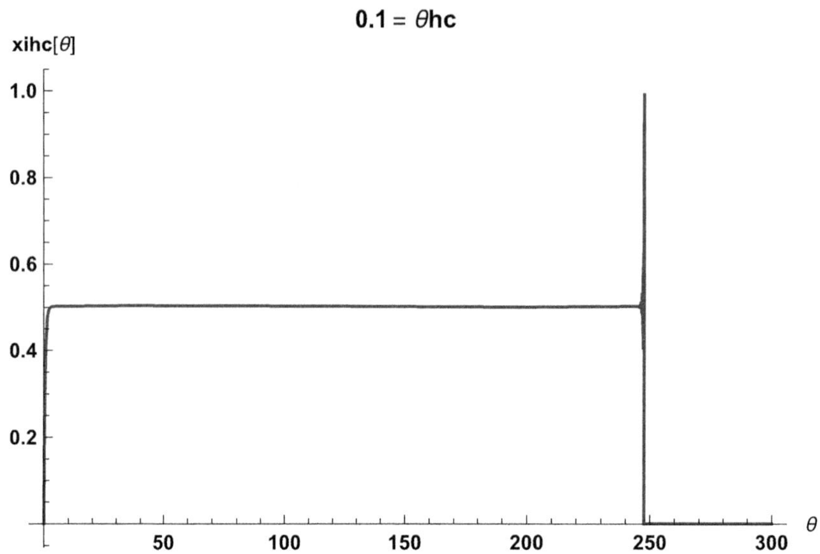

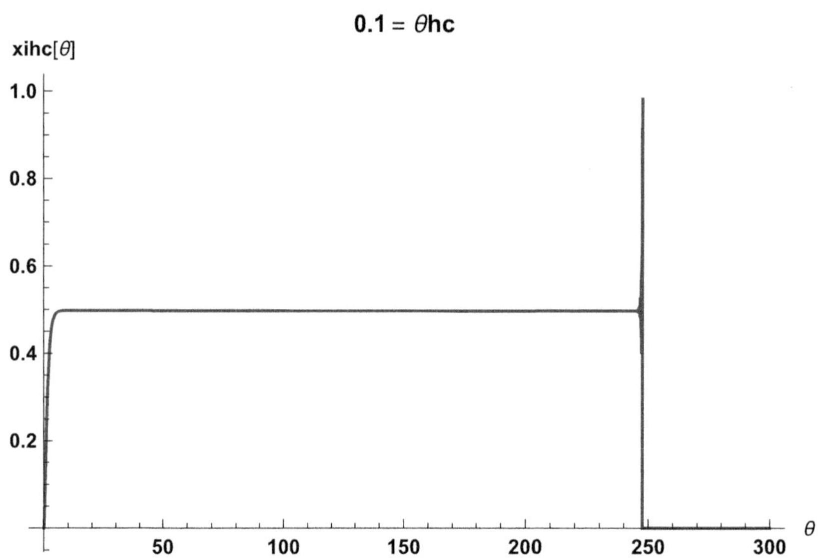

We can modify the Module function one more time to output the value of the concentration at tmax and then set tmax just below and just above the instability in order to test the accuracy of the results. We do this as follows:

iinh3[zf_, zo_, yo_, τhc_, τsc_, θhc_, Kd_, tmax_]:=

Module[

{yy},

yy[θ_]:=

$$\left(e^{-\theta\left(2+\theta hc\left(\frac{1}{\tau hc}+\frac{Kd}{\tau sc}\right)\right)}\right.$$

$$\left(e^{\theta+\theta\theta hc\left(\frac{1}{\tau hc}+\frac{Kd}{\tau sc}\right)}Kd((-1+yo)\tau hc+(-zf+zo)\tau sc)\right.$$

$$(Kd\theta hc\tau hc+(\theta hc+\tau hc)\tau sc)+$$

$$e^{\theta}\tau sc(Kd(yo-Kdzo)\theta hc\tau hc+$$

$$(-\tau hc+Kdzf\tau hc+yo(\theta hc+\tau hc)-$$

$$Kdzo(\theta hc+\tau hc))\tau sc)+$$

$$e^{\theta\left(2+\theta hc\left(\frac{1}{\tau hc}+\frac{Kd}{\tau sc}\right)\right)}(Kd\tau hc+\tau sc)$$

$$(\tau hc\tau sc+Kd\theta hc(\tau hc+zf\tau sc))))/$$

$$((Kd\tau hc+\tau sc)(Kd\theta hc\tau hc+(\theta hc+\tau hc)\tau sc));$$

yy[tmax]

]

Then we can set tmax to a number just below the divergence and test again, and then we can set tmax to a number just above the divergence and test once again.

First look at a value well below the divergence at $\theta = 150$:

iinh3[0, 0, 0, .1, .1, .1, .01, 150]

Accuracy[%]

0.502488

16.2535

There are just over 16 digits to the right of the decimal place. The Accuracy of any number x is the log base ten of the uncertainty (dx) in x that is - Log[10, dx]. Next we can compute concentration at different specific values of θ, and then Map Accuracy onto the list of values and then plot the accuracies.

```
dat = Table[iinh3[0, 0, 0, .1, .1, .1, .01, θ],
{θ, {150, 200, 225, 230, 240, 245, 247, 247.5,
247.55, 247.552, 247.5525, 247.6, 248}}];
Map[Accuracy, dat]
ListPlot[Map[Accuracy, dat], PlotRange → All]
```

{16.2535, 16.2535, 16.2535, 16.2535, 16.2535, 16.2534, 16.1976, 16.0211, 15.9558, 15.9532, 15.9525, 323.607, 323.607}

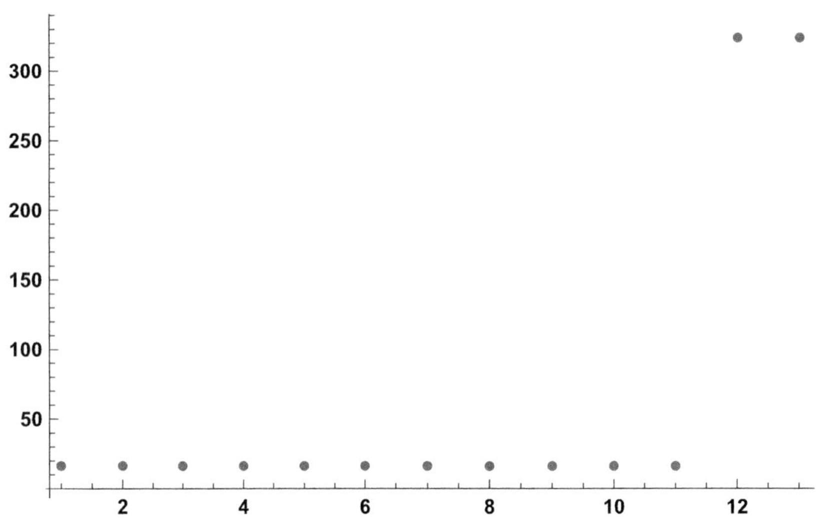

Up to the divergence the accuracy is to 16 places, after the divergence it is 300 places. A pure integer or zero, has infinite accuracy because there is no uncertainty in its value.

```
Accuracy[0]
```

∞

The question is why is this happening. The source of the divergence comes from the lead terms in the solutions that involve the exponential function raised to very high negative and positive powers. We can take the functions apart and test the behavior of just these parts to see this happen:

(* First create functions of the problematic exponential terms *)

$$\textbf{test1}[\theta\_]:=N\left[e^{-\theta\left(2+\theta\text{hc}\left(\frac{1}{\tau\text{hc}}+\frac{\text{Kd}}{\tau\text{sc}}\right)\right)}\right]$$

$$\textbf{test2}[\theta\_]:=N\left[e^{\theta+\theta\text{hc}\left(\frac{1}{\tau\text{hc}}+\frac{\text{Kd}}{\tau\text{sc}}\right)}\right]$$

(* Apply values to the dimensionless parameters *)

$\text{zf}=0;$

$\text{zo}=0;$

$\text{yo}=1.0;$

$\tau\text{hc}=10^{-1};$ **(* Dimensionless Parameters *)**

$\tau\text{sc}=10^{-1}$

$\theta\text{hc}=10;$

$\text{Kd}=.01;$

(* Evaluate the exponents *)

$$\left(\theta\text{hc}\left(\frac{1}{\tau\text{hc}}+\frac{\text{Kd}}{\tau\text{sc}}\right)\right)$$

$$\theta\text{hc}\left(\frac{1}{\tau\text{hc}}+\frac{\text{Kd}}{\tau\text{sc}}\right)$$

(* Evaluate the exponentials *)

$$e^{-\theta\left(2+\theta\text{hc}\left(\frac{1}{\tau\text{hc}}+\frac{\text{Kd}}{\tau\text{sc}}\right)\right)}$$

$$e^{\theta+\theta\theta\text{hc}\left(\frac{1}{\tau\text{hc}}+\frac{\text{Kd}}{\tau\text{sc}}\right)}$$

(* Evaluate the functions *)

test1[6]

test2[6]

$$\left(e^{-\theta\left(2+\theta\text{hc}\left(\frac{1}{\tau\text{hc}}+\frac{\text{Kd}}{\tau\text{sc}}\right)\right)}\right.$$

$$\left(e^{\theta+\theta\theta\text{hc}\left(\frac{1}{\tau\text{hc}}+\frac{\text{Kd}}{\tau\text{sc}}\right)}\text{Kd}((-1+\text{yo})\tau\text{hc}+(-\text{zf}+\text{zo})\tau\text{sc})\right.$$

$(\text{Kd}\theta\text{hc}\tau\text{hc}+(\theta\text{hc}+\tau\text{hc})\tau\text{sc})+$

$e^{\theta}\tau\text{sc}$

$(\text{Kd}(\text{yo}-\text{Kdzo})\theta\text{hc}\tau\text{hc}+$

$(-\tau\text{hc}+\text{Kdzf}\tau\text{hc}+\text{yo}(\theta\text{hc}+\tau\text{hc})-$

$\text{Kdzo}(\theta\text{hc}+\tau\text{hc}))\tau\text{sc})+$

$e^{\theta\left(2+\theta\text{hc}\left(\frac{1}{\tau\text{hc}}+\frac{\text{Kd}}{\tau\text{sc}}\right)\right)}(\text{Kd}\tau\text{hc}+\tau\text{sc})$

$(\tau\text{hc}\tau\text{sc}+\text{Kd}\theta\text{hc}(\tau\text{hc}+\text{zf}\tau\text{sc}))))$

$\frac{1}{10}$

101.

101.

$e^{-103.\theta}$

$e^{102.\theta}$

4.036548581771182`*^-269

6.14077×10^{265}

$e^{-103.\theta}\left(0.+0.101e^{\theta}+0.00202e^{103.\theta}\right)$

This tells the story. We have parameters that vary by several orders of magnitude and they are involved as exponents to the exponential functions. The instability is in the computation of these numbers.

$$\text{Plot}\left[e^{-\theta(100)}\left(\text{“0.”}+\text{“0.101”}e^{\theta}+\frac{101e^{103\theta}}{50000}\right),\right.$$

$\{\theta,0,10\text{^}1\},\text{PlotRange}\to\text{All}]$

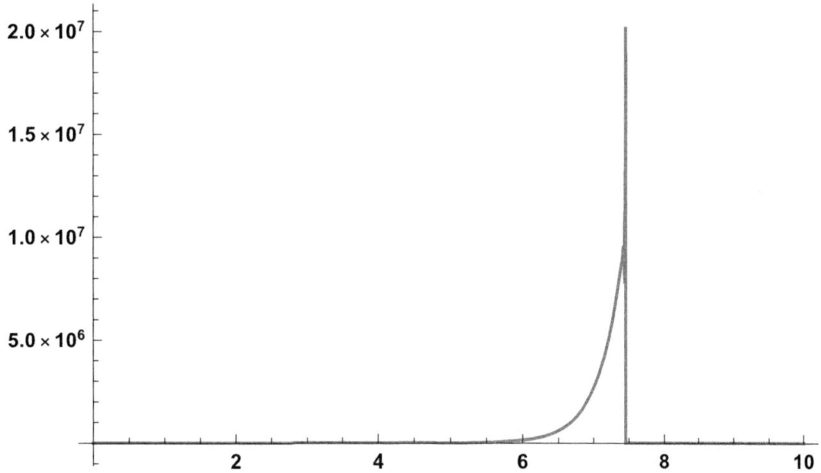

So, we can expect that the time-dependent solutions are accurate in regions well away from the divergence point. But what this really teaches us is the inherent value of the steady state solutions. These were easy to develop, they are simple and not complicated, they are infinitely stable and they are useful. As long as the time it takes to get to a steady state is not long, say 5-10 holding times, then the steady state solutions are exceptionally valuable. It is not surprising that most of the chemical, and petroleum were built by engineers using steady state models and not much more than slide rules.

We will come back to this overall unit scheme later in our studies, when we seek to write models for a group of units, but for now we can use the knowledge we have gained here by applying it to some other seemingly different systems that are actually quite similar at the level of analysis.

Conclusion

In this chapter we covered the approaches that we can take to handling mass transfer between phases using rate processes, as well as equilibrium stage. We also learned how to develop and check complicated and parametrically sensitive solutions to coupled time-dependent equations. The time-dependent solutions we derived would be tedious to derive by hand and in this way *Mathematica* behaves well as an assistant doing the symbolic computations for us. We also have seen how to work back and forth between the straightforward steady state or "long time" solutions and the more complicated time-dependent solutions. We will use all of these new skills as we move forward to analyze and model new physical and chemical phenomena.

CHAPTER 6

Adsorption and permeation

Adsorption

Adsorption is a fundamental physical process that takes place when solids are in contact with liquids, gases, or vapors. It is also a method of separation that is practiced for different purposes. Removal of volatile organics contaminants (VOCs) from air is one, removal of water vapor from nitrogen is another. Hydrogen purification often involves removing trace quantities of methane, carbon monoxide, carbon dioxide, and water vapor especially for applications in the electronics fabrication processes that can require "nine nines" 99.9999999% purity (that is less that one part impurity in 10^9 parts of the gas!). Adsorption combined with permeation through a palladium alloy membrane can give this level of purification for high purity hydrogen. Permeation of hydrogen through a palladium membrane, gives the highest attainable purity, since hydrogen and only hydrogen can be transported through the metal. We will cover permeation after we examine adsorption.

From the perspective of physics and chemistry, adsorption is the process by which molecules in a fluid phase come in contact with a solid and interact with it. The interaction of physical adsorption does not cause reaction, which is to say the adsorbed molecule is not disassembled in the process. The species that is adsorbed is referred to as the "adsorbate." The solid that does the adsorption is termed the "adsorbent." Adsorption and absorption sound so similar that they can be confused, but they are different.

Typically physical adsorption is reversible and reaches equilibrium between the fluid and solid surface rapidly, or even instantaneously. Chemical adsorption, or chemisorption, typically does involve chemistry, bond breaking and making, between the adsorbate and the adsorbent, but physical adsorption does not. Hydrogen is adsorbed on activated carbon, but it is chemisorbed on a metallic surface like platinum or palladium.

Physical adsorption is reversible, that is to say the adsorbate can come back off the surface to the fluid phase with their full molecular integrity intact. The reason for this is that the forces that adsorb the molecule to the surface are relatively weak. These forces arise due to London and dispersion interactions between the adsorbate and the adsorbent. The heats of adsorption often lie in the range from 5-10 kcal/mol or about 20-40 kjoule/mole. Chemical adsorption forces create heats of reaction at the surface that can be ten times larger.

Introduction to Chemical Engineering Analysis Using *Mathematica*
https://doi.org/10.1016/B978-0-12-820051-3.00011-5

Although adsorption can happen on any surface, adsorbents used in engineering processes are porous solids that have sponge-like structures. Naturally-occurring pumice stone provides a mental model of what an adsorbent solid is like. However, there is a significant difference between pumice and an industrial adsorbent. Pumice stone has some pores that are visible to the naked eye. They are in the millimeter range of sizes. Because the pores are large and although the density of pumice is less than water, $\sim 0.2\ g/\mathrm{cm}^3$, the surface area inside the pores is not large - it is only 0.5 - 1.0 m^2/g. An estimate of the average pore size, $\bar{r}_{\text{pore}}$, can be calculated from the reciprocal of the product of the density and surface area:

$$\bar{r}_{\text{pore}} = \left(\frac{1}{\rho * \mathrm{SA}}\right) = \left(0.2\ \frac{g}{\mathrm{cm}^3} * \left(\frac{100\mathrm{cm}}{m}\right)^3 * 1\frac{m^2}{g}\right)^{-1}$$

$$\left(\frac{1}{\rho * \mathrm{SA}}\right) = \left(0.2\ \frac{g}{\mathrm{cm}^3} * \left(\frac{100\mathrm{cm}}{m}\right)^3 * 1\frac{m^2}{g}\right)^{-1}$$

$$5 * 10^{-6} m$$

The average pore size of pumice is estimated to be $5 * 10^{-6}$m or 5 microns. While in everyday life that is a very small number, when it comes to molecules, and adsorption of them into the pores within a solid, that number is very large. Hence the surface area is not large.

By contrast an activated carbon used in water purification will have a density of about $2\ g/\mathrm{cm}^3$ and a surface area of 600 - 1200 m^2/g. The same calculation assuming a surface area of 1000 m^2/g gives:

$$\left(\frac{1}{\rho * \mathrm{SA}}\right) = \left(2.0\ \frac{g}{\mathrm{cm}^3} * \left(\frac{1000\mathrm{cm}}{m}\right)^3 * 1\frac{m^2}{g}\right)^{-1}$$

$$5 * 10^{-10} m$$

This is four orders of magnitude smaller than the average pore size of pumice. Recall that there are $10^9\mathrm{nm}/m$ and that makes this average pore size 0.5 nm. There are 10 angstroms per nanometer, so this corresponds to an average pore size of 5 Å. Molecules such as nitrogen and oxygen are about 3 Å in length. So these are extremely small pores and it is these molecularly-sized pores that give rise to the high internal surface area of the adsorbents.

There are two more points to make. Firstly, this calculation is a crude estimate of average pore size, useful, but crude. In fact there is a range of pore sizes starting at the micron and submicron range down to the nanometer size. The larger pores, do not provide much surface area, but they do provide access to the smaller pores. Think of them as super highways going through cities with many smaller roads (micro and nano pores) and parking spaces off the highway (adsorption sites) that they make accessible. Secondly, activated carbon is but one type of adsorbent used in chemical process technology, others include gamma-alumina, silica gel, zeolites, and diatomaceous earth. A high surface area porous solid, like silica gel, is found packed with many consumer products, often in a small packet, and sometimes in the product itself. It is there to adsorb water vapor.

When the discussion turns to removal of some component from a fluid by an adsorbent, then the term "adsorption" becomes more global and, hence, ambiguous. The reason for this is that mass transfer and diffusion may be combined with adsorption. In other words, the component to be adsorbed must first move through the bulk gas phase to the near vicinity of the adsorbent particle, this is termed external mass transfer. From the near external surface region, the component must then be transported through the pore space of the particles. This is called internal mass transfer or diffusion because it is within the particle. Finally, from the fluid phase within the pores, the component must be adsorbed by the surface in order to be removed from the gas and that is physical adsorption per se. Anyone of these processes, external, internal, or adsorption can, in principle, be the slowest step and, therefore, the process that controls the observed rate. Most often it is not the adsorption that is slow, in fact this step usually comes to equilibrium quickly. Think of how fast frost forms on a beer mug taken from the freezer on a humid summer day. There is instantaneous adsorption of water vapor from the air to the cold glass surface. Once the cold glass surface is covered, additional water vapor adsorbs on the frozen water layer. More typically it is the internal mass transport process through the molecularly-sized that is rate limiting. This is, however, sometimes lumped with the true adsorption process and the overall phenomenon is called "adsorption."

True adsorption is a "mass action" process rather than a mass transfer process. What this means is that it will occur even in the absence of a concentration gradient between the bulk gas and the surface. It comes about due to the rapid and chaotic motion of the fluid phase molecules, and there impingement on the surface. From the elementary kinetic theory of an ideal gas, we can compute the number of molecules impinging upon a surface per unit time per unit area at a given temperature and pressure. It is:

$$\frac{\text{Number of Molecules}}{\text{Area} * \text{Time}} = \frac{1}{4}\,\mathrm{C\,L}\;\bar{v} = \frac{1}{4}\sqrt{\frac{8RT}{\pi \mathrm{MW}}}\,\frac{\mathrm{PL}}{RT} = P\,\mathrm{L}\sqrt{\frac{1}{2\pi RT\mathrm{MW}}} = \frac{P}{\sqrt{2\pi RT\mathrm{MW}}}$$

The number of molecules hitting a surface per unit time per unit area has the dimensions of a flux, and it is proportional to the pressure of the gas, the mean speed of the gas molecules, and that goes as the reciprocal of the square root of temperature. At room temperature and pressure the impingement frequency of nitrogen is:

$$\frac{1}{4}\mathrm{C\,L}\ \bar{v} = \frac{1}{4}\sqrt{\frac{8RT}{\pi \mathrm{MW}}}\ \frac{\mathrm{PL}}{RT} ==$$

NumberForm[

PowerExpand[

$$\left(\frac{0.8\ \mathrm{atm}}{4}\right)\sqrt{\left(\frac{8*8.314*10^7*300}{\pi*28}\right)\left(\frac{Joule}{mole*K}\right)\left(\frac{\frac{g*\mathrm{cm}^2}{s^2}}{Joule}\right)\left(\frac{mole}{g}\right)(K)}$$

$$\left(\frac{6.02*10^{23}}{1000*0.08205*300}\right)\left(\frac{molecule}{mole*K}\right)\left(\frac{L}{\mathrm{cm}^3}\right)\left(\frac{mole*K}{L*atm}\right)]],2]$$

$$\frac{\mathrm{PL}\sqrt{\frac{RT}{\mathrm{MW}}}}{\sqrt{2\pi}\,RT} == \frac{(2.3\times10^{23})\ \mathrm{molecules}}{\mathrm{cm}^2 * s}$$

So about one mole of nitrogen molecules strikes every square centimeter ever second. No wonder the time to equilibration of adsorption is so fast!

Irving Langmuir, the Nobel prize-winning physical chemist (General Electric), built an elegant structure upon this foundation in kinetic theory. He reasoned that not every molecule would adsorb, but only some would. Furthermore, one reason for this was that to be adsorbed, there should be a site for adsorption of the molecule to occur. It stands to reason then, that on the basis of mass action, the rate of adsorption should be proportional to the concentration of molecules in the gas phase and to the number of sites available on the surface. But even more, the rate should be related at any time to the number of sites not covered at that time, rather than to the total number of sites present per unit area. Conversely and again by the principle of mass action, the rate of desorption should be proportional to the number of sites currently occupied at that time. Using **ka** and **kd** as the proportionality constants, that we will call the rate constant for adsorption and desorption, respectively, we can write the net rate of adsorption for gas phase species i as the difference between the rate of adsorption and the rate of desorption:

$$\mathrm{rate}_{i,\mathrm{ads,net}} = k_{\mathrm{ai}} C_{i\,g}(C_{i-\mathrm{sites,total}} - C_{i-\mathrm{sites,\ occupied\ by}\ i}) - k_{d,s} C_{i-\mathrm{sites,\ occupied\ by}\ i}$$

The sites are assumed to all be identical, and the adsorption at one site does not affect that at another site, i.e. they interact with the gas phase independently. In addition to the two rate constants the term $\boldsymbol{C_{i-\mathrm{sites,total}}}$ is also a constant, it is the number of sites available on the solid per unit area. This raises another point, if this and the concentration of occupied sites are written on a per unit area basis, and the gas phase concentration, C_{ig}, is written on a per unit volume basis, then what are the dimensions of the rate constant?

The net rate of adsorption is the number of molecules or moles adsorbed per unit area per unit time, where the area is the area made available for adsorption by a given mass or volume of the adsorbent solid. Therefore the two rates on the right-hand side must also be per unit volume-time. This means that the rate constants must be dimensioned as follows:

$$\frac{\text{mole}}{\text{length}^2\ \text{time}} = \left[\frac{\text{length}^3}{\text{mole time}}\right]\left[\frac{\text{mole}}{\text{length}^3}\right]\left[\frac{\text{mole}}{\text{length}^2}\right] - \left[\frac{1}{\text{time}}\right]\left[\frac{\text{mole}}{\text{length}^2}\right]$$

These are the dimensions we expect from mass action kinetics for second order adsorption and first order desorption.

Consider now an adsorbent which offers no resistance to mass transfer because it is "macroporous." This means that the pores within the solid are large, greater than 20 nm in diameter or width, and that transport of small molecules (0.2 nm) is unhindered and takes place as if in the bulk phase surrounding the solid. This means that the bulk gas phase concentration is the same in the pore spaces within the solid as it is outside the solid.

If the gas around the adsorbent solid occupies some fraction, ϵ, of a volume **V**, and if this volume contains an adsorbing species, **i**, then the rate of adsorption of that species onto the adsorbent and the rate of depletion of that species from the gas phase are coupled batch processes.

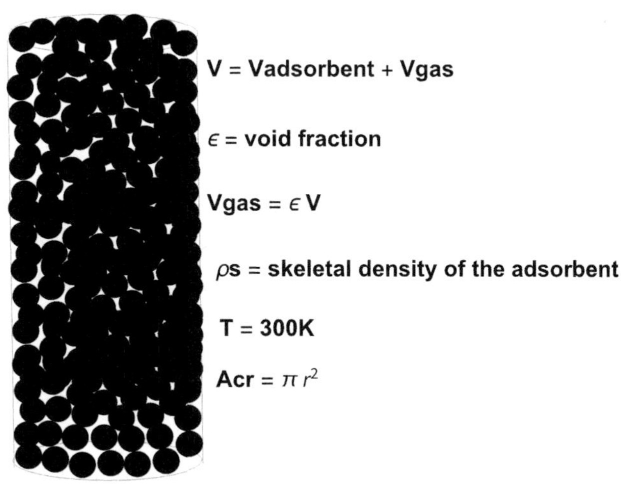

The component mass balances for the gas and solid phases are as follows:

$$\text{Gas phase:} \quad \frac{dC_{i,g}\epsilon V}{dt} = -(1-\epsilon)\,\text{rate}_{i,\text{ads,net}}\, A_s \rho_s \text{V}$$

$$\text{Solid phase:} \quad \frac{dC_{i,s}(1-\epsilon)A_{\text{tot}}}{dt} = (1-\epsilon)\,\text{rate}_{i,\text{ads,net}}\, A_s \rho_s \text{V}$$

$C_{i,g} = \dfrac{g}{\text{volume}}$ gas phase

ϵ = void fraction in the bed of solid and gas

$C_{i,s} = \dfrac{g}{\text{area}}$ surface phase

V = total volume occupied by solid and gas in the vessel

$A_{\text{tot}} = A_s \rho_s V = \text{cm}^2, \text{Length}^2$

$\text{rate}_{i,\text{ads,net}}$ = net rate of adsorption of i on solid

$\text{As} = \dfrac{\text{cm}^2}{g}, \dfrac{\text{Length}^2}{\text{mass}}$

$\rho_s = \dfrac{g}{\text{cm}^3}, \dfrac{\text{mass}}{\text{Length}^3}$

When rearranged and written with the explicit rate of adsorption these become:

$$\text{Gas phase:} \quad \frac{dC_{i,g}}{dt} = -\frac{(1-\epsilon)}{\epsilon}\left(k_a C_{i,g}\left(C_{i,s,\text{tot}} - C_{i,s}\right) - k_{d,s} C_{i,s}\right) A_s \rho_s$$

$$\text{Solid phase:} \quad \frac{dC_{i,s}}{dt} = \left(k_a C_{i,g}\left(C_{i,s,\text{tot}} - C_{i,s}\right) - k_{d,s} C_{i,s}\right)$$

There are two types of experiments that these equations suggest and that actually are done to obtain the rate constants for adsorption of a species, **i**, on a given adsorbent. The first experiment is done gravimetrically. The adsorbent is placed in small container suspended from a balance inside of an evacuated enclosure. After heating the sample under dynamic vacuum to remove any water, or other adsorbed molecules, the sample is cooled to the experimental temperature, and then the adsorbate is admitted in such a way that its pressure remains constant throughout the course of the experiment. This is done either by constant delivery or by connection to a large ballast volume of the adsorbate gas. The mass uptake is measured as a function of time. The parameters, **Ci,s,tot**, and **As** are typically known from separate measurements. So only **ka** and **kd** need to be fitted to the data, either in differential or integral form. To fit to the integral form, we need the expression for mass adsorbed per time, thus we need to integrate the mass balance for the adsorbate over time:

$$\frac{dC_{i,s}A_{\text{tot}}}{dt} = \left(k_a C_{i,g}\left(C_{i,s,\text{tot}} - C_{i,s}\right) - k_{d,s}C_{i,s}\right) A_{\text{tot}}$$

$$\frac{dm_{i,s}}{dt} = \left(k_a C_{i,g}\left(C_{i,s,\text{tot}} - C_{i,s}\right) - k_{d,s}C_{i,s}\right) \rho_s A_s V$$

$$\frac{dm_{i,s}}{dt} = \left(k_a C_{i,g}\rho_s A_s V\left(C_{i,s,\text{tot}} - C_{i,s}\right) - k_{d,s}\rho_s A_s V C_{i,s}\right)$$

$$\frac{dm_{i,s}}{dt} = k_a C_{i,g}\left(m_{i,s,\text{tot}} - m_{i,s}\right) - k_{d,s}m_{i,s}$$

where

$$m_{i,s,\text{tot}} = C_{i,s,\text{tot}}\ \rho_s A_s V \equiv \frac{\text{mass}}{\text{area}}\text{area}$$

Under the conditions of the experiment, the concentration of **i** in the gas phase is a constant, which makes this equation readily integrated:

```
Clear[ka, kd, Cg]
Simplify[DSolve[
{∂_t m_is[t] == kaCig (m_istot − m_is[t]) − kd m_is[t],
m_is[0] == 0}, m_is[t], t]]
```

$$\left\{\left\{m_{\text{is}}[t] \to \frac{\text{Cig}e^{-(\text{Cig ka}+\text{kd})t}\left(-1+e^{(\text{Cig ka}+\text{kd})t}\right)\text{ka}\ m_{\text{istot}}}{\text{Cigka}+\text{kd}}\right\}\right\}$$

We can put some realistic numbers into this equation to see how it would behave. We can take ϵ to be 0.4, which is a reasonable number for a bed of particles. The area per unit volume of adsorbent can be taken as 1000 m^2 per g ($\sim 10^7$ cm^2 per g), the density of the solid is on the order of 2 g cm^{-3}, the number of sites per unit area, $N_{i,s,tot}$, is on the order of $10^{14}cm^2$ making $C_{i,s,tot} \sim 10^{-9}$mole sites cm^2. (On a perfect surface there are $\sim 10^{15}$ atoms per cm^2, so we have taken 10% of this value as an estimate of the number of sites, which corresponds to one site in every 1 nm^2. Finally, the mass concentration of the adsorbate, if the latter is ideal, is $\frac{\text{Pi MW}i}{RT}$. We use these numbers and a value of **ka** which is one order of magnitude larger than **kd**. If the system were to come to equilibrium, then the mass uptake would go to zero.

This would be the same when rate of adsorption was balanced exactly by the rate of desorption. We can compute the mass of i on the solid as follows:

$$\frac{dm_{i,s}}{dt} = 0 = k_a C_{i,g}\left(m_{i,s,\text{tot}} - m_{i,s}\right) - k_{d,s} m_{i,s}$$

Clear$\left[m_{\text{istot}}, m_{\text{iseq}}, \text{ka}, \text{kd}, \text{Cg}\right]$

Solve$\left[\text{kaCg}\left(m_{\text{istot}} - m_{\text{iseq}}\right) - \text{kd}m_{\text{iseq}} == 0, m_{\text{iseq}}\right]$

$$\left\{\left\{m_{\text{iseq}} \to \frac{\text{Cg ka } m_{\text{istot}}}{\text{Cg ka } + \text{kd}}\right\}\right\}$$

The full, time-dependent solution comes from the solution of the material balance equation. Both solutions are presented below:

(* The expression on the right needs to be divided by grams to make it dimensionless for plotting *)

$$m_{\text{is}}[\text{t\_}] := N\left[-\frac{\text{Cg}\left(-1 + e^{-(\text{Cgka}+\text{kd})t}\right)\text{ka}m_{\text{istot}}}{\text{Cgka} + \text{kd}}\right]$$

$$\text{ka} = 10\frac{\text{cm}^3}{\text{sec} * g};$$

$$\text{kd} = .001\frac{1}{\text{sec}};$$

$$\text{pi} = 1 \text{ atm};$$

$$\text{Mwi} = 100\frac{g}{\text{mole}};$$

$$T = 300K;$$

$$R = 82.05\frac{\text{cm}^3\text{ atm}}{\text{mole} * K};$$

$$\text{Cg} = \frac{\text{piMwi}}{R\,T};$$

$$\epsilon = 0.4;$$

$$V = 2\;10^6\text{ cm}^3;$$

$$\rho\text{s} = 2.\;g\text{ cm}^{-3};$$

```
As = 10^7 cm^2 g^-1;

Cistot = 10^14 cm^-2 (1mole / (6.02 10^23)) Mwi;

Print[
"The mass concentration of i on the adsorbent Cistot =", %]

mistot = Cistot V ρs;

Print[
"The total mass of i that could be on the adsorbent assuming 1 molecule per site is
mistot = ", mistot]

(* In the following two expressions the unit of mass needs to be eliminated for plotting *)

miseq = (Cg ka mistot) / (Cg ka + kd);

Print["The total mass of i on the adsorbent at equilibrium = ", %]

miseq / (ρs ϵ V) 1000 mg/g;

Print["The mg of i per g of adsorbent =", %]
```

The mass concentration of i on the adsorbent $C_{\mathrm{istot}} = \dfrac{1.66110^{-8} g}{\mathrm{cm}^2}$

The total mass of i that could be on the adsorbent assuming 1 molecule per site is $m_{\mathrm{istot}} = 664452.g$

The total mass of i on the adsorbent at equilibrium = $648489.g$

The mg of i per g of adsorbent = $\dfrac{405.306\mathrm{mg}}{g}$

Now we can plot the time dependence of the change in mass with time. Remember that *Mathematica* requires pure numbers for plotting. Because of that we will place time units in the argument next to t (sec):

```
tmax = 150;

Plot[ mis[t sec] / (ρs ϵ V) 1000, {t, 0, tmax},
```

$$\text{AxesLabel} \to \left\{\text{“t/s”}, \text{“}\frac{\text{mg}i}{g\ \text{adsorbent}}\text{”}\right\},$$

PlotStyle → Hue[0], PlotRange → All,

$$\text{Epilog} \to \text{Line}\left[\left\{\left\{0, \frac{m_{\text{iseq}}1000}{\rho s\epsilon V}\right\}, \left\{\text{tmax}, \frac{m_{\text{iseq}}1000}{\rho s\epsilon V}\right\}\right\}\right]$$

]

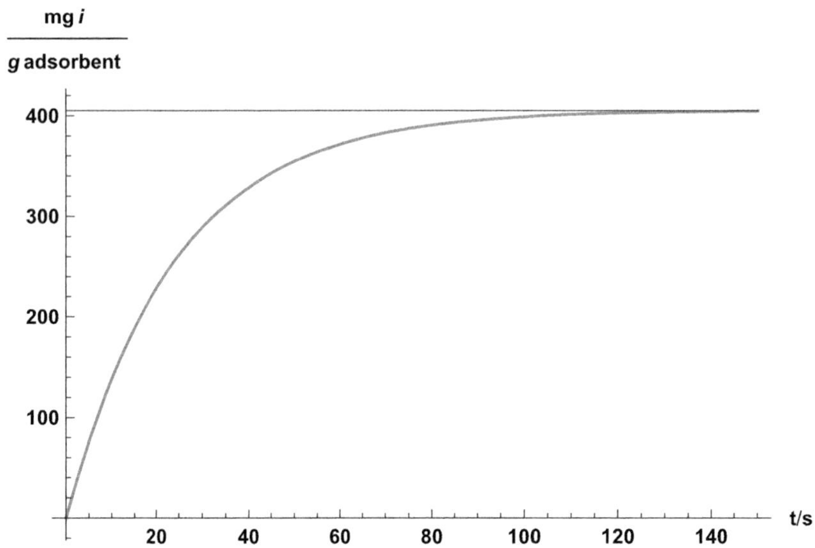

In this case, the equilibrium is reached at a level of 80 mg of adsorbate per gram of adsorbent.

A different experiment that sounds simple is to expose the adsorbent to a volume of gas and then measure the vapor pressure change of species i as a function of time. This has the same aim as the procedure we just analyzed, but it is more complex. A bit of analysis will show us why. Let $\boldsymbol{V_o}$, be the volume which is occupied by the gas at a known pressure and temperature. Once the two volumes are connected, the total volume of the system is **Vtot =** $\boldsymbol{\epsilon}$ **V +** $\boldsymbol{V_o}$ on the basis of a solid that was space occupying, but not adsorbing. The initial pressure, $\boldsymbol{P_1}$, after opening a valve between the two is given by:

$$P_o V_o = P_1\,(\,\epsilon\ \text{V} + V_o\,) \Longrightarrow P_1 = \frac{P_o V_o}{(\epsilon V + V_o)}$$

Is this the correct initial pressure to use, or should we account for the internal void of the adsorbent as well when we compute the initial pressure? To do so would lead to one more term in volume, that of the void fraction within the solid. This is not the void between the solid particles, but that which is within the solid particles. If the unoccupied fraction of the porous adsorbent particles is ξ, then this extra volume is **ξ Vp** where Vp is the volume of the particles and that is (1 - ϵ)V. The corrected initial pressure would be:

$$P_o V_o = P_1 (\epsilon \text{ V} + V_o + \xi \text{ Vp}) \Longrightarrow P_1 = \frac{P_o V_o}{(\epsilon V + V_o + \xi(1-\epsilon)V}$$

Assuming we can compute a reasonable initial pressure, then the pressure is measured as a function of time and the data would be fitted either to the differential, or integral expressions from the mass balance equations for the gas phase between the adsorbent particles and the porous solid, adsorbent phase. The adsorbate is only adsorbed once it has alit on the surface of the adsorbent. Hence we use the total surface area of the adsorbent in the bed. We get this from (1 - ϵ) A_s ρ_g V, where ϵ is once again the void fraction of the bed of adsorbent particles, A_s is the surface area per gram (or kilogram) of the adsorbent, ρ_g is the density of the adsorbent in mass per volume and V is the total volume of the bed of particles, including the void between the particles.

In the differential terms on the left-hand side, we note that for the gas phase the volume of that phase is ϵ V and for the adsorbent phase the volume is (1 - ϵ) V. Obviously ϵ V + (1 - ϵ) V = V the total volume in the unit.

Gas Phase Mass Balance for the Adsorbate:

$$\frac{d\epsilon C_{i,g} V}{dt} = -(1-\epsilon)\rho_s A_s(k_a C_{i,g} \left(C_{i,s,\text{tot}} - C_{i,s}\right) - k_{d,s} C_{i,s}) \text{ V}$$

$$\frac{dC_{i,g}}{dt} = -\frac{(1-\epsilon)}{\epsilon} \rho_s A_s(k_a C_{i,g} \left(C_{i,s,\text{tot}} - C_{i,s}\right) - k_{d,s} C_{i,s})$$

This equation states that the rate of change in the concentration of the adsorbate, i, in the gas phase is equal to the difference between the rate of adsorption onto the surface of the adsorbent and the rate of desorption of species i from that same surface and back to the gas phase.

In some instances, it is convenient to recast this equation in terms of the partial pressure of species i. Recasting looks as follows: if we assume species i behaves ideally in the gas phase, then we know that this is the relationship between the gas phase mass concentration of i and its vapor pressure:

$$P_i \text{ V} = n_i \text{ R T}$$

The partial pressure of i times the volume occupied equals the product of the number of moles of i and R T. This rearranges to give moles per unit volume:

$$\frac{P_i}{RT} = \frac{n_i}{V}$$

To express this in mass (grams) of i per unit volume (cm^3) we multiply through by the molecular weight of i:

$$\frac{P_i}{RT}\text{MW}_i = \frac{n_i}{V}\text{MW}_i$$

$$\frac{P_i}{RT}\text{MW}_i = \frac{n_i \text{MW}_i}{V}$$

$$\frac{P_i}{RT}\text{MW}_i = C_{i,g}$$

$$P_i = C_{i,g}\frac{\text{R T}}{\text{MW}_i}$$

We can manipulate the rate equation in a similar fashion. Firstly, we can multiply the left-hand side by unity in the form of $\left(\frac{\text{RT}}{\text{RT}}\right)\left(\frac{\text{MW}_i}{\text{MW}_i}\right)$.

$$\left(\frac{\text{RT}}{\text{RT}}\right)\left(\frac{\text{MW}_i}{\text{MW}_i}\right)\frac{dC_{i,g}}{dt} = -\frac{(1-\epsilon)}{\epsilon}\rho_s A_s(\, k_a C_{i,g}\left(C_{i,s,\text{tot}} - C_{i,s}\right) - k_{d,s}C_{i,s})$$

Then we bring $\frac{\text{RT}}{\text{MW}_i}$ inside the derivative to convert concentration of i to partial pressure of i and then rearrange:

$$\left(\frac{\text{MW}_i}{\text{RT}}\right)\frac{d\left(\frac{\text{RT}}{\text{MW}_i}C_{i,g}\right)}{dt} = -\frac{(1-\epsilon)}{\epsilon}\rho_s A_s(\, k_a C_{i,g}\left(C_{i,s,\text{tot}} - C_{i,s}\right) - k_{d,s}C_{i,s})$$

Where

$$\left(\frac{\text{RT}}{\text{MW}_i}C_{i,g}\right) = P_{i,g}$$

$$\left(\frac{\text{MW}_i}{\text{RT}}\right)\frac{dP_{i,g}}{dt} = -\frac{(1-\epsilon)}{\epsilon}\rho_s A_s(\, k_a C_{i,g}\left(C_{i,s,\text{tot}} - C_{i,s}\right) - k_{d,s}C_{i,s})$$

$$\frac{dP_{i,g}}{dt} = -\frac{(1-\epsilon)}{\epsilon}\rho_s A_s(k_a \frac{\mathrm{RT}}{\mathrm{MW}_i} C_{i,g} \left(C_{i,s,\mathrm{tot}} - C_{i,s}\right) - k_{d,s}\frac{\mathrm{RT}}{\mathrm{MW}_i} C_{i,s})$$

$$\frac{dP_{i,g}}{dt} = -\frac{(1-\epsilon)}{\epsilon}\rho_s A_s(k_a P_{i,g} \left(C_{i,s,\mathrm{tot}} - C_{i,s}\right) - \frac{\mathrm{RT}}{\mathrm{MW}_i} k_{d,s} C_{i,s})$$

We can re-express this equation in terms of the mole fraction of species i, $x_{i,g}$, in the gas phase where $P_{i,g} = x_{i,g} P_{\mathrm{tot}}$, which is useful because the mole fraction varies from 0 to 1:

$$\frac{dP_{i,g}}{dt} = -\frac{(1-\epsilon)}{\epsilon}\rho_s A_s(k_a P_{i,g} \left(C_{i,s,\mathrm{tot}} - C_{i,s}\right) - \frac{\mathrm{RT}}{\mathrm{MW}_i} k_{d,s} C_{i,s})$$

$$\frac{dX_{i,g}}{dt} = -\frac{(1-\epsilon)}{\epsilon}\rho_s A_s(k_a X_{i,g} \left(C_{i,s,\mathrm{tot}} - C_{i,s}\right) - \frac{\mathrm{RT}}{\mathrm{MW}_i P_{\mathrm{tot}}} k_{d,s} C_{i,s})$$

This equation looks very similar to those we wrote for the earlier case, and so it may seem to be simple to solve, once we also have the mass balance for the adsorbent phase, because we then have two equations for the two unknowns. In fact they are not simple to solve because the pressure of the gas is a function of time, whereas it was a constant in the previous case, and it is multiplied by the change in concentration of i on the adsorbent surface, which also is time dependent. Having the product of both the pressure and the concentration on the surface, both of which are time-dependent functions, makes this a harder problem to solve analytically, i.e. symbolically because it introduces a non-linearity into the equations. The previous experiment had the advantage of being designed around an analysis that was simple to carry out and solve analytically.

We also need to write the mass balance equation for the adsorbate and the adsorbent phase. Every molecule of i that leaves the gas phase appears as a molecule adsorbed on the adsorbent surface. So the net rate of adsorption is the same as the net rate of loss of i from the gas phase and opposite in sign:

Adsorbate Mass Balance in/on the Adsorbent:

$$\frac{dC_{i,s}(1-\epsilon)\rho_s A_s V}{dt} = (1-\epsilon)\rho_s A_s V \left(k_a C_{i,g} \left(C_{i,s,\mathrm{tot}} - C_{i,s}\right) - k_{d,s} C_{i,s}\right)$$

Multiply the first term of the rate of adsorption by unity:

$$\frac{dC_{i,s}}{dt} = \left(\frac{\mathrm{RT}}{\mathrm{RT}}\right)\left(\frac{\mathrm{MW}_i}{\mathrm{MW}_i}\right) \left(k_a C_{i,g} \left(C_{i,s,\mathrm{tot}} - C_{i,s}\right) - k_{d,s} C_{i,s}\right)$$

Rearrange, regroup, and recast:

$$\frac{dC_{i,s}}{dt} = \left(\frac{\mathrm{MW}_i}{\mathrm{RT}}\right) \left(k_a C_{i,g} \left(\frac{\mathrm{RT}}{\mathrm{MW}_i}\right) \left(C_{i,s,\mathrm{tot}} - C_{i,s}\right) - k_{d,s} C_{i,s}\right)$$

$$C_{i,g}\left(\frac{\text{RT}}{\text{MW}_i}\right) \equiv \left[\frac{g}{\text{cm}^3}\right]\left[\frac{\text{cm}^3 - \text{atm}}{K - \text{mol}}\right][\text{K}]\left[\frac{\text{mol}}{g}\right] \equiv \text{atm}$$

$$\frac{dC_{i,s}}{dt} = \left(k_a\left(\frac{\text{MW}_i}{\text{RT}}\right)P_{i,g}\left(C_{i,s,\text{tot}} - C_{i,s}\right) - k_{d,s}C_{i,s}\right)$$

$$\frac{dC_{i,s}}{dt} = \left(k_a P_{\text{tot}}\left(\frac{\text{MW}_i}{\text{RT}}\right)X_{i,g}\left(C_{i,s,\text{tot}} - C_{i,s}\right) - k_{d,s}C_{i,s}\right)$$

Before we try to solve these two new equations using *Mathematica*, we should be sure that we understand all the variables and parameters, that we also check them for unit consistency and then take time to compute some basic numbers that we will need to have in the course of solving this problem.

For the adsorption rate constant, the units are $\frac{\text{cm}^3}{g\text{ sec}}$ while for the desorption rate constant, the units are inverse time, sec^{-1}. We have chosen to express the pressure in the gas phase in terms of atmospheres. Obviously, it is an arbitrary choice, but it is a typical pressure unit in industry and US laboratories. The molecular weight is the molecular weight of the adsorbate in units of $\frac{g}{\text{mole}}$; when we group $\frac{RT}{\text{MWi}}$ the units are $\frac{\text{cm}^3-\text{atm}}{g}$, where R is in $\frac{\text{cm}^3\text{ atm}}{\text{mole}K}$, T is in K, and Mwi^{-1} is in $\frac{\text{mole}}{g}$. The surface concentrations $C_{i,s,\text{tot}}$ and $C_{i,s}$ are in units of $\frac{g}{\text{cm}^2}$. Notice that $C_{i,s,\text{tot}}$ is a constant. The way to calculate this is to know the number of sites available to adsorb one molecule of the adsorbate i per cm^2, the number of $\frac{\text{cm}^2}{g}$ of adsorbent, the density of the adsorbent and the total volume of the adsorbent. What this will give us is the total mass of adsorbate i that can be adsorbed. The product $A_s\rho_s$ is made up of the surface area per gram (or kilogram) of the adsorbent in $\frac{\text{cm}^2}{g}$ and the density, which is actually the skeletal density of the adsorbent in $\frac{g}{\text{cm}^3}$, as opposed to its compacted bulk density. The compacted bulk density is the density of a solid when it is packed in a unit. It is not the intrinsic density of the solid.

On a typical surface there are about 10^{15} atoms per cm^2. Even a small molecule is many atoms in size and will take up more space on the surface. So estimates of 10^{12} to 10^{14} sites per cm^2 are a reasonable range for the number of adsorption sites on the adsorbent depending on the size of the adsorbate and how much space it occupies.

We can examine the equation for the rate of change in the gas phase concentration of i for its unit and dimensional consistency:

$$\frac{dC_{i,g}}{dt} = -\frac{(1-\epsilon)}{\epsilon}\rho_s A_s(\,k_a C_{i,g}\left(C_{i,s,\text{tot}} - C_{i,s}\right) - k_d C_{i,s})$$

Dimensions Analysis

$$\left[\frac{\text{mass}}{L^3\text{time}}\right] \equiv \left(\left[\frac{\text{mass}}{L^3}\right]\left[\frac{L^2}{\text{mass}}\right]\right)\left(\left[\frac{L^3}{\text{mass time}}\right]\left[\frac{\text{mass}}{L^3}\right]\left[\frac{\text{mass}}{L^2}\right] - \left[\frac{1}{\text{time}}\right]\left[\frac{\text{mass}}{L^2}\right]\right)$$

$$\left[\frac{\text{mass}}{L^3\text{ time}}\right] \equiv \left[\frac{\text{mass}}{L^3\text{ time}}\right] - \left[\frac{\text{mass}}{L^3\text{ time}}\right]$$

Units Analysis

$$\left[\frac{g}{\text{cm}^3\text{ sec}}\right] \equiv \left(\left[\frac{g}{\text{cm}^3}\right]\left[\frac{\text{cm}^2}{g}\right]\right)\left(\left[\frac{\text{cm}^3}{g\text{ sec}}\right]\left[\frac{g}{\text{cm}^3}\right]\left[\frac{g}{\text{cm}^2}\right] - \left[\frac{1}{\text{sec}}\right]\left[\frac{g}{\text{cm}^2}\right]\right)$$

$$\left[\frac{g}{\text{cm}^3\text{ sec}}\right] \equiv \left[\frac{1}{\text{cm}}\right]\left(\left[\frac{g}{\text{cm}^2\text{ sec}}\right] - \left[\frac{g}{\text{cm}^2\text{ sec}}\right]\right)$$

$$\left[\frac{g}{\text{cm}^3\text{ sec}}\right] \equiv \left[\frac{g}{\text{cm}^3\text{ sec}}\right] - \left[\frac{g}{\text{cm}^3\text{ sec}}\right]$$

$$\left[\frac{g}{\text{cm}^3\text{ sec}}\right] \equiv \left[\frac{g}{\text{cm}^3\text{ sec}}\right]$$

The equation is consistent both in its general dimensions and its units.

Let's also check the unit consistency of the equation after we have converted the gas phase concentration to pressure (using the ideal gas law).

Units Analysis

$$\frac{dP_{i,g}}{dt} = -\frac{(1-\epsilon)}{\epsilon}A_s\rho_s\left(k_a P_{i,g}\left(C_{i,s,\text{tot}} - C_{i,s}\right) - \frac{\text{RT}}{\text{MW}_i}k_{d,s}C_{i,s}\right)$$

$$\left[\frac{\text{atm}}{\text{sec}}\right] \equiv \left(\left[\frac{g}{\text{cm}^3}\right]\left[\frac{\text{cm}^2}{g}\right]\right)\left(\left[\frac{\text{cm}^3}{g\text{ sec}}\right][\text{atm}]\left[\frac{g}{\text{cm}^2}\right] - \right.$$

$$\left.\left[\frac{\text{cm}^3-\text{atm}}{K-\text{mol}}\right][\text{K}]\left[\frac{\text{mol}}{g}\right]\left[\frac{1}{\text{sec}}\right]\left[\frac{g}{\text{cm}^2}\right]\right)$$

$$\left[\frac{\text{atm}}{\text{sec}}\right] \equiv \left(\left[\frac{1}{\text{cm}}\right]\right)\left(\left[\frac{\text{cm}}{\text{sec}}\right][\text{atm}] - [\text{cm}][\text{atm}]\left[\frac{1}{\text{sec}}\right]\right)$$

$$\left[\frac{\text{atm}}{\text{sec}}\right] \equiv \left[\frac{\text{atm}}{\text{sec}}\right] - \left[\frac{\text{atm}}{\text{sec}}\right]$$

$$\left[\frac{\text{atm}}{\text{sec}}\right] \equiv \left[\frac{\text{atm}}{\text{sec}}\right]$$

And for the version written in terms of mole fraction i, after dividing through by P_{tot}:

$$\frac{dX_{i,g}}{dt} = -\frac{(1-\epsilon)}{\epsilon}\left(k_a X_{i,g}\left(C_{i,s,\text{tot}} - C_{i,s}\right) - \frac{\text{RT}}{\text{MW}_i P_{\text{tot}}} k_{d,s}\rho_s A_s C_{i,s}\right)$$

$$\left[\frac{1}{\text{sec}}\right] = \left(\left[\frac{\text{cm}^2}{g\,\text{sec}}\right][1]\left[\frac{g}{\text{cm}^2}\right] - \left[\frac{\text{cm}^3-\text{atm}}{K-\text{mol}}\right][\text{K}]\left[\frac{\text{mol}}{g}\right]\left[\frac{1}{\text{atm}}\right]\left[\frac{1}{\text{sec}}\right]\left[\frac{g}{\text{cm}^2}\right]\left[\frac{1}{\text{cm}}\right]\right)$$

$$\left[\frac{1}{\text{sec}}\right] = \left[\frac{1}{\text{sec}}\right] - \left[\frac{1}{\text{sec}}\right]$$

The same can be done for the rate of change of species i on the adsorbent surface:

Dimensions Analysis

$$\frac{dC_{i,s}}{dt} = \left(\left(\frac{k_a}{\rho_s A_s}\right) P_{\text{tot}}\left(\frac{\text{MW}_i}{\text{RT}}\right) X_{i,g}\left(C_{i,s,\text{tot}} - C_{i,s}\right) - k_{d,s}C_{i,s}\right)$$

$$\left[\frac{\text{mass}}{L^2\text{time}}\right] = \left[\frac{L^2}{\text{masstime}}\right][L][\text{pressure}]\left[\frac{\text{mass}}{\text{mole}}\right]\left[\frac{\text{mole}-\text{temp}}{L^3-\text{pressure}}\right]\left[\frac{1}{\text{temp}}\right]\left[\frac{\text{mass}}{L^2}\right]$$

$$-\left[\frac{1}{\text{time}}\right]\left[\frac{\text{mass}}{L^2}\right]$$

$$\left[\frac{\text{mass}}{L^2\,\text{time}}\right] \equiv \left[\frac{\text{mass}}{L^2\text{time}}\right] - \left[\frac{\text{mass}}{L^2\text{time}}\right]$$

Units Analysis

$$\frac{dC_{i,s}}{dt} = \left(\left(\frac{k_a}{\rho_s A_s}\right) P_{\text{tot}}\left(\frac{\text{MW}_i}{\text{RT}}\right) X_{i,g}\left(C_{i,s,\text{tot}} - C_{i,s}\right) - k_{d,s}C_{i,s}\right)$$

For this version of the net rate of adsorption in terms of vapor pressure of species i:

$$\left[\frac{g}{\text{cm}^2\,\text{sec}}\right] \equiv \left(\left[\frac{\text{cm}^2}{g\,\text{sec}}\right][\text{cm}]\,[\text{atm}]\left[\frac{g}{\text{mole}}\right]\left[\frac{K-\text{mole}}{\text{cm}^3-\text{atm}}\right]\left[\frac{1}{K}\right]\left[\frac{g}{\text{cm}^2}\right]\right.$$

$$\left.-\left[\frac{1}{\text{sec}}\right]\left[\frac{g}{\text{cm}^2}\right]\right)$$

$$\left[\frac{g}{\text{cm}^2\,\text{sec}}\right] \equiv \left[\frac{g}{\text{cm}^2\,\text{sec}}\right] - \left[\frac{g}{\text{cm}^2\,\text{sec}}\right]$$

These dimensional and unit analyses show that the equations are internally consistent. It is always important to test dimensional and unit consistency before moving forward with solutions.

It is also of great interest to calculate how much surface area in total can be held within a modestly-sized adsorber unit based on bed size of the adsorbent. A small to medium sized adsorber would be one that is cylindrical with a meter or so diameter and 2 m high. The cross sectional area and volume would be as shown here:

Void Fraction in the bed of adsorbent particles = ϵ

Cross Sectional Area = $A_{cr} = \pi\ r^2$ and

Volume = V = A_{cr}

Volume of the Gas = $V_g = \epsilon V$

Volume of adsorbent solid = V_s = (1-ϵ) V

Here are some computed values for these parameters:

```
ϵ = 0.4; (* Void fraction of the adsorbent bed of particles *)
r = 0.564m; (* radius of the vessel *)
h = 2m; (* height of the vessel *)
Acr = π(r)^2;
Print[" Acr = ", Acr]
V = %% * h;
Print[" V = ", V]
Vg = ϵV;
Print["Vg =", Vg]
Vs = (1 − ϵ)V;
Print["Vs =", Vs]
```

Acr = 0.999328 m^2

V = 1.99866 m^3

V_g = 0.799462 m^3

V_s = 1.19919 m^3

Now we need to know the total area available for adsorption within the bed. This will be the product of the surface area per gram of adsorbent, the density of the adsorbent in g/cm^3 and the total volume of the bed in cm^3:

Total Area for Adsorption in the Adsorbent Bed $= A_s\ \rho_g\ V_s$

$$= \left(1200\ \frac{\text{m}^2}{g}\right)\left(100\ \frac{\text{cm}}{m}\right)^2\left(2.2\ \frac{g}{\text{cm}^3}\right)\left(1.2\ m^3\right)\left(100\frac{\text{cm}}{m}\right)^3$$

$$= 3.2 \times 10^{13}\ \text{cm}^2$$

The total area available for adsorption is about 3.2 x 10^{13}cm^2, which is a huge area. One way to understand how much area this is, is to restate it in numbers of football fields. One football field is 57,600 ft^2, thus we have as follows:

$$\mathbf{3.2\ 10^{13}\ cm^2}\left(\frac{\mathbf{1\ in}}{\mathbf{2.25\ cm}}\right)^2\left(\frac{\mathbf{1\ ft}}{\mathbf{12\ in}}\right)^2$$

$$\%\left(\frac{\mathbf{1\text{“football field”}}}{\mathbf{5.76 * 10^4\ ft^2}}\right)$$

4.38957×10^{10} ft^2

762079. football field

The area is equivalent to about 762 thousand football fields and all packed into a volume of about 2 m^3. The numbers are not what we are used to in the macroscopic world of experience. Adsorption is a nanoscale phenomenon because to play on these football fields, the players would need to be less that 6Å in height.

The next logical question to ask is “how many adsorbate molecules could this adsorbent volume or bed hold?” Usually this would be measured in the laboratory for us, but we can estimate the number subject to a few assumptions. Firstly, assume a perfectly flat surface and about 10^{15} atoms per cm^2 and we can assume (guesstimate) that one site for adsorption takes about ten atoms, meaning there might be 10^{14} sites per cm^2 of surface. This also means that each molecule adsorbed would take up 10^{-14}cm^2. So how small is the molecule we are adsorbing? We will assume it takes up a circular space and that area is pi times the molecular radius squared. Then we find this:

$$\left(10^{-14}\ \frac{\text{cm}^2}{\text{molecule adsorbed}}\right)\left(\frac{1m}{100\ \text{cm}}\right)^2\left(10^9\frac{\text{nm}}{m}\right)^2$$

$$= \left(10^{-14} * \frac{1}{10^4} * 10^{18}\right)\ \text{nm}^2$$

$$= 1\ \frac{\text{nm}^2}{\text{molecule adsorbed}}$$

$$= \pi\ {r_m}^2 \Rightarrow r_m = \sim 0.56\ \text{nm or } 5.6\ \text{Å}$$

This tells us that if we have 10^{14} sites per cm^2, then we have adsorbate molecules that are about 10 - 11Å or 1 - 1.1 nm in diameter.

Back to the question of how many molecules can be accumulated on this adsorbent. We will assume that the adsorption process is Langmuirian, meaning that there is one layer of adsorbate molecules, and only one layer, on the surface. Imagine the molecule lying on the surface with a radius r_m. If we draw a line through the molecule and then spin it around its center point while remaining on the plain, a circle results that has an area of $\pi\ {r_m}^2$. We will assume that this is the area occupied by the molecule. If we lay these molecules in a close packed two dimensional array, then the fraction of the area occupied to that unoccupied is $\frac{\pi}{4} = 0.78$. We calculated the area occupied by the molecule to be about 1.0 nm^2 and the total area of the adsorbent to be $3.2 * 10^{13} cm^2$. The number of molecules that can be accommodated then is this:

$$\text{Number of Molecules Adsorbed} = \frac{\frac{\pi}{4}\ \text{Total Adsorbent Area}}{\text{Area Occupied by One Molecule}}$$

$$(\pi/4)\frac{3.2 * 10^{13}\text{cm}^2}{1\frac{\text{nm}^2}{\text{molecules}}\left(\frac{1\text{cm}}{10^7\text{nm}}\right)^2}$$

(* No. Molecules Adsorbed *);

Print["No. of molecules adsorbed = ", %]

$$\%\% \Big/ \left(6.02\ 10^{23}\frac{\text{molecules}}{\text{mole}}\right)\ \text{(* Moles Adsorbed *);}$$

Print["No. of moles adsorbed = ", %]

$$\%\% \left(100\frac{g}{\text{mole}}\right)\ \text{(* Grams Adsorbed *);}$$

Print["No. of grams adsorbed = ", %]

$$\%\% / \left(3.2\ 10^{13}\text{cm}^2\right);$$

Print["Concentration on surface = ", %]

No. of molecules adsorbed = 2.51327×10^{27}molecules

No. of moles adsorbed = 4174.87mole

No. of grams adsorbed = 417487.g

Concentration on surface = $1.3047 \times 10^{-8} \frac{g}{cm^2}$

The results are that in theory 2.5 x 10^{27} molecules, corresponding to about ~4200 moles of adsorbate. If the molecular weight is about 100 g/g-mole, then nearly 420,000 g of species i can be adsorbed. But this does not take into account thermodynamics at the surface. The adsorbent has a density of 2.2 g/cm^3 and it has a volume of 1.2 m^3 or 1.2 x 10^6cm^3. This corresponds to 2640 Kg. So the adsorbent can adsorb about 16% of its own weight. This assumes only monolayer coverage. If the adsorbate were to form multilayers, or if it were to condense in the adsorbent's pore volume, even more could be retained, although technically this would involve different physical processes.

We computed that the volume of gas space between the particles in the vessel was 0.8 m^3. Let's compute the mass of species i that is present in this space initially, assuming it remains at 300K and behaves ideally. The total pressure is 1 atm and the mole fraction is $X_i = 0.01$ and the molecular weight is 100 g/ g-mole. Therefore we have:

$$\frac{X_i P_{\text{tot}} V_{\text{tot}}}{RT}\left(\frac{1}{\text{Mw}_i}\right) = m_i$$

$$\frac{\mathbf{0.01\ 1atm\ 0.8}\ \boldsymbol{m}^3\ \mathbf{10}^6\ \frac{\mathbf{cm}^3}{\boldsymbol{m}^3}}{\mathbf{82.05}\frac{\mathbf{cm}^3\ \mathbf{atm}}{\mathbf{mole}\ \boldsymbol{K}}\ \mathbf{300}\boldsymbol{K}\left(\frac{\mathbf{1mole}}{\mathbf{100}\boldsymbol{g}}\right)}$$

32.5005g

The adsorbent has the capacity to adsorb 42 g of species i and there are about 33 g in the gas phase of the unit at the start of the experiment. In principle, all of product i could be adsorbed, if the equilibrium were favorable. If that were to happen, then the concentration of species i on the adsorbent surface would be 1 x $10^{-8}\frac{g}{\text{cm}^2}$. We will use this as the maximum adsorbate concentration.

We can return to the equations to be solved to understand how the system will behave in time. Remember that in this instance the vapor pressure of species i will vary with time and that makes this a harder problem to solve.

$$\frac{dC_{i,g}}{dt} = -\frac{(1-\epsilon)}{\epsilon}\rho_s A_s\left(k_a C_{i,g}\left(C_{i,s,\text{tot}} - C_{i,s}\right) - k_{d,s} C_{i,s}\right)$$

$$\frac{dC_{i,s}}{dt} = k_a C_{i,g}\left(C_{i,s,\text{tot}} - C_{i,s}\right) - k_{d,s} C_{i,s}$$

Clear[ka, kd, x1, *T*, *R*, co, ϵ, *V*, ρs, As, cist]

Timing[

DSolve[

$$\left\{\partial_t \text{cig}[t] == -\frac{(1-\epsilon)}{\epsilon}\rho\text{s}\right.$$

(ka cig[*t*](cist − cis[*t*]) − kd cis[*t*]),

∂*t*cis[*t*] == (ka cig[*t*](cist − cis[*t*]) − kdcis[*t*]),

cig[0] == cigo,

cis[0] == ciso}, {cig[*t*], cis[*t*]}, *t*]]

{11.4582, {}}

The reply after about 11-12 seconds to our request for a symbolic solution is the empty set { }. We could continue to try and find conditions under which this may be soluble analytically (symbolically), but it is unlikely to produce results. It is more productive to assign parameter values and to try and solve this numerically.

$$\frac{dC_{i,g}}{dt} = -\frac{(1-\epsilon)}{\epsilon}\rho_s A_s\left(k_a C_{i,g}\left(C_{i,s,\text{tot}} - C_{i,s}\right) - k_{d,s} C_{i,s}\right)$$

$$\frac{dC_{i,s}}{dt} = k_a C_{i,g}\left(C_{i,s,\text{tot}} - C_{i,s}\right) - k_{d,s} C_{i,s}$$

Clear[ka, kd, xi, ci, PP1, *T*, *R*, co, ϵ, *V*, ρs, As]

P_{tot}=.

Mw_i=.

C_{istot}=.

ka = 0.1(* $\frac{\text{cm}^3}{g\,\text{sec}}$ *);

kd = .001(* sec^{-1} *);

$x1 = 0.01;$

$P_{tot} = 1$ (* atm *);

$Mw_i = 100(* \frac{g}{mole} *);$

$T = 300(* K *);$

$R = 82.05(* \frac{cm^3 \, atm}{mole K} *);$

$\epsilon = 0.4;$ (* unitless *)

$V = 2\ 10^6(* cm^3 *);$

$\rho s = 2.(* gcm^{-3} *);$

$As = 1.2\ 10^7(* cm^2 g^{-1} *);$

$co = 0;$

$C_{istot} = 1.\ 10^{-8}(* \frac{g}{cm^2} *);$

$$mio = \frac{x1 * P_{tot} * \epsilon * V}{R * T * \frac{1}{Mw_i}};$$

Print["This is the mass of i in the

gas phase at t = 0, mio = ", mio]

tmax = 500;

Print[

"These are the numerical solutions for cig[t] and

cis[t]"]

$a = $ NDSolve[

$$\left\{\partial_t \text{cig}[t] == -\frac{(1-\epsilon)\rho s As}{\epsilon}\right.$$

$(\text{kacig}[t]\,(C_{istot} - \text{cis}[t]) - \text{kd cis}[t])\,,$

$\partial_t \text{cis}[t] ==$

```
(kacig[t] (C_istot − cis[t]) − kd cis[t]),
cig[0] == x1 * P_tot / (R * T * 1/Mw_i),
cis[0] == co},
{cig[t], cis[t]}, {t, 0, tmax}]
ccg[t_]:=Evaluate[cig[t]/.a]
ccs[t_]:=Evaluate[cis[t]/.a]
Print["This is the mass of i on the adsorbent
 at t = tmax, ccs[tmax]ρs As (1-ϵ) V = ",
ccs[tmax]ρsAs(1 − ϵ)V]

Plot[ccg[t]ϵV, {t, 0, tmax},
AxesLabel →
{"t/s", "mass i in the gas phase/g"},
PlotRange → {{0, tmax}, {0, 35}}]

Plot[ccs[t](1 − ϵ)ρsAsV, {t, 0, tmax},
AxesLabel → {"t/s", "mass i adsorbed/g"},
PlotRange → {{0, tmax}, {0, 35}}]
```

What follows are the outputs of the code written above. Included are print statements that identify what each number represents and then the graphs of the change in mass of the gas and adsorbent phases.

This is the mass of i in the gas phase at t = 0, mio = 32.5005

These are the numerical solutions for cig[t] and cis[t]

{{cig[t] → InterpolatingFunction[□][t], cis[t] → InterpolatingFunction[□][t]}}

This is the mass of species i in the gas phase at t = tmax, ccg[tmax]ϵ V = {0.878492}

This is the mass of i on the adsorbent at t = tmax, ccs[tmax]ρs As (1-ϵ)V = {31.622}

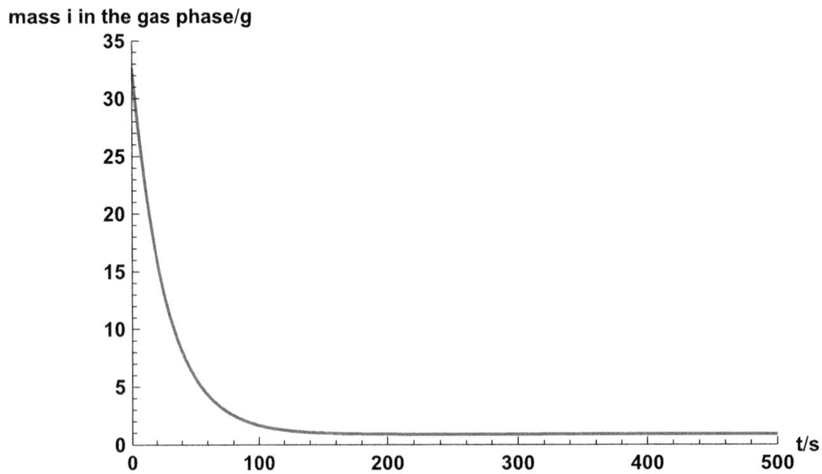

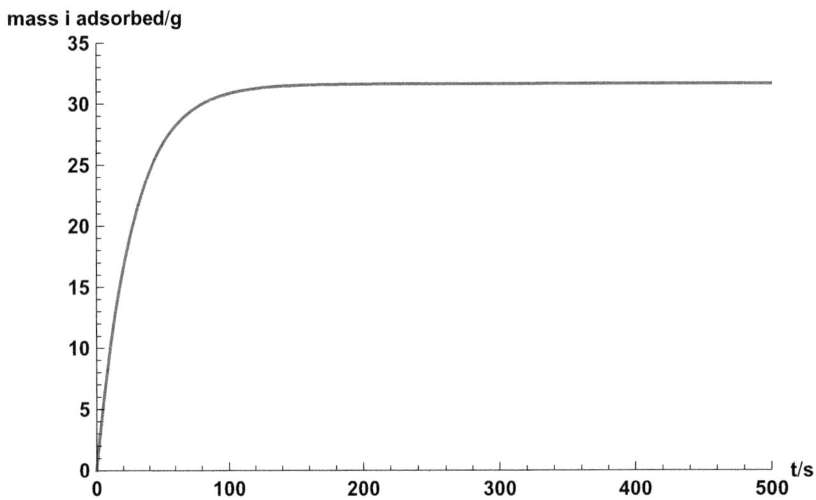

The extent of adsorption is close to complete in the time allotted and more importantly based on the equilibrium constant for adsorption. That constant is the ratio of ka/kd, which is $10^2 cm^3/g$.

As we can see from these solutions this is anything but a simple experiment. This is good illustration of how the analysis can be used to define and, indeed to design, the experiment.

Semi-continuous adsorption

From the laboratory vantage there is another experiment that can be done to obtain adsorption rate parameters and this is the semi-continuous approach. Here the adsorbate is fed at a mass flow rate that is equal to the rate of adsorption. A very sensitive pressure transducer is connected to a mass flow controller for the adsorbate. As this gas is adsorbed, were there no flow into the system the pressure would drop, leading to the complexities we have just analyzed. If, however the mass flow controller is used in such a way that it opens whenever there is a slight δP of pressure drop below the fixed experimental pressure set point, then the pressure can be maintained as a constant. The mass flow rate into the system is the same as the mass rate of adsorption "out" of the gas phase and "onto" the adsorbent. The physical situation corresponding to this is shown here:

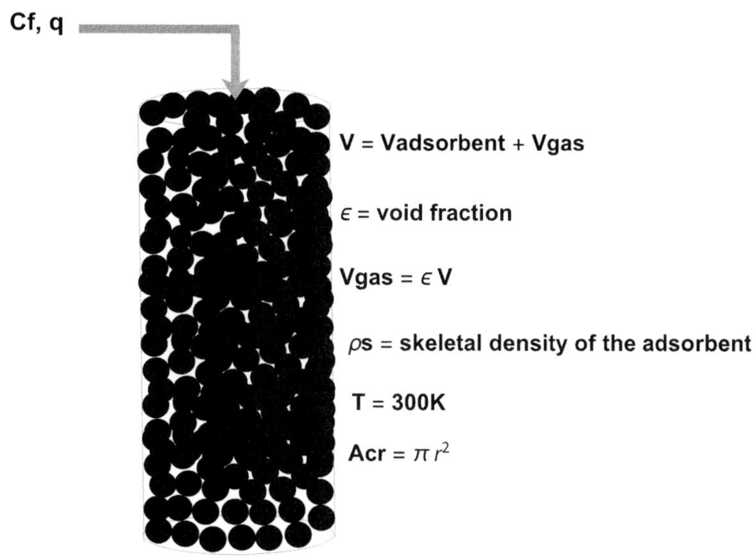

We will work with on gaseous component that partitions between the gas phase in the space within the vessel and the adsorbed phase on the surface of the adsorbent. We will also assume that the gas behaves ideally. Therefore the density of the gas in the vessel, $\rho_g[t]$, changes with the pressure in the vessel, but the density in the supply stream, ρ_f, will be assumed to be constant.

$$\rho_g[t] = \frac{P_g[t]}{RT}\mathrm{Mw}_g \text{ and } \rho_f = \frac{P_f}{RT}(\mathrm{Mw}_g)$$

Now let's think about mass of gas in the vessel at the end of the process, which we will define as the time at which all the sites on the adsorbent that can be filled are filled and the pressure in the vessel is the same as the supply stream pressure. This time can be designated at t_{fin}. All the mass in the gas phase and on the adsorbent entered the vessel through the supply stream. Therefore the total mass of gas in the vessel is the product of the gas density, at t_{fin}, the volumetric flow rate and the time:

Total mass of gas = ρ_f q t_{fin}

This mass is made up of that which is in the empty space in the vessel and the which is adsorbed:

Total mass of gas = mass gas phase + mass of gas adsorbed

$$(\rho_f - \rho_g[t_{\text{fin}}])\,q\,t_{\text{fin}} = \rho_g[t_{\text{fin}}]\,\epsilon\,V + Cs[t_{\text{fin}}]\,(1-\epsilon)V\rho_s A_s$$

At any given time before t final the same expression also holds:

$$(\rho_f - \rho_g[t])\,q\,t = \rho_g[t]\,\epsilon\,V + Cs[t]\,(1-\epsilon)V\rho_s A_s$$

We can derive the rate of change by taking the derivative with respect to time and reverse their positions:

$$\frac{d}{dt}[(\rho_f - \rho_g[t])\,q\,t] = \frac{d}{dt}[\rho_g[t]\,\epsilon\,V + Cs[t]\,(1-\epsilon)V\rho_s A_s]$$

$$\frac{d\left[(\rho_f - \rho_g[t])\epsilon V + Cs[t](1-\epsilon)V\rho_s A_s\right]}{dt} = (\rho_f - \rho_g[t])\,q$$

This is the total material balance for this system.

The next step is to take this into account by examining the gas phase separately from the adsorbent:

Gas phase:

$$\frac{dC_g\epsilon V}{dt} = C_f\,q - (1-\epsilon)\,\text{rate}_{i,\text{ads,net}}\,A_s\rho_s V$$

Notice that we wrote this in terms of the concentration of the gas in the vessel and in the supply stream. When there is only one species involved, the density and the concentration are the same. Also ρf is a constant and drops out upon differentiation. So we can rewrite this as:

$$\frac{d\rho_g[t]\epsilon V}{dt} = (\rho_f - \rho_g[t])\,q - (1-\epsilon)\,\text{rate}_{i,\text{ads,net}}\,A_s\rho_s V$$

The density on the left, $\rho_g[t]$ will increase with time. The density in the first term on the right-hand side, the convective flow term, is the fixed density of the feed, ρ_f, which is at a higher initial pressure and density than the pressure and density in the vessel, $\rho_g[t]$. The other density on the right-hand side is the skeletal density of the adsorbent, ρ_s, and it is also a constant.

On the adsorbent phase the rate of mass change of i is given as the net rate of adsorption:

Solid phase:

$$\frac{dC_s(1-\epsilon)A_s\rho_s V}{dt} = (1-\epsilon)\,\mathrm{rate}_{\mathrm{ads,net}}\,A_s\rho_s \mathrm{V}$$

Note that although the gas phase balance includes a convective flow term for the mass flow of gas into the system, the adsorbent phase mass balance does not. In this mass balance, we use the concentration of the gas adsorbed on the adsorbent surface.

If we add the two balance equations we come back to the total material balance:

$$\frac{d\rho_g[t]\epsilon V}{dt} = \left(\rho_f - \rho_g[t]\right)\mathrm{q} - (1-\epsilon)\,\mathrm{rate}_{i,\mathrm{ads,net}}\,A_s\rho_s \mathrm{V}$$

$$+\frac{dC_s(1-\epsilon)A_s\rho_s V}{dt} = (1-\epsilon)\,\mathrm{rate}_{\mathrm{ads,net}}\,A_s\rho_s \mathrm{V}$$

$$\frac{d\rho_g[t]\epsilon V}{dt} + \frac{dC_s(1-\epsilon)A_s\rho_s V}{dt} = \left(\rho_f - \rho_g[t]\right)]\,\mathrm{q}$$

$$\frac{d\left[\rho_g[t]\epsilon V + \mathrm{Cs}[t](1-\epsilon)V\rho_s A_s\right]}{\mathrm{dt}} = \left(\rho_f - \rho_g[t]\right)\mathrm{q}$$

Let's think about what is happening in this system and what the mathematics is telling us. To do this, look again at the gas phase mass balance equation. Indeed it is a balance, because the rate of change of mass in the gas phase is equal to the difference between the rate of supply of gas to the vessel and the rate of adsorption within the vessel.

$$\frac{d\rho_g[t]\epsilon V}{dt} = \left(\rho_f - \rho_g[t]\right)\mathrm{q} - (1-\epsilon)\,\mathrm{rate}_{\mathrm{ads,net}}\,A_s\rho_s \mathrm{V}$$

If the rate of supply to the vessel is larger than the rate of adsorption, the pressure will rise above the adsorbent. If this is reversed, then pressure would actually fall, though this may be unlikely to happen in an actual system. It is possible that with a controller (human, mechanical or electronic) monitoring the system that it could quickly adjust the flow rate of the supply so that the pressure in the vessel remained nearly constant. Which is to say:

$$\text{if } \rho_g \approx \text{constant, then } \frac{d\rho_g \epsilon V}{dt} \approx 0$$

and

$$(\rho_f - \rho_g)\text{q} = (\ 1 - \epsilon\ )\ \text{rate}_{\text{ads,net}}\ A_s\rho_s\text{V}$$

$$(\rho_f - \rho_g)\ \text{q} = (\ 1 - \epsilon\ )A_s\rho_s\text{V}\ [\ k_a\rho_g\left(C_{s,\text{tot}} - C_s\right) - k_d C_s\ ]$$

$$\frac{(\rho_f - \rho_g)\,q}{(1-\epsilon)A_s\rho_s V} = [\ k_a\rho_g\left(C_{s,\text{tot}} - C_s\right) - k_d C_s\ ]$$

This equation results, if the gas phase concentration does not change. This means that the rate at which mass flows into the unit is exactly balanced by the rate of adsorption. That very nice result can now be used to replace the explicit rate of adsorption on the right side of the mass balance for the adsorbed gas:

$$\frac{dC_s(1-\epsilon)A_s\rho_s V}{dt} = (\rho_f - \rho_g)\ \text{q}$$

We can check this for unit consistency:

$$\frac{dC_s}{dt} = \frac{(\rho_f - \rho_g)\,q}{(1-\epsilon)A_s\rho_s V} \equiv \left(\frac{g}{\text{cm}^3}\right)\left(\frac{\text{cm}^3}{\text{sec}}\right)\left(\frac{g}{\text{cm}^2}\right)\left(\frac{\text{cm}^3}{g}\right)\left(\frac{1}{\text{cm}^3}\right) \equiv \frac{g}{\text{cm}^2\ \text{sec}}$$

Given that the gas phase concentration is constant, the concentration on the surface of the adsorbent is immediately integrated to a linear form in time:

$$C_s[\ \text{t}\ ] = \frac{(\rho_f - \rho_g)\,q}{(1-\epsilon)A_s\rho_s V}\ \text{t} = \text{Constant} * \text{t}$$

So, we have a system that is in a pseudo-steady state, so called because the gas phase pressure remains constant, even as the concentration on the surface of the adsorbent is linearly increasing in time. We also notice that there is no mathematical limit to the mass that can be adsorbed on the surface. To cut this solution off we need additional information.

Using the same parameters as we used earlier and with the addition of the density of the delivery stream and the volumetric flow rate of that stream, we can compute the time that would be required to load the adsorbent to capacity. This will be the ratio of the total mass capacity of the adsorbent given as ($C_{\text{stot}}(1-\epsilon)\rho s\text{As}V$) to the mass delivery rate, which is the product of the density of the gas and the flow rate ($(\rho_f - \rho_g)$ q) with a conversion factor from seconds to hours:

$$\boldsymbol{q = 10000(*\ \frac{\text{cm}^3}{\text{sec}}\ *);}$$

$$\boldsymbol{C_{\text{stot}} = 1.\ 10^{-8}(*\ \frac{g}{\text{cm}^2}\ *);}$$

$\epsilon = 0.4$; (* unitless *)

$V = 2\ 10^6$(* cm^3 *);

$\rho\mathrm{s} = 2.$(* $g\mathrm{cm}^{-3}$ *);

$\mathrm{As} = 1.2\ 10^7$(* $\mathrm{cm}^2 g^{-1}$ *);

$\mathrm{Pf} = 1.5$(* atm *);

$\mathrm{Pg} = 1.0$(* atm *)

$\mathrm{Mw} = 16$;

$$\rho\mathrm{f} = \frac{\mathrm{Pf}}{R * T * \frac{1}{\mathrm{Mw}}}(* \frac{g}{\mathrm{cm}^3} *);$$

$$\rho\mathrm{g} = \frac{\mathrm{Pg}}{R * T * \frac{1}{\mathrm{Mw}}}(* \frac{g}{\mathrm{cm}^3} *);$$

$$\frac{C_{\mathrm{stot}}(1-\epsilon)\rho\mathrm{sAs}V}{((\rho\mathrm{f}-\rho\mathrm{g})q)};$$

$$\frac{C_{\mathrm{stot}}(1-\epsilon)\rho\mathrm{sAs}V}{3600((\rho\mathrm{f}-\rho\mathrm{g})q)};$$

On this basis, the adsorbent could be used for just over 24 hours before it would require regeneration or replacement.

It is tempting to try and do more with this result, but we have to be careful. Look again at this equation:

$$\frac{(\rho\mathrm{f}-\rho\mathrm{g})q}{(1-\epsilon)A_s\rho_s V} = [\ k_a\rho_g\left(C_{s,\mathrm{tot}} - C_s\right) - k_d C_s\ ]$$

Every parameter on the left-hand side is a constant. That implies that the right-hand side should be a constant also. But, we just found that change in surface concentration is a linear function of time. What makes the rate function on the right side so interesting and useful is that it is self limiting. As the surface concentration grows, and becomes closer in magnitude to the maximum surface concentration, the driving force for adsorption is diminished and desorption becomes more important. This is critical behavior in the dynamics of this system. Yet the solution we found has no such dynamics built into it. Making simplifying assumptions was more important when computing power was limited. But now we can compute the solution numerically without making assumptions, so let's do that now.

Without any simplifying assumptions, we may numerically solve the equations for this system. Once again there are a raft of parameters that we need to set. We always begin by clearing the parameter and variable names of value from previous models. We will set the adsorption rate constant to 0.1 $\frac{cm^3}{g\,sec}$, and the desorption rate constant to 10^{-5} sec^{-1}. For the molecular weight we will use that of methane, 16 g per mole. The temperature will be 300 Kelvin. The pressure of the supply stream will be set to 1.1 atm, and the initial pressure of the gas in the vessel will be 1 atm. The total volume of the unit will be 2 x $10^6 cm^3$, skeletal density will be 2 gm per cm^3, the surface area will be 1.2 x 10^7 cm^2 per gram, with a total of 10^{-9} g per cm^2 for possible sites occupied by the adsorbate and the bed of adsorbent will have void fraction of 0.4.

Clear[ka, kd, xi, ci, PP1, T, R, co, ϵ, V, ρs, As]

Pf=.

C_{stot}=.

$$ka = 10^{-1}(* \frac{cm^3}{g\,sec} *);$$

$$kd = 10^{-5}(* \ sec^{-1} \ *);$$

$$Mw = 16;$$

$$T = 300(* \ K \ *);$$

$$R = 82.05(* \frac{cm^3\,atm}{mole K} *);$$

$$Pf = 1.1\ (* \ atm \ *);$$

$$\rho f = \frac{Pf}{R * T * \frac{1}{Mw}};$$

$$Pg = 1;$$

$$\rho go = \frac{Pg}{R * T * \frac{1}{Mw}};$$

$$q = 100000(* \frac{cm^3}{sec} *);$$

$$\epsilon = 0.4;\ (* \ unitless \ *)$$

$$V = 2\ 10^6(* \ cm^3 \ *);$$

$$\rho s = 2.(* \ gcm^{-3} \ *);$$

As = 1.2 10^7(* cm^2g^{-1} *);

C_{stot} = 1. 10^{-9}(* $\frac{g}{\text{cm}^2}$ *);

tmax = 150000

150000

We can set up the solution and representation of the solutions graphically as follows. We clear the key variable names. Then we set up the solution of the differential equations. These are entered as we developed them. The initial conditions are that density of the gas in the vessel is based on the pressure being at 1 atm, which gives us ρgo. We have set the initial surface concentration of the adsorbate on the adsorbent surface to zero.

Clear[ρg, cs, t]

b = NDSolve[

$$\left\{\partial_t \rho g[t] == \frac{(\rho f - \rho g[t])\, q}{\epsilon V} - \frac{(1-\epsilon)\rho s As}{\epsilon}(ka\rho g[t]\,(C_{\text{stot}} - cs[t]) - kd\, cs[t]),\right.$$

$$\partial_t cs[t] == (ka\rho g[t]\,(C_{\text{stot}} - cs[t]) - kd\, cs[t]),$$

ρg[0] == ρgo,

cs[0] == 0},

{ρg[t], cs[t]}, {t, 0, tmax}]

$\rho\rho$g[t_]:=Evaluate[ρg[t]/.b]

ccs[t_]:=Evaluate[cs[t]/.b]

{{ρg[t] → InterpolatingFunction[□] [t], cs[t] → InterpolatingFunction[□] [t]}}

The numerical solutions are given function names, $\rho\rho$g [t] and ccs [t]. Three plots of $\rho\rho$g [t] are given with a decreasing range of the ordinate to see the behavior. We have included a horizontal line corresponding to the density of the feed gas. With the surface concentration, we plot the actual values of ccs [t] and the value ratioed to C_{stot}. In both graphs the maximum concentrations are included as a horizontal line.

b1 = Plot[$\rho\rho$g[t], {t, 0, tmax},

AxesLabel →

{"t (s)", "conc in the gas phase (g/cm^3)"},

```
Epilog → Line[{{0, ρf}, {tmax, ρf}}],

PlotRange → {{0, tmax}, {0, ρf}}, PlotStyle → Hue[0]]

b2 = Plot[ρρg[t], {t, 0, tmax},

AxesLabel →

{"t (s)", "conc i in the gas phase (g/cm^3)"},

Epilog → Line[{{0, ρf}, {tmax, ρf}}],

PlotRange → {{0, tmax}, {0.95ρf, ρf}},

PlotStyle → Hue[0]]

b3 = Plot[ρρg[t], {t, 0, tmax},

AxesLabel →

{"t (s)", "conc in the gas phase (g/cm^3)"},

Epilog → Line[{{0, ρf}, {tmax, ρf}}],

PlotStyle → Hue[0]]

b4 = Plot[ccs[t], {t, 0, tmax},

AxesLabel → {"t (s)", "CCs [ t ] (g/cm^2)"},

Epilog → Line [{{0, Cstot}, {tmax, Cstot}}],

PlotRange → {{0, tmax}, {0, 1.1Cstot}}]

b5 = Plot[ccs[t] / Cstot , {t, 0, tmax},

AxesLabel → {"t (s)", "CCs [ t ]/Cistot (Unitless)"},

Epilog → Line[{{0, 1}, {tmax, 1}}],

PlotRange → {{0, tmax}, {0, 1}}]
```

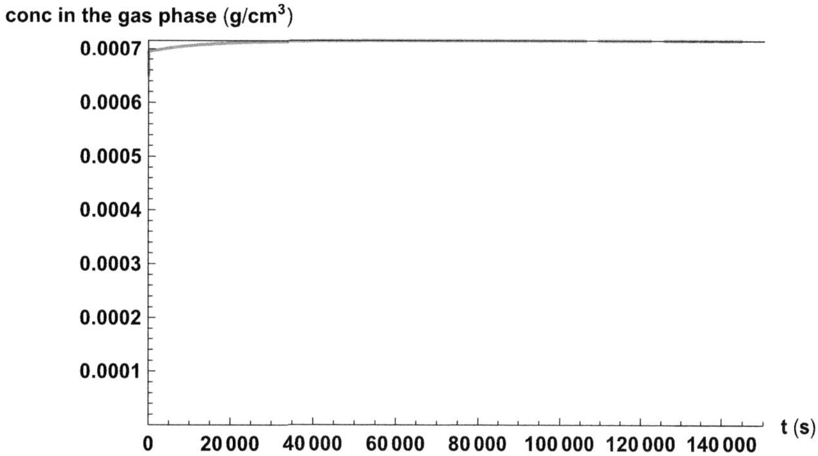
conc in the gas phase (g/cm³)
0.0007
0.0006
0.0005
0.0004
0.0003
0.0002
0.0001
0
20 000
40 000
60 000
80 000
100 000
120 000
140 000
t (s)

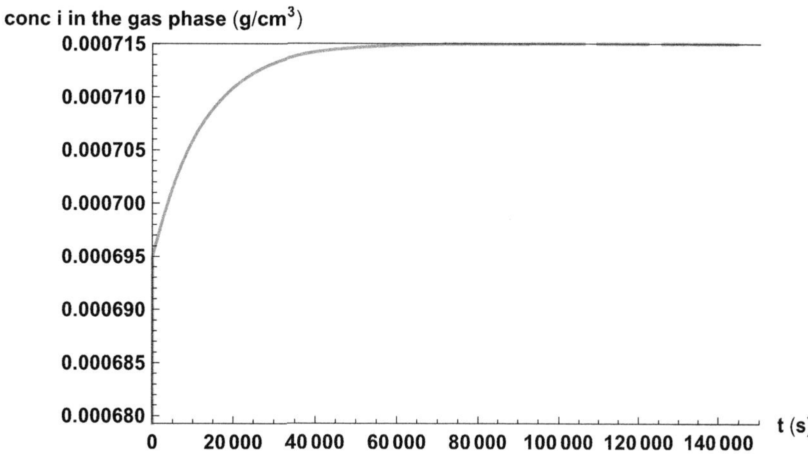
conc i in the gas phase (g/cm³)
0.000715
0.000710
0.000705
0.000700
0.000695
0.000690
0.000685
0.000680
0
20 000
40 000
60 000
80 000
100 000
120 000
140 000
t (s)

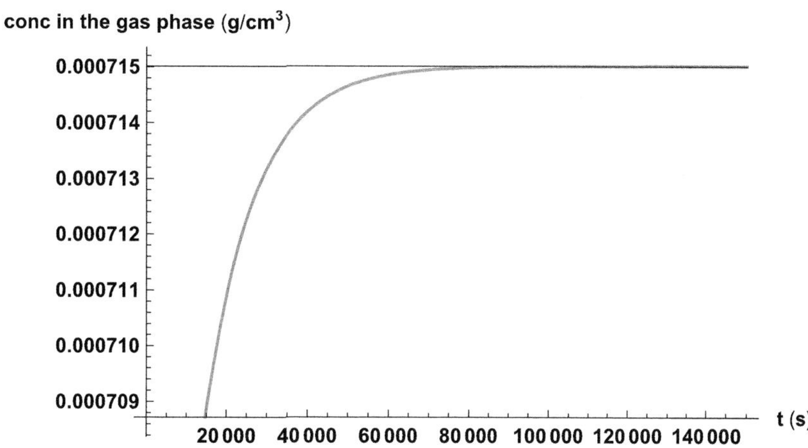
conc in the gas phase (g/cm³)
0.000715
0.000714
0.000713
0.000712
0.000711
0.000710
0.000709
20 000
40 000
60 000
80 000
100 000
120 000
140 000
t (s)

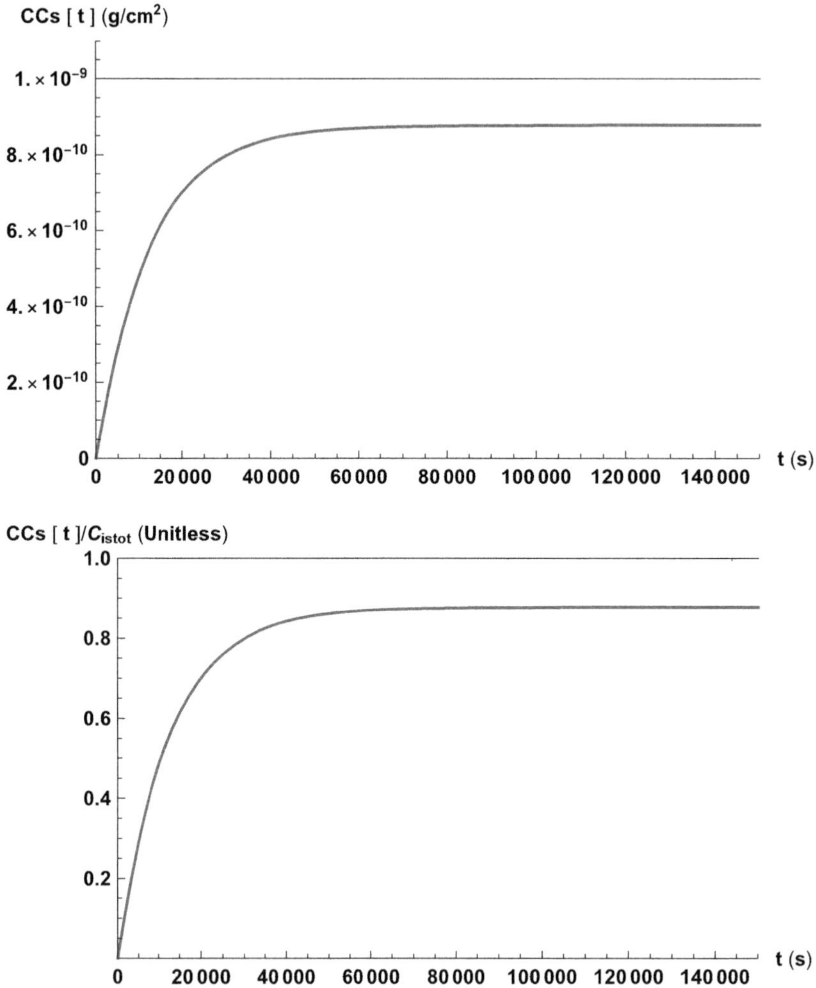

What we observe for this set of parameters is that the density of the gas in the vessel rises for about 16 hours (60,000 seconds) and then it has reached the density of the feed and no longer changes. This is because the pressure in the vessel cannot rise above that in the supply stream. The rise in density is smooth and the solution appears to be stable throughout the computation. The rise time of the adsorption process is shorter and this process stops at about 8 hours (30,000 seconds). The concentration of the adsorbate on the surface of the adsorbent comes to equilibrium at a value that is only about 80% coverage of the sites available, based on a purely surface area argument. This is because the equilibrium is set by the adsorption and desorption rate constants as we will see later. The behavior we see here is what we expect in general for this process.

Now we can return to the question of the pseudo-steady state and results derived therefrom. Recall that the result was that the surface concentration would vary linearly in time and that the slopes of that line would be this expression:

$$\text{Cs}[t] \approx \frac{(\rho\text{f} - \rho\text{go})q}{(1-\epsilon)\text{As}\rho\text{s}V}t$$

We can put a line with this slope on the graph for the surface concentration (b3). First we graph it and then show the two graphs:

tmax = 40000;

b5 = Plot$\left[\frac{(\rho\text{f} - \rho\text{go})q}{(1-\epsilon)\text{As}\rho\text{s}V}t, \{t, 0, \text{tmax}\},\right.$

PlotStyle → Hue[0.3]];

b3 = Plot[ccs[t], {t, 0, tmax},

AxesLabel → {"t (s)", "CCs [t] (g/cm^2)"},

Epilog → Line [{{0, C_{stot}}, {tmax, C_{stot}}}],

PlotRange → {{0, tmax}, {0, 1.1C_{stot}}}];

Show[b3, b5]

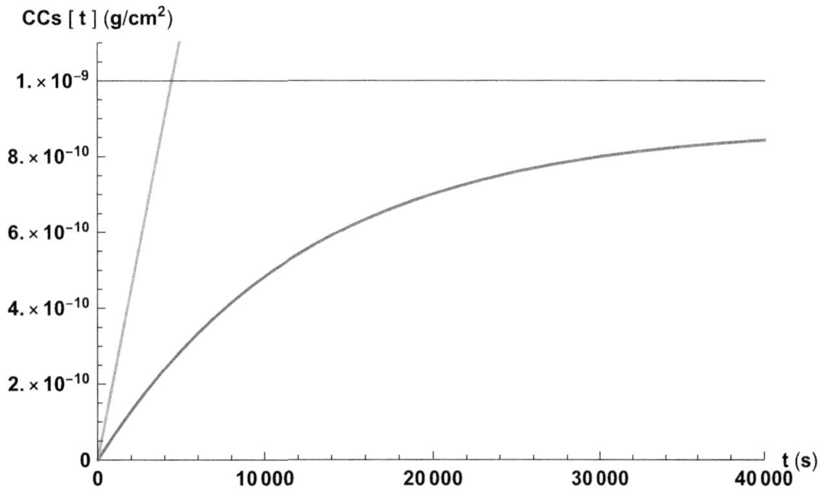

Although the early time behavior of the surface concentration is linear, the prediction from the pseudo-steady state approximation is a very poor fit, it is not close to the correct slope. If we look back to the adsorption rate, we can see that the early rate should be governed by ka ρgo C_{stot} and that this should be the proper slope of the surface concentration. We can add this to the graphs to check.

b7 = Plot [kaC_{stot} * t, {t, 0, 40000},

PlotStyle → Hue[0.1]];

Show[b3, b5, b7]

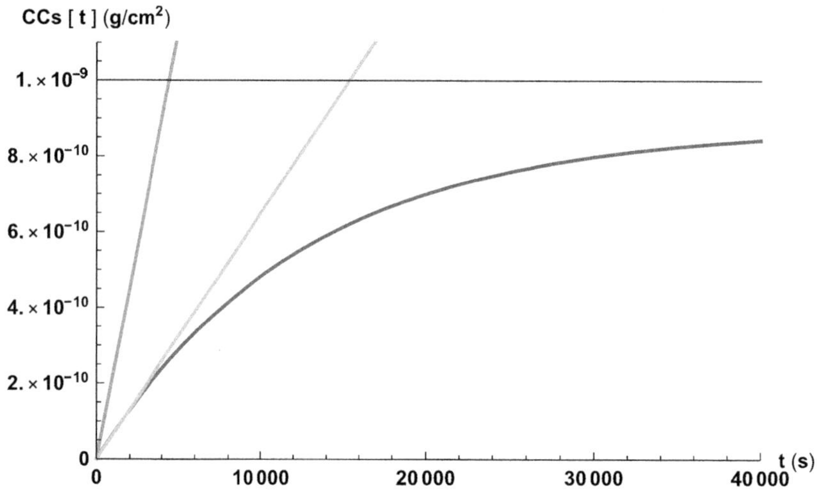

The very early time behavior of the surface concentration is nicely predicted by this result and that is quite logical. Again, if we look at the full expression this makes sense:

$$k_a\rho_g\left(C_{s,\text{tot}}-C_s\right)-k_dC_s$$

The reasoning is that ρ_g will be close to ρ_{go} and constant, and that $C_{s,\text{tot}}$ will be much greater than C_s and that the product of C_s and k_d will be so small as to be negligible, hence:

at t $\simeq$ 0

$$k_a\rho_g\left(C_{s,\text{tot}}-C_s\right)-k_dC_s \;\rightarrow\; k_a\rho_gC_{s,\text{tot}}$$

And this term is close to constant for a brief time early in the process.

Now if we wanted to operate close to the pseudo-steady state, we could do so if we adjust the flow rate q to a lower value so that the rate of delivery is equal to the early rate of adsorption. That would imply that

$$k_a\rho_gC_{s,\text{tot}}\approx\frac{\left(\rho_f-\rho_g\right)q}{(1-\epsilon)A_s\rho_sV}\Rightarrow \text{q}\approx k_a\rho_{\text{go}}C_{s,\text{tot}} * \left(\frac{\left(\rho_f-\rho_{\text{go}}\right)}{(1-\epsilon)A_s\rho_sV}\right)^{-1}$$

We can compute this value for q directly:

ka ρgo $C_{s,\text{tot}}$

$6.500101564086939 \times 10^{-14}$

$$\frac{\rho\text{f} - \rho\text{go}}{\text{As}\rho\text{s}V(1 - \epsilon)}$$

$2.256979709752411 \times 10^{-18}$

%%/%

28800.

According to this, if q = 28 800 cm^3 per second, the supply rate will be balanced by the adsorption rate early on.

$q = 28800$;

tmax = 40000;

$$\text{b8} = \text{Plot}\left[\frac{(\rho\text{f} - \rho\text{go})q}{(1 - \epsilon)\text{As}\rho\text{s}V}t, \{t, 0, \text{tmax}\},\right.$$

PlotStyle → Hue[0.3]];

Show[b3, b5, b8, b7]

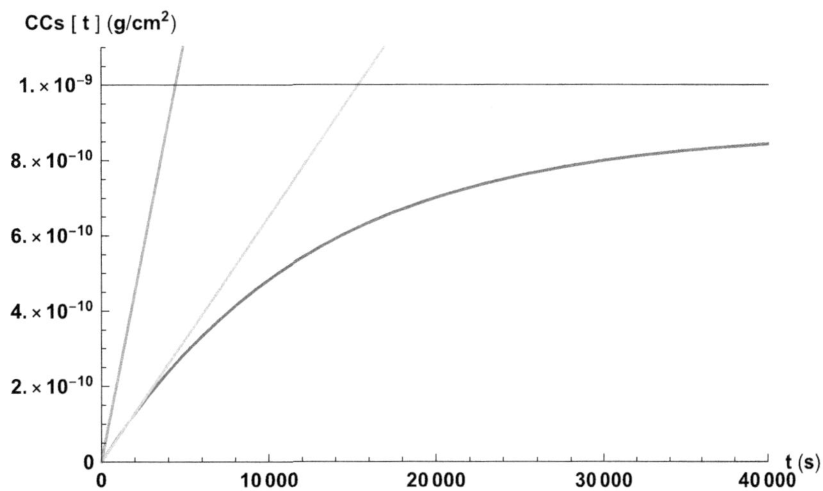

We have four graphs on the plot, but the plots for b8 and b7 now overlap perfectly. If we now go back to the overall model, we can keep every other parameter the same, but reset q to this new smaller value and see what happens.

We put all three of the input cells together into one cell for efficiency. We also rename the graphs with a b at the end of their number name to indicate they correspond to the new condition and we make them dashed. We suppress their output and then show them with the original graphs.

Clear[ka, kd, xi, ci, PP1, T, R, co, ϵ, V, ρs, As]

Pf=.

C_{stot}=.

$$\text{ka} = 10^{-1}(* \frac{\text{cm}^3}{g\,\text{sec}} *);$$

$$\text{kd} = 10^{-5}(* \ \text{sec}^{-1} \ *);$$

$$\text{Mw} = 16;$$

$$T = 300(* \ K \ *);$$

$$R = 82.05(* \frac{\text{cm}^3\ \text{atm}}{\text{mole}K} *);$$

$$\text{Pf} = 1.1 \ (* \ \text{atm} \ *) \ ;$$

$$\rho\text{f} = \frac{\text{Pf}}{R * T * \frac{1}{\text{Mw}}};$$

$$\text{Pg} = 1;$$

$$\rho\text{go} = \frac{\text{Pg}}{R * T * \frac{1}{\text{Mw}}};$$

$$q = 28800(* \frac{\text{cm}^3}{\text{sec}} *);$$

$$\epsilon = 0.4; \ (* \ \text{unitless} \ *)$$

$$V = 2\ 10^6(* \ \text{cm}^3 \ *);$$

$$\rho\text{s} = 2.(* \frac{g}{\text{cm}^3} *);$$

$$\text{As} = 1.2\ 10^7(* \ \frac{\text{cm}^2}{g} \ *);$$

$C_{stot} = 1.\ 10^{-9}(* \frac{g}{cm^2} *);$

tmax = 150000

Clear[ρg, cs, t]

b = NDSolve[

$\left\{\partial_t \rho g[t] == \frac{(\rho f - \rho g[t])\, q}{\epsilon V} - \frac{(1-\epsilon)\rho sAs}{\epsilon}(ka\rho g[t]\,(C_{stot} - cs[t]) - kd\, cs[t]),\right.$

$\partial_t cs[t] == (ka\rho g[t]\,(C_{stot} - cs[t]) - kd\, cs[t]), \rho g[0] == \rho go, cs[0] == 0\},$

{ρg[t], cs[t]}, {t, 0, tmax}];

$\rho\rho$g[t_]:=Evaluate[ρg[t]/.b]

ccs[t_]:=Evaluate[cs[t]/.b]

b1b = Plot[$\rho\rho$g[t], {t, 0, tmax},

AxesLabel →

{"t (s)", "conc in the gas phase (g/cm^3)"},

Epilog → Line[{{0, ρf}, {tmax, ρf}}],

PlotRange → {{0, tmax}, {0, ρf}},

PlotStyle → {Dashed, Hue[0]}];

b2b = Plot[$\rho\rho$g[t], {t, 0, tmax},

AxesLabel →

{"t (s)", "conc i in the gas phase (g/cm^3)"},

Epilog → Line[{{0, ρf}, {tmax, ρf}}],

PlotRange → {{0, tmax}, {0.95ρf, ρf}},

PlotStyle → {Dashed, Hue[0]}];

b3b = Plot[$\rho\rho$g[t], {t, 0, tmax},

AxesLabel →

{"t (s)", "conc in the gas phase (g/cm^3)"},

Epilog → Line[{{0, ρf}, {tmax, ρf}}],

PlotStyle → {Dashed, Hue[0]}];

```
b4b = Plot[ccs[t], {t, 0, tmax},
AxesLabel → {"t (s)", "CCs [ t ] (g/cm²)"},
Epilog → Line [{{0, Cstot}, {tmax, Cstot}}],
PlotRange → {{0, tmax}, {0, 1.1Cstot}},
PlotStyle → Dashed];
b5b = Plot[ccs[t] / Cstot, {t, 0, tmax},
AxesLabel →
{"t (s)", "CCs [ t ]/Cistot (Unitless)"},
Epilog → Line[{{0, 1}, {tmax, 1}}],
PlotRange → {{0, tmax}, {0, 1}},
PlotStyle → Dashed];
Show[b1, b1b]
Show[b2, b2b]
Show[b3, b3b]
Show[b4, b4b, b7]
Show[b5, b5b]
```

150000

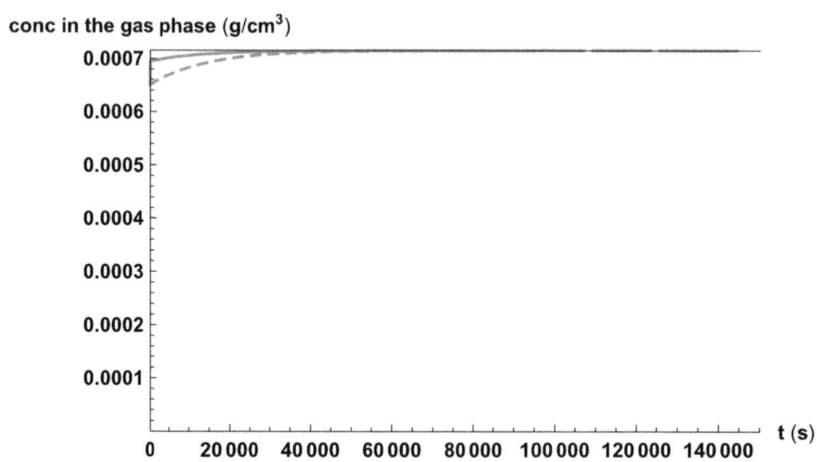

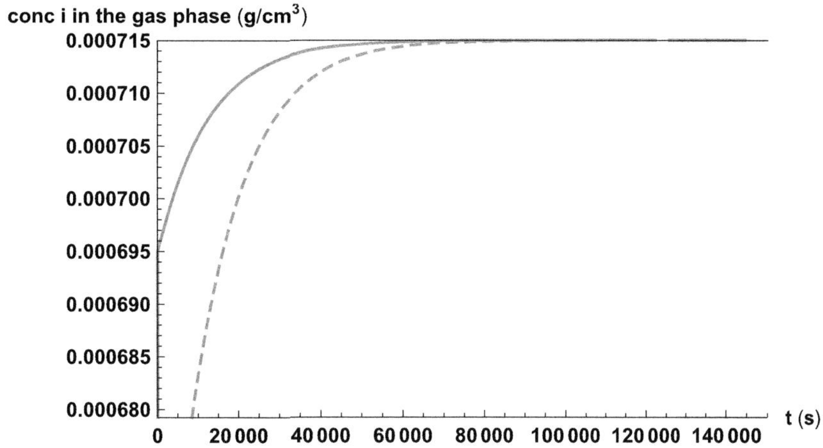
conc i in the gas phase (g/cm³)
0.000715
0.000710
0.000705
0.000700
0.000695
0.000690
0.000685
0.000680
0
20 000
40 000
60 000
80 000
100 000
120 000
140 000
t (s)

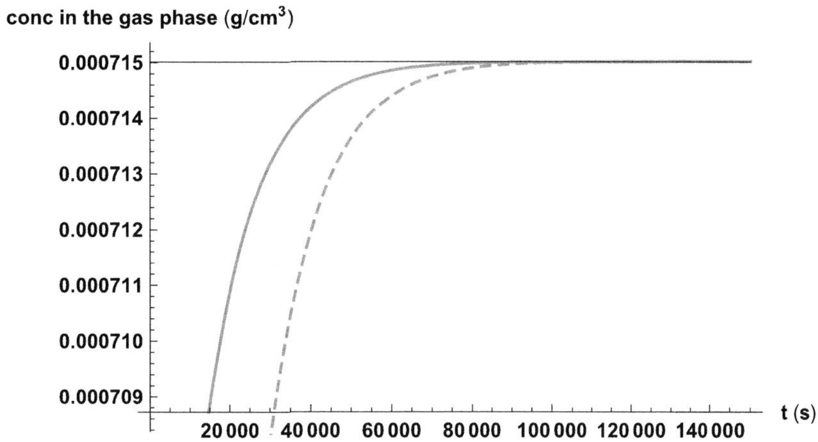
conc in the gas phase (g/cm³)
0.000715
0.000714
0.000713
0.000712
0.000711
0.000710
0.000709
20 000
40 000
60 000
80 000
100 000
120 000
140 000
t (s)

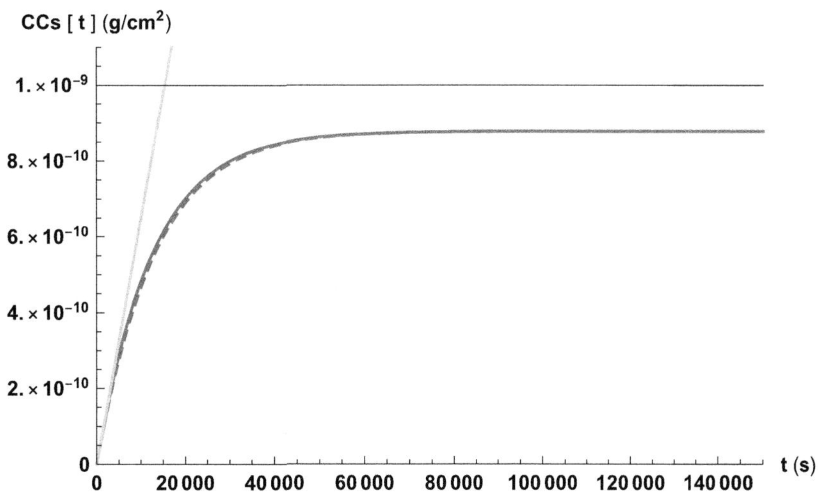
CCs [t] (g/cm²)
1. × 10⁻⁹
8. × 10⁻¹⁰
6. × 10⁻¹⁰
4. × 10⁻¹⁰
2. × 10⁻¹⁰
0
0
20 000
40 000
60 000
80 000
100 000
120 000
140 000
t (s)

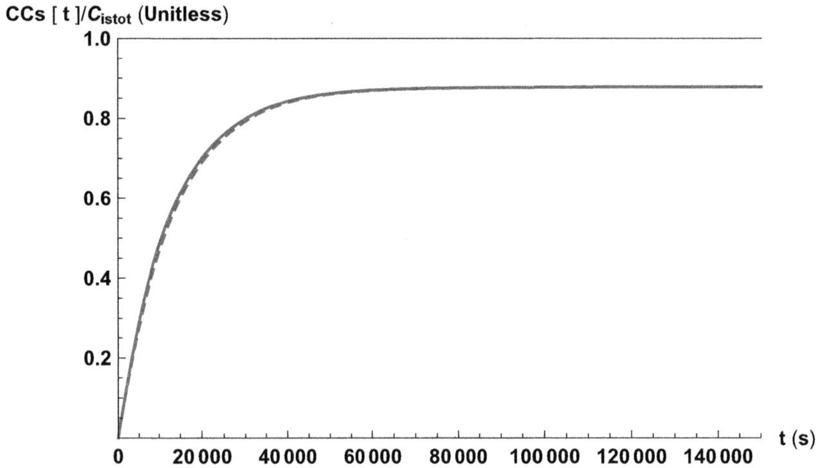

What we notice is that the density of the gas phase does not jump under this condition, but rises smoothly. We also see that the rate of adsorption is not affected by the change to this new condition. And, lastly the early time behavior shows that the flow rate into the unit nicely matches the rate of adsorption. A fit to the early time behavior that comes from this condition would be a nice way to obtain adsorption rate parameters. A fit to the longer time behavior would give the desorption rate constant. That long time behavior is the equilibrium condition.

We also have seen that the equilibrium markedly affects the utilization of the adsorbent. We can get this from consideration of an equilibrium condition. When the concentration of some species i on the surface of the adsorbent is no longer changing, then rate of adsorption and desorption are equal. From this we find:

$$0 = \left(k_a C_{i\mathrm{ge}} \left(C_{\mathrm{istot}} - C_{\mathrm{ise}}\right) - k_d C_{\mathrm{ise}}\right)$$

This can be rearranged to give the ratio of the rate constants on the left-hand side and equilibrium concentrations on the right-hand side and those are equal to the adsorption equilibrium constant:

$$\frac{\mathrm{ka}}{\mathrm{kd}} = \frac{\mathrm{Cise}}{\mathrm{Cige(Cistot - Cise)}} = \mathrm{Kads}$$

Where

Cise = the equilibrium surface concentration of i

Cige = the equilibrium gas phase concentration of i

Cistot = the total number of molecules of i that could be adsorbed if all were adsorbed in a close-packed monolayer using all the area

Kads = the adsorption equilibrium constant, which is the ratio $\frac{\text{ka}}{\text{kd}}$.

Dividing through by the total number of sites provides the fraction of sites that are occupied at equilibrium, and we call this θ in the Langmuir isotherm.

$$\text{Kads} = \frac{(\text{Cise}/\text{Cistot})}{\text{Cige}(\text{Cistot} - \text{Cise})/\text{Cistot}}$$

$$\text{Kads} = \frac{\theta}{\text{Cige}(1-\theta)}$$

Solve$\left[\textbf{Kads==}\frac{\theta}{\textbf{Cige}(1-\theta)}, \theta\right]$

$$\left\{\left\{\theta \to \frac{\text{CigeKads}}{1+\text{CigeKads}}\right\}\right\}$$

If we measure the rate of adsorption at early time in an experiment, paying close attention to the mass flow rate into the system so that it balanced the rate of adsorption, then we can compute ka from the flow and the physical parameters of the adsorbent, the bed, and the vessel:

$$k_a \rho_g C_{\text{is,tot}} \approx \frac{(\rho_f - \rho_g) q}{(1-\epsilon) A_s \rho_s V}$$

and

$$k_a \approx \frac{(\rho_f - \rho_g) q}{(1-\epsilon) A_s \rho_s V} \left(\frac{1}{\rho_g C_{\text{is,tot}}}\right)$$

Because, as we see below, in that time frame, the gas phase concentration (density) changes very little.

tmax = 5000

b1b = Plot[$\rho\rho$**g[** t **], {** t **, 0, tmax},**

AxesLabel →

{“t (s)”, “conc in the gas phase (g/cm^3**)”},**

```
Epilog → Line[{{0, ρf}, {tmax, ρf}}],
PlotRange → {{0, tmax}, {0, ρf}},
PlotStyle → {Dashed, Hue[0]}]
```

5000

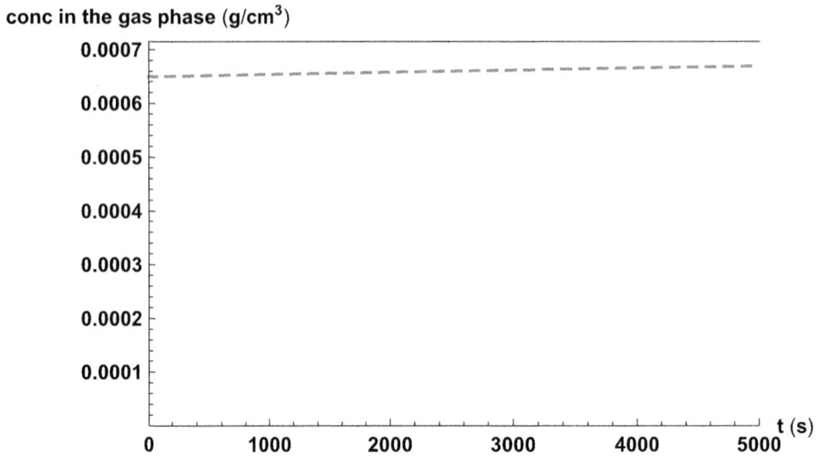


```
tmax = 5000;
q = 28800;
b4 = Plot[ccs[t], {t, 0, tmax},
AxesLabel → {"t (s)", "CCs [ t ] (g/cm²)"},
PlotRange → {{0, tmax}, {0, 1.1ccs[tmax]}}];
b4b = Plot[ccs[t], {t, 0, tmax},
AxesLabel → {"t (s)", "CCs [ t ] (g/cm²)"},
PlotRange → {{0, tmax}, {0, 1.1ccs[tmax]}},
PlotStyle → Dashed];
```

$$b8 = \text{Plot}\left[\frac{(\rho f - \rho go)q}{(1-\epsilon)\text{As}\rho s V}t, \{t, 0, \text{tmax}\},\right.$$

```
PlotStyle → Hue[0.3]];
Show[b4b, b8]
```

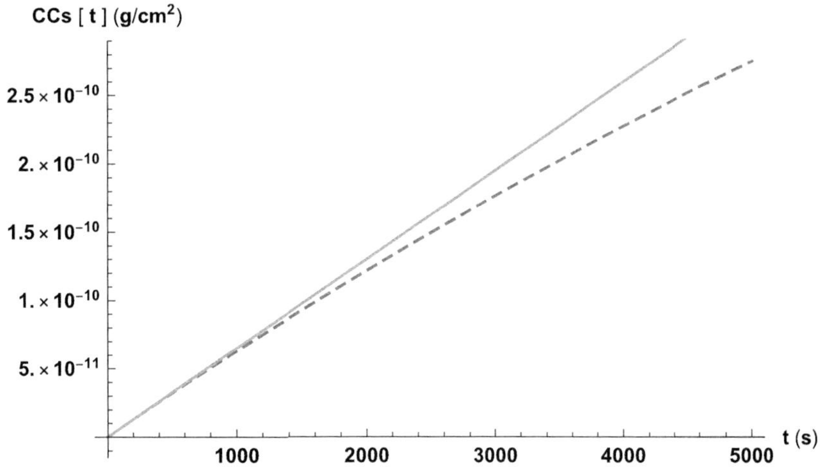

To obtain ka explicitly we need a measure, or reliable calculation of $C_{\mathrm{is,tot}}$. In the absence of that information, we can obtain the product ka $C_{\mathrm{is,tot}}$ = ka', which has the dimensions of $\frac{\mathrm{cm}}{g\ \mathrm{sec}}$. From this and a value for Kads we can compute the desorption rate constant. If we do not have these numbers, then a full non-linear statistical fit to the time dependent data can be used to obtain these parameters.

The essence of the Langmuirian model of adsorption is that it is self-limiting. We see this in the equilibrium expression that we have developed.

$$\theta = \frac{\mathrm{CigeKads}}{1 + \mathrm{CigeKads}}$$

If we assume that Cige corresponds to the concentration or density of an ideal gas, then we can reexpress this in terms of pressure.

$$\theta = \frac{\mathrm{CigeKads}}{1 + \mathrm{CigeKads}} = \frac{\mathrm{Pige/RTMwiKads}}{1 + \mathrm{Pige/RTMwiKads}} = \frac{\mathrm{Pige}K'\mathrm{ads}}{1 + \mathrm{Pige}K'\mathrm{ads}}$$

where the constants have been incorporated into the equilibrium constant so that it has inverse pressure units which it must have since its product with pressure must be unitless:

$$\theta = \frac{\mathbf{PigeKads}}{\mathbf{1 + PigeKads}};$$

Kads = 1;

Plot[θ, {Pige, 0, 100}, PlotRange → {{0, 100}, {0, 1}}]

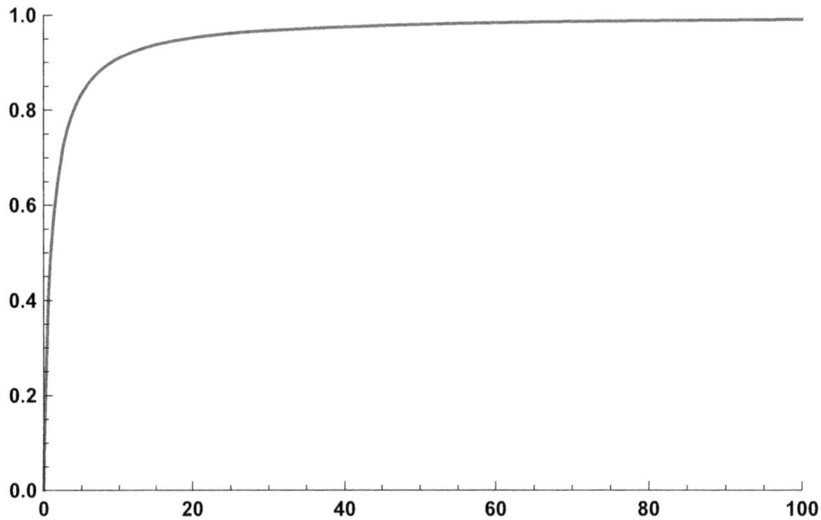

The equilibrium has two regions - the low pressure region and the high pressure region. At low pressure Pige Kads << 1 and the fractional coverage varies linearly with pressure. At high pressure the situation reverses and Pige Kads >> 1, hence the expression goes to unity.

```
Plot[PigeKads/(1 + PigeKads), {Pige, 0, 0.2},
AxesLabel → {P, "θ"},
PlotStyle → Hue[0.4],
PlotLabel->"Low P range",
Epilog → Line[{{0, 0}, {0.2, 0.2 * Kads}}]]
Plot[PigeKads/(1 + PigeKads), {Pige, 20, 100},
PlotRange → {{20, 100}, {0, 1}},
AxesLabel → {P, "θ"},
PlotStyle → Hue[0.6],
PlotLabel->"High P range",
Epilog → Line[{{0, 1}, {100, 1}}]]
```

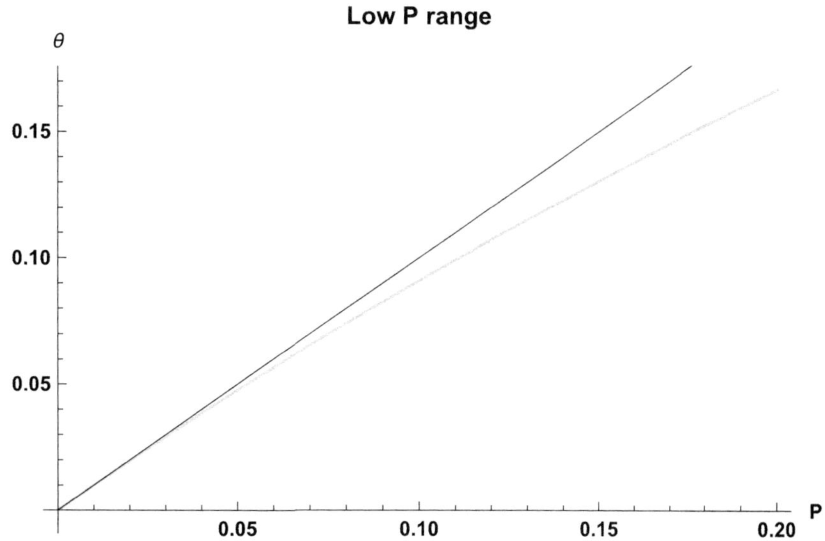

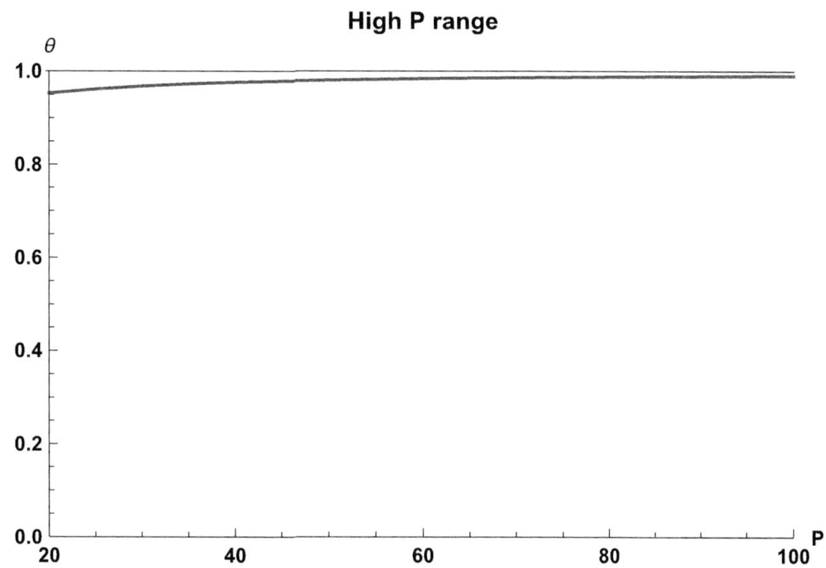

In the low concentration limit the adsorption isotherm is a nearly linear law as was the partition coefficient between two immiscible phases, and this is called the Henry's Law region. The black line remains linear over the same region. The upper pressure region is the surface saturation region, with the black line representing $\theta = 1$.

Continuous adsorption

The first two cases were batch and semi-continuous adsorption. Batch adsorption is more likely to be done in a laboratory than in a process. Semi-continuous adsorption can be either done as a dynamic laboratory experiment the data from which provides essential adsorption parameters, or as process, such as that for the storage of a gas like hydrogen or methane in a vessel containing adsorptive material such as carbon to be used as a fuel. Most adsorption processes of a larger scale are operated on continuous bases with flow in and flow out of the unit. Generation of oxygen or nitrogen by pressure swing adsorption operates as a continuous process with cycles. The removal of color from raw sugar in solution is done continuously. Water treatment is also done continuously. Gas drying and other "polishing" steps are done continuously.

We have done so much with the batch and semi-continuous models that the extension to the continuous case is relatively straightforward, but we have to be careful. In this problem, we have a stream flowing into the unit that consists of a non-adsorbing gas i for inert and an adsorbing gas or vapor j. We will express gas concentrations and densities in terms of mass per unit volume, or g per cm^3. The vessel contains both the adsorbent, which is the "solid phase" abbreviated as s and interparticle gas phase abbreviated as g. The parameters in the stream supplying the vessel will be designated with an f (for feed). The streams leaving the vessel will be assumed to have identical densities and concentrations to those in the vessel. Both in the vessel and in the exit streams, these entities will be designated with an v for vessel. The key characterizing variables are:

ρ^g_{fi} = density of gas i in the feed stream, constant

ρ^g_{fj} = density of gas j in the feed stream, constant

ρ^g_{vi} = density of gas i in the vessel, variable

ρ^g_{vj} = density of gas j in the vessel, variable

C^s_{vj} = concentration of j on the adsorbent in the vessel, variable

q_f = volumetric flow rate into the vessel

q = volumetric flow rate out of the vessel

X^g_{fi} = mole fraction of gas i in the feed stream, constant

X^g_{fj} = mole fraction of gas i in the feed stream, constant

X^g_{vi} = mole fraction of gas i in the vessel

X^g_{vj} = mole fraction of gas j in the vessel

There will be other parameters needed to describe this process that are the same as those used previously.

At any time in the vessel the total masses of species i and j are the sum of the mass of i and j in the interparticle volume, ϵ V, and the mass of species j accumulated on the adsorbent. We can express this sum as follows:

$$\text{Total mass in the adsorber unit} = \rho^g_{vi}\epsilon\ \mathrm{V} + \rho^g_{vj}\ \epsilon\ \mathrm{V} + C^s_{vj}(1 - \epsilon)\ \rho_g A_s \mathrm{V}$$

The mass of inert gas and adsorbate gas that flowed into the vessel up to that time t are given as:

$$\rho^g_{fi} q_f\ \mathrm{t} + \rho^g_{fj} q_f\ \mathrm{t}$$

The mass that has flowed out of the vessel up to time t will be:

$$\rho^g_{vi}\mathrm{q}\ \mathrm{t} + \rho^g_{vj}\mathrm{q}\ \mathrm{t}$$

The differential total mass balance is:

$$\partial_t\left[\rho^g_{vi}\epsilon V + \rho^g_{vj}\epsilon V + C^s_{vj}(1-\epsilon)\rho_g A_s V\right]$$

and the right-hand side is just the sum of the derivatives of the mass flowing to and from the vessel, and that which is adsorbed:

$$(\rho^g_{fi} + \rho^g_{fj0})\ q_f - (\rho^g_{vi} + \rho^g_{vj})\mathrm{q}$$

and

$$\partial_t\left[\rho^g_{vi}\epsilon V + \rho^g_{vj}\epsilon V + C^s_{vj}(1-\epsilon)\rho_g A_s V\right] = (\rho^g_{fi} + \rho^g_{fj0})\ q_f - (\rho^g_{vi} + \rho^g_{vj})\mathrm{q}$$

At this point it is best to indicate which of these are parameters that stay constant and which are variables that may change with time.

$$\partial_t\left[\rho^g_{vi}\epsilon V + \rho^g_{vj}[t]\epsilon V + C^s_{vj}[t](1-\epsilon)\rho_g A_s V\right] = (\rho^g_{fi} + \rho^g_{fj})\ q_f - (\rho^g_{vi} + \rho^g_{vj}[t])\mathrm{q}$$

The density of species j in the gas phase may change and the concentration of species j on adsorbent surface also will change. We will not see much change in the density of the non-adsorbing gas so we will fix it as a parameter. One may ask why isn't the rate of adsorption shown in the right-hand side. This is because this is an instantaneous rate expression. When the rate of adsorption is large it will suppress the density of species i exiting the vessel. Hence the right-hand side would be made up of both the input flow terms of species i and j, but the exit flow term would be reduced to just that of species i. In other words, while species i is adsorbing we may expect there is no flow of species i from the unit. That is how the accumulation will be handled. As the adsorbent capacity is used up, the rate of adsorption will smoothly diminish and the exit density of species i will rise until it is the same as the feed density.

Next, we write the mass balances for each component in each phase within the vessel.

Species i, Gas Phase:

$$\partial_t \left[\rho_{vi}^{g} \, \epsilon V\right] = \rho_{fi}^{g} \, q_f - \rho_{vi}^{g} \, \mathrm{q}$$

If ρ_{vi}^{g} is a constant, then:

$$0 = \rho_{fi}^{g} \, q_f - \rho_{vi}^{g} \, q_f$$

and

$$\mathrm{q} = \frac{\rho_{fi}^{g} q_f}{\rho_{vi}^{g}}$$

further

$$\text{if } \rho_{fi}^{g} = \rho_{vi}^{g} \text{ then } q_f = q_f$$

But what if ρ_{vi}^{g} is not a constant, then:

$$\partial_t \left[\rho_{vi}^{g} \, [t]\right] = \frac{\rho_{fi}^{g} q_f - \rho_{vi}^{g} q}{\epsilon V}$$

Species i, Solid Phase:

NA

There is no adsorption of species i so this, the gas phase mass balance in the vessel, is the only relevant equation. But, we do notice that we now have a relationship for q the exit flow rate. It is the product of the feed flow rate and the ratio of the density of species i in the feed to that in the vessel's gas phase.

Now, we do the same for species j, which will be adsorbed, as specified in the statement of the work.

Species j, Gas Phase:

$$\partial_t \left[\rho^g_{vj}[t]\epsilon V\right] = \rho^g_{fj}\, q_f \ - \ \rho^g_{vj}[t]\, \text{q} \ - r_{\text{ads}}(1 - \epsilon)\, \rho_g A_s V$$

$$\partial_t \left[\rho^g_{vj}[t]\right] = \frac{\rho^g_{fj} q_f - \rho^g_{vj}[t] q}{\epsilon V} - r_{\text{ads}} \frac{(1-\epsilon)}{\epsilon} \rho_g A_s$$

$$\partial_t \left[\rho^g_{vj}[t]\right] = \frac{\rho^g_{fj} q_f - \rho^g_{vj}[t] q}{\epsilon V} - \frac{(1-\epsilon)}{\epsilon} \rho_g A_s \left(k_a \rho^g_{vj}[t] \left(C^{s*}_{vj} - C^s_{vj}[t]\right) - k_d C^s_{vj}[t]\right)$$

Species j, Solid Phase:

$$\partial_t \left[C^s_{vj}[t](1-\epsilon)\rho_g A_s V\right] = r_{\text{ads}}(1-\epsilon)\rho_g A_s$$

$$\partial_t \left[C^s_{vj}[t]\right] = r_{\text{ads}} = \left(k_a \rho^g_{vj}[t] \left(C^{s*}_{vj} - C^s_{vj}[t]\right) - k_d C^s_{vj}[t]\right)$$

The three equations that will govern this system are:

$$\partial_t \left[\rho^g_{vi}[t]\right] = \frac{\rho^g_{fi} q_f - \rho^g_{vi}[t] q}{\epsilon V}$$

$$\partial_t \left[\rho^g_{vj}[t]\right] = \frac{\rho^g_{fj} q_f - \rho^g_{vj}[t] q}{\epsilon V} - \frac{(1-\epsilon)}{\epsilon} \rho_g A_s \left(k_a \rho^g_{vj}[t] \left(C^{s*}_{vj} - C^s_{vj}[t]\right) - k_d C^s_{vj}[t]\right)$$

$$\partial_t \left[C^s_{vj}[t]\right] = \left(k_a \rho^g_{vj}[t] \left(C^{s*}_{vj} - C^s_{vj}[t]\right) - k_d C^s_{vj}[t]\right)$$

In the code that we will write, it is better to do away with superscripts and subscripts. So we will rename the variables this way:

$$\rho^g_{fi} = \rho\text{fig}$$

$$\rho^g_{fj} = \rho\text{fjg}$$

$$\rho^g_{vi} = \rho\text{vig}$$

$$\rho^g_{vj} = \rho\text{vjg}$$

$$C^s_{vj} = \text{cvjs}$$

$$q_f = \text{qf}$$

$$\text{q} = \text{q}$$

$$X^g_{fi} = \text{xfig}$$

$$X^{g}_{fj} = \text{xfjg}$$

$$X^{g}_{vi} = \text{xvig}$$

$$X^{g}_{vj} = \text{xvjg}$$

For this case, we will let the adsorption be highly favorable with ka/kd = $10^6 \frac{\text{cm}^3}{g}$ and with ka = $10 \frac{\text{cm}^3}{g\,\text{sec}}$, and kd = 10^{-5}sec^{-1}. We will use the same molecular weight, 16 g/mole, as in the earlier example, temperature will be set to 300K. The pressure in the supply stream will be 1.1 atmospheres. The mole fraction of the non-adsorbing gas, i, will be 0.9 and the balance will be the adsorbate, j. The flow rates in and out of the unit will be set to 1500 cm^3 per second, which is about 130 m^3 per day, giving the unit a holding time of about 24 minutes. The other parameters are the same as we have used in the other models.

```
Clear[ka, kd, Pfig, ρfig, ρfjg, ρgo, co, ϵ, V,
ρs, As]
ka = 10(* cm^3/(g sec) *);
kd = 10^-5(* sec^-1 *);
Mw = 16;
T = 300(* K *);
R = 82.05(* cm^3 atm/(mole K) *);
Pfig = 1.1 (* atm *);
ρfig = .9Pfig/(R*T*1/Mw); (* g/cm^3 *)
ρfjg = .1Pfig/(R*T*1/Mw); (* g/cm^3 *)
ρgo = Pfig/(R*T*1/Mw)(* g/cm^3 *)
q = 1500(* cm^3/sec *);
qf = 1500(* cm^3/sec *);
ϵ = 0.4; (* unitless *)
V = 2 10^6(* cm^3 *);
ρs = 2.(* gcm^-3 *);
```

$\text{As} = 1.2\ 10^7(*\ \text{cm}^2 g^{-1}\ *);$

$C_{\text{stot}} = 1.\ 10^{-9}(*\ \frac{g}{\text{cm}^2}\ *);$

$\text{tmax} = 500000;$

$b = \text{NDSolve}\left[\left\{\left\{\partial_t \rho\text{vig}[t] == \frac{\rho\text{fig} * \text{qf} - \rho\text{vig}[t] * q}{\epsilon V},\right.\right.\right.$

$\partial_t \rho\text{vjg}[t] == \frac{\rho\text{fjg} * \text{qf} - \rho\text{vjg}[t] * q}{\epsilon V} - \frac{(1-\epsilon)}{\epsilon}\rho\text{sAs}*$

$\frac{\partial \text{cvjs}(t)}{\partial t} = \text{ka}\rho\text{vjg}(t)\,(C_{\text{stot}} - \text{cvjs}(t)) - \text{kdcvjs}(t)\rho\text{vig}[0] == 1\rho\text{go},$

$\rho\text{vjg}[0] == 0\rho\text{go},$

$\text{cvjs}[0] == 0\},$

$\{\rho\text{vig}[t], \rho\text{vjg}[t], \text{cvjs}[t]\}, \{t, 0, \text{tmax}\}];$

$\rho\rho\text{vig}[\text{t\_}] := \text{Evaluate}[\rho\text{vig}[t] /. b]$

$\rho\rho\text{vjg}[\text{t\_}] := \text{Evaluate}[\rho\text{vjg}[t] /. b]$

$\text{ccvjs}[\text{t\_}] := \text{Evaluate}[\text{cvjs}[t] /. b]$

$\text{b1c} = \text{Plot}\left[\rho\rho\text{vig}[t] * R * T * \frac{1}{\text{Mw}}, \{t, 0, \text{tmax}\},\right.$

$\text{AxesLabel} \rightarrow$

{"t (s)", "Pi (atm)"},

$\text{Epilog} \rightarrow$

$\text{Line}\left[\left\{\left\{0, \rho\text{fig} * R * T * \frac{1}{\text{Mw}}\right\},\right.\right.$

$\left.\left.\left\{\text{tmax}, \rho\text{fig} * R * T * \frac{1}{\text{Mw}}\right\}\right\}\right],$

$\text{PlotRange} \rightarrow \left\{\{0, \text{tmax}\}, \left\{0, 1.2 * \rho\text{fig} * R * T * \frac{1}{\text{Mw}}\right\}\right\},$

$\text{PlotStyle} \rightarrow \{\text{Dashed}, \text{Hue}[0]\}]$

$\text{b1c2} = \text{Plot}\left[\rho\rho\text{vig}[t] * R * T * \frac{1}{\text{Mw}}, \{t, 0, \text{tmax}\},\right.$

$\text{AxesLabel} \rightarrow$

{"t (s)", "Pi (atm)"},

```
Epilog →
Line[{{0, ρfig * R * T * 1/Mw},
{tmax, ρfig * R * T * 1/Mw}}],
PlotRange →
{{0, tmax}, {.9ρfig * R * T * 1/Mw,
1.2 * ρfig * R * T * 1/Mw}},
PlotStyle → {Dashed, Hue[0]}]
b2c = Plot[ρρvjg[t] * R * T * 1/Mw, {t, 0, tmax},
AxesLabel →
{"t (s)", "Pj (atm)"},
Epilog →
Line[{{0, ρfjg * R * T * 1/Mw},
{tmax, ρfjg * R * T * 1/Mw}}], PlotRange → All,
PlotStyle → {Dashed, Hue[0]}]
b3c = Plot[ccvjs[t], {t, 0, tmax},
AxesLabel → {"t (s)", "CCs [ t ] (g/cm^2)"},
Epilog → Line [{{0, Cstot}, {tmax, Cstot}}],
PlotRange → {{0, tmax}, {0, 1.1Cstot}},
PlotStyle → Dashed]
b4c = Plot[ccvjs[t] / Cstot , {t, 0, tmax},
AxesLabel → {"t (s)", "CCs [ t ]/Cistot (Unitless)"},
Epilog → Line[{{0, 1}, {tmax, 1}}],
PlotRange → {{0, tmax}, {0, 1}}, PlotStyle → Dashed]
```

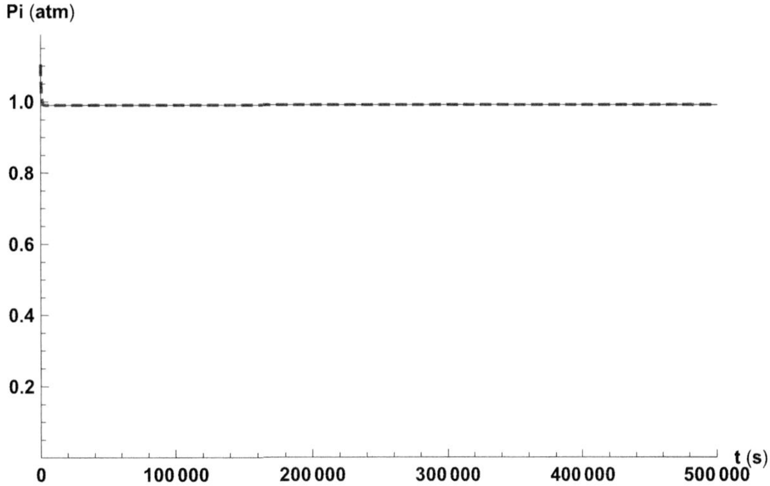
Pi (atm)
1.0
0.8
0.6
0.4
0.2
0
100 000
200 000
300 000
400 000
500 000
t (s)

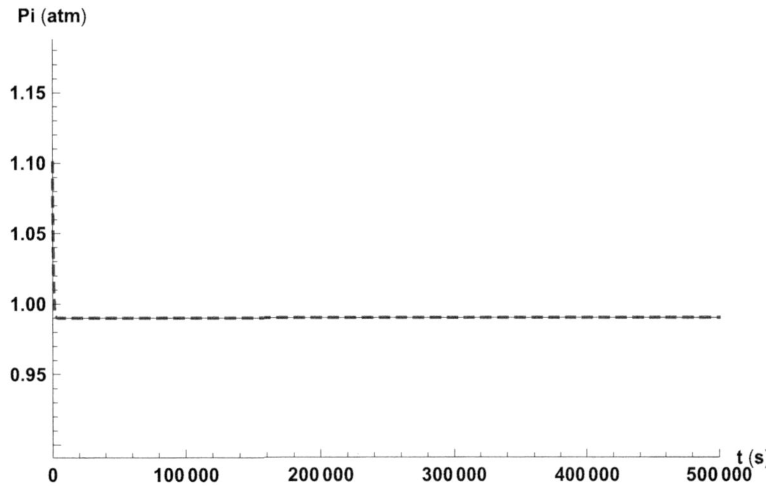
Pi (atm)
1.15
1.10
1.05
1.00
0.95
0
100 000
200 000
300 000
400 000
500 000
t (s)

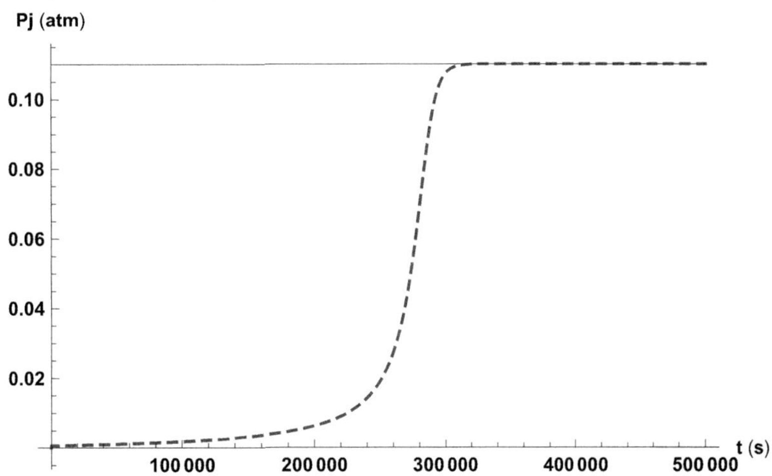
Pj (atm)
0.10
0.08
0.06
0.04
0.02
100 000
200 000
300 000
400 000
500 000
t (s)

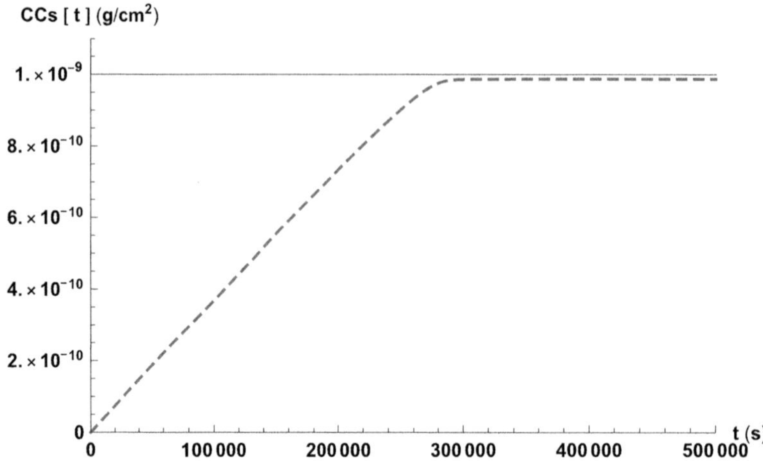

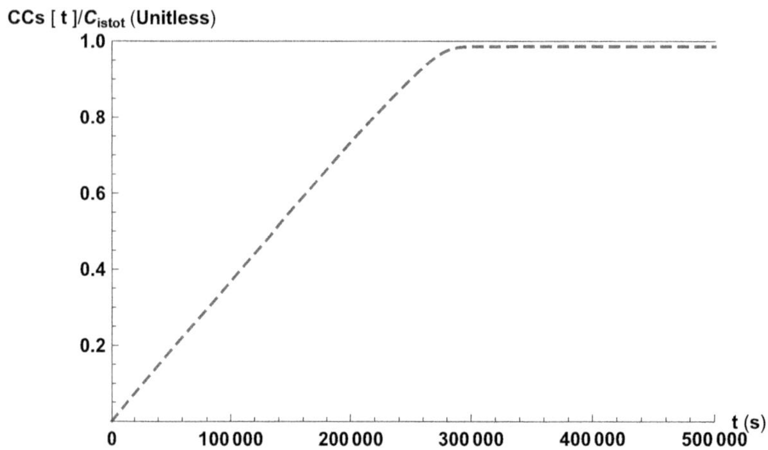

Starting with the first two graphs, we note that the partial pressure of gas i drops very rapidly from 1.1 atm to 1 atm. The partial pressure of the adsorbate, j, shows s-curve behavior. Early on, it exits the system with a very low pressure that begins to rise with time. When the adsorbent nears full capacity, the adsorbate "breaks through," in the exit stream and rises in partial pressure until it matches that in the feed stream. The last two graphs are the change in concentration of the adsorbate on the adsorbent. Notice that the rise is mostly linear until the concentration approaches its limit. This is because we chose a very highly favorable equilibrium and a large rate constant for the adsorption process. While all this behavior is typical, the details are very much a result of our particular choices of parameter values.

To see how the behavior of this system changes with changes in the values of the parameters, it is useful to build a function from the code that we have produced to this point. We will choose to vary the adsorption rate constant and the maximum time and we will wrap of the code in a Module and give the function the name ads. The two parameters are named inside the function this way:

```
ads[ka_,tmax_]:=
```

Then we create the Module to wrap all the other parameters and their values along with the names of all the variables that will be use locally in the code:

$$\text{Module}\left[\left\{\text{kd} = \frac{1}{10^5}, \text{Cstot} = \frac{1}{10^9}, \text{qf} = 1500, q = 1500, \text{Mw} = 16, \text{Pfi} = 1.1,\right.\right.$$

$$\left.\text{R} = 82.05, \text{T} = 300, \epsilon = 0.4, V = 2\ 10^6,\ \rho\text{s} = 2., \text{As} = 1.2\ 10^7\right\}.$$

Parameter values may also be calculated on the fly, and this is useful if they involve one of the input parameters. In this case they don't, but the point is made.

$$\rho\text{go} = \frac{\text{MwPfi}}{\text{RT}} \qquad \rho\text{fi} = \frac{0.9\text{MwPfi}}{\text{RT}} \qquad \rho\text{fj} = \frac{0.1\text{MwPfj}}{\text{RT}}$$

The differential equations can be treated as a set and we will name it eqs inside the code. When we create the numerical solution, we will name this set as the first term, followed by the initial conditions, the functions for which we are solving and the time domain over which the solution will be determined:

```
res=NDSolve[{eqs,ρi(0)=ρgo,ρj(0)=0,cj(0)=0},{ρi(t),ρj(t),cj(t)},{t,0,tmax}];
```

We will create a set of solutions directly by calling them out of the results from NDSolve. We called the results "res". Inside res are the three interpolating functions that we want to use. They are inside a set of brackets; the first 1, the second number is the number of the actual function, of which there are three and, then we want the second term of those functions in each case. Therefore, we have:

```
{res[[1,1,2]],res[[1,2,2]],res[[1,3,2]]};
```

Lastly, to observe behavior, we will observe the graphs of these functions. We create one for each, that is one for the pressure of the non-adsorbed gas, another for the pressure of the adsorbed species in the gas phase and the last for the ratio of the actual surface concentration of the adsorbed species to the maximum possible concentration on the surface.

$$\left\{\text{Plot}\left[\frac{RT\,\text{res}[[1,1,2]]}{\text{Mw}},\ \{t, 0, \text{tmax}\},\right.\right.$$

$$\text{AxesLabel} \to \{\text{t (s), Pi (atm)}\},$$

$$\text{Epilog} \to \text{Line}\left[\left(\left\{\left\{0, \frac{\rho\text{fi}RT}{\text{Mw}}\right\},\right)\right], \left\{\text{tmax}, \frac{\rho\text{fi}RT}{\text{Mw}}\right\}\right\}\Big],$$

$$\text{PlotRange} \to \{\{0, \text{tmax}\}, \left\{0, \frac{1.\rho\text{fi}RT}{\text{Mw}}\right\}\Big\},$$

$$\text{PlotStyle} \to \{\text{Dashed}, Hue[0]\}\Big],$$

$$\text{Plot[res[[1,2,2]]}\ \frac{RT}{\text{Mw}}, \{\text{t}, 0, \text{tmax}\},$$

$$\text{AxesLabel} \to \{\text{t (s), Pj (atm)}\},$$

$$\text{Epilog} \to \text{Line}[\{\{0, \frac{\rho\text{fj}RT}{\text{Mw}}\Big\}, \left\{\text{tmax}, \frac{\rho\text{fj}RT}{\text{Mw}}\right\}\Big\}\Big],$$

$$\text{PlotRange} \to \{\{0, \text{tmax}\}, \left\{0, \frac{1.\rho\text{fj}RT}{\text{Mw}}\right\}\Big\},$$

$$\text{PlotStyle} \to \{\text{Dashed}, Hue[0]\}\Big],$$

$$\text{Plot}\left[\frac{\text{res}[[1,3,2]]}{\text{Cstot}}, \{t, 0, \text{tmax}\},\right.$$

$$\text{AxesLabel} \to \{\text{t (s), CCs[t]/Cstot (g/cm}^2\Big)\},$$

$$\text{Epilog} \to \text{Line}[\{\{0, \frac{\rho\text{fj}RT}{\text{Mw}}\Big\}, \{\text{tmax}, \frac{\rho\text{fj}RT}{\text{Mw}}\Big\}\Big\}\Big],$$

$$\text{PlotRange} \to \{\{0, \text{tmax}\}, \{0, \frac{1.\rho\text{fj}RT}{\text{Mw}}\Big\}\Big\},$$

$$\text{PlotStyle} \to \{\text{Dashed}, \text{Hue}[0.1]\}\Big],$$

$$\text{Plot}\left[\frac{\text{res}[[1,3,2]]}{\text{Cstot}}, \text{Null}\{t, 0, \text{tmax}\},\right.$$

$$\text{AxesLabel} \to \left\{\text{t (s), CCs[t]/Cstot}\ \left(\text{g/cm}^2\right)\right\},$$

$$\text{Epilog} \to \text{NullLine}[\{\{0, 1\}, \{\text{tmax}, 1\}\}],$$

$$\text{PlotRange} \to \{\{0, \text{tmax}\}, \{0, 1.1\}\},$$

$$\text{PlotStyle} \to \text{Dashed}]\}]$$

We also elaborate the graphs for readability. Thus, we add axes labels, horizontal lines for the maximum values of the functions, and we set the plot range rather than having *Mathematica* choose it. Finally, we end the code with a closed square bracket. This closes the whole of the Module:

Module[{... ... ...}]

This is all shown in the following input cell:

ads[ka_, tmax_]:=

Module $\left[\left\{\mathbf{kd} = \frac{1}{10^5}, \mathbf{Cstot} = \frac{1}{10^9}, \mathbf{qf} = \mathbf{1500}, q = \mathbf{1500},\right.\right.$

$\mathbf{Mw = 16, Pfi = 1.1}, R = \mathbf{82.05}, T = \mathbf{300}, \epsilon = \mathbf{0.4},$

$V = 2 * 10^6, \rho\mathbf{s} = 2., \mathbf{As} = 1.2 * 10^7, b, \rho\mathbf{go}, \rho\mathbf{fi},$

$\rho\mathbf{fj}, \rho\mathbf{i}, \rho\rho\mathbf{i}, \rho\mathbf{j}, \rho\rho\mathbf{j}, \mathbf{cj}, \mathbf{ccj}, \mathbf{res}\},$

$\rho\mathbf{go} = \frac{\mathbf{MwPfi}}{RT};$

$\rho\mathbf{fi} = \frac{\mathbf{0.9MwPfi}}{RT};$

$\rho\mathbf{fj} = \frac{\mathbf{0.1MwPfi}}{RT};$

$\mathbf{eqs} = \left\{\partial_t \rho\mathbf{i}[t] == \frac{\rho\mathbf{fiqf} - \rho\mathbf{i}[t]q}{\epsilon V},\right.$

$\partial_t \rho\mathbf{j}[t] == \frac{\rho\mathbf{fjqf} - \rho\mathbf{j}[t]q}{\epsilon V} -$

$\frac{(1-\epsilon)\rho\mathbf{sAs}(\mathbf{ka}\rho\mathbf{j}[t](\mathbf{Cstot} - \mathbf{cj}[t]) - \mathbf{kdcj}[t])}{\epsilon},$

$\partial_t \mathbf{cj}[t] == \mathbf{ka}\rho\mathbf{j}[t](\mathbf{Cstot} - \mathbf{cj}[t]) - \mathbf{kdcj}[t]\};$

$\mathbf{res} = \mathbf{NDSolve}[\{\mathbf{eqs}, \rho\mathbf{i}[0] == \rho\mathbf{go}, \rho\mathbf{j}[0] == 0,$

$\mathbf{cj}[0] == 0\}, \{\rho\mathbf{i}[t], \rho\mathbf{j}[t], \mathbf{cj}[t]\}, \{t, 0, \mathbf{tmax}\}];$

{res[[1, 1, 2]], res[[1, 2, 2]], res[[1, 3, 2]]};

{Plot[res[[1, 1, 2]]*RT*/Mw, {*t*, 0, tmax},

```
AxesLabel → {"t (s)", "Pi (atm)"},
Epilog → Line[{{0, ρfiRT/Mw}, {tmax, ρfiRT/Mw}}],
PlotRange → {{0, tmax}, {0, 1.ρfiRT/Mw}},
PlotStyle → {Dashed, Hue[0]}],

Plot[res[[1, 2, 2]]RT/Mw, {t, 0, tmax},
AxesLabel → {"t (s)", "Pj (atm)"},
Epilog → Line[{{0, ρfjRT/Mw}, {tmax, ρfjRT/Mw}}],
PlotRange → {{0, tmax}, {0, 1.ρfjRT/Mw}},
PlotStyle → {Dashed, Hue[0.6]}],

Plot[res[[1, 3, 2]]/Cstot, {t, 0, tmax},
AxesLabel → {"t (s)", "CCs[t]/Cstot (g/cm^2)"},
Epilog → Line[{{0, 1}, {tmax, 1}}],
PlotRange → {{0, tmax}, {0, 1.1}},
PlotStyle → Dashed]}
]
```

That is the code and we have now created a unique function. We can use this function to produce many different cases and we can learn how this system behaves with variations in parameters. It is as if we can do many different experiments, but they are all simulations. This is the very essence of building learning models. Remember that any or all of the parameters of the system could be made function variables. To use this code, we call the function with the two parameters, ka and tmax specified and hit shift return as we always do. Let's choose ka = 1 and tmax = 500,000 again. This should be similar to the results we generated with ka = 10, but we expect the transitions may be less sharp:

```
ads[1, 500000]
```

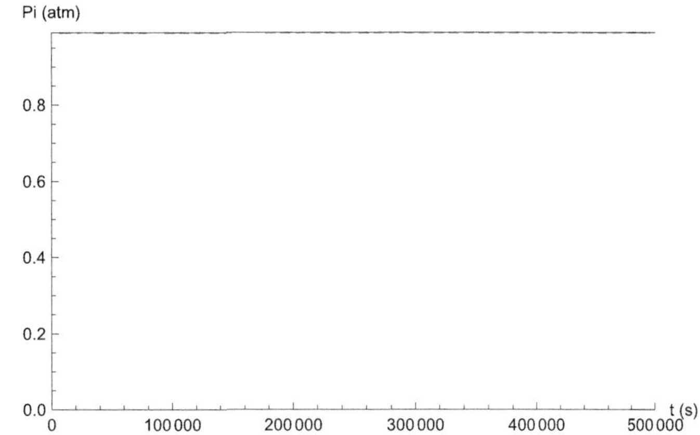

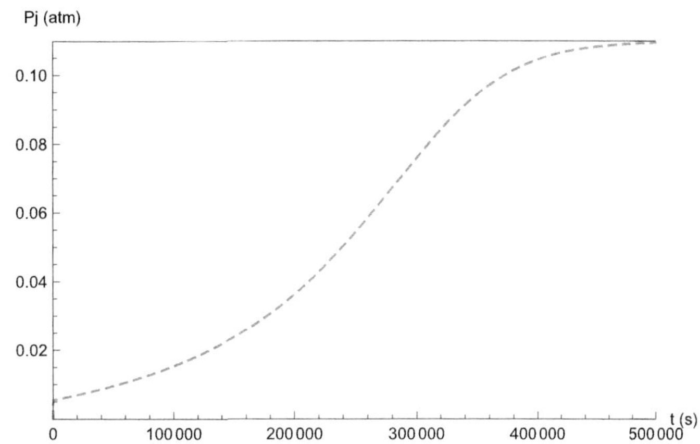

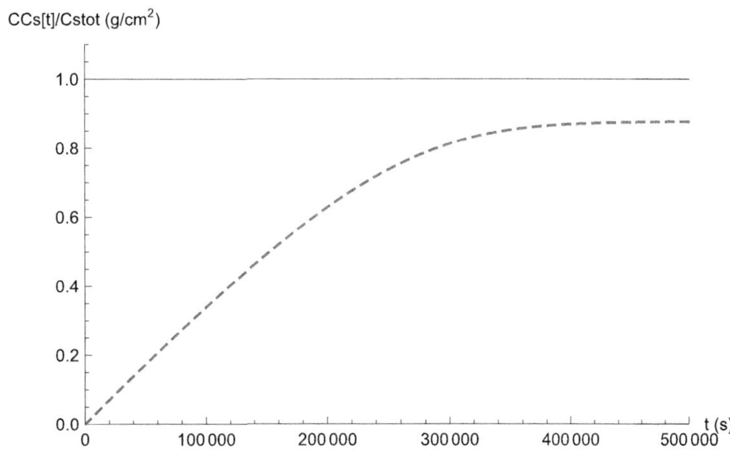

As we expected the change in adsorbate pressure and surface concentration are smoother and the surface equilibrium point is lower, all as it should be. We note that the graph of the pressure of the non-adsorbed gas is not interesting, so let's drop it from the output. Let's also add to the plots the value of ka. We can do that by adding this phrase to the plotting code:

PlotLabel → " = ka" ka

For ease of understanding this phrase has been added to the new code for the function ads2:

ads2[ka_, tmax_]:=

$$\text{Module}\left[\left\{\text{kd} = \tfrac{1}{10^5}, \text{Cstot} = \tfrac{1}{10^9}, \text{qf} = 1500, q = 1500,\right.\right.$$

$$\text{Mw} = 16, \text{Pfi} = 1.1, R = 82.05, T = 300, \epsilon = 0.4,$$

$$V = 2 * 10^6, \rho\text{s} = 2., \text{As} = 1.2 * 10^7, b, \rho\text{go}, \rho\text{fi},$$

$$\rho\text{fj}, \rho\text{i}, \rho\rho\text{i}, \rho\text{j}, \rho\rho\text{j}, \text{cj}, \text{ccj}, \text{res}\},$$

$$\rho\text{go} = \frac{\text{MwPfi}}{RT};$$

$$\rho\text{fi} = \frac{0.9\text{MwPfi}}{RT};$$

$$\rho\text{fj} = \frac{0.1\text{MwPfi}}{RT};$$

$$\text{eqs} = \left\{\partial_t \rho\text{i}[t] == \frac{\rho\text{fiqf} - \rho\text{i}[t]q}{\epsilon V},\right.$$

$$\partial_t \rho\text{j}[t] == \frac{\rho\text{fjqf} - \rho\text{j}[t]q}{\epsilon V} - \frac{(1-\epsilon)\rho\text{sAs}(\text{ka}\rho\text{j}[t](\text{Cstot} - \text{cj}[t]) - \text{kdcj}[t])}{\epsilon},$$

$$\partial_t \text{cj}[t] == \text{ka}\rho\text{j}[t](\text{Cstot} - \text{cj}[t]) - \text{kdcj}[t]\};$$

$$\text{res} = \text{NDSolve}[\{\text{eqs}, \rho\text{i}[0] == \rho\text{go}, \rho\text{j}[0] == 0,$$

$$\text{cj}[0] == 0\}, \{\rho\text{i}[t], \rho\text{j}[t], \text{cj}[t]\}, \{t, 0, \text{tmax}\}];$$

$$\{\text{res}[[1,1,2]], \text{res}[[1,2,2]], \text{res}[[1,3,2]]\};$$

$$\text{Plot}[\text{res}[[1,2,2]]\frac{RT}{\text{Mw}}, \{t, 0, \text{tmax}\},$$

AxesLabel → {"t (s)", "Pj (atm)"},

$$\text{Epilog} \to \text{Line}\left[\left\{\left\{0, \frac{\rho\text{fj}RT}{\text{Mw}}\right\}, \left\{\text{tmax}, \frac{\rho\text{fj}RT}{\text{Mw}}\right\}\right\}\right],$$

$$\text{PlotRange} \to \left\{\{0, \text{tmax}\}, \left\{0, \frac{1.\rho\text{fj}RT}{\text{Mw}}\right\}\right\},$$

PlotStyle → {Dashed, *Hue*[0]},

```
PlotLabel → " = ka"ka],
Plot[res[[1, 3, 2]]/Cstot, {t, 0, tmax},
AxesLabel →
{"t (s)", "CCs [ t ]/Cstot (g/cm²)"},
Epilog → Line[{{0, 1}, {tmax, 1}}],
PlotRange → {{0, tmax}, {0, 1.1}},
PlotStyle → Dashed,
PlotLabel → " = ka"ka]}
]
```

We can use this new function to see how the behavior of the adsorbate changes with variation in the value of ka. We will start with a low value of ka where the equilibrium is not favorable and move to higher values. We also want a neat side-by-side output, so that we can see how the behavior evolves. To do this we use the Table function and in a different way. Instead of letting there be a range of values, we provide a set or list of values to be used. When we do this, it looks as follows:

Table[ads2[k, 500000],{k,{.0001, .001, .01, .1, 1, 10, 100}}]

We wrap this in Grid to provide a array of graphs. There are other ways to do this, but this works pretty well here, but it's better on a computer, where the graphs can be stretched to see the abscissa properly:

```
Grid[Table[ads2[k, 500000],
{k, {.0001, .01, 1, 100}}]]
```

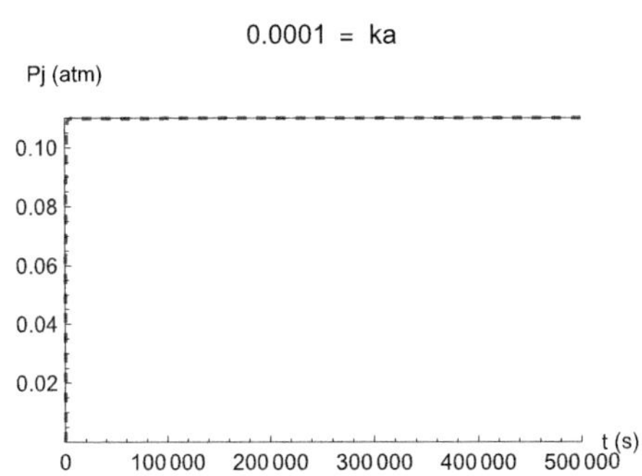

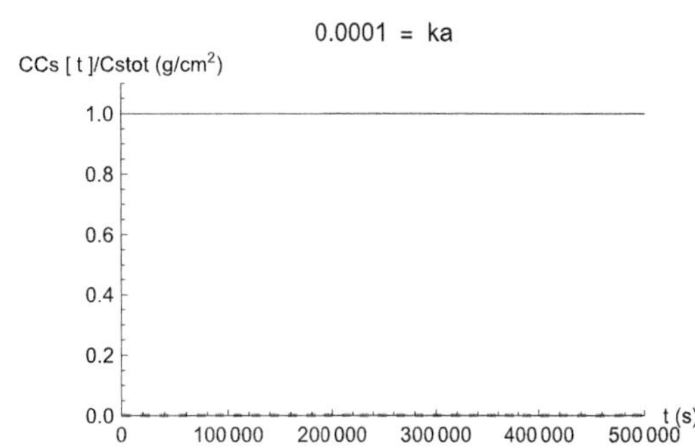

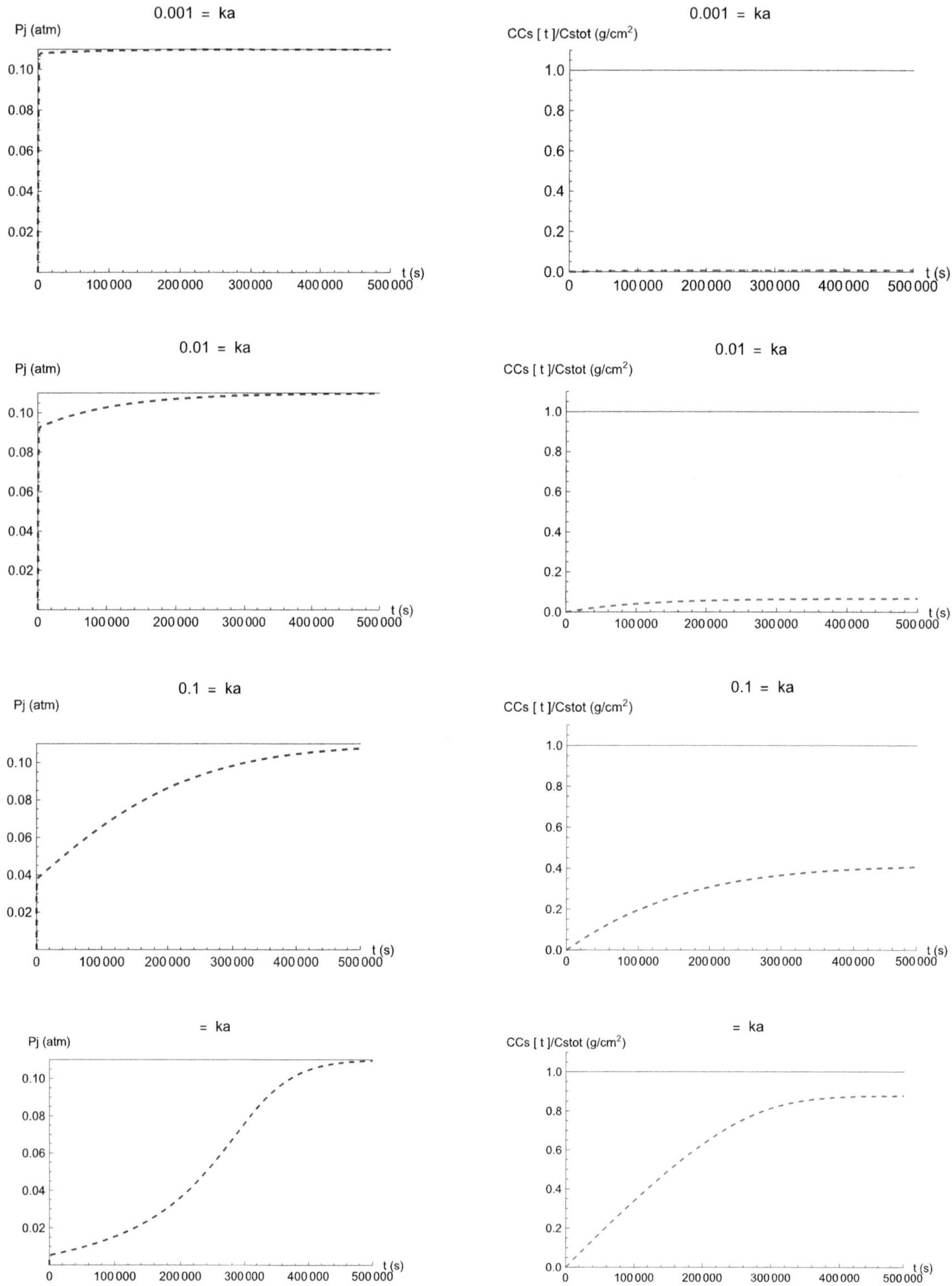
0.001 = ka
Pj (atm)
t (s)
0.001 = ka
CCs [t]/Cstot (g/cm²)
t (s)
0.01 = ka
Pj (atm)
t (s)
0.01 = ka
CCs [t]/Cstot (g/cm²)
t (s)
0.1 = ka
Pj (atm)
t (s)
0.1 = ka
CCs [t]/Cstot (g/cm²)
t (s)
= ka
Pj (atm)
t (s)
= ka
CCs [t]/Cstot (g/cm²)
t (s)

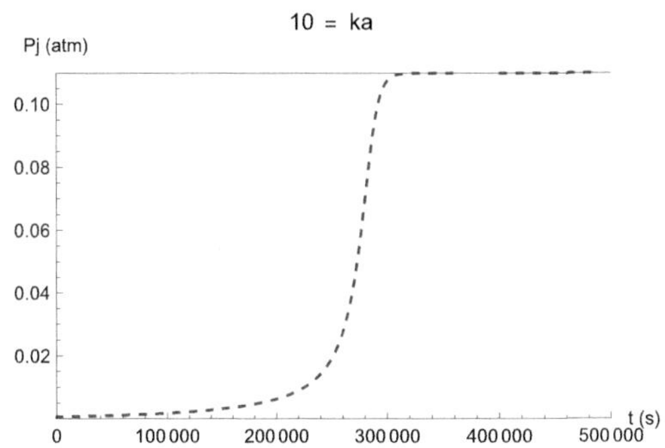

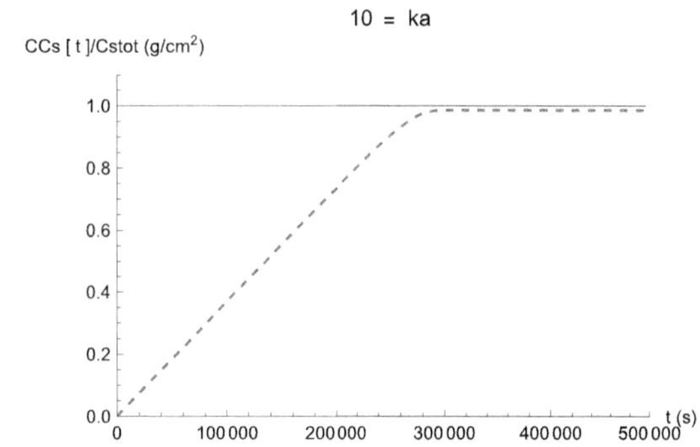

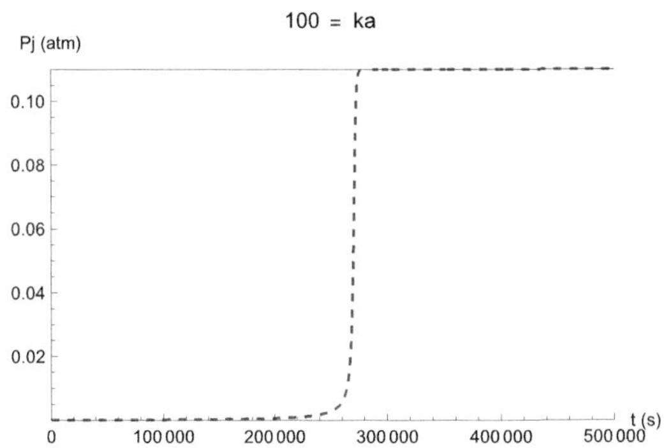

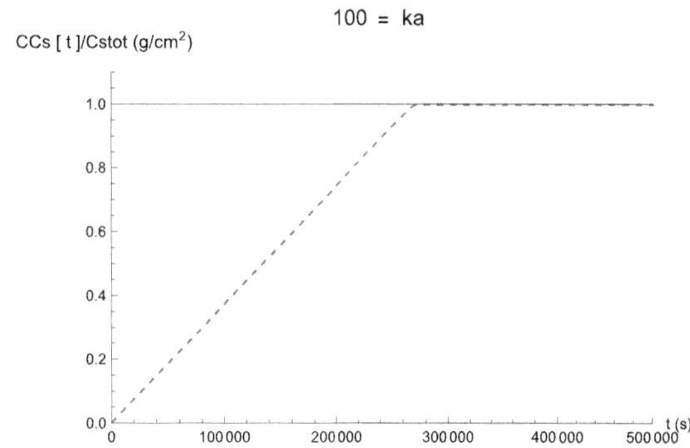

Lastly, we can take one more step and use the power of *Mathematica* even more explicitly. We can redo the code and let it be a function of just ka. This is useful if we want to focus on the effect of this one parameter. We can also choose to use only the graph of the pressure of the adsorbate in the exit stream. This is most useful because this is what we measure. Then we can use this new function ads3 inside the Manipulate function. We set ka as the variable and *Mathematica* generates a "slider" for us from the plot. Using a wide range of ka values (0.001 to 10) and very small steps (0.0001), we compute each of these curves and we can see how the behavior changes. This cannot be seen in the print version of the book, but, hopefully, it is obvious how it works.

ads3[ka_]:=

Module$\left[\left\{\mathbf{tmax} = \mathbf{500000}, \mathbf{kd} = \frac{1}{10^5}, \mathbf{Cstot} = \frac{1}{10^9},\right.\right.$

$\mathbf{qf} = \mathbf{1500}, \boldsymbol{q} = \mathbf{1500}, \mathbf{Mw} = \mathbf{16}, \mathbf{Pfi} = \mathbf{1.1}, \boldsymbol{R} = \mathbf{82.05},$

$\boldsymbol{T} = \mathbf{300}, \boldsymbol{\epsilon} = \mathbf{0.4}, \boldsymbol{V} = \mathbf{2 * 10^6}, \boldsymbol{\rho}\mathbf{s} = \mathbf{2.}, \mathbf{As} = \mathbf{1.2 * 10^7},$

$\boldsymbol{b}, \boldsymbol{\rho}\mathbf{go}, \boldsymbol{\rho}\mathbf{fi}, \boldsymbol{\rho}\mathbf{fj}, \boldsymbol{\rho}\mathbf{i}, \boldsymbol{\rho\rho}\mathbf{i}, \boldsymbol{\rho}\mathbf{j}, \boldsymbol{\rho\rho}\mathbf{j}, \mathbf{cj}, \mathbf{ccj}, \mathbf{res}\},$

$$\rho\text{go} = \frac{\text{MwPfi}}{RT};$$

$$\rho\text{fi} = \frac{0.9\text{MwPfi}}{RT};$$

$$\rho\text{fj} = \frac{0.1\text{MwPfi}}{RT};$$

$$\text{eqs} = \left\{\partial_t \rho\text{i}[t] == \frac{\rho\text{fiqf} - \rho\text{i}[t]q}{\epsilon V},\right.$$

$$\partial_t \rho\text{j}[t] == \frac{\rho\text{fjqf} - \rho\text{j}[t]q}{\epsilon V} - \frac{(1-\epsilon)\rho\text{sAs}(\text{ka}\rho\text{j}[t](\text{Cstot} - \text{cj}[t]) - \text{kdcj}[t])}{\epsilon},$$

$$\left.\partial_t \text{cj}[t] == \text{ka}\rho\text{j}[t](\text{Cstot} - \text{cj}[t]) - \text{kdcj}[t]\right\};$$

$\text{res} = \text{NDSolve}[\{\text{eqs}, \rho\text{i}[0] == \rho\text{go}, \rho\text{j}[0] == 0,$

$\text{cj}[0] == 0\}, \{\rho\text{i}[t], \rho\text{j}[t], \text{cj}[t]\}, \{t, 0, \text{tmax}\}];$

{res[[1, 1, 2]], res[[1, 2, 2]], res[[1, 3, 2]]};

$\text{Plot}[\text{res}[[1,2,2]]\frac{RT}{\text{Mw}}, \{t, 0, \text{tmax}\},$

AxesLabel → {"t (s)", "Pj (atm)"},

$$\text{Epilog} \to \text{Line}\left[\left\{\left\{0, \frac{\rho\text{fj}RT}{\text{Mw}}\right\}, \left\{\text{tmax}, \frac{\rho\text{fj}RT}{\text{Mw}}\right\}\right\}\right],$$

$$\text{PlotRange} \to \left\{\{0, \text{tmax}\}, \left\{0, \frac{1.\rho\text{fj}RT}{\text{Mw}}\right\}\right\},$$

PlotStyle → {Dashed, Hue[0]}, PlotLabel → " = ka"ka]

]

Manipulate[ads3[ka], {ka, 0.001, 10, .00001}]

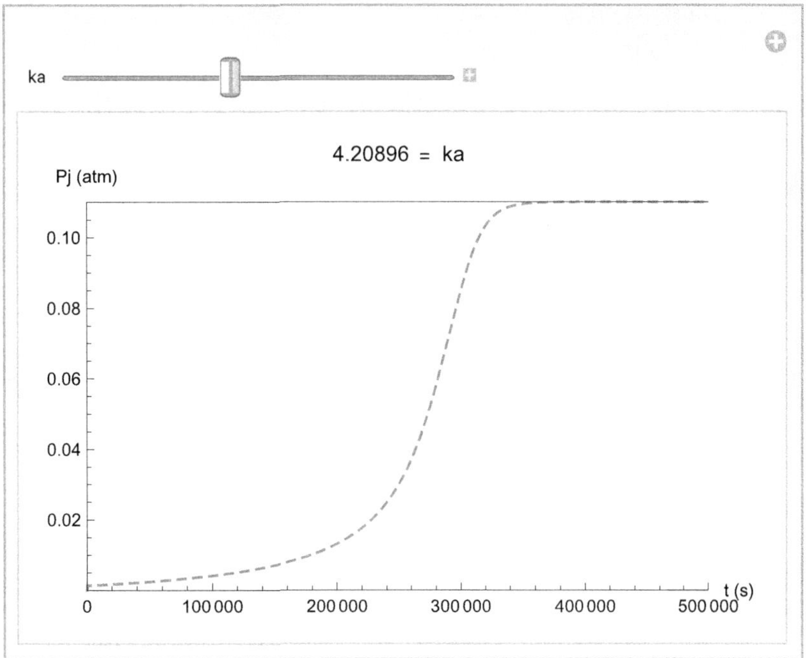

On a computer, you will notice that as you slide the button, the value of ka is automatically updated. It is possible to add as many function variables as you desire and to make sliders of all of them. When done in this way, this code moves from a model to a simulation and even to a simple simulator of sorts.

We can modify the code to show the effect of the variation in ka on the concentration of the adsorbed species:

ads4[ka_]:=

$$\text{Module}\left[\left\{\text{tmax} = 500000, \text{kd} = \frac{1}{10^5}, \text{Cstot} = \frac{1}{10^9},\right.\right.$$

$$\text{qf} = 1500, q = 1500, \text{Mw} = 16, \text{Pfi} = 1.1, R = 82.05,$$

$$T = 300, \epsilon = 0.4, V = 2 * 10^6, \rho\text{s} = 2., \text{As} = 1.2 * 10^7,$$

$$b, \rho\text{go}, \rho\text{fi}, \rho\text{fj}, \rho\text{i}, \rho\rho\text{i}, \rho\text{j}, \rho\rho\text{j}, \text{cj}, \text{ccj}, \text{res}\},$$

$$\rho\text{go} = \frac{\text{MwPfi}}{RT};$$

$$\rho\text{fi} = \frac{0.9\text{MwPfi}}{RT};$$

$$\rho\text{fj} = \frac{0.1\text{MwPfi}}{RT};$$

$$\text{eqs} = \left\{\partial_t \rho\text{i}[t] == \frac{\rho\text{fiqf} - \rho\text{i}[t]q}{\epsilon V},\right.$$

$$\partial_t \rho\text{j}[t] == \frac{\rho\text{fjqf} - \rho\text{j}[t]q}{\epsilon V} - \frac{(1-\epsilon)\rho\text{sAs}(\text{ka}\rho\text{j}[t](\text{Cstot} - \text{cj}[t]) - \text{kdcj}[t])}{\epsilon},$$

$$\left.\partial_t \text{cj}[t] == \text{ka}\rho\text{j}[t](\text{Cstot} - \text{cj}[t]) - \text{kdcj}[t]\right\};$$

```
res = NDSolve[{eqs, ρi[0] == ρgo, ρj[0] == 0,
cj[0] == 0}, {ρi[t], ρj[t], cj[t]}, {t, 0, tmax}];
{res[[1, 1, 2]], res[[1, 2, 2]], res[[1, 3, 2]]};
Plot[res[[1, 3, 2]]/Cstot, {t, 0, tmax},
AxesLabel → {"t (s)", "CCs [ t ]/Cstot (g/cm²)"},
Epilog → Line[{{0, 1}, {tmax, 1}}],
PlotRange → {{0, tmax}, {0, 1.1}},
PlotStyle → Dashed,
PlotLabel → " = ka"ka]
]
Manipulate[ads4[ka], {ka, 0.001, 10, .0000001}]
```

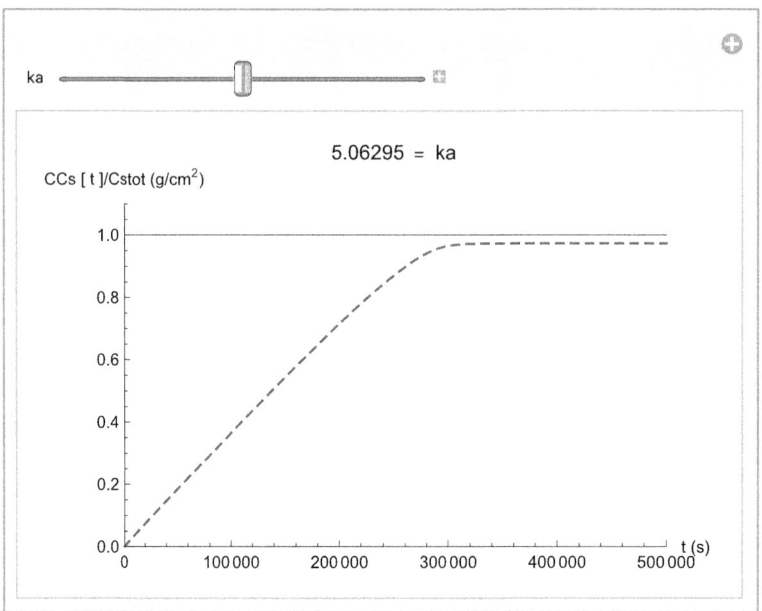

That all works very nicely. Let's move now to a consideration of plug flow.

Plug flow and a simple rate model

To this point, we have assumed that the gas is well-mixed in the adsorber unit. That implies that the gas phase is every where the same, the same density and concentration at every position. It also implies that the gas exiting the unit also has the same density and concentration as is prevalent within the unit. Further, we have said nothing about the shape of the vessel. At this point, we will bring more definition to the shape of the vessel and we will introduce the idea that the gas flows through the unit as a plug, hence the name plug flow.

We can assume that the vessel holding the adsorbent is more like a pipe with a certain diameter and length. The aspect ratio is the ratio of the length to the diameter of the unit.

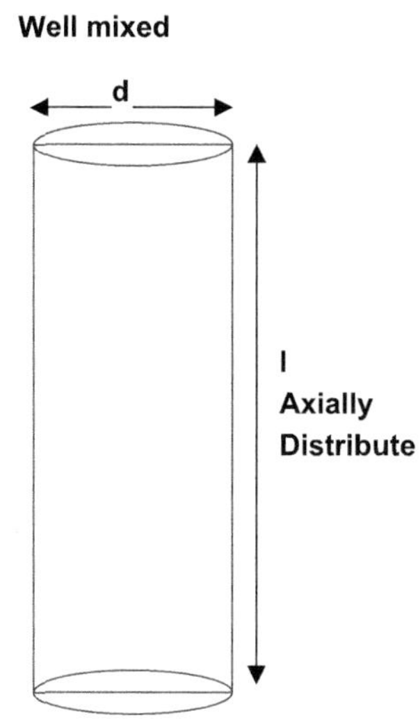

The volume of this unit will be the product of the cross sectional area and the length. If the length is take to be ten times the diameter, then that is the aspect ratio, 10:1 and the volume is:

$$V = \pi \frac{d^2}{4}(10\, d) = 2 \text{ x } 10^6 cm^3$$

$$d = 63 \text{ cm and } l = 630$$

This unit is about two feet in diameter and 20 feet long. If the gas flows as a plug through the unit, then we are assuming that the radial mixing is perfect in the cross sectional area. However, the concentration of the adsorbate will change with position from the top of the bed of adsorbent to the bottom. Put another way, the concentration will be axially distributed and radially well-mixed. At any given point the concentration may be higher just above that point in the axial direction and lower just below it, but at that position the concentration will be the same at any position across the bed in the radial direction.

Imagine a gas that is non-adsorbed that contains a component that is adsorbed, entering at the top of this unit. The concentration of that adsorbate will be as high as it can be at the entrance to the bed of the adsorbent. As the gas progresses downward through the bed, the concentration of the adsorbate will decrease as it is adsorbed. This will continue as the gas travels through the bed toward the exit. Typically, a concentration profile develops in the axial direction. This is what we want to account for in our model.

Consider this unit again, but think of it as a set of smaller units of the same diameter, but shorter in length and connected in series, meaning that the outflow from the unit above is the inflow to the unit below.

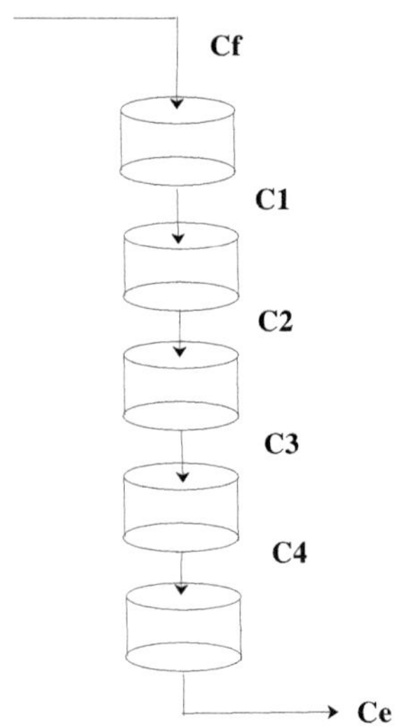

The concentration in each unit is assumed to be well-mixed axially. That sounds counter intuitive, but this will become clearer momentarily. If this were literally the system, then each unit would be well-mixed and the concentration profile would be a series of concentration drops along the series, they would resemble stepwise drops like this:

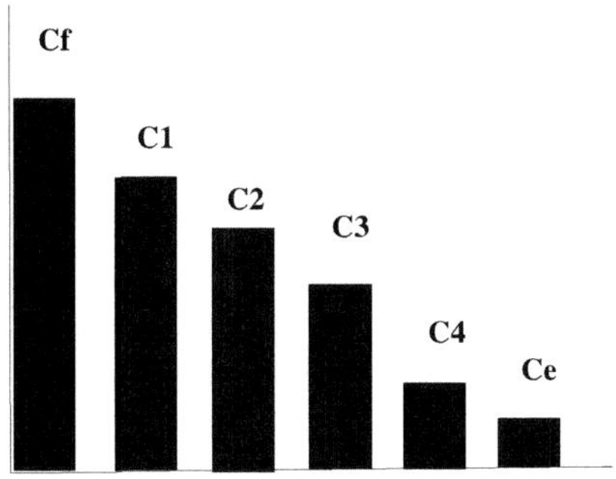

While this may look like a linear drop, it need not be linear and may be quite non-linear. Next, we can imagine pushing these units very close together, with some clever plumbing. In fact they may be so close together that there is no distance between them. Were this the case, then instead of labeling each section as a separate unit, we might just as well consider them to be in series along the original length of the single unit. Furthermore, we need not divide the original volume into just five sections, we can imagine dividing it into ten sections, fifty sections, or an even higher number. This is easy to imagine and it leads us to the idea that the governing equation for this system is a partial differential equation in time and in the axial direction. We can think of it being developed this way. For any one unit we have the governing equation for a well-mixed system in which species B is adsorbed:

C_B = concentration of the adsorbate in the gas space

$C_{B,\mathrm{in}}$ = concentration of adsorbate in the supply stream

q = volumetric flow rate

V = volume of the vessel

A_{cr} = cross sectional area of the vessel

Δz = the axial dimension of the vessel

ρ_s = density of adsorbent

A_s = surface area per gram of adsorbent

$$\frac{dC_B \epsilon V}{dt} = (C_{B,\mathrm{in}} - C_B)\mathrm{q} - (1 - \epsilon)\, r_{\mathrm{ads}}\rho_s\, A_s \mathrm{V}$$

$$\frac{dC_B \epsilon A_{\mathrm{cr}} \Delta \mathrm{z}}{dt} = (C_{B,\mathrm{in}} - C_B)\mathrm{q} - (1 - \epsilon)\, r_{\mathrm{ads}}\rho_s\, A_s \mathrm{V}$$

$$\frac{dC_B}{dt} = \frac{(C_{B,\mathrm{in}} - C_B)\, q}{\Delta \mathrm{z} A_{\mathrm{cr}}} - \frac{(1-\epsilon)}{\epsilon}\, r_{\mathrm{ads}}\rho_s\, A_s \frac{V}{A_{\mathrm{cr}}\Delta \mathrm{z}}$$

$$\frac{dC_B}{dt} = - \frac{\left(C_B - C_{B,\text{in}}\right) q}{\Delta z A_{\text{cr}}} - \frac{(1-\epsilon)}{\epsilon} r_{\text{ads}} \rho_s \, A_s \frac{V}{A_{\text{cr}} \Delta z}$$

$$\frac{dC_B}{dt} = - \frac{\Delta C_B q}{\Delta z A_{\text{cr}}} - \frac{(1-\epsilon)}{\epsilon} r_{\text{ads}} \rho_s \, A_s$$

This is a differential-difference equation as written. If we let Δz approach zero, then it becomes a partial differential equation in time and axial dimension z, and when we divide the volume flow rate by the cross sectional area, we obtain the velocity of the plug of gas flowing through the system in the axial direction:

$$\frac{dC_B}{dt} = - v_z \frac{\Delta C_B}{\Delta z} - \frac{(1-\epsilon)}{\epsilon} r_{\text{ads}} \rho_s \, A_s$$

$$\lim_{\Delta z \to 0} \left[\frac{dC_B}{dt} = - v_z \frac{\Delta C_B}{\Delta z} - \frac{(1-\epsilon)}{\epsilon} r_{\text{ads}} \rho_s \, A_s\right]$$

$$\frac{\partial C_B}{\partial t} = - v_z \frac{\partial C_B}{\partial z} - \frac{(1-\epsilon)}{\epsilon} r_{\text{ads}} \rho_s \, A_s$$

The differential equation that governs the system as we have imagined it, is subject to plug flow. The concentration will be both time and axially dependent. In most instances the time derivative is set to zero and solutions are found that depend only on z, the axial dimension. But, the transient solution can be found analytically, if we let the rate of adsorption be very simple, i.e. first order in the concentration. This is not very realistic, but it is an approximation that could hold very early in the process.

$$\left\{\text{cb}^{(0,1)}[z,t] == -\frac{\text{As ka } (1-\epsilon)\rho\text{scb}[z,t]}{\epsilon} + \text{vz } \text{cb}^{(1,0)}[z,t],\ \text{cb}[0,t] == \text{Cbf}\right\}$$

Because it is first order, we can specify the initial condition and the boundary condition and then determine if *Mathematica* can find a solution. The initial condition is the state of the concentration of the adsorbate in the gas phase of the unit at time equal to zero and across the whole of the axial dimension. We can imagine that this could be zero. However, zeros can lead to mathematical instabilities, so instead we will simply let cb[z,0] == cbo, which can be set to zero or very close to zero, once we have a solution. The boundary condition will be the state of the concentration of the adsorbate at the entry to the unit at all times. Here we will let that concentration be a constant, i.e. the supply stream concentration, and that is given as cb[0,t]==cbf.

```
Clear[ka, ρs, As, vz, z, t, ϵ, cbf, cb, cbo, cstot]

eqns =

{∂tcb[z, t] == −vz ∂zcb[z, t]−
```

$\frac{(1-\epsilon)}{\epsilon}$**ka cb[$z$, t] cstotρs As, cb[z, 0]==cbo,**

cb[0, t] == cbf};

DSolve[eqns, cb[z, t], {z, t}]//FullSimplify

$$\left\{\left\{\text{cb}[z,t]\to e^{\frac{\text{As cstot ka } z(-1+\epsilon)\rho\text{s}}{\text{vz }\epsilon}}\left(\text{cbo } e^{\frac{\text{As cstot ka }(t\text{vz}-z)(-1+\epsilon)\rho\text{s}}{\text{vz }\epsilon}}\left(\text{HeavisideTheta}[t]-\text{HeavisideTheta}\left[t-\frac{z}{\text{vz}}\right]\right)+\text{cbf HeavisideTheta}\left[t-\frac{z}{\text{vz}}\right]\right)\right\}\right\}$$

The result is interesting, the solution includes the HeavisideTheta function. Written in traditional form it looks as follows, where θ stands for that function.

$$\text{cb}(z,t)=e^{-\frac{\text{As cstot ka }\rho\text{s}z(1-\epsilon)}{\text{vz }\epsilon}}\left(\text{cboe}^{-\frac{\text{As cstot ka }\rho\text{s }(1-\epsilon)(t\text{ vz}-z)}{\text{vz }\epsilon}}\left(\theta(t)-\theta\left(t-\frac{z}{\text{vz}}\right)\right)+\text{cbf }\theta\left(t-\frac{z}{\text{vz}}\right)\right)$$

Let's recall what the Heaviside function does; it provides a jump from zero to unity when t = 0 and then it remains equal to unity for all t > 0. When this is multiplied by cbo, at t = 0 the value of this term jumps to initial concentration value.

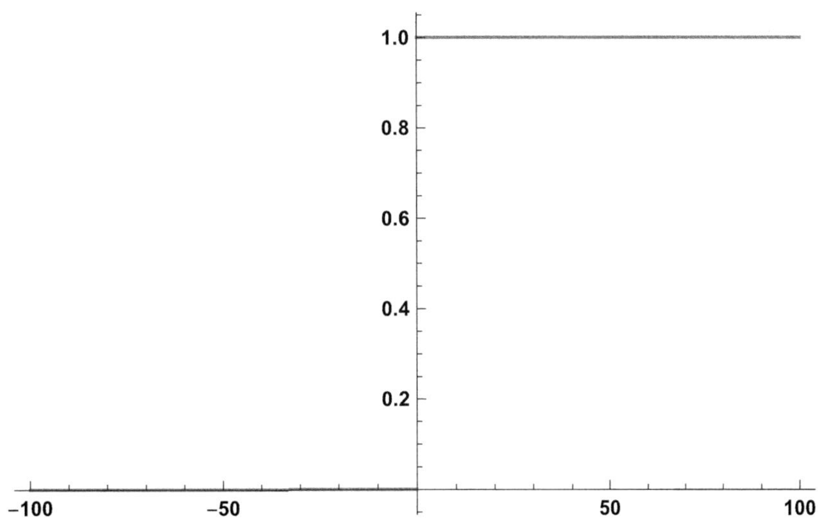

The second HeavisideTheta function that is present has the argument (t - $\frac{z}{vz}$). Using a value of vz = 10 we can examine this function also

vz = 10;

Plot3D[HeavisideTheta[$t-\frac{z}{\text{vz}}$], {t, 0, 10},

{z, 0, 20},

AxesLabel → { "t, time", " z, axial dimension",

"θ[z,t] "}]

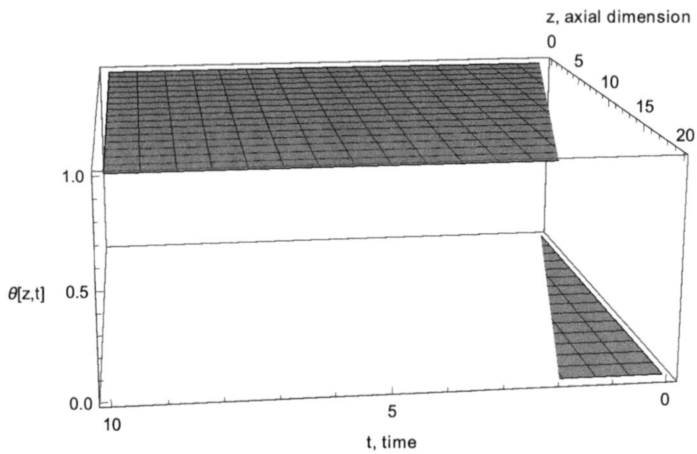

Here we see that when t = 0 and z = 0, the function jumps to unity. Notice that cbf jumps to the feed value at that point. Then t and z/vz are equal the function jumps to unity. Thus, in the concentration function, this becomes equal to the feed concentration at that time and position. This is why we see a triangle in the lower plane. (Time is increasing to the left in this view to make the jump easier to visualize.) Along the hypotenuse of that triangle the concentration jumps from zero to the concentration of the feed stream.

We also notice that there are also exponentials that involve the parameters that specify the nature of the adsorption process. The two exponentials are decreasing functions of time and axial position, which makes good physical sense These exponentials are expected because if ka is favorable (large), then the concentration of adsorbate in the gas phase will diminish with increasing position and time. Hence, the solution makes sense logically. The next step is to use parameter values that we have utilized before to examine how this solution behaves in z and t. We can anticipate that early on we will see jumps in concentration at the inlet and later smooth decreases in concentration with position and time - which is to say we will see the combination of the HeavisideTheta and the diminishing exponentials.

Clear[ka, ρs, As, vz, z, t, ϵ, cbf, cb, cbo, cstot]

As = $1.2 * 10^7$;

cstot = $1 * 10^{-9}$;

ka = 100;

ρs = 2;

ϵ = .5;

vz = 10;

cbf = $4 * 10^{-4}$;

cbo = 10^{-9};

tmax = 2;

cb[z_, t_]:=

$$e^{\frac{\text{Ascstotka}z(-1+\epsilon)\rho\text{s}}{\text{vz}\epsilon}}\left(\text{cbo}e^{\frac{\text{Ascstotka}(t\text{vz}-z)(-1+\epsilon)\rho\text{s}}{\text{vz}\epsilon}}\left(\text{HeavisideTheta}[t]-\text{HeavisideTheta}\left[t-\frac{z}{\text{vz}}\right]\right)+\text{cbfHeavisideTheta}\left[t-\frac{z}{\text{vz}}\right]\right)$$

Plot3D[

cb[z, t], {z, 0, 20}, {t, 0, tmax}, PlotRange → All,

MaxRecursion → 1,

AxesLabel → { "z, axial dimension", " t, time",

"Cb[z,t] "}]

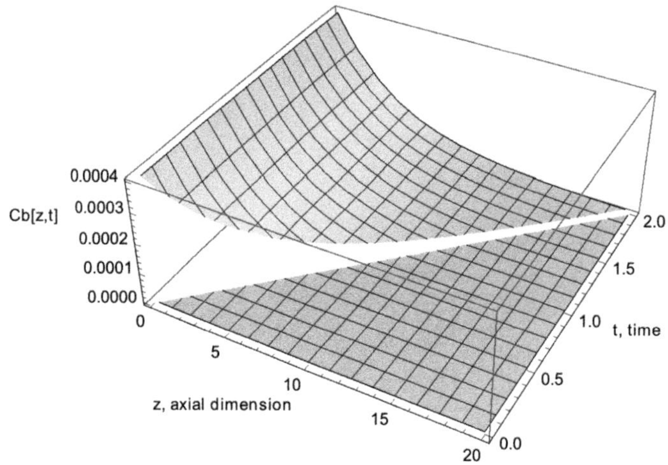

As expected, we see the jump discontinuity, which appears as the white region between the two regions of the graph that has a straight line base from Cb[0, 0] = 0 to Cb[20, 20] = 0 and a curve from just beyond Cb[~0, ~0] = Cbf to Cb[20, 2.0] = 0. Notice that along the left upper edge, the concentration is constant with time and equal to Cbf. This is at the inlet to the unit and is the boundary condition. At longer time, we notice that the concentration is a continuous, downward exponential.

It is best to write some code at this point to create a function with the most important variables using the Module command. Here is that function, which will be called pfrads for plug flow reactor adsorption. (This is not a reactor, but PFR is common terminology.) We make this a function of the adsorption rate constant, the linear velocity and the maximum time.

pfrads[ka_, vz_, tmax_]:=Module[{As $= 1.2 * 10^7$,

cstot $= 1 * 10^{-9}$,

ρ**s** $= 2$,

$\epsilon = .5$,

cbf $= 4 * 10^{-4}$,

cbo $= 10^{-9}$,

cb},

cb[z_, t_]:=

$$e^{\frac{\text{Ascstotka}z(-1+\epsilon)\rho\text{s}}{\text{vz}\epsilon}}$$

$$\left(\text{cbo}e^{\frac{\text{Ascstotka}(t\text{vz}-z)(-1+\epsilon)\rho\text{s}}{\text{vz}\epsilon}}\right.$$

```
(HeavisideTheta[t] − HeavisideTheta[t − z/vz]) +
cbfHeavisideTheta[t − z/vz]);
Plot3D[
cb[z, t], {z, 0, 20}, {t, 0, tmax},
PlotRange → All,
PlotLabel → {ka“ = ka” , vz“ = vz”},
AxesLabel → { “z, axial dimension”, “ t, time”,
“Cb[z,t] ”}]
]
```

We can test this function and see if it does what is expected.

```
pfrads[100, 10, 20]
```

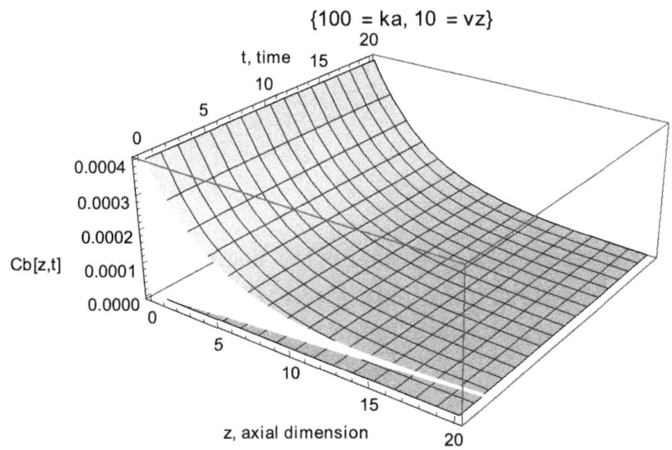

Now, we can go ahead and run a Table of different parameter values to get a fuller sense of the behavior of the system.

```
res = Table[pfrads[ka, vz, tmax],
{ka, { 0.001, 100, 1000}},
{vz, {1., 10, 100}}, {tmax, {100}}];
res[[1, 1, 1]]
res[[2, 1, 1]]
```

res[[3, 1, 1]]

res[[1, 2, 1]]

res[[2, 2, 1]]

res[[3, 2, 1]]

res[[1, 3, 1]]

res[[2, 3, 1]]

res[[3, 3, 1]]

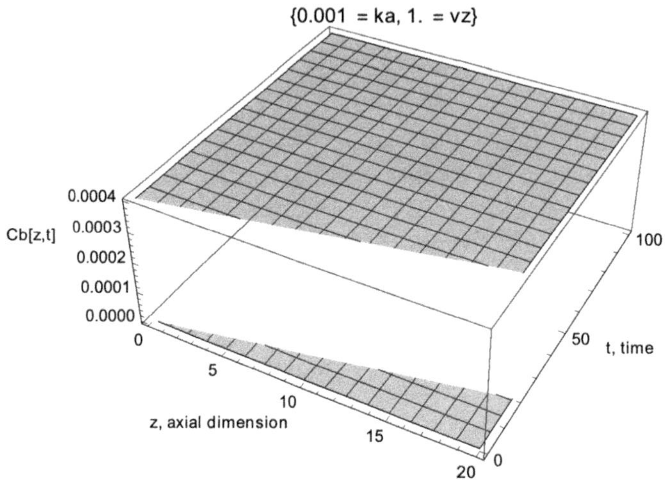

Firstly, in this case the adsorption rate constant is low and so too is the linear velocity. Because the adsorption is not favorable the jump goes to the full feed concentration and remains there. The jump is so pronounced because the linear velocity is low.

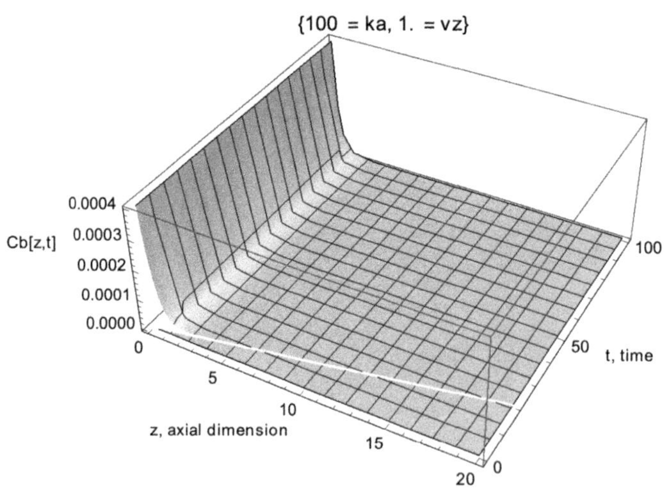

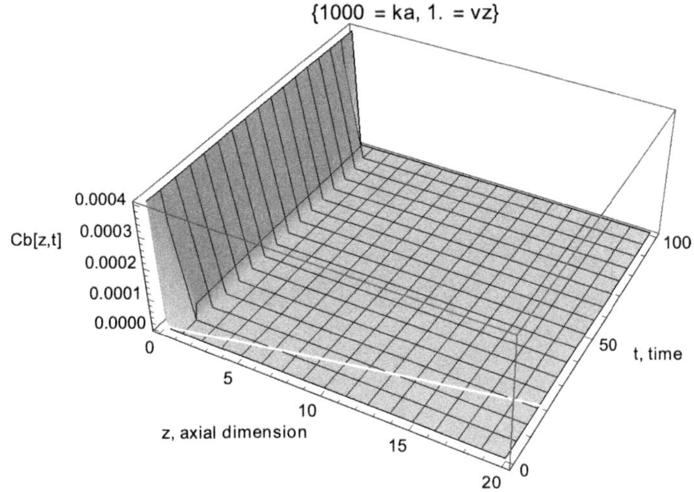

Next, as the adsorption rate constant becomes larger, the adsorption is very pronounced, especially since the linear velocity is low. At such low velocity, the holding time in the unit is quite long and compared to the characteristic time for the adsorption process.

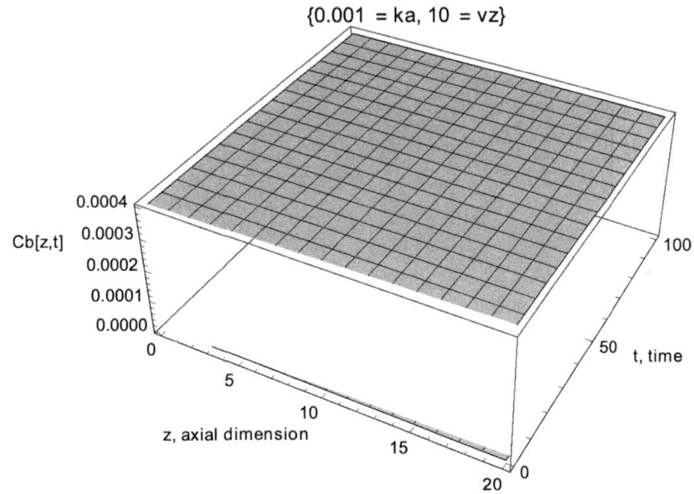

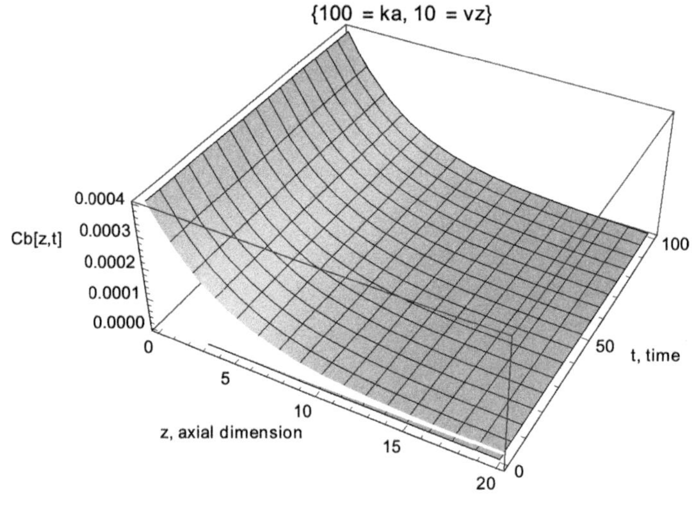

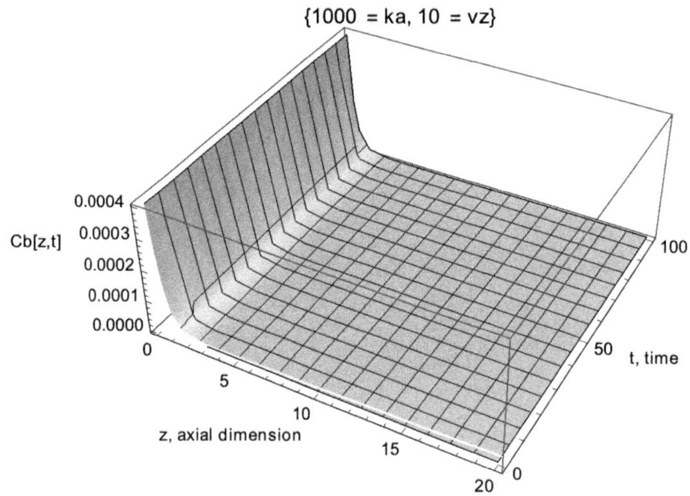

These three graphs show the interplay between the adsorption rate process and rate of flow through the vessel. With a low adsorption rate constant, we again see the jump to the feed concentration, but the whole process ends more quickly. As the adsorption characteristic time becomes shorter with larger ka values the surface goes from curved to waterfall-like.

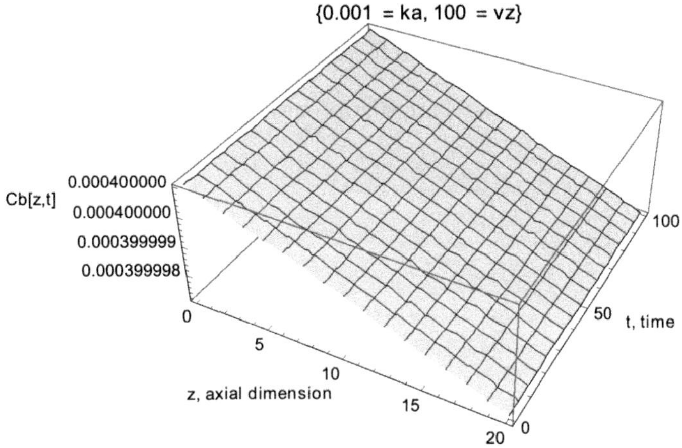
{0.001 = ka, 100 = vz}
Cb[z,t]
0.000400000
0.000400000
0.000399999
0.000399998
0
5
10
15
20
z, axial dimension
0
50
100
t, time

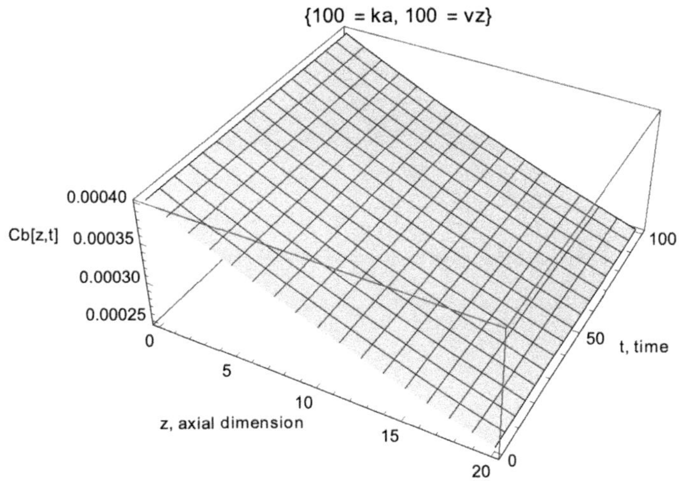
{100 = ka, 100 = vz}
Cb[z,t]
0.00040
0.00035
0.00030
0.00025
0
5
10
15
20
z, axial dimension
0
50
100
t, time

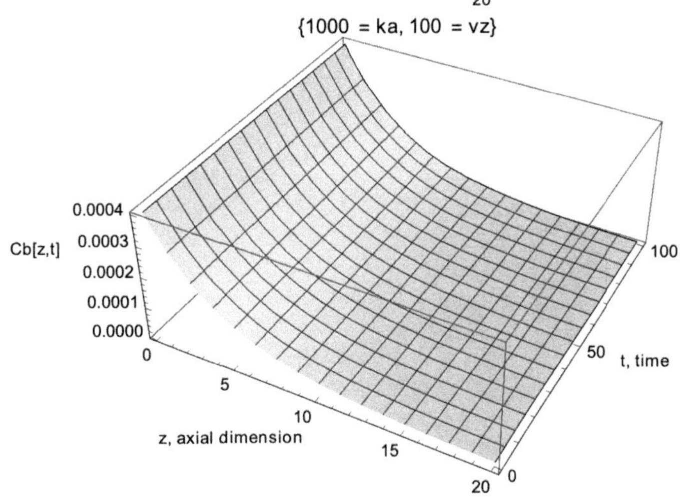
{1000 = ka, 100 = vz}
Cb[z,t]
0.0004
0.0003
0.0002
0.0001
0.0000
0
5
10
15
20
z, axial dimension
0
50
100
t, time

When a large ka, i.e. very short adsorption characteristic times, is combined with high velocity the surface looks curved again, but with lower values of ka it approaches a plane.

This is a very nice system to build a Manipulation around. In this case we can manipulate three of the parameters.

```
Manipulate[
pfrads[ka, vz, tmax],
{ka, 0.01, 100, 1},
{vz, 1., 100, 1},
{tmax, 10, 100, 1}
      ]
```

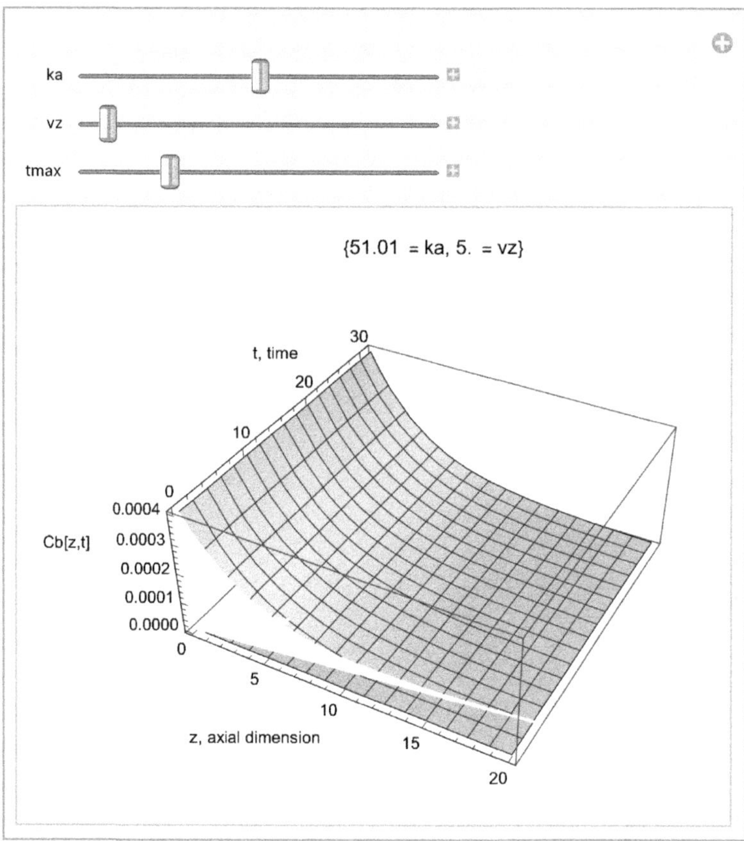

We also should note that with the parameters that we have used, parameters that are of reasonable values, the transients in this system are very short lived and the onset of the steady state is fast. It is no wonder that most treatments of plug flow assume steady state. It is also the case that this was all made possible by utilizing a very simple function for the adsorption rate. Therefore, we should go back now and build a model and simulation with our more realistic Langmuirian adsorption rate expression.

Langmuir and plug flow a numerical approach

The Langmuirian rate expression is:

$$\partial_t \mathrm{cj}[t] == \mathrm{ka}\ \rho\mathrm{j}[t](\mathrm{Cstot} - \mathrm{cj}[t]) - \mathrm{kd}\ \mathrm{cj}[t]$$

Because it involves two concentration variables and their product, it does not lend itself to analytical solutions. So, what we want to do at this point is to combine plug flow with the Langmuirian rate. To do so we will use numerical methods to find the solution.

What led to the partial differential equation in the last section was the mathematical limit we placed on a series of well-mixed control volumes as their axial extent went to zero. This was the diagram that prompted us to take this particular approach:

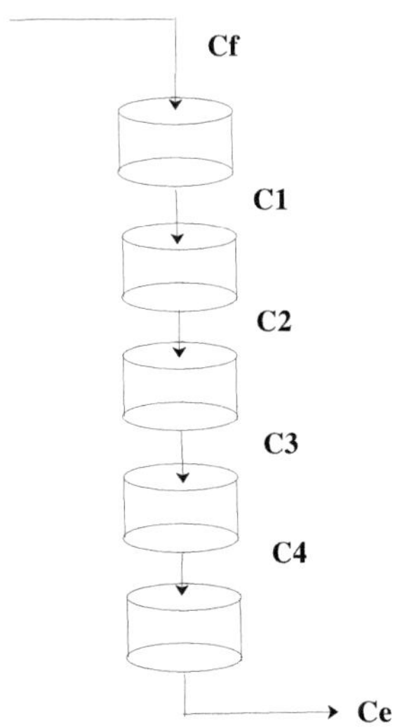

We derived the differential difference equation that described any unit and then limited it to the case of it being vanishingly small axially.

$$\frac{dC_B}{dt} = -\ v_z \frac{\Delta C_B}{\Delta z} - \frac{(1-\epsilon)}{\epsilon} r_{\mathrm{ads}} \rho_s\ A_s$$

$$\lim_{\Delta z \to 0}\left[\frac{dC_B}{dt} = -\, v_z \frac{\Delta C_B}{\Delta z} - \frac{(1-\epsilon)}{\epsilon} r_{\text{ads}} \rho_s\, A_s\right]$$

$$\frac{\partial C_B}{\partial t} = -\, v_z \frac{\partial C_B}{\partial z} - \frac{(1-\epsilon)}{\epsilon} r_{\text{ads}} \rho_s\, A_s$$

In this way, the control volume was well-mixed both in the physical radial direction and in the infinitesimal axial direction. Thus instead of a series of downward steps in concentration, there would be a smooth curvilinear function.

Now, we can go back to the differential difference equation and instead of seeing it as a means to a partial differential equation, we can actually employ it for modeling the system. What we will do is to set up a series of equations that are differential in time, and distributed in space, which is along the axial dimension. The distributed part will be made up of differences rather than a derivative in the axial dimension z. This equation can be used to do this:

$$\frac{dC_B}{dt} = -\, v_z \frac{\Delta C_B}{\Delta z} - \frac{(1-\epsilon)}{\epsilon} r_{\text{ads}} \rho_s\, A_s$$

More explicitly, we will use this form of that equation:

$$\frac{dC_B}{dt} = -\, \frac{\left(C_B - C_{B,\text{in}}\right) q}{\Delta z A_{\text{cr}}} - \frac{(1-\epsilon)}{\epsilon} r_{\text{ads}} \rho_s\, A_s \frac{V}{A_{\text{cr}} \Delta z}$$

Next we will write this equation with the Langmuirian adsorption rate:

$$\frac{dC_{\text{Bg}}}{dt} = \frac{\left(C_{\text{Bgf}} - C_{\text{Bg}}\right) q}{\Delta z A_{\text{cr}}} - \frac{(1-\epsilon)}{\epsilon} \left(\text{ka} C_{\text{Bg}} \left(C_{\text{stot}} - C_{\text{Bs}}\right) - \text{kd} C_{\text{Bs}}\right) \rho_s\, A_s$$

and then express the rate of change in the surface concentration

$$\frac{dC_{\text{Bs}}}{dt} = \left(\text{ka} C_{\text{Bg}} \left(C_{\text{stot}} - C_{\text{Bs}}\right) - \text{kd} C_{\text{Bs}}\right)$$

where,

C_{Bgf} = concentration of B in the gas phase feed stream

C_{Bg} = concentration of B in the gas phase

C_{Bs} = concentration of B on the surface of the adsorbent

We need to write this pair of equations for each control volume and putative "unit," indexing the concentrations for each and letting the exit flow of the nth vessel be the entering flow for the $n^{\text{th}} + 1$ vessel.

Clear[ka, kd, Cg, Cs, Cgo, Cg0, Δz, Acr, *q*, *ρ*s,

As, Cstot , ϵ, nmax]

nmax = 3;

$$\text{Table}\left[\partial_t \text{Cg}[n][t] == \frac{(\text{Cg}[n-1][t] - \text{Cg}[n][t])q}{\Delta\text{z Acr}} - \frac{1}{\epsilon}(1-\epsilon)\rho\text{s As}\right.$$

(ka Cg[*n*][*t*](Cstot − Cs[*n*][*t*]) − kd Cs[*n*][*t*]),

{*n*, 1, nmax}]

Table[

∂*t*Cs[*n*][*t*] == kaCg[*n*][*t*](Cstot − Cs[*n*][*t*])−

kdCs[*n*][*t*], {*n*, 1, nmax}]

$$\left\{\text{Cg}[1]'[t] == \frac{q(\text{Cg}[0][t] - \text{Cg}[1][t])}{\text{Acr}\Delta\text{z}} - \frac{1}{\epsilon}\text{As}(1-\epsilon)\rho\text{s}(\text{kaCg}[1][t](\text{Cstot} - \text{Cs}[1][t]) - \text{kdCs}[1][t]),\right.$$

$$\text{Cg}[2]'[t] == \frac{q(\text{Cg}[1][t] - \text{Cg}[2][t])}{\text{Acr}\Delta\text{z}} - \frac{1}{\epsilon}\text{As}(1-\epsilon)\rho\text{s}(\text{kaCg}[2][t](\text{Cstot} - \text{Cs}[2][t]) - \text{kdCs}[2][t]),$$

$$\left.\text{Cg}[3]'[t] == \frac{q(\text{Cg}[2][t] - \text{Cg}[3][t])}{\text{Acr}\Delta\text{z}} - \frac{1}{\epsilon}\text{As}(1-\epsilon)\rho\text{s}(\text{kaCg}[3][t](\text{Cstot} - \text{Cs}[3][t]) - \text{kdCs}[3][t])\right\}$$

Using the Table function as we have just done is a efficient way to create the set of coupled equations for the gas and adsorbent phases. We are almost ready to solve these equations, but before we do so, let’s set all the parameters, and numerically evaluate all the coefficients in order to see if they are similar in magnitudes or not. If they are, then we expect the numerical solution to operate smoothly. However, if they are not similar in magnitude, then we say the equations are “stiff”, and when that is the case we know that the numerical analysis may be very sensitive to initial conditions and the method of solution.

tmax = 1000;

ρs = 2;

As = $1.2 * 10^7$;

$\epsilon = .5;$

Cstot $= 1 * 10^{-9};$

ka $= .01;$

kd $= 0.001;$

Cgo $= 10^{-}9;$

$q = 1000.;$

Δ**z** $= 633./$**nmax;**

Acr $= 3117.;$

Cg[0][*t*] $= 4 * 10^{-4};$

$q/(\Delta z \text{ Acr});$

$q/(\Delta z\text{Acr})//N;$

nmax $= 1;$

$$\text{Table}\left[\partial_t \text{Cg}[n][t] == \frac{(\text{Cg}[n-1][t] - \text{Cg}[n][t])q}{\Delta z\text{Acr}} - \frac{1}{\epsilon}(1-\epsilon)\rho\text{s As}\right.$$

(kaCg[*n*][*t*](Cstot − Cs[*n*][*t*]) − kdCs[*n*][*t*]),

{*n*, 1, nmax}]//*N*//Simplify

Table[∂_tCs[*n*][*t*] == kaCg[*n*][*t*](Cstot − Cs[*n*][*t*])−

kdCs[*n*][*t*], {*n*, 1, nmax}]//*N*//Simplify

$\{\text{Cg}[1.]'[t] == 2.0273 * 10^{-7} + 24000.\text{Cs}[1.][t] + \text{Cg}[1.][t](-0.000746827 + 240000.\text{Cs}[1.][t])\}$

$\{\text{Cs}[1.]'[t] == \text{Cg}[1.][t](1.0000 * 10^{-11} - 0.01\text{Cs}[1.][t]) - 0.001\text{Cs}[1.][t]\}$

The coefficients are very different - orders of magnitude different. This means that this is a stiff set of equations. We also know from the analytical solution that under some circumstances there are jump discontinuities. So, a solution for these equations may be problematic to implement. Let's see what happens when we have just two equations for one cell.

solns =

NDSolve[

```
{Cg[1]'[t] == 2.0273 * 10^-7 + 24000.Cs[1][t]+ Cg[1][t](-0.000747 + 240000.Cs[1][t]),
Cs[1]'[t] == 1.00 * 10^-11Cg[1][t] - 0.01Cg[1][t]-
0.001Cs[1][t], Cg[1][0] == 0, Cs[1][0] == 0},
{Cg[1][t], Cs[1][t]}, {t, 0, tmax}];
Plot[solns[[1, 1, 2]], {t, 0, tmax}, PlotRange → All]
```

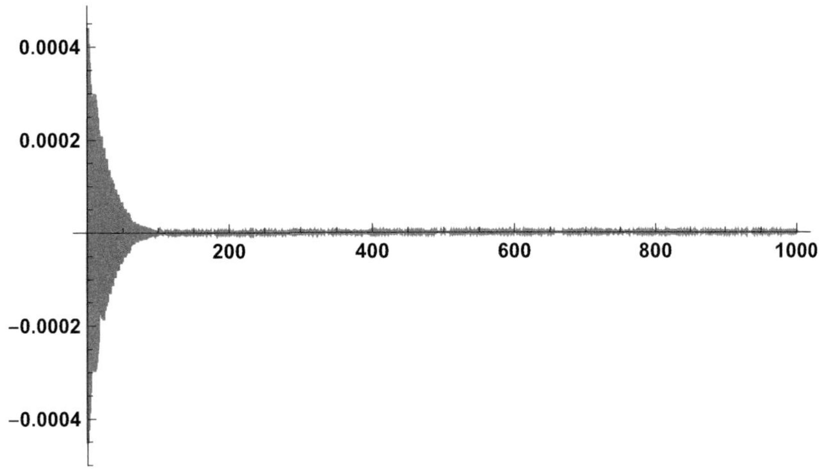

We see numerical instability in the result. We could explore different methods of numerical solution, i.e. extended Runge-Kutta, and others, but we have another way to handle this problem and that is to non-dimensionalize the equations. By non-dimensionalizing, the parameters are lumped and the equations are often less stiff. Also, the values of the dependent variable functions will fall on the domain between zero and one.

We will begin with the equation for the surface concentration in any cell n:

$$\partial_t \mathrm{Cs}[n][t] == \mathrm{ka}(\mathrm{Cstot} - \mathrm{Cg}[n][t])\mathrm{Cg}[1][t] - \mathrm{kdCs}[1][t]$$

In this case we will first divide through by Cstot, that will give us a new variable that we will call the fractional surface concentration,

$$\partial_t \mathrm{frs}[n][t]{==}(\mathrm{ka\ Cg}[1][t](1 - \mathrm{frs}[n][t]) - \mathrm{kd\ frs}[n][t])$$

Next, we convert the gas phase concentration Cg[n][t] in a fractional concentration by multiplying it by unity in the form of $\frac{\mathrm{Cg}[0][t]}{\mathrm{Cg}[0][t]}$, which is the concentration in the supply stream and never changes. This Cg[0][t] is the maximum concentration that the system can attain, hence:

$$\partial_t \text{frs}[n][t]==(\text{ka Cg}[0][t]\text{frg}[n][t](1-\text{frs}[n][t])-\text{kd frs}[n][t])$$

Lastly, instead of keeping time in real units, we will normalize it against the holding time. The holding time, $\theta = \frac{\text{Acr}\Delta z}{q}$ and we will let τ be the dimensionless time unit and it is equal to $\frac{t}{\theta}$:

$$\theta\ \partial_t \text{frs}[n][t] = \partial_\tau \text{frs}[n][\tau]$$

thus

$$\partial_\tau \text{frs}[n][\tau]==(\text{ka}\theta\text{Cg}[0][\tau]\text{frg}[1][\tau](1-\text{frs}[n][\tau])-\text{kd}\theta\text{frs}[n][\tau])$$

This equation is dimensionless. We turn to the gas phase equation. We already know how to transition to frg[n][t], by dividing through term by term with Cg[0] [t], to frs[n][t] by dividing through by Cstot and to τ by multiplying through by θ and this gives us:

$$\partial_\tau \text{frg}[n][\tau]==(1-\text{frg}[n][\tau])-\left(\frac{1-\epsilon}{\epsilon}\right)(\text{ka}\ \theta\rho\text{s As Cstot frg}[n][\tau](1-\text{frs}[n][\tau]) -\frac{\text{kd}\ \theta\rho\text{s As Cstot}}{\text{Cg}[0][t]}\text{frs}[n][\tau]\Big)$$

These are the non-dimensionalized versions of the gas and surface fractional concentrations.

We set up the problem as we did before, examining the results in one cell only first to see how the coefficients look and how the equations behave. Here are the parameter values again:

```
Clear[ka, kd, Cg, Cs, Cgo, Cg0, Δz, Acr, q, ρs,
As, Cstot , ϵ, nmax, solns, frg, frs]
tmax = 100;
nmax = 1;
ρs = 2;
As = 1.2 * 10^7;
ϵ = .5;
Cstot = 10^-9;
ka = 100;
```

kd = 0.001;

$q = 1.;$

Δz = 633./nmax;

Acr = 3160.;

Cg[0][t] = 4 * 10^−4;

Next, we can compute the groups that make up the coefficients of the dependent variable functions. It's the case that all four of these are dimensionless groups. Let's see how they look in terms of magnitudes:

$\Theta = \frac{\text{Acr } \Delta z}{q}$ **(* Capital theta is the overall holding time in the unit *)**

$\theta = \frac{\text{Acr}\Delta z}{q \text{ nmax}}$ **(* small theta is the holding time in each cell or subdivision of the unit as a whole *)**

ka θ ρs As Cstot

$\frac{\text{kd } \theta \rho \text{s As Cstot}}{\text{Cg[0][}t\text{]}}$

kaθ Cg[0][t]

kdθ

2.00028×10^6

2.00028×10^6

4.80067×10^6

120017.

80011.2

2000.28

These groups vary by about three orders of magnitude, but that is much less than the coefficients of the equations in dimensional form. This is encouraging. What happens when we try to solve the equations?

eqns =

{∂_tfrg[1][t]==(1 − frg[1][t])−

$\left(\frac{1-\epsilon}{\epsilon}\right)$ (ka$\theta\rho$sAsCstot frg[1][t](1 − frs[1][t])− $\frac{\text{kd}\ \theta\rho\text{s As Cstot}}{\text{Cg[0][t]}}$ frs[1][t]),

∂_tfrs[1][t]== (ka θ Cg[0][t]frg[1][t](1 − frs[1][t])− kd θfrs[1][t]),

frg[1][0] == 0, frs[1][0] == 0}

vars = {frg[1][t], frs[1][t]}

solns = NDSolve[eqns, vars, {t, 0, tmax}]

Plot[solns[[1, 1, 2]], {t, 0, tmax}, PlotRange → All,

AxesLabel → {t, frg[1]}]

Plot[solns[[1, 2, 2]], {t, 0, tmax}, PlotRange → All,

AxesLabel → {t, frs[1]}, PlotStyle → Hue[.0]]

{frg[1]′[t] == 1 − frg[1][t] − 1.(4.80067 × 10^6frg[1][t](1 − frs[1][t]) − 120017.frs[1][t]),
frs[1]′[t] == 80011.2frg[1][t](1 − frs[1][t]) − 2000.28frs[1][t], frg[1][0] == 0,
frs[1][0] == 0}

{frg[1][t], frs[1][t]}

{{frg[1][t] → InterpolatingFunction[□] [t], frs[1][t] → InterpolatingFunction[□] [t]}}

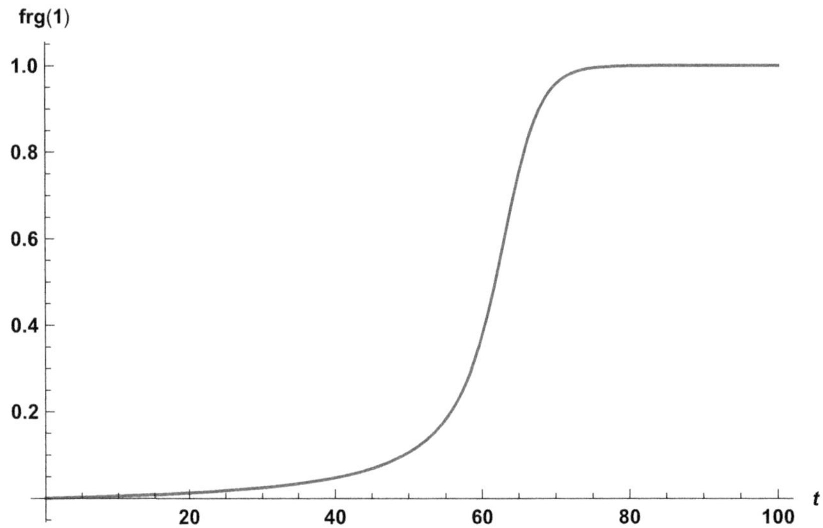

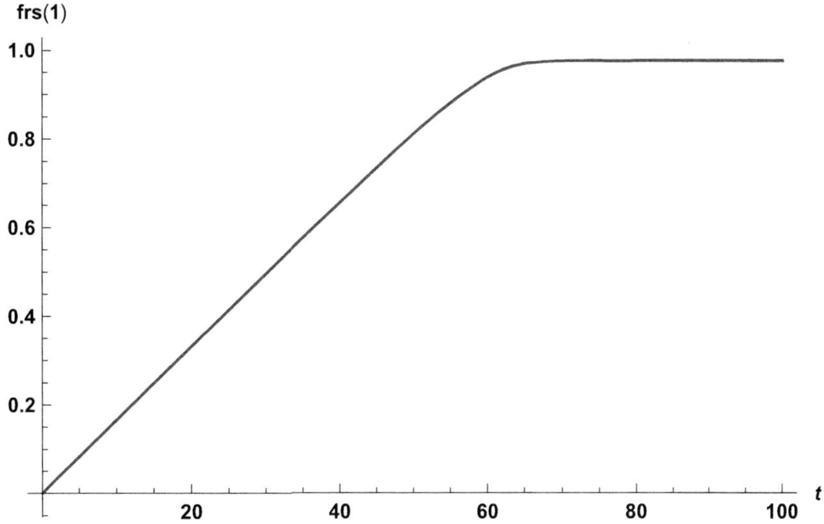

These look as we would expect them to look, it's a good result. Remember the Heaviside functions in the analytical solution? Look at the step in this solution for frg[1]. It is as if we are looking at one spatial slice from that analytical solution. For this choice of adsorption and desorption values the fractional surface concentration rises almost linearly to its maximum value. This is all possible because we non-dimensionalized the equations. And, for our effort, we not only got more soluble equations, we also obtained our first set of dimensionless numbers.

Now, we can redo the code just a little bit, use Table to generate the equations and initial conditions and then look at the breaking this unit up into n cells where n gets large.

```
Clear[ka, kd, Cg, Cs, Cgo, Cg0, Δz, Acr, q, ρs,
As, Cstot , ϵ, nmax, solns, frg, frs]
tmax = 750;
nmax = 10;
ρs = 2;
As = 1.210^7;
ϵ = .5;
Cstot = 10^-9;
ka = 100;
```

kd = 0.001;

$q = 1.;$

Δz = 633./nmax;

Acr = 3160.;

Cg[0][t] = 4 10^{-4};

$$\theta = \frac{\text{Acr } \Delta\text{z}}{q \text{ nmax}};$$

frg[0][t] = 1;

des = Table$\big[\{\partial_t$frg[n][t] == (frg[n − 1][t] − frg[n][t])−

$$\left(\frac{1-\epsilon}{\epsilon}\right)(\text{ka}\theta\rho\text{sAsCstotfrg}[n][t](1-\text{frs}[n][t]) - \frac{\text{kd}\theta\rho\text{sAsCstot}}{\text{Cg}[0][t]}\text{frs}[n][t]\right),$$

∂_tfrs[n][t]==

(kaθ Cg[0][t]frg[n][t](1 − frs[n][t])−

kdθfrs[n][t]),

frg[n][0] == 0, frs[n][0] == 0}, {n, 1, nmax}];

vars = Flatten[Table[{frg[n][t], frs[n][t]},

{n, 1, nmax}]];

cstrs = Flatten[des];

solns = NDSolve[cstrs, vars, {t, 0, tmax}];

gfr = Table[Plot[solns[[1, n, 2]], {t, 0, tmax},

PlotRange → {{0, tmax}, {0, 1.1}},

AxesLabel → {t, frg[n]}], {n, 1, 2nmax, 2}]

sfr = Table[Plot[solns[[1, n, 2]], {t, 0, tmax},

PlotRange → {{0, tmax}, {0, 1.1}},

AxesLabel → {t, frs[n]}, PlotStyle → Hue[0.1]],

{n, 2, 2nmax, 2}]

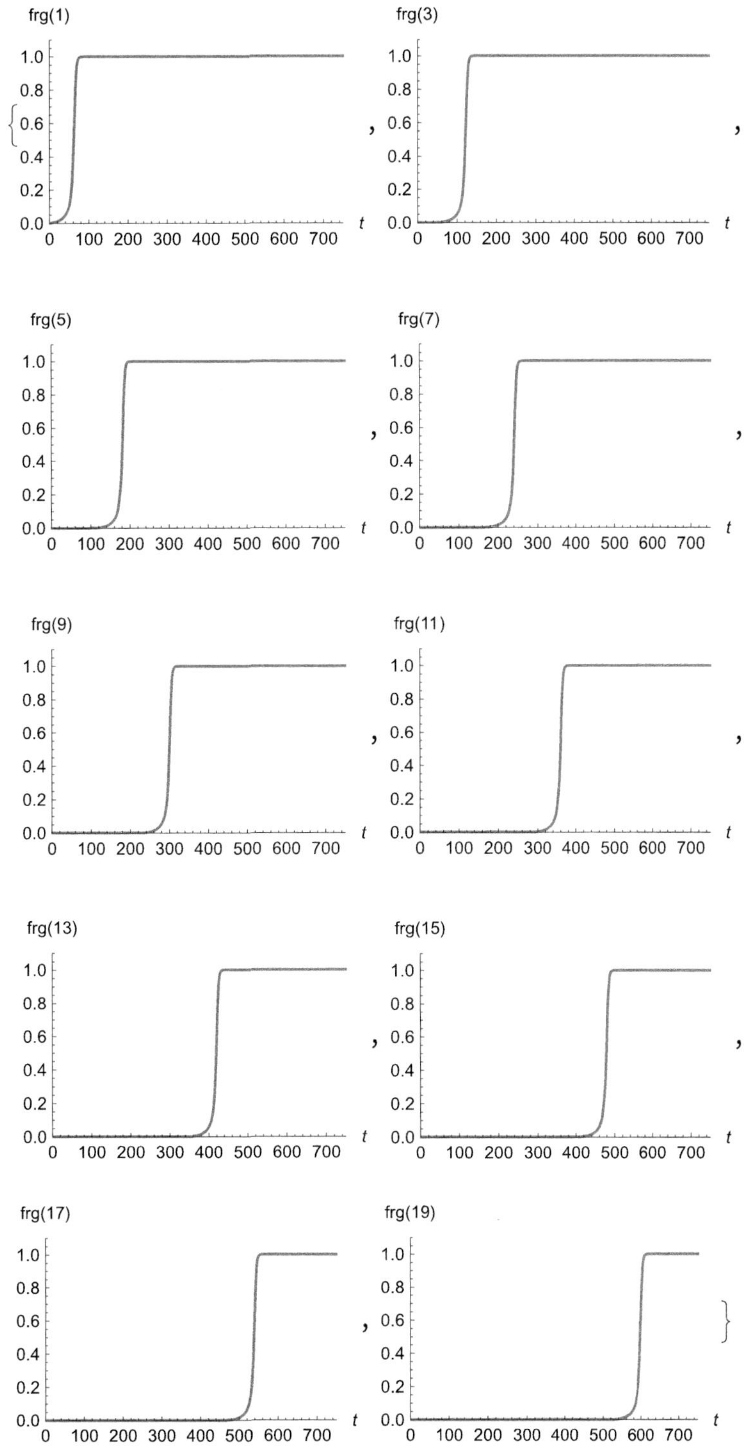
frg(1)
frg(3)
frg(5)
frg(7)
frg(9)
frg(11)
frg(13)
frg(15)
frg(17)
frg(19)
t

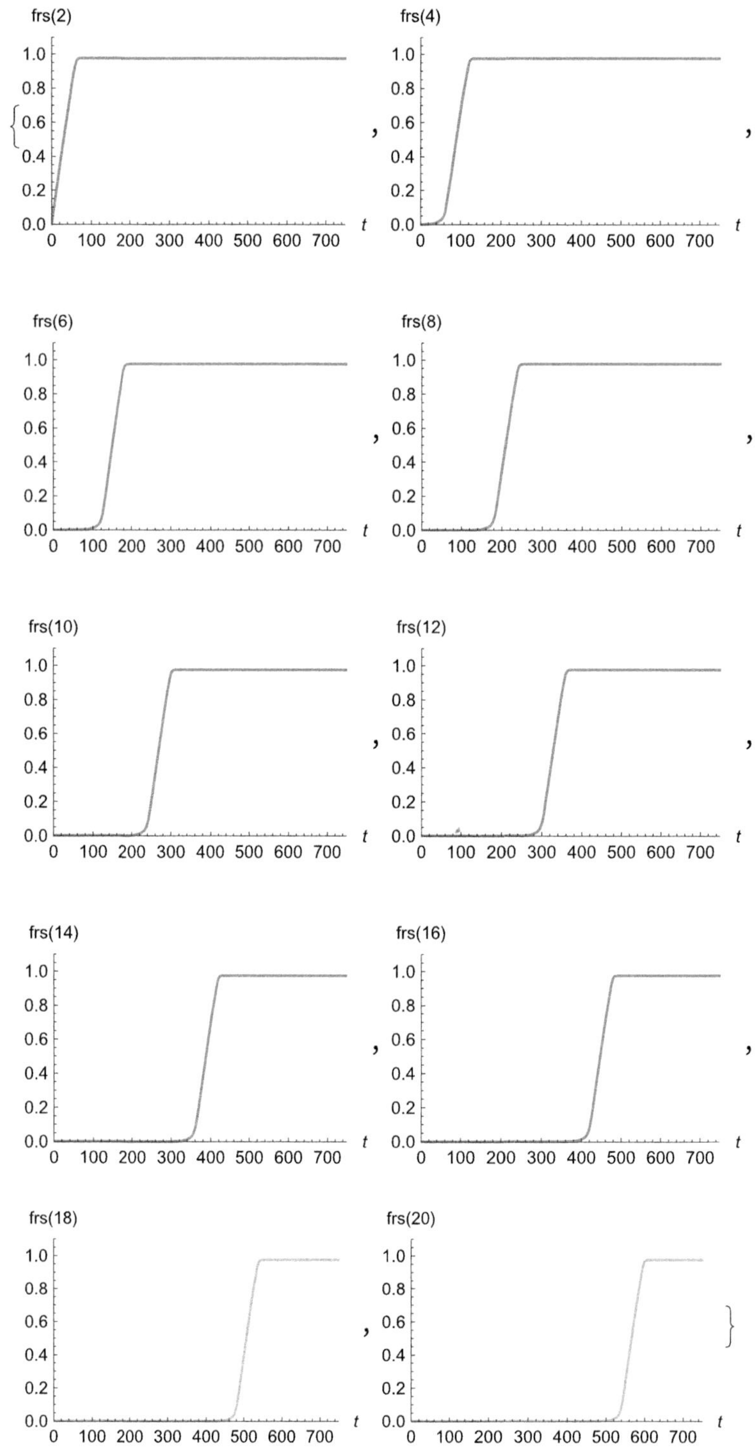
frs(2)
frs(4)
frs(6)
frs(8)
frs(10)
frs(12)
frs(14)
frs(16)
frs(18)
frs(20)
t

These graphics are perfectly acceptable as a means to understanding the dynamics. But, a more elegant way to visualize this information is to put each of the results in a 3D grid that shows the results simultaneously in position and time. We can do that by using ParametricPlot3D, wrapped a round a table function that creates each plot in time at a different value of z, i.e. at a different position:

```
Table[{nΔz, t, First[frg[n][t]/.solns]},
{n, 1, nmax}]

ParametricPlot3D[
Evaluate[
Table[{t/tmax, nΔz/633, First[frg[n][t]/.solns]},
{n, 1, 2nmax, 1}]], {t, 0, tmax},
PlotRange → {{0, tmax/tmax}, {0, nmaxΔz/633},
{0, 1.1}}, AxesLabel → {t, z, "frg[z,t]"},
PlotStyle → Hue[{t/tmax}//N]]
```

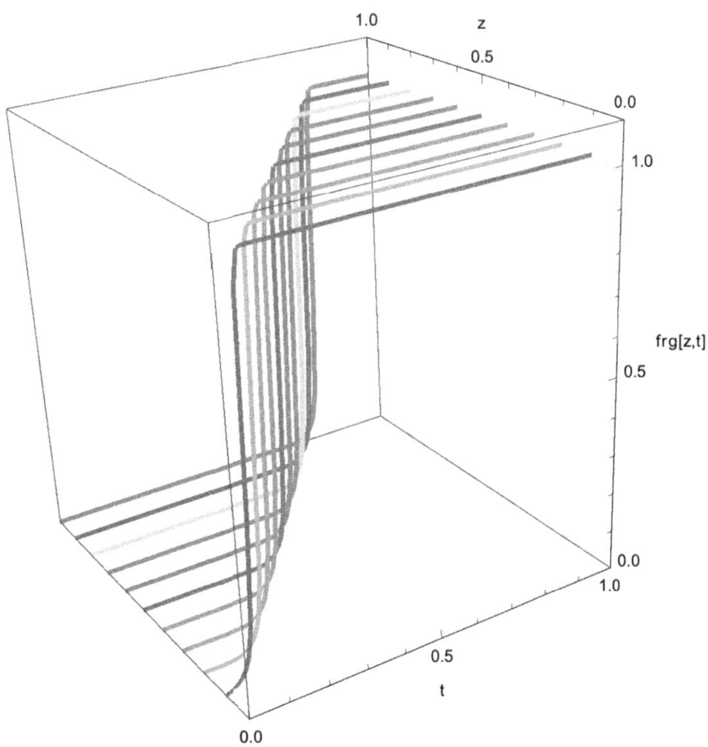

In that example, we set the value of the volumetric flow rate to 1, if we want to know if the analysis is stable, we can test it with a much larger value of q. Here we use a value of 40,000:

Clear[ka, kd, Cg, Cs, Cgo, Cg0, Δz, Acr, q, ρs,

As, Cstot , ϵ, nmax, solns, frg, frs]

tmax = 750;

nmax = 10;

ρs = 2;

As = 1.210^7;

ϵ = .5;

Cstot = 10^{-9};

ka = 100;

kd = 0.001;

q = 40000.;

Δz = 633./nmax;

Acr = 3160.;

Cg[0][t] = 4 10^{-4};

$$\theta = \frac{\text{Acr } \Delta z}{q \text{ nmax}}$$

frg[0][t] = 1;

des = Table[{∂_tfrg[n][t] == (frg[n − 1][t] − frg[n][t])−

$$\left(\frac{1-\epsilon}{\epsilon}\right)\left(\text{ka}\theta\rho\text{sAsCstotfrg}[n][t](1-\text{frs}[n][t]) - \frac{\text{kd}\theta\rho\text{sAsCstot}}{\text{Cg}[0][t]}\text{frs}[n][t]\right),$$

∂_tfrs[n][t]== (kaθ Cg[0][t]frg[n][t](1 − frs[n][t])− kdθfrs[n][t]),

frg[n][0] == 0, frs[n][0] == 0}, {n, 1, nmax}];

vars = Flatten[Table[{frg[n][t], frs[n][t]},

{n, 1, nmax}]];

cstrs = Flatten[des];

solns = NDSolve[cstrs, vars, {t, 0, tmax}];

gfr = Table[Plot[solns[[1, n, 2]], {t, 0, tmax},

```
PlotRange → {{0, tmax}, {0, 1.1}},
AxesLabel → {t, frg[n]}], {n, 1, 2nmax, 2}];
sfr = Table[Plot[solns[[1, n, 2]], {t, 0, tmax},
PlotRange → {{0, tmax}, {0, 1.1}},
AxesLabel → {t, frs[n]}, PlotStyle → Hue[0.1]],
{n, 2, 2nmax, 2}];
ParametricPlot3D[
Evaluate[
Table[{t/tmax, nΔz/633, First[frg[n][t]/.solns]},
{n, 1, 2nmax, 1}]], {t, 0, tmax},
PlotRange → {{0, tmax/tmax}, {0, nmaxΔz/633},
{0, 1.1}}, AxesLabel → {t, z, "frg[z,t]"},
PlotStyle → Hue[{t/tmax}//N]]
```

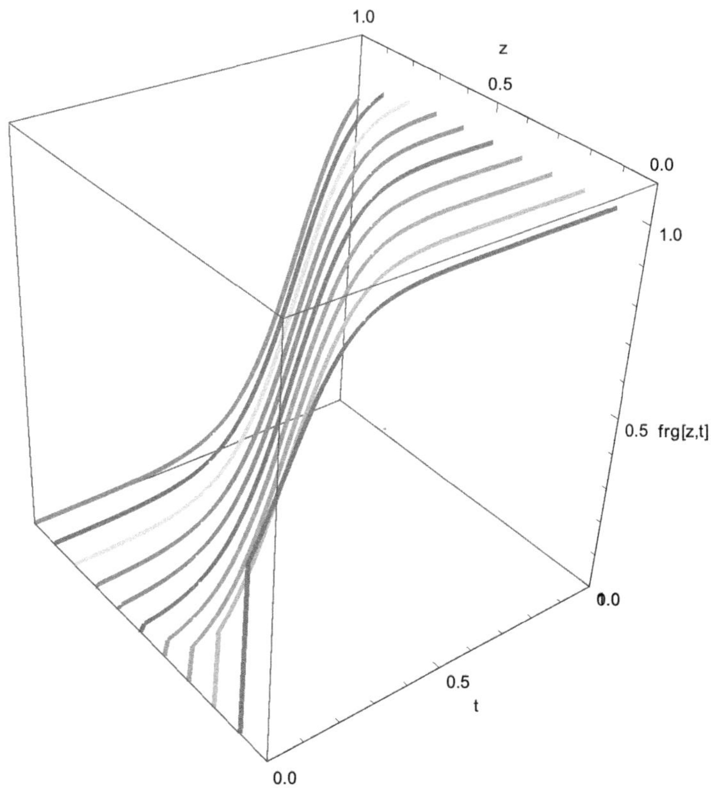

The solutions are fine, and we see the vertical jumps in concentration at very short time in the first five cells. Despite the jump, the results appear stable and physical.

We have covered adsorption in some detail because it is an important process, and understanding the fundamentals is basic and necessary. It is also a preparation for chemical reaction kinetics. There is far more that can be done to build much more realistic models of actual adsorber units. For example beyond the plug flow approach lies the next more sophisticated method called dispersion modeling, which takes into account the radial distribution profile. We also have not included mass transfer effects from the gas to the adsorbent, or diffusion within the adsorbent. There are also in many adsorber units heat of adsorption factors that dominate the dynamics. An example is the adsorptive production of oxygen from air over a lithium-exchanged zeolite. The interaction of nitrogen's quadrupole with the lithium ion is strong enough to set off a heat transfer wave that travels down the bed preceding the mass transfer wave, and raising the temperature along the way. We have literally only touched on the most elementary of concepts here. A full simulation of a adsorber might include a model for the combined mass, heat and fluid dynamical effects, all of which are beyond the scope of this elementary introduction, but are what make chemical engineering models and simulations so rich in physics.

Permeation

Permeation is somewhat like adsorption in that no chemical reaction takes place and because adsorption is a step in the permeation process. It is fundamentally different in that the adsorbed species moves through a membrane or barrier and exits on a different side from where it started. It is a process by which mass is transferred by a linear driving force from a region of higher to lower concentration, or pressure, and, more generally, from higher to lower chemical potential.

The membrane may be polymeric, metallic, or ceramic. Our bodies and all living organisms use membranes as critical structural components of cells. Separating the inside from the outside of the cell provides the cell with its own integrity and, ultimately, specialization. Only certain molecules and ions are allowed to move across these natural membranes, rendering them in some cases highly selective.

Synthetic membranes seek to emulate, or mimic, their natural counterparts, but with much simpler structures, and mechanisms of operation for much less complex separations. Polymeric membranes can be obtained that will separate molecules on the basis of their relative affinities for the interior of the membrane. Those molecules with higher affinities partition themselves to a larger degree into the membrane. With higher concentrations within, they also transport across the membranes faster. A classic example of this is the membrane that is used for hemodialysis. Rendered incapable of clearing the blood of toxins, the patient with renal dysfunction, can be "dialyzed" by passing their blood continuously through a highly specialized membrane unit. The polymers making up the membranes transport these toxins to a dialysate solution in which they are very soluble, and they move from a region of higher chemical potential to one that is lower. Thereby the blood is returned to the patient in refreshed a state.

Ceramic and metallic membranes hold the promise of conducting small molecule separations continuously and possibly with less energy than required by other processes. Ceramic membranes, typically made up of representative and transition metal oxides, offer the opportunity to operate at elevated temperatures. Recent work has focused on using such membranes as a part of a chemical reactor, thus offering enhanced conversions at the same conditions. For example, by selectively transporting one product away from the reaction zone continuously, this partially open system can move to a steady state that is different than that imposed by the equilibrium for a either a closed system, or its fully open equivalent. Metallic membranes of palladium and its alloys are special in that they transport hydrogen and only hydrogen. This makes them particularly interesting for hydrogen purification, recovery and use. They may also play a role in fuel cells. As the world seeks to move away from fossil fuels, hydrogen is interesting because the by-product of its combustion is merely water vapor. Eventually, hydrogen as a fuel, may be used in commercial aviation, and possibly for public transportation. Most hydrogen today is still derived from fossil fuels in processes such as methane reforming, hence, there is still co-production of carbon dioxide. The challenge is to produce hydrogen by other means, such as the electrolysis of water using solar energy or wind energy to drive that reaction.

Before we can begin to work with membranes we must know how to analyze their fundamental behavior and that is the goal of this section of the chapter.

Permeation as adsorption and diffusion

Batch

Permeation involves the transport of molecules across a membrane. The transport process involves either dissolution, or adsorption of the molecule being transported within the substance of the membrane. The driving force for transport is provided by moving the molecule from regions of higher chemical potential to lower chemical potential. In its simplest form this means going from a region of higher concentration toward one of lower concentration.

Permeation is the global process of transport across a membrane, given in terms of the measurable concentrations (or pressures) in the bulk above and below the membrane. Transport within the membrane is diffusion. Partitioning and dissolution are the processes that take a molecule like hydrogen, or oxygen from the bulk gas phase to the surface of the membrane and then within it. At the opposite, low potential side the reverse process takes place.

A flux, **J**, is a measure of the rate of permeation across the membrane and it has units of mass (or moles) per unit time per unit area. The permeation rate is found to be proportional to the difference between the concentrations on either side of the membrane. The proportionality constant is called the permeability, $\boldsymbol{P_m}$, with intrinsic dimensions of $\frac{\textbf{Length}}{\textbf{time}}$, the same as the mass transfer coefficient and the same as velocity.

$$J = P_m \left(C^I - C^{II} \right)$$

Higher fluxes come with higher permeabilities, concentrations, or pressures. The high concentration (chemical potential) side of the membrane, from which mass typically flows, is the termed the retentate, while that side to which mass flows is the permeate.

At relatively typical pressures and concentrations, the permeability is the product of two terms: the adsorption constant (or partition coefficient) and the diffusivity:

$$P_m = K\,D \quad (K = Kd \text{ or } Kads \ldots)$$

Consider the following diagram for a batch permeation.

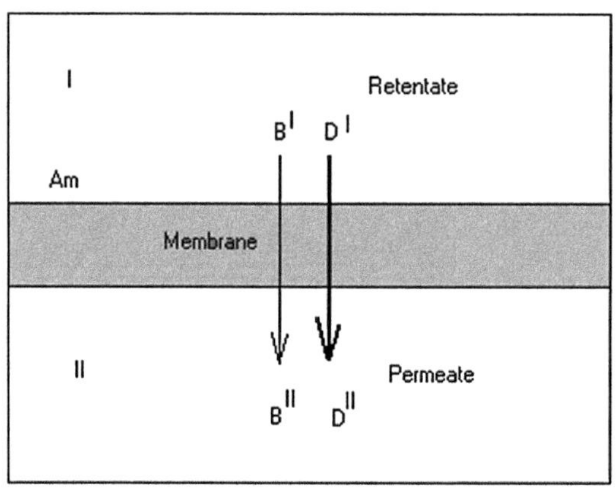

The concentrations of species **B** and **D** are given as $\boldsymbol{C_B^I}$ and $\boldsymbol{C_D^I}$ on the retentate side and as $\boldsymbol{C_B^{II}}$ and $\boldsymbol{C_D^{II}}$ on the permeate side of the membrane. The permeation process is considered to take place at fixed temperature. The membrane has an area **Am** through which the flux is measured. The volumes of the two compartments are $\boldsymbol{V^I}$ and $\boldsymbol{V^{II}}$. Each molecule will have its own permeability and we will assume that they permeate independently of one another. The material balance equations are:

$$\frac{dC_B^I V^I}{dt} = - P_{m,B}\,\mathrm{Am}\,(C_B^I - C_B^{II})$$

$$\frac{dC_D^I V^I}{dt} = - P_{m,D}\,\mathrm{Am}\,(C_D^I - C_D^{II})$$

$$\frac{dC_B^{II} V^{II}}{dt} = P_{m,B}\,\mathrm{Am}\,(C_B^I - C_B^{II})$$

$$\frac{dC_D^{II} V^{II}}{dt} = P_{m,D}\,\mathrm{Am}\,(C_D^I - C_D^{II})$$

There are four equations that describe the system. We see that independent of which side of the membrane is the higher concentration side, these equations still work, as they must if they are to be valid. The concentrations could be in mass per volume or in moles per volume, we will assume the former. If **B** and **D** were gases we could express these equations in terms of the pressure. In all four equations the right-hand side is just the flux of the component times the area of the membrane. The volumes of the compartments are constant, so they can be brought to the right-hand sides:

$$\frac{dC_B^I}{dt} = - \frac{P_{m,B}\mathrm{Am}\left(C_B^I - C_B^{II}\right)}{V^I}$$

$$\frac{dC_D^I}{dt} = - \frac{P_{m,D}\mathrm{Am}\left(C_D^I - C_D^{II}\right)}{V^I}$$

$$\frac{dC_B^{II}}{dt} = \frac{P_{m,B}\mathrm{Am}\left(C_B^I - C_B^{II}\right)}{V^{II}}$$

$$\frac{dC_D^{II}}{dt} = \frac{P_{m,D}\mathrm{Am}\left(C_D^I - C_D^{II}\right)}{V^{II}}$$

This rearrangement provides a time constant that is useful; it is the ratio of $\frac{\mathbf{PmAm}}{\boldsymbol{V}}$, which has dimensions of reciprocal time. Therefore the reciprocal of this group is a time constant or *characteristic time*. Notice that $\frac{\boldsymbol{V}}{\mathbf{Am}}$ is a *characteristic length* and $\frac{\mathbf{1}}{\mathbf{Pm}}$ is a time per length, so their product is a time.

We can solve these equations analytically. For the case in which the permeate side has zero concentrations of **B** and **D** initially and the retentate side is charged with non-zero initial concentrations of **B** and **D**:

Clear[PmB, PmA, Am, VI, VII, C1Bo, C1Do, plI, plII,

plIII, plIV]

VI = VII;

memsol =

Flatten[

Simplify[

DSolve[

$$\left\{\partial_t \mathrm{C1B}[t] == -\frac{\mathrm{PmBAm}}{\mathrm{VI}}(\mathrm{C1B}[t] - \mathrm{C2B}[t]),\right.$$

$$\partial_t \mathrm{C1D}[t] == -\frac{\mathrm{PmDAm}}{\mathrm{VI}}(\mathrm{C1D}[t] - \mathrm{C2D}[t]),$$

$$\partial_t \mathrm{C2B}[t] == +\frac{\mathrm{PmBAm}}{\mathrm{VII}}(\mathrm{C1B}[t] - \mathrm{C2B}[t]),$$

$$\partial_t \mathrm{C2D}[t] == +\frac{\mathrm{PmDAm}}{\mathrm{VII}}(\mathrm{C1D}[t] - \mathrm{C2D}[t]),$$

C1B[0] == C1Bo, C1D[0]==C1Do,

C2B[0] == 0, C2D[0] == 0},

{C1B[*t*], C2B[*t*], C1D[*t*], C2D[*t*]},*t*]

]

];

CIB[t_]:=Evaluate[C1B[*t*]/.memsol[[1]]]

CIIB[t_]:=Evaluate[C2B[*t*]/.memsol[[2]]]

CID[t_]:=Evaluate[C1D[*t*]/.memsol[[3]]]

CIID[t_]:=Evaluate[C2D[*t*]/.memsol[[4]]]

$$C1Bo = 1;$$

$$C1Do = 1;$$

$$PmB = 10^{-4};$$

$$PmD = 10^{-6};$$

$$VII = 1;$$

$$Am = 10;$$

$$plI = Plot\left[CIB[t], \left\{t, 0, \frac{VI}{PmBAm}\right\},\right.$$

DisplayFunction → Identity,

AxesLabel → {t, "C1B[t]"},

PlotStyle → Hue[0], PlotRange → All];

$$plII = Plot\left[CID[t], \left\{t, 0, \frac{VI}{PmBAm}\right\},\right.$$

DisplayFunction → Identity,

AxesLabel → {t, "C1D[t]"},

PlotStyle → Hue[0.8], PlotRange → All];

$$plIII = Plot\left[CIIB[t], \left\{t, 0, \frac{VI}{PmBAm}\right\},\right.$$

DisplayFunction → Identity,

AxesLabel → {t, "C2B[t]"},

PlotStyle → {Hue[0],

Dashing[{0.15, 0.05}]}, PlotRange → All];

$$plIV = Plot\left[CIID[t], \left\{t, 0, \frac{VI}{PmBAm}\right\},\right.$$

DisplayFunction → Identity,

AxesLabel → {t, "C2D[t]"},

PlotStyle → {Hue[0.8],

Dashing[{0.15, 0.05}]}, PlotRange → All];

Show[GraphicsArray[{{plI, plII}, {plIII, plIV}}]]

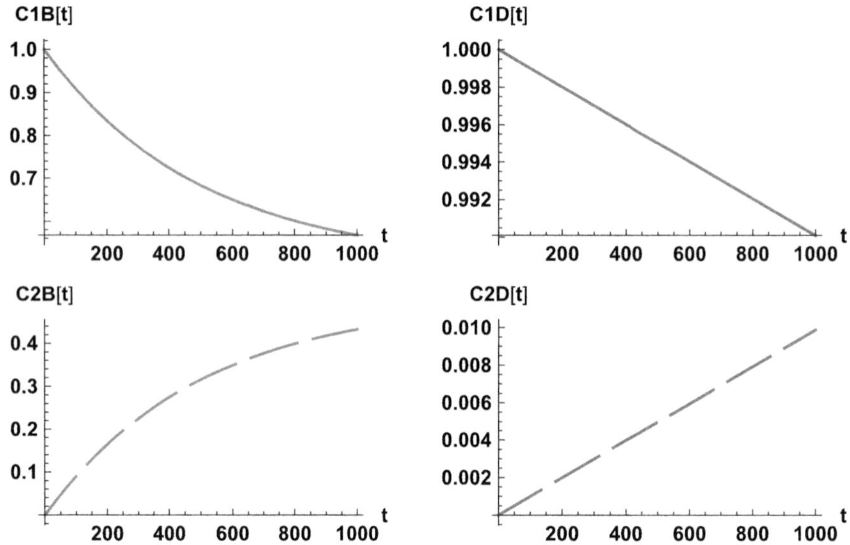

We see that over the period of time equal to $\frac{\mathbf{Vl}}{\mathbf{PmBAm}}$, the concentration of **B** has fallen sharply on the retentate side and has risen as sharply on the permeate side. The two sides are almost at equilibrium with respect to species **B**, with each cell going to 0.5 concentration units. The other species, **D**, has barely begun to transfer across the membrane. It takes on the order of 10^2 times longer to get to equilibrium. Were this kind of separation achievable, the membrane's selectivity would be quite high, since there is so little **D** on the permeate side compared to **B**. It would be very nice to try this continuously to see how well the system would work.

Continuous permeation

The continuous process would have to have feed and exit on the retentate side and at least exit flow on the permeate side. We could have an additional "sweep" flow on the permeate side, but that adds very little to the analysis. The new physical situation is as follows:

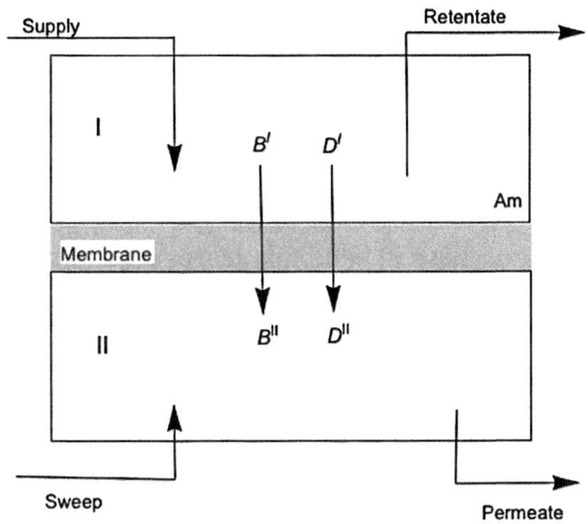

The material balance equations for components **B** and **D** on the retained and permeate sides become:

$$\frac{dC_B^I V^I}{dt} = (C_{\mathrm{Bf}}^I - C_B^I)\, q^I - P_{m,B}\,\mathrm{Am}\,(C_B^I - C_B^{\mathrm{II}})$$

$$\frac{dC_D^I V}{dt} = (C_{\mathrm{Df}}^I - C_D^I)\, q^I - P_{m,D}\,\mathrm{Am}\,(C_D^I - C_D^{\mathrm{II}})$$

$$\frac{dC_B^{\mathrm{II}} V^{\mathrm{II}}}{dt} = -C_B^{\mathrm{II}}\, q^{\mathrm{II}} + P_{m,B}\,\mathrm{Am}\,(C_B^I - C_B^{\mathrm{II}})$$

$$\frac{dC_D^{\mathrm{II}} V^{\mathrm{II}}}{dt} = -C_D^{\mathrm{II}}\, q^{\mathrm{II}} + P_{m,D}\,\mathrm{Am}\,(C_D^I - C_D^{\mathrm{II}})$$

When the system runs at steady state, the derivatives are identically zero. We can divide both sides of both of the equations by the volumes to give:

$$0 = (C_{\mathrm{Bf}}^I - C_B^I)\,\frac{q^I}{V^I} - \frac{P_{m,B}\mathrm{Am}}{V^I}\,(C_B^I - C_B^{\mathrm{II}})$$

$$0 = (C_{\mathrm{Df}}^I - C_D^I)\,\frac{q^I}{V^I} - \frac{P_{m,D}\mathrm{Am}}{V^I}\,(C_D^I - C_D^{\mathrm{II}})$$

$$0 = -C_B^{\mathrm{II}}\,\frac{q^{\mathrm{II}}}{V^{\mathrm{II}}} + \frac{P_{m,B}\mathrm{Am}}{V^{\mathrm{II}}}\,(C_B^I - C_B^{\mathrm{II}})$$

$$0 = -C_D^{\mathrm{II}}\,\frac{q^{\mathrm{II}}}{V^{\mathrm{II}}} + \frac{P_{m,D}\mathrm{Am}}{V^{\mathrm{II}}}\,(C_D^I - C_D^{\mathrm{II}})$$

There are now two characteristic times in the equations, the first is the holding time, $\left(\frac{q}{V}\right)^{-1}$ and the second is the permeation time $\left(\frac{P_m\mathrm{Am}}{V}\right)^{-1}$. We have four equations, two inlet concentrations, four outlet concentrations, two flow rates, two volumes, two permeabilities, and one area for a total of 13 variables and parameters. If we know the two inlet concentrations of **B** and **D**, their two permeabilities, the two volumes, and the area of the membrane, the retentate flow rate, then we have eight of these in hand that leaves four to be calculated, if we measure one. Let's say we measure the permeate flow rate. Then we should be able to compute the four exit concentrations:

```
(* Clear all variable and parameter names *)

Clear[PmB, PmD, Am, VI, VII, qI, qII, CB1ss, CD1ss,

CB2ss, CD2ss, solmem, CBIf]


VI = VII;

PmB = 10^-4;

PmD = 10^-6;

VII = 1;

x = 5; (* Parameter values *)

Am = 10^x;

CBIf = .1;

CDIf = 0.1;

qI = 10;

qII = qI;


(* Coupled steady state equations *)

solmem = Flatten[

Simplify[

Solve[
```

$$\left\{0 == (\mathrm{CBIf} - \mathrm{CBI})\frac{\mathrm{qI}}{\mathrm{VI}} - \frac{\mathrm{PmBAm}}{\mathrm{VI}}(\mathrm{CBI} - \mathrm{CBII}),\right.$$

$$0 == (\text{CDIf} - \text{CDI})\frac{\text{qI}}{\text{VI}} - \frac{\text{PmDAm}}{\text{VI}}(\text{CDI} - \text{CDII}),$$

$$0 == -\text{CBII}\frac{\text{qII}}{\text{VII}} + \frac{\text{PmBAm}}{\text{VII}}(\text{CBI} - \text{CBII}),$$

$$0 == -\text{CDII}\frac{\text{qII}}{\text{VII}} + \frac{\text{PmDAm}}{\text{VII}}(\text{CDI} - \text{CDII})\Big\},$$

```
{CBI, CDI, CBII, CDII}]
    ]
    ];

(* Extraction of solutions from solmem *)

CB1ss = NumberForm[Evaluate[CBI/.solmem], {3, 3}];
CD1ss = NumberForm[Evaluate[CDI/.solmem], {3, 3}];
CB2ss = NumberForm[Evaluate[CBII/.solmem], {3, 3}];
CD2ss = NumberForm[Evaluate[CDII/.solmem], {3, 3}];

Print["The steady state concentrations are: "]
Print[" CB1ss = "CB1ss]
Print[" CD1ss = "CD1ss]
Print[" CB2ss = "CB2ss]
Print[" CD2ss = "CD2ss]
```

The steady state concentrations are:

CB1ss = 0.067

CD1ss = 0.099

CB2ss = 0.033

CD2ss = 0.001

We might want to know how long it would take this system to reach a steady state. To find this we must solve the same set of equations, that we have just examined at the steady state, but in time. We will also include an inlet flow on the permeate side to simulate the effect of a sweep stream on the dynamics of this system. This flow on the permeate side would serve the purpose of continuously sweeping the lower side of the membrane to clear the permeate. If the flow rate is large compared the volume, ideally, we will have near zero concentrations of the permeate and maximal permeation rates. We still need only consider the component balances at this point. Economically, this approach of using a sweep gas may work if the retentate is the valuable product and the permeate can be diluted to very low concentrations. We will use all the same numbers and equations except that **qII** will be fixed at the inlet and outlet of the permeate side of the unit.

```
Clear[PmB, PmA, Am, VI, VII, C1B, C1D, C2B, C2D,
CIBo, CIDo, plI, plII, plIII, plIV, t]
CB1ss = 0.067;
CD1ss = 0.099;
CB2ss = 0.033;
CD2ss = 0.001;
CIBo = 0.000001;
CIDo = 0.000001;
VI = 1;
PmB = 10^-4;
PmD = 10^-6;
VII = VI;
x = 5;
Am = 10^x;
CBIf = .1;
CDIf = 0.1;
qI = 10;
qII = 10;
n = 5;
```

$$\frac{\text{VI}}{\text{PmBAm}}//N$$

eqns =

$$\left\{\partial_t \text{CBI}[t] == (\text{CBIf} - \text{CBI}[t])\frac{\text{qI}}{\text{VI}} - \frac{\text{PmBAm}}{\text{VI}}(\text{CBI}[t] - \text{CBII}[t]),\right.$$

$$\partial_t \text{CDI}[t] == (\text{CDIf} - \text{CDI}[t])\frac{\text{qI}}{\text{VI}} - \frac{\text{PmDAm}}{\text{VI}}(\text{CDI}[t] - \text{CDII}[t]),$$

$$\partial_t \text{CBII}[t] == (-\text{CBII}[t])\,\text{dim}\,\tfrac{\text{qII}}{\text{VII}} + \tfrac{\text{PmBAm}}{\text{VII}}(\text{CBI}[t] - \text{CBII}[t]),$$

$$\partial_t \text{CDII}[t] == (-\text{CDII}[t])\frac{\text{qII}}{\text{VII}} + \frac{\text{PmDAm}}{\text{VII}}(\text{CBI}[t] - \text{CBII}[t]),$$

$$\text{CBI}[0] == \text{CIBo},\ \text{CDI}[0]{=}{=}\text{CIDo}, \text{CBII}[0] == 10^{-10},$$

$$\left.\text{CDII}[0] == 0\right\};$$

numsol =

NDSolve[

eqns,

{CBI[t], CDI[t], CBII[t], CDII[t]},{t, 0, n}   ]

C1B[t_]:=Evaluate[First[CBI[t]/.numsol]]

C1D[t_]:=Evaluate[First[CDI[t]/.numsol]]

C2B[t_]:=Evaluate[First[CBII[t]/.numsol]]

C2D[t_]:=Evaluate[First[CDII[t]/.numsol]]

$$\text{pl1} = \text{Plot}\left[\frac{\text{C1B}[t]}{\text{CBIf}}, \{t, 0, n\},\right.$$

$$\text{AxesLabel} \to \left\{\text{t}, \text{“}\frac{\text{C1B}[t]}{\text{CBIf}}\text{”}\right\},$$

PlotStyle → Hue[0],

PlotRange → {{0, n}, {0, 1}},

Epilog → {RGBColor[0, .5, 1],

$$\left.\text{Line}\left[\left\{\left\{0, \frac{\text{CB1ss}}{\text{CBIf}}\right\}, \left\{n, \frac{\text{CB1ss}}{\text{CBIf}}\right\}\right\}\right]\right\}\Big];$$

```
pl2 = Plot[C1D[t]/CDIf, {t, 0, n},
DisplayFunction → Identity,
AxesLabel → {t, "C1D[t]/CDIf"},
PlotStyle → Hue[0.8],
PlotRange → {{0, n}, {0, 1}}];

pl3 = Plot[C2B[t], {t, 0, n},AxesLabel → {t, "C2B[t]"},
PlotStyle → {Hue[0],
Dashing[{0.15, 0.05}]},
PlotRange → {{0, n}, {0, CBIf}},
Epilog → {RGBColor[0, .5, 1],
Line[{{0, CB2ss }, {n, CB2ss }}]}];

pl4 = Plot[C2D[t], {t, 0, n},AxesLabel → {t, "C2D[t]"},
PlotStyle → {Hue[0.8],
Dashing[{0.15, 0.05}]},
PlotRange → {{0, n}, {0, CDIf}},
AxesOrigin → {0, 0}];

Show[GraphicsGrid[{{pl1, pl2}, {pl3, pl4}}]]
Plot[C2B[t]qII, {t, 0, n},
AxesLabel → {t, "CIIB[t]*qII[t]"},
PlotStyle → Hue[0.5],
PlotRange → {{0, n}, {0, .5}}]

{{CBI[t] → InterpolatingFunction[□][t],
CDI[t] → InterpolatingFunction[□][t],
CBII[t] → InterpolatingFunction[□][t],
CDII[t] → InterpolatingFunction[□][t]}}
```

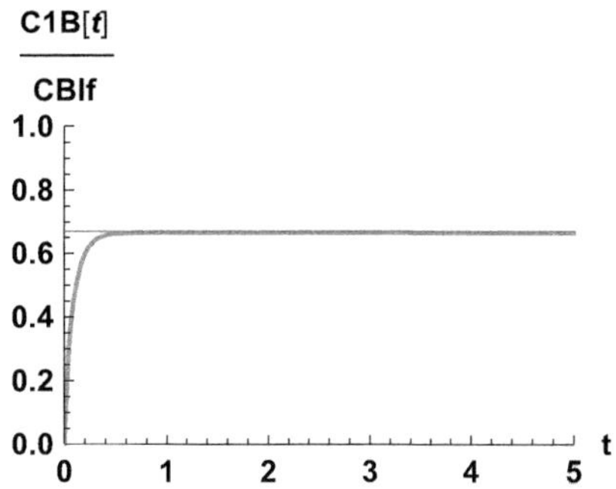
C1B[t]
CBIf
1.0
0.8
0.6
0.4
0.2
0.0
0
1
2
3
4
5
t

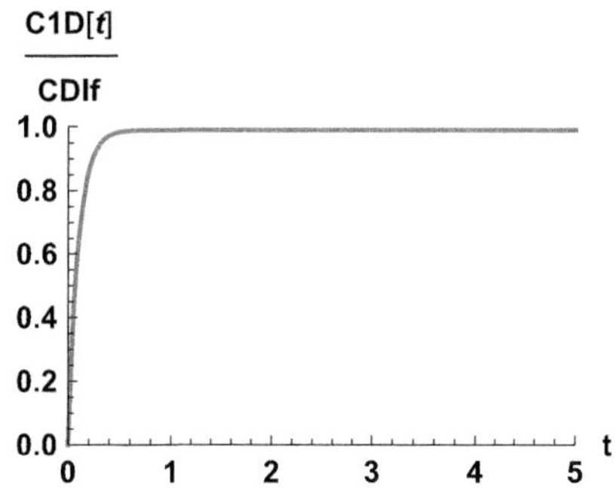
C1D[t]
CDIf
1.0
0.8
0.6
0.4
0.2
0.0
0
1
2
3
4
5
t

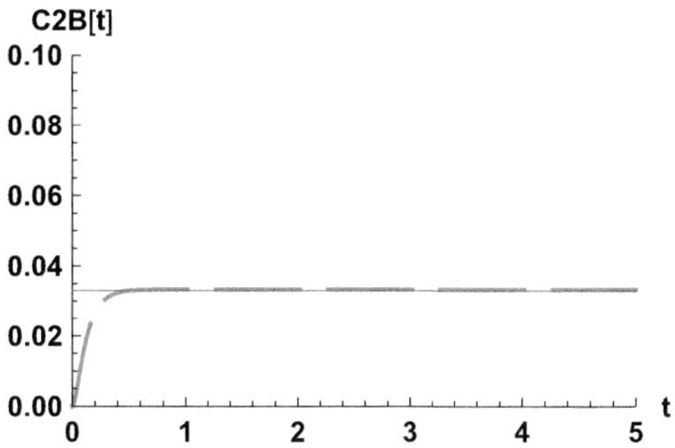
C2B[t]
0.10
0.08
0.06
0.04
0.02
0.00
0
1
2
3
4
5
t

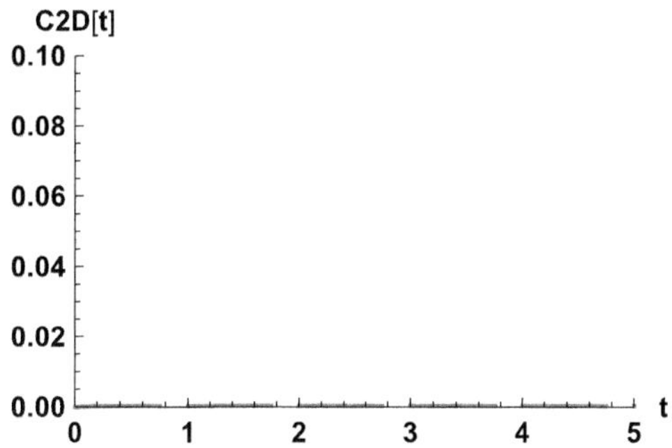
C2D[t]
0.10
0.08
0.06
0.04
0.02
0.00
0
1
2
3
4
5
t

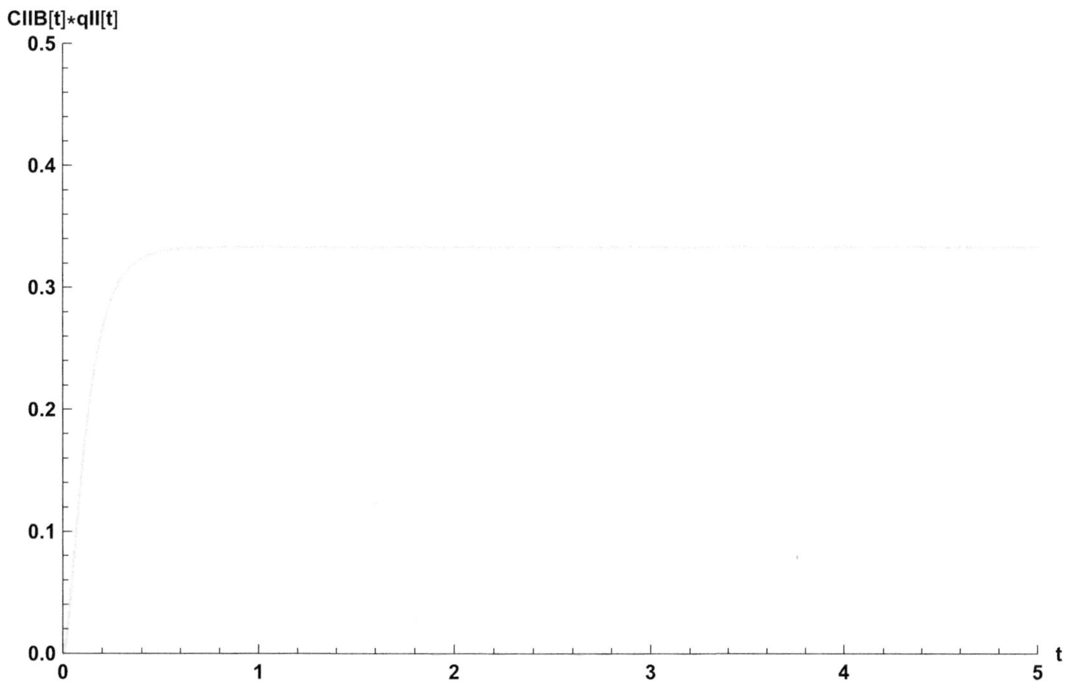
CIIB[t]*qII[t]
0.5
0.4
0.3
0.2
0.1
0.0
0
1
2
3
4
5
t

The transient analysis evolves to the steady state in a time equal one to two multiples of the characteristic permeation time, $\frac{\text{VI}}{\text{PmBAm}}$. Let's think about all of this one more time. How should we change this system if we wish to get the ultimate removal of **B** from the feed stream with the areas, feed flow and permeances all fixed? The answer is to increase the flow rate of the sweep gas. The analysis shows us that if we could reduce the concentration of **B** on the permeate side to near zero values, we could remove nearly 50% of it versus only 33% with these conditions. How could we do this? Increasing the flow rate of the sweep gas holding all else fixed, will have the effect of bringing the concentration of **B** by increasing the driving force for **B** across the membrane. How much higher would the sweep gas flow have to be to accomplish this? The answer is shown below; on the order of a factor of 10 increase will do it!

```
Clear[PmB, PmA, Am, VI, VII, CIBo, CIDo, pII, pIII,
pIIII, pIIV, t]
CB1ss = 0.067;
CD1ss = 0.099;
CB2ss = 0.033;
CD2ss = 0.001;
CIBo = 0.000001;
CIDo = 0.000001;

VI = 1;
PmB = 10^-4;
PmD = 10^-6;
VII = VI;
x = 5;
Am = 10^x;
CBIf = .1;
CDIf = 0.1;
qI = 10;
qII = 100;
```

$n = 10;$

eqns =

$$\left\{\partial_t \text{CBI}[t] == (\text{CBIf} - \text{CBI}[t])\frac{\text{qI}}{\text{VI}} - \frac{\text{PmBAm}}{\text{VI}}(\text{CBI}[t] - \text{CBII}[t]),\right.$$

$$\partial_t \text{CDI}[t] == (\text{CDIf} - \text{CDI}[t])\frac{\text{qI}}{\text{VI}} - \frac{\text{PmDAm}}{\text{VI}}(\text{CDI}[t] - \text{CDII}[t]),$$

$$\partial_t \text{CBII}[t] == (-\text{CBII}[t])\frac{\text{qII}}{\text{VII}} + \frac{\text{PmBAm}}{\text{VII}}(\text{CBI}[t] - \text{CBII}[t]),$$

$$\partial_t \text{CDII}[t] == (-\text{CDII}[t])\frac{\text{qII}}{\text{VII}} + \frac{\text{PmDAm}}{\text{VII}}(\text{CBI}[t] - \text{CBII}[t]),$$

$$\text{CBI}[0] == \text{CIBo},\ \text{CDI}[0]==\text{CIDo}, \text{CBII}[0] == 10^{-10},$$

$$\left.\text{CDII}[0] == 0\right\};$$

```
numsol = Flatten[
NDSolve[
eqns,
{CBI[t], CDI[t], CBII[t], CDII[t]},
{t, 0, n}
    ]
];
C1B[t_]:=Evaluate[CBI[t]/.numsol[[1]]]
C1D[t_]:=Evaluate[CDI[t]/.numsol[[2]]]
C2B[t_]:=Evaluate[CBII[t]/.numsol[[3]]]
C2D[t_]:=Evaluate[CDII[t]/.numsol[[4]]]
```

$$\text{pl1} = \text{Plot}\left[\frac{\text{C1B}[t]}{\text{CBIf}}, \{t, 0, n\},\right.$$

$$\text{AxesLabel} \to \left\{\text{t}, \text{“}\frac{\text{C1B}[t]}{\text{CBIf}}\text{”}\right\},$$

PlotStyle → Hue[0],

PlotRange → {{0, n}, {0, .75}},

Epilog → {RGBColor[0, .5, 1],

$$\text{Line}\left[\left\{\left\{0, \frac{\text{CB1ss}}{\text{CBIf}}\right\}, \left\{n, \frac{\text{CB1ss}}{\text{CBIf}}\right\}\right\}\right]\Big]$$

$$\text{pl2} = \text{Plot}\left[\frac{\text{C1D}[t]}{\text{CDIf}}, \{t, 0, n\},\right.$$

$$\text{AxesLabel} \to \left\{\text{t}, \text{“}\frac{\text{C1D}[t]}{\text{CDIf}}\text{”}\right\},$$

PlotStyle → Hue[0.8],

PlotRange → {{0, n}, {0, 1}}]

pl3 = Plot[C2B[t], {t, 0, n},AxesLabel → {t, “C2B[t]”},

PlotStyle → {Hue[0],

Dashing[{0.15, 0.05}]},

PlotRange → {{0, n}, {0, .5CBIf}},

Epilog → {RGBColor[0, .5, 1],

Line[{{0, CB2ss }, {n, CB2ss }}]}]

pl4 = Plot[C2D[t], {t, 0, n},AxesLabel → {t, “C2D[t]”},

PlotStyle → {Hue[0.8],

Dashing[{0.15, 0.05}]},

PlotRange → {{0, n}, {0, .001CDIf}},

AxesOrigin → {0, 0}]

Plot[C2B[t]qII, {t, 0, n},

AxesLabel → {t, “CIIB[t] qII[t]”},

PlotStyle → Hue[0.5],

PlotRange → {{0, n}, {0, 1}}]

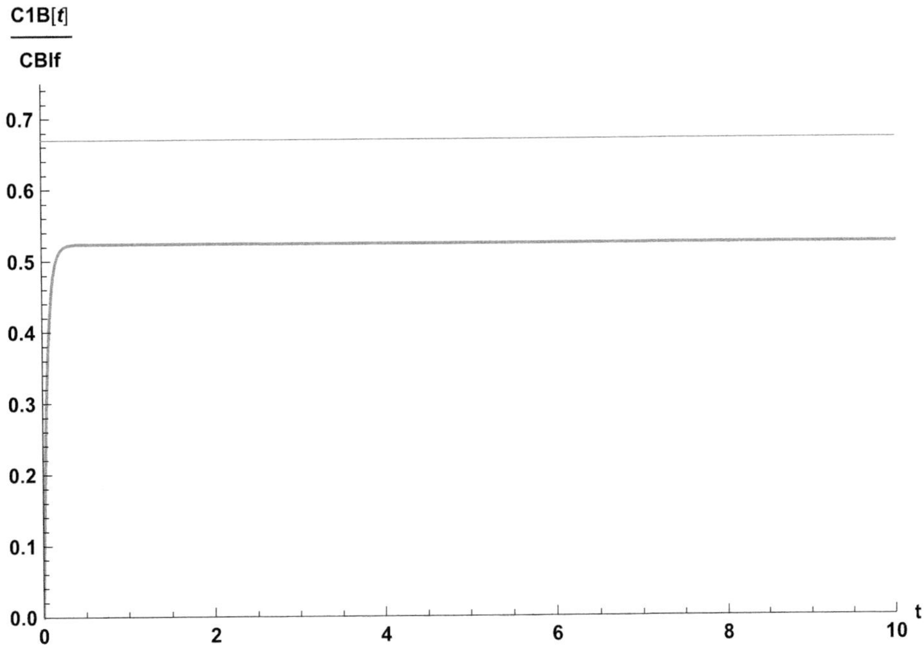
C1B[t]
CBIf
0.7
0.6
0.5
0.4
0.3
0.2
0.1
0.0
0
2
4
6
8
10
t

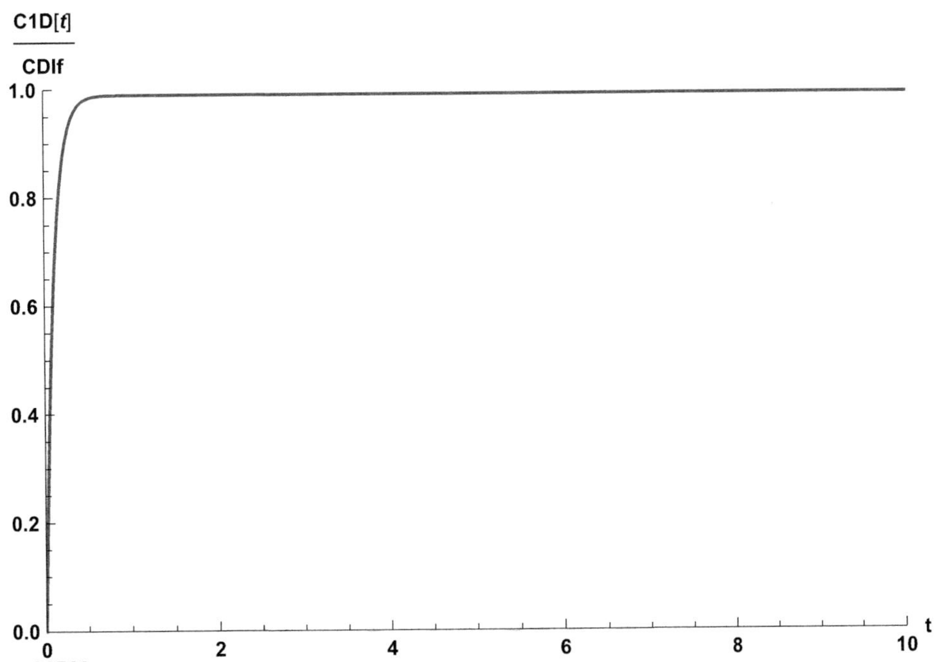
C1D[t]
CDIf
1.0
0.8
0.6
0.4
0.2
0.0
0
2
4
6
8
10
t

C2B[t]

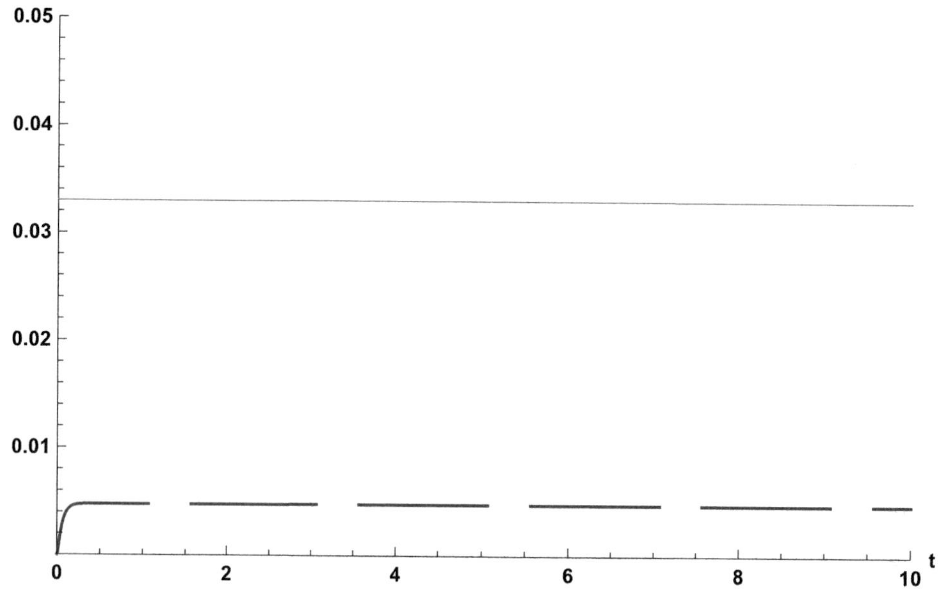
0.05
0.04
0.03
0.02
0.01
0
2
4
6
8
10
t

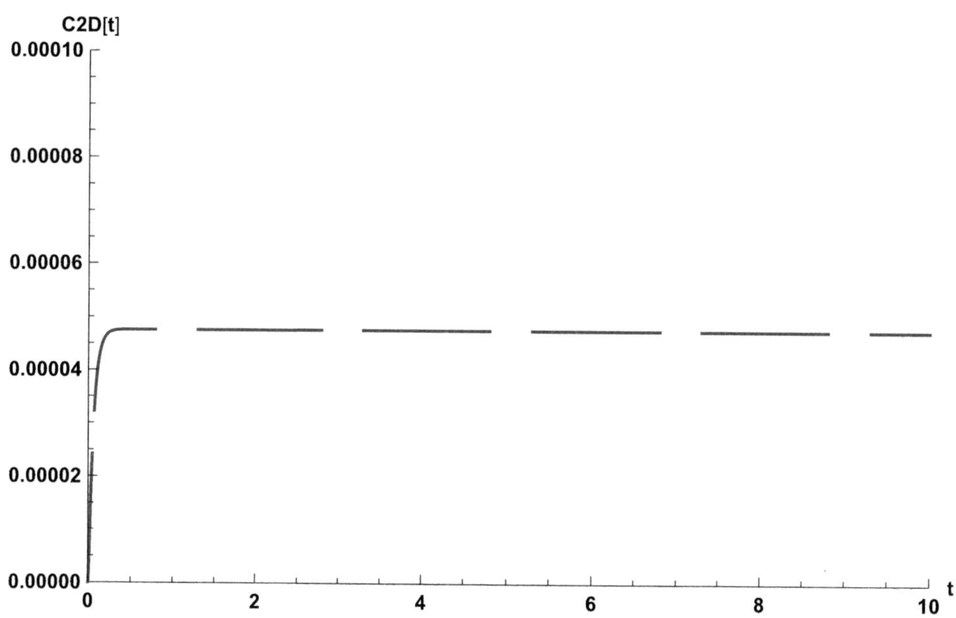
C2D[t]
0.00010
0.00008
0.00006
0.00004
0.00002
0.00000
0
2
4
6
8
10
t

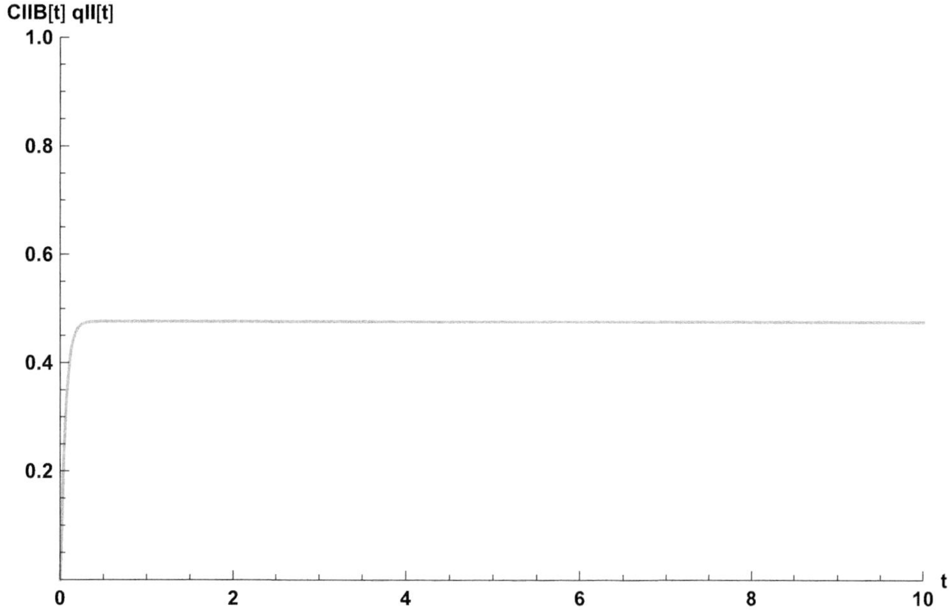

Now we can use these fundamentals about permeation in a new way - to study the swelling of a cell.

Expanding cell

Consider the following problem. A nearly spherical cell consists of a thin membrane surrounding a salt solution. Outside of the cell membrane there is a solution that is isotonic with that within the membrane. Everything is in equilibrium. If the cell is instantaneously removed from its surroundings and placed into a volume of pure water, then the action of osmosis will drive water through the membrane to cause dilution of its contents. The transport across the membrane is a permeation process with a rate of:

$$J_{H2O} = \mathrm{Pm}(C_{H2O}^{Outside} - C_{H2O}^{Inside})$$

The direction of the flow is from the region of lower salt concentration outside to higher salt concentration inside, and from higher water concentration to lower water concentration. The concentration of water on the outside of the cell is taken to be equal to the density of pure water, $\boldsymbol{C_{H2O}^{Outside}} = \boldsymbol{\rho_{H2O}}$. Inside the membrane the concentration of water is increasing as the density of the solution is decreasing. The density of the cellular content follows the linear relationship:

$$\rho_{cell}[t] = \rho_{H2O} + \gamma\ C_{salt}^{Cell}[\mathrm{t}]$$

While water transports osmotically across the cell membrane to dilute the cellular contents, the cell grows. The membrane stretches to accommodate the newly accumulated mass. The geometry of the cell, spherical, remains constant, that is it grows in every direction through the whole of the solid angle by the same amount in the same time period. The physical situation with this cell is sketched below:

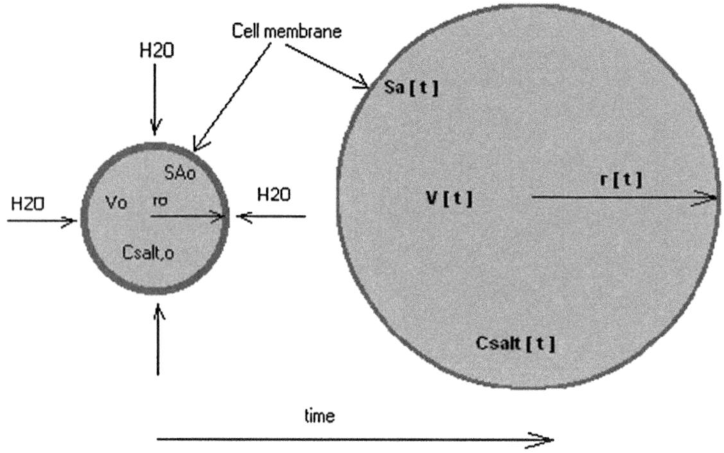

We would like to know how the membrane grows as a function of time, how much the cellular contents are diluted in time and similar information about this process. We will find the answers to these questions by modeling the dynamic process. We begin by writing the material balance equations within the cell in a general form:

$$\frac{dm_{\text{tot}}[t]}{dt} = P_m \text{ SA } (C_{\text{H2O}}^{\text{Outside}} - C_{\text{H2O}}^{\text{Inside}})$$

$$\frac{dm_{\text{salt}}[t]}{dt} = 0$$

$$\frac{dm_{\text{H2O}}[t]}{dt} = P_m \text{ SA } (C_{\text{H2O}}^{\text{Outside}} - C_{\text{H2O}}^{\text{Inside}})$$

The surface area and volume of the cell are functions of the radius:

$$\text{V}[\text{t}] = \frac{4}{3}\pi \text{ r}[\text{t}]^3 \quad \text{SA}[\text{t}] = 4\pi \text{ r}[\text{t}]^2$$

and

$$\frac{dV[t]}{dt} = 4\pi \text{ r}[\text{t}]^2 \frac{dr[t]}{dt} \quad \frac{d\text{SA}[t]}{dt} = 8\pi \text{ r}[\text{t}] \frac{dr[t]}{dt}$$

We will need to relate the water concentration to the salt concentration and the change in density to the water flowing into the cell:

$$\rho_{cell}[t] = \rho_{H2O} + \gamma\, C_{salt}^{Cell}[\,t\,]$$

$$C_{salt}^{Cell}[\,t\,] + C_{H2O}^{Cell}[\,t\,] = \rho_{H2O} + \gamma\, C_{salt}^{Cell}[\,t\,]$$

$$C_{H2O}^{Cell}[\,t\,] = \rho_{H2O} + (\gamma - 1)\, C_{salt}^{Cell}[\,t\,]$$

$$C_{salt}^{Cell}[\,t\,] = \frac{C_{H2O}^{Cell}[t] - \rho_{H2O}}{(\gamma - 1)}$$

$$\rho_{cell}[t] = \rho_{H2O} + \gamma\, \frac{C_{H2O}^{Cell}[t] - \rho_{H2O}}{(\gamma - 1)}$$

$$\rho_{cell}[t] = \frac{\gamma C_{H2O}^{Cell}[t] - \rho_{H2O}}{(\gamma - 1)}$$

With these expressions and the component balances we can now find expression for $C_{salt}^{Cell}[\,t\,]$ and $C_{H2O}^{Cell}[\,t\,]$ as functions of r [t].

$$\frac{dm_{H2O}[t]}{dt} = P_m\, \text{SA}\ (\,C_{H2O}^{Outside} - C_{H2O}^{Inside}\,)$$

$$\frac{dm_{salt}[t]}{dt} = 0 = \frac{dC_{salt}^{Cell}[t]V[t]}{dt} = C_{salt}^{Cell}[t]\frac{dV[t]}{dt} + \text{V}\,[\,t\,]\frac{dC_{salt}^{Cell}[t]}{dt}$$

$$C_{salt}^{Cell}[t] = \frac{C_{salt}^{Cell}[0]V[0]}{V[t]} = \frac{C_{salt}^{Cell}[0]\,{r_o}^3}{r[t]^3}$$

$$C_{H2O}^{Cell}[\,t\,] = \rho_{H2O} + (\gamma - 1)\,\frac{C_{salt}^{Cell}[0]\,{r_o}^3}{r[t]^3}$$

Taking the derivative of this we find that the change in salt concentration as a function of time:

$$\frac{\partial C_{H2O}^{Cell}[t]}{\partial t} = \frac{\partial\left(\frac{(\gamma-1)\left(r_o^3 C_{salt}^{Cell}(0)\right)}{r(t)^3} + \rho_{H2O}\right)}{\partial t}$$

$$C_{H2O}^{Cell\,\prime}[t] = -\frac{3(-1+\gamma)r_o^3 C_{salt}^{Cell}[0]r'[t]}{r[t]^4}$$

Returning to the total material balance, we can use the definitions of all the variables in terms of **r [t]** to solve for the derivative of **r [t]** in terms of just the cell parameters. Once we have this we can then solve for **r [t]** explicitly in time:

$$\frac{dm_{\text{tot}}[t]}{dt} = P_m \text{ SA } (C_{\text{H2O}}^{\text{Outside}} \text{-} C_{\text{H2O}}^{\text{Cell}}[t])$$

$$\frac{d\rho_{\text{cell}}[t]V[t]}{dt} = P_m \text{ SA } (C_{\text{H2O}}^{\text{Outside}} \text{-} C_{\text{H2O}}^{\text{Cell}}[t])$$

The following re-expresses the left- and right-hand sides in terms of **r [t]** and then solves for **r ' [t]**. Finally, we solve for **r [t]** and plot the function:

C_{H2O}**[t_]:=+(**γ**−1)**

ρ_{cell}**[t_]:=+**γ

V[t_]:=$\frac{4}{3}r[t]^3$

lhs =

SA[t_]:=4π $r[t]^2$

rhs = PmSA[t**] (**ρ_{H2O}**−)**

Solve[lhs == rhs, $r'[t]$**]**

$$-\frac{4\gamma r_o^3 C_{\text{salt},o}^{\text{Cell}} r'[t]}{r[t]} + 4r[t]^2 \left(\rho_{\text{H2O}} + \frac{\gamma r_o^3 C_{\text{salt},o}^{\text{Cell}}}{r[t]^3} \right) r'[t]$$

$$-\frac{4\pi \text{Pm}(-1+\gamma) r_o^3 C_{\text{salt},o}^{\text{Cell}}}{r[t]}$$

$$\left\{ \left\{ r'[t] \to -\frac{\pi \text{Pm}(-1+\gamma) r_o^3 C_{\text{salt},o}^{\text{Cell}}}{r[t]^3 \rho_{\text{H2O}}} \right\} \right\}$$

DSolve[

$$\left\{ \partial_t r[t] == -\frac{\pi \text{Pm}(-1+\gamma) \text{ro}^3 C_{\text{salt},o}^{\text{Cell}}}{r[t]^3 \rho_{\text{H2O}}}, \right.$$

$r[0]$ **== ro},**

$r[t]$**,** t**]//PowerExpand//Simplify**

$$\left\{\left\{r[t]\to-\frac{\text{ro}^{3/4}\left(\text{ro}\rho_{\text{H2O}}-4\pi\,\text{Pm}t(-1+\gamma)C_{\text{salt},o}^{\text{Cell}}\right)^{1/4}}{\rho_{\text{H2O}}^{1/4}}\right\},\right.$$

$$\left\{r[t]\to-\frac{i\,\text{ro}^{3/4}\left(\text{ro}\rho_{\text{H2O}}-4\pi\,\text{Pm}t(-1+\gamma)C_{\text{salt},o}^{\text{Cell}}\right)^{1/4}}{\rho_{\text{H2O}}^{1/4}}\right\},$$

$$\left\{r[t]\to\frac{i\,\text{ro}^{3/4}\left(\text{ro}\rho_{\text{H2O}}-4\pi\,\text{Pm}t(-1+\gamma)C_{\text{salt},o}^{\text{Cell}}\right)^{1/4}}{\rho_{\text{H2O}}^{1/4}}\right\},$$

$$\left.\left\{r[t]\to\frac{\text{ro}^{3/4}\left(\text{ro}\rho_{\text{H2O}}-4\pi\,\text{Pm}t(-1+\gamma)C_{\text{salt},o}^{\text{Cell}}\right)^{1/4}}{\rho_{\text{H2O}}^{1/4}}\right\}\right\}$$

There are four solutions to this equation, two are real and two are complex. Of the two real solutions, only the second is physical, since it is a growing function of time:

$$\frac{\text{ro}^{3/4}\left(\text{ro}\rho_{\text{H2O}}-4\pi\,\text{Pm}t(-1+\gamma)C_{\text{salt},o}^{\text{Cell}}\right)^{1/4}}{\rho_{\text{H2O}}^{1/4}}$$

$$\boldsymbol{r[\text{t\_}]:=\frac{\text{ro}^{3/4}\left(\text{ro}\rho_{\text{H2O}}-4\pi\,\text{Pm}t(-1+\gamma)C_{\text{salt},o}\right)^{1/4}}{\rho_{\text{H2O}}^{1/4}}}$$

$\mathbf{ro = 10^{-4};}$

$\mathbf{Pm = 10^{-3};}$

$\boldsymbol{\gamma = 0.9;}$

$\boldsymbol{C_{\text{salt},o} = 0.01;}$

$\boldsymbol{\rho_{\text{H2O}} = 1;}$

$$\mathbf{tmax = 1000\frac{3Pm}{\boldsymbol{r}[0]}}$$

(* We solve for the time when the cell explodes *)

$$\mathbf{Solve}\left[\mathbf{3ro}==\frac{\text{ro}^{3/4}\left(\text{ro}\rho_{\text{H2O}}-4\pi\,\text{Pmtcr}(-1+\gamma)C_{\text{salt},o}\right)^{1/4}}{\rho_{\text{H2O}}^{1/4}},\ \mathbf{tcr}\right]$$

Plot$\boldsymbol{[r[t]10^4, \{t, 0, \text{tmax}\},}$

AxesLabel → {"1000 θ/sec", "r[t]x10⁴/μ"},

PlotStyle → Hue[0.1],

PlotLabel → "Radial Cell Growth"]

30000.

{{tcr → 636.62}}

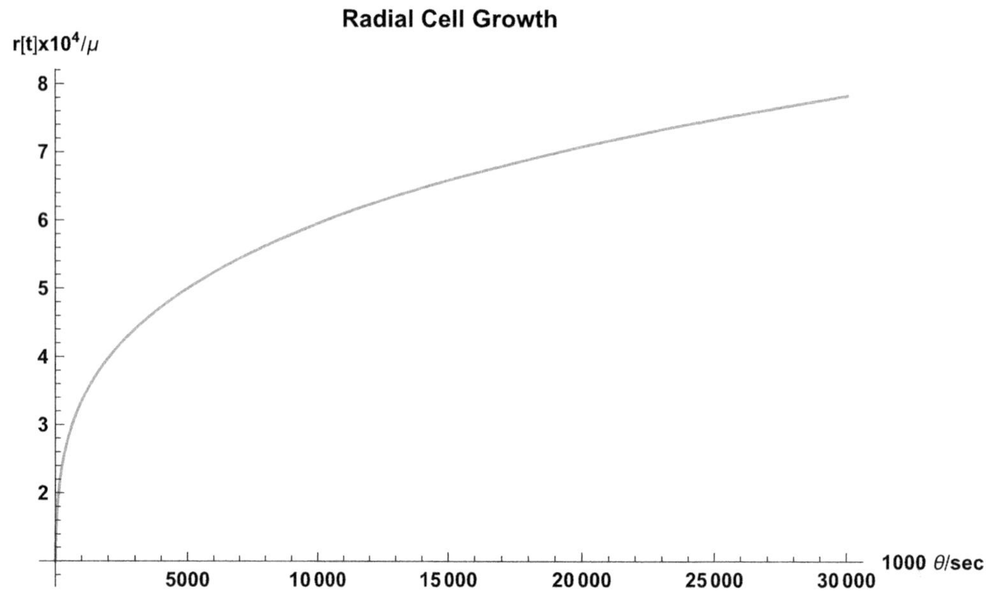

This solution to the problem shows that the cell grows very fast at short time, but then it slows at longer times. There is, however, one noticeable issue that crops up with this solution, and that is the radius continues to grow with increasing time. The critical time is 636.6 seconds because that is the time that the radius goes to three times its initial value. We took three times the initial value to be the critical radius, and beyond which the membrane surface cannot withstand the internal pressure. Hence, the membrane will break at that critical radius. Thus the solution should be indicative of this behavior. (Note: The choice of 3ro is not fundamental. It is completely arbitrary in the absence of any other information. Most bacterial cells that are roundish fall in the range of 0.4 to 2 microns. That is a factor of 5 in size range. So that a smaller cell might be able to grow to three times its initial diameter seems reasonable. In actuality the degree to which a cell may swell is determined by the mechanics of the particular cell's wall.)

We can modify the solution to include this behavior. The critical radius can be expressed as a multiple of the initial radius at time zero. We want the function to literally "blow-up" when we reach this radius. The critical condition can be expressed with the **UnitStep** function. It will have a value of unity until the critical condition is reached, whereupon it will go to zero. Recall the behavior of the **UnitStep** given in terms of **r [t]** and the critical condition of **3 r [0]**.

Plot[1 − 1UnitStep[*r*[*t*] − 3ro], {*t*, 0, 1200},

PlotStyle → Hue[0]]

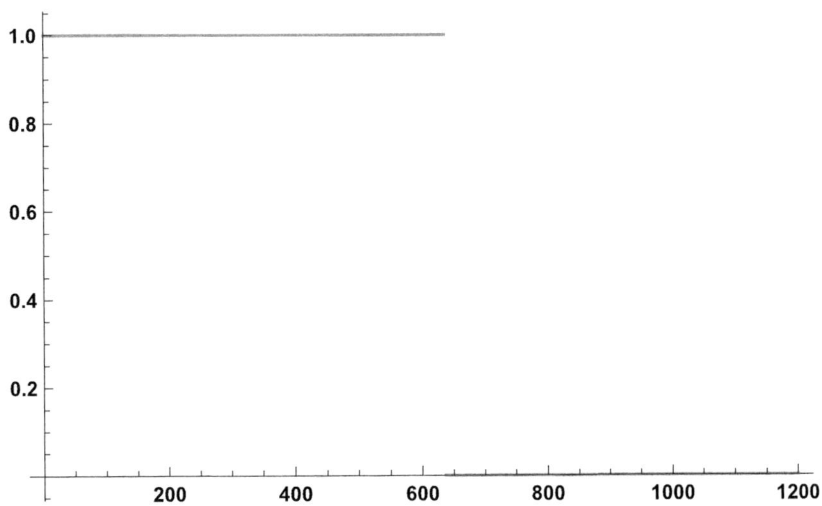

By multiplying **r[t]** by the **UnitStep** function we obtain the correct behavior, the solution is meaningful up to the critical radius, but not beyond. After the critical radius is reached, the solution goes to zero.

Clear[rbl, us]

rbl[t_]:=(1 − 1UnitStep[*r*[*t*] − 3ro])

$$\frac{\mathbf{ro}^{3/4}\left(\mathbf{ro}\rho_{\mathrm{H2O}}-4\pi\mathbf{Pm}t(-1+\gamma)C_{\mathrm{salt},o}\right)^{1/4}}{\rho_{\mathrm{H2O}}^{1/4}}$$

$\mathbf{ro} = 10^{-4};$

$\mathbf{Pm} = 10^{-3};$

$\gamma = 0.9;$

$C_{\text{salt},o} = 0.01;$

$\rho_{\text{H2O}} = 1;$

$\frac{3\text{Pm}}{\text{ro}}$

$\text{tmax} = 25\frac{3\text{Pm}}{\text{ro}}$

$10^4 r[\text{tmax}]$

Plot$\left[\left\{\text{rbl}[t]10^4\right\}, \{t, 0, \text{tmax}\},\right.$

AxesLabel $\rightarrow$ {"1000 θ/sec", "r[t]x$10^4/\mu$"},

PlotStyle $\rightarrow$ {Hue[0], Hue[0.1]},

PlotLabel $\rightarrow$ "Radial Cell Growth",

PlotRange $\rightarrow$ $\left\{\{0, \text{tmax}\}, \left\{0, r[\text{tmax}]10^4\right\}\right\}$,

Epilog $\rightarrow$ $\left\{\text{Line}\left[\left\{\left\{0, 3 \times 10^4\text{ro}\right\}, \left\{\text{tmax}, 3 \times 10^4\text{ro}\right\}\right\}\right],\right.$

$\left.\left.\text{Line}\left[\left\{\{636, 0\}, \left\{636, 3 \times 10^4\text{ro}\right\}\right\}\right]\right\}\right]$

30

750

3.12402

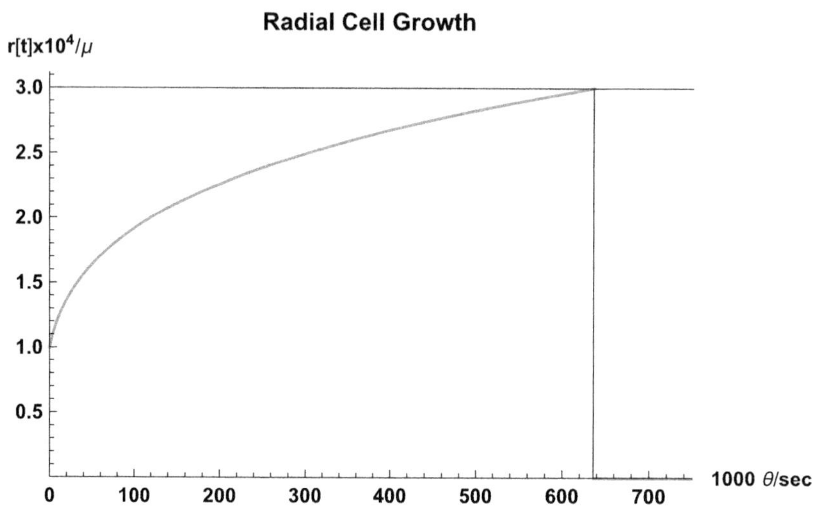

We also might wish to project the growth in the plane. When we do so we can use our expression for the radius as a function of time, and then plot the circular surfaces at integer time steps separated by a constant increment. The following code does this:

```
Show[
Graphics[
Table[{RGBColor[1, 0.2, 0], Circle[{0, 0}, rbl[t]]},
{t, 0, .8tmax, 20}], AspectRatio → Automatic,
Axes → Automatic]]
```

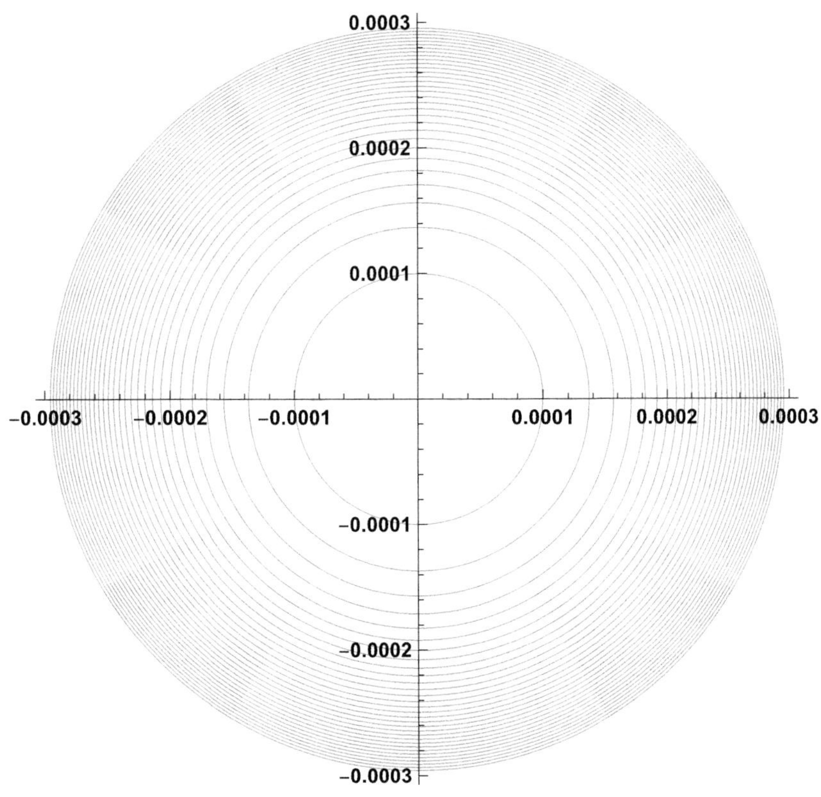

This is a good visualization of the expansion. The closer spacing of each successive ring near the edges corresponds well with the shape of the graph of the radius versus time.

Conclusion

We have covered material in this chapter that is extensive both in terms of the fundamental phenomena of adsorption and permeation. We also have used *Mathematica* at an increasingly sophisticated level. Both the fundamentals and our learning of the Wolfram language will be valuable to us for the rest of this book.

CHAPTER 7

Reacting species - kinetics and batch reactors

Chemical kinetics has been a fascinating subject of inquiry for over 100 years. Chemical kinetics is the mathematical description of chemically reacting systems. Chemical kinetics also is a science that lies at the heart of industrial chemistry and chemical engineering. The concept of *rate* is central to the science of chemical kinetics and it is also a central issue in the design of reactors for chemicals production. Chemical kinetics have another purpose beside establishing the rate of reaction and that is to disclose the mechanism of reaction. By careful experimentation and mathematical analysis, the mechanism of a reaction can be understood and in many cases the propensity to enhance the rate of production to a desired species may be improved. These are the motivations for studying kinetics.

Chemical thermodynamics comes together with chemical kinetics because, thermodynamics provides the extent to which a reaction may take place at any set of conditions. Thermodynamics also provides the heats the will be released, or absorbed, during reaction. For this reason in the analysis of chemically reacting systems, a thermodynamic analysis should always come first, because it provides us with the information necessary to know what is, or is not possible, and how much heat may be released or absorbed. But, thermodynamics cannot inform us about rates of reaction. To find rates of reaction and mechanism, we have to deploy careful experiments and the analysis of data from those experiments. The simulation of chemical reactions by theoretical methods has made rapid advances in the last two decades, but it is not yet a substitute for experiment and kinetics analysis.

For many chemical reactions, the equilibrium thermodynamics are favorable, but the rate of reaction is slow. Therefore a catalyst is used to improve the rate of reaction. Catalysts are critical elements in providing reactions rates that allow the chemistry to be practiced economically. The catalyst is an agent that is not consumed by the reaction; it is not a reactant. Empirically, the catalyst lowers the temperature required for a reaction take place at an acceptable rate. Yet, the extent of reaction is still dictated by the equilibrium thermodynamics, even in the presence of a catalyst. The catalyst accelerates the rate to equilibrium. As powerful as thermodynamics is, it does not provide any information about how fast, or slow the rate of approach to equilibrium will be.

Introduction to Chemical Engineering Analysis Using *Mathematica*
https://doi.org/10.1016/B978-0-12-820051-3.00012-7

This chapter is devoted to the kinetics of chemical reactions as they take place in a closed system. This closed system is referred to as a batch reactor. The use of the term batch connotes one at a time, and not continuous. In such a reactor, the reactants are mixed, the vessel is closed, and the reactions take place for a certain length of time at a specific temperature. Then the reactor is opened, the products are removed and that batch is complete. Discontinuous, intermittent, and on-off could also be used to describe this kind of reactor. Batch reactors are used in laboratories, but they are also used in the industrial production of low-volume, high-value chemicals.

How chemical reactions take place

We know from chemistry that reactions take place when molecules collide with one another. We also know that reactions often take place faster at higher temperature and that a catalyst often improves rate further. Enzymes are natural catalysts, they work by orienting molecules along specific directions that are preferred for reaction. Industrial catalysts are often solids. Raney nickel is a porous metal catalyst. Alumina with transition metals such as platinum are used in petroleum refining. High surface area carbons with nanoparticles of transition metals like palladium or rhodium dispersed over the inner surfaces of their pores are used in food and pharmaceutical production. Catalysts supported in ceramic monoliths are used in pollution control, including in automobiles. Catalysts are similar to the adsorbents that we discussed in the last chapter. We know that reactions have strong temperature and orientational dependence, and that collisions alone are not enough for reaction to take place, but let's ground our thinking about chemical kinetics in the kinetic theory of gases.

From kinetic theory of gases, the number of collisions in the gas phase that take place per unit volume and time at fixed temperature and pressure, is given by the collision number, Zab, between gas molecules, A and B:

$$
\begin{aligned}
\mathrm{Zab} &= \pi \frac{(\mathrm{da}+\mathrm{db})^2}{4} \sqrt{\frac{8RT}{\pi \mu}} \frac{\mathrm{NaNbNav}^2}{V^2} \\
&= \pi \frac{(\mathrm{da}+\mathrm{db})^2}{4} \sqrt{\frac{8RT}{\pi \mu}} \frac{\mathrm{Pa\,Pb}}{(RT)^2} \\
&= A_c\, \bar{v}\, C_a\, C_b
\end{aligned}
$$

where

R = ideal gas constant, [J/mole-K]

T = absolute temperature, [K]

$\mu = \frac{\text{MaMb}}{(\text{Ma} + \text{Mb})}$, mean molecular weight

M = molecular weight, [g/mole]

$A_c = \frac{\pi(\text{da} + \text{db})^2}{4}$, reaction cross section, [cm^2]

da, db = molecular diameters of A and B, [cm]

Na, Nb = [number]

V = volume, [cm^3]

Nav = Avogadro's number, [number per mole]

This is the product of the reaction cross section at collision, $\pi\frac{(\text{da} + \text{db})^2}{4}$, the mean speed $\sqrt{\frac{8RT}{\pi\mu}}$ and the product of the concentrations of A and B, $\frac{\text{Pa}\,\text{Pb}}{(RT)^2}$. We can compute this value for standard conditions:

Clear[Zab]

da = 2.5 10^{-8}cm;

db = 3 10^{-8}cm;

R1 = 8.314 $10^7 \frac{\text{gcm}^2}{s^2\text{mol}K}$;

$T = 300K$;

$\mu = 30\frac{g}{\text{mol}}$;

Pa = .8atm;

Pb = .2atm;

R2 = 82.05$\frac{\text{cm}^3\text{atm}}{\text{mol}K}$;

Nav = 6.02 $10^{23}\frac{1}{\text{mol}}$;

Zab == π

PowerExpand[%]

$$\text{Zab} == \frac{1.04617 \times 10^{28}\sqrt{\frac{\text{cm}^2}{s^2}}}{\text{cm}^4}$$

$$\text{Zab} == \frac{1.04617 \times 10^{28}}{\text{cm}\ ^3 s}$$

This shows that we have $O(10^{28})$ collisions per cm^3 per second between molecules such as oxygen and nitrogen in air at room temperature. The collision number is also notable because, in the absence of any other information it has the form that we would expect for the rate of reaction between two molecules.

$$\left(\pi\frac{(\text{da}+\text{db})^2}{4}\sqrt{\frac{8\text{R}1T}{\pi\,\mu}}\right)\left(\frac{\text{Pa}\,\text{PbNav}^2}{(\text{R}2T)^2}\right) \longrightarrow k_{\text{ab}}\ \text{Ca Cb}$$

The first term in the collision number is akin to a rate constant at any given temperature, and the second parenthetical term is the product of the reactant concentrations. If we convert to deciliters, the rate constant-like number is the quotient of the collision number and the concentration products:

$$\frac{\mathbf{1.046\ 10^{28}}}{\mathbf{cm^3\ sec}}\left(\frac{\mathbf{10^3 cm^3}}{\mathbf{dm^3}}\right)\frac{\mathbf{1mole}}{\mathbf{6.02\ 10^{23}}}$$

$$\left(\frac{\mathbf{1000cm^3}}{\mathbf{1dm^3}}\right)^{\mathbf{2}}\left(\frac{\mathbf{1mole}}{\mathbf{6.02\ 10^{23}}}\right)^{\mathbf{2}}$$

%%/%

$$\frac{1.73754 \times 10^7\text{mole}}{\text{dm}^3\ \text{sec}}$$

$$\frac{0.000264071\text{mole}^2}{\text{dm}^6}$$

$$\frac{6.57983 \times 10^{10}\text{dm}^3}{\text{mole sec}}$$

This magnitude of 10^{10} to 10^{11} $\frac{\text{dm}^3}{\text{mole sec}}$ is important to consider. It is not so high as to be completely out of the realm of actual rate constants, but it is higher than that for many reactions. Perhaps, even more important to notice, is that this pseudo-rate constant has temperature dependence of $T^{\frac{1}{2}}$:

$$\pi\frac{(\text{da}+\text{db})^2}{4}\sqrt{\frac{8\text{R}T}{\pi\,\mu}}$$

From basic chemistry, we know empirically many rates constants double with each increase of ten degrees and so the $T^{\frac{1}{2}}$ is a weaker dependence on temperature than expected.

At the beginning of the twentieth century, Arrhenius articulated a simple, but elegant mathematical statement for the rate constant that was consistent with experiment:

$$k[T] = A\, e^{\frac{-Ea}{RT}}$$

Here the A is a pre-exponential factor – the A factor – which he ascribed to geometric effects while the temperature dependence is accounted for in the exponential. The term Ea is the activation energy. This is a threshold energy that must be surmounted for a collision to lead to reaction. We can rewrite this statement in terms of the reaction temperature, $T_{rxn} = \frac{Ea}{R}$.

$$k[T] = A\, e^{\frac{-Trxn}{T}}$$

If we take the threshold energy to be 40,000 cal/mol, then given R = 1.98 $\sim 2\frac{cal}{mole\ K}$, then the "threshold reaction temperature" is ~20,000 K. At this threshold temperature of reaction we can see that adding ten degrees to an initial temperature will indeed double the rate constant and, if all else is the same, the rate of reaction. We can also see that if the threshold temperature is 40,000 K then a ten degree temperature rise will nearly quintuple the rate, whereas at 10,000 K, the rate is barely raised by 1.5 times its initial value. This is strong temperature dependence

T2 = T1 + 10;

T1 = 500;

Trxn = 20000;

$$\textbf{Solve}\left[\left\{\frac{k1}{k2}==\frac{e^{-\frac{Trxn}{T1}}}{e^{-\frac{Trxn}{T2}}}\right\}, k2\right] //N$$

{{k2 → 2.1909k1}}

We have left the issue of the dimensions of the rate constant not discussed. The reason is that the dimensions change with the dependence upon concentration, so we post-pone a deeper consideration of dimensions until we reach that point. However, for a reaction that involves the collision of two species, i.e. a bimolecular reaction, the dimensions are $\frac{Length^3}{mole-time}$.

Reactions take place in a localized region of space, i.e. a system defined by a control volume. The control volume can be real, such as a cell, or it can be an abstraction such as an infinitesimal region of space within a cell. We choose the control volume according to the problem we are solving and the appropriateness of the level of analysis to be undertaken.

We will begin with the case of the batch reactor. In this case the vessel defines the control volume. We will move to systems with flow in, and with both flow in and out. The former is the case of semi-batch operation while the latter will be treated as the continuous stirred tank reactor (CSTR) and the plug flow reactor (PFR). All the rate analyses that we will need can be introduced within the context of these four different kinds of reactors.

No-flow/batch system

We know from the conservation of mass that when we run a reaction in a batch reactor the mass of products must be equal to the mass of reactants, so long as nothing escaped from the reactor. This holds absolutely and is independent of the chemical reaction type, mechanism, or stoichiometry. All that chemical reactions do is to rearrange the atoms and mass in and between the molecules. In essence the labels on the mass change, but the mass remains the same. If the reaction in solution leads to a gas such as the reaction of baking soda with vinegar, that is sodium bicarbonate with dilute acetic acid, then a mass change can seem to take place because one of the products is a gas and can escape the vessel:

$$Na_2CO_3 + 2\ CH_3COOH \rightarrow 2\ Na\,(CH_3COO) + CO_2 \uparrow\ + H_2O$$

But if we trap the escaping gas, we will find that the total mass of products is the same as was the total mass of reactants.

What is the proper expression for a batch reactor? We know that the total mass balance will be equal to zero based on the conservation of mass and the assumption that nothing escapes the vessel. This means that the net accumulation will be zero:

$$\frac{d\rho V}{dt} = 0$$

If the net accumulation is zero for the overall mass in the vessel, then what can we say about the component balances? If we know that the reaction proceeds forward to completion, then the concentrations of the species in the reaction vessel change with time. Therefore even though the overall mass does not change, the mass of any given species, or component does change and this is what we measure. Consider the reaction:

$$A + B \rightarrow D + E$$

This is simple stiochiometrically and we can assume that it is highly favorable, meaning that chemical equilibrium predicted by thermodynamics lies far toward the right side and that product, D, returns to A and B very slowly. The *component mass balances* become:

$$\frac{dCaV}{dt} = \frac{dCbV}{dt} = -r_a V = -r_b V$$

$$\frac{dCdV}{dt} = \frac{dCeV}{dt} = r_d V = r_e V$$

$$r_a = r_b = -r_d = -r_e = \mathrm{r}$$

The reactants are considered to be decreasing in concentration, and the products are increasing in concentration. Thus, the rates of reaction for A and B are taken to be negative and the product rate is positive. The important point is that the rates are oppositely signed as is shown by the last expression. If the reaction does not produce a change in volume, then the control volume does not change, and the volume term is a constant that can be canceled across all the equations:

$$\frac{dCa}{dt} = \frac{dCb}{dt} = -r$$

$$\frac{dCd}{dt} = \frac{dCe}{dt} = r$$

The next step is to find kinetics for the rate of reaction that can be used as a constitutive relationship to replace the rate, **r**, on the right-hand side. When we say we look for a constitutive relationship for the kinetics to replace the right-hand side, we are seeking to make the equation *autonomous*. This means that we are seeking a function for the *RHS* that is explicit in the concentration of one or more of the components.

Simple irreversible reactions - first to N^{th} order

First order kinetics

The simplest case to consider by far is that of first order or linear kinetics in a constant volume batch reactor. If the rate of reaction is directly proportional to the concentration, then we refer to this rate as being first order in the concentration of reactant, and the right-hand side becomes:

$$\mathrm{r} = \mathrm{k\ Ca}$$

$$\frac{dCa}{dt} = - k\,Ca$$

$$Ca(t) = Cao\, e^{-kt}$$

This is an expression for the rate of decay of the concentration of species A. It should remind us of the expression we derived for the change in level of the draining tank, for which we used a linear constitutive relationship between level and rate of flow or the simplified rate of adsorption we used in the last chapter. The dimensions of **k** in this case are reciprocal time, i.e. sec^{-1} or min^{-1} etc. The reason for this is that the rate of reaction is given in dimensions of $\frac{moles}{volume\ time}$. Therefore to be dimensionally consistent the first order rate constant must have dimensions of inverse time.

Since the stoichiometry for the rate of reaction of component B is the same we can show that:

$$\frac{dCa}{dt} = \frac{dCb}{dt}$$

$$Ca - Cao = Cb - Cbo$$

$$Cb(t) = Ca(t) - (Cao + Cbo)$$

From which we find:

$$Cb(t) = Cao\, e^{-kt} - Cao + Cbo$$

$$= Cao\, [\, e^{-kt} - 1] + Cbo$$

In a similar fashion we find that the rate of appearance of component D is:

$$\frac{dCa}{dt} = - \frac{dCd}{dt}$$

$$Ca - Cao = - (Cd - Cdo)$$

$$Cd(t) = (Cao + Cdo) - Ca(t)$$

$$= (Cao + Cdo) - Cao\, e^{-kt}$$

$$= Cao[1 - e^{-kt}\,] + Cdo$$

The change in concentration of component E at any time would follow the same form with the substitution of Ceo for Cdo.

We can now plot these concentration functions so that we can see how a reaction system of this kind would behave in time. To do this we will assume that the concentrations of the products are both zero at time zero and that the initial concentrations of the two reactants are both equal. Also in order to make the behavior general, we will plot the change in the *ratio of the concentrations of the reactants to an initial concentration* of one of the reactants, Cao. We can go one step further and normalize the time coordinate with the "inverse reaction time." What is that in this case? For a first order reaction rate constant, its dimension is the reciprocal of time, inverse time. So, for the first order case, the rate constant is the inverse of the characteristic time for the chemical reaction. Therefore, if we multiply the rate constant, k, by real time, t, the result is dimensionless time which we shall refer to as τ. In fact we already had this result in hand. Look back at the expression for the change in concentration of A with time. We notice that the *RHS* has an exponential term, the argument of which is the product, **k t**. Since the exponential is a transcendental function, like sine, cosine etc., the argument must be a pure number which is dimensionless. Thus the solution of the differential equation that leads to this result naturally generates the dimensionless time, τ, simply as an outcome of the solution procedure.

We plan to plot Ca(τ)/Cao, Cb(τ)/Cao, Cd(τ)/Cdo, and Ce(τ)/Ceo against the dimensionless time, τ. The beauty of doing this is that the product, kt = τ, varies as natural numbers. This curve is dimensionless on both the ordinate and the abscissa, and that makes it general. In other words all first order rates when plotted in this dimensionless form will fall on this curve. We will come back to this shortly.

To emphasize the non-dimensional nature of the concentrations, we can introduce a new variable, the Greek letter Φ for dimensionless concentrations. When the initial concentration of B is divided by that of A, we will call this Φbo, and likewise for the other two species. The new expressions in dimensionless form will be:

$$\Phi a = e^{-\tau}$$

$$\Phi b = [\, e^{-\tau} - 1] + \Phi bo$$

$$\Phi d = [1 - e^{-\tau}] - \Phi do$$

$$\Phi e = [1 - e^{-\tau}] - \Phi eo$$

Since the initial concentrations of the products, D and E, are taken to be zero, the corresponding dimensionless initial concentrations are also zero. Also, we can see that Φd = Φe, so we will examine only Φd, and if Φbo =1, then Φa = Φb and we need only consider Φa:

```
firstordpl1 = Plot[
{N[Exp[−τ]], N[(1 − Exp[−τ])]}, {τ, 0, 5},
AxesLabel → {"τ", "Φa,Φd"},
PlotStyle → {Hue[0.0], Hue[0.5]},
Epilog → {Line[{{1, 0}, {1, N[Exp[−1]]}}],
Line[{{0, N[Exp[−1]]}, {1, N[Exp[−1]]}}]}
]
N[Exp[−1]]
```

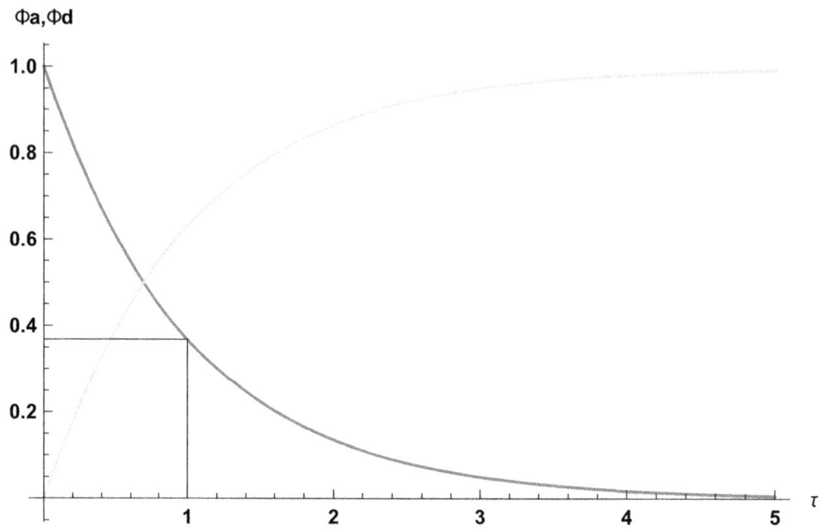

0.367879

The plots show that after time passes five times **k t**, the first order reaction will be near completion and Φa will asymptotically approach zero, just as Φd approaches 1. Also, the point on the curve above $\tau = 1$, is at Φa = 0.37. With raw concentration versus time experimental data, if we plot the dimensionless concentration against regular time, then we can extend a horizontal line from the point 0.37 on the ordinate to the curve and then extend a vertical line to the abscissa, then that time when multiplied by **k**, the rate constant, will be equal to unity. Therefore the reciprocal of that time is the value of the first order rate constant for the process. To test how well the first order rate fits the experimental data, one can plot the experimental (τ, Φa) points on this curve to determine the goodness of fit.

We can use the following data to execute this procedure. The initial concentration of species a was 3 x 10^{-2} mole/L:

t(sec)	Ca(mole/L)	t(sec)	Ca(mole/L)
0.	0.0300	1860.	0.0031
60.	0.0277	1920.	0.0027
120.	0.024	1980.	0.0019
180.	0.0234	2040.	0.0027
240.	0.0207	2100.	0.001
300.	0.0212	2160.	0.0005
360.	0.017	2220.	0.0005
420.	0.0171	2280.	0.0017
480.	0.0162	2340.	0.0015
540.	0.014	2400.	0.0009
600.	0.013	2460.	0.0005
660.	0.0124	2520.	0.0007
720.	0.0107	2580.	0.0007
780.	0.0108	2640.	0.0009
840.	0.0089	2700.	0.0018
900.	0.0086	2760.	0.0012
960.	0.0085	2820.	0.0011
1020.	0.0074	2880.	0.0007
1080.	0.0074	2940.	0.0004
1140.	0.0063	3000.	0.0007
1200.	0.0059	3060.	0.0004
1260.	0.0046	3120.	0.0001
1320.	0.0043	3180.	−0.0002
1380.	0.0047	3240.	0.0001
1440.	0.0037	3300.	0.0013
1500.	0.0046	3360.	−0.0002
1560.	0.0034	3420.	0.0003
1620.	0.0026	3480.	0.0006
1680.	0.0025	3540.	0.0002
1740.	0.0027	3600.	−0.0006
1800.	0.0028		

The data from the experiment are placed into a list of data points and named dat:

```
dat = {{0., 0.0300}, {60., 0.0277}, {120., 0.024},
{180., 0.0234}, {240., 0.0207}, {300., 0.0212},
{360., 0.017}, {420., 0.0171}, {480., 0.0162},
{540., 0.014}, {600., 0.013}, {660., 0.0124},
{720., 0.0107}, {780., 0.0108}, {840., 0.0089},
{900., 0.0086}, {960., 0.0085}, {1020., 0.0074},
{1080., 0.0074}, {1140., 0.0063}, {1200., 0.0059},
{1260., 0.0046}, {1320., 0.0043}, {1380., 0.0047},
{1440., 0.0037}, {1500., 0.0046}, {1560., 0.0034},
{1620., 0.0026}, {1680., 0.0025}, {1740., 0.0027},
{1800., 0.0028}, {1860., 0.0031}, {1920., 0.0027},
{1980., 0.0019}, {2040., 0.0027}, {2100., 0.001},
{2160., 0.0005}, {2220., 0.0005}, {2280., 0.0017},
{2340., 0.0015}, {2400., 0.0009}, {2460., 0.0005},
{2520., 0.0007}, {2580., 0.0007}, {2640., 0.0009},
{2700., 0.0018}, {2760., 0.0012}, {2820., 0.0011},
{2880., 0.0007}, {2940., 0.0004}, {3000., 0.0007},
{3060., 0.0004}, {3120., 0.0001}, {3180., −0.0002},
{3240., 0.0001}, {3300., 0.0013}, {3360., −0.0002},
{3420., 0.0003}, {3480., 0.0006}, {3540., 0.0002},
{3600., −0.0006}};
```

A plot of this data looks as follows:

```
ListPlot[dat,
AxesLabel → {"t (sec)", " Ca (moles/L)"}]
```

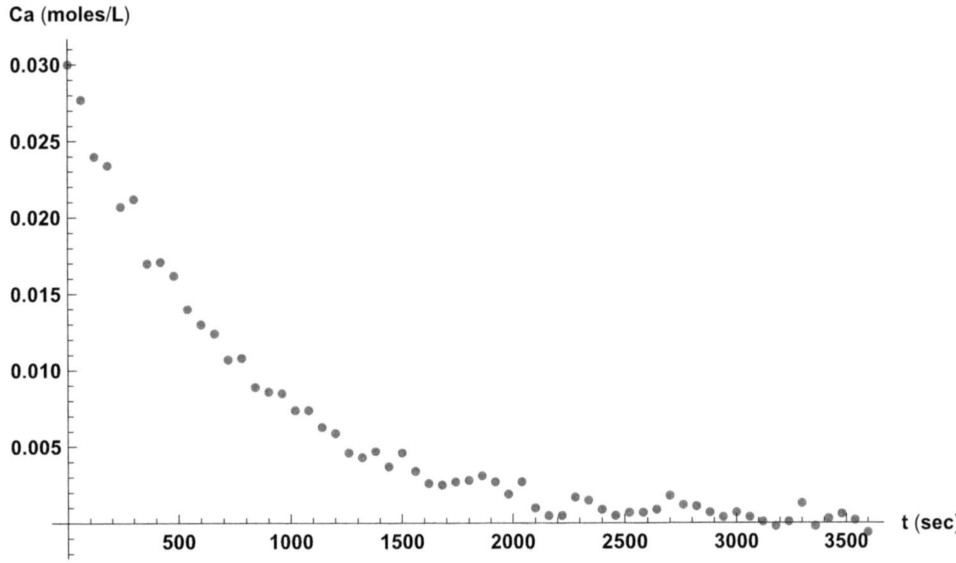

The data are somewhat noisy, notice that toward the end of the experiment, a few concentration points fall below zero, but we can tolerate that at this point. Next we will divide every concentration point by the initial concentration and replot the points and we will add a horizontal line at y = 0.37

$$\textbf{dat2} = \textbf{Table}\left[\left\{\textbf{dat[[}n\textbf{, 1]]}, \frac{\textbf{dat[[}n\textbf{, 2]]}}{\textbf{.03}}\right\}, \{n, 1, 60\}\right];$$

ListPlot[%, Epilog → Line[{{0, 0.37}, {1000, 0.37}}]]

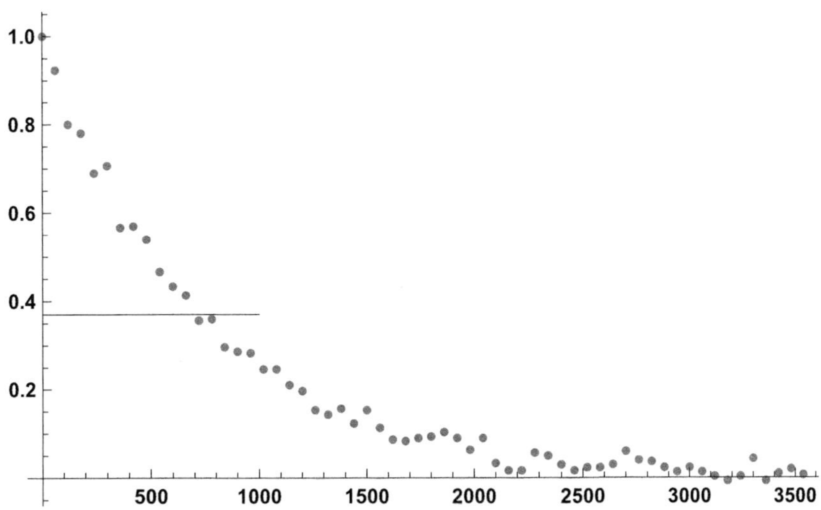

From the plot it looks as though the line cuts through at the 13th or 14th data point. These are:

dat2[[13]]

dat2[[14]]

$$\frac{1}{\text{dat2[[13, 1]]}}$$

$$\frac{1}{\text{dat2[[14, 1]]}}$$

{720., 0.356667}

{780., 0.36}

0.00138889

0.00128205

Hence, the data indicates that the rate constant is in the range from 0.0013 and 0.0014 $\sec^{-1}$. We can choose the value between these two and that corresponds to a time of 750 seconds. We can add a vertical line from the that point on the ordinate through the data.

$$\textbf{dat2} = \textbf{Table}\left[\left\{\textbf{dat[[}n\textbf{, 1]]}, \frac{\textbf{dat[[}n\textbf{, 2]]}}{\textbf{.03}}\right\}, \{n, 1, 60\}\right];$$

ListPlot[%,

Epilog → {Line[{{0, 0.37}, {900, 0.37}}],

Line[{{750, 0}, {750, 0.4}}]}]

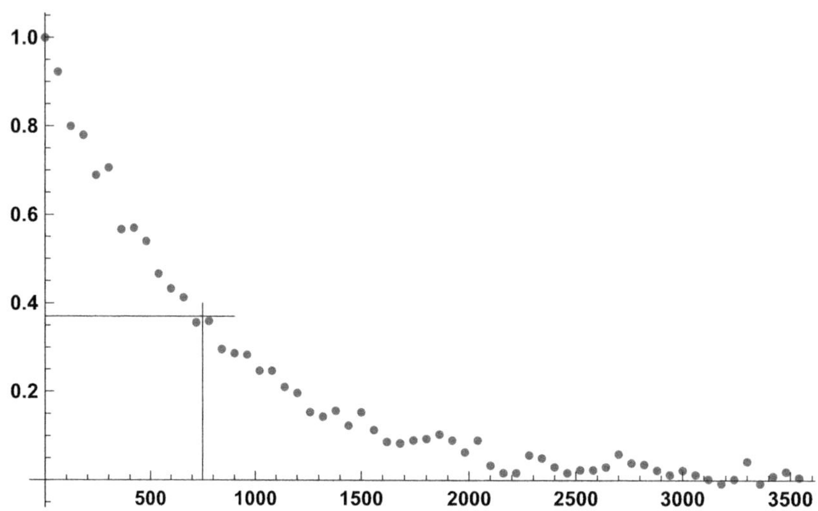

Lastly, we can put the data on the general graph for first order rates by normalizing all the time points with this rate constant, k = $\frac{1}{750}\text{sec}^{-1}$, and we can add the earlier graph as well:

$$\text{dat3} = \text{Table}\left[\left\{\frac{\text{dat[[}n\text{, 1]]}}{750}, \frac{\text{dat[[}n\text{, 2]]}}{.03}\right\}, \{n, 1, 60\}\right];$$

```
dat3pl =
ListPlot[%,
Epilog → {Line[{{0, 0.37}, {1.2, 0.37}}],
Line[{{1, 0}, {1, 0.4}}]}];
firstordpl1 = Plot[
N[Exp[−τ]], {τ, 0, 5},
AxesLabel → {"τ", "Φa"},
PlotStyle → {Hue[0.0]}
];
Show[%%, %]
```

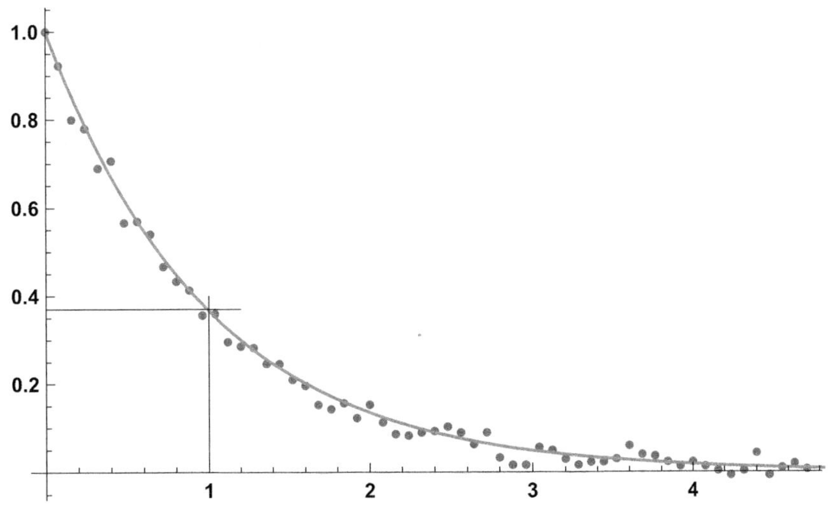

Given the error in the data, the goodness of fit appears to be acceptable. We need not rely on the eye to judge the fit, we can get the goodness of fit numerically. If we fit this to Exp[- a τ] then the value of "a" should be unity:

```
nlinm = NonlinearModelFit[dat3, Exp[−aτ], a, τ]
nlinm["FitResiduals"];
ListPlot[%, Filling->Axis, PlotLabel->FitResiduals]
nlinm["EstimatedVariance"]
nlinm["ANOVATable"]
nlinm["ANOVATableSumsOfSquares"]
```

FittedModel []

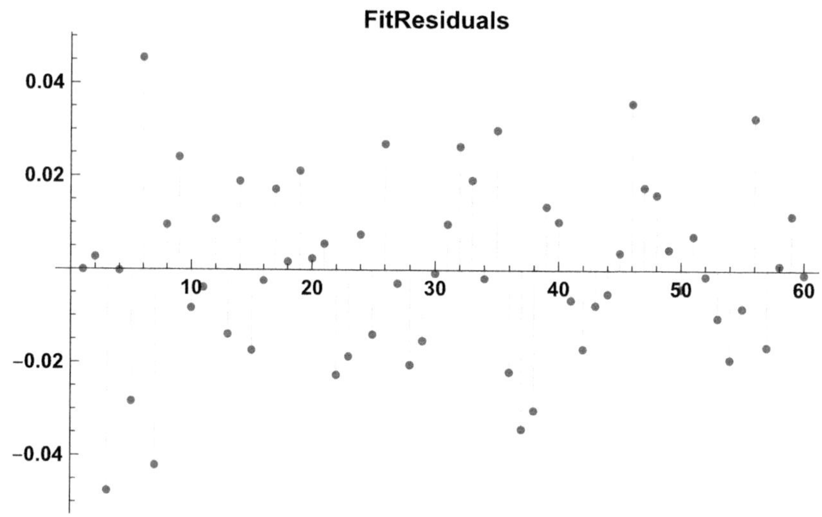

0.000359697

	DF	SS	MS
Model	1	6.48434	6.48434
Error	59	0.0212221	0.000359697
Uncorrected Total	60	6.50557	
Corrected Total	59	3.90279	

{6.48434, 0.0212221, 6.50557, 3.90279}

The statistics confirm what we saw by eye for this system that the fit is acceptable.

If the reaction that leads to the consumption of A is simply A alone converting to products, then first order kinetics are expected. On the other hand if the reaction is $A + B \rightarrow D + E$, then under the conditions of the experiment, the kinetics may appear to be first order, when in fact they are second order. This kind of observation leads kineticists to call the kinetics pseudo-first order. In the next section we will cover this explicitly.

Second order kinetics overall

For the reaction of A and B to produce C and D, it is likely that the kinetics would be second order overall, with first order in the concentration of both A and B, rather than just first order in A. If this were the case, then the solution would be different than that which we found above and would be derived as follows:

$$r = k\ \mathrm{Ca}\ \mathrm{Cb}$$

$$\frac{d\mathrm{Ca}}{\mathrm{dt}} = -\ k\ \mathrm{Ca}\ \mathrm{Cb}$$

This cannot be solved as written because we need an expression for Cb in terms of Ca in order to do the integration. This can be obtained by going back to the stoichiometric statement:

$$\frac{d\mathrm{Ca}}{\mathrm{dt}} = \frac{d\mathrm{Cb}}{\mathrm{dt}}$$

$$\int_{\mathrm{Cao}}^{\mathrm{Ca}} d\mathrm{Ca} == \int_{\mathrm{Cbo}}^{\mathrm{Cb}} d\mathrm{Cb}$$

$$\mathrm{Ca} - \mathrm{Cao} = \mathrm{Cb} - \mathrm{Cbo}$$

$$\mathrm{Cb} = \mathrm{Ca} - \mathrm{Cao} + \mathrm{Cbo}$$

$$\mathrm{Cb} = \mathrm{Ca} - (\mathrm{Cao} - \mathrm{Cbo})$$

Now, we can substitute this expression for Cb in terms of Ca into the differential equation describing the change in Ca, make it autonomous, and derive an expression for the time dependence of Ca:

$$\frac{d\mathrm{Ca}}{\mathrm{dt}} = -\ k\ \mathrm{Ca}\ \mathrm{Cb}$$

$$= -\ k\ \mathrm{Ca}\ [\mathrm{Ca} - (\mathrm{Cao} - \mathrm{Cbo})]$$

$$= -\ k\ \mathrm{Ca}^2 + k\ \mathrm{Ca}\ (\mathrm{Cao} - \mathrm{Cbo})$$

$$= -k\,[Ca^2 - Ca\,(Cao - Cbo)]$$

Notice that if Cao = Cbo, then the right-most term goes to zero and the rate expression just reduced to - k Ca^2. Using *Mathematica* we have two primary choices on how to proceed with the solution to this equation – we can rearrange it into its separable components and then integrate both sides of the equation, or we can solve it directly with DSolve, we will do the latter.

```
Clear[k, Ca, Cb, Cao, Cbo]

DSolve[{Ca'[t]== - k + kCa[t](Cao - Cbo),
Ca[0]==Cao}, Ca[t], t]
```

$$\left\{\left\{\mathrm{Ca}[t] \to \frac{(\mathrm{Cao}-\mathrm{Cbo})e^{\mathrm{Cao}kt+\frac{\mathrm{Cbo}(-\mathrm{Log}[\mathrm{Cao}]+\mathrm{Log}[\mathrm{Cbo}])}{\mathrm{Cao}-\mathrm{Cbo}}}}{-e^{\mathrm{Cbo}kt+\frac{\mathrm{Cao}(-\mathrm{Log}[\mathrm{Cao}]+\mathrm{Log}[\mathrm{Cbo}])}{\mathrm{Cao}-\mathrm{Cbo}}}+e^{\mathrm{Cao}kt+\frac{\mathrm{Cbo}(-\mathrm{Log}[\mathrm{Cao}]+\mathrm{Log}[\mathrm{Cbo}])}{\mathrm{Cao}-\mathrm{Cbo}}}}\right\}\right\}$$

This is a remarkable looking result. We find, however, that if we work through all the algebra, it is not complicated. As we see in the following:

```
DSolve[{Ca'[t]== - k + kCa[t](Cao - Cbo),
Ca[0]==Cao}, Ca[t], t];
%//FullSimplify
```

$$\left\{\left\{\mathrm{Ca}[t] \to \frac{\mathrm{Cao}(\mathrm{Cao}-\mathrm{Cbo})}{\mathrm{Cao}-\mathrm{Cbo}\,e^{(-\mathrm{Cao}+\mathrm{Cbo})kt}}\right\}\right\}$$

We check the solution by replacing it in the differential equation and simplifying.

$$\mathbf{Ca[t\_]:=\frac{Cao(Cao-Cbo)}{Cao-Cbo\,e^{(-Cao+Cbo)kt}}}$$

```
Simplify[∂tCa[t]== - k + kCa[t](Cao - Cbo)]
```

True

The solution is valid as LHS and RHS are equal. But notice that Ca[t] will go to zero, if Cao = Cbo. It is because, we implicitly assumed they were not equal when we solved the differential equation. If they were zero, then the right most term would disappear, but since we retained it, we made an implicit assumption that they were not equal.

This solution is one that we would like to explore as we did with the previous solution for the first order rate equation. To do so we could at this point convert the solution for Ca[t] into one that is non-dimensionalized. In fact it would be easier to have non-dimensionalized the equation to be solved in the first place, and we will learn more by doing so. Hence, we will re-work the differential equation and resolve it.

$$\frac{d\mathrm{Ca}}{\mathrm{dt}} = -\mathrm{k}\ \mathrm{Ca}^2 + \mathrm{k}\ \mathrm{Ca}\ (\mathrm{Cao} - \mathrm{Cbo})$$

Firstly, we will let Φa = Ca/Cao and Φb = Cb/Cao, once again. In this case we can multiply through on both sides of the equation by Cao/Cao. In the case of the first term on the *RHS*, we will multiply by $\mathrm{Cao}^2/\mathrm{Cao}^2$. This yields:

$$\mathrm{Cao}\ \frac{d\Phi\mathrm{a}}{\mathrm{dt}} = -\mathrm{k}\ \mathrm{Cao}^2\Phi\mathrm{a}^2 + \mathrm{k}\ \mathrm{Cao}\ \Phi\mathrm{a}\ (\mathrm{Cao} - \mathrm{Cbo})$$

If we divide each side by the residual Cao on the *LHS*, then we find:

$$\frac{d\Phi\mathrm{a}}{\mathrm{dt}} = -\mathrm{k}\ \mathrm{Cao}\Phi\mathrm{a}^2 + \mathrm{k}\ \Phi\mathrm{a}\ (\mathrm{Cao} - \mathrm{Cbo})$$

Instead of kt = τ from the first order rate, for the second order rate we have found that k Cao t = τ. We can group the rate constant and the initial concentration of A parenthetically to give:

$$\mathrm{k}\ \mathrm{Cao}\ \mathrm{t} = \tau$$

$$(\mathrm{k}\ \mathrm{Cao})\ \mathrm{t} = \tau$$

$$\mathrm{k'}\ \mathrm{t} = \tau$$

The product of k and Cao has units of inverse time, the same as the first order rate constant. Thus, we can identify the product (k Cao) as the new rate constant k' and this is now a *pseudo-first order rate constant*. Of course the choice of Cao was arbitrary; if we had chosen to non-dimensionalize in terms of Cbo, then k' would still be pseudo-first order, but it would be the product of Cbo and k. We can immediately see that if we had run a kinetics experiment with B in such great excess over A that its concentration change would have been undetectable, we would observe first order kinetics rather than second order overall kinetics.

Returning to the non-dimensionalizing of the equation; we had the following form:

$$\text{Cao}\,\frac{d\Phi a}{\text{dt}} = \text{-k Cao}^2\Phi a^2 + \text{k Cao }\Phi a\text{ (Cao - Cbo)}$$

By dividing both sides of this equation by (k Cao^2), we get the result that we were seeking, which is properly dimensionless on both sides:

$$\frac{d\Phi a}{k\text{Caodt}} = -\ \Phi a^2 + \ \Phi a\frac{(\text{Cao} - \text{Cbo})}{\text{Cao}}$$

The last step we take is to recognize that (k Cao dt) is the same as $d\tau$, rendering the equation as:

$$\frac{d\Phi a}{d\tau} = -\ \Phi a^2 + \ \Phi a\frac{(\text{Cao} - \text{Cbo})}{\text{Cao}}$$

Following the procedure from the latter half of the last section, we can now experiment with the behavior of this equation by solving it and plotting the dimensionless results. To do so let M replace the ratio involving the initial concentrations.

DSolve$\left[\left\{\Phi_a'[\tau] == -\Phi_a[\tau]^2 + M\Phi_a[\tau], \Phi_a[0] == \Phi_{a0}\right\},\right.$

$\left.\Phi_a[\tau], \tau\right]$

$$\left\{\left\{\Phi_a[\tau] \rightarrow \frac{e^{M\tau}M\Phi_{a0}}{M - \Phi_{a0} + e^{M\tau}\Phi_{a0}}\right\}\right\}$$

Thus, we find that:

$$\Phi_a[\tau] \rightarrow \frac{e^{M\tau}M\Phi_{a0}}{M - \Phi_{a0} + e^{M\tau}\Phi_{a0}}$$

This can be plotted to determine how the dimensionless concentration of A changes with time. We know that the concentration change for B will follow that of A, based on the stoichiometry. The change in concentration of the products D and E will also track each other, so we need only solve for one. To solve for the change in the dimensionless concentration of D, we recall that:

$$\frac{d\text{Ca}}{\text{dt}} = -\ \frac{d\text{Cd}}{\text{dt}}$$

$$\text{Ca - Cao} = \text{Cdo - Cd}$$

$$\therefore \text{Cd (t)} = \text{Cdo - Ca (t)} + \text{Cao}$$

If we divide every term by Cao to render this expression dimensionless, we find:

$$\Phi_d(\tau) = \Phi_{d0} - \Phi_a(\tau) + 1$$

Taking the initial concentration of D as zero and replacing for $\Phi a(\tau)$ we have:

$$\Phi_d(\tau) = 1 - \frac{e^{M\tau} M \Phi_{a0}}{M - \Phi_{a0} + e^{M\tau}\Phi_{a0}}$$

These two equations can now be plotted to determine their behavior after we assign initial values to Φao, Cao and Cbo and to M. Let Φao = 1, and Cbo = 1.05:

$$\Phi_{a0} = 1;$$

Cao = 1;

Cbo = 1.05;

$$M = \frac{(\text{Cao} - \text{Cbo})}{\text{Cao}};$$

secordpl1 =

$$\text{Plot}\left[\left\{N\left[\frac{e^{M\tau} M \Phi_{a0}}{M - \Phi_{a0} + e^{M\tau}\Phi_{a0}}\right], N\left[1 - \frac{e^{M\tau} M \Phi_{a0}}{M - \Phi_{a0} + e^{M\tau}\Phi_{a0}}\right]\right\},\right.$$

$\{\tau, 0, 5\}$,

AxesLabel → {"τ", "Φa,Φd"},

PlotStyle → {Hue[0.0], Hue[0.5]}]

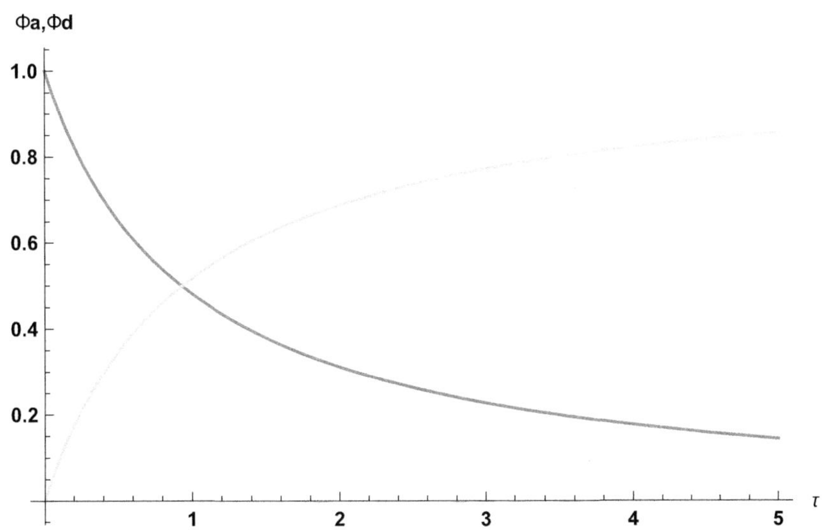

What happens if Cbo is equal to Cao? Then, as we stated in the beginning of this development, we will find that M = 0, and this solution is meaningless, because if they are equal, then the equation for this system changes as follows:

$$\frac{d\text{Ca}}{dt} = -\text{ k Ca Cb}$$

$$= -\text{ k Ca}^2 + \text{k Ca (Cao - Cbo)}$$

$$= -\text{ k Ca}^2$$

or in dimensionless form:

$$\frac{d\Phi\text{a}}{d\tau} = -\ \Phi\text{a}^2 +\ \Phi\text{a}\frac{(\text{Cao} - \text{Cbo})}{\text{Cao}}$$

$$\frac{d\Phi\text{a}}{d\tau} = -\ \Phi\text{a}^2$$

Working with the dimensionless form we have:

DSolve[{Φa′[τ]== − Φa[τ]², Φa[0] == Φao}, Φa[τ],

τ]

$$\left\{\left\{\Phi\text{a}[\tau] \to \frac{\Phi\text{ao}}{1 + \tau\Phi\text{ao}}\right\}\right\}$$

By the same stoichiometric relationship between D and A, but using this new solution for dimensionless A concentration we find:

$$\Phi_d(\tau) = \Phi_{\text{d0}} - \Phi_a(\tau) + 1$$

$$= \Phi_{\text{d0}} - \frac{\Phi_{\text{a0}}}{1 + \tau\Phi_{\text{a0}}} + 1$$

Using the same procedure as we used before, we can solve for and plot the new solution subject to the condition that Cao = Cbo and we obtain:

$\Phi_{a0} = 1$;

$\Phi_{d0} = 0$;

secordpl2 = Plot[

$$\left\{N\left[\frac{\Phi_{\text{a0}}}{1 + \tau\Phi_{\text{a0}}}\right], N\left[\Phi_{\text{d0}} - \frac{\Phi_{\text{a0}}}{1 + \tau\Phi_{\text{a0}}} + 1\right]\right\}, \{\tau, 0, 10\},$$

AxesLabel → {“τ”, “Φa,Φd”},

```
PlotStyle → {{Dashing[{0.15, 0.05}], Hue[0.0]},
{Dashing[{0.15, 0.05}], Hue[0.5]}}]
```

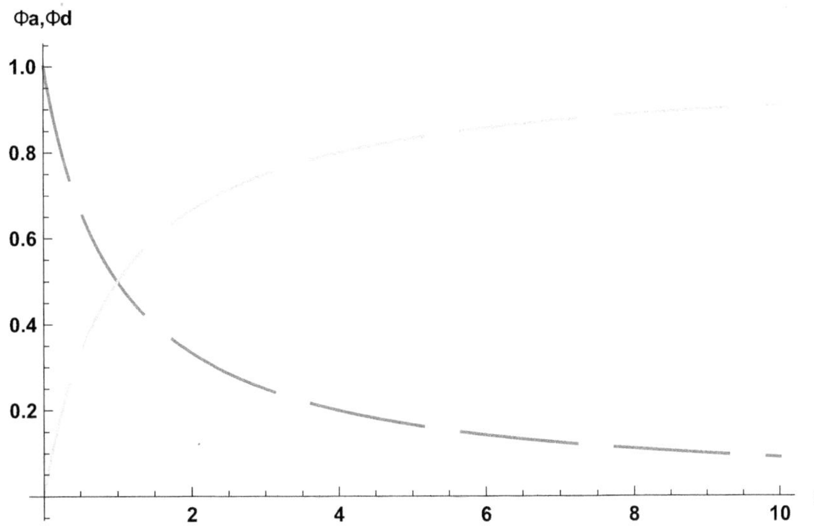

This set of two solutions can be compared to the set of two solutions that we obtained earlier with Cao ≠ Cbo.

```
Show[secordpl1, secordpl2,
PlotLabel → "Dashing for Cao=Cbo"]
```

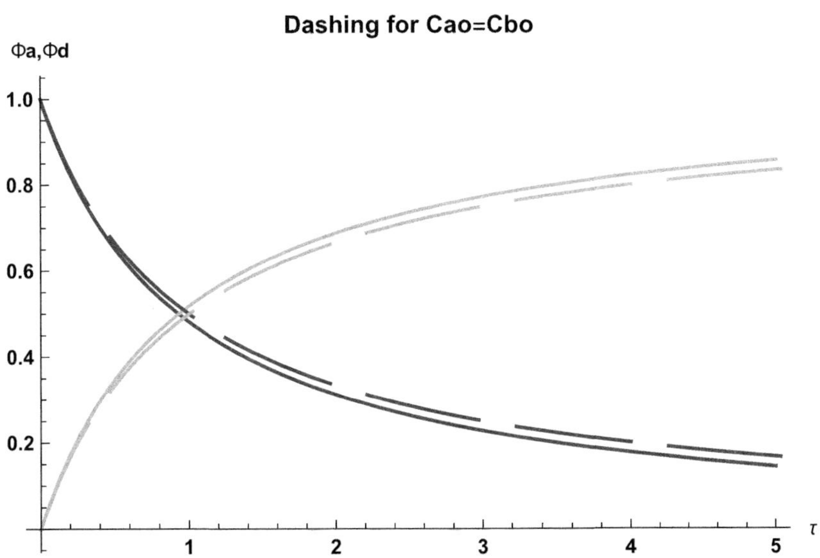

We see that with M = - 0.05, the first solution is almost the same as the second solution. As M gets even smaller the two solutions become close enough to consider them practically the same, but they are never exactly the same.

Given the form of the rate expression with k Ca^2, it is natural to wonder if it applies elsewhere and it does. The solution we derived for the case of A reacting with B and with equal initial concentrations to produce D and E, is also a description of the similar case in which A reacts with itself to give D:

$$A + A \rightarrow D$$

If this reaction happens to follow second order kinetics, then for every two moles of A reacted we get one mole of D, then the resultant analysis will lead to the same result we have just seen:

$$\frac{d\text{Ca}}{\text{dt}} = - \frac{d\text{Cd}}{\text{dt}}$$

$$\text{Ca - Cao = Cdo - Cd}$$

$$\text{Cd = Cdo + Cao - Ca}$$

$$\frac{d\text{Ca}}{\text{dt}} = - \text{k Ca}^2$$

$$\text{Ca}[t] = \frac{\text{Cao}}{1 + \text{Cao}kt}$$

and

$$\text{Cd }[\text{t}] = \text{Cdo} + \text{Cao}\left(1 - \frac{1}{1 + \text{Cao }kt}\right)$$

These are just the same expressions that we had already obtained for the second order case in which two different species were involved, and they had the same initial concentrations.

N^{th} order

The order of a reaction may not be as simple as first or second order. We often find non-integral order or "power-law" kinetics. This typically indicates that the "reaction" rate we have measured is not for a single reaction, that it is not for just one elementary step, but it is a rate for several elementary steps taken in their entirety. Normally, we refer to rate expressions such as these as *global rates or kinetics*. They are global in the sense of overall or measurable, as opposed to *intrinsic or fundamental rates and kinetics*. Consider the reaction of A to B:

$$A \rightarrow B$$

$$r_{A-} = \mathrm{k}\, C_A{}^n$$

$$\frac{dC_A}{dt} = -\mathrm{k}\, C_A{}^n$$

$$\frac{dC_B}{dt} = +\mathrm{k}\, C_A{}^n$$

When we non-dimensionalize, these become:

$$\frac{d\Phi_A}{d\tau} = -\, \Phi_A{}^n$$

$$\frac{d\Phi_B}{d\tau} = +\, \Phi_A{}^n$$

Solving for the concentrations of A and B we find:

```
Clear[Ca, Cao, Cbo, ca, k, τ, solnord]
solnthA = Simplify[
DSolve[
{Ca'[t] == -kCa[t]^n, Ca[0] == Cao},
{Ca[t]}, t]
  ];
ca[t]:=First[Evaluate[Ca[t]/.solnthA]]
ca[t]
```

$$\left(\mathrm{Cao}^{1-n} + k(-1+n)t\right)^{\frac{1}{1-n}}$$

```
solnthB = Simplify[
DSolve[
```

```
{Cb'[t] == +k(ca[t])^n, Cb[0] == Cbo},
{Cb[t]}, t]
  ];
cb[t]:=First[Evaluate[Cb[t]/.solnthB]]
cb[t]
```

$$\mathrm{Cao}^{1-n}\left(\left(\left(\mathrm{Cao}^{1-n}\right)^{\frac{1}{1-n}}\right)^{n}-\left(\left(\mathrm{Cao}^{1-n}+k(n-1)t\right)^{\frac{1}{1-n}}\right)^{n}\right)-k(n-1)t\left(\left(\mathrm{Cao}^{1-n}+k(n-1)t\right)^{\frac{1}{1-n}}\right)^{n}+\mathrm{Cbo}$$

Repeating, but on the basis of the non-dimensionalized equations we find:

```
Clear[Φa, Φb, Φao, Φbo, n, τ]
nda = DSolve[
{Φa'[τ] == −Φa[τ]^n, Φa[0] == Φao}, Φa[τ], τ];
Φa1[τ_]:=First[Evaluate[Φa[τ]/.nda]]//
FullSimplify;
Print["Φa1[τ]=", Φa1[τ]]

ndb = Simplify[DSolve[
{Φb'[τ] == Φa1[τ]^n, Φb[0] == 0}, Φb[τ], τ]];
Φb1[τ_]:=First[Evaluate[Φb[τ]/.ndb]]//
FullSimplify;
Print["Φb1[τ]=", Φb1[τ]]
```

$$\Phi\mathrm{a1}[\tau]=\left((n-1)\tau+\Phi\mathrm{ao}^{1-n}\right)^{\frac{1}{1-n}}$$

$$\Phi\mathrm{b1}(\tau)=-(n-1)\tau\left(\left((n-1)\tau+\Phi\mathrm{ao}^{1-n}\right)^{\frac{1}{1-n}}\right)^{n}+\Phi\mathrm{ao}^{1-n}\left(\left(\left(\Phi\mathrm{ao}^{1-n}\right)^{\frac{1}{1-n}}\right)^{n}-\left(\left((n-1)\tau+\Phi\mathrm{ao}^{1-n}\right)^{\frac{1}{1-n}}\right)^{n}\right)$$

```
n = 1.5;
```

```
Φao = 1;
Plot[{Φa1[τ], Φb1[τ]}, {τ, 0, 10},
AxesLabel → {"τ", "Φa,Φd"},
PlotStyle → {Hue[0.0], Hue[0.5]},
PlotLabel → n"order"]
```

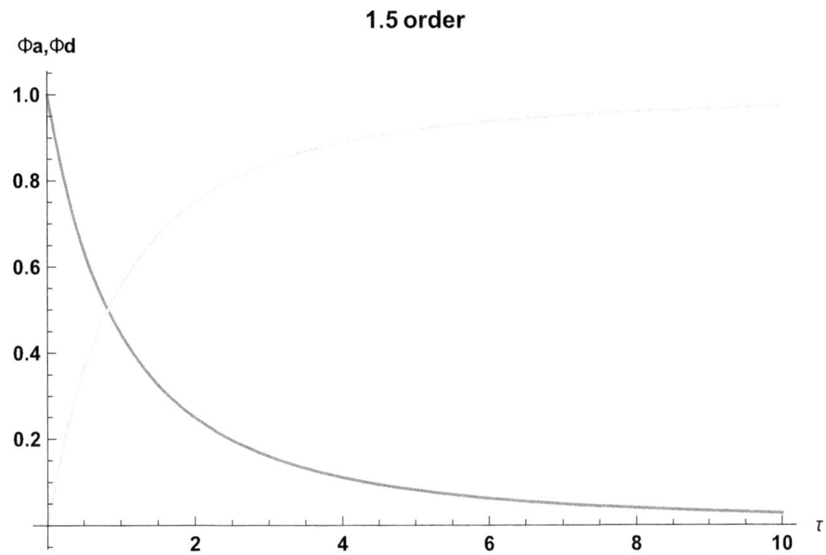

This was a fast excursion through simple, empirical chemical kinetics of highly favorable reactions, for which we can ignore the back reaction and its rate. We can also see how non-dimensionalizing, even in systems of equations such as these, is helpful.

Reversible reactions - chemical equilibrium

Chemical reactions do not move in the forward direction only, but move in either direction and come to position of chemical equilibrium. Our goal in this section is to understand how we analyze such a common situation and at the same time seek the interrelationships between kinetics and thermodynamics, as they apply to chemical systems.

Take as a starting point the most simple, reversible reaction:

$$A \Longleftrightarrow B$$

This is "simple" because the stoichiometry is one mole of reactant goes to one mole of product, and because the conversion of A to B follows first order kinetics, as does the conversion of B back to A. Thus, when we assemble the two component mass balance equations in a constant volume batch reactor, we find:

$$\frac{d\text{Ca}}{\text{dt}} = -r_a + r_b$$

$$\frac{d\text{Cb}}{\text{dt}} = r_a - r_b$$

These expressions reflect the fact that the overall rate of "accumulation" of either species will be the difference between their rates of formation and depletion, that is the net rate. If we take both r_a and r_b as first order in Ca and Cb, respectively, then we have:

$$\frac{d\text{Ca}}{\text{dt}} = -k_a\text{Ca} + k_b\text{Cb}$$

$$\frac{d\text{Cb}}{\text{dt}} = k_a\text{Ca} - k_b\text{Cb}$$

These two equations can be solved simultaneously to give Ca [t] and Cb [t] for any arbitrary initial concentrations of A and B.

```
Clear[Ca, Cb, Φa, Φb]
rvrsol1 = Simplify[DSolve[
{Ca'[t] == −kaCa[t] + kbCb[t],
Cb'[t] == +kaCa[t] − kbCb[t],
Ca[0] == Cao, Cb[0] == Cbo},
{Ca[t], Cb[t]}, t]];
Φa[t_]:=rvrsol1[[1]][[1]][[2]]/Cao
Φb[t_]:=rvrsol1[[1]][[2]][[2]]/Cao
Φa[t]
Φb[t]
```

$$\frac{\text{Cbo}\left(1 - e^{-(\text{ka}+\text{kb})t}\right)\text{kb} + \text{Cao}\left(e^{-(\text{ka}+\text{kb})t}\text{ka} + \text{kb}\right)}{\text{Cao}(\text{ka} + \text{kb})}$$

$$\frac{\text{Cao}\left(\text{ka} - e^{-(\text{ka}+\text{kb})t}\text{ka}\right) + \text{Cbo}\left(\text{ka} + e^{-(\text{ka}+\text{kb})t}\text{kb}\right)}{\text{Cao}(\text{ka}+\text{kb})}$$

We could have non-dimensionalized these equations completely, as we have done for other cases, but then we would lose the individual contributions of ka and kb, so instead we have referenced to the initial concentration of A, but we have retained the real time, t.

ka = 1.;

kb = 1.;

Cao = 1.;

Cbo = 0.;

Plot[{N[Φa[t]], N[Φb[t]]}, {t, 0, 5},

PlotRange → All,

AxesLabel → {t, "Φa,Φd"},

PlotStyle → {Hue[0.0], Hue[0.5]},

PlotLabel → {ka"=ka", kb"=kb", Cao"=Cao",

Cbo"=Cbo"}]

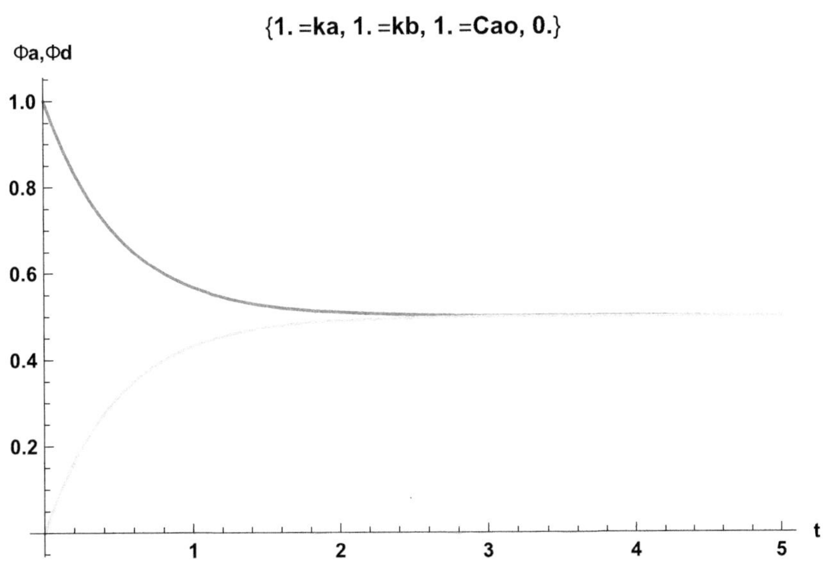

What we observe is that the concentrations of A and B move to a value of 0.5 by t = 3 and, then, they remain unchanged. This is the *equilibrium point* for the parameters that we set. We could have calculated this before solving the differential equations explicitly. The reason is that at equilibrium the rates of the forward and reverse reaction are equal, that is why the system appears unchanging. Given that this is the case, then we can reason that the accumulation terms, that is the differentials on the *LHS* are zero-valued, because their arguments are no longer time dependent. Thus, after fully non-dimensionalizing, we can see that:

$$\frac{d\Phi a}{d\tau} = -\Phi a + \frac{k_b}{(k_a + k_b)}(1 + \Phi bo)$$

$$\frac{d\Phi a}{d\tau}\Big|_{eq} = 0$$

$$\therefore \quad \Phi a\,|_{eq} = \frac{k_b}{(k_a + k_b)}(1 + \Phi bo)$$

Since we chose ka = kb and Φbo = 0, we find that:

$$\Phi a\,|_{eq} = 0.5$$

We can also note that the kinetics relate directly to the thermodynamics (equilibrium) in this manner:

$$0 = -\Phi a + \frac{k_b}{(k_a + k_b)}(1 + \Phi bo)$$

$$0 = -\Phi a\,(k_a + k_b) + k_b(1 + \Phi bo)$$

$$0 = -\Phi a\,k_a - \Phi a k_b + k_b(1 + \Phi bo)$$

$$0 = -\Phi a\,k_a + k_b(1 + \Phi bo - \Phi a)$$

But we have already shown that:

$$\Phi b = 1 + \Phi bo - \Phi a$$

$$\therefore \quad 0 = -\Phi a\,k_a + \Phi b\,k_b$$

$$\frac{\Phi b|_{eq}}{\Phi a|_{eq}} = \frac{k_a}{k_b} = K_{eq}$$

This well-known result provides the kinetics definition of chemical equilibrium and relates the rate constant from thermodynamics to the ratio of the forward and reverse rate constants.

One more example of mixed order is worth analyzing in the same way. Let us take as an example the case of one molecule dividing into two different molecules. Examples abound – PCl_5 reacts to give PCl_3 and Cl_2, ethylbenzene ($C_6H_5CH_2CH_3$) reacts to give styrene ($C_6H_5CH = CH_2$) and dihydrogen (H_2). We can generalize this type of reaction to:

$$A \longrightarrow B + D$$

For this case, we will assume that in the direction from A to B and D, the rate is first order in A and in the opposite direction, we will take it as second order overall, first order in B and D each.

$$r_A = k_A\, C_A$$

$$r_B = r_D = k_B\, C_B C_D$$

These are constitutive kinetics that we need to complete the model for this type of reaction taking place in a constant volume batch reactor. The component equations are:

$$\frac{dC_A}{\mathrm{dt}} = -\ k_A C_A + k_B C_B C_D$$

$$\frac{dC_B}{\mathrm{dt}} = \ k_A C_A - k_B C_B C_D$$

$$\frac{dC_D}{\mathrm{dt}} = \ k_A C_A - k_B C_B C_D$$

If we try to solve the three simultaneous equations in their initial form, here is the result we get back:

```
Remove[Ca, Cb, Cd, ka, kb]

DSolve[

{Ca'[t] == −kaCa[t] + kbCb[t]Cd[t],

Cb'[t] == +kaCa[t] − kbCb[t]Cd[t],

Cd'[t] == +kaCa[t] − kbCb[t]Cd[t],

Ca[0] == Cao, Cb[0] == Cbo, Cd[0] == Cdo},

{Ca[t], Cb[t], Cd[t]}, t]
```

Solve::tdep	The equations appear to involve the variables to be solved for in an essentially non-algebraic way.
DSolve::dsing	Unable to fit initial/boundary conditions {Ca[0] == 1, Cb[0] == 0, Cd[0] == Cdo }.

{ }

Alternatively, if we set the initial concentrations of B and D equal, then we can re-express Cd in terms of Cb. Now, we can solve analytically as follows:

Remove[ca, cb, Ca, Cb, Cd, ka, kb, Cao, Cbo, *t*]

Clear $[k_A, k_B, C_{\mathrm{Ao}}, C_{\mathrm{Bo}}, \Phi_{\mathrm{Ao}}, \Phi_{\mathrm{Bo}}, \Phi_{\mathrm{Do}}]$

```
rvrsol2 = FullSimplify[DSolve[
{Ca'[t] == −kaCa[t] + kbCb[t]^2,
Cb'[t] == +kaCa[t] − kbCb[t]^2,
Ca[0] == Cao, Cb[0] == Cbo},
{Ca[t], Cb[t]}, t]]
cb1[t_]:=Evaluate[Cb[t]/.rvrsol2[[1, 1]]]
ca1[t_]:=Evaluate[Ca[t]/.rvrsol2[[1, 2]]]
cb2[t_]:=Evaluate[Cb[t]/.rvrsol2[[2, 1]]]
ca2[t_]:=Evaluate[Ca[t]/.rvrsol2[[2, 2]]]
```

$$\left\{\left\{\mathrm{Cb}[t] \to \frac{1}{2\mathrm{kb}}\left(-\mathrm{ka} + \sqrt{\mathrm{ka}}\sqrt{\mathrm{ka} + 4(\mathrm{Cao} + \mathrm{Cbo})\mathrm{kb}}\right.\right.\right.$$

$$\mathrm{Tanh}\left[\tfrac{1}{2}\sqrt{\mathrm{ka}}\sqrt{\mathrm{ka} + 4(\mathrm{Cao} + \mathrm{Cbo})\mathrm{kb}}\,t\right.$$

$$\left.\left.\left. -\mathrm{ArcTanh}\left[\frac{\mathrm{ka} + 2\mathrm{Cbokb}}{\sqrt{\mathrm{ka}}\sqrt{\mathrm{ka} + 4(\mathrm{Cao} + \mathrm{Cbo})\mathrm{kb}}}\right]\right]\right),$$

$$\mathrm{Ca}[t] \to \frac{1}{2\mathrm{kb}}\left(\mathrm{ka} + 2(\mathrm{Cao} + \mathrm{Cbo})\mathrm{kb} - \sqrt{\mathrm{ka}}\sqrt{\mathrm{ka} + 4(\mathrm{Cao} + \mathrm{Cbo})\mathrm{kb}}\right.$$

$$\mathrm{Tanh}\left[\tfrac{1}{2}\sqrt{\mathrm{ka}}\sqrt{\mathrm{ka} + 4(\mathrm{Cao} + \mathrm{Cbo})\mathrm{kb}}\,t\right.$$

$$-\text{ArcTanh}\left[\frac{\text{ka}+2\text{Cbokb}}{\sqrt{\text{ka}}\sqrt{\text{ka}+4(\text{Cao}+\text{Cbo})\text{kb}}}\right]\Big]\Big)\Big\},$$

$$\Big\{\text{Cb}[t]\to\frac{1}{2\text{kb}}\Big(-\text{ka}+\sqrt{\text{ka}}\sqrt{\text{ka}+4(\text{Cao}+\text{Cbo})\text{kb}}$$

$$\text{Tanh}\Big[\tfrac{1}{2}\sqrt{\text{ka}}\sqrt{\text{ka}+4(\text{Cao}+\text{Cbo})\text{kb}}t$$

$$+\text{ArcTanh}\left[\frac{\text{ka}+2\text{Cbokb}}{\sqrt{\text{ka}}\sqrt{\text{ka}+4(\text{Cao}+\text{Cbo})\text{kb}}}\right]\Big]\Big),$$

$$\text{Ca}[t]\to\frac{1}{2\text{kb}}\Big(\text{ka}+2(\text{Cao}+\text{Cbo})\text{kb}-\sqrt{\text{ka}}\sqrt{\text{ka}+4(\text{Cao}+\text{Cbo})\text{kb}}$$

$$\text{Tanh}\Big[\tfrac{1}{2}\sqrt{\text{ka}}\sqrt{\text{ka}+4(\text{Cao}+\text{Cbo})\text{kb}}t$$

$$+\text{ArcTanh}\left[\frac{\text{ka}+2\text{Cbokb}}{\sqrt{\text{ka}}\sqrt{\text{ka}+4(\text{Cao}+\text{Cbo})\text{kb}}}\right]\Big]\Big)\Big\}\Big\}$$

We have four solutions, two for Ca[t] and two for Cb[t]. The solutions involve the hyperbolic tangent function. That may seem surprising, but it's reasonable as we will see shortly. For this case, it is not immediately obvious which of the pairs are the physically meaningful solutions. The easiest approach is to apply regular parameter values and to graph them:

```
ka = 0.01;

kb = 30;

Cao = .001;

Cbo = 0;

Plot[{ca1[t], ca2[t]}, {t, 0, 150},

PlotStyle → Hue[0.6], PlotLabel → "ca1, ca2"]

Plot[{cb1[t], cb2[t]}, {t, 0, 150},

PlotStyle → Hue[0.1], PlotLabel → "cb1, cb2"]
```

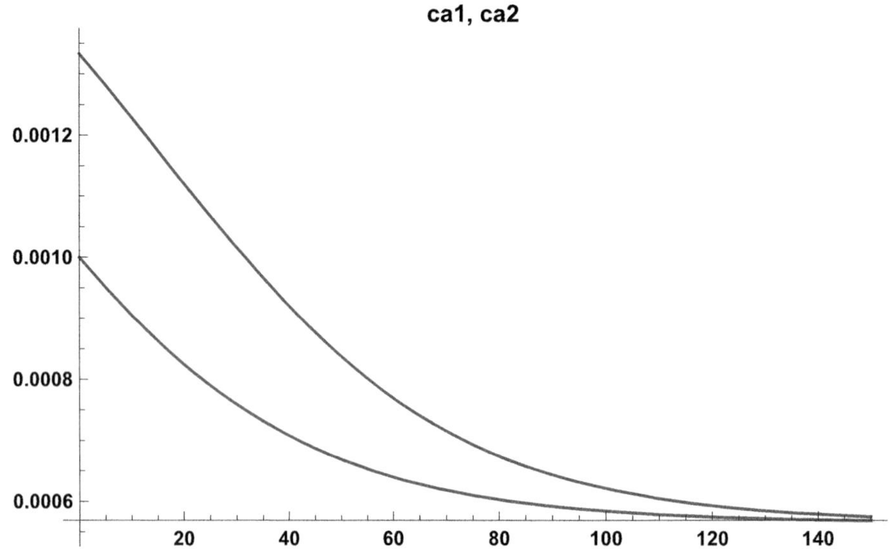

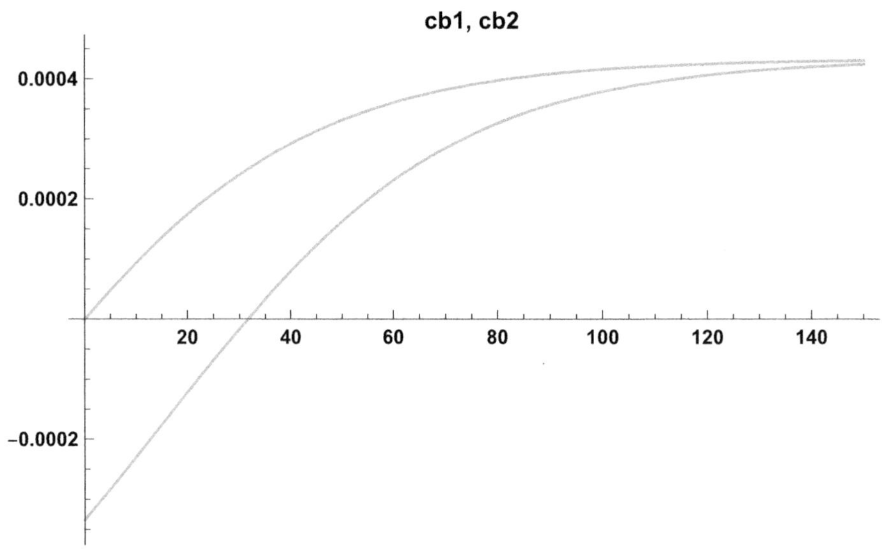

It is now clear that the second solution in each pair is the physically realistic solution as negative concentrations are frowned upon. When we take these two solutions and graph them with these parameter values, we see smooth behavior and the onset of chemical equilibrium.

```
cnc = Plot[{ca2[t], cb2[t]}, {t, 0, 150},
PlotStyle → {Hue[0.6], Hue[0.1]},
AxesLabel → {t, "Cᵢ[t]"}, PlotRange → All]
```

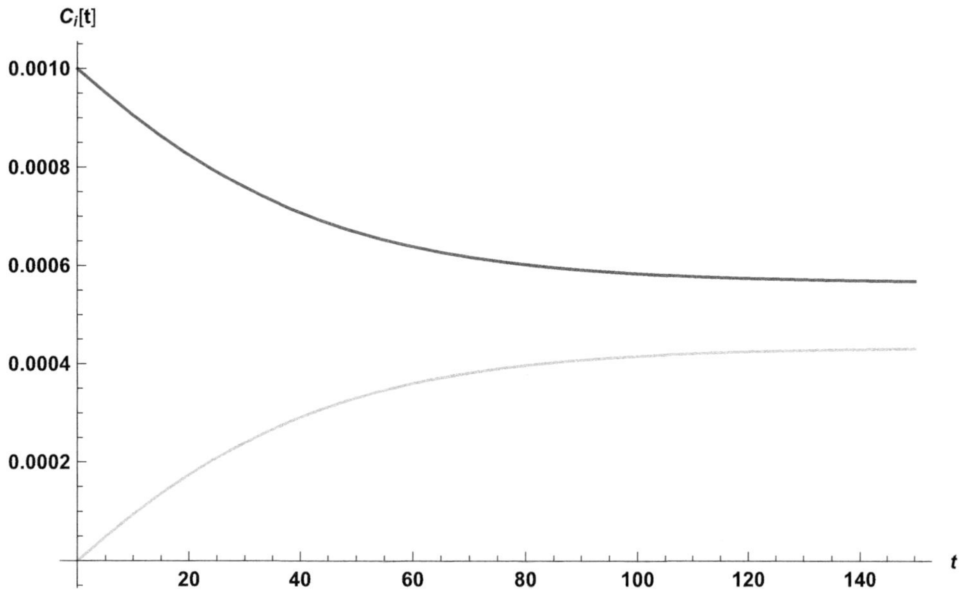

Now let's go back to the equations and non-dimensionalize them:

$$\partial_t \mathrm{Ca}[t] == -\mathrm{kaCa}[t] + \mathrm{kbCb}[t]^2$$

$$\partial_t \mathrm{Cb}[t] == +\mathrm{kaCa}[t] - \mathrm{kbCb}[t]^2$$

We begin with the first equation:

$$\frac{\mathrm{Cao}}{\mathrm{Cao}}\partial_t \mathrm{Ca}[t] == -\mathrm{ka}\frac{\mathrm{Cao}}{\mathrm{Cao}}\mathrm{Ca}[t] + \mathrm{kb}\frac{\mathrm{Cao}^2}{\mathrm{Cao}^2}\mathrm{Cb}[t]^2$$

$$Cao\partial_t \phi a[t] == -\mathrm{kaCao}\phi a[t] + kb\mathrm{Cao}^2\phi b[t]^2$$

Recognizing that $k_B C_{\mathrm{Ao}}{}^2$ is in every term we can divide through by the product of this concentration parameter and the rate constant for the reverse reaction. This product has units of inverse time since the rate constant is second order. Therefore on the left-hand side we have $\frac{1}{C_{\mathrm{Ao}}k_B}\frac{d\Phi_A}{\mathrm{dt}}$ which is just the same as $\frac{d\Phi_A}{d\tau}$ where $\mathrm{d}\tau = C_{\mathrm{Ao}}k_B\mathrm{dt}$:

$$\frac{\partial_t \phi a[t]}{\mathrm{kbCao}} == -\frac{\mathrm{ka}}{\mathrm{kbCao}}\phi a[t] + \phi b[t]^2$$

$$\partial_t \phi a[\tau] == -\frac{\mathrm{ka}}{\mathrm{kbCao}}\phi a[\tau] + \phi b[\tau]^2$$

and for B we have:

$$\partial_t \phi b[\tau] == \frac{ka}{kbCao} \phi a[\tau] - \phi b[\tau]^2$$

This makes the equations in complete dimensionless form. The group of constants, $\frac{k_A}{k_B C_{Ao}}$ that make up the coefficient of the linear term in Φ_A are worth looking at in more detail. Recall that k_B is a second order rate constant with dimensions of vol/ mol/ time, when this is multiplied by initial concentrations, C_{Ao}, the resultant dimensions are 1/time, the same dimensions as that of k_A, which is the other term in the numerator. These are also the dimensions of the product $C_{Ao}k_B$ seen in the denominator. This makes sense and the overall group is dimensionless. When we provide the relative magnitudes of the rate constants and the initial concentration, then this term can be evaluated. We can solve these equations in their dimensionless concentrations and then compare these results to those obtained from the solutions of the fully dimensional equations that we already have derived. Here are those solutions:

Remove[ca, cb, Ca, Cb, Cd, ka, kb, Cao, Cbo, t]

Clear [k_A, k_B, C_{Ao}, C_{Bo}, ϕa, ϕb, ϕao, ϕbo, Φ_{Ao},

Φ_{Bo}, Φ_{Do}, ca, cb, Ca, Cb, Cd, ka, kb, Cao, Cbo, t]

ka=.

kb=.

C_{Ao}=.

ϕao=.

ϕbo=.

ndrvrsol3 = FullSimplify[DSolve[

$$\left\{\phi a'[t] == -\frac{ka}{kbC_{Ao}} \phi a[t] + \phi b[t]^2,\right.$$

$$\phi b'[t] == +\frac{ka}{kbC_{Ao}} \phi a[t] - \phi b[t]^2,$$

ϕa[0] == ϕao, ϕb[0] == ϕbo},

{ϕa[t], ϕb[t]}, t]]

Φb1[t_]:=Evaluate[ϕb[*t*]/.ndrvrsol3[[1, 1]]]

Φa1[t_]:=Evaluate[ϕa[*t*]/.ndrvrsol3[[1, 2]]]

Φb2[t_]:=Evaluate[ϕb[*t*]/.ndrvrsol3[[2, 1]]]

Φa2[t_]:=Evaluate[ϕa[*t*]/.ndrvrsol3[[2, 2]]]

$$\left\{\left\{\phi\mathrm{b}[t]\to-\frac{1}{2\mathrm{kb}C_{\mathrm{Ao}}}\left(\mathrm{ka}+\sqrt{\mathrm{ka}}\sqrt{\mathrm{ka}+4\mathrm{kb}(\phi\mathrm{ao}+\phi\mathrm{bo})C_{\mathrm{Ao}}}\right.\right.\right.$$
$$\left.\mathrm{Tanh}\left[\mathrm{ArcTanh}\left[\frac{\mathrm{ka}+2\mathrm{kb}\phi\mathrm{bo}C_{\mathrm{Ao}}}{\sqrt{\mathrm{ka}}\sqrt{\mathrm{ka}+4\mathrm{kb}(\phi\mathrm{ao}+\phi\mathrm{bo})C_{\mathrm{Ao}}}}\right]-\frac{\sqrt{\mathrm{ka}}t\sqrt{\mathrm{ka}+4\mathrm{kb}(\phi\mathrm{ao}+\phi\mathrm{bo})C_{\mathrm{Ao}}}}{2\mathrm{kb}C_{\mathrm{Ao}}}\right]\right),$$
$$\phi\mathrm{a}[t]\to\frac{1}{2\mathrm{kb}C_{\mathrm{Ao}}}\left(\mathrm{ka}+2\mathrm{kb}(\phi\mathrm{ao}+\phi\mathrm{bo})C_{\mathrm{Ao}}+\sqrt{\mathrm{ka}}\sqrt{\mathrm{ka}+4\mathrm{kb}(\phi\mathrm{ao}+\phi\mathrm{bo})C_{\mathrm{Ao}}}\right.$$
$$\left.\left.\mathrm{Tanh}\left[\mathrm{ArcTanh}\left[\frac{\mathrm{ka}+2\mathrm{kb}\phi\mathrm{bo}C_{\mathrm{Ao}}}{\sqrt{\mathrm{ka}}\sqrt{\mathrm{ka}+4\mathrm{kb}(\phi\mathrm{ao}+\phi\mathrm{bo})C_{\mathrm{Ao}}}}\right]-\frac{\sqrt{\mathrm{ka}}t\sqrt{\mathrm{ka}+4\mathrm{kb}(\phi\mathrm{ao}+\phi\mathrm{bo})C_{\mathrm{Ao}}}}{2\mathrm{kb}C_{\mathrm{Ao}}}\right]\right)\right\},$$
$$\left\{\phi\mathrm{b}[t]\to\tfrac{1}{2\mathrm{kb}C_{\mathrm{Ao}}}\left(-\mathrm{ka}+\sqrt{\mathrm{ka}}\sqrt{\mathrm{ka}+4\mathrm{kb}(\phi\mathrm{ao}+\phi\mathrm{bo})C_{\mathrm{Ao}}}\right.\right.$$
$$\left.\mathrm{Tanh}\left[\mathrm{ArcTanh}\left[\frac{\mathrm{ka}+2\mathrm{kb}\phi\mathrm{bo}C_{\mathrm{Ao}}}{\sqrt{\mathrm{ka}}\sqrt{\mathrm{ka}+4\mathrm{kb}(\phi\mathrm{ao}+\phi\mathrm{bo})C_{\mathrm{Ao}}}}\right]+\frac{\sqrt{\mathrm{ka}}t\sqrt{\mathrm{ka}+4\mathrm{kb}(\phi\mathrm{ao}+\phi\mathrm{bo})C_{\mathrm{Ao}}}}{2\mathrm{kb}C_{\mathrm{Ao}}}\right]\right),$$
$$\phi\mathrm{a}[t]\to\tfrac{1}{2\mathrm{kb}C_{\mathrm{Ao}}}\left(\mathrm{ka}+2\mathrm{kb}(\phi\mathrm{ao}+\phi\mathrm{bo})C_{\mathrm{Ao}}-\sqrt{\mathrm{ka}}\sqrt{\mathrm{ka}+4\mathrm{kb}(\phi\mathrm{ao}+\phi\mathrm{bo})C_{\mathrm{Ao}}}\right.$$
$$\left.\left.\left.\mathrm{Tanh}\left[\mathrm{ArcTanh}\left[\frac{\mathrm{ka}+2\mathrm{kb}\phi\mathrm{bo}C_{\mathrm{Ao}}}{\sqrt{\mathrm{ka}}\sqrt{\mathrm{ka}+4\mathrm{kb}(\phi\mathrm{ao}+\phi\mathrm{bo})C_{\mathrm{Ao}}}}\right]+\frac{\sqrt{\mathrm{ka}}t\sqrt{\mathrm{ka}+4\mathrm{kb}(\phi\mathrm{ao}+\phi\mathrm{bo})C_{\mathrm{Ao}}}}{2\mathrm{kb}C_{\mathrm{Ao}}}\right]\right)\right\}\right\}$$

These are the solutions of the dimensionless equations. Notice that these four solutions contain TangentHyperbolic and ArcTangentHyperbolic functions. That is expected and is consistent with the solutions we derived from the dimensionalized equations. When we plot these solutions, we see that they mimic the dimensionalized solution:

ka = 0.01;

kb = 30;

C_{Ao} = .001;

ϕao = 1;

```
φbo = 0;
tmax = 150;
Table[{Φa1[τ], Φa2[τ]}, {τ, 0, kbCAotmax, 1}]
Plot[{Φa1[τ], Φa2[τ]}, { τ, 0, kbCAotmax},
PlotStyle → Hue[0.6], PlotLabel → "Φa1, Φa2",
PlotRange → All]
Table[{Φb1[τ], Φb2[τ]}, {τ, 0, kbCAotmax}]
Plot[{Φb1[τ], Φb2[τ]}, {τ, 0, kbCAotmax},
PlotStyle → Hue[0.1], PlotLabel → "Φb1, Φb2"]
Plot[{Φa2[τ], Φb2[τ]}, {τ, 0, kbCAotmax},
PlotStyle → {Hue[0.6], Hue[0.1]},
PlotLabel → "Φa2, Φb2", PlotRange → All]
```

{{1.33333, 1.}, {0.982789, 0.740442}, {0.731299, 0.624207}, {0.620821, 0.583937}, {0.582849, 0.57127}}

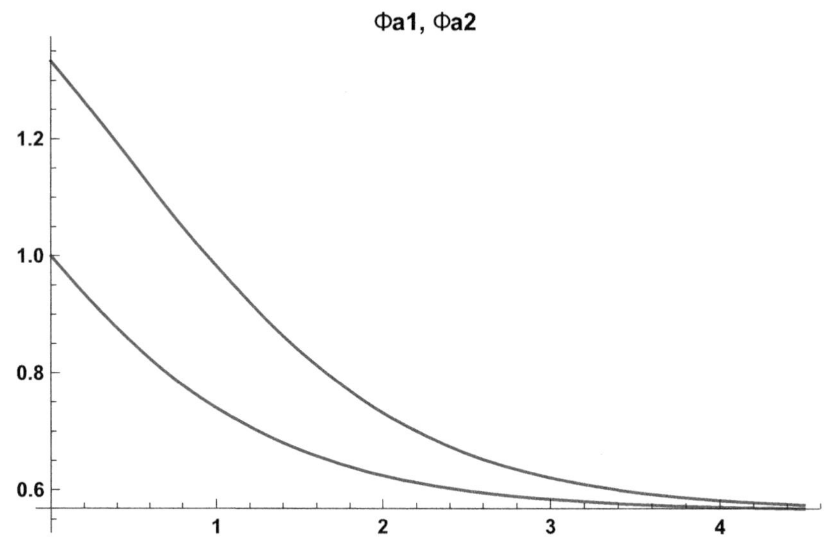

{{−0.333333, $2.891205793294679 \times 10^{-17}$}, {0.017211, 0.259558}, {0.268701, 0.375793}, {0.379179, 0.416063}, {0.417151, 0.42873}}

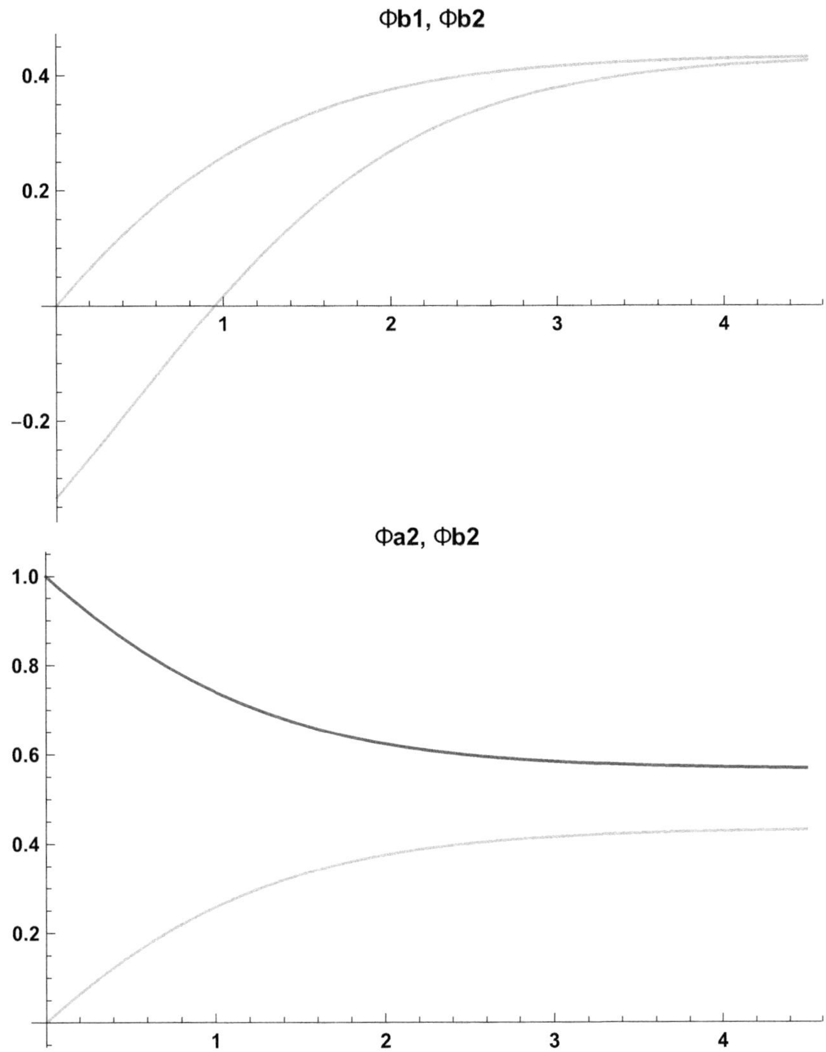

It is again obvious after plotting the solutions that the second set is the set that is physically realistic.

Though the shapes of the solutions look the same as the solutions derived originally, we can plot them in the following way to be sure. We compare the dimensional solutions to the non-dimensionalized solutions directly. We will convert those dimensionalized solutions to their non-dimensionalized forms. To do this we will convert ca2[t] and cb2[t] to Φa[τ] and Φb[τ] by dividing by Cao. We will compute the value of this function at 15 points from 0 to 150 in steps of ten. Then we will compute the corresponding dimensionless times from the product

$k_B C_{Ao} t$ also from 0 to 150 in steps of ten. Then we merge the two dimensionless concentrations with the dimensionless times to create a set of points. Once we have done this for both Ca and Cb, we can overlay these points on the continuous plots of $\Phi a[\tau]$ and $\Phi b[\tau]$. Here it is:

ka = 0.01;

kb = 30;

Cao = C_{Ao} = 0.001;

Cbo = 0;

ϕao = 1;

ϕbo = 0;

ca2[t_]:=

$$\frac{1}{2kb}$$

$$\left(ka + 2(Cao + Cbo)kb - \sqrt{ka}\sqrt{ka + 4(Cao + Cbo)kb}\ \text{Tanh}\left[\frac{1}{2}\sqrt{ka}\sqrt{ka + 4(Cao + Cbo)kb}\,t + \text{ArcTanh}\left[\frac{ka + 2Cbokb}{\sqrt{ka}\sqrt{ka + 4(Cao + Cbo)kb}}\right]\right]\right)$$

cb2[t_]:=

$$\frac{1}{2kb}$$

$$\left(-ka + \sqrt{ka}\sqrt{ka + 4(Cao + Cbo)kb}\ \text{Tanh}\left[\frac{1}{2}\sqrt{ka}\sqrt{ka + 4(Cao + Cbo)kb}\,t + \text{ArcTanh}\left[\frac{ka + 2Cbokb}{\sqrt{ka}\sqrt{ka + 4(Cao + Cbo)kb}}\right]\right]\right)$$

Φb2[t_]:=

$$+\frac{1}{2kbC_{Ao}}$$

$$\left(-\text{ka}+\sqrt{\text{ka}}\sqrt{\text{ka}+4\text{kb}(\phi\text{ao}+\phi\text{bo})C_{\text{Ao}}}\right.$$

$$\text{Tanh}\left[\text{ArcTanh}\left[\frac{\text{ka}+2\text{kb}\phi\text{bo}C_{\text{Ao}}}{\sqrt{\text{ka}}\sqrt{\text{ka}+4\text{kb}(\phi\text{ao}+\phi\text{bo})C_{\text{Ao}}}}\right]+\right.$$

$$\left.\left.\frac{\sqrt{\text{ka}}t\sqrt{\text{ka}+4\text{kb}(\phi\text{ao}+\phi\text{bo})C_{\text{Ao}}}}{2\text{kb}C_{\text{Ao}}}\right]\right)$$

Φa2[t_]:=

$$\frac{1}{2\text{kb}C_{\text{Ao}}}$$

$$(\text{ka}+2\text{kb}(\phi\text{ao}+\phi\text{bo})C_{\text{Ao}}-$$

$$\sqrt{\text{ka}}\sqrt{\text{ka}+4\text{kb}(\phi\text{ao}+\phi\text{bo})C_{\text{Ao}}}$$

$$\text{Tanh}\left[\text{ArcTanh}\left[\frac{\text{ka}+2\text{kb}\phi\text{bo}C_{\text{Ao}}}{\sqrt{\text{ka}}\sqrt{\text{ka}+4\text{kb}(\phi\text{ao}+\phi\text{bo})C_{\text{Ao}}}}\right]+\right.$$

$$\left.\left.\frac{\sqrt{\text{ka}}t\sqrt{\text{ka}+4\text{kb}(\phi\text{ao}+\phi\text{bo})C_{\text{Ao}}}}{2\text{kb}C_{\text{Ao}}}\right]\right)$$

a = Plot[{Φa2[τ]}, {τ, 0, $k_B C_{\text{Ao}}$150},
AxesLabel → {t, "Φ_{Ao}"}];

b = Plot[{Φb2[τ]}, {τ, 0, $k_B C_{\text{Ao}}$150},
AxesLabel → {t, "Φ_{Bo}"}, PlotStyle → Hue[0.8]];

cnca = Table[ca2[t] / C_{Ao} , {t, 0, 150, 10}];
tao = Table [$k_B C_{\text{Ao}} t$, {t, 0, 150, 10}];
Transpose[Partition[Join[%, %%], Length[tao]]];
capl = ListPlot[%, PlotRange → All,
PlotStyle → Hue[0.3]];

cncb = Table[cb2[t] / C_{Ao} , {t, 0, 150, 10}];
tao = Table [$k_B C_{\text{Ao}} t$, {t, 0, 150, 10}];

```
Transpose[Partition[Join[%, %%], Length[tao]]];
cbpl = ListPlot[%, PlotRange → All,
PlotStyle → Hue[0.1]];

Show[{a, capl}, PlotRange → All]
Show[{b, cbpl}, PlotRange → All]
```

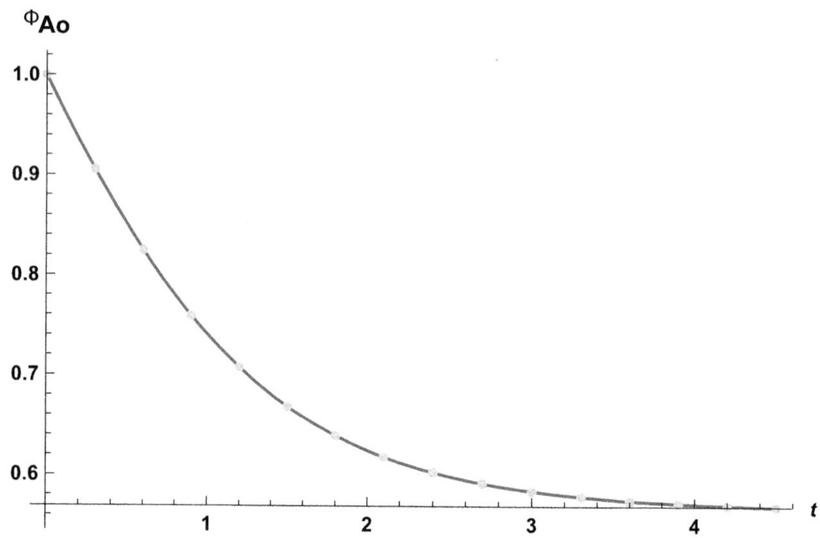

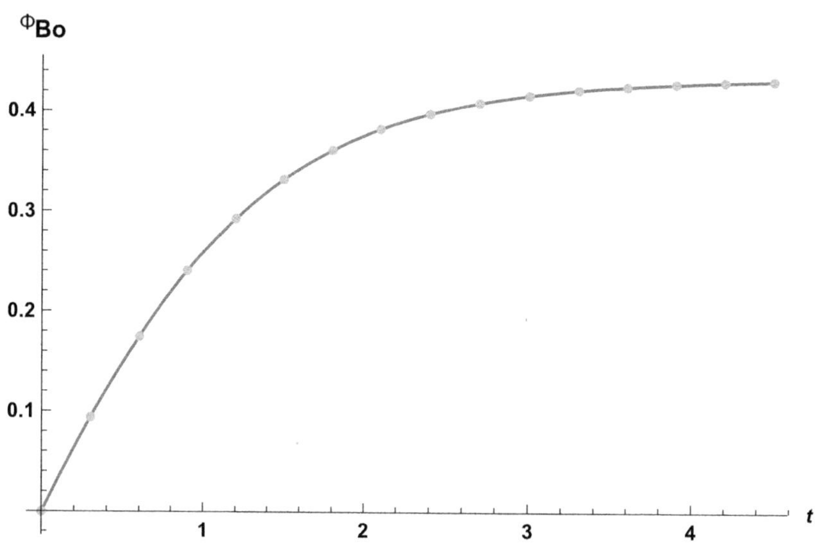

It is obvious that the two sets of solutions are in agreement. The lesson in this is that when we develop symbolic solutions, although they may appear to be the same, it is important to reduce those solutions to numbers and to be sure that they are.

Complex reactions

Series

Reactions rarely take place in isolation of other reactions. Reversibility is one example of the simultaneity of reaction chemistries. Another is that which arises when a reaction is immediately proceeded by another reaction. When reactions occur in series, they are also referred to as being consecutive. Consider:

$$A \rightarrow B \rightarrow D$$

If each of these proceed via a first-order rate process, then this can be analyzed readily. Higher order reaction rates follow the same analysis, but they require a bit more mathematical effort. We can begin by writing the key material balance equations:

$$\frac{dC_A}{dt} = - k_A C_A$$

$$\frac{dC_B}{dt} = k_A C_A - k_B C_B$$

$$\frac{dC_D}{dt} = k_B C_B$$

We can see that the first of these equations can be integrated immediately, to give:

$$Ca = Cao \exp(-k_A t)$$

This can be substituted into the second equation and the integration can be done for the concentration of B. Subsequently, we substitute this into the equation for the rate of change of D and integrate once more for the full solution. Alternatively, we can let the DSolve algorithm do for us all at once:

```
Clear[cA, cB, ka, kb, cAo, cBo, cDo]

Simplify[

DSolve[
```

```
{cA'[t] == -kacA[t],
cB'[t] == kacA[t] - kbcB[t],
cD'[t] == kbcB[t],
cA[0] == cAo,
cB[0] == 0,
cD[0] == 0},
{cA[t], cB[t], cD[t]},
t]
    ]
```

$$\{\{\mathrm{cA}[t] \to \mathrm{cAo}e^{-\mathrm{ka}t},$$

$$\mathrm{cB}[t] \to -\frac{\mathrm{cAo}\left(e^{-\mathrm{ka}t} - e^{-\mathrm{kb}t}\right)\mathrm{ka}}{\mathrm{ka} - \mathrm{kb}},$$

$$\mathrm{cD}[t] \to \frac{\mathrm{cAo}\left(\mathrm{ka} - e^{-\mathrm{kb}t}\mathrm{ka} + \left(-1 + e^{-\mathrm{ka}t}\right)\mathrm{kb}\right)}{\mathrm{ka} - \mathrm{kb}}\}\}$$

We could also have chosen to non-dimensionalize the differential equations before solving them in order to find a general solution in fewer absolute parameters. We can divide all by $\boldsymbol{k_A}$ $\boldsymbol{C_{Ao}}$, which will give us:

$$\left(\frac{1}{k_A C_{Ao}}\right)\frac{dC_A}{\mathrm{dt}} = -\ \Phi_A$$

$$\left(\frac{1}{k_A C_{Ao}}\right)\frac{dC_B}{\mathrm{dt}} = \Phi_A - \frac{k_B}{k_A}\Phi_B$$

$$\left(\frac{1}{k_A C_{Ao}}\right)\frac{dC_D}{\mathrm{dt}} = \frac{k_B}{k_A}\Phi_B$$

But k_A dt = dτ since $\boldsymbol{k_A}$ is an inverse time constant associated with the rate of the first chemical reaction and τ is "reduced" time. This gives us the following three equations:

$$\frac{d\Phi_A}{d\tau} = -\ \Phi_A$$

$$\frac{d\Phi_B}{d\tau} = \Phi_A - \frac{k_B}{k_A}\Phi_B$$

$$\frac{d\Phi_D}{d\tau} = \frac{k_B}{k_A}\Phi_B$$

Now the ratio of $\frac{k_B}{k_A}$ is not a reciprocal of an equilibrium constant since both reactions are considered to be irreversible. It is simply the ratio of the rate constants, and we will leave it as such.

```
Clear["Global`* "]

ka=.

kb=.

Φao=.

sersol1 = Simplify[

DSolve[

{Φa'[τ]== − Φa[τ],

Φb'[τ] == +Φa[τ] − kb/ka Φb[τ],

Φd'[τ] == +kb/ka Φb[τ],

Φa[0] == Φao,

Φb[0] == 0,

Φd[0] == 0},

{Φa[τ], Φb[τ], Φd[τ]},

τ]

    ];

ΦA[τ_]:=sersol1[[1]][[1]][[2]]

ΦB[τ_]:=sersol1[[1]][[2]][[2]]

ΦD[τ_]:=sersol1[[1]][[3]][[2]]

Φa[τ]==ΦA[τ]

Φb[τ]==ΦB[τ]

Φd[τ]==ΦD[τ]
```

```
ka = 2.;

kb = 1.;

Φao = 1.;

Plot[

{ΦA[τ], ΦB[τ], ΦD[τ]},

{τ, 0, 10},

PlotRange → All,

AxesLabel → {"τ", "Φa,Φb,Φd"},

PlotStyle →

{Hue[0.3], Hue[0.5], Hue[0.8]},

PlotLabel → {ka"= ka", kb"= kb", Φao"= Φao"}]
```

$$\Phi a[\tau] == e^{-\tau}\Phi ao$$

$$\Phi b[\tau] == -\frac{\left(e^{-\tau} - e^{-\frac{kb\tau}{ka}}\right) ka\Phi ao}{ka - kb}$$

$$\Phi d[\tau] == \frac{\left(ka - e^{-\frac{kb\tau}{ka}} ka + \left(-1 + e^{-\tau}\right) kb\right) \Phi ao}{ka - kb}$$

Applying parameter values, we see that A diminishes smoothly, B rises and falls in concentration and D ultimately dominates:

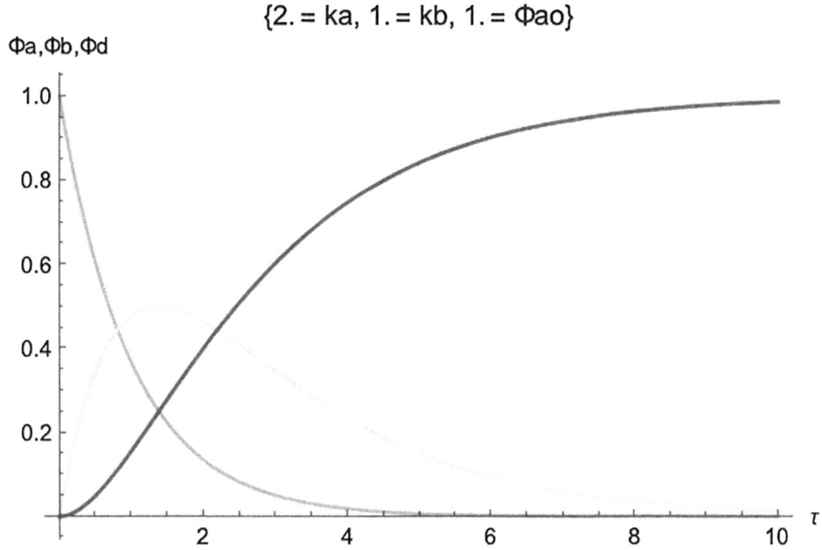

This is a common enough scenario that we may want to experiment with the parameter values, so we create a function sersol2 to let us vary ka, kb, Φao, and τmax:

Clear["Global`*"]

sersol2[ka_, kb_, Φao_, τmax_]:=

Module[{Φa, Φb, Φd},

Φa[τ_]:= $e^{-\tau}\,\Phi\mathrm{ao}$**;**

Φb[τ_]:= $\dfrac{\left(e^{-\tau}-e^{-\frac{\mathrm{kb}\tau}{\mathrm{ka}}}\right)\mathrm{ka}\Phi\mathrm{ao}}{-\mathrm{ka}+\mathrm{kb}}$**;**

Φd[τ]:= $\dfrac{\left(\mathrm{ka}-e^{-\frac{\mathrm{kb}\tau}{\mathrm{ka}}}\,\mathrm{ka}+\left(-1+e^{-\tau}\right)\mathrm{kb}\right)\Phi\mathrm{ao}}{\mathrm{ka}-\mathrm{kb}}$**;**

Plot[

{Φa[τ], Φb[τ], Φd[τ]},

{τ, 0, τmax},

```
PlotRange → {{0, τmax}, {0, 1}},
AxesLabel → {"τ", "Φi"},
PlotStyle →
{Hue[0.3], Hue[0.5], Hue[0.8]},
PlotLabel → {ka"= ka", kb"= kb"}]
]
Show[sersol2[1., .5, 1., 10]]
```

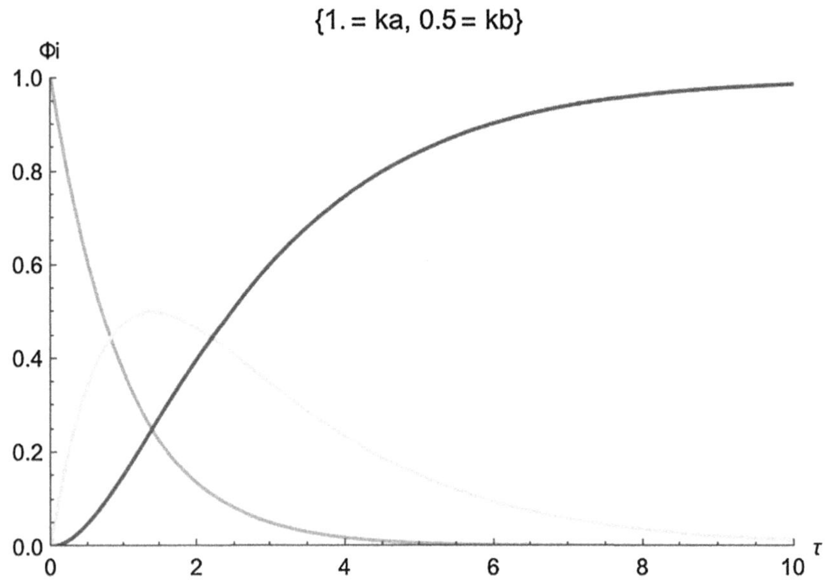

Now one can run as many cases as one should like to in order to compare the effects of different parameters. We have left the semi-colon out after the **Plot** routine in the **Module** function, so that **Graphics** are a bona-fide output. This allows us to use sersol as part of **GraphicsArray**.

```
Show[
GraphicsGrid[
{Table[sersol2[1., n, 1., 10], {n, 0.01, 1.01, .5}],
Table[sersol2[1., n, 1., 10], {n, 1.51, 2.52, .5}],
Table[sersol2[1., n, 1., 10], {n, 3.01, 4.02, .5}],
```

```
Table[sersol2[1., n, 1., 10], {n, 4.51, 24.51, 10}]

}]]
```

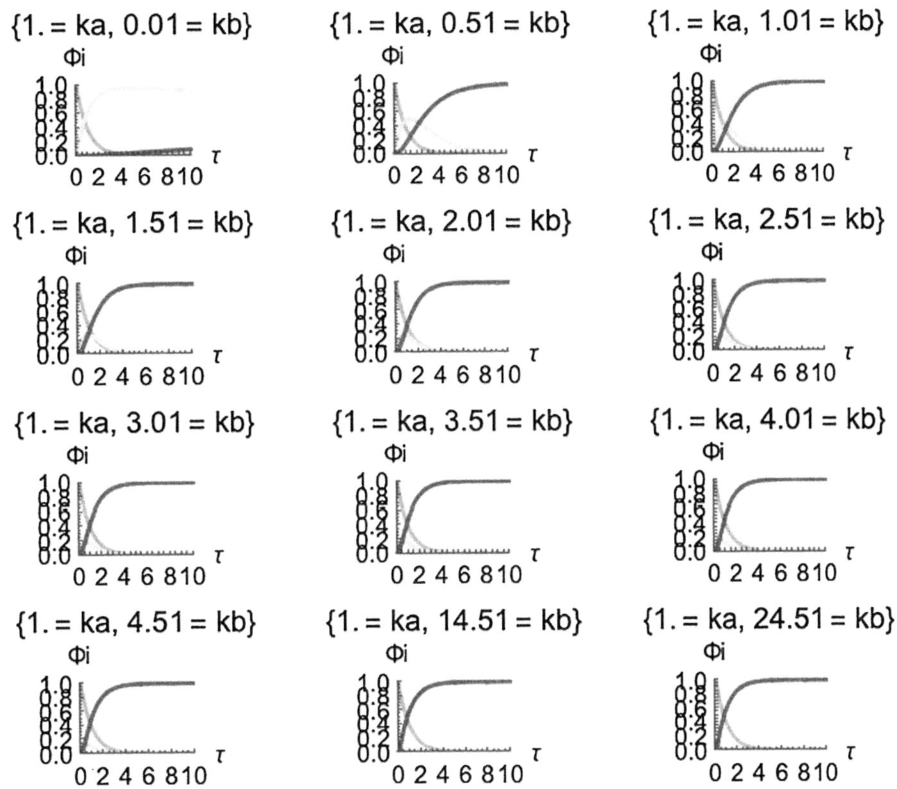

This array shows the full gamut of the effects that the magnitude of kb at fixed ka has upon the chemistry. We see that when kb is 10^2 smaller than ka, the reaction appears to be that of A $\rightarrow$ B. When we begin to increase the magnitude of kb we see that the prevalence of B grows, as does the final amount of D at 10τ. With longer times, the amount of B would grow to be equal to the original amount of A, but we are concerned here with a fixed batch holding time of 10τ. As kb increases and overtakes ka, the maximum amount of B continuously diminishes and shifts to earlier τ. Finally, with kb at $\sim$25ka, the maximum in B is shifted to very short τ and to a value of less than 0.1.

The better way to get a sense for this is to use the Manipulate function again. In what follows, we fixed kb and Φao and chose to vary ka and τmax.

Manipulate[sersol2[ka, .2, 1, τmax],

{ka, 0.01, 100, .001}, {τmax, 10, 1500, 1}]

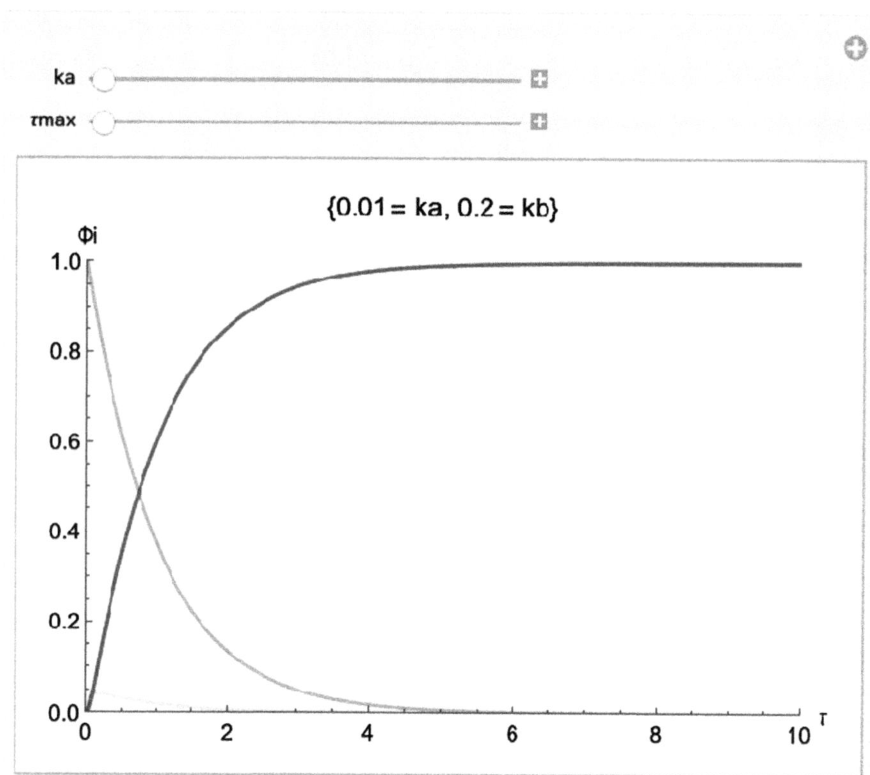

If we were to change the kinetics so that the first reaction was second order in A and the second reaction was first order in B, then we would see largely the same picture emerging in the graphs of dimensionless concentration versus time. There would of course be differences, but not large departures in the trends from what we have observed for this all-first-order case. But, what if the reactions have rate expressions that are not so readily integrable? What if we have widely differing, mixed-order concentration dependencies? In some cases one can develop fully analytical (closed-form) solutions like the ones we have derived for the first order case, but in other cases this not possible, we must instead turn to numerical methods for efficient solution.

Suppose that the following reaction is a series network with square kinetics for the first reaction and half-order kinetics for the second:

$$2A \rightarrow B \rightarrow D$$

Then accounting for the stoichiometry of two going to one we have the following set of equations to solve:

$$\frac{d\mathrm{Ca}}{dt} = -\ \mathrm{ka}\ \mathrm{Ca}^2$$

$$\frac{d\mathrm{Cb}}{dt} = +\frac{\mathrm{kaCa}^2}{2} -\ \mathrm{kb}\ \sqrt{\mathrm{Cb}}$$

$$\frac{d\mathrm{Cd}}{dt} = +\ \mathrm{kb}\sqrt{\mathrm{Cb}}$$

$$\frac{d\mathrm{Cb}}{dt} = -\frac{1}{2}\frac{d\mathrm{Ca}}{dt} - \frac{d\mathrm{Cd}}{dt}$$

Because of the mix in orders, non-dimensionalizing must be done carefully. Looking at the equations for the rate of change of B or D, we have the rate term kb $\sqrt{\mathrm{Cb}}$. This must have dimensions of $\frac{\mathrm{mol}}{\mathrm{Volume}*\mathrm{time}}$. That means that the rate constant kb must have units of $\sqrt{\frac{\mathrm{mol}}{L*s^2}}$ so that when multiplied by $\sqrt{\frac{\mathrm{mol}}{L}}$ the dimensions are correct.

We begin with the first equation:

$$\frac{\mathrm{Cao}}{\mathrm{Cao}}\frac{d\mathrm{Ca}}{dt} = -\ \frac{\mathrm{Cao}^2}{\mathrm{Cao}^2}\mathrm{ka}\ \mathrm{Ca}^2$$

$$\mathrm{Cao}\ \frac{d\Phi a}{dt} = -\ \mathrm{Cao}^2\ \mathrm{ka}\ \Phi a^2$$

$$\frac{d\Phi a}{dt} = -\ \mathrm{Cao}\ \mathrm{ka}\ \Phi a^2$$

$$\frac{d\Phi a}{d\tau} = -\ \ \Phi a^2$$

Since we are going to need the non-dimensionalized form for the rate of change of **D** to obtain **B** we proceed with this equation next:

$$\frac{d\mathrm{Cd}}{dt} = +\ \mathrm{kb}\ \sqrt{\mathrm{Cb}}$$

$$\frac{\mathrm{kaCao}^2}{\mathrm{kaCao}^2}\frac{d\mathrm{Cd}}{dt} = +\ \mathrm{kb}\ \sqrt{\mathrm{Cb}}$$

$$\mathrm{kaCao}^2\ \frac{d\mathrm{Cd}/\mathrm{Cao}}{\mathrm{kaCao}dt} = +\ \mathrm{kb}\ \sqrt{\mathrm{Cb}}$$

$$\frac{d\Phi d}{d\tau} = + \frac{\text{kb}}{\text{kaCao}^2}\sqrt{\text{Cb}}$$

$$\frac{d\Phi d}{d\tau} = + \frac{\text{kb}}{\text{kaCao}^{3/2}\text{Cao}^{1/2}}\sqrt{\text{Cb}}$$

$$\frac{d\Phi d}{d\tau} = + \frac{\text{kb}}{\text{kaCao}^{3/2}}\frac{\sqrt{\text{Cb}}}{\sqrt{\text{Cao}}}$$

$$\frac{d\Phi d}{d\tau} = + \frac{\text{kb}}{\text{kaCao}^{3/2}}\sqrt{\Phi b}$$

Notice that to be properly dimensionless, we divide kb by kaCao$^{3/2}$. Finally, we take these two results and combine them to derive the non-dimensionalized form for the rate of change of B. We first show that the overall non-dimensionalized equation is parallel in form to the fully dimensional equations and then make the appropriate substitutions etc.:

$$\frac{d\text{Cb}}{dt} = -\frac{1}{2}\frac{d\text{Ca}}{dt} - \frac{d\text{Cd}}{dt}$$

$$\frac{1}{\text{kaCao}^2}\frac{d\text{Cb}}{dt} = \frac{1}{\text{kaCao}^2}\left[-\frac{1}{2}\frac{d\text{Ca}}{dt} - \frac{d\text{Cd}}{dt}\right]$$

$$\frac{d\Phi b}{d\tau} = -\frac{1}{2}\frac{d\Phi a}{d\tau} - \frac{d\Phi d}{d\tau}$$

$$\frac{d\Phi b}{d\tau} = \frac{\Phi a^2}{2} - \frac{\text{kb}}{\text{ka}\sqrt{\text{Cao}^3}}\sqrt{\Phi b}$$

If we try to solve this analytically, we find that we cannot do it, at least not directly with **DSolve**:

Clear["Global`*"]

DSolve[

$\{\Phi\text{a}'[\tau] == -\Phi\text{a}[\tau]^2,$

$\Phi\text{b}'[\tau] == +\frac{\Phi\text{a}[\tau]^2}{2} - \frac{\text{kb}}{2\text{ka}\sqrt{\text{Cao}^3}}\sqrt{\Phi\text{b}[\tau]},$

$\Phi\text{d}'[\tau] == +\frac{\text{kb}}{2\text{ka}\sqrt{\text{Cao}^3}}\sqrt{\Phi\text{b}[\tau]},$

$\Phi\text{a}[0] == \Phi\text{ao},$

```
Φb[0] == 0,
Φd[0] == 0},
{Φa[τ], Φb[τ], Φd[τ]},
τ]
```

$$\text{DSolve}\left[\left\{\Phi a'[\tau] == -\Phi a[\tau]^2,\ \Phi b'[\tau] == \frac{\Phi a[\tau]^2}{2} - \frac{kb\sqrt{\Phi b[\tau]}}{2\sqrt{Cao^3}ka},\ \Phi d'[\tau] == \frac{kb\sqrt{\Phi b[\tau]}}{2\sqrt{Cao^3}ka},\ \Phi a[0] == \Phi ao,\ \Phi b[0] == 0,\ \Phi d[0] == 0\right\},\ \{\Phi a[\tau], \Phi b[\tau], \Phi d[\tau]\},\ \tau\right]$$

We turn then to numerical methods in **NDSolve** and find the solution readily, so long as we specify parameters.

```
Clear["Global`*"]
ka = 10.;
kb = .5;
Cao = 1;
τmax = 40;
sersol3 = NDSolve[
{Φa'[τ] == −Φa[τ]^2,
```

$$\Phi b'[\tau] == +\frac{\Phi a[\tau]^2}{2} - \frac{kb}{ka\sqrt{Cao^3}}\sqrt{\Phi b[\tau]},$$

$$\Phi d'[\tau] == +\frac{kb}{ka\sqrt{Cao^3}}\sqrt{\Phi b[\tau]},$$

```
Φa[0] == 1,
Φb[0] == 0,
Φd[0] == 0},
{Φa[τ], Φb[τ], Φd[τ]},
```

```
{τ, 0, τmax}];
ΦA[τ_]:=sersol3[[1]][[1]][[2]]
ΦB[τ_]:=sersol3[[1]][[2]][[2]]
ΦD[τ_]:=sersol3[[1]][[3]][[2]]
Plot[{ΦA[τ], ΦB[τ], ΦD[τ]},
{τ, 0, τmax},
PlotRange → {{0, τmax}, {0, 1}},
AxesLabel → {"τ", "Φi"},
PlotStyle →
{Hue[0.3], Hue[0.5], Hue[0.8]},
PlotLabel → {ka"= ka", kb"= kb"}]
```

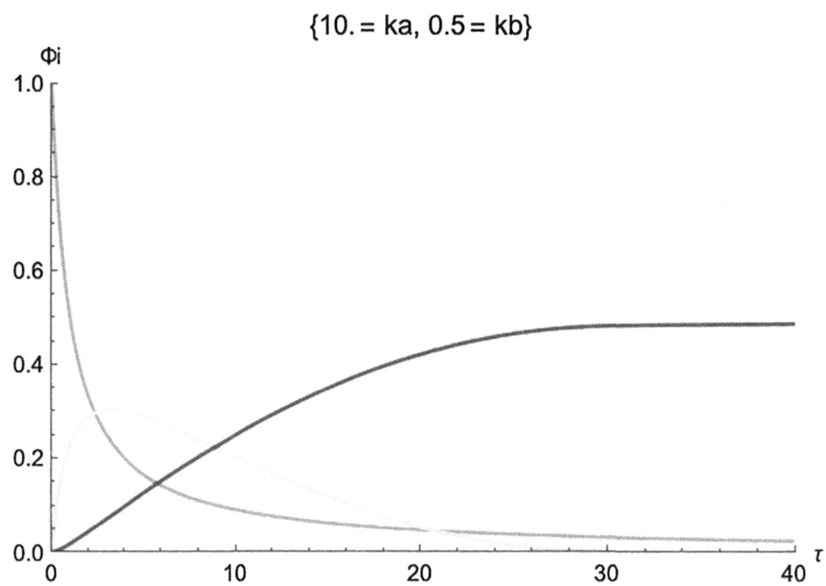

Mathematica allows us to approach this problem by creating a Module function of the numerical routine. With this Module we can use a Table loop to find how the solutions vary with different parameters. We can do this as follows:

```
Clear["Global`* "]

mixord[ka_, kb_, Φao_, Cao_, τmax_]:=

Module[{sersol3, Φa, Φb, Φd, A, B, D, τ},

sersol = NDSolve[

{Φa'[τ] == −Φa[τ]^2,

Φb'[τ] == +Φa[τ]^2/2 − kb/(ka Sqrt[Cao^3]) Sqrt[Φb[τ]],

Φd'[τ] == +kb/(ka Sqrt[Cao^3]) Sqrt[Φb[τ]],

Φa[0] == Φao,

Φb[0] == 0,

Φd[0] == 0},

{Φa[τ], Φb[τ], Φd[τ]},

{τ, 0, τmax}];

A[τ] = Evaluate[Φa[τ]/.sersol];

B[τ] = Evaluate[Φb[τ]/.sersol];

D[τ] = Evaluate[Φd[τ]/.sersol];


Plot[{A[τ], B[τ], D[τ]},

{τ, 0, τmax},

PlotRange → {{0, τmax}, {0, 1}},

AxesLabel → {"τ", "Φi"},

PlotStyle →

{Hue[0.3], Hue[0.5], Hue[0.8]},

PlotLabel → {ka"= ka", kb"= kb"}]

]

Show[mixord[30., .5, 1, 1, 100]]
```

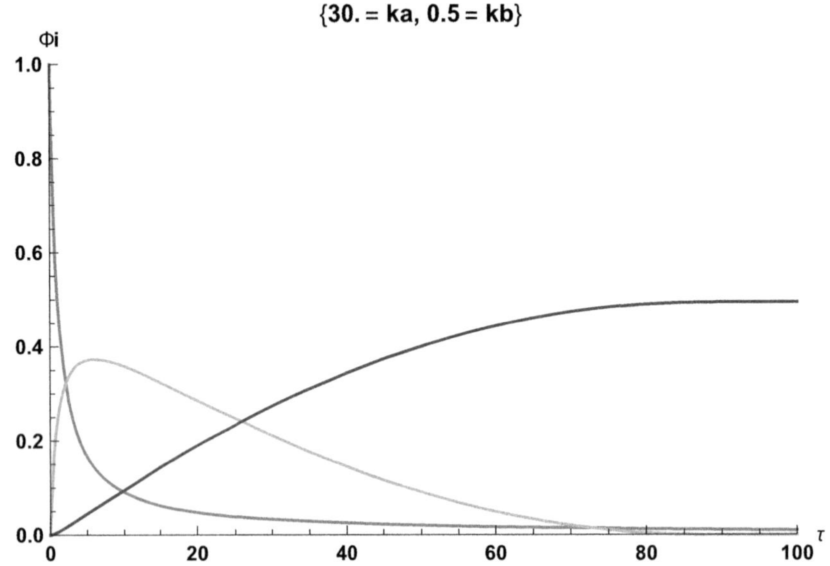

Show[

GraphicsGrid[

{Table[mixord[*n*, .5, 1., 1., 20],

{*n*, 0.1, 20.1, 10}],

Table[mixord[5., *m*, 1., 1., 20], {*m*, .1, 2.1, 1}]}

]]

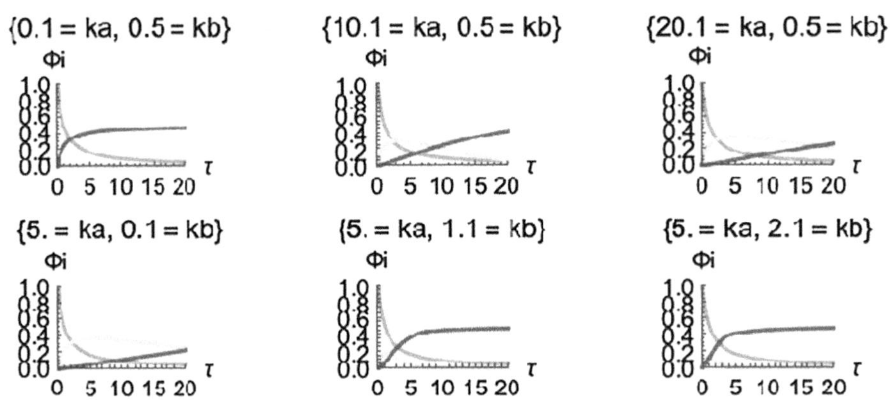

Notice that at a constant value of kb, as the value of ka increases, we observe more of the concentration of B at shorter times, which makes sense. But, also notice that the concentrations of

A and B, have more sensitivity to the magnitude of these rate constants than we observed in previous cases. For example, even with a small value of ka = 0.1, we find that with kb = 0.5, nearly all of A is converted to D in about 15τ and the presence of B is near zero throughout. In simpler terms every time a molecule of A is converted to B, that B molecule is immediately converted to D. Thus we see very little B. At constant ka, as kb increases the trend is reversed. We also note that these equations are numerically "stiff" for some values of their parameters. For example if we choose ka = 0.001 and kb = 0.5, the integration becomes unstable after about 6τ, see the gaps below.

Show[mixord[.001, .5, 1, 1, 15]]

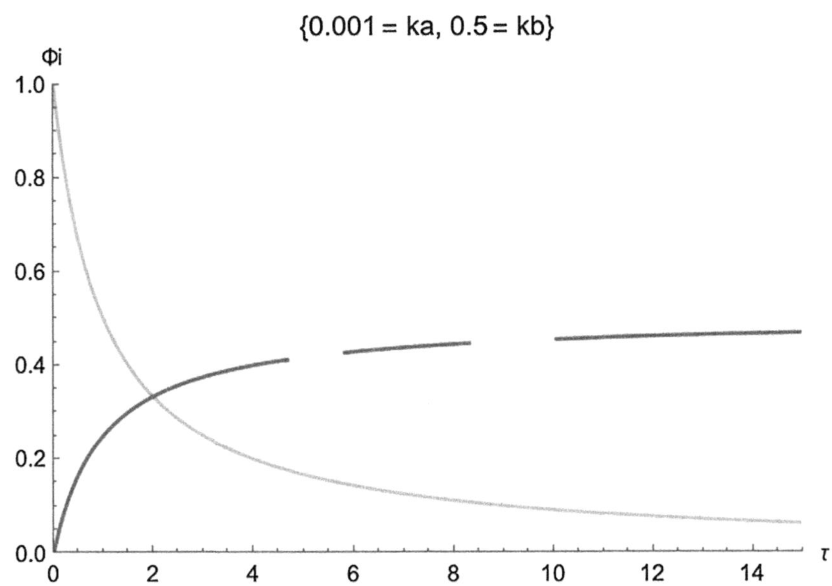

Series-parallel reactions

The next case that we shall consider is that of the series-parallel reaction system with all first order rates. This keeps things more straightforward and lets us observe the general behavior of such a group of reactions. If the order of the rates of reaction becomes higher, or non-integral, then numerical methods such as those used in the last section may be employed.

If instead of the one species reacting to one other species, we will look at the situation in which there can be two products formed by competing reactions. We can also let one of the two primary products react to produce one other product. Thus, this set of reactions, or reaction *network* will involve four components and three rate constants as follows:

$$\begin{array}{l} \quad k_1 \quad k_2 \\ A \rightarrow B \rightarrow D \\ \quad \searrow k_3 \\ \quad\quad E \end{array}$$

This network of reactions that can be solved readily by using the same analysis methods that we have used to this point.

$$\frac{dC_A}{\mathrm{dt}} = -\,\mathrm{k1}\ C_A - \mathrm{k3}\ C_A = -\,(\mathrm{k1} + \mathrm{k3})\ C_A$$

$$\frac{dC_B}{\mathrm{dt}} = \mathrm{k1}\ C_A - \mathrm{k2}\ C_B$$

$$\frac{dC_D}{\mathrm{dt}} = \mathrm{k2}\ C_B$$

$$\frac{dC_E}{\mathrm{dt}} = \mathrm{k3}\ C_A$$

We note that in the first equation the rate constants are the proportionality factor that determine how much of A proceeds to B and E. Also the rate of depletion of A follows an observed rate constant that is the sum of the two rate constants for the parallel forward reactions. The other equations are much as we would expect. We can non-dimensionalize using the sum, k1 + k3 and, of course $\boldsymbol{C_{Ao}}$.

$$\frac{d\Phi_A}{\mathrm{d}\tau} = -\,\Phi_A$$

$$\frac{d\Phi_B}{\mathrm{d}\tau} = \kappa 1\ \Phi_A - \kappa 2\ \Phi_B$$

$$\frac{d\Phi_D}{\mathrm{d}\tau} = \kappa 2\ \Phi_B$$

$$\frac{d\Phi_E}{\mathrm{d}\tau} = \kappa 3\ \Phi_A$$

where

$$\kappa 1 = \frac{\mathrm{k1}}{(\mathrm{k1} + \mathrm{k3})}, \quad \kappa 2 = \frac{\mathrm{k2}}{(\mathrm{k1} + \mathrm{k3})}, \quad \kappa 3 = \frac{\mathrm{k3}}{(\mathrm{k1} + \mathrm{k3})}$$

Here is the DSolve routine for this network of reactions. We have used Φi to denote the dimensionless concentration of component i. One more routine is added here, we have nested the DSolve routine in the Simplify function "Simplify[DSolve[<>]]". This makes sure that the output statement corresponds to an algebraically reduced form.

```
Clear["Global`* "]
Simplify[
DSolve[{Φa'[τ]== − Φa[τ],
Φb'[τ]==κ1Φa[τ] − κ2Φb[τ],
Φd'[τ]==κ2Φb[τ], Φe'[τ]==κ3Φa[τ],
Φa[0]==Φao,
Φb[0]==Φbo, Φd[0]==Φdo, Φe[0]==Φeo}, {Φa[τ],
Φb[τ], Φd[τ], Φe[τ]}, τ]]
```

$$\Big\{\Big\{\Phi a[\tau] \to e^{-\tau}\Phi ao,\ \Phi b[\tau] \to \frac{e^{-\tau}\kappa 1\Phi ao - e^{-\kappa 2\tau}(\kappa 1\Phi ao + \Phi bo - \kappa 2\Phi bo)}{-1+\kappa 2},\ \Phi d[\tau] \to -\frac{\kappa 1\left(-1+e^{-\kappa 2\tau}+\kappa 2 - e^{-\tau}\kappa 2\right)\Phi ao}{1-\kappa 2} + \Phi bo - e^{-\kappa 2\tau}\Phi bo + \Phi do,\ \Phi e[\tau] \to \kappa 3\left(\Phi ao - e^{-\tau}\Phi ao\right) + \Phi eo\Big\}\Big\}$$

Note that the expression for the loss of A from the system goes as a typical exponential decay, but recall that τ is non-dimensional, it is the product of real time and the group of the rate constants for the two reactions that consume A. If there were three or more first order, A-consuming reactions, then we would use the sum of all their rate constants. If there were reactions consuming A and reactions producing A, simultaneously, then we would still take the sums of the rate constants, but the signs would be positive and negative, so we would have a sum and difference in the argument leading to τ.

It is also noteworthy that the stoichiometry will be controlled by the rate constants k1 and k3. This is clear and evident in the expression for Φe [τ]. If Φeo is zero, then at large τ, Φe [τ]→ $\kappa 3$ Φao, where $\kappa 3 = \frac{\text{k1}}{(\text{k1+k3})}$, the ratio of k3 to the sum of k1 and k3. This ratio, $\kappa 3$, is also a measure of the *selectivity* of the reaction network.

Below we have assigned values to the parameters of the system. The rate constant, k1, has been set to unity and all the others are set in relation to it. In this case, the rate constant of the third step, which leads to E, is set at twice the value of that of the step leading to B. The rate constant between B and D is taken as half the magnitude of k1. The initial concentration of A is unity and zero for the other species. The solutions derived from DSolve are implemented as local functions Φi [τ_].

Clear["Global`*"]

k1 = 1.;

k2 = 0.5k1;

k3 = 2k1;

τmax = 10;

κ1 =;

κ2 = ;

κ3 = ;

Φao = 1;

Φbo = Φdo = Φeo = 0;

Φa[τ_]:= $E^{-\tau}\Phi ao$

Φb[τ_]:=

$$\frac{1}{-1+\kappa 2}\left(E^{-(1+\kappa 2)\tau}\left(-\left(E^{\tau}-E^{\kappa 2\tau}\right)\kappa 1\Phi ao+E^{\tau}(-1+\kappa 2)\Phi bo\right)\right)$$

Φd[τ_]:=

$$\frac{1}{-1+\kappa 2}$$

$$\left(\kappa 1\left(-1+E^{-\kappa 2\tau}+\kappa 2-E^{-\tau}\kappa 2\right)\Phi ao+\right.$$

$$\left.E^{-\kappa 2\tau}(-1+\kappa 2)\left(\left(-1+E^{\kappa 2\tau}\right)\Phi bo+E^{\kappa 2\tau}\Phi do\right)\right)$$

Φe[τ_]:= $\kappa 3\left(\Phi ao-E^{-\tau}\Phi ao\right)+\Phi eo$

Plot[{Φa[τ], Φb[τ], Φd[τ], Φe[τ]}, {τ, 0, τmax},

PlotRange → {{0, τmax}, {0, 1}},

AxesLabel → {"τ", "Φi"},

PlotStyle →

{Hue[0.2], Hue[0.4], Hue[0.6], Hue[0.8]},

PlotLabel → {k1"= k1", k2"= k2", k3"= k3",

A : Yellow + Green, *B* : Green, *D* : Blue, *E* : Magenta}]

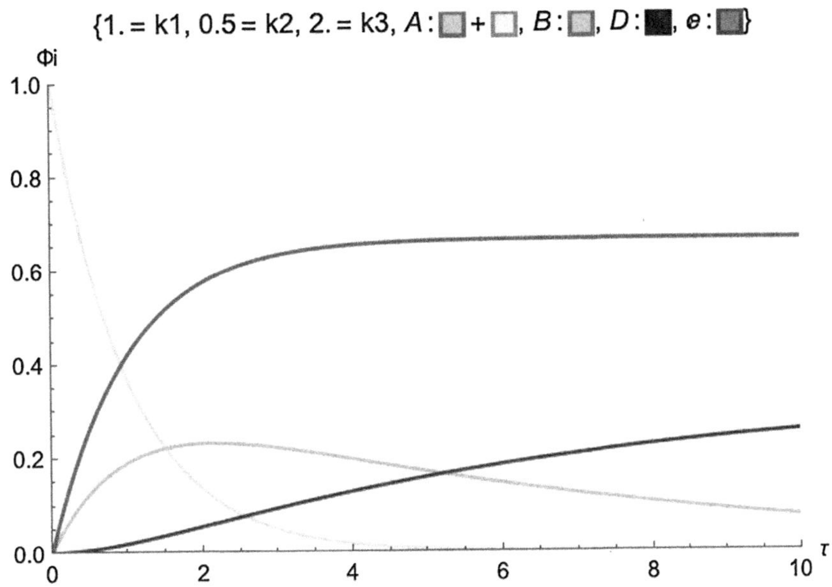

Since A (yellow-green - light gray-mid gray in the print version) is consumed rapidly by the two pathways to B (dark green - mid gray in the print version) and E (magenta - semi-dark gray in the print version), its concentration profile drops sharply with time. The concentration of E rises very rapidly in response to the drop in A, but B lags behind. The reason is that B is not only formed more slowly, but as it is formed it is depleted by reaction to produce D (blue - dark gray in the print version), albeit at a comparably slower rate than the other reactions. The ultimate products, E and D, finally approach constant values at longer time and their magnitudes are 0.66 and 0.33, each corresponding to the rate constant ratios as is predicted.

This is an illustration of the intricacies that can develop even in a very simple reaction network. Imagine the kinds of complexity that arises in some petroleum processing steps which involve numerous reactant molecules and many potential pathways for reaction (thermal, acid-catalyzed, metal catalyzed...).

High dilution

Running a reaction with the reactants are high dilution is both a solution to practical problem and an instance of fundamental interest in kinetics. From a practical standpoint, many macrocyclic compounds may be synthesized using the method of high dilution. Macrocyclic compounds are used as chelates for metals, hence they are used in very basic process such extractive metallurgy and also in advanced technologies, especially medicine. The point is that their value is manifold and that makes the use of high dilution worthwhile to execute. But what is it that high dilution condition of reaction provides? And why does it work? Consider the following reactions and their kinetics:

The simplified reactions are that molecule A can either react with itself to undergo ring formation, think of it has the head reacting with the tail of the molecule, or it can react with another A molecule to form oligomers and then polymers. The oligomers and polymer forming reactions do not lead to the desired product, only the ring closure reaction does. But, we notice that the reaction to form the ring is first order in A, and the reaction to oligomer is second order. This is critical to know or to intuit because we see immediately that if the concentration of A is less than 1, the squared term will always be much smaller than the linear term. Hence, by being at high dilution, we favor the formation of the ring compound. In simple terms the mathematics are showing that at high dilution, for any given period of time, the molecule A is more likely to react with itself, in a head to tail fashion than it is to collide with another A to form oligomer or polymer. That is really interesting concept and the outcome is well known. We can "prove" this by assembling a small code for the kinetics of this process and then we can examine the results at low and high dilution. In essence we can do a simulation of an experiment.

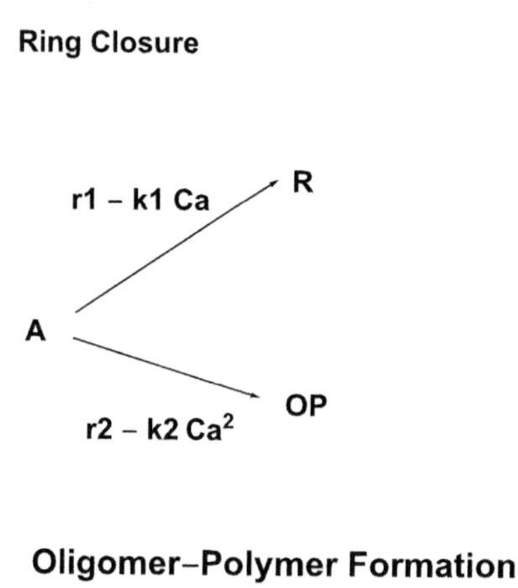

We can call the Module "dileffect" for dilution effect. We will supply the volume of the reaction solution, the initial concentration of A, and the final time. Remember we are simulating a simple batch reactor, so the reactant is charged into the reactor at initial time. Hence, Ao is parameter that will affect the outcome of the calculation, as it would also affect the outcome of an experiment. In one instance, we will do a series of comparisons by changing the volume of the reaction solution and we will also change the concentration Ao when we do so in order to simulate a dilution of the reactant at constant moles. Suppose we wanted to generate a certain number of moles and mass of product. By diluting the concentration, we also reduce the total moles. The counter to the dilution would be to increase volume. We will also run a series in which the volume is held constant and the concentration is reduced. Both will show the dilution effect.

We begin the Module with the rate constant parameters. We have made the rate constant for oligomer formation much larger than that for the ring closure, keeping in mind that it is a second order rate constant. Then we specify the rate equations for the diminution of A, and the formation of R and OP. These results are assigned to functions. Next we compute the selectivities to R and to OP. Lastly we plot the concentrations singly and on one graph. The Module keeps every variable local and this makes it easier to do multiple calculations with the variation in the parameters.

```
Clear["Global`*"]

dileffect[V_, Ao_, tf_]:=

Module[

{k1 = 0.04,
k2 = 1},

sln = NDSolve[{∂t A[t]V == -k1A[t]V - k2A[t]^2 V,
∂t R[t]V == +k1A[t]V,
∂t OP[t]V == +k2A[t]^2 V,
A[0] == Ao, R[0] == 0, OP[0] == 0},
{A[t], R[t], OP[t]}, {t, 0, tf}];
```

```
CA[t_]:=Evaluate[A[t]/.sln];

CR[t_]:=Evaluate[R[t]/.sln];

COP[t_]:=Evaluate[OP[t]/.sln];


{Print["Selectivity to R =",

Round[(CR[tf]/(CR[tf] + COP[tf])), .01]100"%"];

Print["Selectivity to OP =",

Round[COP[tf]/(CR[tf] + COP[tf]), 0.01]100"%"];


a = Plot[{CA[t]}, {t, 0, tf}, PlotRange → All,

PlotStyle → {Dashed, Hue[1]},

PlotLabel → "CA[t]"],


r = Plot[{CR[t]}, {t, 0, tf}, PlotRange → All,

PlotStyle → {Hue[0.25]}, PlotLabel → "CR[t]"],


op = Plot[{COP[t]}, {t, 0, tf}, PlotRange → All,

PlotStyle → {Hue[0.6]}, PlotLabel → "COP[t]"],

Show[a, r, op]

}]
```

We begin with a high initial concentration of A and low volume:

```
dileffect[100, 1, 200]
```

Selectivity to R ={13.%}

Selectivity to OP ={87.%}

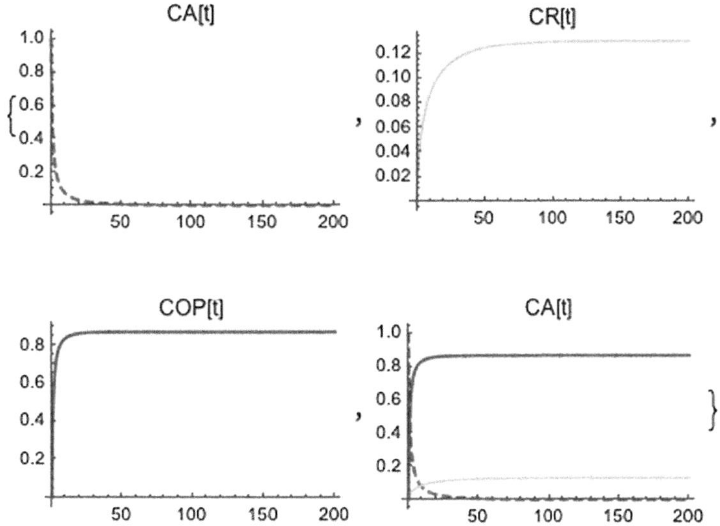

The results show that under these conditions the kinetics are such that they dramatically favor the formation of oligomer-polymer (OP - Blue - dark gray in the print version). The selectivity is 87% in favor of the less desired product. We notice that this reaction would be over in short time.

Let's reduce the concentration by ten and increase the volume by ten. The preserves the initial number of moles at 100.

dileffect[1000, 0.1, 200]

Selectivity to R ={50.%}

Selectivity to OP ={50.%}

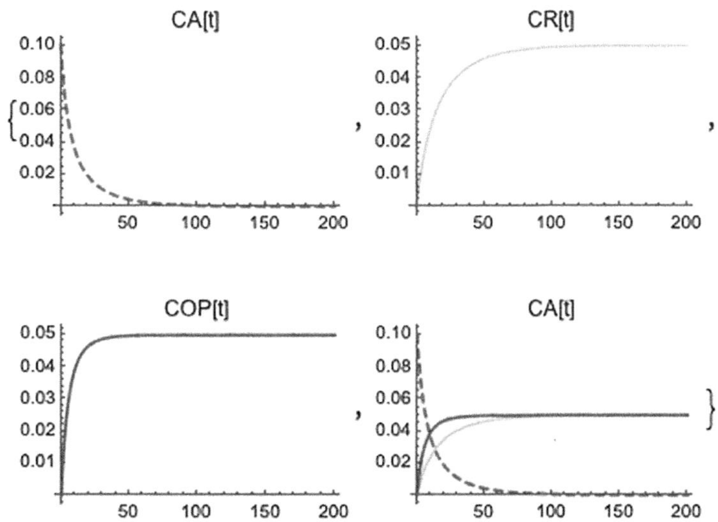

Already we see the effect of dilution - the selectivities to the ring closed product, R, and to the oligomer-polymer, OP, are equal at 50%.

Next, we set the initial concentration at another decade lower and raise the volume by another factor of ten and then we do it again.

dileffect[10000, 0.01, 200]

Selectivity to R ={89.%}

Selectivity to OP ={11.%}

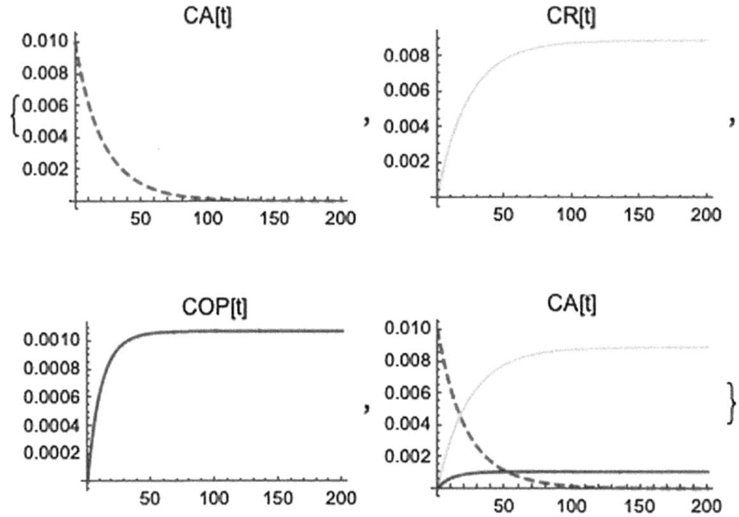

dileffect[100000, 0.001, 200]

Selectivity to R ={99.%}

Selectivity to OP ={1.%}

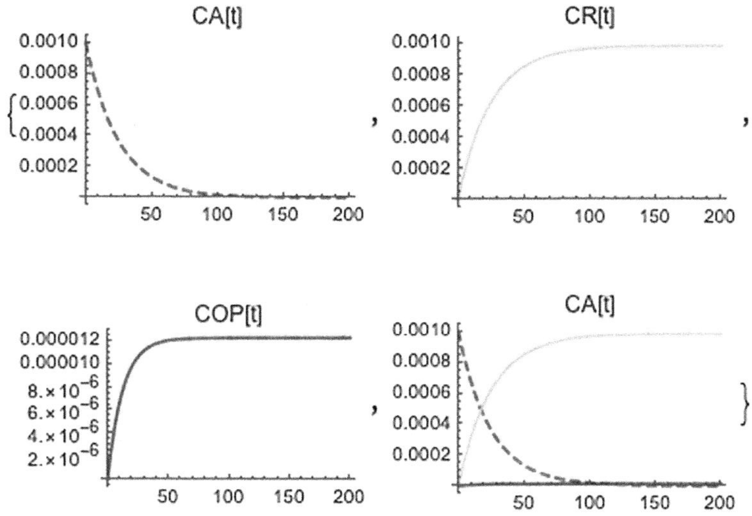

We can see the selectivity to the desired product goes to 89% and then to 99% with each decadal change in concentration. This is the effect of high dilution.

Would it have mattered if we had not varied the volume and had simply reduced the concentration. The answer is no, as we can see by running cases from 1.0 to 0.001 mole per liter of A in 100 liters of solution:

Table[dileffect[100, cao, 200],

{cao, {1, 0.1, 0.01, 0.001}}]

Selectivity to R ={13.%}

Selectivity to OP ={87.%}

Selectivity to R ={50.%}

Selectivity to OP ={50.%}

Selectivity to R ={89.%}

Selectivity to OP ={11.%}

Selectivity to R ={99.%}

Selectivity to OP ={1.%}

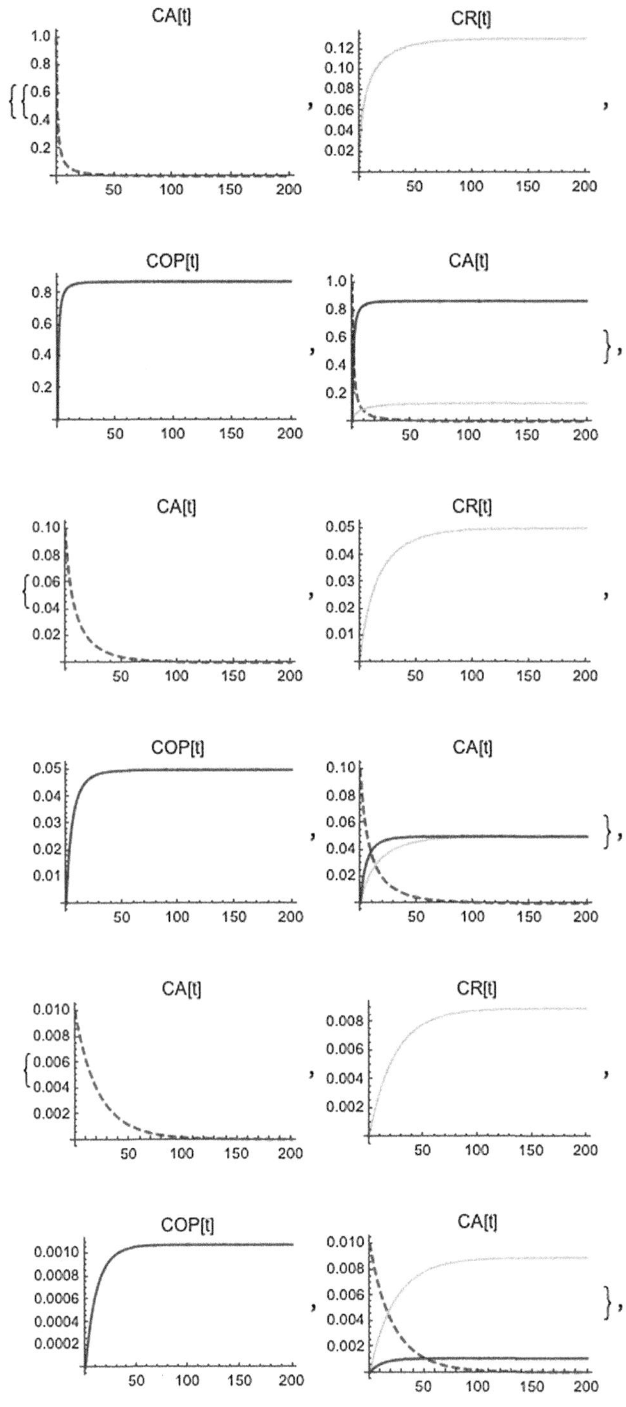
CA[t]
CR[t]
COP[t]
CA[t]
CA[t]
CR[t]
COP[t]
CA[t]
CA[t]
CR[t]
COP[t]
CA[t]

This is an interesting effect and the idea can be observed in various other instances, that can combine thermodynamic and effects to drive reaction and selectivity both.

Langmuir-Hinshelwood-Hougen-Watson kinetics (LHHW)

In heterogeneous catalysis, the kinetics must account for the reaction that takes place on the surface of the solid. Hence heterogeneous catalysis is also referred to as contact catalysis in the older literature. Reaction takes place in combination with adsorption, which we have already studied. Now we need to couple adsorption with the mass action kinetics for the surface reaction. To do this we will assume that the rates of adsorption and desorption are fast compared to the rates of surface chemical reaction. This is a good assumption for many cases, but not all.

We will consider first the case of a surface reaction that takes A into B, for example an isomerization. The reactant A adsorbs onto a site reacts at that site to form B and then B desorbs to the gas phase, relinquishing the site for another round of reaction. This is the picture on two equivalent sites in the schematic:

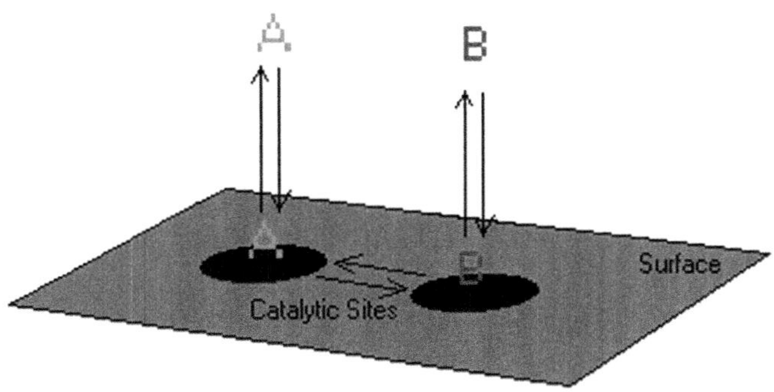

Given that adsorption and desorption of A and B are at the same site, they are competing for the sites. We account for this is in the adsorption rate term, as shown for A below:

$$r_{A,\text{ads}} = k_{A,\text{ads}} C_A [\, C_{\text{tot}} - C_{A,\text{surf}} - C_{B,\text{surf}} \,] - k_{A,\text{des}} C_{A,\text{surf}}$$

This says that the rate of adsorption of species A is proportional to the gas phase concentration of species A, and the difference between the total number of sites available and those occupied by species A or B. This accounts for the competition for the same sites by species A and species B. The rate of desorption for species A is as always first order in the surface concentration of A. At adsorption-desorption equilibrium this rate goes to zero so we have:

$$k_{A,\text{ads}} C_A [\, C_{\text{tot}} - C_{A,\text{surf}} - C_{B,\text{surf}} \,] = k_{A,\text{des}} C_{A,\text{surf}}$$

$$C_A \frac{k_{A,\text{ads}}}{k_{A,\text{des}}} = \frac{C_{A,\text{surf}}}{\left(C_{\text{tot}} - C_{A,\text{surf}} - C_{B,\text{surf}}\right)}$$

$$C_A K_A = \frac{C_{A,\text{surf}}}{\left(C_{\text{tot}} - C_{A,\text{surf}} - C_{B,\text{surf}}\right)} \frac{C_{\text{tot}}}{C_{\text{tot}}}$$

$$C_A K_A = \frac{C_{A,\text{surf}}}{C_{\text{tot}}} \frac{C_{\text{tot}}}{\left(C_{\text{tot}} - C_{A,\text{surf}} - C_{B,\text{surf}}\right)}$$

$$C_A K_A = \theta_A \frac{1}{(1 - \theta_A - \theta_B)}$$

Where θ_A stands for the fraction of the total sites occupied by A on the surface, $\frac{C_{A,\text{surf}}}{C_{\text{tot}}}$ and the same meaning is attributed to θ_B for species B. Repeating this analysis for species B we find:

$$C_B K_B = \theta_B \frac{1}{(1 - \theta_A - \theta_B)}$$

We can solve the two equations simultaneously for θ_A and for θ_B to get them both in terms of the gas phase concentrations and the adsorption constants for each:

Clear["Global` * "]

***θ*=.**

***K*=.**

Solve [{C_A==,

C_B==},]

$$\left\{\left\{\theta_A \to \frac{C_A K_A}{1 + C_A K_A + C_B K_B}, \theta_B \to \frac{C_B K_B}{1 + C_A K_A + C_B K_B}\right\}\right\}$$

The surface reaction is reversible and is first order in the surface concentrations of A and of B.

$$r_{A \to B,\text{surf}} = k_{A \to B,\text{surf}} C_{A,\text{surf}} - k_{B \to A,\text{surf}} C_{B,\text{surf}}$$

Multiplying through by $\frac{C_{\text{tot}}}{C_{\text{tot}}}$ provides these expressions in terms of the fractional surface concentrations:

$$r_{A \to B,\text{surf}} = k_{A \to B,\text{surf}}\, C_{\text{tot}}\, \theta_A - k_{B \to A,\text{surf}}\, C_{\text{tot}}\, \theta_B$$

Replacing with the expressions for the fractional surface concentrations:

$$r_{A\to B,\text{surf}} = \frac{k_{A\to B,\text{surf}} C_{\text{tot}} C_A K_A - k_{B\to A,\text{surf}} C_{\text{tot}} C_B K_B}{1 + C_A K_A + C_B K_B}$$

The reversible surface reaction has associated with it an equilibrium constant, which is just the ratio of the forward to the reverse surface rate constants:

$$K_{A\Leftrightarrow B,\text{surf}} = \frac{k_{A\to B,\text{surf}}}{k_{B\to A,\text{surf}}}$$

$$\therefore k_{B\to A,\text{surf}} = \frac{k_{A\to B,\text{surf}}}{K_{A\Leftrightarrow B,\text{surf}}}$$

Making this substitution and factoring out the product of the forward surface rate constant and the total surface concentration of sites:

$$r_{A\to B,\text{surf}} = \frac{k_{A\to B,\text{surf}} C_{\text{tot}} \left(C_A K_A - \dfrac{C_B K_B}{K_{A\Leftrightarrow B,\text{surf}}} \right)}{1 + C_A K_A + C_B K_B}$$

On the far right of the numerator, we have two parameters that may be difficult to obtain independently, they are the adsorption equilibrium constant for B and the surface equilibrium constant $K_{A\Leftrightarrow B,\text{surf}}$ for the reaction $A_{\text{surf}} \Leftrightarrow B_{\text{surf}}$. Our goal is to clear these by re-expressing them in terms of something that is unchanging, after all both of these may be strong functions of the catalyst structure and composition. The overall reaction, A$\Leftrightarrow$ B, is, however, one which is fixed at any temperature and pressure by the overall equilibrium constant. *This is independent of the catalyst.* Therefore we want to use this in the reaction rate expression. Here is how we do it:

$$K_{\text{eq}} = \frac{C_B}{C_A} = \frac{C_B}{C_{B,\text{surf}}} \frac{C_{B,\text{surf}}}{C_{A,\text{surf}}} \frac{C_{A,\text{surf}}}{C_A} \quad \text{(This is multiplication by unity)}$$

$$= \frac{1}{K_B} K_{A\Leftrightarrow B,\text{surf}} K_A \quad \text{(Each term is an equilibrium constant)}$$

$$\therefore \quad \frac{K_B}{K_{A\Leftrightarrow B,\text{surf}}} = \frac{K_A}{K_{\text{eq}}} \quad \text{(Rearrangement)}$$

and

$$r_{A\to B,\text{surf}} = \frac{k_{A\to B,\text{surf}} C_{\text{tot}} \left(C_A K_A - \dfrac{C_B K_A}{K_{\text{eq}}} \right)}{1 + C_A K_A + C_B K_B} \quad \left(\text{Replace } \frac{K_B}{K_{A\Leftrightarrow B,\text{surf}}} \text{ with } \frac{K_A}{K_{\text{eq}}}\right)$$

$$r_{A\to B,\text{surf}} = \frac{k_{A\to B,\text{surf}} K_A C_{\text{tot}} \left(C_A - \frac{C_B}{K_{\text{eq}}}\right)}{1 + C_A K_A + C_B K_B} \quad \text{(Bring } K_A \text{ out to the front)}$$

$$r_{A\to B,\text{surf}} = \frac{k_{f,\text{AB}} \left(C_A - \frac{C_B}{K_{\text{eq}}}\right)}{1 + C_A K_A + C_B K_B} \quad (k_{A\to B,\text{surf}} K_A = k_{f,\text{AB}})$$

The product $\boldsymbol{k_{A\to B,\text{surf}} K_A C_{\text{tot}}}$ is usually taken as the "global" forward rate constant on the surface, $k_{f,\text{AB}}$. Now we can proceed to see how this equation behaves.

Consider a batch reactor of volume V, into which the catalyst that does the conversion of A to B has been placed. The catalyst occupies a fraction (1 - ϵ) of the reactor volume. The component balances for A and B are:

$$\frac{d\left[C_A \epsilon V\right]}{dt} = \text{-}\,(1 - \epsilon)\, \frac{k_{f,\text{AB}} \left(C_A - \frac{C_B}{K_{\text{eq}}}\right)}{1 + C_A K_A + C_B K_B} \text{V}$$

$$\frac{d\left[C_B \epsilon V\right]}{dt} = +\,(1 - \epsilon)\, \frac{k_{f,\text{AB}} \left(C_A - \frac{C_B}{K_{\text{eq}}}\right)}{1 + C_A K_A + C_B K_B} \text{V}$$

In the following cell we compute the two concentrations as functions of time, and in the Epilog we compute the equilibrium levels of A and B and graph horizontal lines corresponding to each. The equilibrium level of reaction in this case is give by:

$$\text{Keq} = \frac{\alpha}{1 - \alpha}$$

```
Clear["Global`*"]

ϵ = 0.4;

kf = 10^-1;

Ka = 1; (* Parameters *)

Kb = 10;

Keq = .5;

Cao = 1;

Cbo = 0;


tmax = 50;
```

LHHW1 = NDSolve[

$$\left\{Ca'[t]==-\frac{(1-\epsilon)}{\epsilon}\frac{kf\left(Ca[t]-\frac{Cb[t]}{Keq}\right)}{1+KaCa[t]+KbCb[t]},\right.$$

$$Cb'[t]==+\frac{(1-\epsilon)}{\epsilon}\frac{kf\left(Ca[t]-\frac{Cb[t]}{Keq}\right)}{1+KaCa[t]+KbCb[t]},$$

Ca[0] == Cao,

Cb[0] == Cbo},

{Ca[t], Cb[t]},

{t, 0, tmax}];

CA[t_]:=Evaluate[Ca[t]/.LHHW1]

CB[t_]:=Evaluate[Cb[t]/.LHHW1]

Plot[{CA[t], CB[t]},

{t, 0, tmax},

AxesLabel → {t, "Ca[t],Cb[t]"},

PlotStyle → {Hue[0.5], Hue[0.6]},

Epilog →

{(* This computes the equilibrium levels *)

{RGBColor[0, 0, 1], Dashed,

Line[

$$\left\{\left\{0, \text{Flatten}\left[\text{NSolve}\left[\text{Keq}==\frac{\alpha}{1-\alpha},\alpha\right]\right][[1]][[2]]+\right.\right.$$

.002},

$$\left\{\text{tmax}, \text{Flatten}\left[\text{NSolve}\left[\text{Keq}==\frac{\alpha}{1-\alpha},\alpha\right]\right][[1]][[\right.$$

2]] + .002}}

]},

```
{RGBColor[0, 1, 1], Dashed,
Line[
{{0,
(1 - Flatten[NSolve[Keq==α/(1-α), α]][[1]][[2]]+
.002)},
{tmax,
(1 - Flatten[NSolve[Keq==α/(1-α), α]][[1]][[2]]+
.002)}}
  ]}
    }
    ]
```

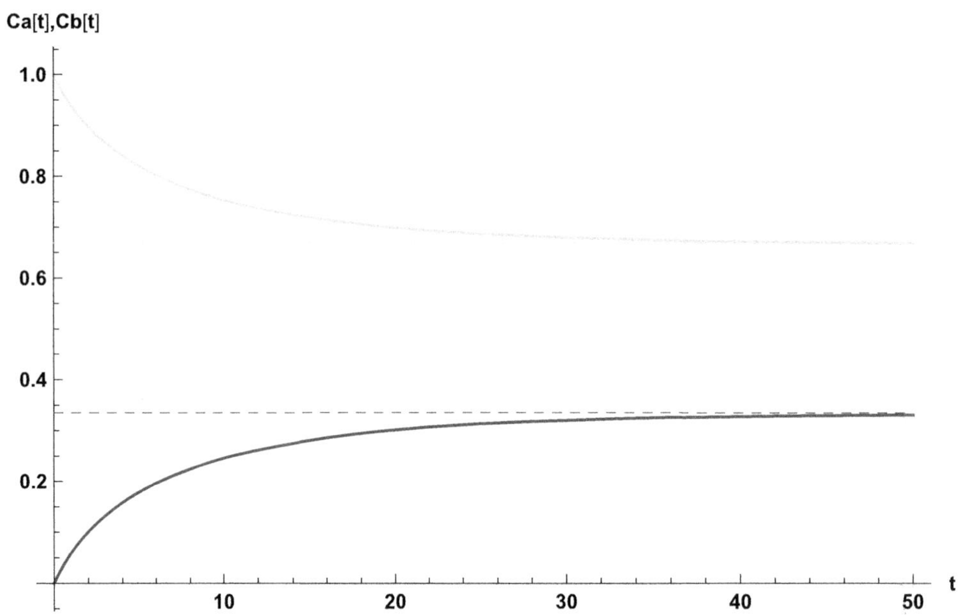

The hydrogenation and dehydrogenation of alkenes and alkanes are reversible processes that favor the alkane and involve multiple sites. For this reason it is worth considering a prototypical general case to see how we progress with the LHHW analysis.

The reaction proceeding from alkene to alkane can be taken as the forward direction, since it is the thermodynamically favored direction. The overall reaction is:

$$A + H_2 \Longleftrightarrow B$$

The individual steps including the sites ($\otimes$) are as follows:

$$A + \otimes \Longleftrightarrow A\otimes \quad \text{Adsorption/Desorption}$$

$$H_2 + 2\otimes \Longleftrightarrow 2\,H\otimes \quad \text{Dissociative Adsorption/Desorption}$$

$$A\otimes + 2\,H\otimes \Longleftrightarrow B\otimes + 2\otimes \quad \text{Surface Reaction}$$

$$B\otimes \Longleftrightarrow B + \otimes \quad \text{Adsorption/Desorption}$$

In this mechanism the dihydrogen molecule must dissociatively adsorb prior to reacting with the alkene, A. This requires two sites. When the surface reaction takes place to convert the alkene and two hydrogen atoms into one alkane, those two sites are liberated. So we need to examine how the dissociative adsorption step is handled and what ramification this has upon the rate expression, assuming all the adsorption-desorption steps are at equilibrium.

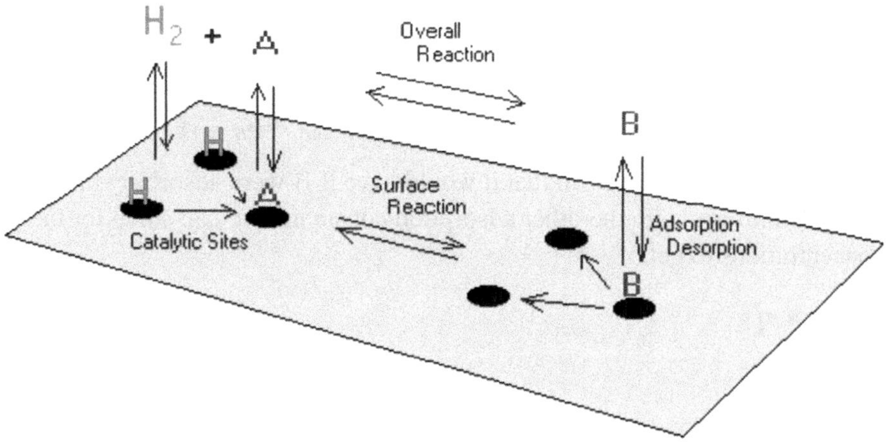

The dissociative adsorption-desorption of hydrogen follows this rate expression:

$$r_{\mathrm{H2,ads}} = k_{\mathrm{H2,ads}} C_{\mathrm{H2}} (C_{\mathrm{tot}} - C_{H,\mathrm{surf}} - C_{A,\mathrm{surf}} - C_{B,\mathrm{surf}})^2 - k_{H,\mathrm{des}} C_{H,\mathrm{surf}}{}^2$$

The adsorption includes the difference between the total concentration of sites and the sites occupied by hydrogen atoms, A and B. Notice also that this term is squared because there are two sites involved in the adsorption process. On the desorption side of the expression, we see that the rate depends upon the square of the concentration of surface hydrogen atoms. Once again we assume this step and the other adsorption-desorption steps to be at equilibrium:

$$k_{H2,ads}C_{H2}(\ C_{tot} - C_{H,surf} - C_{A,surf} - C_{B,surf}\ )^2 = k_{H,des}C_{H,surf}{}^2$$

$$\frac{k_{H2,ads}}{k_{H,des}}C_{H2} = \frac{C_{H,surf}{}^2}{(C_{tot} - C_{H,surf} - C_{A,surf} - C_{B,surf})^2}$$

$$K_{H2,ads}C_{H2} = \frac{C_{H,surf}{}^2}{(C_{tot} - C_{H,surf} - C_{A,surf} - C_{B,surf})^2}\frac{C_{tot}{}^2}{C_{tot}{}^2}]$$

$$K_{H2,ads}C_{H2} = \left(\frac{C_{H,surf}{}^2}{C_{tot}{}^2}\right)\left(\frac{C_{tot}{}^2}{(C_{tot} - C_{H,surf} - C_{A,surf} - C_{B,surf})^2}\right)$$

$$K_{H2,ads}C_{H2} = \frac{\theta_{H,surf}{}^2}{(1 - \theta_{H,surf} - \theta_{A,surf} - \theta_{B,surf})^2}$$

$$\sqrt{K_{H2,ads}C_{H2}} = \frac{\theta_{H,surf}}{(1 - \theta_{H,surf} - \theta_{A,surf} - \theta_{B,surf})}$$

From the same analyses of the adsorption-desorption processes for A and B we find:

$$K_{A,ads}C_A = \frac{\theta_{A,surf}}{(1 - \theta_{H,surf} - \theta_{A,surf} - \theta_{B,surf})}$$

$$K_{B,ads}C_B = \frac{\theta_{B,surf}}{(1 - \theta_{H,surf} - \theta_{A,surf} - \theta_{B,surf})}$$

The term for B is written in the form that it would have if B were adsorbing in order to keep the meaning K_B uniform with the other adsorption constants. We can solve for the fractional surface concentrations to get:

Clear["Global`*"]

θ=.

K=.

Simplify[Solve$\left[\left\{C_A K_A = \frac{\theta_A}{-\theta_A-\theta_B-\theta_H+1},\right.\right.$

$C_B K_B = \frac{\theta_B}{-\theta_A-\theta_B-\theta_H+1},$

$\left.\sqrt{C_{H2}K_{H2}} = \frac{\theta_H}{-\theta_A-\theta_B-\theta_H+1}\right\},$

$\{\theta_A, \theta_B, \theta_H\}$]]

$$\left\{\left\{\theta_A \to \frac{C_A K_A}{C_A K_A + C_B K_B + \sqrt{C_{H2}K_{H2}} + 1},\right.\right.$$

$$\theta_B \to \frac{C_B K_B}{C_A K_A + C_B K_B + \sqrt{C_{H2} K_{H2}} + 1},$$

$$\theta_H \to \frac{\sqrt{C_{H2} K_{H2}}}{C_A K_A + C_B K_B + \sqrt{C_{H2} K_{H2}} + 1}\}\}$$

The surface reaction rate begins to take shape:

$$r_{A+2H \Leftrightarrow B,\text{surf}} = k_{A+2H \to B,\text{surf}} C_{A,\text{surf}}\ C_{H,\text{surf}}{}^2\ - k_{B \to A+2H,\text{surf}} C_{B,\text{surf}} C_{\text{empty,sites}}{}^2$$

$$r_{A+2H \Leftrightarrow B,\text{surf}} = k_{A+2H \to B,\text{surf}} C_{A,\text{surf}}\ C_{H,\text{surf}}{}^2 \left(\frac{C_{\text{tot}}{}^3}{C_{\text{tot}}{}^3}\right) - k_{B \to A+2H,\text{surf}} C_{B,\text{surf}} C_{\text{empty,sites}}{}^2 \left(\frac{C_{\text{tot}}{}^3}{C_{\text{tot}}{}^3}\right)$$

$$r_{A+2H \Leftrightarrow B,\text{surf}} = k_{A+2H \to B,\text{surf}} C_{\text{tot}}{}^3 \theta_A\ \theta_H{}^2\ -\ k_{B \to A+2H,\text{surf}} C_{\text{tot}}{}^3 \theta_B \theta^2$$

$$r_{A+2H \Leftrightarrow B,\text{surf}} = k_{A+2H \to B,\text{surf}} C_{\text{tot}}{}^3 \theta_A\ \theta_H{}^2\ - k_{B \to A+2H,\text{surf}} C_{\text{tot}}{}^3 \theta_B\ (1 - \theta_H - \theta_A - \theta_B)^2$$

$$r_{A+2H \Leftrightarrow B,\text{surf}} = k_{A+2H \to B,\text{surf}} C_{\text{tot}}{}^3 (\theta_A\ \theta_H{}^2\ - \frac{\theta_B\ (1 - \theta_H - \theta_A - \theta_B)^2}{K_{A+2H \Leftrightarrow B,\text{surf}}})$$

This begins to look a bit formidable because of the algebra that would be involved in manipulating this expression. We will let *Mathematica* do most of the algebraic manipulations by following these steps that use PowerExpand to expand the higher order terms, Together, which brings separate terms over the same denominator and FullSimplify, which does just that. Here is the result of this approach to the parenthetical expression of the right-hand side of the rate expression:

Clear[K]

$$\boldsymbol{\theta_A = \frac{C_A K_A}{1 + C_A K_A + C_B K_B + \sqrt{C_{H2} K_{H2}}};}$$

$$\boldsymbol{\theta_B = \frac{C_B K_B}{1 + C_A K_A + C_B K_B + \sqrt{C_{H2} K_{H2}}};}$$

$$\boldsymbol{\theta_H = \frac{C_{H2} K_{H2}}{C_{H2} K_{H2} + (1 + C_A K_A + C_B K_B)\sqrt{C_{H2} K_{H2}}};}$$

FullSimplify[

Together[

PowerExpand$\boldsymbol{\left[\theta_A \theta_H^2 - \frac{\theta_B(-\theta_A - \theta_B - \theta_H + 1)^2}{K_{A+2H \Leftrightarrow B,\text{surf}}}\right]}$

]

]

$$\frac{-C_B K_B + C_A C_{H2} K_A K_{H2} K_{A+2H \Leftrightarrow B,\text{surf}}}{\left(1 + C_A K_A + C_B K_B + \sqrt{C_{H2}}\sqrt{K_{H2}}\right)^3 K_{A+2H \Leftrightarrow B,\text{surf}}}$$

We can manipulate this into the form that we are interested in:

$$r_{A+2H\Leftrightarrow B,\mathrm{surf}} = k_{A+2H\rightarrow B,\mathrm{surf}} C_{\mathrm{tot}}{}^{3} \frac{\left(C_A C_{\mathrm{H2}} K_A K_{\mathrm{H2}} - \frac{C_B K_B}{K_{A+2H\Leftrightarrow B,\mathrm{surf}}}\right)}{\left(1 + C_A K_A + C_B K_B + \sqrt{C_{\mathrm{H2}} K_{\mathrm{H2}}}\right)^3}$$

$$K_{\mathrm{eq}} = \frac{C_B}{C_A C_{\mathrm{H2}}} = \frac{C_B}{C_{B,\mathrm{surf}}} \frac{C_{B,\mathrm{surf}}}{C_{A,\mathrm{surf}}} \frac{C_{A,\mathrm{surf}}}{C_A} \frac{C_{H,\mathrm{surf}}{}^2}{C_{H,\mathrm{surf}}{}^2} \frac{1}{C_{\mathrm{H2}}}$$

$$= \frac{C_B}{C_{B,\mathrm{surf}}} \frac{C_{B,\mathrm{surf}}}{C_{A,\mathrm{surf}}} \frac{C_{A,\mathrm{surf}}}{C_A} \frac{1}{C_{H,\mathrm{surf}}{}^2} \frac{C_{H,\mathrm{surf}}{}^2}{C_{\mathrm{H2}}}$$

$$= \frac{C_B}{C_{B,\mathrm{surf}}} \frac{C_{B,\mathrm{surf}}}{C_{A,\mathrm{surf}} C_{H,\mathrm{surf}}{}^2} \frac{C_{A,\mathrm{surf}}}{C_A} \frac{C_{H,\mathrm{surf}}{}^2}{C_{\mathrm{H2}}}$$

$$= \frac{K_{A+2H\Leftrightarrow B,\mathrm{surf}} K_A K_{\mathrm{H2}}}{K_B}$$

$$\therefore \frac{K_B}{K_{A+2H\Leftrightarrow B,\mathrm{surf}}} = \frac{K_A K_{\mathrm{H2}}}{K_{\mathrm{eq}}}$$

$$r_{A+2H\Leftrightarrow B,\mathrm{surf}} = k_{A+2H\rightarrow B,\mathrm{surf}} K_A K_{\mathrm{H2}} C_{\mathrm{tot}}{}^{3} \frac{\left(C_A C_{\mathrm{H2}} - \frac{C_B}{K_{\mathrm{eq}}}\right)}{\left(1 + C_A K_A + C_B K_B + \sqrt{C_{\mathrm{H2}} K_{\mathrm{H2}}}\right)^3}$$

$$r_{A+2H\Leftrightarrow B,\mathrm{surf}} = k_{\mathrm{global}} \frac{\left(C_A C_{\mathrm{H2}} - \frac{C_B}{K_{\mathrm{eq}}}\right)}{\left(1 + C_A K_A + C_B K_B + \sqrt{C_{\mathrm{H2}} K_{\mathrm{H2}}}\right)^3}$$

where

$$k_{\mathrm{global}} = k_{A+2H\rightarrow B,\mathrm{surf}} K_A K_{\mathrm{H2}} C_{\mathrm{tot}}{}^{3}$$

Now we can write some code that will evaluate the kinetics, and the equilibrium, and graph the relevant gas and surface phase concentrations for us all-at-once. The equilibrium extent of reaction can be computed as follows for any given value of the equilibrium constant:

```
Keq = .5;
NSolve[Keq == α/(Cao(1 − α)^2), α]
```

$$\left\{\left\{\alpha \to \frac{\mathrm{Cao} - 1.\sqrt{2.\mathrm{Cao} + 1.} + 1.}{\mathrm{Cao}}\right\}, \left\{\alpha \to \frac{\mathrm{Cao} + \sqrt{2.\mathrm{Cao} + 1.} + 1.}{\mathrm{Cao}}\right\}\right\}$$

We see that the second of the two evaluations is the correct one. We embed this into the **Epilog** as the y-coordinate of a line for the equilibrium concentration of the product at time zero and tmax, and we take one minus this value to get the line corresponding to the equilibrium concentration of reactant. By doing it this way, we can set the magnitude of the equilibrium constant and the code will automatically compute these concentrations for the graph. This allows us to visualize immediately, where the concentrations are at any time relative to the equilibrium concentrations.

Clear["Global`* "]

$\epsilon = .4;$

$\text{kglo} = .01;$

$\text{Ka} = .1;$

$\text{Kb} = .01;$

$\text{KH2} = .5;$

$\text{Keq} = .5;$

$\text{Cao} = 1;$

$\text{CH2o} = 1;$

$\text{Cbo} = 0;$

$\text{tmax} = 500;$

LHHW2 = NDSolve[

$$\{\partial_t \text{Ca}[t] == -\frac{(1-\epsilon)}{\epsilon}\frac{\text{kglo}\left(\text{Ca}[t]\text{CH2}[t] - \frac{\text{Cb}[t]}{\text{Keq}}\right)}{\left(1 + \text{KaCa}[t] + \text{KbCb}[t] + \sqrt{\text{KH2 CH2}[t]}\right)^3},$$

$$\partial_t \text{CH2}[t] == -\frac{(1-\epsilon)}{\epsilon}\frac{\text{kglo}\left(\text{Ca}[t]\text{CH2}[t] - \frac{\text{Cb}[t]}{\text{Keq}}\right)}{\left(1 + \text{KaCa}[t] + \text{KbCb}[t] + \sqrt{\text{KH2 CH2}[t]}\right)^3},$$

$$\partial_t \text{Cb}[t] == +\frac{(1-\epsilon)}{\epsilon}\frac{\text{kglo}\left(\text{Ca}[t]\text{CH2}[t] - \frac{\text{Cb}[t]}{\text{Keq}}\right)}{\left(1 + \text{KaCa}[t] + \text{KbCb}[t] + \sqrt{\text{KH2 CH2}[t]}\right)^3},$$

Ca[0] == Cao,

CH2[0] == CH2o,

Cb[0] == Cbo},

{Ca[t], CH2[t], Cb[t]},

{t, 0, tmax}];

cA[t_]:=Evaluate[Ca[t]/.LHHW2]

cH2[t_]:=Evaluate[CH2[t]/.LHHW2]

cB[t_]:=Evaluate[Cb[t]/.LHHW2]

$$\theta_A[\text{t\_}]:=\frac{\text{Ka cA}[t]}{1+\text{Ka cA}[t]+\text{Kb cB}[t]+\sqrt{\text{KH2 cH2}[t]}}$$

$$\theta_H[\text{t\_}]:=\frac{\text{KH2 cH2}[t]}{1+\text{Ka cA}[t]+\text{Kb cB}[t]+\sqrt{\text{KH2 cH2}[t]}}$$

Plot[{cA[t], cB[t]}, {t, 0, tmax},

AxesLabel → {t, "Ca[t],Cb[t]"},

PlotStyle → {Hue[0.5], Hue[0.6]},

PlotLabel → Keq"= Keq",

Epilog → {

{RGBColor[0, 0, 1], Dashed,

Line[

$$\left\{\left\{0, \text{Flatten}\left[\text{NSolve}\left[\text{Keq}==\frac{\alpha}{(1-\alpha)^2}, \alpha\right]\right]\text{[[2]][[}\right.\right.$$

2]] + .002},

$$\left\{\text{tmax, Flatten}\left[\text{NSolve}\left[\text{Keq}==\frac{\alpha}{(1-\alpha)^2}, \alpha\right]\right]\text{[[2]][[}\right.$$

2]] + .002}}

]},

{RGBColor[0, 1, 1], Dashed,

Line[

{{0,

```
(1 - Flatten[NSolve[Keq== α/(1 - α)^2, α]][[2]][[
2]] + .002)},
{tmax,
(1 - Flatten[NSolve[Keq== α/(1 - α)^2, α]][[2]][[
2]] + .002)}}
  ]}
     }
     ]

Plot[{θA[t], θH[t]}, {t, 0, tmax},
AxesLabel → {t, "θA[t],θH[t]"},
PlotStyle → {Hue[0.3], Hue[0.5]},
PlotStyle → {{Dashing[{0.05, 0.025}], Hue[0.0]}}]
```

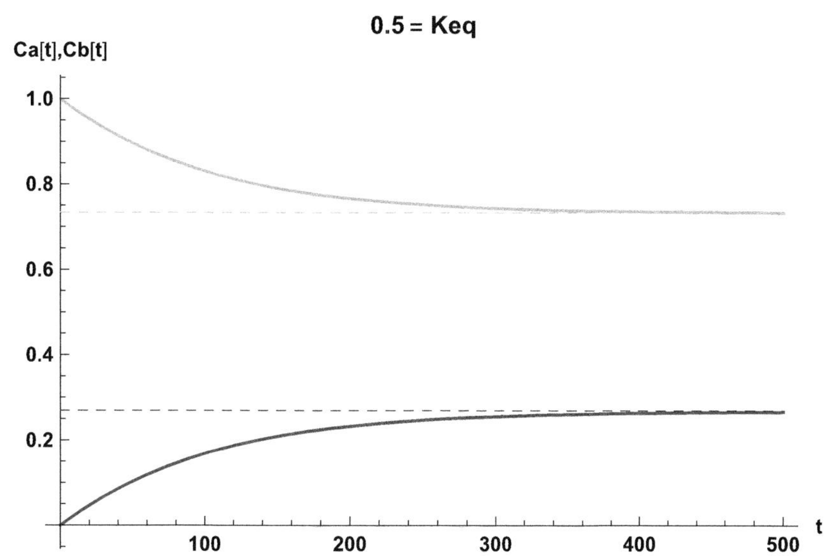

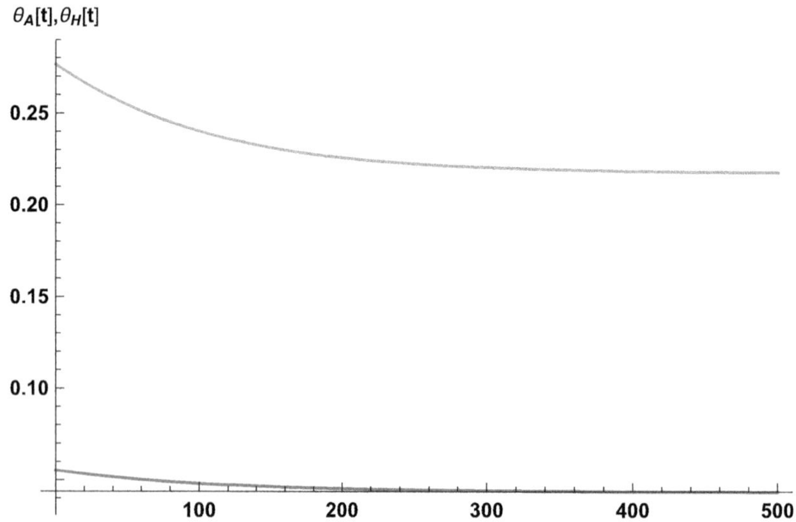

The next step is to make this into a **Module** so that we can test parametric sensitivity readily.

LH[*K*eq_, kglo_, tmax_]:=Module[{$\epsilon = .4$,

Ka = .1,

Kb = .01,

KH2 = .5,

Cao = 1,

CH2o = 1,

Cbo = 0},

LHHW2 = NDSolve[

$$\{\partial_t \mathrm{Ca}[t] == -\frac{(1-\epsilon)}{\epsilon}\frac{\mathrm{kglo}\left(\mathrm{Ca}[t]\mathrm{CH2}[t]-\frac{\mathrm{Cb}[t]}{\mathrm{Keq}}\right)}{\left(1+\mathrm{KaCa}[t]+\mathrm{KbCb}[t]+\sqrt{\mathrm{KH2\ CH2}[t]}\right)^3},$$

$$\partial_t \mathrm{CH2}[t] == -\frac{(1-\epsilon)}{\epsilon}\frac{\mathrm{kglo}\left(\mathrm{Ca}[t]\mathrm{CH2}[t]-\frac{\mathrm{Cb}[t]}{\mathrm{Keq}}\right)}{\left(1+\mathrm{KaCa}[t]+\mathrm{KbCb}[t]+\sqrt{\mathrm{KH2\ CH2}[t]}\right)^3},$$

$$\partial_t \mathrm{Cb}[t] == +\frac{(1-\epsilon)}{\epsilon}\frac{\mathrm{kglo}\left(\mathrm{Ca}[t]\mathrm{CH2}[t]-\frac{\mathrm{Cb}[t]}{\mathrm{Keq}}\right)}{\left(1+\mathrm{KaCa}[t]+\mathrm{KbCb}[t]+\sqrt{\mathrm{KH2\ CH2}[t]}\right)^3},$$

```
Ca[0] == Cao,
CH2[0] == CH2o,
Cb[0] == Cbo},
{Ca[t], CH2[t], Cb[t]},
{t, 0, tmax}];

cA[t_]:=Evaluate[Ca[t]/.LHHW2];
cH2[t_]:=Evaluate[CH2[t]/.LHHW2];
cB[t_]:=Evaluate[Cb[t]/.LHHW2];
θA[t_]:=(Ka cA[t])/(1 + Ka cA[t] + Kb cB[t] + Sqrt[KH2 cH2[t]]);
θH[t_]:=(KH2 cH2[t])/(1 + Ka cA[t] + Kb cB[t] + Sqrt[KH2 cH2[t]]);
Plot[{cA[t], cB[t]}, {t, 0, tmax},
AxesLabel → {t, "Ca[ t ],Cb[ t ]"},
PlotStyle → {Hue[0.5], Hue[0.6]},
PlotLabel → Keq"= Keq",
Epilog → {
{RGBColor[0, 0, 1], Dashed,
Line[
{{0, Flatten[NSolve[Keq==α/(1 − α)^2, α]][[2]][[
2]] + .002},
{tmax,
Flatten[NSolve[Keq==α/(1 − α)^2, α]][[2]][[
2]] + .002}}
   ]},
{RGBColor[0, 1, 1], Dashed,
```

```
Line[
{{0,
(1 - Flatten[NSolve[Keq==α/(1 - α)^2, α]][[2]][[
2]] + .002)},
{tmax,
(1 - Flatten[NSolve[Keq==α/(1 - α)^2, α]][[2]][[
2]] + .002)}}
  ]}
    }
    ]]

LH[.1, .05, 50]
```

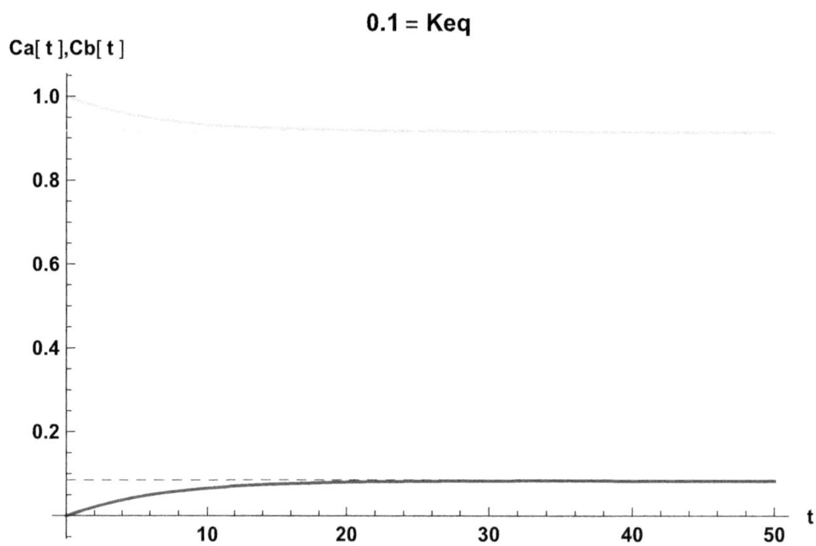

When the equilibrium constant is small (0.01), then even with the forward rate constant at its maximum value there is no reaction as can be seen below:

Manipulate[LH[Keq, kglo, 500], {Keq, .01, 100, .01},

{kglo, .0001, .1, .00001}]

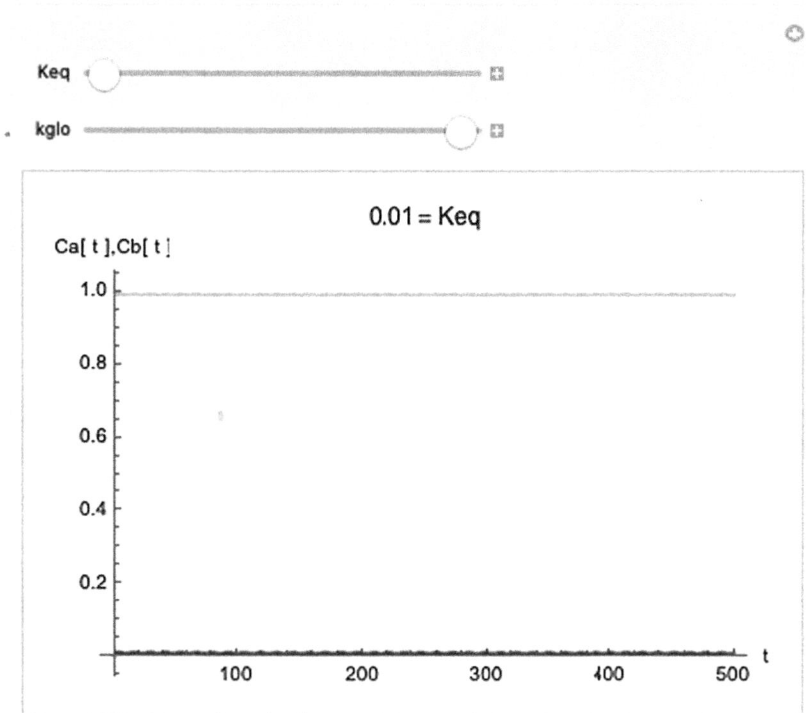

When the equilibrium is favorable (~46), as long as the rate constant is large enough for reaction to occur, it does.

Manipulate[LH[Keq, kglo, 500], {Keq, .01, 100, .01},

{kglo, .0001, .1, .00001}]

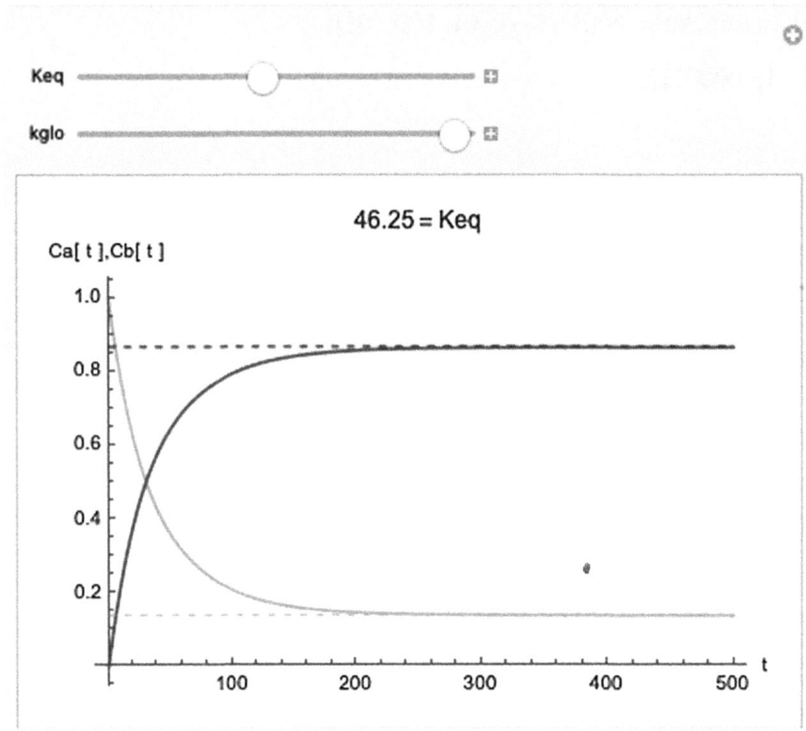

Finally, with the equilibrium constant at a maximum (100), if the rate constant is too low, then the system remains dormant chemically and exceptionally long times would be required to see an appreciable extent of reaction.

```
Manipulate[LH[Keq, kglo, 500], {Keq, .01, 100, .01},
{kglo, .0001, .1, .00001}]
```

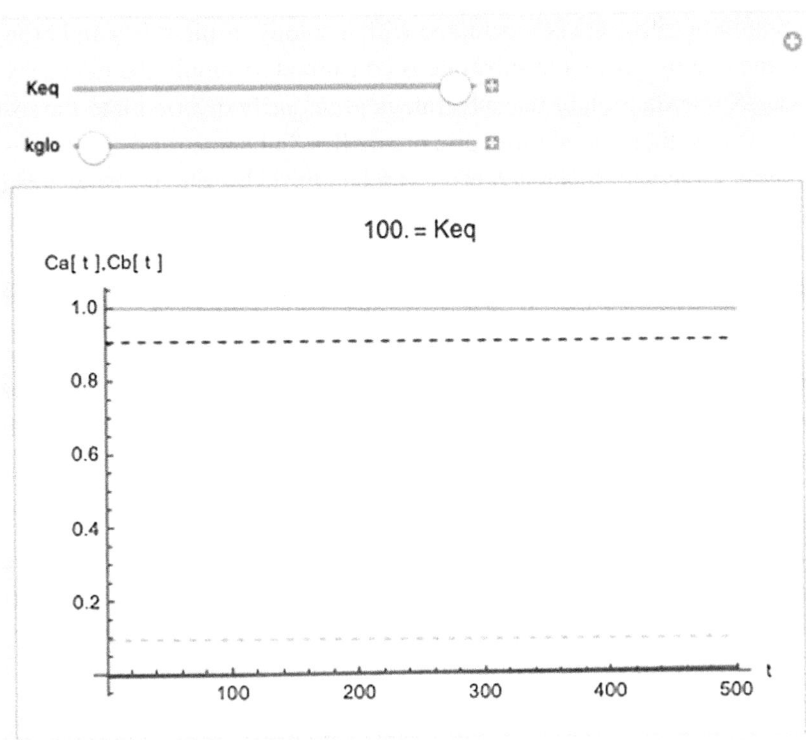

Microbial population dynamics

At the present there is a research explosion in the biological sciences that is unprecedented. The breakthroughs in the basic sciences of genomics and related disciplines have brought us to the threshold of a new era in biological technology. Paramount to this new technology is the use of microbes, that is cellular organisms as reactors. Organisms have evolved mechanisms for dealing with environmental stress, such as the presence of a new substrate chemical in their surroundings, by rerouting their metabolic pathways. Metabolic engineers can take advantage of this through a procedure of accelerated adaptation in order to generate new microbes that consume a given substrate and produce a specific target chemical.

Microbes use enzymes as catalysts to make desired or beneficial reactions takes place, and typically under mild conditions. Brewing of beer and fermentation of fruit and vegetable mass high in starches to produce consumable ethanol, is the oldest and most familiar instance of using microbial action to fulfill a desired end. But now much more has been demonstrated from the production of essential human hormones to the synthesis of specialty chemicals.

If a reactor containing substrate is inoculated with a colony of microbes and brought to maturity, then as the colony grows the substrate is consumed to supply the microbes with their building blocks. Some fraction of the substrate is necessarily diverted into the formation of bio-mass, that is cells–their membranes and organelles–but some other fraction is used to produce the target molecule. In a batch process when the substrate has been consumed, the microbial colony either dies rapidly, or if the process is to be stopped prior to complete substrate consumption, it is killed by a rapid change in conditions. (For example by raising the temperature as is done in the pasteurization of raw milk.) From this point the problem of recovering the target molecule is one of separating it from the biomass and aqueous medium.

The basis of life is molecular, therefore we can describe the rates of substrate consumption, product formation, and even microbe population growth in much the same way that we would describe the rates of molecular-level chemical processes.

We will take the microbe, substrate, and product concentrations to be a[t], b[t], c[t], respectively. The ways in which these kinetics are written look somewhat different, but they are actually equivalent mathematically to the Langmuir-Hinshelwood-Hougen-Watson kinetics. The equations that describe the rates of change of each of these are shown below:

$$\text{a'}[\,t\,] = (\mu\text{max}\,\frac{b[t]}{b[t]+\text{Ks}} \text{ - k) a}[\,t\,]$$

$$\text{b'}[\,t\,] = (\frac{\mu\text{max}}{\text{ys}}\,\frac{b[t]}{b[t]+\text{Ks}} \text{ - k) a}[\,t\,]$$

$$\text{c'}[\,t\,] = (\alpha + \beta\mu\text{max}\frac{b[t]}{b[t]+\text{Ks}} \text{ - k) a}[\,t\,]$$

Where μmax is a maximum rate constant, Ks is a saturation concentration, and ys is a dimensionless parameter that is similar to a stoichiometric coefficient. Similarly, α and β are dimensionless numbers that are also similar to stoichiometric coefficients–they relate the rate of production of the desired molecule to the rate of growth of microbial cell mass.

In the cell that follows, we build a model for these kinetics to examine how they behave:

```
Clear["Global`*"]

μ = .15; "μmax";

K = .04; "Ks";

y = 1; "ys";

α = 10^-n;
```

```
n = 2;
β = .1;
tmax = 300;
k = 0.1;
"a[t] is the change in microbial concentration; decreasing";
"b[t] is the change in the substrate concentration; increasing";
"c[t] is the change in product concentration; increasing";
bugs1 = NDSolve[{
        a'[t]==μ(b[t]/(K + b[t]) - k) a[t],
        b'[t]== - μ/y b[t]/(K + b[t]) a[t],
        c'[t]==(α + βμ b[t]/(K + b[t])) a[t],
        a[0]==.01,
        b[0]==100,
        c[0]==0
        },
      {a[t], b[t], c[t]},
      {t, 0, tmax}];
a1[t_]:=Evaluate[a[t]/.bugs1];
b1[t_]:=Evaluate[b[t]/.bugs1];
c1[t_]:=Evaluate[c[t]/.bugs1];
pa1 = Plot[a1[t], {t, 0, tmax},
PlotStyle → Hue[0.0],
PlotRange → {{0, tmax}, {0, b1[0][[1]]}}];
pb1 = Plot[b1[t], {t, 0, tmax},
```

PlotStyle → Hue[.3]];

pc1 = Plot[c1[*t*], {*t*, 0, tmax},];

Show[{pa1, pb1, pc1}, PlotLabel → tmax"= tmax,a[t]:red, b[t]:grn, c[t]:blk"]

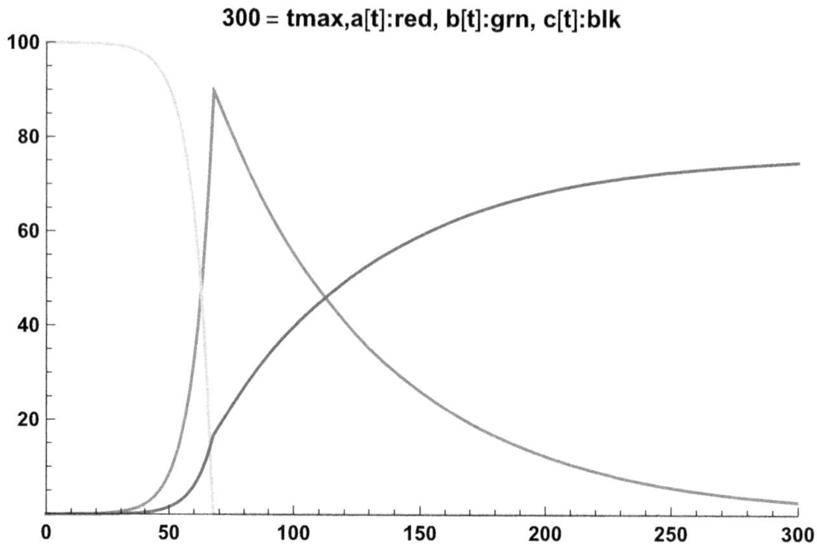

How do we interpret these results? Remember a[t] is the change in microbial concentration (red - mid gray in the print version), b[t] is the change in the substrate concentration (green - light gray in the print version), and c[t] is the change in product concentration (blue - dark gray in the print version). The substrate concentration (green curve - light gray curve in the print version) falls slowly at very short time, just after inoculation of the microbial colony. This is an induction period (at the top of the green curve - light gray curve in the print version) over which the colony grows slowly (see the bottom of the red curve - mid gray curve in the print version). After this induction time, the colony of microbes (red - mid gray curve in the print version) suddenly grows, "explosively," and reaches a maximum (see the peak in the red curve - mid gray curve in the print version). At the same time, that the explosive growth occurs, the substrate is diminished at precipitous rate (see the near vertical drop in the green curve - light gray curve in the print version). Once the substrate is used up (see the bottom of the green curve - light gray curve in the print version), the colony begins to diminish in number (see the falloff on the red curve - mid gray curve in the print version). This occurs in the present case at a much slower rate than did their growth. During the period of explosive growth and shortly after the maximum is attained in microbial population, the product concentration increases (see the blue curve - dark gray curve in the print version) and then levels to a constant with time as the colony finally expires.

This is another very interesting subject to explore with continuous variation of parameters. Here is the function, microbs, that will allow us to simulate and explore these kinetics. We will include {μ, n, K, k, tmax} as the parameters to vary:

```
microbs[μ_, n_, K_, k_, tmax_]:=Module[{
y = 1,
α = 10^-n,
β = .1},
"a[t] is the change in microbial concentration;
decreasing";
"b[t] is the change in the substrate concentration;
increasing";
"c[t] is the change in product concentration;
increasing";
bugs1 = NDSolve[{
      a'[t]==μ(b[t]/(K + b[t]) - k) a[t],
      b'[t]== -(μ/y) b[t]/(K + b[t]) a[t],
      c'[t]==(α + βμ b[t]/(K + b[t])) a[t],
      a[0]==.01,
      b[0]==100,
      c[0]==0
      },
    {a[t], b[t], c[t]},
    {t, 0, tmax}];
a1[t_]:=Evaluate[a[t]/.bugs1];
b1[t_]:=Evaluate[b[t]/.bugs1];
```

```
c1[t_]:=Evaluate[c[t]/.bugs1];
{pa1 = Plot[a1[t], {t, 0, tmax},
PlotStyle → Hue[0.0],
PlotRange → {{0, tmax}, {0, b1[0][[1]]}}];
pb1 = Plot[b1[t], {t, 0, tmax},
PlotStyle → Hue[.3]];
pc1 = Plot[c1[t], {t, 0, tmax}];
Show[{pa1, pb1, pc1},
PlotLabel → tmax"= tmax,a[t]:red, b[t]:grn, c[t]:blk"]}]
```

Here are the parameters that can be varied in microbs[μ, n, K, k, tmax] and we test the function to be sure it provides the expected output:

```
microbs[.1, 2, .05, .1, 500][[1]]
```

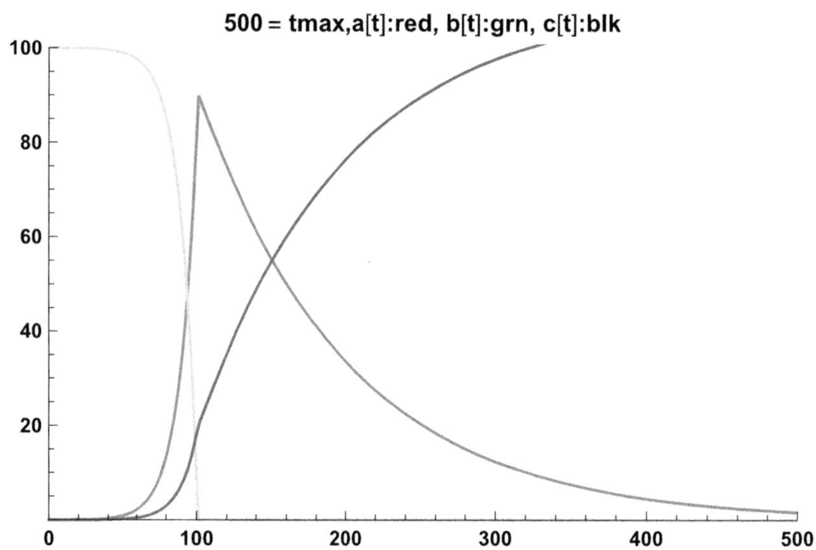

The Manipulate function allows us to vary any of the parameters in microbs. Here we manipulate small k, the rate constant.

Manipulate[microbs[.1, 2, .05, *k*, *t*][[1]],

{*k*, 0.0001, 1, 0.0001}, {*t*, 0.001, 1500, 1}]

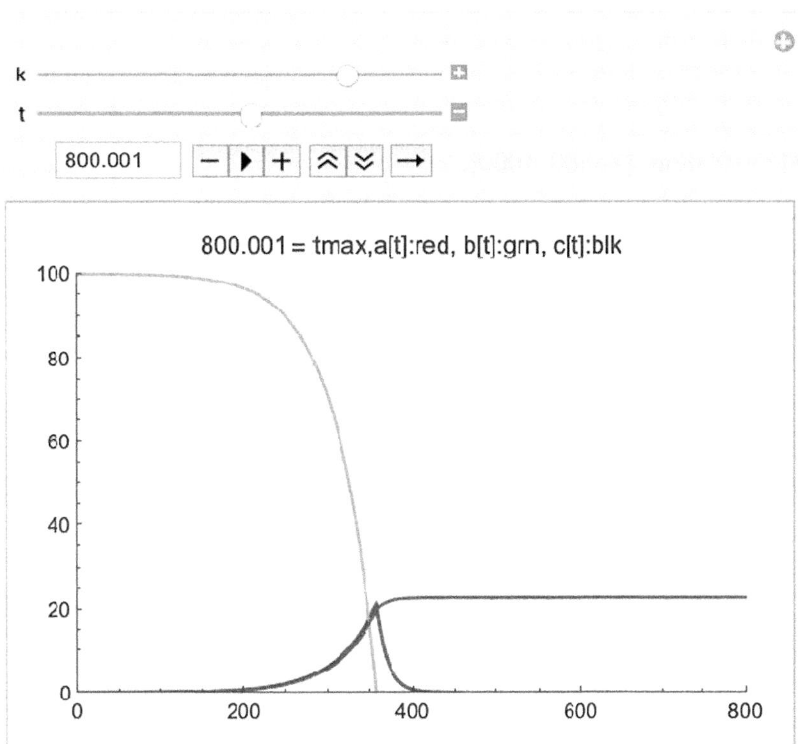

Temperature dependence of reaction rates

A century ago Arrhenius (Nobel Prize Chemistry 1903) organized a considerable body of anecdotal information and experimental data about the temperature dependence of chemical reaction rates into one simple, but elegant, mathematical statement:

$$k\,[\,T\,] = A\ \mathrm{Exp}\,[\,-\frac{Ea}{RT}\,]$$

Most importantly, this statement tells us that the temperature dependence is exponential. Although this is nothing more than a simple exponential, let's explore it and think about its chemical implications.

The activation energy, or Ea, is the energy that must be imparted to the molecule for the reaction to take place. Thinking in energy units of kiloJoules (kJ), the following graphs show the rate constant values vary continuously from 300K to 1000K for activation energies between 10 and 50 kJ per mole:

```
k[T_, Ea_]:=AExp[−Ea/(RT)]
R = 8.314;
A = 10^13;
Table[Plot[k[T, 1000Ea], {T, 300, 1000},
PlotStyle → Hue[Ea/50]],
{Ea, {10, 15, 20, 30, 40, 50}}];
Show[%, PlotRange → All]
```

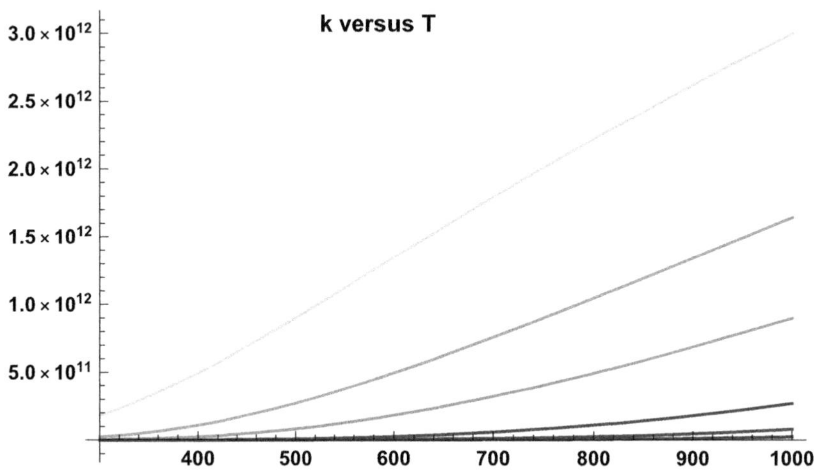

We can take the log of the rate constant and plot that as a function of temperature and parametric in the activation energy as seen below. With zero activation energy and constant pre-exponential factor, there is no temperature dependence (red - dark gray in the print version, top), and with high activation energy (80 kJ/mole, blue - mid gray in the print version, bottom), there is the most curvature and, hence, the most dependence on temperature.

```
R = 8.314;
A = 10^13;
Table[Plot[Log[k[T, 1000Ea]], {T, 1000, 300},
PlotStyle → Hue[Ea/50]],
```

```
{Ea, {0, 5, 10, 15, 20, 25, 30, 35, 40, 45, 50,
60, 70, 80}}];
Show[%, PlotRange → All,
Epilog → {Line[{{310, 0}, {310, 35}}],
Line[{{320, 0}, {320, 35}}]}]
```

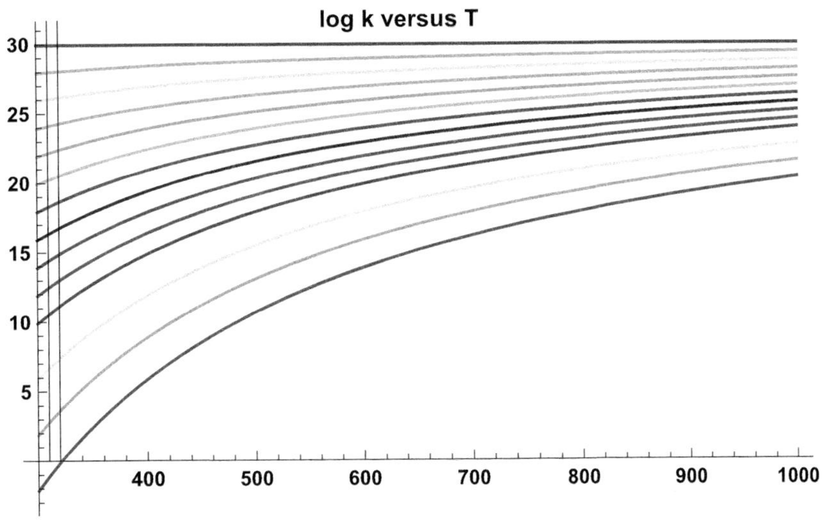

The most typical way to plot and show the temperature dependence of the rate constant as predicted by the Arrhenius expression, is to show the log of the rate constant against inverse temperature. This results in a linear function of temperature as follows:

```
A = 10^13;
R = 1.99;
Ea=.
Table[Table[{1/T//N, N[−Ea/(RT) + Log[A]]},
{T, 200, 100000, 10}],
{Ea, {0, 5, 10, 15, 20, 25, 30, 35, 40, 45, 50, 60}}];
ListPlot[%, AxesLabel → {"T(K)", "Log[ k [ T ] ]"},
PlotRange → All]
```

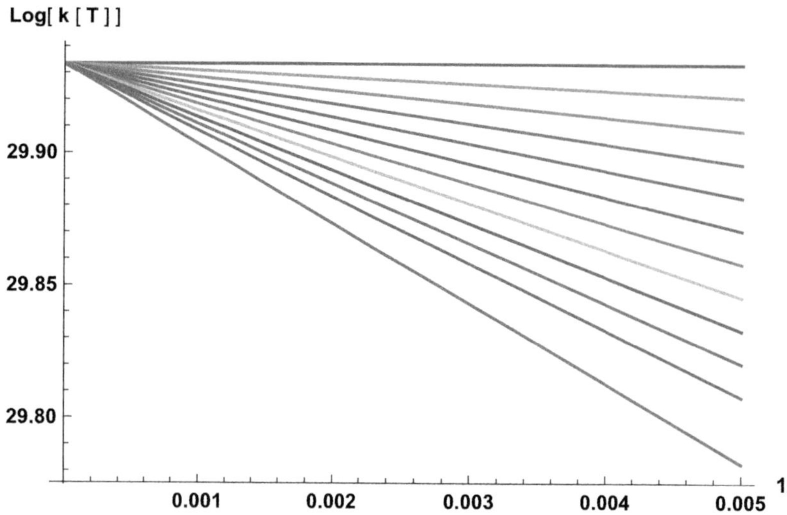

Each of these lines cuts through the abscissa at 29.9336, which corresponds to the Log_e of the constant A-factor, as it should:

Log[*A*]//*N*

Exp[%]

29.9336

$1. \times 10^{13}$

We can find the ratio of two rate constant values that have the same activation energy at different temperatures as follows:

$$\textbf{ratio[T1\_, T2\_, Ea\_]:= Exp}\left[-\frac{\text{Ea}}{R}\left(\frac{1}{\text{T2}}-\frac{1}{\text{T1}}\right)\right]$$

At an activation energy of 55 kJ per mole, we see the ratio is just at a value of two between 300 and 310K:

$R = 8.314$;

ratio$\left[300, 310, 55 * 10^3\right]$

2.03669

We can also see that as the magnitude of the temperatures rises, then for the same ten degree difference, the ratio is diminished for the same activation energy:

```
ratio[600, 610, 50 * 10^3]
```

1.17859

When we plot the ratio of two rate constants that are different by ten degrees as a continuous function of temperature and parametric in the activation energy in a range from 0 to 60 kJ/mole, we obtain this family of curves:

```
R = 8.314;
Table[Plot[ratio[T, T + 10, 10^3Ea], {T, 1000, 300},
PlotStyle → Hue[Ea/50]],
{Ea, {0, 5, 10, 15, 20, 25, 30, 35, 40, 45, 50,
60, 70, 80}}];
Show[%, PlotRange → All,
Epilog → {Line[{{310, 0}, {310, 3.5}}],
Line[{{320, 0}, {320, 3.5}}]},
AxesLabel → {"T(K)", k[T2]/k[T1]},
PlotLabel → "Rule of Ten Degrees"
]
```

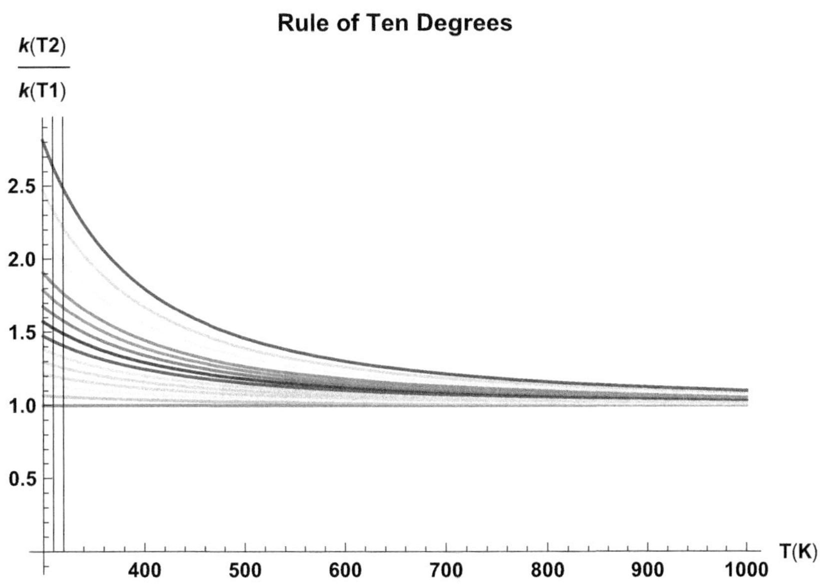

The curves are ordered from the highest activation energy (80 kJ/mole) to the lowest (0 kJ/mole). Note that the two vertical lines correspond to the temperatures of 310 and 320 K. This makes clear the general tendency for the ratios of the exponential factors to converge as the temperature increases and that the rule of ten leading to a doubling is only valid at low overall temperatures and higher activation energies. This also treats the A-factor as an absolute constant.

The second point is that although the Arrhenius expression takes the A-factor to be a constant and temperature independent, this factor does show some weak temperature dependence. The A-factor is a geometric constant that accounts for the fact that even if the molecules have sufficient energy to react, they may also need to be oriented properly for reaction to take place. In the case of a first order reaction, the dimensions of the A-factor are inverse time. When the reaction is second order, its dimensions are $\frac{\text{Volume}}{\text{mole}-\text{time}}$ and for nth order the dimensions are $\left(\frac{1}{\text{time}}\right)\left(\frac{\text{volume}}{\text{mole}}\right)^{n-1}$.

In the chapter on adsorption, we introduced the collision factor for two molecules, A and B as:

$$\text{Zab} = \pi\frac{(\text{da}+\text{db})^2}{4}\sqrt{\frac{8RT}{\pi\mu}}\frac{\text{NaNbNav}^2}{V^2}$$

$$= \pi\frac{(\text{da}+\text{db})^2}{4}\sqrt{\frac{8RT}{\pi\mu}}\frac{\text{Pa Pb}}{(RT)^2}$$

$$= A_c\bar{v}C_aC_b$$

R = ideal gas constant, [j/mol-K]

T = absolute temperature, [K]

$\mu = \frac{\text{MaMb}}{(\text{Ma}+\text{Mb})}$, mean molecular weight

M = molecular weight, [Kg/mol]

$A_c = \frac{\pi(\text{da}+\text{db})^2}{4}$, reaction cross section, [cm^2]

da, db = molecular diameters of A and B, [cm]

Na, Nb = [number of molecules per unit volume]

V = volume, [cm^3]

Nav = Avogadro's number, [number per mole]

If every collision were to lead to a reaction, then this expression would be taken as a rate expression, and the first two parts would be analogous to the rate constant for a chemical reaction.

$$k_{\text{collision}} = \pi \frac{(\text{da} + \text{db})^2}{4} \sqrt{\frac{8RT}{\pi \mu}}$$

The first term, $\pi \frac{(\text{da}+\text{db})^2}{4}$, is taken to be the collision cross section. It is easier to picture as $\pi (\text{ra} + \text{rb})^2$ and the physical picture is this one:

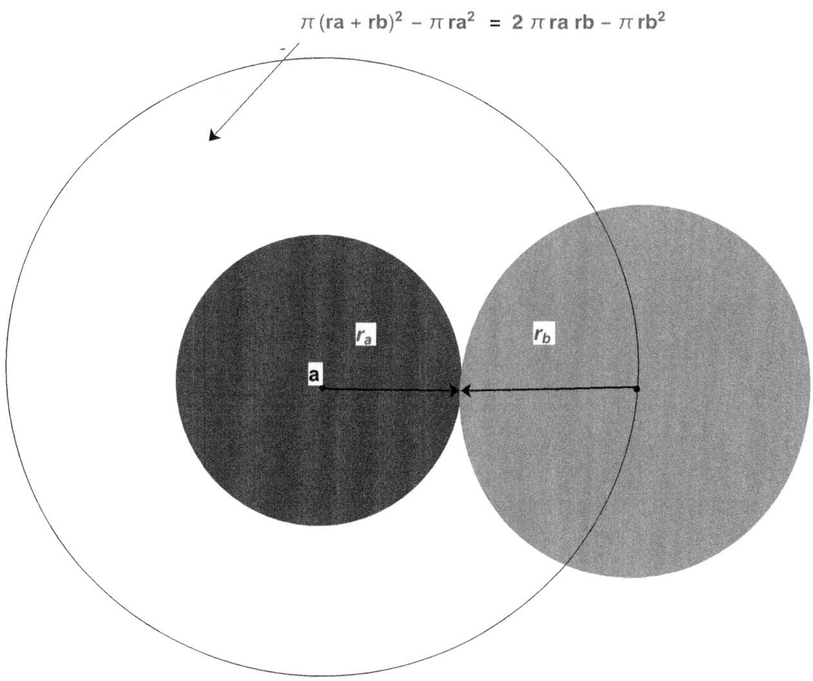

Any encounter between A and B that brings the two spherical centers within the sum (ra + rb) is a collision. The area given by (2π ra rb - πrb^2) takes into account that this can happen from any direction around the circle centered at a. Hence this gives rise to the collision cross section as an area per molecule. The last point is important, to remember.

The second term, $\sqrt{\frac{8RT}{\pi \mu}}$, is the average or mean speed of A and B. Remember R is given in Joules per mole per degree Kelvin and the basic SI units of a Joule are $\frac{\text{kg}-\text{meter}^2}{\text{sec}^2}$. Thus the molar masses must be given in terms of kg/mole. After working through the units the expression under the radical sign is meter^2/ sec^2 and the overall units are meter/ sec.

When we take the product of the molecular cross section and the mean speed the dimensions of this group are Length^3/molecular collision/ or volume per molecule per time. If we convert the number molecular collisions into moles of collisions we have the proper units for a second order rate constant - namely volume/mole/time:

$$k_{\text{collision}} = \pi \frac{(\text{da} + \text{db})^2}{4} \sqrt{\frac{8RT}{\pi \mu}}$$

$$= \left(\frac{\text{nm}}{1\text{molecularcollision}}\right)^2 \left(\frac{6.02 \times 10^{23}}{\text{mole}}\right) \left(\frac{1m}{10^9\text{nm}}\right)^2 \left(\frac{m}{\text{sec}}\right)$$

$$= \left(\frac{m^3}{\text{mol sec}}\right)$$

$$= \left(6.02 \times 10^5 \frac{m}{\text{mole nm}}\right) \left(\pi \frac{(\text{da} + \text{db})^2}{4} \text{nm}^2\right) \left(\left(\sqrt{\frac{8RT}{\pi \mu}} \frac{m}{\text{sec}}\right)\right)$$

Consider that at high enough temperatures, O2 and N2 will react to produce NO, which is oxidized further to NO2.

$$\text{N2} + \text{O2} \longrightarrow 2\ \text{NO}$$

Every reaction begins with a collision. The number of collisions will be given by Zab.

$$\left(\pi \frac{(\text{da} + \text{db})^2}{4}\right) \left(\sqrt{\frac{8RT}{\pi \mu}}\right) \left(\frac{P_{O_2}}{\text{R1}T}\right) \left(\frac{P_{N_2}}{\text{R1}T}\right)$$

The first term is the cross sectional area of collision. Since the diameters of molecular nitrogen and oxygen are the sum of the atomic diameters of nitrogen (.96Å) and oxygen (1.12Å), and these are given in Angstroms, the cross sectional area will be in Å^2. Inside the radical, when computing the mean speed, the value of R will be 8.314 J/mol/K. The molecular weights are given in kg/mol in order to match the mass units in Joules. The units of the mean speed will be meter/sec. For the calculation of the concentrations of N_2 and O_2, the value of R will be 0.08205 liter-atm/mol/K. Assume this reaction takes place at 1 atmosphere total pressure, and that the mole fractions are 0.8 and 0.2 respectively. The concentrations will be in units of mol/liter. The units so far look like this:

$$\text{area} * \text{speed} * \text{concentration}^2$$
$$\text{Å}^2 \quad \left(\frac{m}{\text{sec}}\right) \quad \left(\frac{\text{mol}}{\text{liter}}\right)^2$$

Typical units for a reaction rate are $\frac{\text{mol}}{\text{liter sec}}$. Therefore we need to clear nanometers, meters and reduce the exponent of moles from two to one. And recall that:

$$\text{Å}^2\left(\frac{1\text{nm}}{10\text{Å}}\right)^2\left(\frac{1m}{10^9\text{nm}}\right)^2\left(\frac{m}{\text{sec}}\right)\left(\frac{100\text{cm}}{m}\right)^3\left(\frac{1\text{liter}}{1000\text{cm}^3}\right)\left(\frac{6.02\times10^{23}}{\text{mol}}\right)\left(\frac{\text{mol}^2}{\text{liter}^2}\right)$$

$$\left(\frac{100^3\times6.02\times10^{23}}{10^2\times10^{18}\times1000}\right)\left(\frac{\text{Å}^2\text{nm}^2m^3\text{cm}^3\text{liter}}{\text{Å}^2\text{nm}^2\,\text{sec}\,m^3\text{cm}^3\text{mol}}\right)\left(\frac{\text{mol}^2}{\text{liter}^2}\right)$$

$$\left(\frac{6.02\times10^{29}}{10^{21}}\right)\left(\frac{\text{mol}}{\text{liter sec}}\right)$$

$$\left(6.02\times10^{6}\right)\left(\frac{\text{mol}}{\text{liter sec}}\right)$$

We can check our work:

$$\left(\frac{\mathbf{100^3\times6.02\times10^{23}}}{\mathbf{10^2\times10^{18}\times1000}}\right)\left(\frac{\mathbf{Å^2nm^2}\boldsymbol{m}^3\mathbf{cm^3liter}}{\mathbf{Å^2nm^2\,sec}\,\boldsymbol{m}^3\mathbf{cm^3mol}}\right)\left(\frac{\mathbf{mol^2}}{\mathbf{liter^2}}\right)$$

$$\frac{6.02\times10^{6}\text{mol}}{\text{liter sec}}$$

Thus we can put all the parameters into the mathematical statement in their "natural units", multiply by 6.02 x 10^6 and the result will be in moles per liter per second.

da = .96;

db = 1.12;

***R* = 8.314;**

***T* = 773;**

ma = 3.2 10^{-2};

mb = 2.8 10^{-2};

$$\mu = \mathbf{ma} * \mathbf{mb}/(\mathbf{ma} + \mathbf{mb});$$

$$R = 8.314;$$

$$\mathbf{R1} = \mathbf{0.08205};$$

$$\mathbf{PO2} = \mathbf{0.2};$$

$$\mathbf{PN2} = \mathbf{0.8};$$

$$(\mathbf{6.02}\ )\,(\pi)\,()\ (\mathbf{PO2}/\mathbf{R1}/T)$$

$$(\mathbf{PN2}/\mathbf{R1}/T)$$

$$\frac{851733.\text{mol}}{\text{liter sec}}$$

Under these conditions there would be about 852,000 mole per liter per second and if every collision led to reaction. This would be a very large rate of reaction. Let's see why. If the gas were ideal, one liter of it at this temperature would be about moles 0.016 moles:

$$P = 1\mathbf{atm};$$

$$V = 1L;$$

$$\mathbf{R1} = \mathbf{0.08205}((L\mathbf{atm})/(\mathbf{mol}K));$$

$$T = 773K;$$

$$\frac{PV}{\mathbf{R1}T}$$

$$\left(\frac{P}{\mathbf{R1}T}\right)/\left(\frac{\mathbf{851732.7mol}}{L\,\mathbf{sec}}\right)$$

$$\%10^9\ \mathbf{nsec}/\mathbf{sec}$$

0.0157667mol

$1.8511371623804126 \times 10^{-8}$ sec

18.5114nsec

At a constant rate of 8.52 x 10^5moles per liter per second, a liter of air at 500c (773K) would be converted completely to NO in less than 19 nanoseconds. This is clearly wrong. Why?

The reason it is wrong is that it does not account for the energetics of the collisions. When the average temperature of the gas is at 773K, there are molecules at lower "temperature" and others at "high temperature". But, we don't think of molecules as having a temperature; they have speed and energy. In fact, the Maxwell-Boltzmann distribution describes the distributions of energy for such molecules at any given temperature. Let's look at this now. Written in terms of the scalar speed in three dimensions, for a given set of molecules totaling N_o in number, the fraction of molecules, $\frac{dN}{N_o}$, with a speed between v and v + dv is:

$$\frac{dN}{N_o} = \left(\frac{m}{2\pi k_B T}\right)^{3/2} 4\pi v^2 \text{Exp}\left[-\frac{mv^2}{2k_B T}\right] dv$$

Here k_B is Boltzmann's constant, the magnitude of which is the gas constant in Joules divided by Avogadro's number. We can plot this distribution as a function of speed and at different temperatures.

f[v_, T_]:= $\left(\frac{m}{2\pi k_B T}\right)^{3/2} 4\pi v^2 \text{Exp}\left[-\frac{mv^2}{2k_B T}\right]$

$k_B = 1.38\ 10^{-23}$; (* Joule/K *)

$m = .028/(6.02\ 10^{23})$; (* kg/ molecule *)

Table[Plot[$f[v, T]$, {v, 0, 2500}, PlotRange → All,

PlotStyle → Hue[T/500]], {T, {300, 500, 700, 900}}];

Show[%,

AxesLabel →

{"v,(m/s)", "probability density, (s/m)"}]

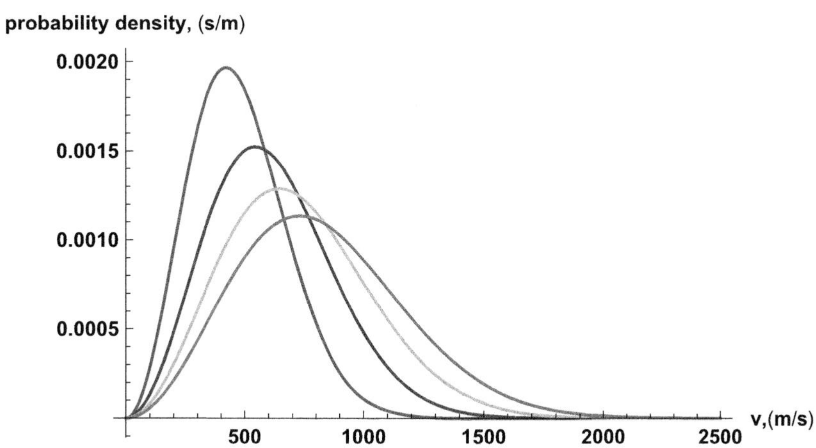

We see that as the temperature parameter is increased, (blue to lavender - semi-dark gray to mid gray in the print version), the breadth of the distribution increases, and the maximum, shifts to higher speeds. We can recast the distribution in terms of energy by making a change of the variable. We do so by observing that the kinetic energy is equal to $\frac{mv^2}{2}$. We need to find the derivative, $\partial_{\mathrm{KE}} v$, and then we need to make the replacements in the distribution function:

$$\boldsymbol{v = \left(\frac{2\mathrm{KE}}{m}\right)^{\frac{1}{2}}}$$

$$\sqrt{2}\sqrt{\frac{\mathrm{KE}}{m}}$$

$\partial_{\mathrm{KE}} v$//PowerExpand

$$\frac{1}{\sqrt{2}\sqrt{\mathrm{KE}}\sqrt{m}}$$

Thus $\partial_{\mathrm{KE}} v = \frac{1}{\sqrt{2}\sqrt{\mathrm{KE}}\sqrt{m}} \partial E$ and with the change of variable we have:

$$\left(\frac{M}{2\pi RT}\right)^{3/2} 4\pi \frac{2\mathrm{KE}}{M} \mathrm{Exp}\left[-\frac{Mv^2}{2RT}\right] \frac{1}{\sqrt{2}\sqrt{\mathrm{KE}}\sqrt{m}} \mathrm{dE}$$

which we can expand and simplify to:

$$\frac{2e^{-\frac{\mathrm{En}M}{mRT}}\sqrt{\mathrm{En}}}{\sqrt{\pi}R^{3/2}T^{3/2}}$$

$$\boldsymbol{\left(\frac{M}{2\pi RT}\right)^{3/2} 4\pi \frac{2\mathrm{KE}}{M} \mathrm{Exp}\left[-\frac{Mv^2}{2RT}\right] \frac{1}{\sqrt{2}\sqrt{\mathrm{KE}}\sqrt{m}} \mathrm{dE}}$$

$$\frac{2\,\mathrm{dE}\, e^{-\frac{Mv^2}{2RT}}\sqrt{\mathrm{KE}}\left(\frac{M}{RT}\right)^{3/2}}{\sqrt{m}M\sqrt{\pi}}$$

$$\boldsymbol{\frac{2e^{-\frac{\mathrm{KE}\,M}{mRT}}\sqrt{\mathrm{KE}}\ \mathrm{PowerExpand}\left[M^{-3/2}\left(\frac{M}{RT}\right)^{3/2}\right]}{\sqrt{\pi}}}$$

$$\frac{2e^{-\frac{\mathrm{KE}M}{mRT}}\sqrt{\mathrm{KE}}}{\sqrt{\pi}R^{3/2}T^{3/2}}$$

Now, we plot the distribution as a function of energy and having converted the mass per molecule to the mass per mole we find:

fE[KE_, T_]:= $\frac{2}{\sqrt{\pi}R^{3/2}T^{3/2}}\sqrt{\mathrm{KE}}\mathrm{Exp}\left[-\frac{\mathrm{KE}}{N_{\mathrm{Av}}k_B T}\right]$

$k_B = 1.38\ 10^{-23}$; (* Joule/K *)

$m = .028/(6.02\ 10^{23})$; (* kg/ molecule *)

$N_{\mathrm{Av}} = 6.02\ 10^{23}$;

$R = N_{\mathrm{Av}} k_B$;

Table[Plot[fE[KE, T], {KE, 0, 100000},

PlotRange → All, PlotStyle → Hue[T/350]],

{T, {200, 400, 600, 1000, 1200}}];

Show[%,

AxesLabel →

{"E(Joule)", "probability density 1/E (Joule)$^{-1}$"},

Epilog → Line[{{85000, 0}, {85000, 10^{-4}}}]]

Table[Plot[fE[KE, T], {KE, 80000, 100000},

PlotRange → All, PlotStyle → Hue[T/350]],

{T, {200, 400, 600, 1000, 1200}}];

Show[%,

AxesLabel →

{"E(Joule)", "probability density 1/E (Joule)$^{-1}$"},

Epilog → Line[{{85000, 0}, {85000, 10^{-7}}}]]

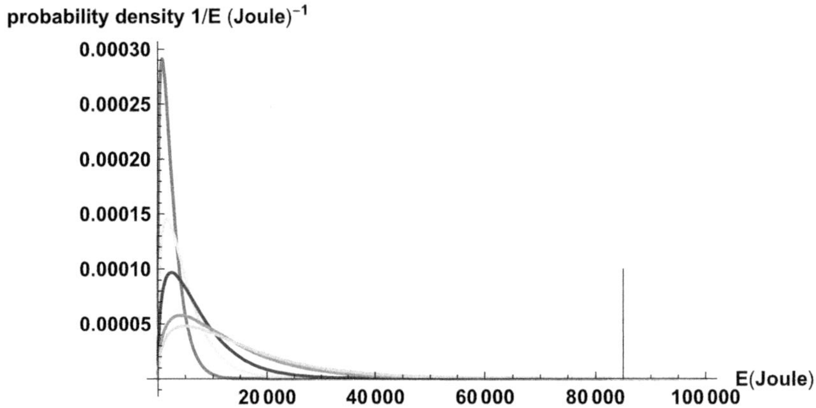

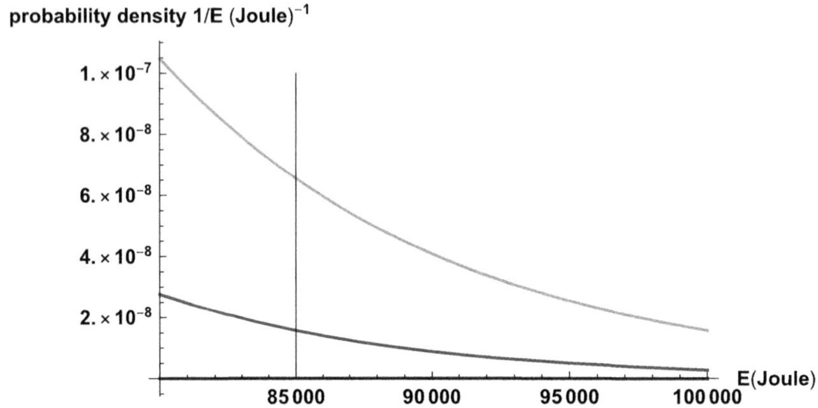

We have taken the plots out to 100,000 Joule, which corresponds to about 25 Kcal per mole. We note that the distribution broadens as the temperature parameter is increased, and that the shift in the maximum is much less pronounced than in the case of the velocity distribution. The breadth of the distribution indicates that there are more higher energy molecules at higher temperature. The vertical line is marked at 85,000 Joule, or 85 KJ. Although the functions appear to be at zero in this region, they are not. They show that for reactions with higher activation energies, the probability of having molecules of adequate energy is small. This illustrates why catalysts are so important. By reducing the activation energy for a reaction, they increase the population of molecules with sufficient energy to be reactive, though the actual effect is more complex than this implies.

Coming back then to the rate of reaction and the rate constant for reaction, we can feel justified on the basis of the Maxwell-Boltzmann distributions and the requirement that a collision happens for a reaction to take place to combine the collision frequency with the Arrhenius statement's of the rate constant and its exponential dependence on temperature. The whole rate expression for a gas phase bimolecular reaction between A and B then becomes:

$$\text{rt} = \left(6.02\ 10^5\right)\left(\pi\,\frac{(\text{da}+\text{db})^2}{4}\right)\left(\sqrt{\frac{8RT}{\pi\,\mu}}\right)\text{Exp}\left[\frac{-1000\ \text{Ea}}{R\ T}\right]\left(\frac{\text{PO2}}{\text{R1}\ T}\right)\left(\frac{\text{PN2}}{\text{R1}\ T}\right)\left(\frac{\text{mol}}{L\,\text{sec}}\right)$$

where the rate constant is:

$$\left(6.02\ 10^5\right)\left(\pi\,\frac{(\text{da}+\text{db})^2}{4}\right)\left(\sqrt{\frac{8RT}{\pi\,\mu}}\right)\text{Exp}\left[\frac{-1000\ \text{Ea}}{R\ T}\right]$$

In this model the A-factor is not simply an empirical constant, it is developed fundamentally from kinetic theory. We can evaluate both the magnitude of the rate constant and the rate of reaction at a given temperature and pressure, with an activation energy of 85 KJoule per mole as follows:

da = .96;

db = 1.12;

R = 8.314;

T = 600;

ma = 3.2 10^{-2};

mb = 2.8 10^{-2};

μ = ma * mb/(ma + mb);

R = 8.314;

R1 = 0.08205;

PO2 = 0.2;

PN2 = 0.8;

Ea = 85;

Print["The magnitude of the rate constant is"]

$$6.02\times 10^5\left(\frac{\pi(\text{da}+\text{db})^2}{4}\right)\sqrt{\frac{8RT}{\pi\,\mu}}\exp\left(-\frac{1000\ \text{Ea}}{RT}\right)$$

(* rate constant *)

Print["The magnitude of the rate is"]

$$\text{rt} = 6.02\times 10^5\left(\frac{\pi(\text{da}+\text{db})^2}{4}\right)\sqrt{\frac{8RT}{\pi\,\mu}}\exp\left(-\frac{1000\ \text{Ea}}{RT}\right)\left(\frac{PO2}{R1\ T}\right)\left(\frac{PN2}{R1\ T}\right)$$

$$\mathrm{rt} = \left(\frac{mol}{L\ sec}\right)$$

The magnitude of the rate constant is

75.0777

The magnitude of the rate is

$$\frac{0.00495645\mathrm{mol}}{L\ \mathrm{sec}}$$

At this rate, how long would it take to convert one liter of these gases at 1 atm pressure to products completely and how does this compare to the time required if every collision were reactive? Firstly, we compute the number of moles:

$P = 1\mathrm{atm};$

$V = 1L;$

$\mathrm{R1} = 0.08205((L\ \mathrm{atm})/(\mathrm{mol}\ K));$

$T = 773K;$

$$\frac{PV}{\mathrm{R1}\ T}$$

0.0157667mol

Next, the ratio of the concentration over the rate (rt), is the time required to convert one liter of reactants to products:

$$\frac{\frac{P}{\mathrm{R1}\ T}}{\mathrm{rt}}$$

3.18105 sec

This is the ratio of the time, based on kinetic theory, to the time we just computed that is based on a combination of kinetic theory and exponential temperature dependence:

$$\frac{3.18105\,\mathrm{sec}}{(1.8 \times 10^{-8}\,\mathrm{sec})}$$

1.76725×10^{8}

By incorporating the exponential temperature dependence, the estimated time for complete reaction is almost 2 x 10^8 longer. We can see the profound effect of the magnitude of the activation energy, by redoing the calculations with an Ea of 20 KJoule per mole.

$\text{da} = .96;$

$\text{db} = 1.12;$

$R = 8.314;$

$T = 600;$

$\text{ma} = 3.2\ 10^{-2};$

$\text{mb} = 2.8\ 10^{-2};$

$\mu = \text{ma} * \text{mb}/(\text{ma} + \text{mb});$

$R = 8.314;$

$\text{R1} = 0.08205;$

$\text{PO2} = 0.2;$

$\text{PN2} = 0.8;$

$\text{Ea} = 20$

Print["The magnitude of the rate constant is"]

$(6.02\ )\ (\pi)\ ()\ \text{Exp}[-1000\text{Ea}/R/T]$

Print["The magnitude of the rate is"]

$\text{rt} = (6.02\ )\ (\pi)\ ()$

$\text{Exp}[-1000\text{Ea}/R/T](\text{PO2}/\text{R1}/T)(\text{PN2}/\text{R1}/T)\left(\frac{\text{mol}}{L\,\text{sec}}\right)$

$P = 1\text{atm};$

$V = 1L;$

$\text{R1} = 0.08205((L\text{atm})/(\text{mol}K));$

$T = 773K;$

$$\frac{PV}{\text{R1}T}$$

Print[

"This is the time required to convert one

liter of reactants to products"]

$$\frac{P}{\text{R1}T} \Big/ \text{rt}$$

Print[

"This is the ratio of the time based on kinetic

theory to the rate estimated based on a

combination of kinetic theory and exponential

temperature dependence"]

$\%\% / (1.8\ 10^{-8}\ \text{sec})$

20

The magnitude of the rate constant is

3.42348×10^{7}

The magnitude of the rate is

$$\frac{2260.1\ \text{mol}}{L\ \text{sec}}$$

0.0157667mol

This is the time required to convert one liter of reactants to products

6.9761×10^{-6} sec

This is the ratio of the time based on kinetic theory to the rate estimated based on a combination of kinetic theory and exponential temperature dependence

387.562

The rate is several orders of magnitude faster.

Another consequence of the exponential dependence of the rate of reaction on temperature, is the possibility of "thermal runway." In this condition, an exothermic reaction drives itself to higher and higher rates such that the system cannot dissipate the heat of reaction fast enough.

This leads to a temperature rise, which increases the rate of reaction. That increase in rate of reaction in turn causes an increase in the rate of heat released. Just like mass, the heat will accumulate inside the system and the temperature will continue to rise, leading to another increase in rate and more heat released that is not dissipated. If the vessel holding the contents is mechanically able to contain the products at the highest temperature attained, then the reaction rate will diminish as soon as all of the reactant is used up. Then heat can dissipate and flow away from the system. On the other hand if the temperature rises to a high enough value to compromise the vessel, then an explosion may occur. Most pressure vessels are engineered with a safety valve that will rupture at a certain pressure in order to prevent and explosion.

To see how this happens, imagine a system with a first order reaction that takes A to B and releases 100,000 Joule/mole. Let the rate constant's preexponential factor be 5 $\sec^{-1}$, the activation energy will be 10,000 Joule/mole. The heat capacity of the contents will be constant for the sake of this calculation and it will have a value of 50 Joule/Mole. The system will have no inputs or outputs, but is closed and the reaction is initiated instantaneously. (As in the ignition of the fuel-air mixture in the combustion chamber of an internal combustion engine.)

Energy, like mass is conserved. Hence, the energy in the system will accumulate if the walls are insulated and if there is an internal or external heat source in this case the internal heat source is the reaction. The mass balance for species A will be a first order decline in concentration that is explicitly dependent on temperature through the rate constant:

$$\partial_t \mathrm{Ca}[t] == -k\mathrm{Ca}[t]$$

$$\partial_t \mathrm{Ca}[t] == -\mathrm{koExp}[-\mathrm{Ea}/(RT[t])]\mathrm{Ca}[t]$$

The rate of heat released by the reaction will be given by the rate of reaction times the heat of reaction:

$$\text{rate of heat released} = -\ \mathrm{k\ Ca[t]}\ \Delta H_{\mathrm{rxn}}$$

This will raise the temperature in the system. The contents of the vessel have a specific heat capacity. It will be dependent on the components of the mixture. We will assume that the reactant and product have heat capacities and densities that are similar enough to call equal and that we do not have to incorporate this into the analysis. We will also assume that the heat capacity is a weak function of temperature and its temperature dependence can be left out of the analysis at this stage. The accumulation of energy in the system will be given as the time derivative of the product of the heat capacity and the temperature at any time and that will be equal to the rate of heat generation from the reaction:

$$\partial_t \rho V T[t]\mathrm{cp} = -\ \mathrm{k\ Ca[t]}\ \Delta H_{\mathrm{rxn}}$$

$$\partial_t \mathrm{Cp}T[t] = \text{ - k Ca[t] } \Delta H_{\mathrm{rxn}}$$

$$\partial_t \mathrm{Cp}T[t] = \text{ - ko Exp[-Ea/(R T[t])] Ca[t] } \Delta H_{\mathrm{rxn}}$$

$$\partial_t T[t] = -\frac{\mathrm{ko}}{\mathrm{Cp}} \text{ Exp[-Ea/(R T[t])] Ca[t] } \Delta H_{\mathrm{rxn}}$$

By dividing Ca[t] by Cao and taking ko to the left-hand side, we can non-dimensionalize the one equation to give:

$$\partial_\tau \phi\mathrm{a}[\tau] = -\mathrm{Exp}[-\mathrm{Ea}/(RT[\tau])]\phi\mathrm{a}[\tau]$$

$$\partial_\tau T[\tau] = H\frac{\mathrm{Cao}}{\mathrm{Cp}}\mathrm{Exp}[-\mathrm{Ea}/(RT[\tau])]\phi\mathrm{a}[\tau]$$

Where ϕa$[\tau]$ is the dimensionless concentration and τ is dimensionless time. Choosing a set of parameters consistent with a gas phase reaction, we can solve these equations numerically:

```
ko = 5; (* sec^-1 *)
Ea = 10000;  (* Joule *)
R = 8.3; (* Joule/mole/K *)
Cp = 50; (* Joule/K *)
H = 100000; (* Joule/mole *)
Cao = 1; (* mole/L *)
tmax = 8(*  sec  *)
sols =
NDSolve[
{∂τ ϕa[τ]== −  Exp[−Ea/(R T[τ])]ϕa[τ],
∂τ T[τ] == H Cao/Cp Exp[−Ea/(R T[τ])]ϕa[τ],
ϕa[0] == 1, T[0] == 500}, {ϕa[τ], T[τ]},
{τ, 0, tmax}];
Φa[τ]:=Evaluate[ϕa[τ]/.sols[[1]]]
Tmp[τ]:=Evaluate[T[τ]/.sols[[1]]]
Φa[τ];
```

```
Tmp[τ];
Plot[Φa[τ], {τ, 0, tmax}, PlotRange → All,
AxesLabel → {τ, "Φa[τ]"}]
Plot[Tmp[τ], {τ, 0, tmax}, PlotRange → All,
AxesLabel → {τ, "Tmp[τ]"}, PlotStyle → Hue[0.1]]
```

8

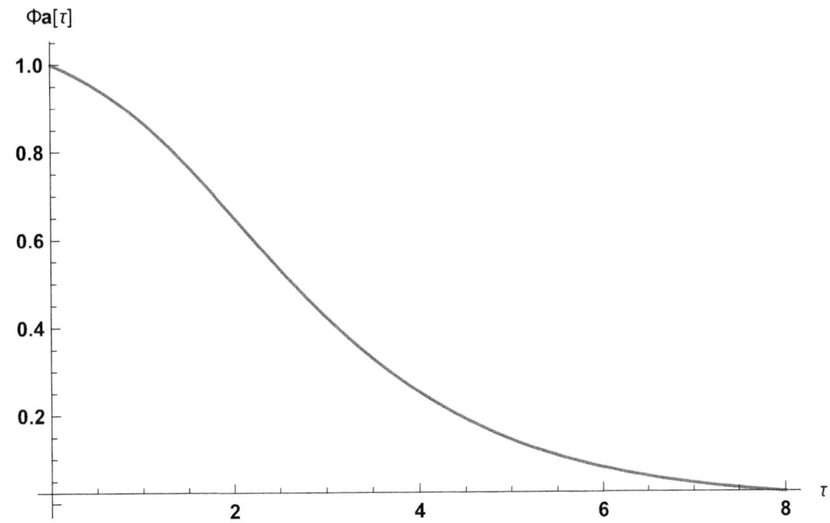

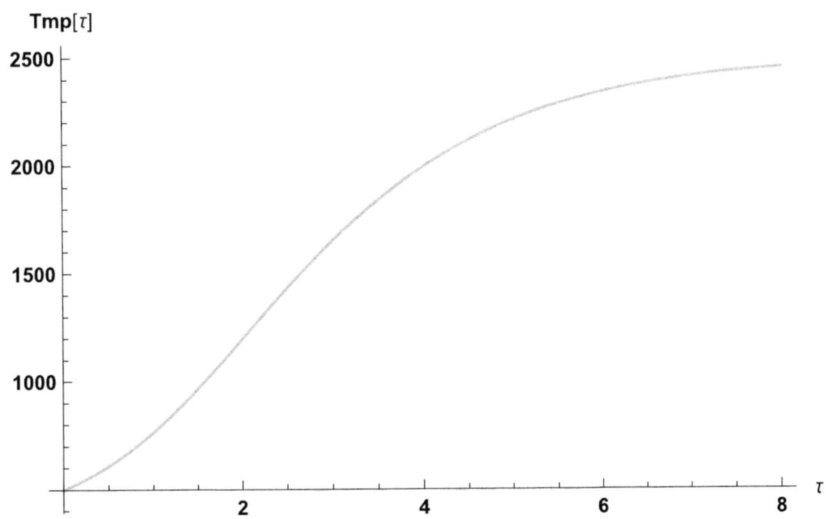

Here we see that for this very exothermic reaction, with a relatively low activation energy, and the assumption of no loss of heat from the system (adiabatic conditions), the temperature climbs rapidly to over 2000K in just $\tau = 4$. This is an extreme analysis, but it shows why the consideration of thermodynamic effects in reaction engineering must be given so much attention and much more detailed attention than we have provided here. Still, even this simplified analysis makes the point that exothermic reactions can "runaway" thermally without proper heat management.

In the case of catalytic reactions that use a solid as the catalyst, thermodynamics can play out in yet another way. Remember the case of A $\Leftrightarrow$ B in the section on Langmuir-Hinshelwood kinetics. We noted that the reactant A must come in contact with the catalyst surface as an adsorbate before reaction would take place to form B. The overall rate of reaction looked as follows:

$$r_{A\to B,\text{surf}} = \frac{k_{f,\text{AB}}\left(C_A - \dfrac{C_B}{K_{\text{eq}}}\right)}{1 + C_A K_A + C_B K_B}$$

and we noted that $k_{f,\text{AB}}$ was the product of the rate constant for the reaction on the surface and the adsorption equilibrium constant that determined the extent to which A was present on that surface:

$$r_{A\to B,\text{surf}} = \frac{k_{A\to B,\text{surf}} K_A C_{\text{tot}}\left(C_A - \dfrac{C_B}{K_{\text{eq}}}\right)}{1 + C_A K_A + C_B K_B}$$

$$k_{f,\text{AB}} = k_{A\to B,\text{surf}} K_A$$

Both $k_{A\to B,\text{surf}}$ and K_A have exponential dependencies on temperature as does $k_{f,\text{AB}}$:

$$k_{f,\text{AB},o}\text{Exp}\left[-\frac{E_{A,\text{app}}}{RT}\right] = k_{A\to B,\text{surf},o}\text{Exp}\left[-\frac{E_{A\to B,\text{surf}}}{RT}\right] K_{\text{A}o}\ \text{Exp}\left[-\frac{\Delta H_{\text{rxn}}}{RT}\right]$$

Here $E_{A,\text{app}}$ is an apparent activation energy. In fact when we look at the right-hand side of the equation, taking into account the law of exponents, it is the sum of the activation energy for the surface reaction and the heat of reaction for A $\longrightarrow$ B.

$$k_{f,\text{AB},o}\text{Exp}\left[-\frac{E_{A,\text{app}}}{RT}\right] = k_{A\to B,\text{surf},o} K_{\text{A}o}\text{Exp}\left[-\frac{(E_{A\to B,\text{surf}} + \Delta H_{\text{ads}})}{RT}\right]$$

$$\therefore E_{A,\text{app}} = (E_{A\to B,\text{surf}} + \Delta H_{\text{ads}})$$

In every case, (that we care about) the activation energy is a positive quantity. However, the heat of adsorption is almost always negative, meaning that heat is released upon adsorption. Therefore if $E_{A\to B,\mathrm{surf}} > 0$ and $\Delta H_{\mathrm{ads}} < 0$, the sum of the two, $E_{A,\mathrm{app}}$, is less than $E_{A\to B,\mathrm{surf}}$. In such a case the effect of temperature on the rate of the reaction may not be as strong as would be anticipated. In the limiting case wherein $E_{A\to B,\mathrm{surf}} \approx |\Delta H_{\mathrm{ads}}|$, there would be almost no temperature dependence.

Also remember that the rate is also dependent upon the surface concentration of A. As the temperature rises, given a negative heat of adsorption, the concentration of surface species will drop and that will bring the rate back down again. Thus at lower temperatures as the temperature is raised, the rate increases, as the rate constant increases, and this increases the rate. But as the temperature is increased further, the rate decreases, even as the rate constant increases because the surface concentration of reactant becomes so small.

$$\frac{k_o \mathrm{Exp}[-E_{A,app}/RT]\, K_{\mathrm{Ao}}\mathrm{Exp}[-\Delta H_{ads,A}/RT]\, C_{tot}\left(C_A - \frac{C_B}{K_{\mathrm{eq}}}\right)}{1 + C_A K_{\mathrm{Ao}}\mathrm{Exp}[-\Delta H_{ads,A}/RT] + C_B K_{\mathrm{Bo}}\mathrm{Exp}[-\Delta H_{ads,B}/RT]}$$

$$\frac{k'_o \mathrm{Exp}[-E_{A,app}/RT]\, K_{\mathrm{Ao}}\mathrm{Exp}[-\Delta H_{ads,A}/RT]\left(C_A - \frac{C_B}{K_{\mathrm{eq}}}\right)}{1 + C_A K_{\mathrm{Ao}}\mathrm{Exp}[-\Delta H_{ads,A}/RT] + C_B K_{\mathrm{Bo}}\mathrm{Exp}[-\Delta H_{ads,B}/RT]}$$

$$r[\mathrm{Ea}, \Delta H_{ads,A}, T] = \frac{k'_o \mathrm{Exp}[-E_{A,app}/RT]\, K_{\mathrm{Ao}}\mathrm{Exp}[-\Delta H_{ads,A}/RT]\left(C_A - \frac{C_B}{K_{\mathrm{eq}}}\right)}{1 + C_A K_{\mathrm{Ao}}\mathrm{Exp}[-\Delta H_{ads,A}/RT] + C_B K_{\mathrm{Bo}}\mathrm{Exp}[-\Delta H_{ads,B}/RT]}$$

With the proper selection of parameters we can make the effect of temperate evident. In the following, we first plot the surface coverage of rate a low conversion of A to B. Thus we leave out the term for B and compute the coverage of the surface by A and examine its temperature dependence. The surface coverage drops in a reverse s-shaped curve to approach zero asymptotically. Next we can plot the rate constant for the surface reaction also as a function of temperature and it increases exponentially. Lastly, we can put the two together to see the cross over between the rate constant and the surface coverage with increasing temperature. This illustrates how the two functional dependencies on temperature counter act in the overall rate. The parameters used herein were chosen specifically to illustrate the overall effect and the results may not be so evident with different parameter values.

$\boldsymbol{R = 1.99;}$

$\mathbf{Ca = 1;}$

$\mathbf{Kao = 10^{-4};}$

```
θ[100, −10000]//N;
a = Plot[θ[T, −10000], {T, 100, 1000},
AxesLabel → {T, θ[T]}]
θ[T_, ΔHads_]:= KaoExp[−ΔHads/(RT)]Ca / (1 + KaoExp[−ΔHads/(RT)]Ca)
k[T_, Ea_]:=ko Exp[−Ea/(RT)]
ko = 300;
R = 1.99;
b = Plot[k[T, 5000], {T, 100, 2000},
AxesLabel → {T, "k[ T ]"}, PlotStyle → Hue[1]]
Show[{a, b}, AxesLabel → {T, "k[ T ], θ[ T ]"}]
Plot[k[T, 5000]θ[T, −10000], {T, 100, 2000},
AxesLabel → {T, r_initial[T]}, PlotStyle → Hue[0.8]]
```

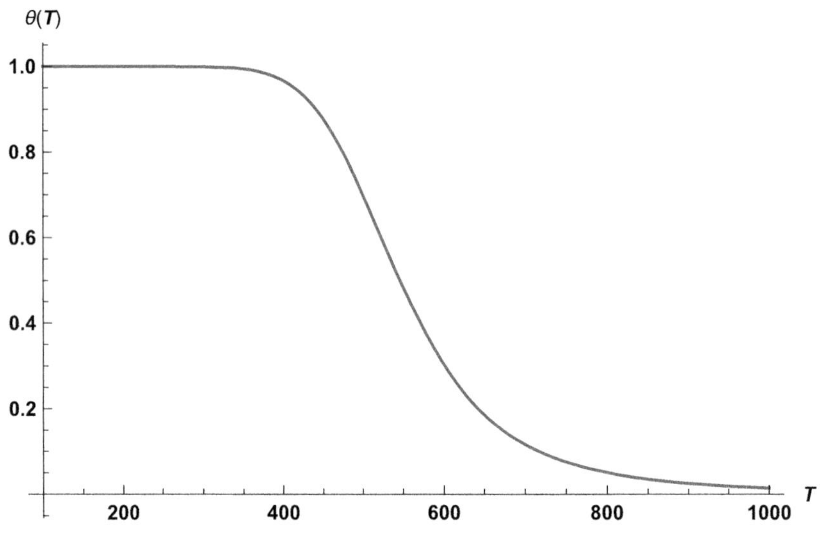

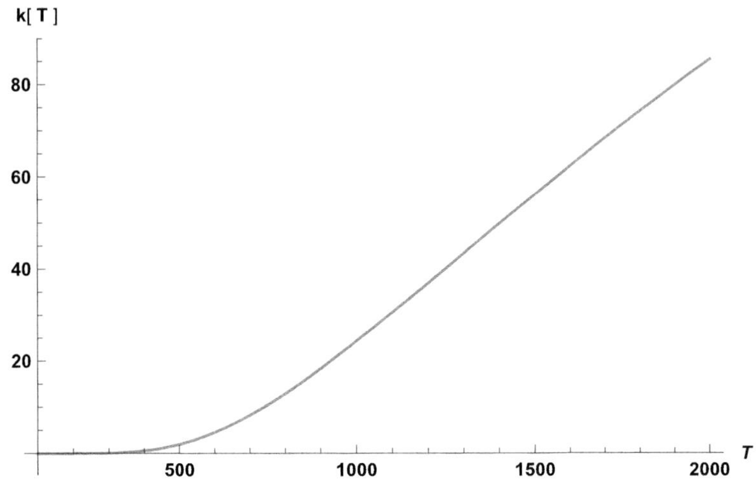
k[T]
80
60
40
20
500
1000
1500
2000
T

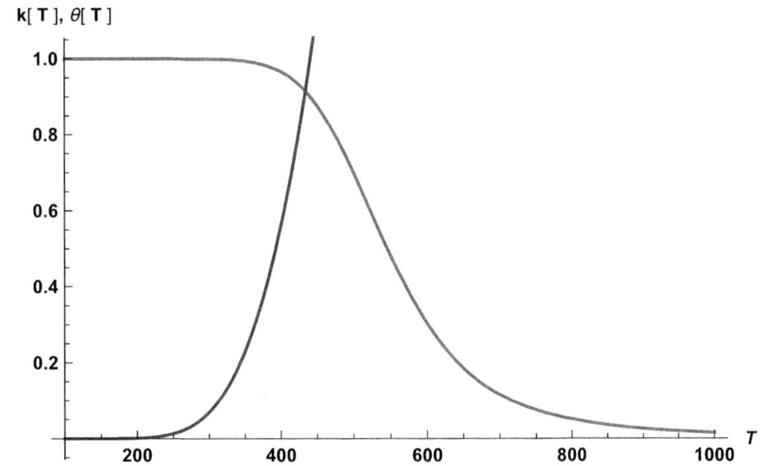
k[T], θ[T]
1.0
0.8
0.6
0.4
0.2
200
400
600
800
1000
T

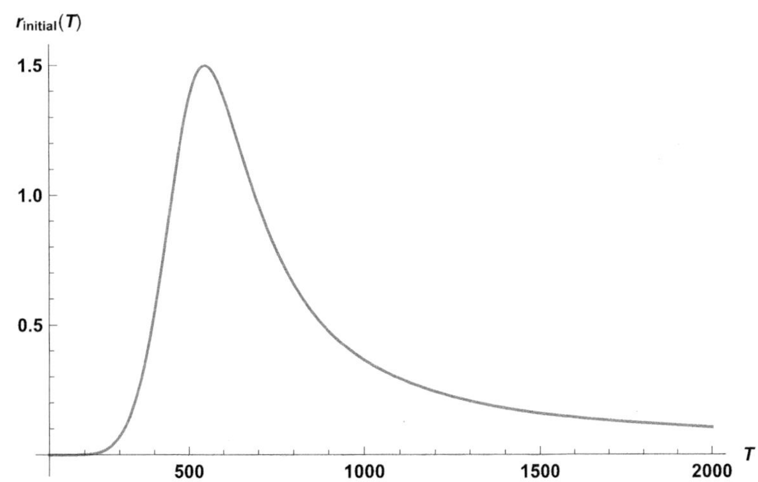
$r_{initial}(T)$
1.5
1.0
0.5
500
1000
1500
2000
T

The last graph shows how the initial rate increases with temperature as the rate constant increases and then decreases with further increases in temperature as the surface concentration diminishes.

Conversion and selectivity

Up to this point we have focused on the fundamental aspects of chemical kinetics. The use of kinetics to understand mechanisms and to make improvements in reaction outcomes is a critical application of this science. Kinetics are also used to design reactors, especially to determine the size needed for a given production goal and they are used to improve the performance of catalysts. Catalysts are used in the majority of chemical processes. Kinetics are used to improve the rate of reaction and the selectivity to the desired product. In this regard, rates are often expressed in terms of conversion and selectivity. We take that up now.

Consider the case of a reactant A that is converted into products B and D. Both reactions occur in parallel:

$$A \longrightarrow B + D$$

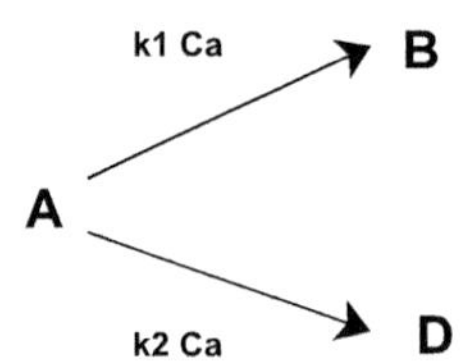

We want to know the extent of reaction in terms of the percent conversion of A and the selectivity to the desired product B and the undesired product D. We can go back to the rate relationships to find these. The differential rates are equated as follows:

$$- \partial_t \mathrm{Ca}[t] = \partial_t \mathrm{Cb}[\,t\,] + \partial_t \mathrm{Cd}[t]$$

Upon integration we obtain these expressions:

$$- (\,\mathrm{Ca}[t] - \mathrm{Cao}) = (\mathrm{Cb}\,[\,t\,] - \mathrm{Cbo}) + (\mathrm{Cd}\,[\,t\,] - \mathrm{Cdo})$$

Assuming that the initial concentrations of B and D are zero, we find:

$$(\mathrm{Cao} - \mathrm{Ca}[\,t\,]) = (\mathrm{Cb}\,[\,t\,]) + (\mathrm{Cd}\,[\,t\,])$$

We can non-dimensionalize these by dividing through by Cao:

$$\frac{(\text{Cao} - \text{Ca}[t])}{\text{Cao}} = \left(\frac{\text{Cb}[t]}{\text{Cao}}\right) + \left(\frac{\text{Cd}[t]}{\text{Cao}}\right)$$

The first term is the fractional conversion of A at any time t. In a batch reactor, the ultimate conversion of A is that measured at the end of the time period for reaction, tf, for time final. The conversion percentage is:

$$\%\ \text{Conversion of A} \ = \frac{(\text{Cao} - \text{Ca}[\text{tf}])}{\text{Cao}} * 100$$

This is often expressed in moles of A, N_a and the initial moles of A, N_{ao}:

$$\%\ \text{Conversion of A} = \frac{V}{V}\frac{(\text{Cao} - \text{Ca}[\text{tf}])}{\text{Cao}} * 100$$

$$\%\ \text{Conversion of A} = \frac{(\text{Nao} - \text{Na}[\text{tf}])}{\text{Nao}} * 100$$

At the end of the reaction period, at time tf, there will be residual A, if it is not used up completely, in addition to the presence of the products B and D. The selectivity to B is the amount of A that was converted to B and the selectivity to D was that converted to D.

$$\%\ \text{Selectivity to B} = \left(\frac{\text{Cb}[t]}{\text{Cao}}\right) * 100$$

$$\%\ \text{Selectivity to D} = \left(\frac{\text{Cb}[t]}{\text{Cao}}\right) * 100$$

Lastly, the yields of B and D are given as the product of the conversion and the selectivity to each product.

$$\%\ \text{Yield of B} = \left(\frac{\text{Cao} - \text{Ca}[\text{tf}]}{\text{Cao}}\right) * \left(\frac{\text{Cb}[t]}{\text{Cao}}\right) * 100$$

$$\%\ \text{Yield of D} = \left(\frac{\text{Cao} - \text{Ca}[\text{tf}]}{\text{Cao}}\right) * \left(\frac{\text{Cd}[t]}{\text{Cao}}\right) * 100$$

So these are all the definitions that we need. Let's re-express the rate in terms of the fractional conversion, X at any time t.

$$\text{X}[\ \text{t}\ ] = \left(\frac{\text{Cao} - \text{Ca}[t]}{\text{Cao}}\right)$$

$$\text{Ca}[\ t\ ] = \text{Cao}\ (\ 1 - \text{X}[\ t\ ])$$

and

$$\partial_t \text{Ca}[t] = -\ (\text{r1} + \text{r2}) == \ -\ \text{r1} - \text{r2}$$

$$\partial_t \text{Ca}[t] = -\text{k1 Ca}[\ t\ ] - \text{k2 Ca}[\ t\ ]$$

$$\text{Ca}[t] = \text{Cao}e^{-(\text{k1}+\text{k2})t}$$

But we can also choose to re-express Ca[t] in terms of the conversion, X[t]:

$$\partial_t \text{Cao}(1 - X[t]) = \ -\ (\text{k1} + \text{k2})\ \text{Cao}\ (\ 1 - \text{X}[\ t\ ])$$

$$\partial_t (1 - X[t]) = \ -\ (\text{k1} + \text{k2})\ (\ 1 - \text{X}[\ t\ ])$$

$$\therefore \partial_t X[t] == \ (\text{k1} + \text{k2})\ (\ 1 - \text{X}[\ t\ ])$$

This is the rate expression in terms of the fractional conversion. Letting X[0] = 0 and integrating, the fraction conversion varies as a function of time according to this expression:

$$X[t] == \left(1 - e^{-(\text{k1}+\text{k2})t}\right)$$

$$\boldsymbol{\int_0^{X[t]} \frac{1}{(1 - x[t])} dx[t]}$$

$$\text{ConditionalExpression}[-\text{Log}[1 - X[t]], \text{Re}[X[t]] \leq 1 \| X[t] \notin \mathbb{R}]$$

k1 = 0.1;

k2 = 0.05;

Plot$\left[\left(1 - e^{-(\text{k1}+\text{k2})t}\right)\right.$, {$t$, 0, 50}, PlotRange → All,

Epilog → {Hue[0.4], Line[{{0, 1}, {50, 1}}]},

AxesLabel → {t, "X[t]"}]

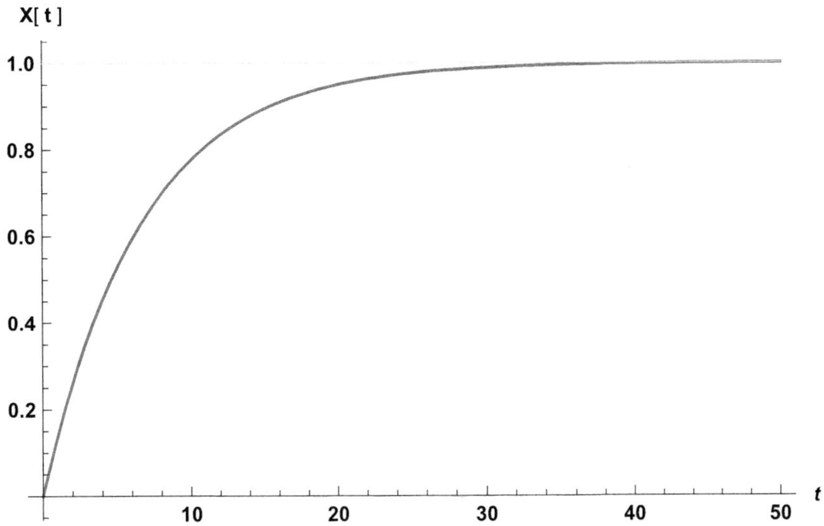

We also want to know the concentrations of A, B, and D as a function of the fractional conversion. Going back to the overall stoichiometry we have:

$$\left(\frac{\text{Cao} - \text{Ca}[t])}{\text{Cao}}\right) = \left(\frac{\text{Cb}[t]}{\text{Cao}}\right) + \left(\frac{\text{Cd}[t]}{\text{Cao}}\right)$$

$$1 - \left(\frac{\text{Ca}[t]}{\text{Cao}}\right) = \left(\frac{\text{Cb}[t]}{\text{Cao}}\right) + \left(\frac{\text{Cd}[t]}{\text{Cao}}\right)$$

$$\text{X}[\text{ t }] = \left(\frac{\text{Cb}[t]}{\text{Cao}}\right) + \left(\frac{\text{Cd}[t]}{\text{Cao}}\right)$$

But we also have:

$$\partial_t \text{Cb}[t] = \text{r1} = \text{k1 Ca}[\text{ t }] = \text{k1 Cao (1 - X[t])}$$

$$= \text{k1 Cao (1 - } \left(1 - e^{-(\text{k1}+\text{k2})t}\right)) = \text{ k1 Cao } e^{-(\text{k1}+\text{k2})t}$$

$$\int d\text{Cb}[t] = \int \text{k1Cao}e^{-(\text{k1}+\text{k2})t}dt$$

$$\text{Cb}[\text{ t }] = \frac{\text{k1Cao}\left(1 - e^{-(\text{k1}+\text{k2})t}\right)}{\text{k1}+\text{k2}} = \left(\frac{\text{k1}}{\text{k1}+\text{k2}}\right)\text{Cao}\left(1 - e^{-(\text{k1}+\text{k2})t}\right)$$

$$= \left(\frac{\text{k1}}{\text{k1}+\text{k2}}\right)(\text{Cao - Ca[t]})$$

And

$$\partial_t \mathrm{Cd}[t] = \mathrm{r2} = \mathrm{k2\ Ca\ [\ t\ ]} = \mathrm{k2\ Cao\ (\ 1\ -\ X[\ t\ ])}$$

$$= \mathrm{k2\ Cao\ 1\ -}\left(1 - e^{-(\mathrm{k1}+\mathrm{k2})t}\right) = \mathrm{k2\ Cao}e^{-(\mathrm{k1}+\mathrm{k2})t}$$

$$\int d\mathrm{Cd}[t] = \int \mathrm{k2Cao}e^{-(\mathrm{k1}+\mathrm{k2})t}dt$$

$$\mathrm{Cd[\ t\ ]} == \frac{\mathrm{k2Cao}\left(1 - e^{-(\mathrm{k1}+\mathrm{k2})t}\right)}{\mathrm{k1}+\mathrm{k2}} = \left(\frac{\mathrm{k2}}{\mathrm{k1}+\mathrm{k2}}\right)\mathrm{Cao}\left(1 - e^{-(\mathrm{k1}+\mathrm{k2})t}\right)$$

$$= \left(\frac{\mathrm{k2}}{\mathrm{k1}+\mathrm{k2}}\right)(\mathrm{Cao\ -\ Ca[t]})$$

If we go back to the expression for the stoichiometry, then we find the following:

$$1 - \left(\frac{\mathrm{Ca}[t]}{\mathrm{Cao}}\right) == \left(\frac{\mathrm{Cb}[t]}{\mathrm{Cao}}\right) + \left(\frac{\mathrm{Cd}[t]}{\mathrm{Cao}}\right)$$

$$1 - \left(\frac{\mathrm{Cao}\ e^{-(\mathrm{k1}+\mathrm{k2})t}}{\mathrm{Cao}}\right) = \left(\frac{\left(\frac{\mathrm{k1}}{\mathrm{k1}+\mathrm{k2}}\right)\mathrm{Cao}\left(1 - e^{-(\mathrm{k1}+\mathrm{k2})t}\right)}{\mathrm{Cao}}\right) + \left(\frac{\left(\frac{\mathrm{k2}}{\mathrm{k1}+\mathrm{k2}}\right)\mathrm{Cao}\left(1 - e^{-(\mathrm{k1}+\mathrm{k2})t}\right)}{\mathrm{Cao}}\right)$$

$$\left(1 - e^{-(\mathrm{k1}+\mathrm{k2})t}\right) = \left(\frac{\mathrm{k1}}{\mathrm{k1}+\mathrm{k2}}\right)\left(1 - e^{-(\mathrm{k1}+\mathrm{k2})t}\right) + \left(\frac{\mathrm{k2}}{\mathrm{k1}+\mathrm{k2}}\right)\left(1 - e^{-(\mathrm{k1}+\mathrm{k2})t}\right)$$

$$1 = \left(\frac{\mathrm{k1}}{\mathrm{k1}+\mathrm{k2}}\right) + \left(\frac{\mathrm{k2}}{\mathrm{k1}+\mathrm{k2}}\right)$$

Hence, the fractions going to products B and D are determined by their fractional rate constants:

$$\%\ \mathrm{Selectivity\ to\ B} = \left(\frac{\mathrm{Cb}[t]}{\mathrm{Cao}}\right) * 100$$

$$= \frac{\left(\frac{\mathrm{k1}}{\mathrm{k1}+\mathrm{k2}}\right)\mathrm{Cao}\left(1-e^{-(\mathrm{k1}+\mathrm{k2})t}\right)}{\mathrm{Cao}} * 100$$

$$= \left(\frac{\mathrm{k1}}{\mathrm{k1}+\mathrm{k2}}\right)\left(1-e^{-(\mathrm{k1}+\mathrm{k2})t}\right) * 100$$

$$= \left(\frac{\mathrm{k1}}{\mathrm{k1}+\mathrm{k2}}\right) * 100 \text{ as } \mathrm{t} \longrightarrow \infty$$

$$\% \text{ Selectivity to D} = \left(\frac{\mathrm{Cd}[t]}{\mathrm{Cao}}\right) * 100$$

$$= \frac{\left(\frac{\mathrm{k2}}{\mathrm{k1}+\mathrm{k2}}\right)\mathrm{Cao}\left(1-e^{-(\mathrm{k1}+\mathrm{k2})t}\right)}{\mathrm{Cao}} * 100$$

$$= \left(\frac{\mathrm{k2}}{\mathrm{k1}+\mathrm{k2}}\right)\left(1-e^{-(\mathrm{k1}+\mathrm{k2})t}\right) * 100$$

$$= \left(\frac{\mathrm{k2}}{\mathrm{k1}+\mathrm{k2}}\right) * 100 \text{ as } \mathrm{t} \longrightarrow \infty$$

The Yield of B is as follows:

$$\% \text{ Yield of B} = (\text{Fractional Conversion} * \text{Fractional Selectivity}) * 100$$

$$\% \text{ Yield of B} = \left(1-e^{-(\mathrm{k1}+\mathrm{k2})t}\right)\left(\frac{\mathrm{k1}}{\mathrm{k1}+\mathrm{k2}}\right)\left(1-e^{-(\mathrm{k1}+\mathrm{k2})t}\right) * 100$$

$$= \left(\frac{\mathrm{k1}}{\mathrm{k1}+\mathrm{k2}}\right)\left(1-e^{-(\mathrm{k1}+\mathrm{k2})t}\right)^2 * 100$$

$$\rightarrow \left(\frac{\mathrm{k1}}{\mathrm{k1}+\mathrm{k2}}\right) * 100 \text{ as } \mathrm{X[t]} \rightarrow 1 \text{ and as } \mathrm{t} \rightarrow \infty$$

The expression for D follows the same form.

Time required for a specified conversion

If the kinetics are known for a given reaction, then the question becomes how long should the reaction be allowed to proceed at the specified conditions to obtain the desired conversion of species A. Recall that:

$$\partial_t \text{Ca}[t] V = \text{- r V}$$

For the first order case:

$$\partial_t \text{Cao}(1 - X[t])\text{V} = \text{ - k Cao (1 - X[t])V}$$

$$\partial_t \text{Cao} V(1 - X[t]) = \text{ - k Cao (1 - X[t])V}$$

Remember that Cao V = Nao and therefore:

$$\text{Cao V } \partial_t (1 - X[t]) = \text{ - k Cao (1 - X[t])V}$$

$$\text{Nao } \partial_t (-X[t]) = \text{ - k Cao (1 - X[t])V}$$

$$\text{- Nao} \frac{dX[t]}{dt} = \text{ - k Cao (1 - X[t])V}$$

Which rearranges to:

$$\text{Nao} \int_0^{X[t]} \frac{1}{k\text{Cao}(1 - X[t])V} dX[t] = \text{t}$$

and which integrates to this:

$$\frac{\text{NaoLog}[k\text{Cao}V]}{k\text{Cao}V\text{Log}[k\text{Cao}V(X[t] - 1)]} = \text{t}$$

and Nao=Cao V

$$\frac{\text{Log}[k\text{Nao}]}{k\text{Log}[k\text{Nao}(X[t] - 1)]} = \text{t}$$

or

$$\frac{\text{Log}[k\text{Nao}]}{\text{Log}[k\text{Nao}(X[t] - 1)]} = \text{k t}$$

This is for the case of first order kinetics. In general we can see that the time required to obtain a given conversion fraction is:

$$\text{t} = \text{Nao} \int_0^{X[t]} \frac{1}{rV} dX[t]$$

This is the "master" equation that most reaction engineers commit to memory and use when they need to make a estimate of reactor volume for a given conversion and rate. Let's see how it works.

If the rate is second order in A, then:

$$\mathrm{t} = \mathrm{Nao}\int_0^{X[t]} \frac{1}{k\,\mathrm{Cao}^2 V(1-X[t])^2} dX[t]$$

$$\mathrm{t} = \frac{\mathrm{Nao}}{\mathrm{Cao}^2 V}\int_0^{X[t]} \frac{1}{k(1-X[t])^2} dX[t]$$

$$\mathrm{t} = \frac{X[t]}{\mathrm{Cao}\,k(1-X[t])}$$

and if the rate is nth ($n > 1$) order in A, then:

$$\mathrm{t} = \mathrm{Nao}\int_0^{X[t]} \frac{1}{k\mathrm{Cao}^n V(1-X[t])^n} dX[t]$$

$$\mathrm{t} = \frac{\mathrm{Nao}}{k\mathrm{Cao}^n V}\int_0^{X[t]} \frac{1}{(1-X[t])^n} dX[t]$$

$$\mathrm{t} = \frac{\mathrm{Nao}}{k\mathrm{Cao}^n V}\frac{(1-X[t])^{1-n}-1}{n-1}$$

$$\mathrm{t} = \frac{1}{k\mathrm{Cao}^{n-1}}\frac{(1-X[t])^{1-n}-1}{n-1}$$

If we apply reasonable values for the parameters, to further see how this works. Let's assume that the order in A is n = 2.5, that the rate constant, k = $0.01\ L^{1.5}\mathrm{mol}^{-1.5}\,\mathrm{sec}^{-1}$ with its attendant units and that Cao = 1 molL^{-1}. Then the time to 0.9 fractional conversion is calculated as follows:

$\boldsymbol{k = 0.01L^{1.5}\mathrm{mol}^{-1.5}\,\mathrm{sec}^{-1};}$

$\boldsymbol{n = 2.5;}$

$\boldsymbol{\mathrm{Cao} = 1\mathrm{mol}L^{-1};}$

$\boldsymbol{X[t] = 0.9;}$

$\boldsymbol{\frac{(1-X[t])^{1-n}-1}{n-1}}$**//PowerExpand**

%/(60 sec / min)

2041.52 sec

34.0253 min

In this case the time required to reach 90% conversion is 34 minutes. Notice that to get to 95% conversion the time required is just over 1.6 h, but to get to 99.99% requires over 18,500 h.

$$\textbf{Table}\left[\frac{(1-X[t])^{1-n}-1}{n-1}\Big/3600\Big/(\sec/h)\text{//PowerExpand},\{X[t],\{0.9,0.95,0.99,0.999,0.9999\}\}\right]$$

{0.567088 h, 1.63783 h, 18.5 h, 585.588 h, 18518.5 h}

This is because the conversion reaches the value of 1 asymptotically, as can be seen in the plot below:

***k*=.;**

$n = 2.5;$

Cao=.;

***t*=.;**

Clear[*X*[*t*]]

$$\textbf{Solve}\left[t==\frac{(1-X[t])^{1-n}-1}{(n-1)}, X[t]\right]\text{//}$$

FullSimplify

$$\left\{\left\{X[t]\to 1.-\frac{0.763143}{(0.666667+\text{Cao}^{1.5}kt)^{0.6666666666666667}}\right\}\right\}$$

$$k = 0.01L^{1.5}\text{mol}^{-1.5}\sec^{-1};$$

$$n = 2.5;$$

$$\text{Cao} = 1\text{mol}L^{-1};$$

$$X[\text{t\_}]:=1.-0.7631\Big/(0.6667+\text{Cao}^{1.5}kt\sec)^{0.6667}\text{//PowerExpand}$$

```
Plot[X[t], {t, 0, 100500}, PlotRange → All,
Epilog → {Hue[0.4], Line[{{0, 1}, {100500, 1}}]},
AxesLabel → {t, "X[ t ]"}]
```

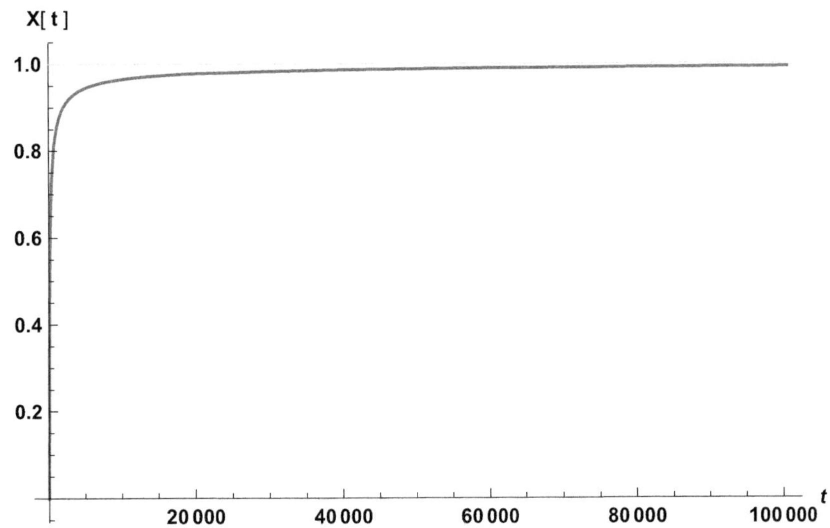

Effect of stoichiometry

Consider the following reaction - one in which one molecule fragments into four, A ⟶ 3 B + D. Imagine that all the species are gases and that we can treat them as ideal. The rate may be a first order decay, r_{A-} = k Ca, as might happen in a pyrolysis, or photochemical process. The rate equations are as follows:

$$\text{A} \longrightarrow 3\,\text{B} + \text{D}$$

$$-\frac{d\text{Ca}}{dt} = \frac{1}{3}\frac{d\text{Cb}}{dt} = \frac{d\text{Cd}}{dt}$$

which upon integration give us:

$$\text{Cao - Ca[t]} = \frac{(\text{Cb}[t] - \text{Cbo})}{3} = \text{Cd [t] - Cdo}$$

Since the gases are ideal we can re-express the concentration in terms of the partial pressure of each species in terms of the ideal gas law:

$$\frac{\text{Pao}}{RT} - \frac{\text{Pa}[t]}{RT} = \frac{(\text{Pb}[t] - \text{Pbo})}{3\text{RT}} = \frac{\text{Pdo}}{RT} - \frac{\text{Pd}[t]}{RT}$$

And assuming the initial pressures of B and D are zero:

$$\text{Pao} - \text{Pa}[t] = \frac{\text{Pb}[t]}{3} = \text{Pd}[t]$$

We also can say that:

$$\text{Ptot}[\,0\,] = \text{Pao}$$

and

$$\text{Ptot}[\,t\,] = \text{Pa}[\,t\,] + \text{Pb}[\,t\,] + \text{Pd}[\,t\,]$$

The rate expression for Pa [t] is:

$$\frac{d\,\text{Pa}[t]}{dt} = -\,k\ \text{Pa}[\,t\,] \Longrightarrow \text{Pa}[\,t\,] = \text{Pao Exp}[\,-\,k\,t\,]$$

$$\frac{d\,\text{Pb}[t]}{dt} = 3\,k\ \text{Pa}[\,t\,] = 3\,k\ \ \text{Pao Exp}[\,-\,k\,t\,]$$

$$\Longrightarrow \text{Pb}[\,t\,] = 3\ \text{Pao}\ (1 - \text{Exp}[\,-\,k\,t\,])$$

$$\frac{d\,\text{Pd}[t]}{dt} = k\ \text{Pa}[\,t\,] = k\ \text{Pao Exp}[\,-\,k\,t\,]$$

$$\Longrightarrow \text{Pd}[\,t\,] = \text{Pao}\ (1 - \text{Exp}[\,-\,k\,t\,])$$

And therefore:

$$\text{Ptot}[\,t\,] = \text{Pao Exp}[\,-\,k\,t\,] + 3\ \text{Pao}\ (1 - \text{Exp}[\,-\,k\,t\,]) + \text{Pao}\ (\,1 - \text{Exp}[\,-\,k\,t\,])$$

$$\text{Ptot}[\,t\,] = \text{Pao}\ (4 - 3\ \text{Exp}[\,-\,k\,t])$$

If the reaction goes to completion, in this closed vessel the pressure will increase fourfold. If the initial pressure of species A was 1000 pounds-per-square-inch-gauge (psig), then at the end of the process the pressure will be 4000 psig. Thus, one has to be very careful in planning reactions in closed vessels.

Ptot[t_]:= Pao(4 − 3Exp[−*kt*])

k **= 0.01**

Pao = 1

Plot[Ptot[*t*], {*t*, 0, 1000},

AxesLabel → {*t*, "P [t]"}]

0.01

1

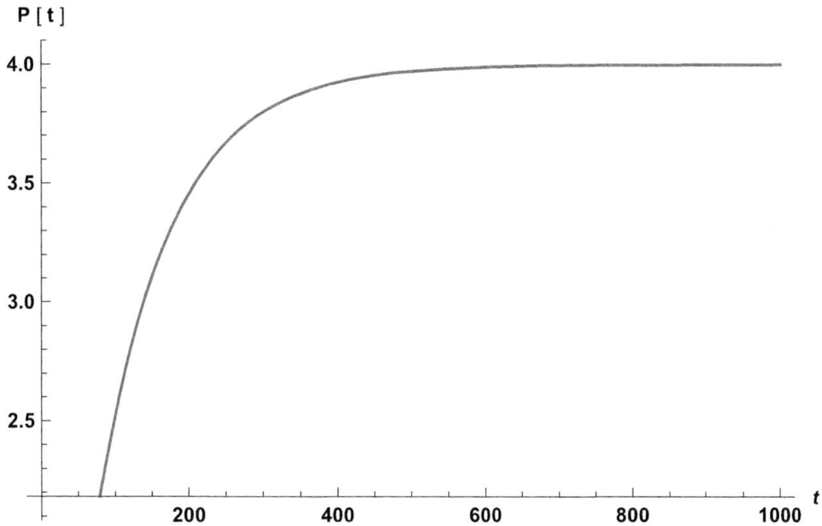

We will see that this stoichiometric effect, becomes particularly important in plug flow reactor, because as the reaction proceeds along the length of the reactor, the production of more moles of gas actually accelerates the gas flow rate and thereby reduces the holding time, meaning that less conversion will take place for a given volume of reactor than if expansion did not occur. In the batch reactor this is manifested simply as a rise in the pressure.

For the general reaction:

$$\mathrm{a\,A + b\,B \longrightarrow d\,D + e\,E}$$

The rate expressions as follows provide the stoichiometric relationships:

$$-\frac{1}{a}\frac{dCa}{dt} = -\frac{1}{b}\frac{dCb}{dt} = \frac{1}{d}\frac{dCd}{dt} = \frac{1}{e}\frac{dCe}{dt}$$

Notice that the rates for A and B are opposite in sign compared to those for D and E and notice that the reciprocals of the stoichiometric coefficients normalize the rates.

Complex reactions and the steady state approximation

Lindemann

Not all reactions are as simple as their stoichiometry may imply. For example various gas phase reactions involve A $\longrightarrow$ B, and from the stoichiometry they may be expected to be unimolecular and to follow first order kinetics. In fact the situation is often more complicated than that. Remember that a reaction follows a collision and that the collision imparts the energy needed for species A to overcome the activation energy needed for reaction. Lindemann, and then Hinshelwood described the kinetics for the following types of reactions:

$$\mathrm{A + M \Longleftrightarrow A^* + M} \quad (1)$$

$$\mathrm{A^* \longrightarrow B} \quad (2)$$

The first step is a bimolecular collision in which A collides with M and is put into an activated state denoted by the asterisk. Once in this state it can relax back to its unactivated state or undergo a unimolecular reaction to produce B. Think of an isomerization as such a reaction. A could be the cis isomer and B would be the trans isomer.

What do the kinetics show us? Let's examine this and see how such a mechanism is developed. First the rate of the bimolecular reaction is as follows:

$$\frac{d\mathrm{Ca}}{dt} = -\,\mathrm{kf1\ Ca\ Cm + kr1\ Ca^*\ Cm}$$

$$\frac{d\mathrm{Ca^*}}{dt} = \mathrm{kf1\ Ca\ Cm - kr1\ Ca^*\ Cm - k2f\ Ca^*}$$

$$\frac{d\mathrm{Cb}}{dt} = \mathrm{k2f\ Ca^*}$$

The notion of the steady state is introduced in order to simplify the analysis and because it is a reasonable approximation. Namely it is assumed that the concentration of A^* is constant, which is to say that it is produced as fast as it is consumed. If correct then the second rate equation is equal to zero:

If $\mathrm{Ca^*} \approx$ constant, then $\dfrac{d\mathrm{Ca^*}}{dt} \approx 0$

and then:

$$0 = \mathrm{kf1\ Ca\ Cm - kr1\ Ca^*\ Cm - k2f\ Ca^*}$$

From which we can find that:

$$Ca^* \rightarrow \frac{\mathrm{kf1CaCm}}{\mathrm{k2f} + \mathrm{kr1Cm}}$$

$$\therefore \frac{d\mathrm{Cb}}{dt} = \frac{\mathrm{kf1k2fCaCm}}{\mathrm{k2f} + \mathrm{kr1Cm}}$$

and

$$\frac{d\mathrm{Ca}}{dt} = -\ \mathrm{kf1\ Ca\ Cm} + \frac{\mathrm{kf1kr1CaCm}^2}{\mathrm{k2f} + \mathrm{kr1Cm}}$$

The species M may be just an inert gas, or it could be any non-reactive gas. Its only role in the mechanism is to impart energy via a collision to A so that it can react to give product B. Therefore, and in keeping with the ideas of kinetic theory, we can consider the two limiting cases firstly, a low pressure of M and secondly, a high pressure of M. At low pressures of M we can take k2f $\gg$ kr1 Cm hence:

$$\frac{d\mathrm{Cb}}{dt} = \frac{\mathrm{kf1k2fCaCm}}{\mathrm{k2f}} \longrightarrow \mathrm{kf1\ Ca\ Cm}$$

The product will appear to be generated by second order kinetics in A and M. Here the rate determining step is the initial collision, which is intuitively clear given the pressure is low. The rate of depletion of A will look as follows:

$$\frac{d\mathrm{Ca}}{dt} = -\ \mathrm{kf1\ Ca\ Cm} + \frac{\mathrm{kf1kr1CaCm}^2}{\mathrm{k2f}}$$

$$= -\mathrm{kf1\ Ca\ Cm}\ (\ 1 - \frac{\mathrm{kr1}}{\mathrm{k2f}}\mathrm{Cm})$$

which is not a simple result to interpret.

On the other hand at high pressure of M, the reverse is true, k2f $\ll$ kr1 Cm:

$$\frac{d\mathrm{Cb}}{dt} = \frac{\mathrm{kf1k2fCaCm}}{\mathrm{kr1Cm}} \longrightarrow \mathrm{k'' \ Ca}$$

and the kinetics appear to be first order in A. In this instance at high pressure, collisions are plentiful and the second reaction, the unimolecular decay prevails as the rate determining step. And we find that the rate of change in A is predicted to be zero.

$$\frac{d\mathrm{Ca}}{dt} = -\ \mathrm{kf1CaCm} + \mathrm{kf1CaCm} = 0$$

which is an odd result that might give us pause.

What if we do not make the steady state approximation? What if, instead, we solve the three equations outright and examine the solutions. In the code below we do just that - we solve the equations, we assign the results to functions and then we test the results in the equations.

```
Clear[Cao, Cbo, Ca, Cb, Cm, ca, cb, ca*, kf1,
kr1, kf2, solns]
Ca*=.
solns = FullSimplify[
DSolve[{∂tCa[t] == −kf1Ca[t]Cm + kr1Ca*[t]Cm,
∂tCa*[t] == kf1Ca[t]Cm − kr1Ca*[t]Cm−kf2Ca*[t],
∂tCb[t]==kf2Ca*[t],
Ca[0] == Cao, Ca*[0] == 0, Cb[0] == 0},
{Ca[t], Ca*[t], Cb[t]}, t]]
ca[t_]:=Evaluate[Ca[t]/.solns[[1]]]
ca*[t_]:=Evaluate[Ca*[t]/.solns[[1]]]
cb[t_]:=Evaluate[Cb[t]/.solns[[1]]]
FullSimplify[{∂tca[t] == −kf1ca[t]Cm + kr1ca*[t]Cm,
∂tca*[t] == kf1ca[t]Cm − kr1ca*[t]Cm − kf2ca*[t],
∂tcb[t]==kf2ca*[t]}]
```

$$\{\{\mathrm{Ca}[]t] \to \left(\mathrm{Cao}\, e^{\left(-\frac{1}{2}t\left(\sqrt{(\mathrm{Cm}(\mathrm{kf1}+\mathrm{kr1})+\mathrm{kf2})^2-4\mathrm{Cmkf1kf2}}+\mathrm{Cm}(\mathrm{kf1}+\mathrm{kr1})+\mathrm{kf2}\right)\right)}\right.$$

$$\left(e^{t\sqrt{(\mathrm{Cm}(\mathrm{kf1}+\mathrm{kr1})+\mathrm{kf2})^2-4\mathrm{Cmkf1kf2}}}+\mathrm{Cmkf1}-\mathrm{Cmkr1}-\mathrm{kf2}\right.$$

$$\mathrm{Cm}(\mathrm{kr1}-\mathrm{kf1})+\mathrm{kf2}+\sqrt{(\mathrm{Cm}(\mathrm{kf1}+\mathrm{kr1})+\mathrm{kf2})^2-4\mathrm{Cmkf1kf2}}$$

$$\left.\left.+e^{t\sqrt{(\mathrm{Cm}(\mathrm{kf1}+\mathrm{kr1})+\mathrm{kf2})^2-4\mathrm{Cmkf1kf2}}}\sqrt{(\mathrm{Cm}(\mathrm{kf1}+\mathrm{kr1})+\mathrm{kf2})^2-4\mathrm{Cmkf1kf2}}\right)\right)\Big/$$

$$2\sqrt{(\mathrm{Cm}(\mathrm{kf1}+\mathrm{kr1})+\mathrm{kf2})^2-4\mathrm{Cmkf1kf2}},$$

$$\mathrm{Cb}[t] \;\rightarrow\; \mathrm{Cao} \;-\; \mathrm{Cao}\, e^{-\frac{1}{2}t(\mathrm{Cm}(\mathrm{kf1}+\mathrm{kr1})+\mathrm{kf2})}$$
$$\mathrm{Cosh}[\tfrac{1}{2}t\sqrt{\mathrm{Cm}^2(\mathrm{kf1}+\mathrm{kr1})^2+2\mathrm{Cmkf2}(\mathrm{kr1}-\mathrm{kf1})+\mathrm{kf2}^2}\,] \;-$$
$$\Bigg(\mathrm{Cao}\, e^{-\frac{1}{2}t(\mathrm{Cm}(\mathrm{kf1}+\mathrm{kr1})+\mathrm{kf2})}\,\mathrm{Cm}(\mathrm{kf1}+\mathrm{kr1})$$
$$+\mathrm{kf2}\,\mathrm{Sinh}[\tfrac{1}{2}t\sqrt{(\mathrm{Cm}(\mathrm{kf1}+\mathrm{kr1})+\mathrm{kf2})^2-4\mathrm{Cmkf1kf2}}\,]\Bigg)\Big/$$
$$\sqrt{(\mathrm{Cm}(\mathrm{kf1}+\mathrm{kr1})+\mathrm{kf2})^2-4\mathrm{Cmkf1kf2}},$$

$$\mathrm{Ca}^*[t] \;\rightarrow\; \Bigg(2\mathrm{CaoCmkf1}\, e^{-\frac{1}{2}t(\mathrm{Cm}(\mathrm{kf1}+\mathrm{kr1})+\mathrm{kf2})}$$
$$\mathrm{Sinh}\left(\frac{1}{2}t\sqrt{(\mathrm{Cm}(\mathrm{kf1}+\mathrm{kr1})+\mathrm{kf2})^2-4\mathrm{Cmkf1kf2}}\right)\Bigg)\Big/$$
$$\sqrt{(\mathrm{Cm}(\mathrm{kf1}+\mathrm{kr1})+\mathrm{kf2})^2-4\mathrm{Cmkf1kf2}}\}\}$$

{True, True, True}

Now let's go back to the steady state analysis and solve explicitly for the concentrations of A and B for a comparison. To distinguish these solutions from the more complete solutions, we add the letter s to signify steady state.

Clear[Cao, Cbo, Ca, Cb, solns, kf1, kr1, kf2, Cm]

sssolns =

FullSimplify[

DSolve[

$$\left\{\partial_t \mathbf{Cas}[t] == -\mathbf{kf1Cas}[t]\mathbf{Cm} + \mathbf{kr1}\frac{\mathbf{kf1Cas}[t]\mathbf{Cm}}{\mathbf{kf2}+\mathbf{kr1Cm}}\mathbf{Cm},\right.$$

$$\partial_t \mathbf{Cbs}[t] == \mathbf{kf2}\frac{\mathbf{kf1Cas}[t]\mathbf{Cm}}{\mathbf{kf2}+\mathbf{kr1Cm}},$$

Cas[0] == Cao, Cbs[0] == 0}, {Cas[*t*], Cbs[*t*]},*t*]]

cas[t_]:=Evaluate[Cas[t]/.sssolns[[1]]]

cas*[t_]:=$\dfrac{\text{kf1cas}[t]\text{Cm}}{\text{kf2}+\text{kr1Cm}}$

cbs[t_]:=Evaluate[Cbs[t]/.sssolns[[1]]]

FullSimplify[

{∂_tcas[t] == −kf1cas[t]Cm + kr1cas*[t]Cm,

∂_tcbs[t]==kf2cas*[t]}]

$$\left\{\left\{\text{Cas}[t]\to \text{Cao}e^{-\frac{\text{Cmkf1kf2}t}{\text{kf2}+\text{Cmkr1}}},\ \text{Cbs}[t]\to \text{Cao}-\text{Cao}e^{-\frac{\text{Cmkf1kf2}t}{\text{kf2}+\text{Cmkr1}}}\right\}\right\}$$

{True, True}

Immediately, we see the tremendous simplification that the steady state approximation rendered. As expected, because the equations are simpler, and because we have substituted for Ca* in the equations, the resulting expressions are also much simpler that are the complete solutions.

Neither set of solutions is difficult to evaluate computationally, now that we have them in explicit form. In fact we can compare the results from the steady state approximation to the complete solutions. To do so, we need to specify the forward and reverse rate constants for the first reaction and that of the second reaction, along with the initial concentration of the reactant A and the background reactant M. We can begin with a concentration of M that is equal to the concentration of A. These can be taken as equivalent to 0.2 mol/L:

kf1 = 0.01;

kr1 = 0.001;

kf2 = 0.02;

Cao = 0.2;

Cm = .2;

tmax = 2000;

```
Print[t, " Ca[t]", " Ca*[t]",
" Cb[t]"]
Table[{t, ca[t], ca*[t], cb[t]}, {t, 0, tmax, 400}]//
TableForm
Print[t, " Cas[t]", " Cas*[t]",
" Cbs[t]"]
Table[{t, cas[t], cas*[t], cbs[t]},
{t, 0, tmax, 400}]//TableForm

complete = Plot[{ca[t], ca*[t], cb[t]}, {t, 0, tmax},
AxesLabel → {t, "Ci[t]"}];

steadystate = Plot[{cas[t], cas*[t], cbs[t]},
{t, 0, tmax}, PlotStyle → Dashed,
AxesLabel → {t, "Ci[t]"}];

Show[complete, steadystate]
```

t	Ca[t]	Ca*[t]	Cb[t]
0	0.2	0.	0.
400	0.0905493	0.00993175	0.0995189
800	0.0410452	0.00450503	0.15445
1200	0.0186055	0.00204209	0.179352
1600	0.0084337	0.000925664	0.190641
2000	0.00382293	0.000419596	0.195757

t	Cas[t]	Cas*[t]	Cbs[t]
0	0.2	0.019802	0.
400	0.0905804	0.00896836	0.10942
800	0.0410241	0.00406179	0.158976
1200	0.0185799	0.00183959	0.18142
1600	0.00841487	0.000833156	0.191585
2000	0.00381111	0.000377338	0.196189

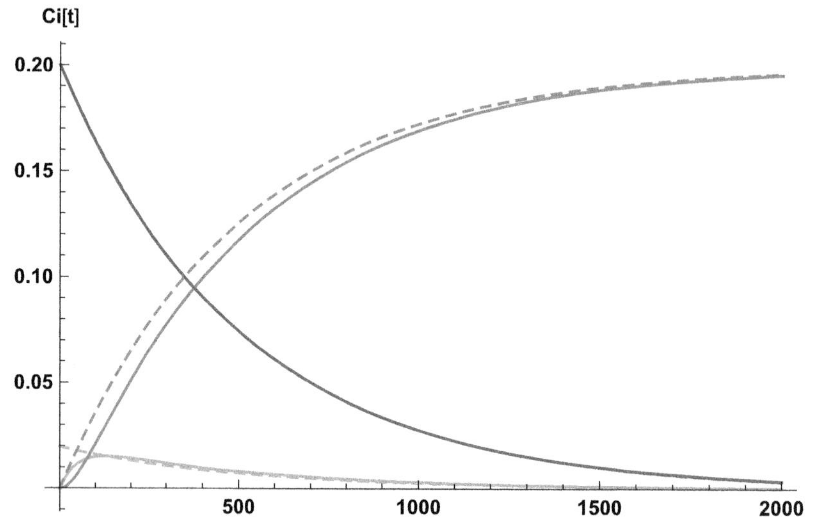

Both the tabular data and the graphs show that at a low value of Cm, there is actually good but not perfect agreement between the complete solution functions (solid) and those derived from the steady state (dashed). Notice that the results for the diminution of A (green - mid gray in the print version) with time are nearly identical in both cases, whereas the steady state result for the product concentration (blue - dark gray in the print version) leads that from the complete solution. The activated complex A^* results are in gold and we see that for the steady state approximation, there is an apparent non-zero starting concentration of this species. What happens when the value of Cm is increased by a factor of 10?

```
kf1 = 0.01;

kr1 = 0.001;

kf2 = 0.02;

Cao = 0.2;

Cm = 2;

tmax = 200;

Print[t, " Ca[t]", " Ca*[t]",

" Cb[t]"]

Table[{t, ca[t], ca*[t], cb[t]}, {t, 0, tmax, 40}]//TableForm

Print[t, " Cas[t]", " Cas*[t]",

" Cbs[t]"]

Table[{t, cas[t], cas*[t], cbs[t]}, {t, 0, tmax, 40}]//TableForm
```

complete = Plot[{ca[t], ca*[t], cb[t]}, {t, 0, tmax},

AxesLabel → {t, "Ci[t]"}];

steadystate = Plot[{cas[t], cas*[t], cbs[t]},

{t, 0, tmax}, PlotStyle → Dashed,

AxesLabel → {t, "Ci[t]"}];

Show[complete, steadystate]

t	Ca[t]	Ca*[t]	Cb[t]
0	0.2	0.	0.
40	0.0926812	0.0698314	0.0374874
80	0.0453872	0.0622823	0.0923304
120	0.0232074	0.0425347	0.134258
160	0.0122396	0.0263287	0.161432
200	0.00659117	0.0155551	0.177854

t	Cas[t]	Cas*[t]	Cbs[t]
0	0.2	0.181818	0.
40	0.096645	0.0878591	0.103355
80	0.0467013	0.0424557	0.153299
120	0.0225672	0.0205157	0.177433
160	0.0109051	0.00991369	0.189095
200	0.0052696	0.00479054	0.19473

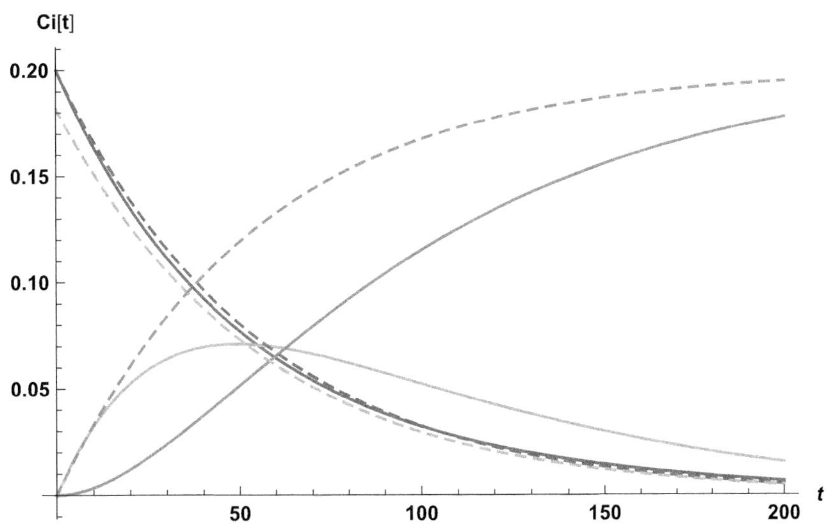

At high pressure, the deviation between the rate of diminution of A is small but evident, the rate of production of the activated species A^* is significant across the time domain, and the rate of product formation shows a short but real induction period for the complete solution that is missed completely in the steady state result. The bottom line is that with symbolic and/or numerical computing power, there is no reason to make the steady state approximation. What is gained in mathematical simplicity is lost in accuracy of the kinetic model.

Another complex reaction is that which takes place with hydrocarbons during pyrolysis. The key to the reaction is formation of radical species that can produce a chain reaction that propagates autocatalytically after an initiation step.

Rice-Herzfeld

The radical reactions of hydrocarbons also submit to complex kinetics and mechanisms made up of many elementary steps. Ethylene is produced by the pyrolysis of ethane according to this rather simple overall all reaction:

$$C_2H_6 \rightarrow C_2H_4 + H_2$$

The actual mechanism of reaction is actually many steps and includes the formation of other side products, namely methane. A less simplified representation of the pyrolysis reaction is:

$$3C_2H_6 \rightarrow 2C_2H_4 + 2\mathrm{CH}_4 + H_2$$

or

$$\frac{3}{2}C_2H_6 \rightarrow C_2H_4 + \mathrm{CH}_4 + \frac{1}{2}H_2$$

We could envision this to involve a very simple mechanism and kinetics. Although the stoichiometry would suggest a termolecular reaction, the probability of three molecules colliding simultaneously is infinitely small and we can be sure that molecularity and stoichiometric coefficients do not have to be the same and are not the same, unless the reaction represented is elementary. Therefore we will assume that the reaction is a unimolecular decay when an ethane molecule absorbs sufficient energy to cause the carbon-carbon and carbon-hydrogen bond lyses. The component rate equations including proper stoichiometric coefficients are as follows:

$$\frac{2}{3}\partial_t C_{C_2H_6}[t] == -\mathrm{k1}C_{C_2H_6}[t]$$

$$\partial_t C_{\mathrm{CH}_4}[t] == \mathrm{k1}C_{C_2H_6}[t]$$

$$\partial_t C_{C_2H_4}[t] == \mathrm{k1}C_{C_2H_6}[t]$$

$$2\partial_t C_{H_2}[t] == \mathrm{k1}C_{C_2H_6}[t]$$

because

$$-\frac{1}{3}\partial_t C_{C_2H_6}[t] = \frac{1}{2}\partial_t C_{C_2H_4}[t] = \frac{1}{2}\partial_t C_{\mathrm{CH}_4}[t] = \partial_t C_{H_2}[t]$$

therefore

$$-\frac{2}{3}\partial_t C_{C_2H_6}[t] = \partial_t C_{C_2H_4}[t] = \partial_t C_{\text{CH}_4}[t] = 2\partial_t C_{H_2}[t]$$

The second order rate constant can be estimated from theoretical considerations that are beyond the scope of this text. Consulting "Kinetics and Mechanism," by J.W. Moore and R.G. Pearson, pg. 173 the A-factor for this constant (k1o) should be about 10^{13} per second and the activation energy can be set equal to the bond energy of a carbon-carbon bond. Hence, this very simple model will look as follows:

```
tmax = 4000;
k1o = 10^13;
Co = 3;
k1 = k1oExp[−243000/8.314/800];
(* Ea ≈ BondEnergy *)

eqns = {2/3 ∂t CC2H6[t] == −k1 CC2H6[t],
∂t CCH4[t] == k1 CC2H6[t],
∂t CC2H4[t] == k1 CC2H6[t] ,
2∂t CH2[t] == k1 CC2H6[t],
CC2H6[0] == Co, CCH4[0] == 0, CC2H4[0] == 0,
CH2[0] == 0};

vars = {CC2H6[t], CCH4[t], CC2H4[t], CH2[t]} ;

ethane = NDSolve[eqns, vars, {t, 0, tmax}];
cC2H6[t_]:=Evaluate[CC2H6[t]/.ethane]
cC2H4[t_]:=Evaluate[CC2H4[t]/.ethane]
cCH4[t_]:=Evaluate[CCH4[t]/.ethane]
cH2[t_]:=Evaluate[CH2[t]/.ethane]

Plot[cC2H6[t], {t, 0, tmax}, PlotRange → All,
PlotLabel → "CC2H6[t]"]
Plot[cC2H4[t], {t, 0, tmax}, PlotRange → All,
```

PlotLabel → "$C_{C_2H_4}$[t]"]

Plot[$c_{CH_4}[t]$, {t, 0, tmax}, PlotRange → All,

PlotLabel → "C_{CH_4}[t]"]

Plot[$c_{H_2}[t]$, {t, 0, tmax}, PlotRange → All,

PlotLabel → "C_{H_2}[t]"]

Plot[{$c_{C_2H_6}[t]$, $c_{C_2H_4}[t]$, $c_{CH_4}[t]$, $c_{H_2}[t]$},

{t, 0, tmax}, PlotRange → All,

PlotLabel → "$c_{C_2H_6}$[t],$c_{C_2H_4}$[t],c_{CH_4}[t],c_{H_2}[t]"]

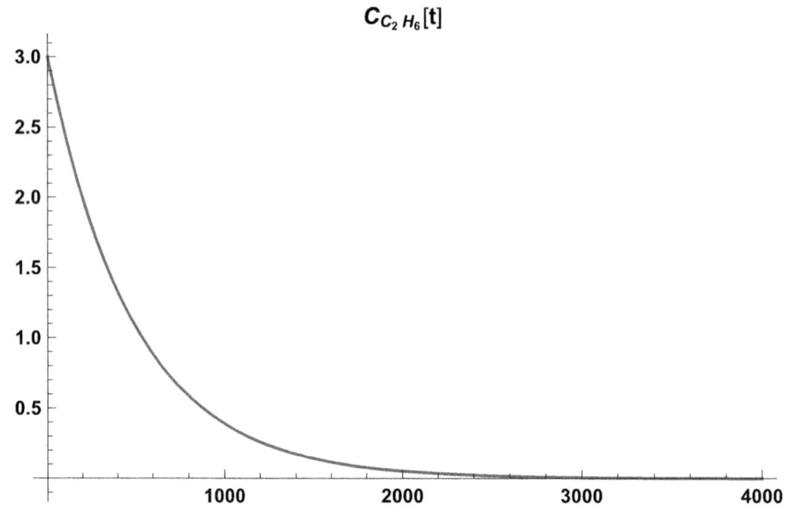

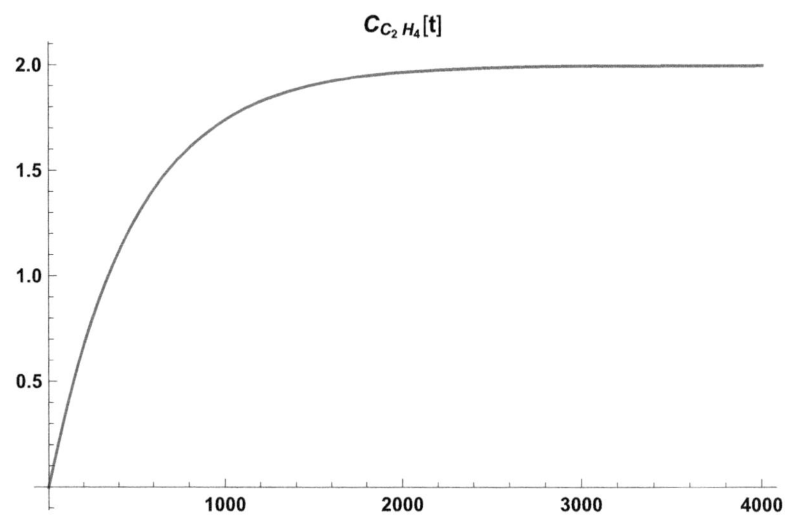

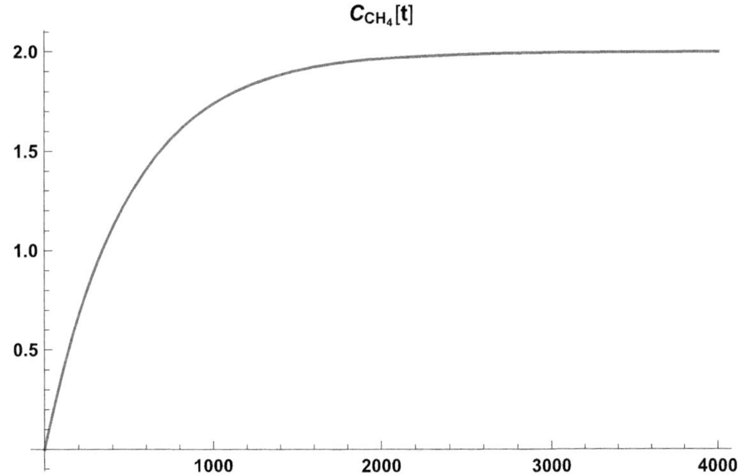
$c_{CH_4}[t]$
2.0
1.5
1.0
0.5
1000
2000
3000
4000

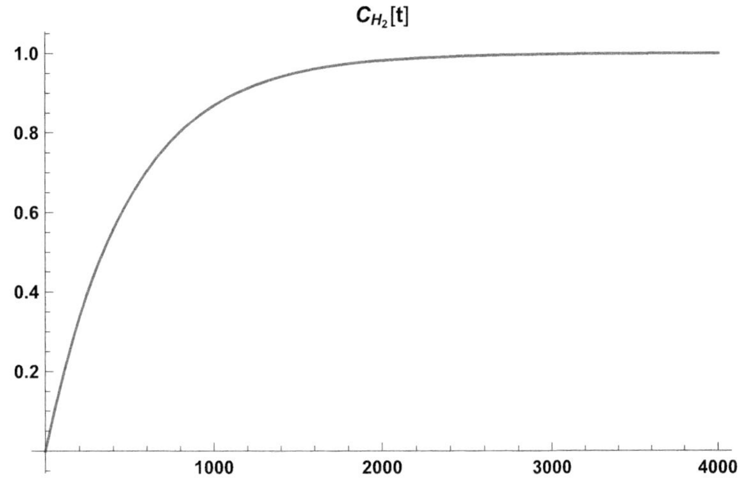
$c_{H_2}[t]$
1.0
0.8
0.6
0.4
0.2
1000
2000
3000
4000

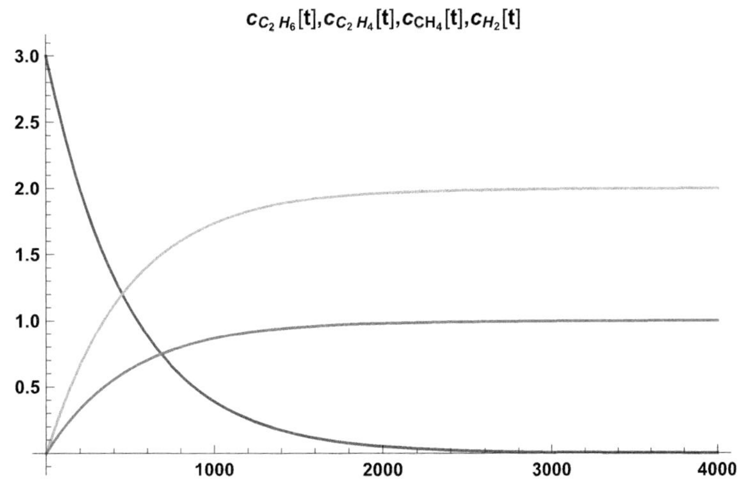
$c_{C_2H_6}[t], c_{C_2H_4}[t], c_{CH_4}[t], c_{H_2}[t]$
3.0
2.5
2.0
1.5
1.0
0.5
1000
2000
3000
4000

In this instance, we set the initial concentration of ethane to 3 moles per liter. With this rate constant and with T at 800K (527C) the reaction is complete in under one hour. If we move the temperature up to 1000K (727C) the rate is much faster as we see:

tmax = 4;

$k1o = 10^{13}$;

Co = 3;

k1 = k1oExp[−243000/8.314/1000];

(* Ea ≈ BondEnergy *)

$$eqns = \left\{\frac{2}{3}\partial_t C_{C_2H_6}[t] == -k1 C_{C_2H_6}[t],\right.$$
$$\partial_t C_{CH_4}[t] == k1 C_{C_2H_6}[t],$$
$$\partial_t C_{C_2H_4}[t] == k1 C_{C_2H_6}[t],$$
$$2\partial_t C_{H_2}[t] == k1 C_{C_2H_6}[t],$$
$$C_{C_2H_6}[0] == Co, C_{CH_4}[0] == 0, C_{C_2H_4}[0] == 0,$$
$$\left. C_{H_2}[0] == 0\right\};$$

$$vars = \{C_{C_2H_6}[t], C_{CH_4}[t], C_{C_2H_4}[t], C_{H_2}[t]\};$$

ethane = NDSolve[eqns, vars, {t, 0, tmax}];

$c_{C_2H_6}$[t_]:=Evaluate[$C_{C_2H_6}[t]$/.ethane]

$c_{C_2H_4}$[t_]:=Evaluate[$C_{C_2H_4}[t]$/.ethane]

c_{CH_4}[t_]:=Evaluate[$C_{CH_4}[t]$/.ethane]

c_{H_2}[t_]:=Evaluate[$C_{H_2}[t]$/.ethane]

Plot[$c_{C_2H_6}[t]$, {t, 0, tmax}, PlotRange → All,
PlotLabel → "$C_{C_2H_6}$[t]"]

Plot[$c_{C_2H_4}[t]$, {t, 0, tmax}, PlotRange → All,
PlotLabel → "$C_{C_2H_4}$[t]"]

Plot[$c_{CH_4}[t]$, {t, 0, tmax}, PlotRange → All,
PlotLabel → "C_{CH_4}[t]"]

Plot[$c_{H_2}[t]$, {t, 0, tmax}, PlotRange → All,
PlotLabel → "C_{H_2}[t]"]

Plot[{$c_{C_2H_6}[t]$, $c_{C_2H_4}[t]$, $c_{CH_4}[t]$, $c_{H_2}[t]$},
{t, 0, tmax}, PlotRange → All,
PlotLabel →
"$c_{C_2H_6}$[t],$c_{C_2H_4}$[t],c_{CH_4}[t],c_{H_2}[t]",
AxesLabel → {t(sec), "Ci [t]"}]

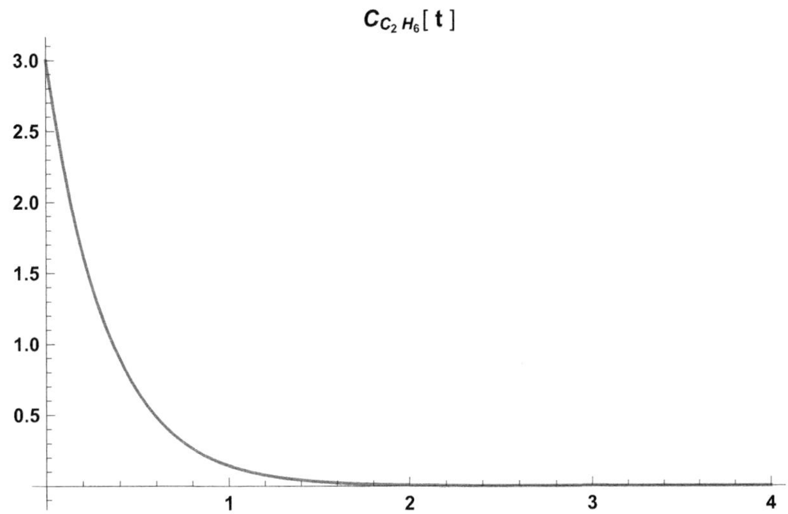

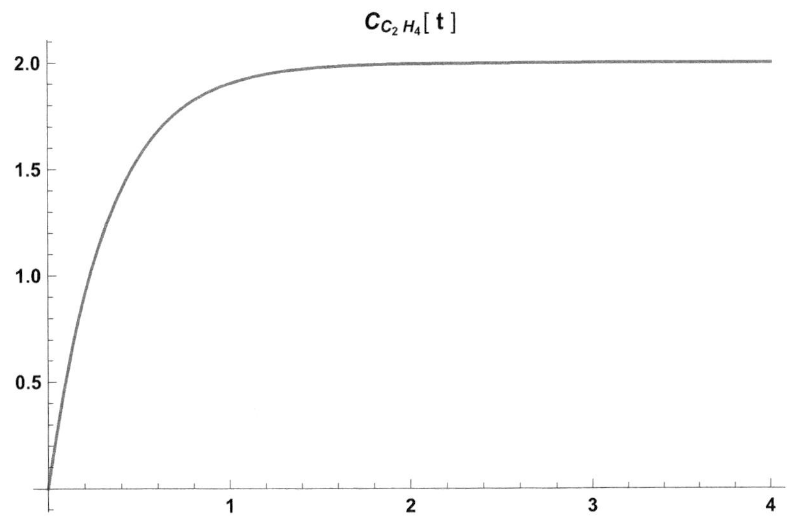

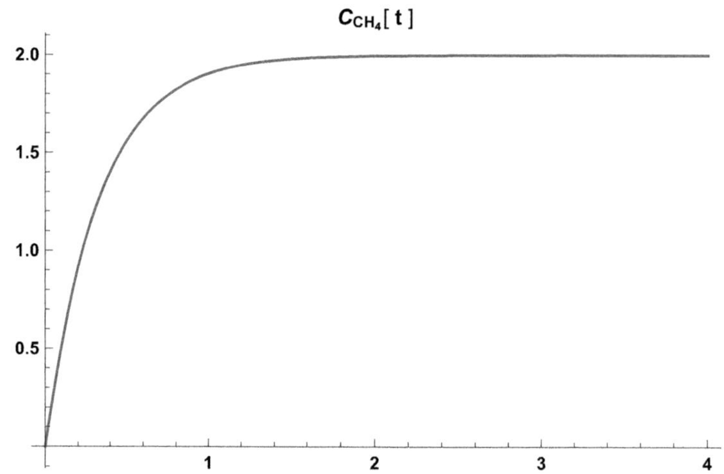
C_{CH_4}[t]
2.0
1.5
1.0
0.5
1
2
3
4

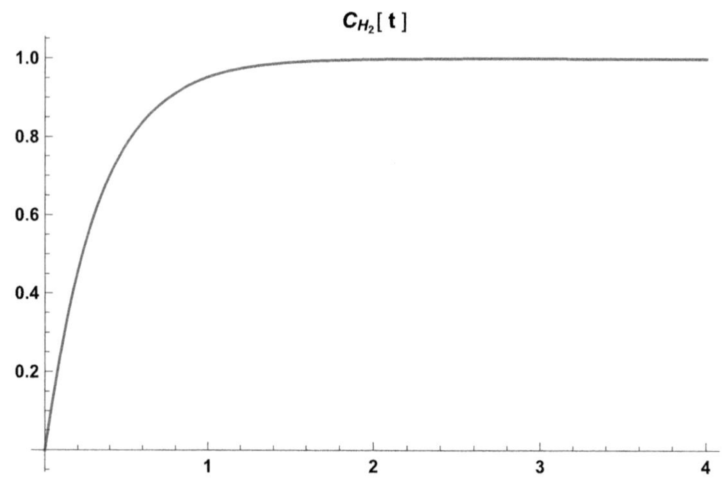
C_{H_2}[t]
1.0
0.8
0.6
0.4
0.2
1
2
3
4

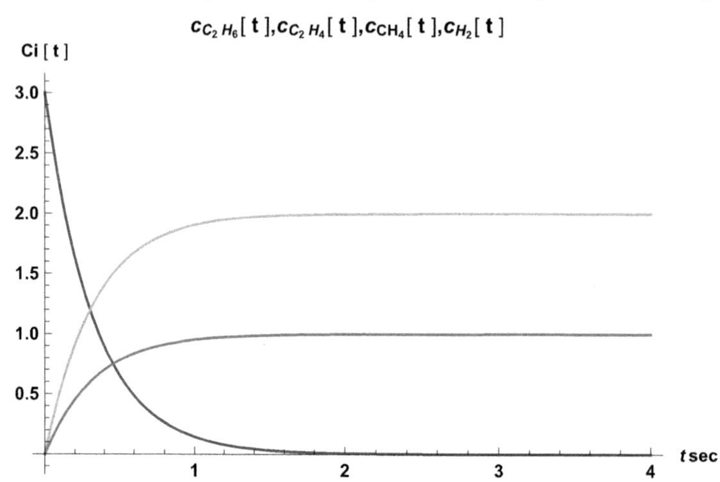
$c_{C_2H_6}$[t],$c_{C_2H_4}$[t],c_{CH_4}[t],c_{H_2}[t]
Ci [t]
3.0
2.5
2.0
1.5
1.0
0.5
1
2
3
4
t sec

At the higher temperature, the rate constant increases by over 1000 and the time to consume three moles of ethane is less than two seconds.

This is acceptable as far as it goes, but it does not go far enough. We know from considerable research that the mechanism of this reaction is considerably more complex than that which we just examined. In fact a mechanism that is still simple, but more realistic involves at least six elementary steps. These include dissociation of the ethane molecule, a bimolecular hydrogen abstraction that generates methane and an ethyl radical, a unimolecular disproportionation of the ethyl radical into ethylene and a hydrogen radical, another bimolecular hydrogen abstraction step to produce dihydrogen and ethyl radical, another disproportionation of the ethyl radical and, finally methyl radical and hydrogen atom association to produce more methane:

$C_2H_6 \rightarrow 2\,CH_3$	$r1 = k1\, C_{C_2H_6}$
$CH_3 + C_2H_6 \rightarrow CH_4 + C_2H_5$	$r2 = k2\, C_{CH_3} C_{C_2H_6}$
$C_2H_5 \rightarrow C_2H_4 + H$	$r3 = k3\, C_{C_2H_5}$
$H + C_2H_6 \rightarrow H_2 + C_2H_5$	$r4 = k4\, C_H C_{C_2H_6}$
$C_2H_5 \rightarrow C_2H_4 + H$	$r5 = k3\, C_{C_2H_5}$
$CH_3 + H \rightarrow CH_4$	$r6 = k6\, C_H C_{CH_3}$

--- --- --- --- --- --- --- ---

$$3\,C_2H_6 \rightarrow 2\,C_2H_4 + 2\,CH_4 + H_2$$

or

$$\tfrac{3}{2}\,C_2H_6 \rightarrow C_2H_4 + CH_4 + \tfrac{1}{2}\,H_2$$

where the rate for each step is also shown. Notice that the overall stoichiometry is identical to that for the over simplified mechanism. We could add more reactions to this mechanism, but this enough for now.

The rate equations for each component of this mechanism are:

$$\text{eqns} = \{\partial_t C_{C_2H_6}[t] = -\text{k1}C_{C_2H_6}[t] - \text{k2}C_{\text{CH}_3}[t]C_{C_2H_6}[t] - \text{k4}C_H[t]C_{C_2H_6}[t],$$

$$\partial_t C_{\text{CH}_3}[t] = +\text{k1}C_{C_2H_6}[t] - \text{k2}C_{\text{CH}_3}[t]C_{C_2H_6}[t] - \text{k6}C_H[t]C_{\text{CH}_3}[t],$$

$$\partial_t C_{\text{CH}_4}[t] = +\text{k2}C_{\text{CH}_3}[t]C_{C_2H_6}[t] + \text{k6}C_H[t]C_{\text{CH}_3}[t],$$

$$\partial_t C_{C_2H_5}[t] = +\text{k2}C_{\text{CH}_3}[t]C_{C_2H_6}[t] - \text{k3}C_{C_2H_5}[t],$$

$$\partial_t C_{C_2H_4}[t] = +\text{k3}C_{C_2H_5}[t],$$

$$\partial_t C_H[t] = +2\text{k3}C_{C_2H_5}[t] - \text{k4}C_H[t]C_{C_2H_6}[t],$$

$$\partial_t C_{H_2}[t] = +\text{k4}C_H[t]C_{C_2H_6}[t]\}$$

Using reasonable values of the rate constants that are based on theoretical principles, (see "Kinetics and Mechanism," by J.W. Moore and R.G. Pearson, pg. 171) we can build a model for the pyrolysis of ethane by employing NDSolve and Module:

C2Pyrol[k1_, k2_, k3_, k4_, k6_, Co_, tmax_]:=

Module[

{eqns =

$\{\partial_t C_{C_2H_6}[t] == -\text{k1}C_{C_2H_6}[t] - \text{k2}C_{\text{CH}_3}[t]C_{C_2H_6}[t] - \text{k4}C_H[t]C_{C_2H_6}[t],$

$\partial_t C_{\text{CH}_3}[t] == +\text{k1}C_{C_2H_6}[t] - \text{k2}C_{\text{CH}_3}[t]C_{C_2H_6}[t] - \text{k6}C_H[t]C_{\text{CH}_3}[t],$

$\frac{3}{8}\partial_t C_{\text{CH}_4}[t] == +\text{k2}C_{\text{CH}_3}[t]C_{C_2H_6}[t] + \text{k6}C_H[t]C_{\text{CH}_3}[t],$

$\partial_t C_{C_2H_5}[t] == +\text{k2}C_{\text{CH}_3}[t]C_{C_2H_6}[t] - \text{k3}C_{C_2H_5}[t],$

$\frac{3}{8}\partial_t C_{C_2H_4}[t] == +\text{k3}C_{C_2H_5}[t],$

$\partial_t C_H[t] == +\text{k3}C_{C_2H_5}[t] - \text{k4}C_H[t]C_{C_2H_6}[t] + \text{k3}\ C_{C_2H_5}[t] - \text{k6}C_H[t]C_{\text{CH}_3}[t],$

$\frac{8}{6}\partial_t C_{H_2}[t] == +\text{k4}C_H[t]C_{C_2H_6}[t],$

$C_{C_2H_6}[0] == \text{Co}, C_{\text{CH}_3}[0] == 0, C_{\text{CH}_4}[0] == 0,$
$C_{C_2H_5}[0] == 0, C_{C_2H_4}[0] == 0, C_H[0] == 0,$
$C_{H_2}[0] == 0\},$

$\text{vars} = \{C_{C_2H_6}[t], C_{\text{CH}_3}[t], C_{\text{CH}_4}[t], C_{C_2H_5}[t],$
$C_{C_2H_4}[t], C_H[t], C_{H_2}[t]\}\ \},$

ethane = NDSolve[eqns, vars, {*t*, 0, tmax}];

$c_{C_2H_6}[\text{t\_}]\text{:=Evaluate}\left[C_{C_2H_6}[t]\text{/.ethane}\right];$
$c_{C_2H_4}[\text{t\_}]\text{:=Evaluate}\left[C_{C_2H_4}[t]\text{/.ethane}\right];$
$c_{\text{CH}_4}[\text{t\_}]\text{:=Evaluate}\left[C_{\text{CH}_4}[t]\text{/.ethane}\right];$

```
cH2[t_]:=Evaluate[CH2[t]/.ethane];
cCH3[t_]:=Evaluate[CCH3[t]/.ethane];
cC2H5[t_]:=Evaluate[CC2H5[t]/.ethane];
cH[t_]:=Evaluate[CH[t]/.ethane];

{Plot[{cCH3[t] * 10^8, cC2H5[t] * 10^8, cH[t] * 10^8},
{t, 0, tmax}, PlotRange → All,
PlotLabel → {"cCH3[ t ]", Hue[0], "cC2H5[ t ]",
Hue[0.3], "cH[ t ]", Hue[0.6]},
PlotStyle → {Hue[0], Hue[0.3], Hue[0.6]},
AxesLabel → {"t(sec)", "(mol/L)*10^8"}],

Plot[{cC2H6[t], .98cC2H4[t], cCH4[t], cH2[t]},
{t, 0, tmax}, PlotRange → All,
PlotLabel → {"cC2H6[ t ]," Hue[0], "cC2H4[ t ]",
Hue[0.6], "cCH4[t]", Hue[0.5], "cH2[ t ]",
Hue[0.8]},
PlotStyle → {Hue[0], Hue[0.6], Hue[0.5],
Hue[0.8]},
AxesLabel → {"t(sec)", "(mol/L)*10^8"}],

Plot[{Co * 30, cC2H6[t] * 30, .98cC2H4[t] * 28,
cCH4[t] * 16, cH2[t] * 2, cCH3[t] * 15, cC2H5[t] * 29,
cH[t] * 1}, {t, 0, tmax}, PlotRange → All,
PlotLabel →
"mC2H6[t],mC2H4[t],mCH4[t],mH2[t],mCH3[t],mC2H5[
t],mH[t]",
AxesLabel → {"t(sec)", "mi (g)"},
```

```
PlotStyle → {{Dashed, Hue[0.6]}, Hue[0],
Hue[0.6], Hue[0.5], Hue[0.8]}],

Plot[{.995Co * 30,
{(cC2H6[t] * 30) + (cC2H4[t] * 28) + (cCH4[t] * 16) +
cH2[t] * 2}}, {t, 0, tmax},
PlotRange → {{0, tmax}, {0, 95}},
PlotLabel →
"mC2H6f + mC2H4f + mCH4f + mH2f + mCH3f + mC2H5f + mHf",
AxesLabel → {"t(sec)", "mi (g)"},
PlotStyle → {{Dashed, Hue[0.6]}, Hue[0.9]}]
}
]
T = 800
k1o = 10^13;
k1 = k1oExp[−243000/8.314/T]
(*Ea ≈ BondEnergy*)
k2o = 10^−5k1o; (*A ≈ 10^8*)
k2 = k2oExp[−50000/8.314/T]
(* Ea ≈ Low *)
k3o = 1k1o; (* A ≈ 10^13 *)
k3 = k3o(* No activation energy *)
k4o = 10^−5k1o; (* A ≈ 10^8 *)
k4 = k4oExp[−50000/8.314/T]
k5o = 1k1o; (* A ≈ 10^13 *)
k5 = k3(* No activation energy *)
k6o = 10^−5k1o; (* A ≈ 10^8 *)
k6 = k6o(* No activation energy *)
```

Co = 3;

tmax = 3500;

C2Pyrol[k1, k2, k3, k4, k6, Co, tmax]

800

0.0013588

54352.2

10000000000000

54352.2

10000000000000

100000000

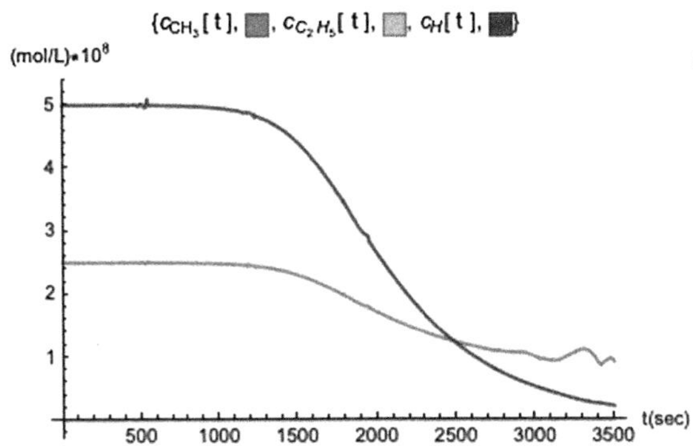

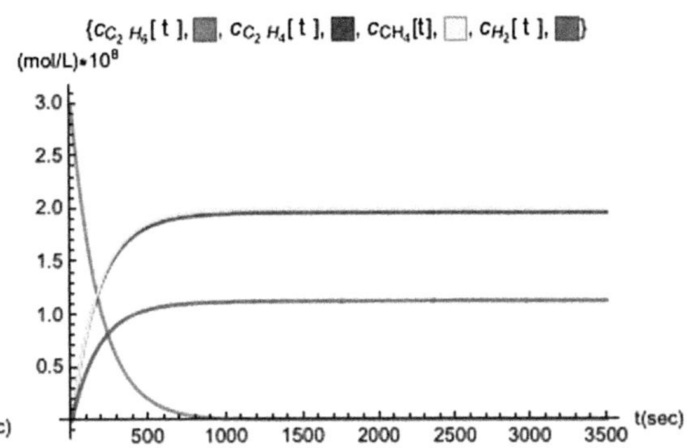

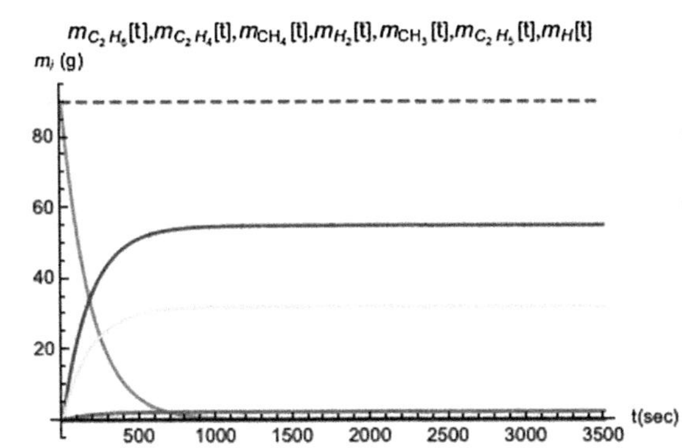

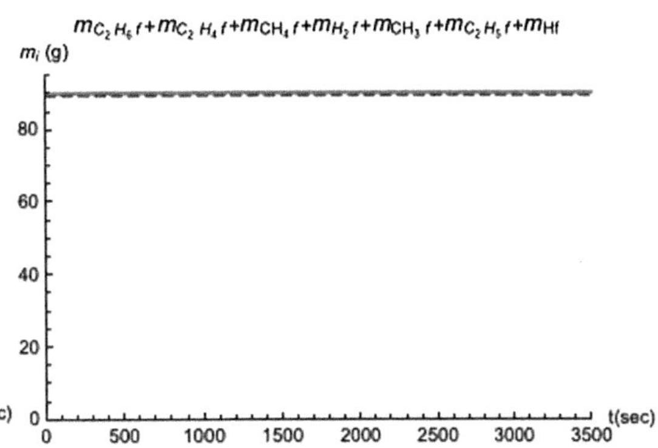

From the output we notice that the total mass balances during and at the end of reaction with the mass of ethane that was present at the initial time (3 mole = 90 g). The concentrations of the radical intermediates are all very small throughout the reaction and they show some mathematical instabilities. The products are two moles of ethylene and methane along with one mole of hydrogen.

Conclusion

In this chapter we have moved from simple kinetics of elementary steps to complex kinetics based upon mechanisms made up of multiple elementary steps that take place simultaneously. We have also used *Mathematica* in more intricate solutions to problems involving many species. These results can now be applied to well-known flow reactors - the fed-batch, or semi-continuous reactor, and two versions of flow reactors - the continuously-stirred-tank reactor (CSTR) and the plug-flow reactor (PFR).

CHAPTER 8

Flow reactors

Introduction to flow reactors

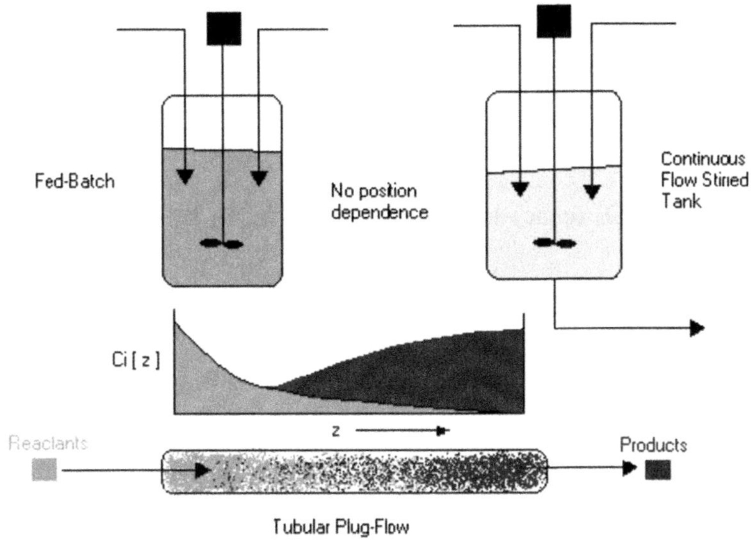

Semi-continuous systems, fed-batch reactors

The fed-batch reactor is a special system that can be used whenever the need arises to carefully control the reaction rate in a batch system. For example if a reaction is highly exothermic, then mixing the reactants at their full stoichiometric ratios initially, can lead to uncontrollable temperature rises that we referred to in the previous chapter as thermal excursion, or runaway. In simple terms, the reaction produces heat at a rate that is faster than the rate at which heat can be transferred away from the vessel. As a result the temperature in the vessel rises. However, by adding one of the reactants slowly, or intermittently, we can control the system and maintain good heat transfer away from the vessel, even if the heat transfer rate is slow. Since the rate of addition of the reactant is limited, the rate of reaction proceeds at a much reduced average rate, which allows the rate of heat transfer to keep pace with the rate of reactions.

A fed-batch is also used when the desired product is among the first to be formed in series of products. In such instances, the rates and the thermodynamics of the reactions are such that

Introduction to Chemical Engineering Analysis Using *Mathematica*
https://doi.org/10.1016/B978-0-12-820051-3.00013-9

the desired product is consumed in subsequent reactions. One way to deal with this is to run the reaction at very high dilution and by adding one of the two reactants intermittently to the other reactant. This has the effect of keeping the product concentration low and that reduces the rates of the subsequent reactions. This approach is well-known at the bench, but it is also used at a larger scale for production of high value chemicals.

If the kinetics are fast, then the concentration changes are nearly instantaneous when one reactant is added intermittently to the solution of the other reactant. An example of this is the dilution of a strong acid such as H_2SO_4 in pure water to make a less concentrated acid. The dilution process must be done slowly to manage the heat of solution that will be dissipated through the reactor walls and to the surroundings. Remember, we were all taught to add acid to water and not the other way around for this very reason. The main reaction is the hydrolysis of the acid to form the aquated proton and the bisulfate (hydrogensulfate) anion:

$$H_2SO_4\ (\mathrm{conc}) + H_2O \longrightarrow H_3O^+\ (\mathrm{aq}) + HSO_4^-\ (\mathrm{aq})$$

$$\Delta H_{\mathrm{rxn}} \approx 72\ \mathrm{kJoule/mole}$$

Although sulfuric acid is diprotic, the second dissociation has a pKa = 2, whereas the first dissociation has a pKa = -3. Therefore, for the purpose of this model, we will consider the first dissociation.

The rate of this reaction is very high, so we can call it instantaneous, especially when the acid concentration is low. By instantaneous, we mean that as soon as the concentrated acid contacts with the water it reacts to give the hydrolyzed product. Put another way, we cannot detect the reactant in solution without highly specialized techniques that are able to make measurements of concentration on the microsecond or less time scale. Hence, for a first order reaction, if the rate constant is about 10^6 seconds^{-1} or larger then the reaction rate is so high that we think of it as being instantaneous. Hydrolyses of acids are reactions that fall in this category. Generally, reactions of inorganics are very fast - examples are acid hydrolyses and precipitations whereas reactions of organics are much slower and more complex. The reasons for this generalization are that inorganic reactions are typically the reactions of ions in aqueous solution, which are fast, the reactions of organics involve the breaking and making of very strong covalent bonds. The strength of these covalent bonds makes the barrier to reaction high compared to reactions involving ionic bonds.

We can look more in depth at the hydrolysis of sulfuric acid. Let's assume we want to make batches of 10 M sulfuric acid and we have concentrated sulfuric acid. The motivation for doing this could be to dilute the acid to use as a catalyst for the hydrolysis of cellulosics, or it could be that there is a demand to be filled for small volume quantities of the weaker form of the acid. There could be other reasons as well, but these are sufficient motivation.

We also will assume that we have a large fed-batch reactor in which to do this. In fact the dilution of sulfuric acid at scale is so common that there are specialized continuous flow units available that have been designed for this purpose, but we will assume that we do not have such a unit. Instead we have a vessel, a good mixer, and a feed pump. We will also assume that the reactor is made of glass that is one quarter of an inch thick. Borosilicate glass is inert to the acid, so it will not be contaminated and the glass is thick enough to make the vessel quite robust.

Concentrated sulfuric acid is 18 Molar, i.e. 18 moles per liter. This will be the feed to the reactor. In the reactor, we will have a premeasured amount of pure water. If we consult the handbooks, we find that "chamber acid," is the product name for sulfuric acid produced by the lead chamber process, and it is about 62% sulfuric acid by weight. At room temperature this has a density of 1.52 kilogram per liter and a molarity of 9.6 moles per liter. If we mix 1 liter of concentrated sulfuric acid with 1 liter of water, the mass fraction of the acid is 0.647, or 64.7% and the molarity will be about 10 mole per liter and the mass fraction will be almost 65%:

$$65\%\ H_2SO_4 \text{ in } 100 \text{ g of diluted acid} \Rightarrow 65 \text{ g H2SO4}$$

$$\text{density of the diluted acid at } 25\ {}^{o}C = 1.54\ g \ / \ \text{cm}^3$$

$$\text{Volume of diluted acid} = \frac{100g}{1.54g\ /\text{cm}^3} = 64.9\ \text{cm}^3 = 0.065\ \text{L}$$

$$\text{Molarity of diluted acid} = \frac{65g}{98.8g/\text{mole} * 0.065L} \approx 10.1 \text{ mole per liter}$$

This is close to chamber acid and it will be the goal for our product simulation.

We need to know the total heat that will be released by this dilution process. The total heat produced is not the rate of heat production - that will be controlled by the rate of addition of the acid to the water - it is the integral over time of the product of the rate of hydrolysis and the heat of reaction. The heat of reaction is 72 kilojoules per mole.

An equal volume of concentrated sulfuric acid at 18 M and water produces 10 M acid. As we saw, this has a density of 1.54 g/cm^3 and it is 65% by weight sulfuric acid. The molecular weight of sulfuric acid is 98.8 g/mole. Let's assume that we want to produce about 2000 liters of 10 M acid. We have 1000 Liters (L) of concentrated H_2SO_4 and 1000 L of pure water. A one-to-one mixture will produce the 10 M acid. We say about 2000 liters, because volume is not conserved. In fact, when we add the concentrated acid and the water together, once they have equilibrated, the volume will be less than the sums of the two volumes.

We can show this by assuming the opposite; what if the two volumes were conserved - what would the density be upon perfect mixing?

$$1000 \text{ L of water} \Rightarrow 1000 \text{ L} \times 1000 \text{ cm}^3/\text{L} \times 1 \, g \Big/ \text{cm}^3 = 1 \times 10^6 \text{ g}$$

$$1000 \text{ L of 18M con } H_2SO_4 \Rightarrow 1000 \text{ L} \times 1.83 \text{ kg/L} = 1830 \text{ kg} = 1.83 \times 10^6 \text{ g}$$

If volume were conserved, the density of this solution would be:

$$(\text{mass of } H_2O + \text{mass of Acid})/(\text{Vol } H_2O + \text{Vol Conc. Acid}) =$$

$$(2.83 \times 10^6 \text{g})/(2000 \text{ L} \times 1000 \text{ cm}^3/\text{L}) = 1.42 \, g \Big/ \text{cm}^3$$

This is almost an 8% difference below the actual density. Working in the other direction we find the volume must be about 1820 L to have a density of 1.54 g/cm^3.

What about the heat released? The reaction is as follows:

$$H_2SO_4 \text{ (conc)} + H_2O \longrightarrow H_3O^+ + HSO_4^- + 72 \text{ kj}$$

We are diluting 1000 liters of 18 M solution, and that corresponds to 18000 moles and each mole generates 72 kj. Therefore, this process will generate 1.296 * 10^6 kj of heat, or 3.09 * 10^{8}cal which is quite a bit of heat to dissipate. The specific heat of the final solution will be about 0.5 cal/g/^oC. The mass of the solution will be 2.84 * 10^6 g. From these numbers and the total heat of reaction, we can calculate what the temperature would be, if all the heat were contained in the reactor.

$$q = \text{sp.ht.} * \text{mass} * (T - 25) \Rightarrow T = 242 \text{ C}$$

That is a much higher temperature than the system would actually get to unless it were very heavily insulated and fully adiabatic. In fact heat will dissipate naturally from the reactor by convection to the air surrounding it. But how fast will it dissipate?

The rate of heat transfer away from the vessel is $\dot{q}$, that is a q for heat with a dot over it to designate it as a rate or time derivative. It is given in joules per second, or calories per second, or watts. This rate is equal to the overall heat transfer coefficient for the system times the surface area through which heat may be dissipated, and the temperature difference between the inside of the vessel and the outside of it.

$$\dot{q} = U \, A \, \Delta T \equiv \text{watts} \equiv \text{joule/sec} \equiv \text{cal/sec...}$$

We are not dealing with the science of heat transfer in this text, but we can estimate the overall heat transfer coefficient. It turns out that heat will move from the bulk of the solution to the wall of the vessel and in so doing it will encounter a resistance at a film that forms between the bulk solution and the wall, even with good agitation. This leads to an "inner heat transfer coefficient" h_i or the internal "film" coefficient. Then the wall of the reactor will offer its own resistance to heat transfer. This resistance will be given by the thickness of the vessel wall divided by the thermal conductivity of the material comprising that wall. At the outer surface of the vessel there will develop another resistance between the surface and the bulk atmosphere that we will call h_o or the external heat transfer coefficient. There is an analogy to this in electrical engineering, which is three resistances in series that ultimately can be described as one cumulative resistance. Based on that analogy, the overall heat transfer coefficient, U, will be comprised as follows:

$$\frac{1}{U} = \frac{1}{h_i} + \frac{L}{\lambda} + \frac{1}{h_o}$$

Let's assume the vessel is made of a 0.25 in glass (borosilicate). On the inside we have the water and acid solution that is being agitated by a mixer. On the outside, we will assume the vessel is not jacketed and is standing in air that is not flowing. These conditions are important to specify. We will be conservative in our estimates of heat transfer coefficient, and a lower rate of heat transfer in an abundance of caution. Consulting the handbooks, or on-line sources, we find that h_i can be about 1000 joules/$m^2/^oC$, which is good, but the thermal conductivity of the glass wall is only 1.14 joule/sec/m/^oC and the outer film heat transfer coefficient corresponds to stagnant air and so it may be as low as 5 joules/$m^2/^oC$. This makes the overall heat transfer coefficient relatively small:

$$\frac{1}{U} = \frac{1}{1000} + \frac{0.25 * \frac{1}{39.4}}{1.14} + \frac{1}{5} = 0.2065 \frac{m^2 * {}^oC}{\text{joule}} = 4.85$$

$$\Longrightarrow \text{U} \simeq 5 \frac{\text{joule}}{m^2 * {}^oC * \text{sec}}$$

We could increase this external heat transfer coefficient, if we assumed that air would be forced over the walls of the reactor. A water jacket would act as an effective heat exchanger and would take heat away even more efficiently, but we will use this worst case scenario because that is how we have pictured the system.

Now we can compute the rate of heat dissipation for a given temperature difference between the inside of the vessel and the surroundings, and from that rate, assuming it remains fixed, we can compute the approximate time required to dissipate all the heat that this reaction will generate. We will assume that the temperature differential will be fixed at 30^oC, and the surroundings will be at 25^oC; the interior of the reactor will be at about 55^oC. The area of the walls of the vessel area:

$$A_w = \text{C} \times \text{H} = \pi \times \text{d} \times \text{H} = 3.14 \times 5 \times 10 = 157 \text{ ft}^2$$

We will also assume that the floor of the reactor also transfers heat through its area in the same way as does the wall.

$$A_f = \frac{\pi d^2}{4} = \frac{\pi * 5^2}{4} = 19 \text{ ft}^2$$

Hence, the total area for heat transfer is:

$$A_{tot} = Aw + Af = 157 + 19 = 176 \text{ ft}^2$$

The rate of heat dissipation is $\dot{q}$ and that is equal to the product of the overall heat transfer coefficient, the total area through which heat may be dissipated and the temperature difference between the inside core of the reactor contents and the outside air in the bulk. We will allow the temperature rise to be 30^oC between the inside and the outside:

$$\dot{q} = \text{U A } \Delta\text{T} = 5 \text{ joule}\Big/m^2\Big/{}^oC/\text{ sec } * (176 \text{ ft}^2 * (39.4\text{in}/m * 1 \text{ ft}/12 \text{ in})^2 (\ 30^oC\ )$$

$$= 285000 \text{ joule/sec}$$

So the system can dissipate 285 000 joules per second. If the heat were generated all at once, the system would require about 76 minutes to come back to room temperature at a constant rate.

$$\text{t} = (1.3 \times 10^9) \,/\, (2.85 \times 10^5) \text{ sec} = 4561 \text{ sec} = 76 \text{ min}$$

We can add this heat dissipation rate back to the heat to be generated to compute the temperature that will be reached inside the reactor as a check on the work we have done assuming the heat generated is removed:

$$\Delta\text{q} = q_{\text{rxn}} - q_{\text{dissipated}} = \text{sp.ht. * mass * (T - 25) - } \dot{q}\ \Delta\text{t} = \text{sp.ht. * mass * (T - 25) - U A (T - 25) } \Delta\text{t}$$

NSolve[

$3.1 * 10^8 -$

$\left(5 * (1.98/8.314) * 176 * (39.4/12)^2 * (T - 25)\right) \frac{4561}{n}$,

T]/.$n \to 1$

$\{\{T \to 55.0839\}\}$

The check shows us that we get the same temperature back, as we should, when we assume that the heat dissipated will match the heat generated, i.e. Δq = 0.

If we allow the system to go to 55^oC, then we should plan on the mixing process to take at least 76 minutes. That means the flow rate into the system should be predicated upon adding the total acid volume over a time of 76-100 minutes, and that gets to the inlet flow rate.

We can assume that the reaction rate is pseudo-first order in the concentration of acid, since for much of the process water is in great excess over the acid. Therefore the rate equation for the acid is:

$$\partial_t C_{\text{H2SO4}}[t]\text{V}[t] = C_{\text{H2SO4},f}\ q_f \text{ - k1 } C_{\text{H2SO4}}[\text{t}]\ C_{\text{H2O}}[\text{t}]\text{V}[\text{t}]$$

$$= C_{\text{H2SO4},f}\,q_f - \text{k1}' C_{\text{H2SO4}}[\text{t}]\text{V}[\text{t}]$$

The volume will change with time and so too will the density. Interestingly, as we noted earlier, the density will increase and the volume will contract as we add the acid to water. Whichever way these change, the total mass balance is the governing equation. And we will need a relationship that ties the change in density to the concentration of the acid in the water:

$$\partial_t \rho[t]\mathrm{V}[\mathrm{t}] = \rho_{\mathrm{H2SO4},f}\ q_f$$

$$\rho[\mathrm{t}] = \rho_{\mathrm{H2O}} + \mathrm{b}\ C_{\mathrm{H2SO4}}[\mathrm{t}]$$

Initially, the density in the system is the density of water. At the end of the process, we expect to be close to 10 M. So we need to find the relationship in the range from zero to nearly 10 M sulfuric acid in order to fit the constant b. Going to Perry's Chemical Engineers' Handbook, we have the following data for increasing weight percentages that we can fit to a polynomial using NonlinearModelFit:

```
adat = {{0, 1.000}, {10, 1.064}, {20, 1.1365},
{30, 1.2150}, {40, 1.2991}, {50, 1.3911},
{60, 1.4940}, {70, 1.60559}, {80, 1.7221},
{90, 1.8091}, {98, 1.8310}}

ListPlot[%, AxesLabel → {"wt %", " ρ"}]
```

{{0, 1.}, {10, 1.064}, {20, 1.1365}, {30, 1.215}, {40, 1.2991}, {50, 1.3911}, {60, 1.494}, {70, 1.60559}, {80, 1.7221}, {90, 1.8091}, {98, 1.831}}

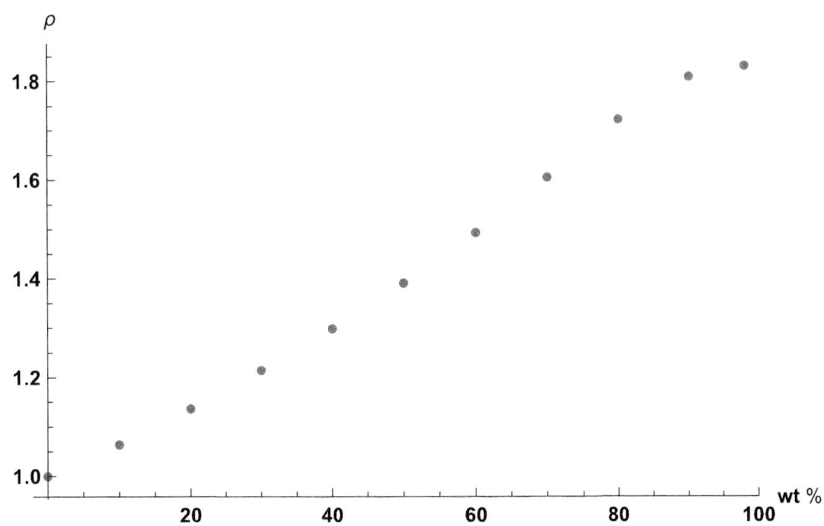

Our calculations should produce concentrations in mole per liter not weight percent. We need to convert the data to molarity and density. To do this, we will multiply the density by its corresponding weight fraction, and then convert to liters and moles, e.g.:

$$1.1365\ \frac{g}{\text{cm}^3} \times \frac{20\%}{100} \times \frac{1000\text{cm}^3}{1L}\ \frac{1\text{mol}}{98.1} = 2.317 \text{ mol/L}$$

in terms of the data points this is in general:

$$\frac{y \times x \times 10}{98.8} \text{ or } \frac{\text{adat}[[n, 2]] * \text{adat}[[n, 1]] * 10}{98.8}$$

which we can enable as follows in a Table function:

$$\textbf{bdat} = \textbf{Table}\left[\left\{\frac{\textbf{adat}[[n, 2]] * \textbf{adat}[[n, 1]] * \textbf{10}}{\textbf{98.8}}, \textbf{adat}[[n, 2]]\right\}, \{n, 1, 10\}\right]$$

ListPlot[%, AxesLabel → {" mol/L", " ρ"}]

{{0., 1.}, {1.07692, 1.064}, {2.30061, 1.1365}, {3.68927, 1.215}, {5.25951, 1.2991}, {7.03998, 1.3911}, {9.07287, 1.494}, {11.3756, 1.60559}, {13.9441, 1.7221}, {16.4797, 1.8091}}

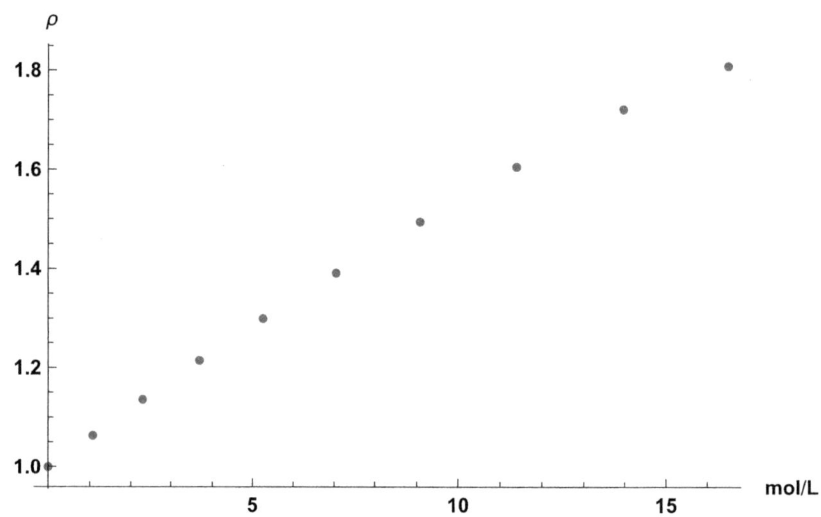

m=.

b=.

dens = LinearModelFit[bdat, *x*, *x*];

Normal[dens]

Show[ListPlot[bdat, AxesLabel → {" mol/L", " ρ"}],

Plot[dens[x], {x, 0, 20}],

Epilog → {Line[{{0, 1.52}, {10, 1.52}}],

Line[{{10, 0}, {10, 1.52}}]}]

This is the linear fit to the density data:

$$1.02376 + 0.0498143x$$

and this plot shows that fit:

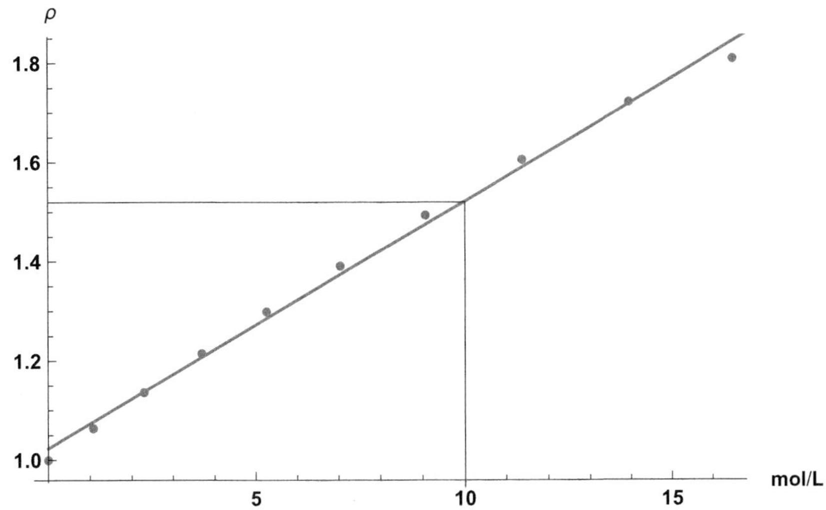

We can assemble the mass balance equations into a model and find the time required at which the density maximizes:

Clear["Global\`* "]

CHf = 18;

k1 = 10^5;

ρo = 1.0237;

ρHf = 1.83;

```
Vo = 1000;
qf = .22;
b = 0.0498;
tf = 10000;
slns = NDSolve[
{ρ[t] == ρo + bCH[t],
∂t(CHS[t]V[t]) == CHfq[t] − k1CHS[t]V[t],
∂t(ρ[t]V[t]) == ρHfq[t],
∂t(CH[t]V[t]) == +k1CHS[t]V[t],
q[t] == If[CH[t] ⩽ 10, qf, 0], ρ[0]==ρo,
CHS[0] == 0, CH[0] == 10^-7, V[0] == Vo, q[0] == qf},
{ρ[t], CHS[t], CH[t], V[t]}, {t, 0, tf}];
ρsol[t_]:= Evaluate[ρ[t]/.slns]
CH2SO4[t_]:=Evaluate[CHS[t]/.slns]
CH+[t_]:=Evaluate[CH[t]/.slns]
Vsol[t_]:=Evaluate[V[t]/.slns]
NSolve [∂t ρsol[t] == 0, t]
```

{{t → 5200.38}}

This provides the functional solutions and the time at which the feed goes to zero, i.e. when the density maximizes at 5200 seconds. Now we need to plot these results to see what they look like.

```
Plot[ρsol[t], {t, 0, tf}, PlotRange → All,
AxesLabel → {"t (sec)", " ρ[ t ] (g/cm^3)"},
Epilog → {{Dashed, Line[{{0, 1.525}, {tf, 1.525}}]},
{Dashed, Line[{{5200, 1.02}, {5200, 1.52}}]}}]
Plot[Vsol[t], {t, 0, tf}, PlotRange → All,
```

```
AxesLabel → {"t (sec)", " V[ t ] (L)"},
Epilog → {{Dashed, Line[{{5200, 1000}, {5200, 2020.}}]}}]
Plot[CH2SO4[t], {t, 0, tf}, PlotRange → All,
 AxesLabel → {"t (sec)", " CH2SO4[ t ] (mol/L)"}]
Plot[CH+[t], {t, 0, tf}, PlotRange → All,
 AxesLabel → {"t (sec)", " CH+[ t ] (L)"},
 Epilog → {{Dashed,Line[{{0, 10.05},{tf, 10.05}}],
 {Dashed, Line[{{5200, 0}, {5200, 10.}}]}}]
```

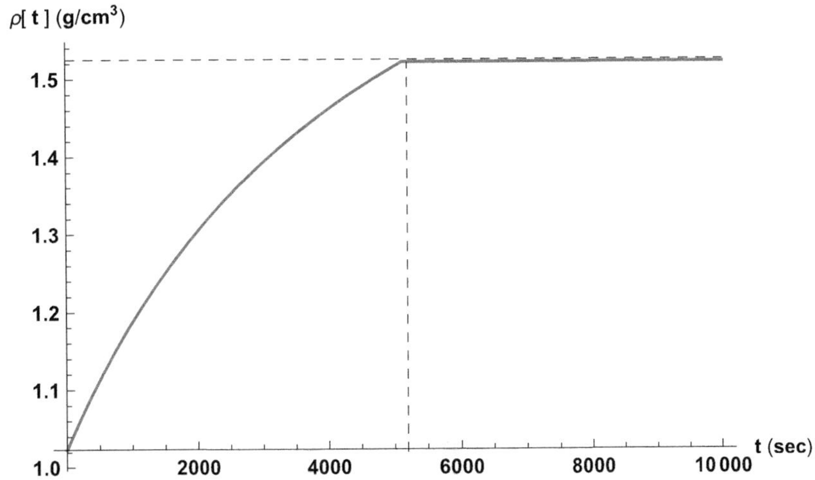

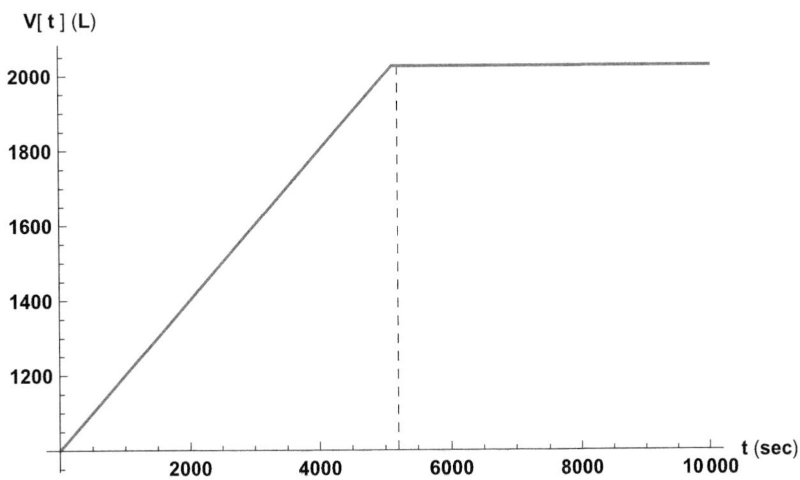

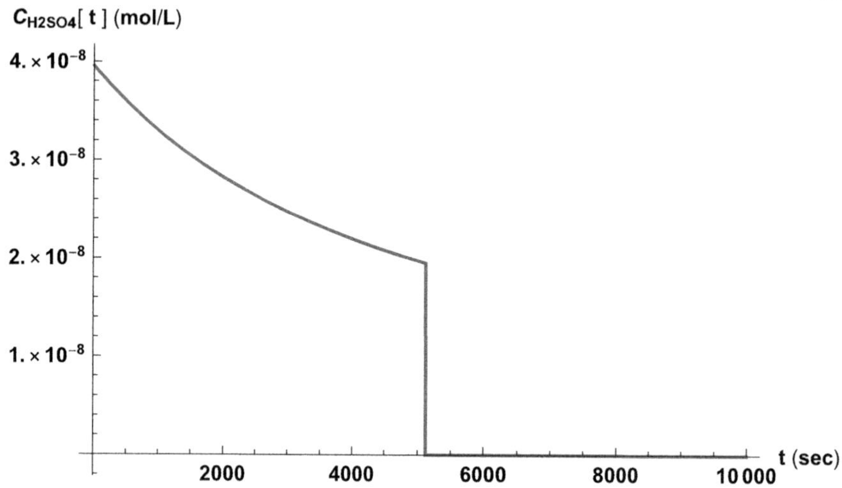

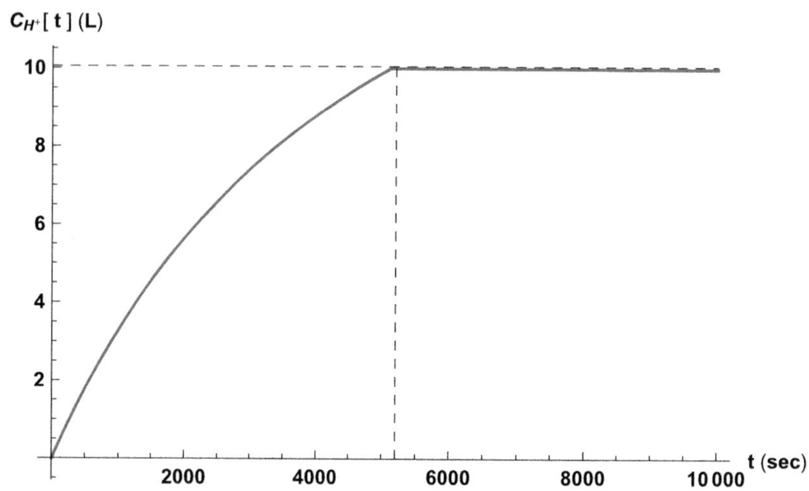

As we anticipated, the calculated density grows smoothly from 1.0 for water to 1.52 for the 10 M solution at 60 ^oC, the volume linearly increases from 1000 L to about 1500 L, which is less than the sum of the volumes of the ingredients. The concentration of H_2SO_4 is very small and drops slowly at first and then abruptly when the feed stops, this makes good sense. Lastly, the concentration of protons in solution rises to 10 M and stops at that target level. Even though we computed a final time well beyond the time needed to reach the product specification, we see that in this case the variables stop changing with longer time because we added the IF statement to the functional definition of the flow rate. The critical time is 5120 seconds. This is when the concentration of protons reaches 10 M. At this point the computed density is 1.52 and the volume is 2027 L which is based on a starting water volume of 1000 L and added volume of acid of 1126 L. The sum of these is 2126 L, if volumes were conserved, this would be the volume at the end of the addition of acid, but volume is not conserved, so the total volume is less than this value. This is contraction in volume is presumably caused by the addition of the protonic, positive point charges into solution, which exert electrocoulombic forces on the dipole of water, thus drawing the waters of hydration surrounding the protons closer together than they are in pure water where the concentration of protons is just 10^{-7}M.

It is useful to repeat the analysis and to add in the heat balance, so that we can see what happens to the temperature of the system during the hydrolysis process. The rate of accumulation of heat energy will be the sum of all the energy fluxes and flows entering and leaving the system because energy is treated like mass since it too is conserved. We need to take into account that the volume will change, as will the density and specific heat of the solution. The fifth variable will be the temperature as a function of time, T[t], and the heat balance equation expressing its rate of change will be:

$$\partial_t[\mathrm{sph}[t](T[t]-\mathrm{To})\rho[t]V[t]\mathrm{MW}] = \rho\ \mathrm{Hf}\ q[t]\ \mathrm{sphf\ Tf}\ +\ \mathrm{k1\ CHS}[t]\ V[t]\Delta\mathrm{Hr}$$
$$-UA(T[t]-\mathrm{To})$$

Once again the rate of heat accumulation is the derivative, and this is equal to the rate of heat entering the reactor with the feed stream, the rate of heat generated by reaction and the flow of heat away from the system through the walls and to the surroundings. We will let the specific heat, sph[t], vary linearly between the value for water and that for sulfuric acid. The specific heat of the feed, sphf, will be a constant, as will the feed stream density, ρHf, and its temperature, Tf. The overall set of equations is as follows:

$$\mathrm{sph}[t] = \mathrm{spho} - .0331\ \mathrm{CHS}[t]$$
$$\rho[t] = \rho\mathrm{o} + b\ \mathrm{CH}[t]$$
$$\partial_t(\mathrm{CHS}[t]\ V[t]) = \mathrm{CHf}\ q[t] - \mathrm{k1\ CHS}[t]\ V[t]$$
$$\partial_t(\rho[t]\ V[t]) = \rho\ \mathrm{Hf}\ q[t]$$
$$\partial_t(\mathrm{CH}[t]\ V[t]) = +\mathrm{k1\ CHS}[t]\ V[t]$$
$$\partial_t(\mathrm{sph}[t](T[t]-\mathrm{To})\rho[t]V[t]\mathrm{MW}) = \rho\ \mathrm{Hf}\ q[t]\ \mathrm{sphf\ Tfd} + \mathrm{k1\ CHS}[t]\ V[t]\ \Delta\mathrm{Hr}$$
$$-UA(T[t]-\mathrm{To})$$
$$q[t] = \mathrm{If}\ [\mathrm{CH}[t] \leqslant 10, \mathrm{qf}, 0]$$

The calculation of the heat balance along with the mass balances and the constitutive relationships for the specific heat and the density is done as follows:

```
Clear["Global`*"]

CHf = 18;

k1 = 10^5;

ρo = 1.0;

ρHf = 1.52;

Vo = 1000;

qf = .22;

b = 0.052;

CHf = 18;

tf = 8000;

sphf = .395 * (8.314/1.98);

spho = .991 * (8.314/1.98);

Tfd = 25;

To = 25;

ΔHr = 72000;

U = 5;

A = 176(39.4/12)^2;

MW = 98;

slns = NDSolve[

{
sph[t] == spho - .0331CHS[t],

ρ[t] == ρo + b CH[t],

∂t(CHS[t] V[t]) == CHf q[t] - k1 CHS[t] V[t],

∂t(ρ[t]V[t]) == ρHf q[t],

∂t(CH[t]V[t]) == +k1 CHS[t] V[t],

∂t(sph[t](T[t] - To)ρ[t] V[t] MW) ==
```

```
ρ Hf q[t] sphf Tfd + k1 CHS[t] V[t] ΔHr −U A(T[t] − To),
q[t] == If[CH[t] ⩽ 10, qf, 0],
ρ[0]==ρo,
CHS[0] == 0,
CH[0] == 10^-7,
V[0] == Vo,
q[0] == qf,
T[0] == 25, sph[0]==spho},
{ρ[t], CHS[t], CH[t], V[t], T[t], sph[t]},
{t, 0, tf}
];
ρsol[t_]:= Evaluate[ρ[t]/.slns]
CH2SO4[t_]:=Evaluate[CHS[t]/.slns]
CH+[t_]:=Evaluate[CH[t]/.slns]
Vsol[t_]:=Evaluate[V[t]/.slns]
Tv[t_]:=Evaluate[T[t]/.slns]

Table[{t, Tv[t]}, {t, 0, tf, 1000}]

Plot[Tv[t], {t, 0, tf}, PlotRange → {{0, tf}, {0, 60}},
 AxesLabel → {"t (sec)", " T[ t ] (°C)"}]
```

{{0, {24.9999}}, {1000, {54.6304}}, {2000, {54.6304}}, {3000, {54.6304}}, {4000, {26.9745}}, {5000, {25.0001}}, {6000, {25.}}, {7000, {25.}}, {8000, {25.}}}

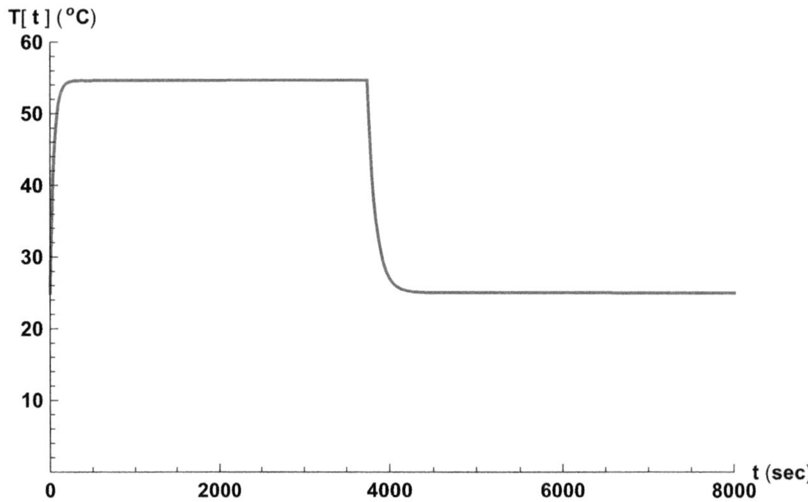

Here we see that the simulated temperature rises sharply to about 55^oC with the addition of the concentrated acid. Then it remains at 55^oC while the acid is added continuously. For this calculation, we have let the heat of reaction be a constant; we could adjust it downward with higher concentrations of acid in solution. When the acid flow stops, the temperature drops back to that of the surroundings. The other functions look as they did in the earlier calculation and they are not shown.

The production of more dilute sulfuric acid is a good example of how the fed-batch reactor works.

```
Clear["Global`*"]

Tsulfuric[qf_, tf_]:=Module[

{CHf = 18,

k1 = 10^5,

ρo = 1.0,

ρHf = 1.52,

Vo = 1000,

b = 0.052,

sphf = .395 * (8.314/1.98),
```

```
spho = .991 * (8.314/1.98),
Tfd = 25,
To = 25,
ΔHr = 72000,
U = 5,
A = 176(39.4/12)^2,
MW = 98},
slns = NDSolve[
{
sph[t] == spho - .0331CHS[t],
ρ[t] == ρo + bCH[t],
∂t(CHS[t]V[t]) == CHfq[t] - k1CHS[t]V[t],
∂t(ρ[t]V[t]) == ρHfq[t],
∂t(CH[t]V[t]) == +k1CHS[t]V[t],
∂t(sph[t](T[t] - To)ρ[t]V[t]MW) ==
ρHfq[t]sphfTfd + k1CHS[t]V[t]ΔHr-
UA(T[t] - To),
q[t] == If[CH[t] ≤ 10, qf, 0],
ρ[0]==ρo,
CHS[0] == 0,
CH[0] == 10^-7,
V[0] == Vo,
q[0] == qf,
T[0] == 25, sph[0]==spho},
{ρ[t], CHS[t], CH[t], V[t], T[t], sph[t]},
{t, 0, tf}
];
```

```
ρsol[t_]:= Evaluate[ρ[t]/.slns];
CH2SO4[t_]:=Evaluate[CHS[t]/.slns];
CH+[t_]:=Evaluate[CH[t]/.slns];
Vsol[t_]:=Evaluate[V[t]/.slns];
Tv[t_]:=Evaluate[T[t]/.slns];

Table[{t, Tv[t]}, {t, 0, tf, 1000}];

Plot[Tv[t], {t, 0, tf}, PlotRange → All,
AxesLabel → {"t (sec)", " T[ t ] (°C)"},
PlotLabel → {qf" = qf"}]
]

Tsulfuric[.5, 5000]
```

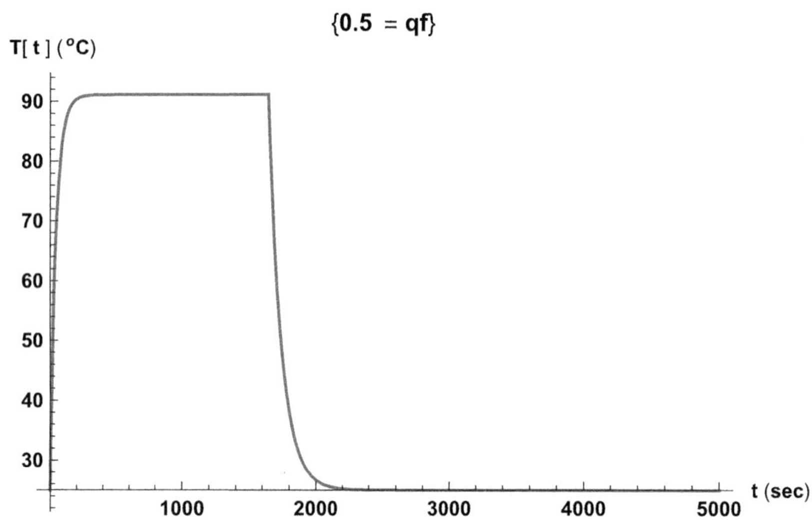

And we can make this into a two parameter continuous simulation with Manipulate:

```
Manipulate[Tsulfuric[qf, tf], {qf, 0.01, .5, .01},
{tf, 1000, 20000, 10}]
```

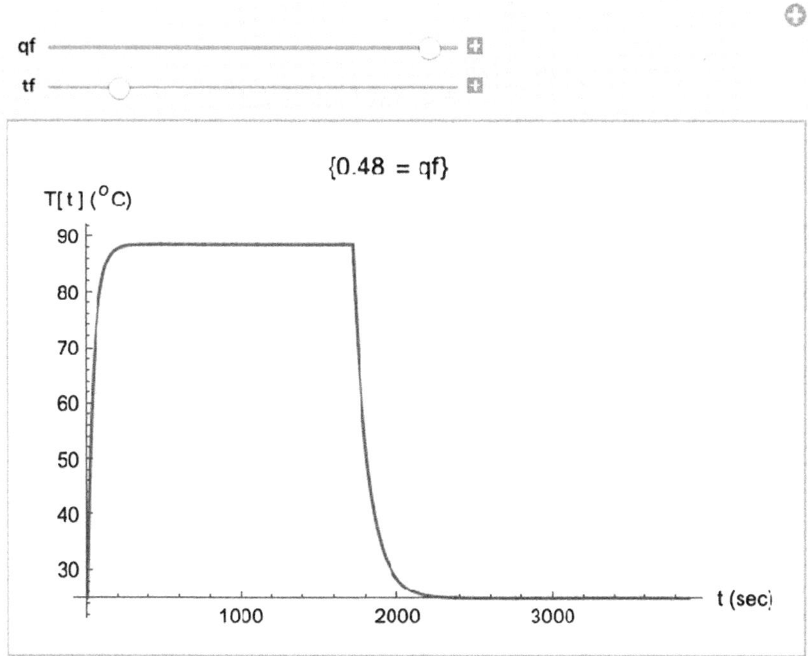

In other instances, one reactant may be added continuously and the product accumulates at a rate that is proportional or equal to the rate of admission of the reactant. For example bioreactors, are fed-batch systems, in that oxygen may be fed continuously to the microbial colony, and the substrate may also be added continuously or it may be charged at the beginning of the batch. Depending on what is to be achieved with any given reaction, the volume of solution within the fed-batch reactor may or may not change significantly.

Let's look at the case of more complex kinetics.

Second order kinetics and negligible volume change

In some cases, the volume change may be negligible from the start to the finish of the batch. For example one reagent may be added in a very concentrated form to a solution of the second reactant already in the reactor. This can lead to a situation in which the volume of added reactant is quite small compared to that of the initial volume of solution, so that considering the overall volume to be constant leads only to a small error. The equations for this are:

$$\frac{dC_A[t]}{dt} = \frac{C_{\mathrm{Af}}q_A}{V} - k_{\mathrm{AB}}C_A[t]C_B[t]$$

$$\frac{dC_B[t]}{dt} = -k_{AB}C_A[t]C_B[t]$$

$$\frac{dC_D[t]}{dt} = +k_{AB}C_A[t]C_B[t]$$

We can seek a complete or analytical solution for this system of equations:

```
Clear["Global`*"]
(* CVFB stands for constant volume fed batch *)
CVFB = DSolve[
{CA'[t] ==_V − CA[t]CB[t],
== − CA[t]CB[t] ,
CD'[t] == CA[t]CB[t],
  CA[0] == CAo,
  CB[0] == CBo,
  CD[0] == CDo},
{CA[t], CB[t], CD[t]},
    t
  ]
```

Mathematica is able to find analytical solutions even though these equations are non-linear.

$$\left\{\left\{\mathrm{CA}[t] \to \left(e^{-\frac{\mathrm{kab}(\mathrm{CAfqA}t-(-\mathrm{CAo}+\mathrm{CBo})V)^2}{2\mathrm{CAfqA}V}}\right.\right.\right.$$

$$\left(\mathrm{CAf}^{3/2}e^{\frac{\mathrm{kab}(\mathrm{CAfqA}t-(-\mathrm{CAo}+\mathrm{CBo})V)^2}{2\mathrm{CAfqA}V}}\sqrt{2\pi}\,\mathrm{qA}^{3/2}t+\right.$$

$$\sqrt{\mathrm{CAf}}\mathrm{CBo}e^{\frac{(-\mathrm{CAo}+\mathrm{CBo})^2\mathrm{kab}V}{2\mathrm{CAfqA}}}\sqrt{2\pi}\sqrt{\mathrm{qA}}V\ +$$

$$\sqrt{\mathrm{CAf}}\mathrm{CAo}e^{\frac{\mathrm{kab}(\mathrm{CAfqA}t-(-\mathrm{CAo}+\mathrm{CBo})V)^2}{2\mathrm{CAfqA}V}}\sqrt{2\pi}\sqrt{\mathrm{qA}}V\ -$$

$$\sqrt{\mathrm{CAf}}\,\mathrm{CBo}\,e^{\frac{\mathrm{kab}(\mathrm{CAfqA}t-(-\mathrm{CAo}+\mathrm{CBo})V)^2}{2\mathrm{CAfqA}V}}\sqrt{2\pi}\sqrt{\mathrm{qA}}V\ +$$

$$\mathrm{CAfCBo}\,e^{\frac{(-\mathrm{CAo}+\mathrm{CBo})^2\mathrm{kab}V}{2\mathrm{CAfqA}}+\frac{\mathrm{kab}(\mathrm{CAfqA}t-(-\mathrm{CAo}+\mathrm{CBo})V)^2}{2\mathrm{CAfqA}V}}$$

$$\sqrt{\mathrm{kab}}\pi\,\mathrm{qA}t\sqrt{V}$$

$$\mathrm{Erf}\left[\frac{(-\mathrm{CAo}+\mathrm{CBo})\sqrt{\mathrm{kab}}\sqrt{V}}{\sqrt{2}\sqrt{\mathrm{CAf}}\sqrt{\mathrm{qA}}}\right]\ +$$

$$\mathrm{CAoCBo}\,e^{\frac{(-\mathrm{CAo}+\mathrm{CBo})^2\mathrm{kab}V}{2\mathrm{CAfqA}}+\frac{\mathrm{kab}(\mathrm{CAfqA}t-(-\mathrm{CAo}+\mathrm{CBo})V)^2}{2\mathrm{CAfqA}V}}$$

$$\sqrt{\mathrm{kab}}\pi\,V^{3/2}\mathrm{Erf}\left[\frac{(-\mathrm{CAo}+\mathrm{CBo})\sqrt{\mathrm{kab}}\sqrt{V}}{\sqrt{2}\sqrt{\mathrm{CAf}}\sqrt{\mathrm{qA}}}\right]\ -$$

$$\mathrm{CBo}^2 e^{\frac{(-\mathrm{CAo}+\mathrm{CBo})^2\mathrm{kab}V}{2\mathrm{CAfqA}}+\frac{\mathrm{kab}(\mathrm{CAfqA}t-(-\mathrm{CAo}+\mathrm{CBo})V)^2}{2\mathrm{CAfqA}V}}$$

$$\sqrt{\mathrm{kab}}\pi\,V^{3/2}\mathrm{Erf}\left[\frac{(-\mathrm{CAo}+\mathrm{CBo})\sqrt{\mathrm{kab}}\sqrt{V}}{\sqrt{2}\sqrt{\mathrm{CAf}}\sqrt{\mathrm{qA}}}\right]\ +$$

$$\mathrm{CAfCBo}\,e^{\frac{(-\mathrm{CAo}+\mathrm{CBo})^2\mathrm{kab}V}{2\mathrm{CAfqA}}+\frac{\mathrm{kab}(\mathrm{CAfqA}t-(-\mathrm{CAo}+\mathrm{CBo})V)^2}{2\mathrm{CAfqA}V}}$$

$$\sqrt{\mathrm{kab}}\pi\,\mathrm{qA}t\sqrt{V}$$

$$\mathrm{Erf}\left[\frac{\sqrt{\mathrm{kab}}(\mathrm{CAfqA}t-(-\mathrm{CAo}+\mathrm{CBo})V)}{\sqrt{2}\sqrt{\mathrm{CAf}}\sqrt{\mathrm{qA}}\sqrt{V}}\right]\ +$$

$$\mathrm{CAoCBo}\,e^{\frac{(-\mathrm{CAo}+\mathrm{CBo})^2\mathrm{kab}V}{2\mathrm{CAfqA}}+\frac{\mathrm{kab}(\mathrm{CAfqA}t-(-\mathrm{CAo}+\mathrm{CBo})V)^2}{2\mathrm{CAfqA}V}}$$

$$\sqrt{\mathrm{kab}}\pi\,V^{3/2}$$

$$\mathrm{Erf}\left[\frac{\sqrt{\mathrm{kab}}(\mathrm{CAfqA}t-(-\mathrm{CAo}+\mathrm{CBo})V)}{\sqrt{2}\sqrt{\mathrm{CAf}}\sqrt{\mathrm{qA}}\sqrt{V}}\right]\ -$$

$$\mathrm{CBo}^2 e^{\frac{(-\mathrm{CAo}+\mathrm{CBo})^2\mathrm{kab}V}{2\mathrm{CAfqA}}+\frac{\mathrm{kab}(\mathrm{CAfqA}t-(-\mathrm{CAo}+\mathrm{CBo})V)^2}{2\mathrm{CAfqA}V}}$$

$$\sqrt{\mathrm{kab}}\pi\,V^{3/2}\mathrm{Erf}[$$

$$\left.\left.\left.\frac{\sqrt{\mathrm{kab}}(\mathrm{CAfqA}t-(-\mathrm{CAo}+\mathrm{CBo})V)}{\sqrt{2}\sqrt{\mathrm{CAf}}\sqrt{\mathrm{qA}}\sqrt{V}}\right]\right)\right)\Big/$$

$$\left(V\left(\sqrt{\text{CAf}}\sqrt{2\pi}\sqrt{\text{qA}}+\text{CBo}e^{\frac{(-\text{CAo}+\text{CBo})^2\text{kab}V}{2\text{CAfqA}}}\right.\right.$$

$$\sqrt{\text{kab}}\pi\sqrt{V}\text{Erf}\left[\frac{(-\text{CAo}+\text{CBo})\sqrt{\text{kab}}\sqrt{V}}{\sqrt{2}\sqrt{\text{CAf}}\sqrt{\text{qA}}}\right]+$$

$$\text{CBo}e^{\frac{(-\text{CAo}+\text{CBo})^2\text{kab}V}{2\text{CAfqA}}}\sqrt{\text{kab}}\pi\sqrt{V}\text{Erf}[$$

$$\left.\left.\frac{\sqrt{\text{kab}}(\text{CAfqA}t-(-\text{CAo}+\text{CBo})V)}{\sqrt{2}\sqrt{\text{CAf}}\sqrt{\text{qA}}\sqrt{V}}]\right)\right),$$

$$\text{CB}[t]\quad\rightarrow$$

$$\left(\sqrt{\text{CAf}}\text{CBo}e^{\frac{(-\text{CAo}+\text{CBo})^2\text{kab}V}{2\text{CAfqA}}-\frac{\text{kab}(\text{CAfqA}t-(-\text{CAo}+\text{CBo})V)^2}{2\text{CAfqA}V}}\sqrt{2\pi}\right.$$

$$\left.\sqrt{\text{qA}}\right)/\left(\sqrt{\text{CAf}}\sqrt{2\pi}\sqrt{\text{qA}}+\text{CBo}e^{\frac{(-\text{CAo}+\text{CBo})^2\text{kab}V}{2\text{CAfqA}}}\right.$$

$$\sqrt{\text{kab}}\pi\sqrt{V}\text{Erf}\left[\frac{(-\text{CAo}+\text{CBo})\sqrt{\text{kab}}\sqrt{V}}{\sqrt{2}\sqrt{\text{CAf}}\sqrt{\text{qA}}}\right]+$$

$$\text{CBo}e^{\frac{(-\text{CAo}+\text{CBo})^2\text{kab}V}{2\text{CAfqA}}}\sqrt{\text{kab}}\pi\sqrt{V}$$

$$\left.\text{Erf}\left[\frac{\sqrt{\text{kab}}(\text{CAfqA}t-(-\text{CAo}+\text{CBo})V)}{\sqrt{2}\sqrt{\text{CAf}}\sqrt{\text{qA}}\sqrt{V}}\right]\right),$$

$$\text{CD}[t]\quad\rightarrow$$

$$\left(e^{-\frac{\text{CAfkabqA}t^2}{2V}}\left(-\sqrt{\text{CAf}}\text{CBo}e^{(-\text{CAo}+\text{CBo})\text{kab}t}\sqrt{2\pi}\right.\right.$$

$$\sqrt{\text{qA}}+\sqrt{\text{CAf}}\text{CBo}e^{\frac{\text{CAfkabqA}t^2}{2V}}\sqrt{2\pi}\sqrt{\text{qA}}+$$

$$\sqrt{\text{CAf}}\text{CDo}e^{\frac{\text{CAfkabqA}t^2}{2V}}\sqrt{2\pi}\sqrt{\text{qA}}+$$

$$\mathrm{CBo}^2 e^{\frac{\mathrm{CAfkabqA}t^2}{2V}+\frac{(-\mathrm{CAo}+\mathrm{CBo})^2\mathrm{kab}V}{2\mathrm{CAfqA}}}\sqrt{\mathrm{kab}\pi}$$

$$\sqrt{V}\,\mathrm{Erf}\left[\frac{(-\mathrm{CAo}+\mathrm{CBo})\sqrt{\mathrm{kab}}\sqrt{V}}{\sqrt{2}\sqrt{\mathrm{CAf}}\sqrt{\mathrm{qA}}}\right]\quad+$$

$$\mathrm{CBoCDo}e^{\frac{\mathrm{CAfkabqA}t^2}{2V}+\frac{(-\mathrm{CAo}+\mathrm{CBo})^2\mathrm{kab}V}{2\mathrm{CAfqA}}}\sqrt{\mathrm{kab}}$$

$$\pi\sqrt{V}\,\mathrm{Erf}\left[\frac{(-\mathrm{CAo}+\mathrm{CBo})\sqrt{\mathrm{kab}}\sqrt{V}}{\sqrt{2}\sqrt{\mathrm{CAf}}\sqrt{\mathrm{qA}}}\right]\quad+$$

$$\mathrm{CBo}^2 e^{\frac{\mathrm{CAfkabqA}t^2}{2V}+\frac{(-\mathrm{CAo}+\mathrm{CBo})^2\mathrm{kab}V}{2\mathrm{CAfqA}}}\sqrt{\mathrm{kab}\pi}\sqrt{V}\,\mathrm{Erf}[$$

$$\left.\frac{\sqrt{\mathrm{kab}}(\mathrm{CAfqA}t-(-\mathrm{CAo}+\mathrm{CBo})V)}{\sqrt{2}\sqrt{\mathrm{CAf}}\sqrt{\mathrm{qA}}\sqrt{V}}\right]\quad+\quad\mathrm{CBo}$$

$$\mathrm{CDo}e^{\frac{\mathrm{CAfkabqA}t^2}{2V}+\frac{(-\mathrm{CAo}+\mathrm{CBo})^2\mathrm{kab}V}{2\mathrm{CAfqA}}}\sqrt{\mathrm{kab}\pi}\sqrt{V}\,\mathrm{Erf}[$$

$$\left.\left.\left.\frac{\sqrt{\mathrm{kab}}(\mathrm{CAfqA}t-(-\mathrm{CAo}+\mathrm{CBo})V)}{\sqrt{2}\sqrt{\mathrm{CAf}}\sqrt{\mathrm{qA}}\sqrt{V}}\right]\right)\right)\Big/$$

$$\left(\sqrt{\mathrm{CAf}}\sqrt{2\pi}\sqrt{\mathrm{qA}}+\mathrm{CBo}e^{\frac{(-\mathrm{CAo}+\mathrm{CBo})^2\mathrm{kab}V}{2\mathrm{CAfqA}}}\sqrt{\mathrm{kab}}\right.$$

$$\pi\sqrt{V}\,\mathrm{Erf}\left[\frac{(-\mathrm{CAo}+\mathrm{CBo})\sqrt{\mathrm{kab}}\sqrt{V}}{\sqrt{2}\sqrt{\mathrm{CAf}}\sqrt{\mathrm{qA}}}\right]\quad+$$

$$\mathrm{CBo}e^{\frac{(-\mathrm{CAo}+\mathrm{CBo})^2\mathrm{kab}V}{2\mathrm{CAfqA}}}\sqrt{\mathrm{kab}\pi}\sqrt{V}$$

$$\left.\left.\left.\mathrm{Erf}\left[\frac{\sqrt{\mathrm{kab}}(\mathrm{CAfqA}t-(-\mathrm{CAo}+\mathrm{CBo})V)}{\sqrt{2}\sqrt{\mathrm{CAf}}\sqrt{\mathrm{qA}}\sqrt{V}}\right]\right)\right\}\right\}$$

These functions for the concentration look somewhat formidable in their "raw" form.

We notice that in the solution for concentration of A, there are exponential decay functions and ErrorFunctions both with constant and with time-dependent arguments. Although still complex, the solution for concentration of B is less complex than those for A or D. The solution for concentration of D includes exponentials in time squared. We can simplify these solutions and then assign them to functions:

FullSimplify[ca[t]]

$$\left\{\left(e^{-\frac{\text{kab}(\text{CAf qA }t+(\text{CAo}-\text{CBo})V)^2}{2\text{CAf qA }V}}\left(\sqrt{\text{CAf}}\sqrt{2\pi}\sqrt{\text{qA}}\right.\right.\right.$$

$$\left(\text{CBo}e^{\frac{(\text{CAo}-\text{CBo})^2\text{kab }V}{2\text{CAf qA}}}V+e^{\frac{\text{kab}(\text{CAf qA }t+(\text{CAo}-\text{CBo})V)^2}{2\text{CAf qA }V}}\right.$$

$$(\text{CAf qA }t+(\text{CAo}-\text{CBo})V))+$$

$$\text{CBo}e^{(\text{CAo}-\text{CBo})\text{kab }t+\frac{\text{CAf kab qA }t^2}{2V}+\frac{(\text{CAo}-\text{CBo})^2\text{kab}V}{\text{CAf qA}}}$$

$$\sqrt{\text{kab}}\pi\sqrt{V}(\text{CAfqA }t+(\text{CAo}-\text{CBo})V)$$

$$\left(\text{Erf}\left[\frac{(-\text{CAo}+\text{CBo})\sqrt{\text{kab}}\sqrt{V}}{\sqrt{2}\sqrt{\text{CAf}}\sqrt{\text{qA}}}\right]+\text{Erf}\right[$$

$$\left.\left.\left.\left.\frac{\sqrt{\text{kab}}(\text{CAf qA }t+(\text{CAo}-\text{CBo})V)}{\sqrt{2}\sqrt{\text{CAf}}\sqrt{\text{qA}}\sqrt{V}}\right]\right)\right)\right)\Big/$$

$$\left(\sqrt{\text{CAf}}\sqrt{2\pi}\sqrt{\text{qA}}V+\text{CBo}e^{\frac{(\text{CAo}-\text{CBo})^2\text{kab}V}{2\text{CAf qA}}}\sqrt{\text{kab}}\right.$$

$$\pi V^{3/2}\left(\text{Erf}\left[\frac{(-\text{CAo}+\text{CBo})\sqrt{\text{kab}}\sqrt{V}}{\sqrt{2}\sqrt{\text{CAf}}\sqrt{\text{qA}}}\right]+\right.$$

$$\left.\left.\left.\text{Erf}\left[\frac{\sqrt{\text{kab}}(\text{CAfqA}t+(\text{CAo}-\text{CBo})V)}{\sqrt{2}\sqrt{\text{CAf}}\sqrt{\text{qA}}\sqrt{V}}\right]\right)\right)\right\}$$

FullSimplify[cb[t]]

$$\left\{\left(\left(\sqrt{\text{CAf}}\text{CBo}e^{-\frac{\text{kab}t(\text{CAfqA}t+2(\text{CAo}-\text{CBo})V)}{2V}}\sqrt{2\pi}\sqrt{\text{qA}}\right)\Big/\right.\right.$$

$$\left(\sqrt{\text{CAf}}\sqrt{2\pi}\sqrt{\text{qA}}+\text{CBo}e^{\frac{(\text{CAo}-\text{CBo})^2\text{kab}V}{2\text{CAfqA}}}\sqrt{\text{kab}}\right.$$

$$\pi\sqrt{V}\left(\text{Erf}\left[\frac{(-\text{CAo}+\text{CBo})\sqrt{\text{kab}}\sqrt{V}}{\sqrt{2}\sqrt{\text{CAf}}\sqrt{\text{qA}}}\right]+\right.$$

$$\left.\left.\left.\text{Erf}\left[\frac{\sqrt{\text{kab}}(\text{CAfqA}t+(\text{CAo}-\text{CBo})V)}{\sqrt{2}\sqrt{\text{CAf}}\sqrt{\text{qA}}\sqrt{V}}\right]\right)\right)\right\}$$

FullSimplify[cd[t]]

$$\{\text{CBo}+\text{CDo}-$$

$$\left(\sqrt{\text{CAf}}\text{CBo}e^{-\frac{\text{kab}t(\text{CAfqA}t+2(\text{CAo}-\text{CBo})V)}{2V}}\sqrt{2\pi}\sqrt{\text{qA}}\right)\Big/$$

$$\left(\sqrt{\text{CAf}}\sqrt{2\pi}\sqrt{\text{qA}}+\text{CBo}e^{\frac{(\text{CAo}-\text{CBo})^2\text{kab}V}{2\text{CAfqA}}}\sqrt{\text{kab}}\right.$$

$$\pi\sqrt{V}\left(\text{Erf}\left[\frac{(-\text{CAo}+\text{CBo})\sqrt{\text{kab}}\sqrt{V}}{\sqrt{2}\sqrt{\text{CAf}}\sqrt{\text{qA}}}\right]+\right.$$

$$\left.\left.\left.\text{Erf}\left[\frac{\sqrt{\text{kab}}(\text{CAfqA}t+\text{CAo}V-\text{CBo}V)}{\sqrt{2}\sqrt{\text{CAf}}\sqrt{\text{qA}}\sqrt{V}}\right]\right)\right)\right\}$$

The simplification has made these equations somewhat less unwieldy, but not much. Thankfully, we have *Mathematica* to manipulate and compute these functions. Notice that in the function for the concentration of D that the arguments of the exponentials are first order in time. Now, we can assign them to functions:

ca[t_]:=

$$\left(e^{-\frac{\textbf{kab}(\textbf{CAf qA }t+(\textbf{CAo}-\textbf{CBo})V)^2}{\textbf{2CAf qA }V}}\right.$$

$$\left(\sqrt{\textbf{CAf}}\sqrt{2\pi}\sqrt{\textbf{qA}}\right.$$

$$\left(\mathrm{CBo}e^{\frac{(\mathrm{CAo}-\mathrm{CBo})^2\mathrm{kab}\,V}{2\mathrm{CAfqA}}}V+\right.$$

$$\left.e^{\frac{\mathrm{kab}(\mathrm{CAf\,qA}\,t+(\mathrm{CAo}-\mathrm{CBo})V)^2}{2\mathrm{CAf\,qA}\,V}}(\mathrm{CAf\,qA}\,t+(\mathrm{CAo}-\mathrm{CBo})V)\right)+$$

$$\mathrm{CBo}\,e^{(\mathrm{CAo}-\mathrm{CBo})\,\mathrm{kab}\,t+\frac{\mathrm{CAf\,kab\,qA}\,t^2}{2V}+\frac{(\mathrm{CAo}-\mathrm{CBo})^2\mathrm{kab}\,V}{\mathrm{CAf\,qA}}}\sqrt{\mathrm{kab}}$$

$$\pi\sqrt{V}(\mathrm{CAfqA}t+(\mathrm{CAo}-\mathrm{CBo})V)$$

$$\left(\mathrm{Erf}\left[\frac{(-\mathrm{CAo}+\mathrm{CBo})\sqrt{\mathrm{kab}}\sqrt{V}}{\sqrt{2}\sqrt{\mathrm{CAf}}\sqrt{\mathrm{qA}}}\right]+\right.$$

$$\left.\mathrm{Erf}\left[\frac{\sqrt{\mathrm{kab}}(\mathrm{CAf\,qA}\,t+(\mathrm{CAo}-\mathrm{CBo})V)}{\sqrt{2}\sqrt{\mathrm{CAf}}\sqrt{\mathrm{qA}}\sqrt{V}}\right]\right)\Bigg)\Bigg)\Bigg/$$

$$\left(\sqrt{\mathrm{CAf}}\sqrt{2\pi}\sqrt{\mathrm{qA}}V+\right.$$

$$\mathrm{CBo}e^{\frac{(\mathrm{CAo}-\mathrm{CBo})^2\mathrm{kab}V}{2\mathrm{CAf\,qA}}}\sqrt{\mathrm{kab}}\pi V^{3/2}$$

$$\left(\mathrm{Erf}\left[\frac{(-\mathrm{CAo}+\mathrm{CBo})\sqrt{\mathrm{kab}}\sqrt{V}}{\sqrt{2}\sqrt{\mathrm{CAf}}\sqrt{\mathrm{qA}}}\right]+\right.$$

$$\left.\left.\mathrm{Erf}\left[\frac{\sqrt{\mathrm{kab}}(\mathrm{CAfqA}t+(\mathrm{CAo}-\mathrm{CBo})V)}{\sqrt{2}\sqrt{\mathrm{CAf}}\sqrt{\mathrm{qA}}\sqrt{V}}\right]\right)\right)$$

$$\mathrm{cb[t\_]:=}\left(\sqrt{\mathrm{CAf}}\mathrm{CBo}e^{-\frac{\mathrm{kab}t(\mathrm{CAf\,qA}\,t+2(\mathrm{CAo}-\mathrm{CBo})V)}{2V}}\sqrt{2\pi}\sqrt{\mathrm{qA}}\right)\Bigg/$$

$$\left(\sqrt{\mathrm{CAf}}\sqrt{2\pi}\sqrt{\mathrm{qA}}+\mathrm{CBo}e^{\frac{(\mathrm{CAo}-\mathrm{CBo})^2\mathrm{kab}V}{2\mathrm{CAfqA}}}\sqrt{\mathrm{kab}}\pi\sqrt{V}\right.$$

$$\left(\mathrm{Erf}\left[\frac{(-\mathrm{CAo}+\mathrm{CBo})\sqrt{\mathrm{kab}}\sqrt{V}}{\sqrt{2}\sqrt{\mathrm{CAf}}\sqrt{\mathrm{qA}}}\right]+\right.$$

$$\left.\left.\mathrm{Erf}\left[\frac{\sqrt{\mathrm{kab}}(\mathrm{CAfqA}t+(\mathrm{CAo}-\mathrm{CBo})V)}{\sqrt{2}\sqrt{\mathrm{CAf}}\sqrt{\mathrm{qA}}\sqrt{V}}\right]\right)\right)$$

cd[t_]:=

$$\mathrm{CBo}+\mathrm{CDo}-$$

$$\left(\sqrt{\mathrm{CAf}}\,\mathrm{CBo}\,e^{-\frac{\mathrm{kab}\,t(\mathrm{CAf}\,\mathrm{qA}\,t+2(\mathrm{CAo}-\mathrm{CBo})V)}{2V}}\sqrt{2\pi}\sqrt{\mathrm{qA}}\right)\Big/$$

$$\left(\sqrt{\mathrm{CAf}}\sqrt{2\pi}\sqrt{\mathrm{qA}}+\mathrm{CBo}\,e^{\frac{(\mathrm{CAo}-\mathrm{CBo})^2\mathrm{kab}\,V}{2\mathrm{CAfqA}}}\sqrt{\mathrm{kab}}\,\pi\sqrt{V}\right.$$

$$\left(\mathrm{Erf}\left[\frac{(-\mathrm{CAo}+\mathrm{CBo})\sqrt{\mathrm{kab}}\sqrt{V}}{\sqrt{2}\sqrt{\mathrm{CAf}}\sqrt{\mathrm{qA}}}\right]+\right.$$

$$\left.\left.\mathrm{Erf}\left[\frac{\sqrt{\mathrm{kab}}(\mathrm{CAf}\,\mathrm{qA}\,t+\mathrm{CAo}\,V-\mathrm{CBo}\,V)}{\sqrt{2}\sqrt{\mathrm{CAf}}\sqrt{\mathrm{qA}}\sqrt{V}}\right]\right)\right)$$

As we have learned, the next step should be to test these solutions for their validity by placing them back in the rate equations, and simplifying the results:

$$\mathrm{Simplify}\left[\partial_t\mathrm{ca}[t]==\frac{C_{\mathrm{Af}}q_A}{V}-k_{\mathrm{AB}}\mathrm{ca}[t]\mathrm{cb}[t]\right]$$

True

$$\mathrm{Simplify}\left[\partial_t\mathrm{cb}[t]==-k_{\mathrm{AB}}\mathrm{ca}[t]\mathrm{cb}[t]\right]$$

True

$$\mathrm{Simplify}\left[\partial_t\mathrm{cd}[t]==+k_{\mathrm{AB}}\mathrm{ca}[t]\mathrm{cb}[t]\right]$$

True

The solutions for the concentrations, though intricate and lengthy, are shown to be valid. We can proceed to check them further by determining if they go back to the proper initial values of concentration when time goes to zero:

```
Limit[ca[t], t → 0]//Simplify
```

CAo

```
Limit[cb[t], t → 0]//Simplify
```

CBo

```
Limit[cd[t], t → 0]//Simplify
```

CDo

We see that they do indeed revert to initial values as they should. Finding the limit from the purely symbolic expressions as time goes to infinity is indeterminate even after ten minutes.

Overall, what we see is that this set of seemingly simple equations is actually not simple and although it appears that they are analytically soluble, the combination of the second order kinetics plus the convective flow term raises the complexity of the solutions. However, they are solutions, because when they are placed back into the rate equations, the two sides of the equations are equal. The next best approach to examining their behavior is to apply a set of reasonable parameter values and to examine them with time.

The concentration of A in the feed stream is taken to be 1 mole per liter. We set the initial concentration of B in the vessel to 0.1 mole per liter, while the initial concentrations of A and D are both set to zero. The flow of A into the unit is taken as a constant equal to 10^{-3} liters per second and the rate constant is fixed at 0.1 liter per mole per second:

```
CAf = 1.;

CAo = 0.;

CBo = .1;

CDo = 0.;

kab = .1;

qA = .001;

V = 100;

tmax = 20000;

Table[{t, ca[t], cb[t], cd[t]},

{t, 0.0, tmax, 2000}]

Plot[{ca[t], cb[t], cd[t]},

{t, 0.0, tmax},
```

PlotStyle → {Hue[0.3], Hue[0.6], Hue[0.9]},

AxesLabel → {t, "Ci[t]"},

PlotLabel → "blue=Cb, purple=Cd, green=Ca"]

These are the concentrations every 2000 seconds.

$\{\{0., 1.45616 \times 10^{-15}, 0.1, -1.38778 \times 10^{-17}\},$
$\{2000., 0.00273966, 0.0827397, 0.0172603\},$
$\{4000., 0.00158483, 0.0615848, 0.0384152\},$
$\{6000., 0.00225607, 0.0422561, 0.0577439\},$
$\{8000., 0.00373216, 0.0237322, 0.0762678\},$
$\{10000., 0.00797885, 0.00797885, 0.0920212\},$
$\{12000., 0.0205525, 0.000552479, 0.0994475\},$
$\{14000., 0.0400013, 1.33834 \times 10^{-6}, 0.0999987\},$
$\{16000., 0.06, 6.07588 \times 10^{-11}, 0.1\},$
$\{18000., 0.08, 5.05227 \times 10^{-17}, 0.1\},$
$\{20000., 0.1, 7.6946 \times 10^{-25}, 0.1\}\}$

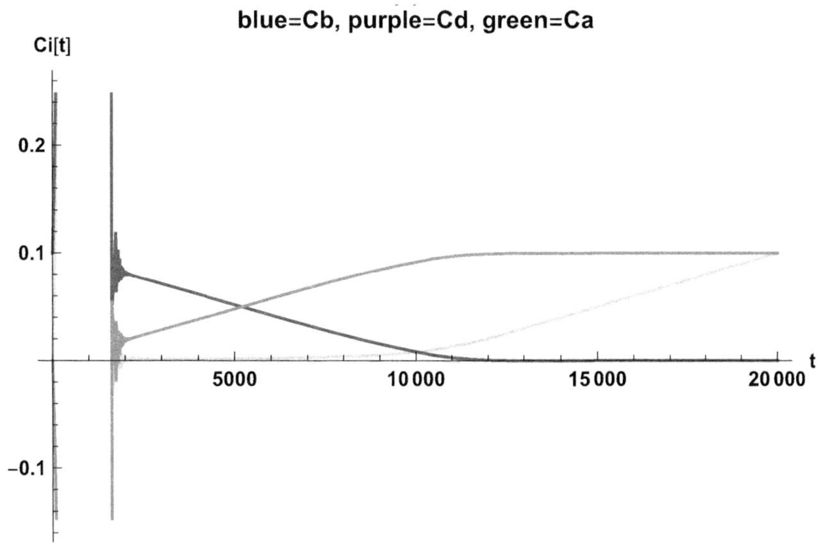

We note that up to a time of about 2000 to 2500 seconds, the numerical evaluation of these functions is very unstable, and then after t ≈ 2000 second, the instability stops and the solutions are stable. We can plot the initial times to see just how unstable they are:

```
CAf = 1.;
CAo = 0.0;
CBo = 0.1;
CDo = 0.0;
kab = .1;
qA = .001;
V = 100;
tmax = 2500;
Table[{t, ca[t], cb[t], cd[t]},
{t, 0.0000000001, tmax, 200}]
Plot[{ca[t], cb[t], cd[t]},
{t, 0.0, tmax},
PlotStyle → {Hue[0.3], Hue[0.6], Hue[0.9]},
AxesLabel → {t, "Ci[t]"},
PlotLabel → "blue=Cb, purple=Cd, green=Ca",
PlotRange → {{0, tmax}, {0, 0.3}}]
Plot[{ca[t], cb[t], cd[t]},
{t, 0.0, tmax},
PlotStyle → {Hue[0.3], Hue[0.6], Hue[0.9]},
AxesLabel → {t, "Ci[t]"},
PlotLabel → "blue=Cb, purple=Cd, green=Ca",
PlotRange → All]
```

$\{\{1. \times 10^{-10}, 1.01915 \times 10^{-13}, 0.1, -1.00017 \times 10^{-13}\},$

{200., 0.626274, 0.724274, −0.624274},

{400., 4.94404, 5.04004, −4.94004},

{600., 33.6032, 33.6972, −33.5972},

{800., 216.37, 216.462, −216.362},

{1000., 1335.88, 1335.97, −1335.87},
{1200., 7922.04, 7922.13, −7922.03},
{1400., 45135., 45135.1, −45135.},
{1600., 247067., 247067., −247067.},
{1800., 0.00805738, 0.0900574, 0.00994262},
{2000., 0.00273966, 0.0827397, 0.0172603},
{2200., 0.000904844, 0.0789048, 0.0210952},
{2400., 0.00119585, 0.0771958, 0.0228042}}

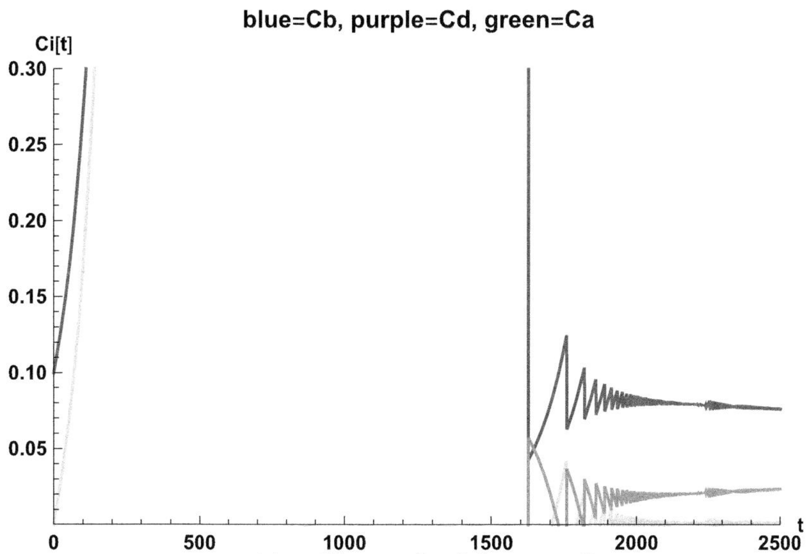

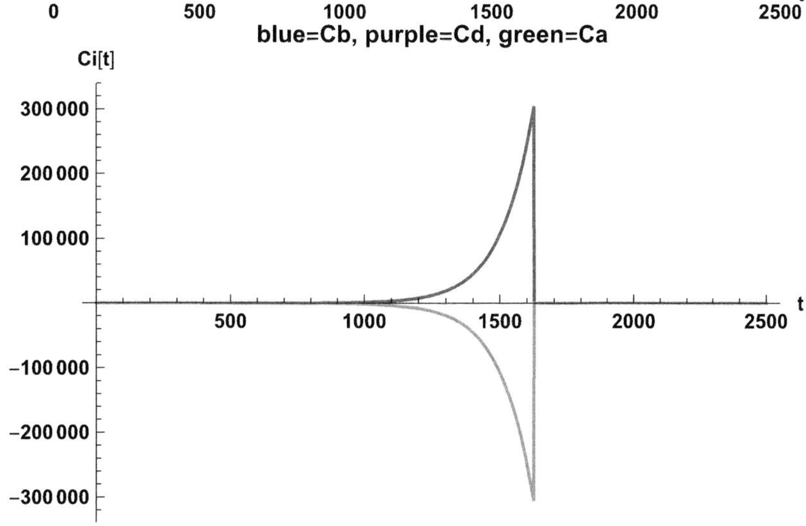

The instability in the calculations is manifest. We note that the instabilities are mostly between $0 < t \lesssim 2500$. To understand what is happening, we can compute the numerator and denominators of Ca[t] separately, then look at their ratio. In what follows, the numerator function is numCa[t] and the denominator is denomCa[t] and their ratio is Ca[t]:

CAf = 1.;

CAo = 0.0;

CBo = 0.1;

CDo = 0.0;

kab = .1;

qA = .001;

V = 100;

numCa[t_]:=

$$\left(e^{-\frac{\text{kab}(\text{CAfqA}t+(\text{CAo}-\text{CBo})V)^2}{2\text{CAfqA}V}}\right.$$

$$\left(\sqrt{\text{CAf}}\sqrt{2\pi}\sqrt{\text{qA}}\right.$$

$$\left(\text{CBo}e^{\frac{(\text{CAo}-\text{CBo})^2\text{kab}V}{2\text{CAfqA}}}V+\right.$$

$$\left.e^{\frac{\text{kab}(\text{CAfqA}t+(\text{CAo}-\text{CBo})V)^2}{2\text{CAfqA}V}}(\text{CAfqA}t+(\text{CAo}-\text{CBo})V)\right)+$$

$$\text{CBo}e^{(\text{CAo}-\text{CBo})\text{kab}t+\frac{\text{CAfkabqA}t^2}{2V}+\frac{(\text{CAo}-\text{CBo})^2\text{kab}V}{\text{CAfqA}}}\sqrt{\text{kab}}$$

$$\pi\sqrt{V}(\text{CAfqA}t+(\text{CAo}-\text{CBo})V)$$

$$\left(\text{Erf}\left[\frac{(-\text{CAo}+\text{CBo})\sqrt{\text{kab}}\sqrt{V}}{\sqrt{2}\sqrt{\text{CAf}}\sqrt{\text{qA}}}\right]+\right.$$

$$\mathrm{Erf}\left[\frac{\sqrt{\mathrm{kab}}(\mathrm{CAfqA}t+(\mathrm{CAo}-\mathrm{CBo})V)}{\sqrt{2}\sqrt{\mathrm{CAf}}\sqrt{\mathrm{qA}}\sqrt{V}}\right]\Bigg)\Bigg)\Bigg)$$

denomCa[t_]:=

$$\Big(\sqrt{\mathrm{CAf}}\sqrt{2\pi}\sqrt{\mathrm{qA}}V+$$

$$\mathrm{CBo}e^{\frac{(\mathrm{CAo}-\mathrm{CBo})^2\mathrm{kab}V}{2\mathrm{CAfqA}}}\sqrt{\mathrm{kab}}\pi V^{3/2}$$

$$\left(\mathrm{Erf}\left[\frac{(-\mathrm{CAo}+\mathrm{CBo})\sqrt{\mathrm{kab}}\sqrt{V}}{\sqrt{2}\sqrt{\mathrm{CAf}}\sqrt{\mathrm{qA}}}\right]+\right.$$

$$\mathrm{Erf}\left[\frac{\sqrt{\mathrm{kab}}(\mathrm{CAfqA}t+(\mathrm{CAo}-\mathrm{CBo})V)}{\sqrt{2}\sqrt{\mathrm{CAf}}\sqrt{\mathrm{qA}}\sqrt{V}}\right]\Bigg)\Bigg)$$

```
Table[{t, numCa[t], denomCa[t],
(numCa[t]/denomCa[t])}, {t, 0, 2000, 10}];
Table[{t, numCb[t], denomCb[t],
(numCa[t]/denomCa[t])}, {t, 0, 2000, 10}];

ListPlot[Table[{t,  numCa[t]}, {t, 0, 20000, 10}],
PlotLabel → "numCa[t]", PlotStyle → Hue[0]]
ListPlot[Table[{t,  denomCa[t]}, {t, 0, 20000, 10}],
PlotLabel → " denomCa[t]"]
ListPlot[Table[{t,  (numCa[t]/denomCa[t])},
{t, 0, 20000, 10}], PlotLabel → "Ca[t]",
PlotStyle → Hue[0.8]]
Show[%%%, %%, %,
```

PlotLabel →

"numCa[t](red), denomCa[t](blue),

Ca[t](purple)"]

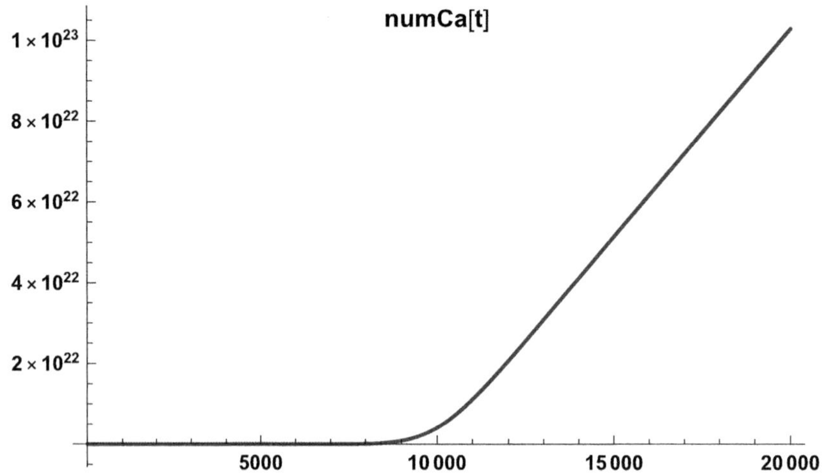

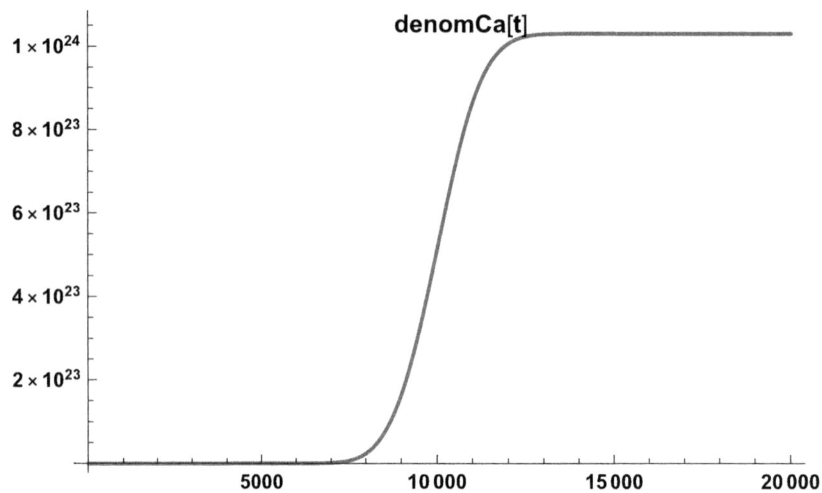

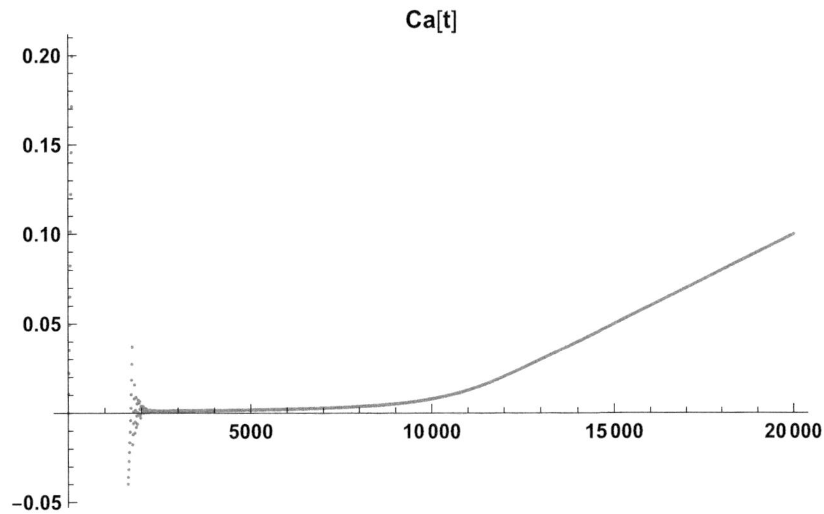

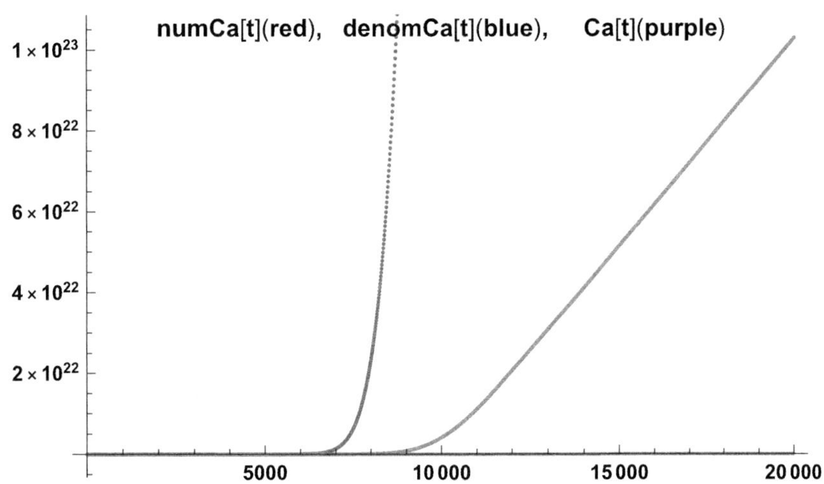

Notice that the numerator and the denominator functions diverge to enormous values from 10^{23} to 10^{24} at longer times. We also notice that denominator gets larger than the numerator faster at times over 5000. In that larger time range the division gives reasonable numbers. At shorter times, this plot shows a region where the value of the denominator is quite small and the numerator is diverging to very large values. But, plot range is so large that it makes seeing the instabilities at the shorter times difficult.

Instead of plotting, we can compute the values for the numerator, denominator, and their ratio between zero and 5000 in steps of 250. Remember the concentration of A coming into the system within the calculation is 1 mole per liter and should always be a fraction between zero and one. In what follows, we see that the ratio diverges to large numbers and then at around t = 1700, it becomes reasonable again.

Table[{*t*, numCa[*t*], denomCa[*t*],

(numCa[*t*]/denomCa[*t*])}, {*t*, 0, 5000, 250}]

$\{\{0, 1.15425 \times 10^{-14}, 7.92665, 1.45616 \times 10^{-15}\},$

{250, 8.58669, 7.92665, 1.08327},

{500, 103.066, 7.92665, 13.0024},

{750, 1081.08, 7.92665, 136.386},

{1000, 10589.1, 7.92665, 1335.88},

{1250, 97381.2, 7.92665, 12285.3},

{1500, 841251., 7.92665, 106129.},

$\{1750, 2.10921 \times 10^{6}, 5.71853 \times 10^{7}, 0.0368839\},$

$\{2000, 1.72335 \times 10^{6}, 6.29038 \times 10^{8}, 0.00273966\},$

$\{2250, 4.89734 \times 10^{6}, 4.74638 \times 10^{9}, 0.0010318\},$

$\{2500, 4.16206 \times 10^{7}, 3.28815 \times 10^{10}, 0.00126577\},$

$\{2750, 2.85547 \times 10^{8}, 2.14674 \times 10^{11}, 0.00133014\},$

$\{3000, 1.81363 \times 10^{9}, 1.31841 \times 10^{12}, 0.00137562\},$

$\{3250, 1.08339 \times 10^{10}, 7.61519 \times 10^{12}, 0.00142266\},$

$\{3500, 6.09416 \times 10^{10}, 4.13711 \times 10^{13}, 0.00147305\},$

$\{3750, 3.22823 \times 10^{11}, 2.11416 \times 10^{14}, 0.00152696\},$

$\{4000, 1.61072 \times 10^{12}, 1.01634 \times 10^{15}, 0.00158483\},$

$\{4250, 7.57105 \times 10^{12}, 4.59674 \times 10^{15}, 0.00164705\},$

$\{4500, 3.35317 \times 10^{13}, 1.95623 \times 10^{16}, 0.0017141\},$

$\{4750, 1.39964 \times 10^{14}, 7.83431 \times 10^{16}, 0.00178655\},$

$\{5000, 5.5074 \times 10^{14}, 2.95296 \times 10^{17}, 0.00186504\}\}$

A plot of the numerator, denominator, and the ratio (below) shows the numerator diverging and taking the ratio, Ca[t] with it.

ListPlot[Table[{*t*, numCa[*t*]}, {*t*, 0, 1700, 10}],

PlotLabel → "numCa[t]", PlotStyle → Hue[0]];

```
ListPlot[Table[{t, denomCa[t]}, {t, 0, 1700, 10}],
PlotLabel → " denomCa[t]"];
ListPlot[Table[{t, (numCa[t]/denomCa[t])},
{t, 0, 1700, 10}], PlotLabel → "Ca[t]",
PlotStyle → Hue[0.8]];
Show[%%%, %%, %,
PlotLabel →
"numCa[t](red), denomCa[t](blue), Ca[t](purple)"]
```

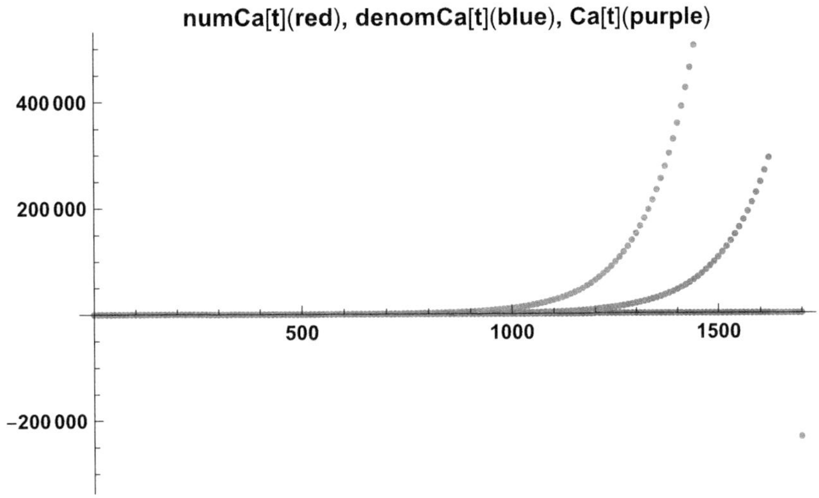

Thus, we have analytical solutions, but they are unusable at early times. If we drop the points before t = 3000, then the shapes of the functions for concentration are as follows:

```
CAf = 1.;
CAo = 0.;
CBo = .1;
CDo = 0.;
kab = .1;
qA = .001;
```

```
V = 100;

tmax = 20000;


Plot[{ca[t], cb[t], cd[t]},

{t, 3000, tmax},

PlotRange → All,

PlotStyle → {Hue[0.3], Hue[0.6], Hue[0.9]},

AxesLabel → {t, "Ci[t]"},

PlotLabel → "green=Ca, blue=Cb, purple=Cd"]
```

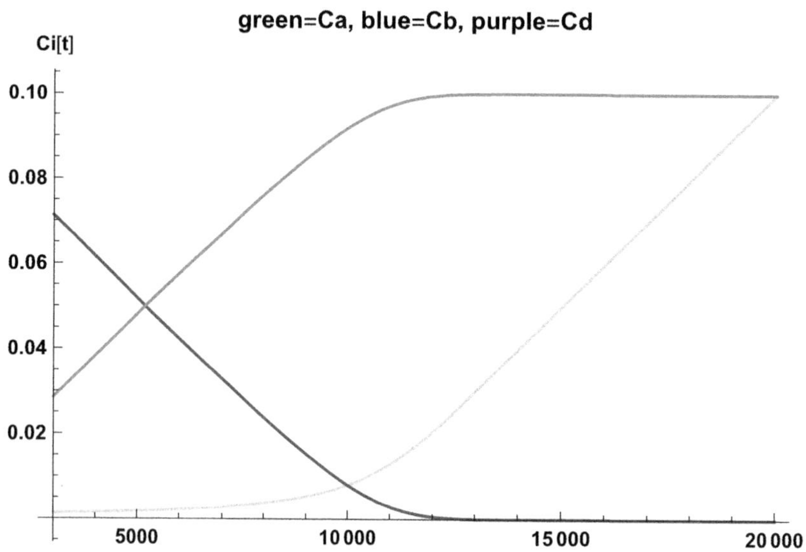

The concentration of B drops nearly linearly with time to zero and the concentration of D, the product, increases linearly over the same time and rises to a concentration of 0.1 mole per liter, which is exactly the same as the concentration of B initially. The concentration of A rises slowly until B is consumed and then it rises linearly. Given the apparent complexity of the analytical solutions, the behavior at sufficient time is almost trivial. Why? Because under the conditions of the fed-batch reactor, at early times while B is being consumed, the concentration of A is small and almost constant. That means the rate of change in A is close to zero and the rate of reaction is controlled by the rate of admission of A to the vessel. Mathematically this becomes:

$$\partial_t \text{ca}[t] = 0 = \frac{\text{CAfqA}}{V} - \text{kabca}[t]\text{cb}[t]$$

$$\therefore \frac{\text{CAfqA}}{V} \approx \text{kabca}[t]\text{cb}[t]$$

$$\partial_t \text{cb}[t] = -\frac{\text{CAfqA}}{V}$$

$$\partial_t \text{cd}[t] = +\frac{\text{CAfqA}}{V}$$

Because $\frac{\text{CAfqA}}{V}$ is a constant, the changes in B and D during this time period are linear. We can see this by superimposing the linear functions for CB[t] (blue, dashed - dark gray in the print version) and CD[t] (purple, dashed - mid gray in the print version) on the overall plot:

```
Plot[(-CAf qA/V t + CBo), {t, 0, 10000},
PlotStyle → {Dashed, Hue[0.6]}];
Plot[(CAf qA/V t), {t, 0, 10000},
PlotStyle → {Dashed, Hue[0.9]}];
Plot[{ca[t], cb[t], cd[t]},
{t, 3000, tmax},
PlotRange → All,
PlotStyle → {Hue[0.3], Hue[0.6], Hue[0.9]},
AxesLabel → {t, "Ci[t]"},
PlotLabel → "blue=Cb, purple=Cd, green=Ca
with linear approximations (dashed)"
];
Show[%, %%, %%%]
```

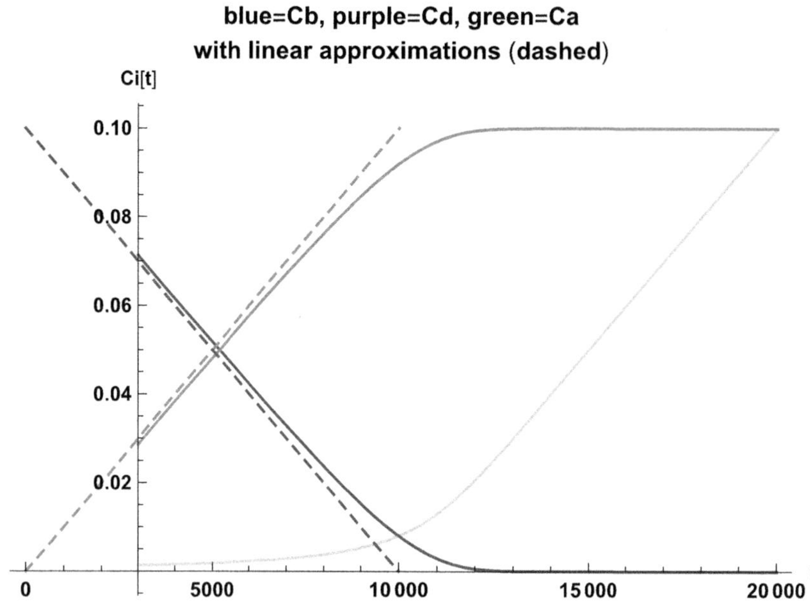

This leads logically to the pseudo-steady state analysis in the next section.

Before leaving this section, let's build a Module function for the analytical solutions, so that we can keep the variable names local within the function, and not have them clash with the same variable names subsequent models. Because these solutions are analytical, we will call the Module "analytical".

analytical[tmax_]:=Module[{CAf = 1.,

CAo = 0.0,

CBo = 0.1,

CDo = 0.0,

kab = .1,

qA = .001,

V = 100},

ca[t_]:=

$$\left(e^{-\frac{\text{kab}(\text{CAfqA}t + (\text{CAo} - \text{CBo})V)^2}{2\text{CAfqA}V}}\right.$$

$$\left(\sqrt{\text{CAf}}\sqrt{2\pi}\sqrt{\text{qA}}\right.$$

$$\left(\text{CBo}e^{\frac{(\text{CAo}-\text{CBo})^2\text{kab}V}{2\text{CAfqA}}}V+\right.$$

$$e^{\frac{\text{kab}(\text{CAfqA}t+(\text{CAo}-\text{CBo})V)^2}{2\text{CAfqA}V}}$$

$$(\text{CAfqA}t+(\text{CAo}-\text{CBo})V))+$$

$$\text{CBo}\,e^{(\text{CAo}-\text{CBo})\text{kab}t+\frac{\text{CAfkabqA}t^2}{2V}+\frac{(\text{CAo}-\text{CBo})^2\text{kab}V}{\text{CAfqA}}}$$

$$\sqrt{\text{kab}\pi}\sqrt{V}(\text{CAfqA}t+(\text{CAo}-\text{CBo})V)$$

$$\left(\text{Erf}\left[\frac{(-\text{CAo}+\text{CBo})\sqrt{\text{kab}}\sqrt{V}}{\sqrt{2}\sqrt{\text{CAf}}\sqrt{\text{qA}}}\right]+\right.$$

$$\left.\left.\left.\text{Erf}\left[\frac{\sqrt{\text{kab}}(\text{CAfqA}t+(\text{CAo}-\text{CBo})V)}{\sqrt{2}\sqrt{\text{CAf}}\sqrt{\text{qA}}\sqrt{V}}\right]\right)\right)\right)\Big/$$

$$\left(\sqrt{\text{CAf}}\sqrt{2\pi}\sqrt{\text{qA}}V+\right.$$

$$\text{CBo}e^{\frac{(\text{CAo}-\text{CBo})^2\text{kab}V}{2\text{CAfqA}}}\sqrt{\text{kab}\pi}V^{3/2}$$

$$\left(\text{Erf}\left[\frac{(-\text{CAo}+\text{CBo})\sqrt{\text{kab}}\sqrt{V}}{\sqrt{2}\sqrt{\text{CAf}}\sqrt{\text{qA}}}\right]+\right.$$

$$\left.\left.\text{Erf}\left[\frac{\sqrt{\text{kab}}(\text{CAfqA}t+(\text{CAo}-\text{CBo})V)}{\sqrt{2}\sqrt{\text{CAf}}\sqrt{\text{qA}}\sqrt{V}}\right]\right)\right);$$

cb[t_]:=

$$\left(\sqrt{\text{CAf}}\text{CBo}e^{-\frac{\text{kab}t(\text{CAfqA}t+2(\text{CAo}-\text{CBo})V)}{2V}}\sqrt{2\pi}\sqrt{\text{qA}}\right)\Big/$$

$$\left(\sqrt{\text{CAf}}\sqrt{2\pi}\sqrt{\text{qA}}+\text{CBo}e^{\frac{(\text{CAo}-\text{CBo})^2\text{kab}V}{2\text{CAfqA}}}\sqrt{\text{kab}\pi}\right.$$

$$\left.\sqrt{V}\left(\text{Erf}\left[\frac{(-\text{CAo}+\text{CBo})\sqrt{\text{kab}}\sqrt{V}}{\sqrt{2}\sqrt{\text{CAf}}\sqrt{\text{qA}}}\right]+\text{Erf}\left[\frac{\sqrt{\text{kab}}(\text{CAfqA}t+(\text{CAo}-\text{CBo})V)}{\sqrt{2}\sqrt{\text{CAf}}\sqrt{\text{qA}}\sqrt{V}}\right]\right)\right);$$

cd[t_]:=

$$\text{CBo}+\text{CDo}-$$

$$\left(\sqrt{\text{CAf}}\text{CBo}e^{-\frac{\text{kab}t(\text{CAfqA}t+2(\text{CAo}-\text{CBo})V)}{2V}}\sqrt{2\pi}\sqrt{\text{qA}}\right)\Big/$$

$$\left(\sqrt{\text{CAf}}\sqrt{2\pi}\sqrt{\text{qA}}+\text{CBo}e^{\frac{(\text{CAo}-\text{CBo})^2\text{kab}V}{2\text{CAfqA}}}\sqrt{\text{kab}\pi}\right.$$

$$\left.\sqrt{V}\left(\text{Erf}\left[\frac{(-\text{CAo}+\text{CBo})\sqrt{\text{kab}}\sqrt{V}}{\sqrt{2}\sqrt{\text{CAf}}\sqrt{\text{qA}}}\right]+\text{Erf}\left[\frac{\sqrt{\text{kab}}(\text{CAfqA}t+\text{CAo}V-\text{CBo}V)}{\sqrt{2}\sqrt{\text{CAf}}\sqrt{\text{qA}}\sqrt{V}}\right]\right)\right);$$

Plot[{ca[t], cb[t], cd[t]},

{t, 0, tmax},

PlotStyle → {Hue[0.3], Hue[0.6], Hue[0.9]},

AxesLabel → {t, "Ci[t]"},

PlotLabel → "blue=Cb, purple=Cd, green=Ca"]]

analytical[20000]

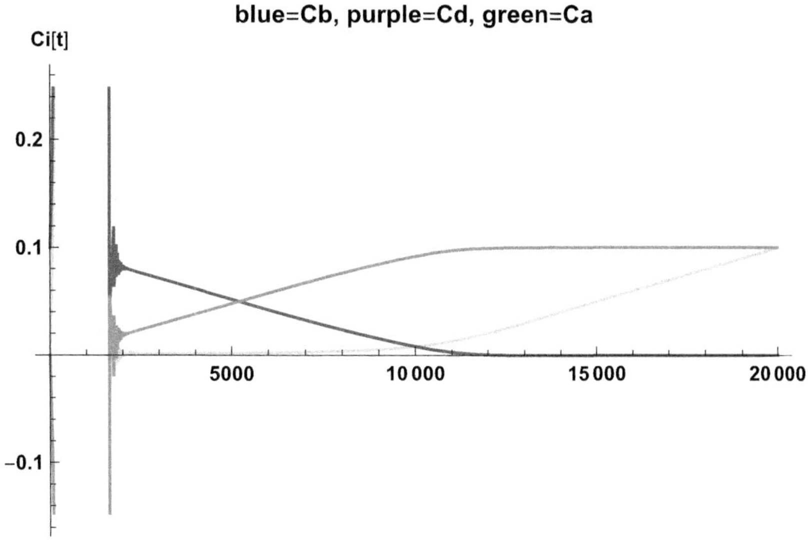

Pseudo-steady state analysis in FBR

As we have seen, the situation that arises in the fed-batch reactor is that the rate of consumption of the added component is balanced exactly by its rate of addition. This is desirable in many cases and is a reason for using a fed-batch reactor. But as we also have seen the inclusion of non-linear kinetics, with the convective flow terms, causes some problems mathematically. But we can simplify the analysis with some logic.

In that case, wherein the rate of change of the mass of A in the reactor is effectively zero, so as long as there is sufficient B present for reaction to take place, there is a period of operation that is referred to as a *pseudo-steady state*. It is called this because at the same time that the concentration of A is nearly constant and its rate of change is close to zero, the rate of change of B is not zero and its concentration changes considerably. Hence it is a pseudo-steady state, or partly at steady state. Let's go back to the equations of change to understand this:

$$\partial_t \mathrm{Ca}[t]V[t] = \mathrm{Caf\ qaf - kab\ Ca[\ t\ ]Cb[\ t\ ]\ V[\ t\ ]}$$

$$\partial_t \mathrm{Cb}[t]V[t] = \mathrm{-\ kab\ Ca[\ t\ ]Cb[\ t\ ]\ V[\ t\ ]}$$

$$\partial_t \mathrm{Cd}[t]V[t] = \mathrm{+\ kab\ Ca[\ t\ ]Cb[\ t\ ]\ V[\ t\ ]}$$

If the rate of change of the concentration of A is zero, then the following simplifications apply:

$$\partial_t \mathrm{Ca}[t]V[t] = 0$$

$$\therefore \text{Caf qaf} \approx \text{kab Ca[t]Cb[t] V[t]}$$

$$\partial_t \text{Cb}[t]V[t] = \text{ - Caf qaf}$$

$$\partial_t \text{Cd}[t]V[t] = \text{ + Caf qaf}$$

If the feed flow rate and concentration of A are constant, then we would find that the changes in the concentrations of B and D are linear in time and with oppositely signed slopes. We use "ss" to designate the steady state, time-dependent concentrations of B and D, here is that simple solution:

Clear["Global\` * "]

Simplify[DSolve[

{$\partial_t(V[t])$ == qa,

∂_t(Cbss[t]$V[t]$) == −Ca qa,

∂_t(Cdss[t]$V[t]$) == +Ca qa,

V[0] == Vo,

Cbss[0] == Cbo,

Cdss[0] == Cdo},

{$V[t]$, Cbss[t], Cdss[t]},

t]]

$$\left\{\left\{V[t] \to \text{qa } t + \text{Vo}, \text{Cbss}[t] \to \frac{-\text{Caqa}t + \text{CboVo}}{\text{qa}t + \text{Vo}}, \text{Cdss}[t] \to \frac{\text{Caqa}t + \text{CdoVo}}{\text{qa}t + \text{Vo}}\right\}\right\}$$

We can write a new Module function that computes and graphs the solutions that we just derived for the steady state.

semi[tmax_]:=

Module[{CAf = 1., CAo = 0., CBo = .1, CDo = 0.,

kab = .1, qA = .001, V = 100},

cbss[t_]:=$\frac{\text{CBo}V - \text{CAfqA}t}{\text{qA}t + V}$;

$$\text{cdss[t\_]:=}\frac{\text{CAfqA}t+\text{CDo}V}{\text{qA}t+V};$$

```
Plot[{cbss[t], cdss[t]},
{t, 0, tmax},
PlotRange → All,
PlotStyle → {{Dashed, Hue[0.6]},
{Dashed, Hue[0.9]}},
AxesLabel → {t, "Ci[t]"},
PlotLabel → "Blue=Cb,Purple=Cd"]
]
semi[20000]
```

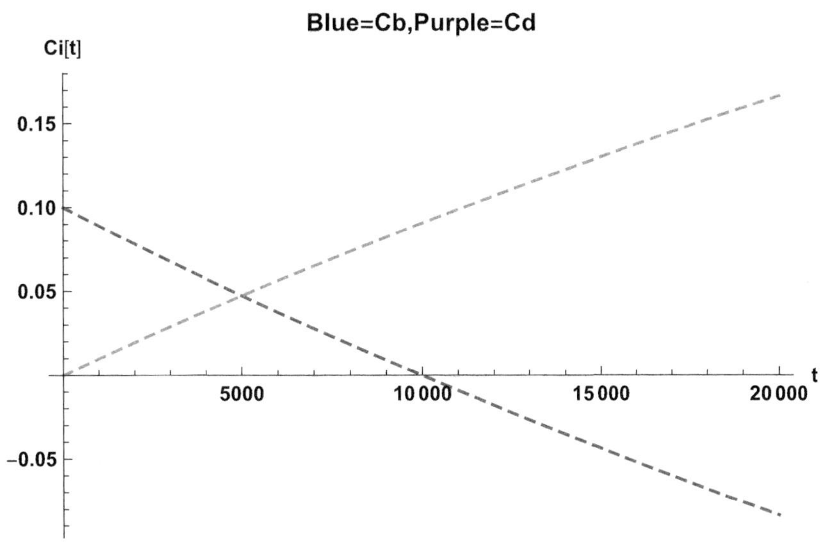

And now we can compare the steady state to the full analytical solutions that we derived before.

```
Show[semi[20000], analytical[20000]]
```

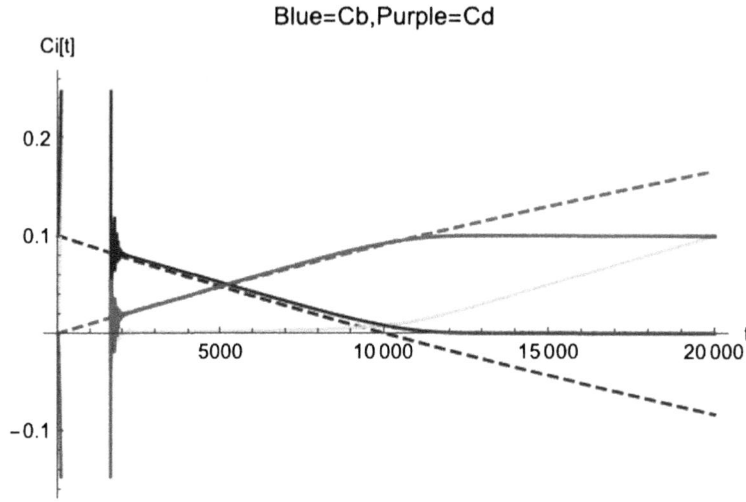

The results show that the steady state solutions (dashed) map well onto the transient solutions for the concentrations of **B** and **D** at early time. Beyond approximately 10 000 time units, the steady state concentrations begin to deviate noticeably from the full solutions. This is also the time at which the concentration of **A** begins to rise above near-zero values. So the usefulness of the steady state solutions is that they allow us to compute the flow rate of reagent **A** and the time dependence of the systems with very simple equations, but we cannot push such an analysis too far beyond its region of applicability. From the perspective of analysis, the pseudo-steady state is important to us in that it explains the behavior of the more complex and complete model in a very straightforward way and it is much easier mathematically, which is why it was used in the past.

Numerical analysis of the fed-batch reactor

Instead of seeking an analytical solution, or imposing the pseudo-steady state, we can analyze the system of equations using numerical methods and NDSolve. Here is the code for doing so:

```
CAf = 1.;

CAo = 0;

CBo = .1;

CDo = 0;

kab = .1;
```

```
qAf = .001;

tmax = 20000;

Vr = 100;

fbsol1 = NDSolve[
{CAn'[t] == − CAn[t]CBn[t],
CBn'(t) = −kabCAn[t]CBn[t] ,
CDn'[t] == CAn[t]CBn[t],
   CAn[0] == CAo,
   CBn[0] == CBo,
   CDn[0] == CDo},
{CAn[t], CBn[t], CDn[t]},
      {t, 0, tmax}];
cAn[t_]:=Evaluate[CAn[t]/.fbsol1]
cBn[t_]:=Evaluate[CBn[t]/.fbsol1]
cDn[t_]:=Evaluate[CDn[t]/.fbsol1]
numerical = Plot[{cAn[t], cBn[t], cDn[t]},
{t, 0, tmax},
PlotRange → All,
PlotStyle → {{Dashed, Hue[0.2]}, {Dashed, Hue[0.5]},
{Dashed, Hue[.8]}},
AxesLabel → {t, "Ci[t]"},
PlotLabel → "blue=Cb, purple=Cd, green=Ca"
]
```

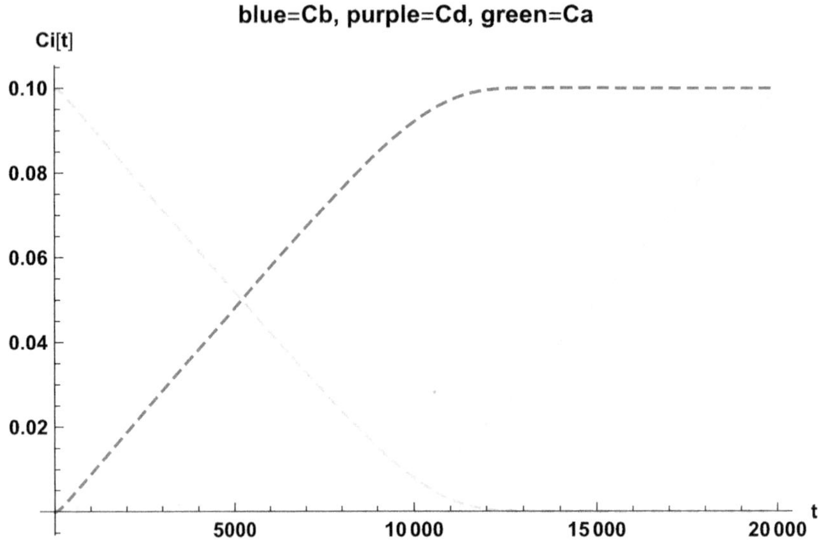

By contrast with the analytical solution, this solution is stable throughout the ranges of times. The two sets of solutions are shown together. It is usually the case that an analytical solution is preferred for its stability over a numerical solution, but that is not the case for this problem.

Show[analytical, numerical]

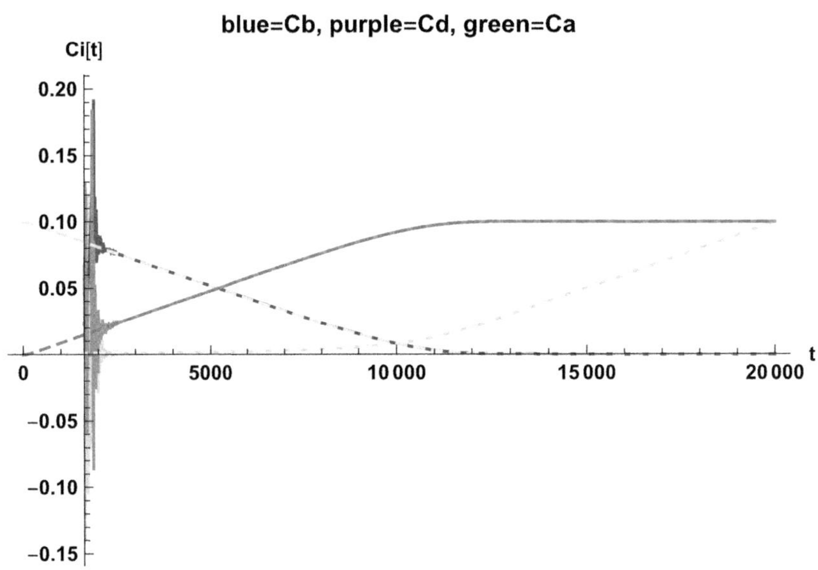

Constant volume batch reactor (CVBR) versus the fed-batch reactor (FBR)

Here is another schematic of a fed-batch reactor with constant volume. The concentration of **A** and **D** in the reactor at time zero will be zero; the concentration of **B** is non-zero initially. As **A** is introduced into the reactor **D** will be formed. If the rate of reaction is very fast, as it is in many reactions of inorganic chemicals, such as a precipitation, then the concentration of **A** will be very small and close to zero until all of **B** is used. On the other hand, if the kinetics are not nearly instantaneous, as is the case with many reactions of organic molecules, there may be **A** present shortly after it is admitted, and then that level will go to zero over time as **D** is formed. This may lead to the addition of **A** intermittently being preferred.

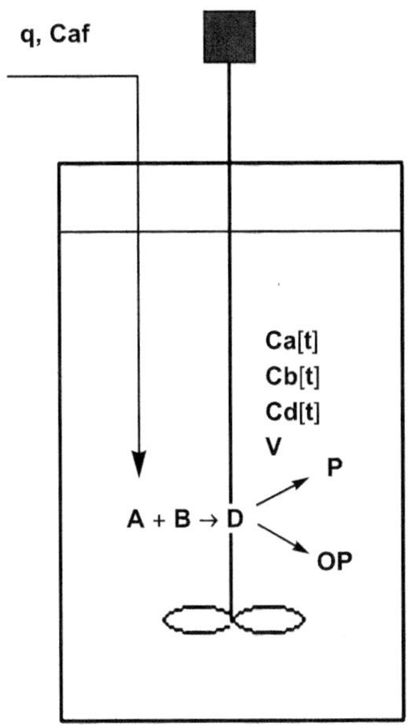

Firstly, we can build a model for these kinetics as if they were taking place in a constant volume batch reactor (CVBR). We have second order kinetics for **A** and **B** reacting to generate **D**. The species **D** reacts along two pathways, the first leads to **P** by first order kinetics and the second leads to **OP** by second order kinetics. The reaction to **P** could be envisioned as a ring-closure reaction, while that to **OP** is oligomer (chain) formation. Product **P** would be the preferred product and the other product, **OP** would be not preferred. We have made this into a Module function with the initial concentrations of **A** and **B**, the volume and the final time as the parameters. Outputs include the relevant graphs, as well as the conversion, the selectivities and the yields of **P** and **OP**.

```
Clear["Global`*"]

(* CVBR stands for constant volume batch reactor *)

CVBR[Cao_, Cbo_, V_, tf_]:=

Module[{k1 = .01, k2 = 0.02,

k3 = 1},

sln = NDSolve[

{CA'[t]V == -CA[t]CB[t]V,

CB'[t]V == -CA[t]CB[t]V,

CD'[t]V == +k1CA[t]CB[t]V - k2CD[t]V-k3CD[t]^2V,

CP'[t]V == +k2CD[t]V,

COP'[t]V==k3CD[t]^2V,

CA[0] == Cao,CB[0] == Cbo, CD[0] == 0., CP[0] == 0.,

COP[0] == 0.},

{CA[t], CB[t], CD[t], CP[t], COP[t]},

    {t, 0, tf}];


ca[t_]:=Evaluate[CA[t]/.sln];

cb[t_]:=Evaluate[CB[t]/.sln];

cd[t_]:=Evaluate[CD[t]/.sln];

cp[t_]:=Evaluate[CP[t]/.sln];

cop[t_]:=Evaluate[COP[t]/.sln];
```

{Print["Conversion =", $\text{Round}\left[\left(1-\left(\frac{\text{cb[tf]}}{\text{Cbo}}\right)\right), 0.001\right] 100\text{"\%"}\right]$;

Print["Selectivity to D =", $\text{Round}\left[\left(\frac{\text{cd[tf]}}{\text{cd[tf]} + \text{cp[tf]} + \text{cop[tf]}}\right), 0.0001\right] 100\text{"\%"}$];

Print["Selectivity to P =", $\text{Round}\left[\left(\frac{\text{cp[tf]}}{\text{cd[tf]} + \text{cp[tf]} + \text{cop[tf]}}\right), 0.01\right] 100\text{"\%"}$];

$$\text{Print}[\text{"Selectivity to OP ="}, \text{Round}\left[\left(\frac{\text{cop[tf]}}{\text{cd[tf]} + \text{cp[tf]} + \text{cop[tf]}}\right), 0.01\right]100\text{"\%"}];$$

$$\text{Print}[\text{"Yield of P ="}, \text{Round}\left[\left(1 - \left(\frac{\text{ca[tf]}}{\text{Cao}}\right)\right)\left(\frac{\text{cp[tf]}}{\text{cd[tf]} + \text{cp[tf]} + \text{cop[tf]}}\right), 0.01\right]100\text{"\%"}];$$

$$\text{Print}[\text{"Yield of OP ="}, \text{Round}\left[\left(1 - \left(\frac{\text{ca[tf]}}{\text{Cao}}\right)\right)\left(\frac{\text{cop[tf]}}{\text{cd[tf]} + \text{cp[tf]} + \text{cop[tf]}}\right), 0.01\right]100\text{"\%"}];$$

```
Plot[{ca[t]V, cb[t]V, cd[t]V, cp[t]V, cop[t]V},
{t, 0, tf}, PlotRange → All,
PlotStyle → {Hue[0.1], Hue[0.1], Hue[0.4],
Hue[0.5], {Dashed, Hue[0.6]}},
PlotLabel → {"ca", Hue[0], "cb", Hue[0.1],
"cd", Hue[0.4], "cp", Hue[0.5], "cop", Hue[0.6]},
AxesLabel → {"t,sec", "N_i[t]"}],

Plot[{ca[t]V, cb[t]V}, {t, 0, tf}, PlotRange → All,
PlotStyle → {Hue[0.1]},
PlotLabel → {"ca", Hue[0], "cb"Hue[0.1]},
AxesLabel → {"t,sec", "N_a,b[t]"}],

Plot[{cd[t]V}, {t, 0, tf}, PlotRange → All,
PlotStyle → {Hue[0.4]},
PlotLabel → {"cd", Hue[0.4]},
AxesLabel → {"t,sec", "N_d[t]"}],

Plot[{cp[t]V}, {t, 0, tf}, PlotRange → All,
PlotStyle → {Hue[0.5]},
PlotLabel → {"cp", Hue[0.5]},
```

```
AxesLabel → {"t,sec", "N_p[t]"}],

Plot[{cop[t]V}, {t, 0, tf}, PlotRange → All,
PlotStyle → {Dashed, Hue[0.6]},
PlotStyle → {Dashed, Hue[0.6]},
PlotLabel → {"cop", Hue[0.6]},
AxesLabel → {"t,sec", "N_op[t]"}]]
]
```

The first run of this model will be done with the concentrations of **A** and **B** fixed at 1 M and the volume at 1000 L. The conversion is 95.2% the yield of **P** is 38% and of **OP** its 57%. If **P** is preferred, then this is not a good result (see below).

CVBR[1, 1, 1000, 2000]

Conversion ={95.2%}

Selectivity to **D** ={0.12%}

Selectivity to **P** ={40.%}

Selectivity to **OP** ={60.%}

Yield of **P** ={38.%}

Yield of **OP** ={57.%}

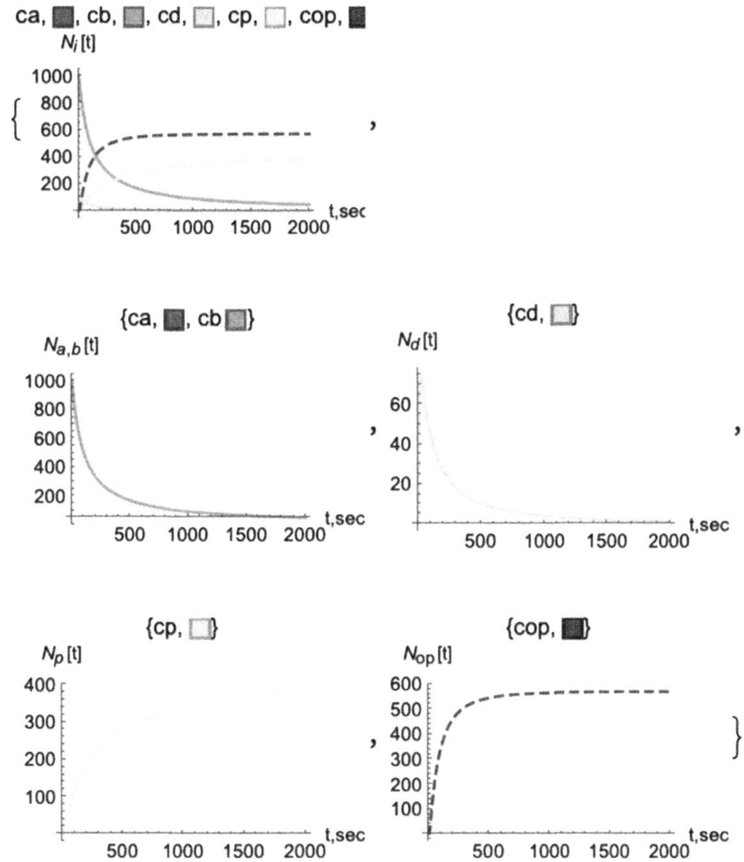

Now, let's examine the dilution effect. If we dilute the concentrations of **A** and **B** in the feed and in the vessel, and increase the overall volume, then we will be using the same number of moles. Here the conversion is lower, but the yields to **P** are much better. Remember yield is the product of conversion and selectivity. We also have much less yield loss to **OP** (see below).

CVBR[.1, .1, 10000, 2000]

Conversion ={66.7%}

Selectivity to **D** ={0.84%}

Selectivity to **P** ={90.%}

Selectivity to **OP** ={9.%}

Yield of **P** ={60.%}

Yield of **OP** ={6.%}

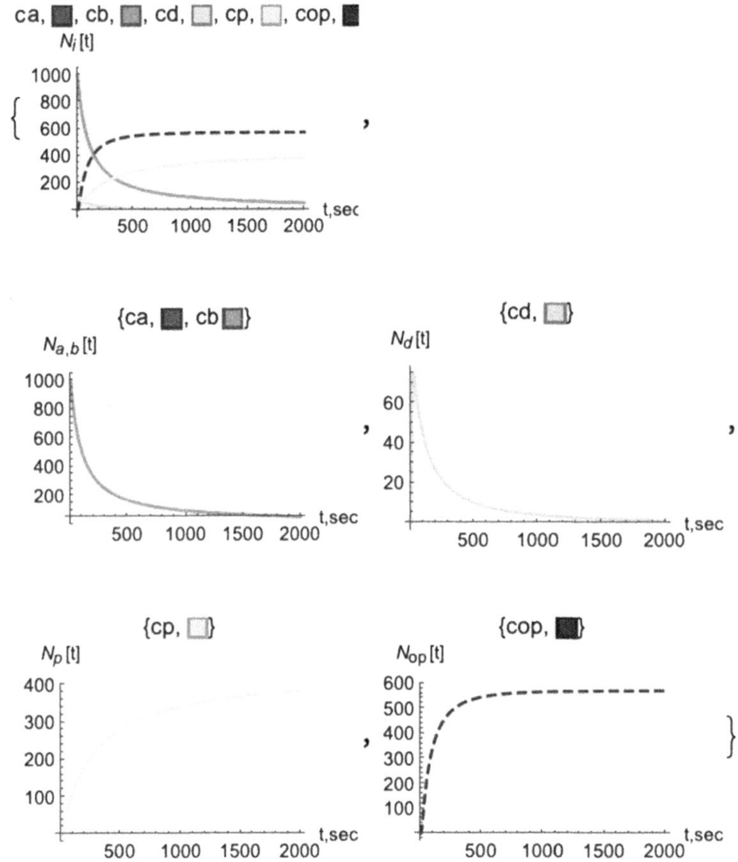

It is not valid to compare yields and selectivities at different levels of conversion. So, if we re-run the high concentration, low volume case, and stop at t = 200 seconds, then the conversion is 66.7%, but the yield to **P** is just 18% and to **OP** it is 46%. So the dilution effect is very real and advantageous (see below):

CVBR[1, 1, 1000, 200]

Conversion ={66.7%}

Selectivity to **D** ={3.96%}

Selectivity to **P** ={27.%}

Selectivity to **OP** ={69.%}

Yield of **P** ={18.%}

Yield of **OP** ={46.%}

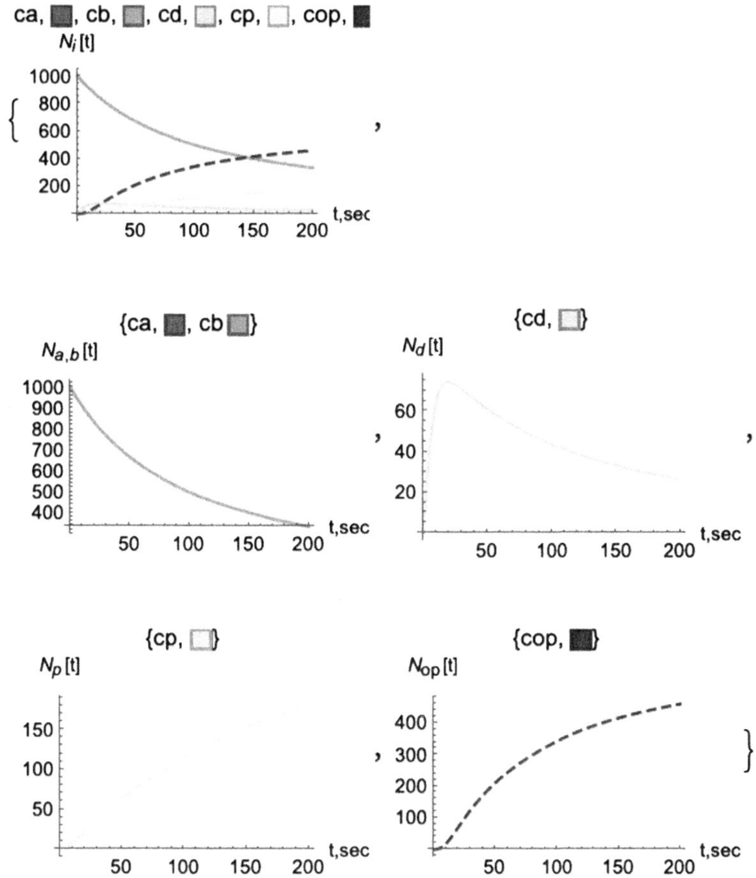

Next, let's reconsider the situation and run the same kinetics in a simulated FBR. We create Module function for this case also so that we can keep variables local and repeat the computation for comparisons.

```
FBR[Cao_, Cbo_, CAf_, q_, V_, tf_]:=Module[
{k1 = .01,
k2 = 0.02,
k3 = 1},
sln = NDSolve[
{CA'[t]V == CAfq − CA[t]CB[t]V,
CB'[t]V == −CA[t]CB[t]V,
```

```
CD'[t]V == +k1CA[t]CB[t]V − k2CD[t]V−
k3CD[t]^2V,
CP'[t]V == +k2CD[t]V,
COP'[t]V==k3CD[t]^2V,
   CA[0] == Cao,
   CB[0] == Cbo, CD[0] == 0., CP[0] == 0.,
COP[0] == 0.},
{CA[t], CB[t], CD[t], CP[t], COP[t]},
      {t, 0, tf}];

ca[t_]:=Evaluate[CA[t]/.sln];
cb[t_]:=Evaluate[CB[t]/.sln];
cd[t_]:=Evaluate[CD[t]/.sln];
cp[t_]:=Evaluate[CP[t]/.sln];
cop[t_]:=Evaluate[COP[t]/.sln];

{Print["Conversion =",
Round[(1 − (cb[tf]/Cbo)), 0.001]100"%"];

Print["Selectivity to D =",
Round[(cd[tf]/(cd[tf] + cp[tf] + cop[tf])),
0.0001]100"%"];
Print["Selectivity to P =",
Round[(cp[tf]/(cd[tf] + cp[tf] + cop[tf])), 0.01]
100"%"];
Print["Selectivity to OP =",
Round[(cop[tf]/(cd[tf] + cp[tf] + cop[tf])), 0.01]
100"%"];
```

```
Print["Yield of P =",
Round[(1 - (cb[tf]/Cbo))
(cp[tf]/(cd[tf] + cp[tf] + cop[tf])), 0.01]
100"%"];
Print["Yield of OP =",
Round[(1 - (cb[tf]/Cbo))
(cop[tf]/(cd[tf] + cp[tf] + cop[tf])), 0.01]
100"%"];

Plot[{ca[t]V, cb[t]V, cd[t]V, cp[t]V, cop[t]V},
{t, 0, tf}, PlotRange → All,
PlotStyle → {Hue[0.], Hue[0.1], Hue[0.4],
Hue[0.5], {Dashed, Hue[0.6]}},
PlotLabel → {"ca", Hue[0], "cb", Hue[0.1],
"cd", Hue[0.4], "cp", Hue[0.5], "cop", Hue[0.6]},
AxesLabel → {"t,sec", "N_i[t]"},
Ticks → {{0, tf/2, tf}, Automatic}],

Plot[{ca[t]V}, {t, 0, tf}, PlotRange → All,
PlotStyle → {Hue[0.0]}, PlotLabel → {"ca", Hue[0]},
AxesLabel → {"t,sec", "N_a[t]"},
Ticks → {{0, tf/2, tf}, Automatic}],

Plot[{cb[t]V}, {t, 0, tf}, PlotRange → All,
PlotStyle → {Hue[0.1]},
PlotLabel → {"cb", Hue[0.1]},
AxesLabel → {"t,sec", "N_b[t]"},
```

```
Ticks → {{0, tf/2, tf}, Automatic}],

Plot[{cd[t]V}, {t, 0, tf}, PlotRange → All,
PlotStyle → {Hue[0.4]},
PlotLabel → {"cd", Hue[0.4]},
AxesLabel → {"t,sec", "Nd[t]"},
Ticks → {{0, tf/2, tf}, Automatic}],

Plot[{cp[t]V}, {t, 0, tf}, PlotRange → All,
PlotStyle → {Hue[0.5]},
PlotLabel → {"cp", Hue[0.5]},
AxesLabel → {"t,sec", "Np[t]"},
Ticks → {{0, tf/2, tf}, Automatic}],

Plot[{cop[t]V}, {t, 0, tf}, PlotRange → All,
PlotStyle → {Dashed, Hue[0.6]},
PlotLabel → {"cop", Hue[0.6]},
AxesLabel → {"t,sec", "Nop[t]"},
Ticks → {{0, tf/2, tf}, Automatic}]
}
]
```

We will let the concentration of **B** in the reactor be 0.1 M, the volume will be 10000L and the feed of **A** will be 0.1 M also. We will feed 2 L per second. Here we see that the conversion is almost 100% of **B** and the yield of **P** is 97%.

```
Cao = 10^-10;
CAf = 0.1;
q = 2;
Cbo = .1;
```

$V = 10000$;

tf = 12000;

FBR[Cao, Cbo, CAf, q, V, tf]

Conversion ={99.9%}

Selectivity to **D** ={0.01%}

Selectivity to **P** ={97.%}

Selectivity to **OP** ={3.%}

Yield of **P** ={97.%}

Yield of **OP** ={3.%}

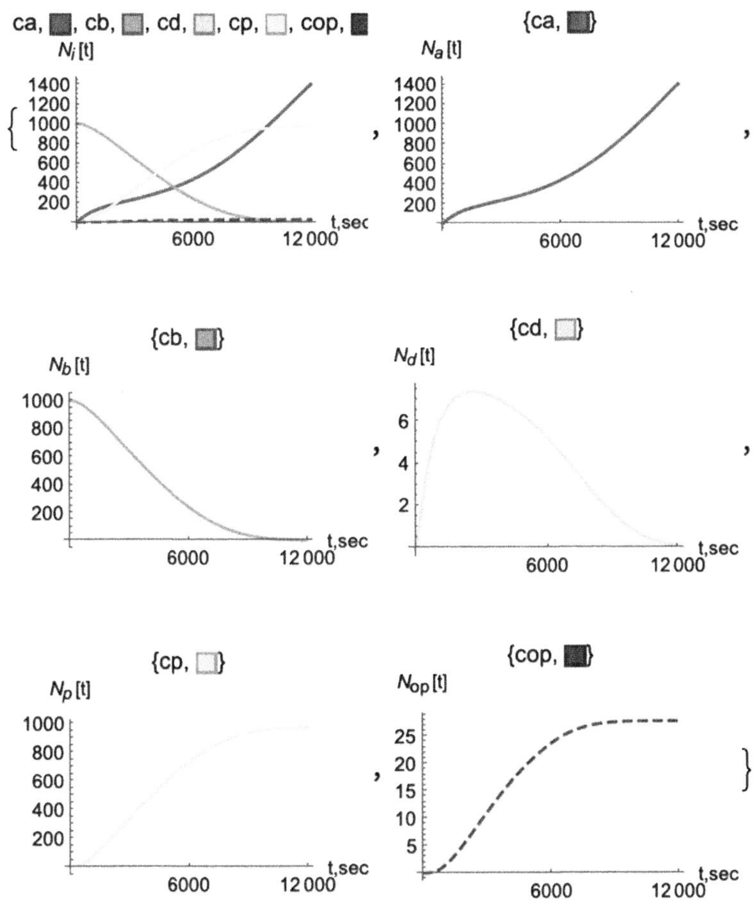

We can repeat the calculation, using V = 1000 L and the concentrations of **A** and **B** being 1 M. By keeping the rate of addition of **A** slow, we can get to above 99% conversion and 84% yield and in the smaller vessel.

Cao = 10^{-10};

CAf = 1;

q = .1;

Cbo = 1;

V = 1000;

tf = 12000;

FBR[Cao, Cbo, CAf, q, V, tf]

Conversion ={99.4%}

Selectivity to **D** ={0.06%}

Selectivity to **P** ={84.%}

Selectivity to **OP** ={16.%}

Yield of **P** ={84.%}

Yield of **OP** ={16.%}

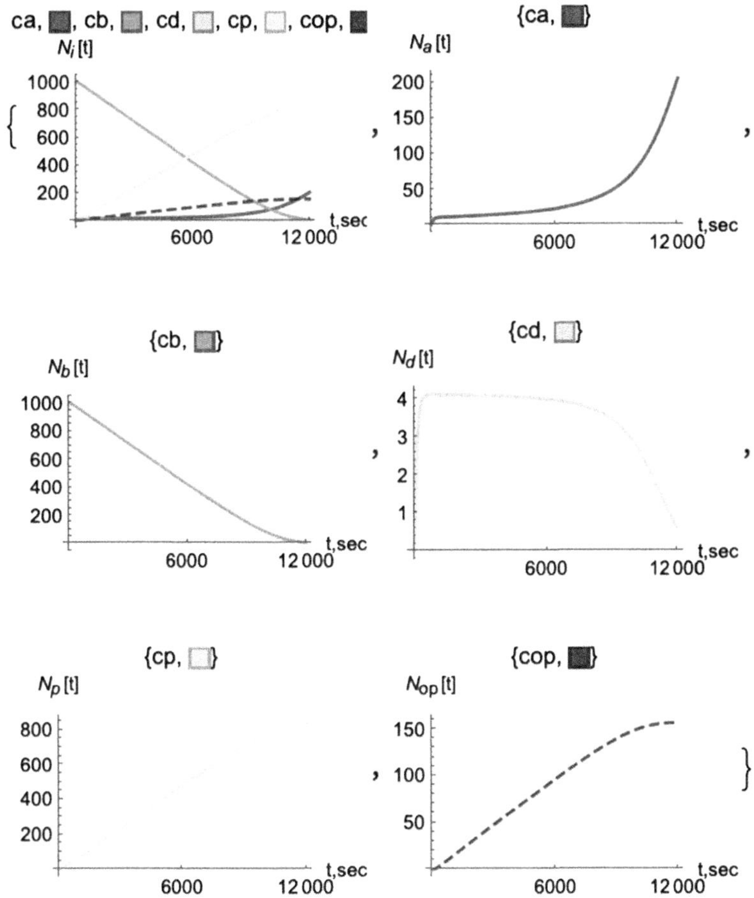

This is interesting because it illustrates why the fed-batch reactor with its semi-continuous operation can be so versatile and useful.

Accounting for larger volume change

The equations for variable volume are the following:

$$\partial_t \mathrm{Ca}[t]V[t] = \mathrm{Caf\ qaf - kab\ Ca[\ t\ ]Cb[\ t\ ]\ V[\ t\ ]}$$

$$\partial_t \mathrm{Cb}[t]V[t] = \mathrm{-\ kab\ Ca[\ t\ ]Cb[\ t\ ]\ V[\ t\ ]}$$

$$\partial_t \mathrm{Cd}[t]V[t] = \mathrm{+\ kab\ Ca[\ t\ ]Cb[\ t\ ]\ V[\ t\ ]}$$

There are now four time-dependent variables and only three equations. We need another equation and that is the total material balance. The mass in the control volume increases only by the additional mass admitted through the feed stream:

$$\partial_t \rho V[t] = \rho \text{ qaf}$$

If the density in the system is unchanging, and if the flow rate in is a constant, then the volume change is linear in time:

$$V[t] = Vo + qaf\ t$$

We need to add this equation to the rate equations to derive the full time-dependent solutions. In other words, we can mix algebraic and differential equations inside of NDSolve. We can repeat the calculations using all the same values and then compare to the case of volume change left constant.

```
Clear["Global`*"]

CAf = 1.;

CAo = 0;

CBo = .1;

CDo = 0;

kab = .1;

qAf = .001;

tmax = 20000;

Vr = 100;


fbsol2 = NDSolve[
{V[t] == qAf t + Vr,
∂t (CA[t]V[t]) == − CA[t]CB[t]V[t],
∂t (CB[t]V[t]) == − CA[t]CB[t]V[t],
∂t (CD[t]V[t]) == CA[t]CB[t]V[t],
  V[0] == Vr,
  CA[0] == CAo,
```

```
  CB[0] == CBo,
  CD[0] == CDo},
{V[t], CA[t], CB[t], CD[t]},
    {t, 0, tmax}];

v[t_]:=Evaluate[V[t]/.fbsol2]
ca[t_]:=Evaluate[CA[t]/.fbsol2]
cb[t_]:=Evaluate[CB[t]/.fbsol2]
cd[t_]:=Evaluate[CD[t]/.fbsol2]

numerical2 = Plot[{ca[t], cb[t], cd[t]},
{t, 0, tmax},
PlotRange → All,
PlotStyle → {Hue[0.2], Hue[0.5], Hue[0.8]},
AxesLabel → {t, "Ci[t]"},
PlotLabel → "Aq=Cb,Ppl=Cd,Gr=Ca"]

Plot[{(v[t])}, {t, 0, tmax},
AxesLabel → {t, "V[t]"},
PlotStyle → Hue[0.6],
PlotRange → {{0, tmax}, {0, 120}}]
```

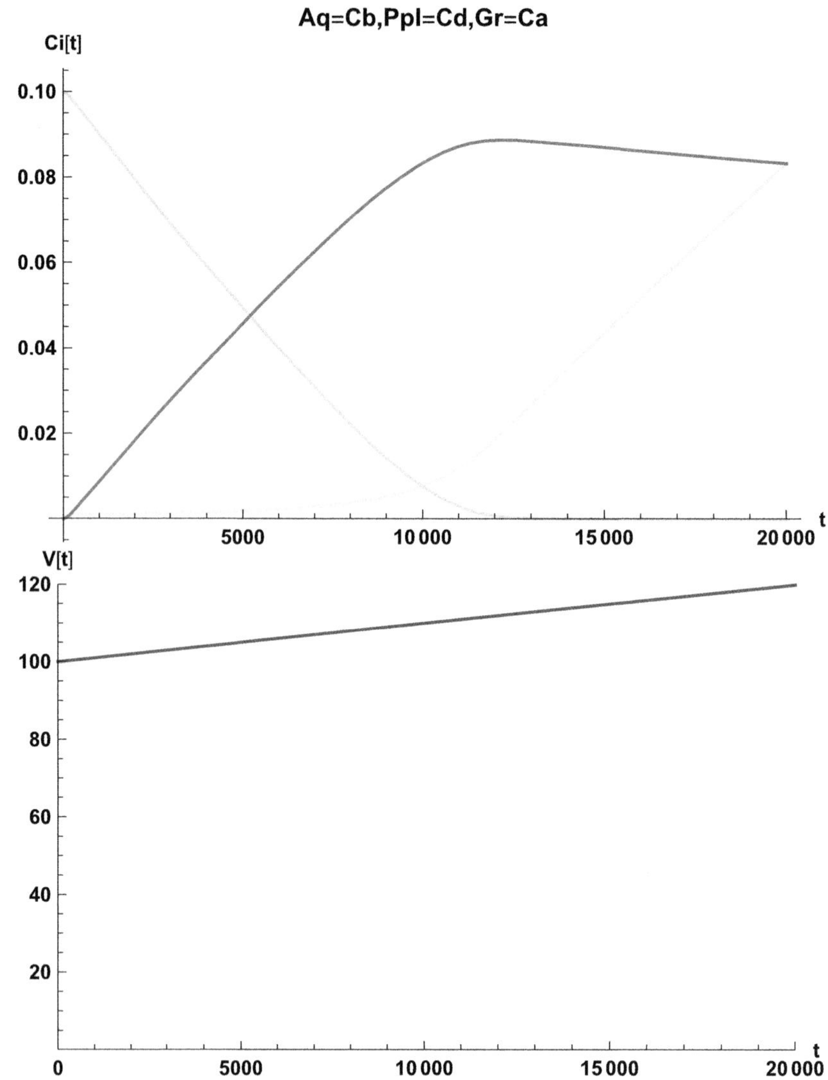

This is about a 20% volume change over the course of the reaction. If the results look familiar, they should. They look like the constant volume case except that as time increases the concentration decreases.

What follows is the comparison that shows the effect of including the volume change in the solutions. The solid lines include the volume change. As we can see in the graphs, the change in the concentration of **B** is almost the same and so too is its rate of change. The major differences we see are in the dilution effects on **D** and **A** after the reaction is complete.

Show[numerical, numerical2]

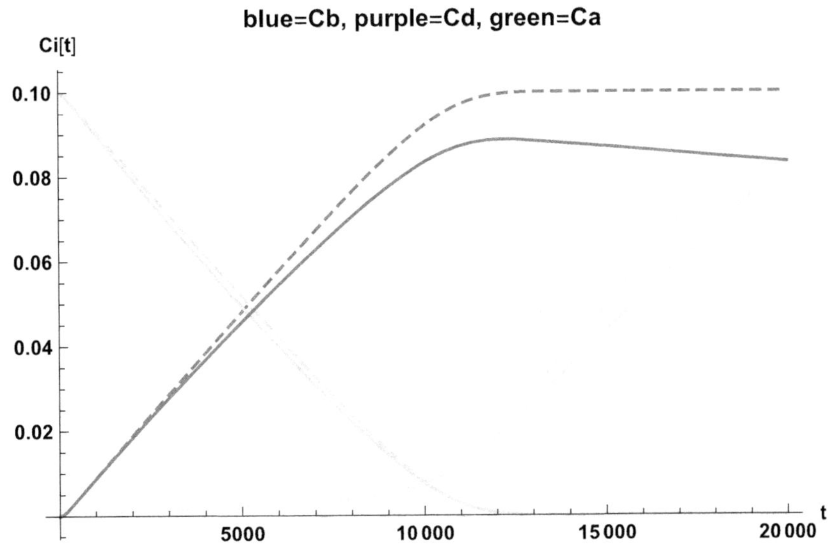

The first graph shows the time dependent concentrations of each component. The profile for **B** drops nearly linearly with time and that of product **D** rises the same way.

Economic potential

To this point we have not discussed any aspect of economics, so let's begin to do so now. Batch and fed-batch or semi-batch reactors are typically used for the production of chemical, pharmaceutical, or biological entities. The reason is that the markets for these chemicals tend to tolerate relatively high prices per unit of mass or volume and total production levels are not large in a relative sense, compared to bulk chemicals. Good examples of such entities are also flavors and fragrances. These are small volume chemicals with high added value. Added value means that because the conversion of starting materials to the product may involve multiple steps, the value after these steps are completed is much higher than the cost of the starting materials and production.

You may ask why is that the case? The reason is that these entities are "value priced," which means they are priced based on the value that they provide, or are perceived to provide, by the customer. In the case of fragrances, a few ounces of a perfume may cost hundreds of dollars at the high end of the market, and yet the essential oils from which they are derived may be two orders of magnitude less. The value is perceived by the customer and is imparted by a proprietary blend of ingredients that has been carefully developed and tested for effect. From this is deducted the cost of advertising, the cost of shipping and the cost of production (synthesis, purification, and blending of the ingredients), testing, and so forth. There are capital costs invested in the production equipment, operating costs due to the cost of people to make analyze and ship the product, and so much more. Business of this kind is fascinating. Highly successful companies that are in this business include Firmenich (Geneva, Switzerland), Givaudan (Vernier, Switzerland), International Flavors & Fragrances (New York, USA), Symrise (Holzminden, Germany), and Takasago (Tokyo, Japan). With the exception of Symrise, each of these companies is over 100 years old. It is a stable business. The key to their success and why they can use batch and semi-batch reactors is that the value of their products is so high that they can make them in batches and fill the world's demand doing so. Some of these products have a total production of say one kilogram per year!

So we need to start to understand this. For products that are low volume, high margin products, efficiency of production is much less of an issue, or not an issue at all. By contrast, production costs are critical to the economics of producing and selling high volume, low margin products. Gasoline and diesel are high volume and low margin products, and they are remarkably inexpensive. (Gasoline is ∼ $2.50-$3.00 per gallon. By contrast, depending on brand and perception thereof, it is not unusual to pay $2.50 per liter for still drinking water, that is ∼ $9.50 per gallon). The low cost of these fuels arises because production is very (extremely) efficient. Petroleum refineries are efficient because they are basically large robots with high capital expenditures (CAPEX) but low (relatively) operational expenditures (OPEX)). The feedstock, crude oil, is cheap, typically between $50 - $100 per barrel (42 gallons) and very available. The product (∼50% gasoline) is not expensive, but the volume of sales is enormous (∼145 x 10^9gallons per year in just the US).

By contrast to the cost of gasoline, worn by only the most self-confident and hardiest of men as a fragrance, Dolce & Gabbana K sells for about $20 per ounce, which is about $2560 per gallon. Even a Ferrari owner would have to think twice about fueling up with D&G K, but they might not think twice about wearing it to cover the smell of gasoline, after they fueled up on the way to the office. Everything else that is a chemical for consumption or use probably lies between the poles of gasoline and D&G K. (There are many other fine fragrances by Valentino, Gucci and others with similar prices that could have been used in this example, none of which have been deployed by the author.)

We need to start incorporating this in our thinking when we go from reactions that are hypothetical like A + B → C to those that are real like the production of L-DOPA, 3-(3,4-dihydroxyphenyl)-L-alanine (L-DOPA), a therapeutic for Parkinson's disease from α-acetamido-4-hydroxy-3-alkoxy-cinnamic acid acetate in the presence of hydrogen and a soluble rhodium catalyst. This reaction led to the Nobel Prize for Dr. William Knowles of Monsanto, who invented the reaction and process.

Chemical thermodynamics gives us the ultimate disposition of a reaction, while saying nothing of the rate of such a reaction to equilibrium. We are looking for something similar that can provide the economic potential of such a reaction. Whereas in thermodynamics, the quantities involved with the thermodynamic potential of a reaction are fixed, namely ΔG, ΔH, and ΔS, in the case of the economic potential, they are subject to change. The constituent quantities are not objects of nature that remain fixed, but, rather, are objects of markets, and markets change. Nonetheless, we can proceed with the calculation of an economic potential, if we realize that the potential is likely to be variable with time or era. Even though it is variable, if it is computed at a given time and reasonable precautions are in place to recognize these fluctuations over time, the idea is still very useful.

The economic potential, or more properly, the maximum economic potential (MEP) of a given product, is the difference between the price of the product per mass (or volume) and the starting materials. Every other consideration, will be a consideration of total cost of production, marketing and distribution, etc., and will deduct from the MEP of the product. So what is the use of the MEP? It tells us how wide the maximum margin is between the market price of the products and the cost of starting materials. It is an idealized margin on the product. It gives a measure of whether the production of this chemical or material has the marginal "width" to be worth pursuing. The narrower the MEP, the more difficult will be the business associated with it, ceteris paribus, which means assuming all else is the same.

We can evaluate the maximum economic potential for the general reaction of A with B to produce D. It will be given by the value per unit mass for the ingredients deducted from the value of the product.

$$\text{MEP} = \left(\frac{\$}{\text{mass product}}\right)(\text{mass of product}) - \left(\frac{\$}{\text{mass reactants}}\right)(\text{mass of reactant})$$

What one seeks to do is to convert reactants with relatively low value into the products with high added value. All the costs for people, capital, plants, energy, special equipment, packaging, sales, marketing and shipping etc. must be paid for from this difference in value.

Process information can be added to this equation to refine it further. If the process leads to 99.99% conversion and 100% selectivity, that means all of the reactant mass is converted to the product mass. This not typical. Usually some fraction of the reactants are converted, but not all of it, and there will be other products as well as the desired product. This is why yield is so valuable as a measure of process efficiency - yield is the product of reactant conversion and the selectivity to a specific product. Yield is sometimes quite high. For example, in the production of metal oxides like alumina, Al_2O_3, a salt of aluminum is added to a base like NaOH, and the products are a hydrated form of alumina and a by-product salt:

$$Al_2(SO_4)_3(aq) + 6\ NaOH(aq) \longrightarrow 2\ Al(OH)_3(ppt) + 3Na_2(SO_4)(aq)$$

$$2\ Al(OH)_3\ (ppt) \longrightarrow Al_2O_3(s) + 3\ H_2O\ (v)$$

$$Al_2(SO_4)_3(aq) + 6\ NaOH(aq) \longrightarrow Al_2O_3(s) + 3Na_2(SO_4)(aq) + 3\ H_2O\ (v)$$

Consulting on-line sources, the following are approximate prices molecular weights and solubilities for these chemicals:

$Al_2(SO_4)_3$ ~ \$200/ton = \$.22/kg; 342.2 g/g-mol; 36.4g/100ml

$Al(OH)_3$ ~ \$300/ton = \$.30/kg; 78 g/g-mol; $1x10^{-4}$g/100ml

γ - Al_2O_3 ~ \$6000/ton = \$6.6/kg; 102 g/g-mol; NA

NaOH ~ \$300/ton = \$.30/kg; 40 g/g-mol; 1000g/100ml

$Na_2(SO_4)$ ~ \$250/ton = \$.25/kg; 142 g/g-mol; 28.1g/100ml

The solubilities are given for 20-25 ^oC. If the reaction were conducted closer to the boiling point of water they would increase. But, note the solubility of $Al(OH)_3$; it is five orders of magnitude less than the corresponding sulfate. Thus, this is a precipitation reaction. After the precipitation is done, the mixture is filtered, washed multiple times and dried. Then, the $Al(OH)_3$, is heat treated in air above 500°C. Each of these steps, the precipitation and the heat treatment (called calcination) are carefully controlled in order to obtain the desired surface area, pore volume, and pore size in the alumina.

Next, look again at the prices of the reactants and the products. The final product, gamma-alumina, which is used as a support for precious metal catalyst particles, is valued at about 20 times more than the other chemicals involved in this reaction. That gives this process a wide MEP. We can run the numbers to see what the MEP comes to.

According to this reaction, 342.2 g of aluminum sulfate reacts with 240 g of sodium hydroxide to produce 156 g of aluminum trihydroxide and 426 g of sodium sulfate. We can always scale these number up. Then 156 g of aluminum trihydroxide is calcined to produce 102 g of gamma alumina and 54 g of water. The MEP of the first reaction is not large enough to make it the basis for an economical way to produce $Al(OH)_3$.

MEP (1) = [(0.156 kg * \$0.3/kg $Al(OH)_3$) + (.426 kg * \$0.25/kg $Na_2(SO_4)$)] - [(0.342 kg * \$.22/ kg $Al_2(SO_4)_3$) + 0.240 kg * \$ 0.30/kg NaOH)] = \$ 0.0061

But the MEP for the second reaction is quite large:

MEP(2) = [(0.102 kg * \$6.6/kg $Al(OH)_3$)] - [(0.156 kg * \$0.3/kg $Al(OH)_3$)] = \$ 0.626

And the MEP for the overall reaction to produce gamma-alumina and sodium sulfate is quite positive:

MEP(overall) = (0.102 kg * \$6.6/kg $Al(OH)_3$) + (.426 kg * \$0.25/kg $Na_2(SO_4)$) - (0.342 kg * \$.22/ kg $Al_2(SO_4)_3$) - (0.240 kg * \$ 0.30/kg NaOH)= \$.632

This says that for every mole of gamma alumina produced by this reaction the MEP overall is about \$.63 per mole of alumina produced. One mole of alumina is 0.102 kg so this corresponds to an MEP of about \$1.77 per pound of alumina. That sounds promising, but it is probably on the edge of being economically viable. What that means is that the costs of manufacturing, shipping, sales, and marketing could easily consume 90% or more of this value potential. When used as a base for precious metal, or even base metals, the specifications on the alumina may be quite stringent. It must have the correct porosity, pore size, surface area, shape (spheres, trilobes, etc.) mechanical strength, and chemical purity. If the base is to be used to support platinum for hydrocarbon reforming, it must be very high purity, whereas if it is to be used to support cobalt and molybdenum for hydrodesulfurization catalysis, the purity may be lower. All these factors will affect the cost of manufacturing of the alumina and the ultimate extent to which any of the MEP is realized. Costs of manufacturing are lower in China and elsewhere in Asia than in the United States and Europe, but the costs of shipping are higher and if rigorous standards of manufacturing must be met, that may factor into the decision of where to produce the material.

$$Al_2(SO_4)_3(aq) + 6\ NaOH(aq) \longrightarrow 2\ Al(OH)_3(ppt) + 3Na_2(SO_4)(aq)$$

$$2\ Al(OH)_3\ (ppt) \longrightarrow Al_2O_3(s) + 3\ H_2O\ (v)$$

$$Al_2(SO_4)_3(aq) + 6\ NaOH(aq) \longrightarrow Al_2O_3(s) + 3Na_2(SO_4)(aq) + 3\ H_2O\ (v)$$

$$A + 6\ B \longrightarrow D + 3\ E + 3F$$

A = $Al_2(SO_4)_3$,

B = NaOH

D = $Al(OH)_3$

E = $Na_2(SO_4)$

F = H_2O

G = Al_2O_3

r = kab CA[t] CB[t]

We simulate the reaction taking place in a vessel large enough to accommodate 1000 L of sodium hydroxide base at 6.4 moles per Liter. To this we will envision adding a solution of

1 M aluminum sulfate with vigorous agitation. We will let the input flow rate be a constant until the target concentration ($< 5 \times 10^{-3} molar$) is reached and then it will go to zero. We do this with the following line of code:

$$qA(t) = \text{If}\left[CB[t] < 5 \times 10^{-3}, 0, qAf\right]$$

This is a way to simulate in the code the automatic shutoff of the feed after the concentration of the base in the unit drops to a number low enough to approximate zero.

```
Clear["Global`*"]

CAf = 1.;

CAo = 0;

CBo = 6.4;

CDo = 0;

CEo = 0;

CFo = 0;

kab = 100;

qAf = 1;

Vr = 1000;

$A = .075;

$B  = .012;

$G = .673;

$E = .036;

$F = 0;

tmax = 1250;

fbsol1 = NDSolve[

{qA[t] == If[CB[t] < 5 * 10^-3, 0, qAf],

V[t]==(Vr + qA[t]t),

∂t(CA[t]V[t]) ==qA[t] - CA[t]CB[t]V[t],

∂t(CB[t]V[t])== - CA[t]CB[t]V[t],

∂t(CD[t]V[t]) == CA[t]CB[t]V[t],
```

$\partial_t(\text{CE}[t]V[t]) == \text{CA}[t]\text{CB}[t]V[t]$,

$\partial_t(\text{CF}[t]V[t]) == 3\text{CA}[t]\text{CB}[t]V[t]$,

CA[0] == CAo,

CB[0] == CBo,

CD[0] == CDo,

CE[0] == CEo,

CF[0] == CFo,

qA[0] == qAf,

V[0] == Vr},

{CA[t], CB[t], CD[t], CE[t], CF[t], qA[t],

V[t]}, {t, 0, tmax}];

ca[t_]:=Evaluate[CA[t]/.fbsol1]

cb[t_]:=Evaluate[CB[t]/.fbsol1]

cd[t_]:=Evaluate[CD[t]/.fbsol1]

ce[t_]:=Evaluate[CE[t]/.fbsol1]

cf[t_]:=Evaluate[CF[t]/.fbsol1]

qa[t_]:=Evaluate[qA[t]/.fbsol1]

Vv[t_]:=Evaluate[V[t]/.fbsol1]

vmax = Max[Table[Vv[t], {t, 0, tmax}]];

vmax

$$\text{ep[t\_]:=}\left(\$\text{G}\frac{\text{cd}[t]}{2}(\text{Vr}+\text{qa}[t]*t)\right)\frac{\text{qa}[t]}{\text{qAf}}+$$

$$(\$\text{E ce}[t](\text{Vr}+\text{qa}[t]*t))\frac{\text{qa}[t]}{\text{qAf}}-$$

$$(\$\text{B cb}[t](\text{Vr}+\text{qa}[t]*t))\frac{\text{qa}[t]}{\text{qAf}}-$$

$$(\$\text{A CAf qa}[t]t)\frac{\text{qa}[t]}{\text{qAf}}$$

```
epm = Max[Table[ep[t], {t, 0, tmax}]];
epm
Plot[Vv[t], {t, 0, tmax},
PlotStyle → Hue[0.35],
Epilog → {Dashed, Hue[0.35],
Line[{{0, vmax}, {tmax, vmax}}]},
PlotLabel → {vmax" = Vmax"}]
Plot[{ca[t], cb[t], cd[t], ce[t]},
{t, 0, tmax},
PlotRange → All,
PlotStyle → {Hue[0.4], Hue[0.6], Hue[0], Hue[0.1]},
AxesLabel → {t, "Ci[ t ]"},
PlotLabel →
"blue = Cb, green = Ca, red = Cd, gold = Ce "]

Show[
Plot[epm, {t, 0, tmax},
PlotRange → {{0, tmax}, {−200, 800}},
PlotStyle → {Dashed, Hue[.6]},
AxesLabel → {t, "$[ t ]"},
PlotLabel → {epm" = $MEP "}],
Plot[ep[t], {t, 0, tmax},
PlotRange → All,
PlotStyle → {Dashed, Hue[0.6]},
AxesLabel → {t, "$[ t ]"}]
]
```

This is vmax: 2066.

This is epm: 751.471

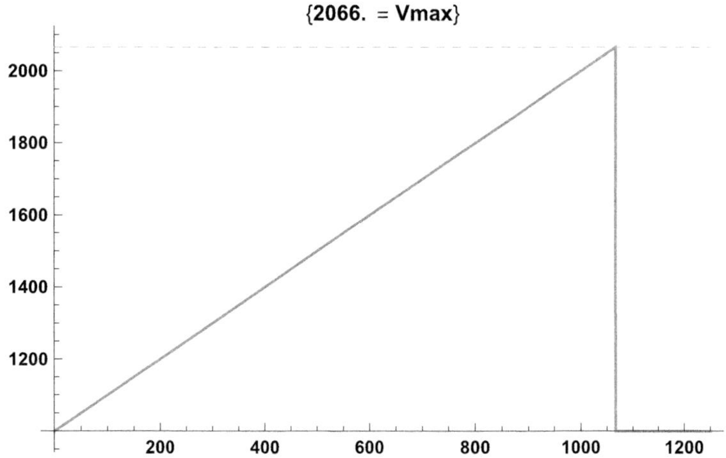
{2066. = Vmax}
2000
1800
1600
1400
1200
200
400
600
800
1000
1200

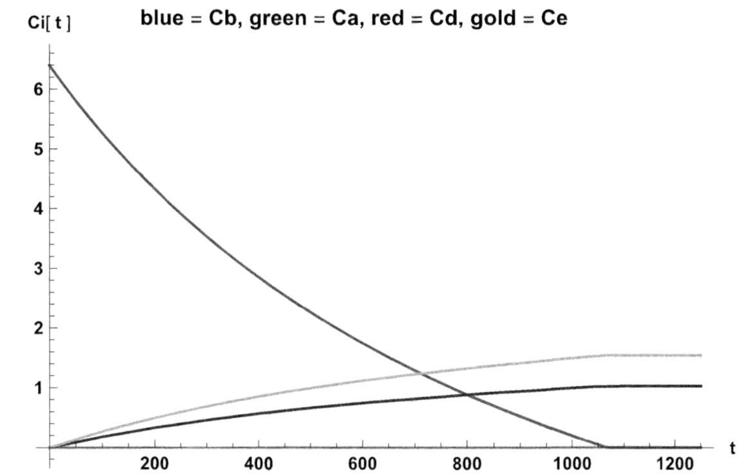
Ci[t]
blue = Cb, green = Ca, red = Cd, gold = Ce
6
5
4
3
2
1
t
200
400
600
800
1000
1200

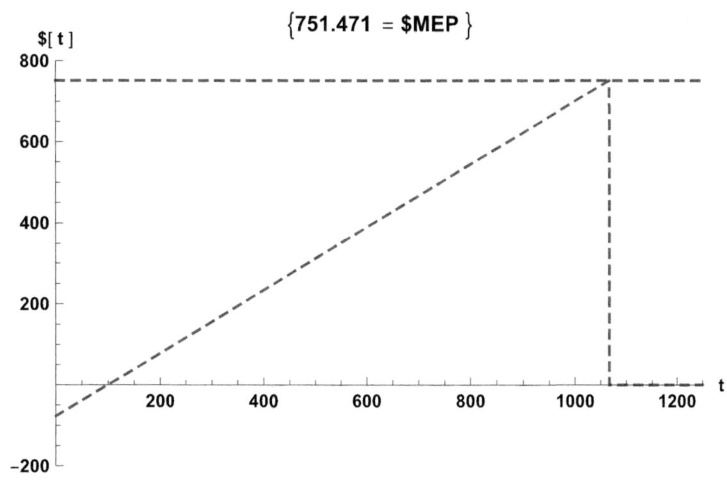
$[t]
{751.471 = $MEP }
800
600
400
200
−200
t
200
400
600
800
1000
1200

The concentration of the base drops in a curvilinear manner with time and the products, $Al(OH)_3$ (red - dark gray in the print version) and $Na_2(SO_4)$ (gold - light gray in the print version) rise curvilinearly. The concentration of $Al_2(SO_4)_3$ (green - semi-gray in the print version) stays close to zero throughout the process. The economic potential starts at a slightly negative value because we defined the value of the reactants as a cost (negative value) and rises sharply to a maximum. It remains at the maximum level. We have specified the economic potential as a time-dependent function:

$$\text{ep}[\text{t\_}]{:=}\left(\$\text{G}\frac{\text{cd}[t]}{2}(\text{Vr}+\text{qa}[t]*t)\right)\frac{\text{qa}[t]}{\text{qAf}}+(\$\text{Ece}[t](\text{Vr}+\text{qa}[t]*t))\frac{\text{qa}[t]}{\text{qAf}}-(\$\text{Bcb}[t](\text{Vr}+\text{qa}[t]*t))\frac{\text{qa}[t]}{\text{qAf}}-(\$\text{ACAfqa}[t]t)\frac{\text{qa}[t]}{\text{qAf}}$$

The first term is the value of the gamma-alumina (the finished product. We compute it by calculating the product of the volume of the solution at any time (Vr+ qa[t] * t)) and the concentration of $Al(OH)_3$ (D) also at any time. This gives us the number of moles of $Al(OH)_3$ (D), but it takes two moles of aluminum hydroxide to make one mole of alumina, hence division by two. This is multiplied by the value per mole of the alumina. The last term in this product is $\frac{\text{qa}[t]}{\text{qAf}}$ and it is included to make the term go to zero after the maximum is reached. At all time before the maximum, this fraction equals unity. The other terms are similarly put together.

We can move from this particular case to the more general. The next step we take is to create a Module function, semi2, which is a code that we may run many different cases of a semi-batch reactor and the reaction of A + B ⟶ D with second order kinetics. We have put every parameter that can be of interest to explore into curly bracketed groups:

```
semi2[{CAf_, CAo_, CBo_, CDo_}, kab_, qAf_,
{$D_, $B_, $A_}, {Mwd_, Mwb_, Mwa_}, Vr_, tmax_]:=
Module[
{V, qA, CA, CB, CD, v, ca, cb, cd, fbsol2, ep, t},
fbsol2 = NDSolve[
{qA[t] == If[CB[t] < 5 * 10^-3, 0, qAf],
∂t V[t] == qA[t],
∂t (CA[t]V[t])==−CA[t]CB[t]V[t],
∂t (CB[t]V[t])== − CA[t]CB[t]V[t],
```

```
∂t(CD[t]V[t]) == CA[t]CB[t]V[t],
  V[0] == Vr,
  CA[0] == CAo,
  CB[0] == CBo,
  CD[0] == CDo,
qA[0] == qAf},
{V[t], CA[t], CB[t], CD[t], qA[t]},
    {t, 0, tmax}];

v[t] = Evaluate[V[t]/.fbsol2];
ca[t] = Evaluate[CA[t]/.fbsol2];
cb[t] = Evaluate[CB[t]/.fbsol2];
cd[t] = Evaluate[CD[t]/.fbsol2];
qa[t] = Evaluate[qA[t]/.fbsol2];

ep[t] = ($Dcd[t]v[t]Mwd − $Bcb[t]v[t] Mwb−
$ACAfqa[t]Mwat);
{
{Graphics[Plot[{ca[t], cb[t], cd[t]},
{t, 0, tmax},
PlotRange → All,
PlotStyle → {},
AxesLabel → {t, "Ci[t]"},
PlotLabel → "Gold = CB, Green = CD, Blue = CA"
  ]
  ]
  },
```

```
{Graphics[
Plot[{(CAfqAfMwa tmax) + CBoVr Mwb,
cb[t]v[t] Mwb, cd[t]v[t]Mwd}, {t, 0, tmax},
AxesLabel → {t, "mi[t]"},
PlotStyle → {},
PlotLabel → "Blue=m[A+B], Green=mD, Gold=mB"]
    ]
    },
{Graphics[Plot[ep[t], {t, 0, tmax},
PlotRange → All, PlotStyle → {},
AxesLabel → {t, "$[t]"},
PlotLabel → "Value"]
    ]
    },
{Graphics[Plot[{(v[t])}, {t, 0, tmax},
PlotRange → All,
AxesLabel → {t, "V[t]"},
PlotStyle → Hue[0, .5, .5]]]}}  ]
```

In this case we have written the economic potential in a way that uses computed masses of reactants and products and their value per unit mass as well. Hence, we go from concentration to moles and from moles to mass with the molecular weight.

ep[t] = ($D cd[t] v[t] Mwd - $B cb[t]v[t] Mwb - $A CAf qAf Mwa t)

Here is the function with all the parameters labeled:

semi2[{CAf_,CAo_,CBo_,CDo_},kab_,qAf_,{$D_,$B_,$A_},{ Mwd_, Mwb_, Mwa_}, Vr_, tmax_]

```
semi2[{1, 0, .1, 0}, .1, .001, {20, 5, 4}, {60, 40, 20},100, 15000]
```

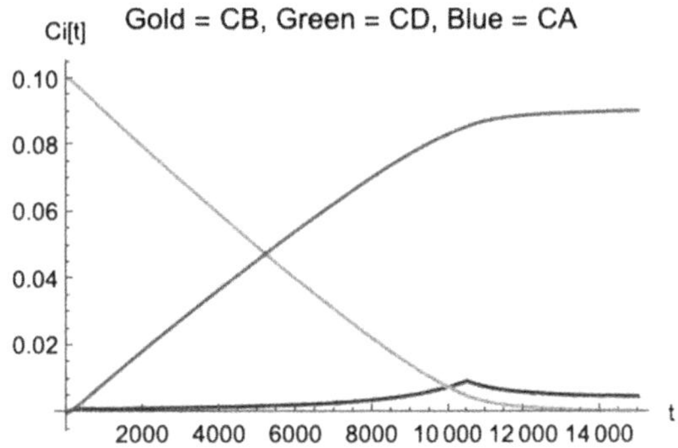
Gold = CB, Green = CD, Blue = CA
Ci[t]
0.10
0.08
0.06
0.04
0.02
2000 4000 6000 8000 10 000 12 000 14 000
t

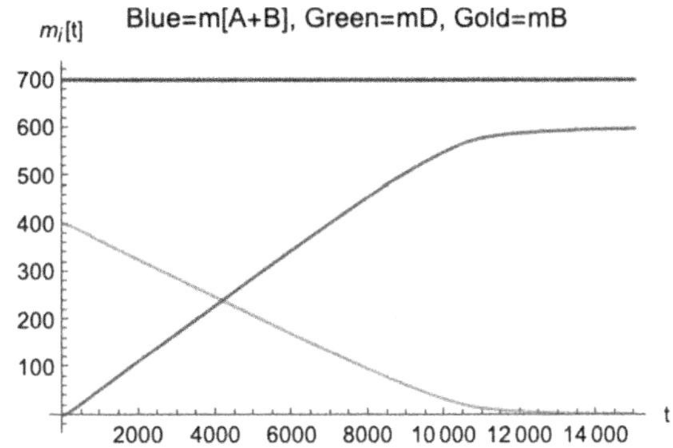
Blue=m[A+B], Green=mD, Gold=mB
m_i[t]
700
600
500
400
300
200
100
2000 4000 6000 8000 10 000 12 000 14 000
t

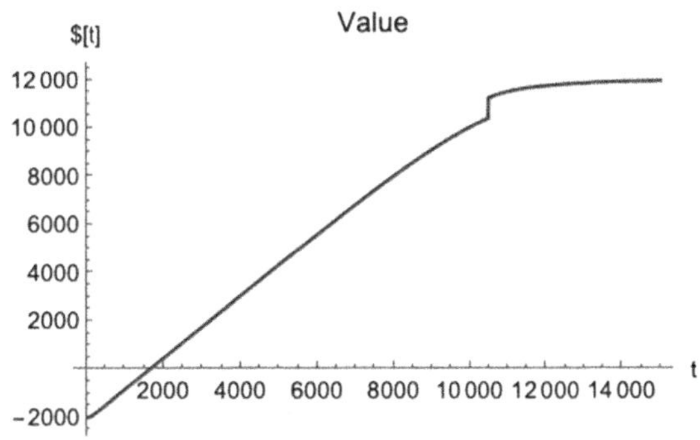
Value
$[t]
12 000
10 000
8000
6000
4000
2000
−2000
2000 4000 6000 8000 10 000 12 000 14 000
t

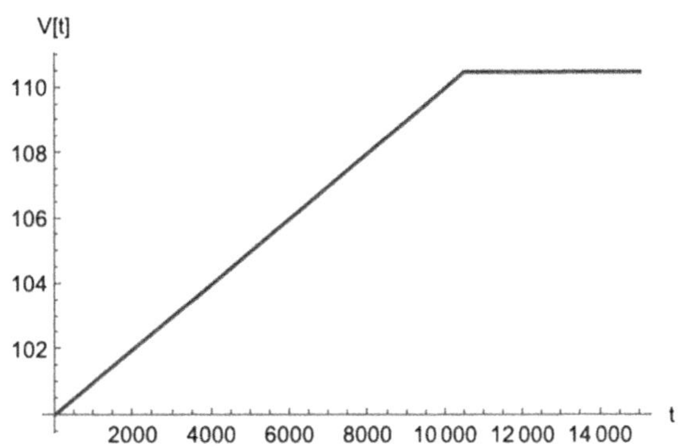
V[t]
110
108
106
104
102
2000 4000 6000 8000 10 000 12 000 14 000
t

The overall value of the product mixture increases significantly with the increased value of the product.

How could one raise the price that the market would bear for alumina? By engineering into it chemical and physical properties that are specialized, or "tuned" for specific uses. If the benefit to the user, the customer, is significant enough, then they may pay extra for this product. Another strategy is to use the same technology to serve a different market. So, instead of making high surface area, narrow-pored gamma-alumina for certain petroleum processes (reforming and hydrotreating), one could shift to making low surface area, large pored alpha-alumina to support silver, which is used as a catalyst for the epoxidation of ethylene. Perhaps, here, the value is higher. From there, one might begin to make low surface area, high density alumina for the ceramics sector and perhaps purity would also be a significant consideration. In this way the business would have a diversified portfolio of different alumina products that are made by the same methods using the same plant, machinery, and people. Diversification allows a business to manage the risk associated with downturns in different sectors of the market. For example, if the petroleum market is down, maybe the chemical and ceramics markets are up and vice versa. Therefore assessing the values of the products versus those of the starting materials is a first step in considering a business and it is far from the last.

The following is a Module function for a semi-batch reactor for

$$A + B \longrightarrow D$$

and using the masses of reactants and product to compute the economic potential of the reactant mixture continuously:

```
semi3[{CAf_, CAo_, CBo_, CDo_}, kab_, qAf_,
{$D_, $B_, $A_}, {Mwd_, Mwb_, Mwa_}, Vr_, tmax_]:=

Module[
{V, qA, CA, CB, CD, v, ca, cb, cd, fbsol2, ep, t},

fbsol2 = NDSolve[
{qA[t] == If[CB[t] < 5 * 10^-3, 0, qAf],
∂t V[t] == qA[t],
∂t (CA[t] V[t]) ==
-CA[t] CB[t] V[t],
```

```
∂t(CB[t]V[t])== − CA[t]CB[t]V[t],
∂t(CD[t]V[t]) == CA[t]CB[t]V[t],
  V[0] == Vr,
  CA[0] == CAo,
  CB[0] == CBo,
  CD[0] == CDo,
qA[0] == qAf},
{V[t], CA[t], CB[t], CD[t], qA[t]},
    {t, 0, tmax}];

v[t] = Evaluate[V[t]/.fbsol2];
ca[t] = Evaluate[CA[t]/.fbsol2];
cb[t] = Evaluate[CB[t]/.fbsol2];
cd[t] = Evaluate[CD[t]/.fbsol2];
qa[t] = Evaluate[qA[t]/.fbsol2];

ep[t] = ($Dcd[t]v[t]Mwd − $Bcb[t]v[t] Mwb−
$ACAfqa[t]Mwat);

Graphics[Plot[ep[t], {t, 0, tmax},
PlotRange → All, PlotStyle → {},
Ticks → Automatic,
AxesLabel → {t, "$[t]"},
PlotLabel → "Value"]]]
```

Using Manipulate, we can make the value of the product continuously varying in order to see the way that this affects the value at the end of the reaction. All the other parameters and the time variable can be changed as well.

```
Manipulate[semi3[{1, 0, .1, 0}, .1, .001, {n, 5, 4},
```

{60, 40, 20}, 100, 15000], {n, 0, 50, .1}]

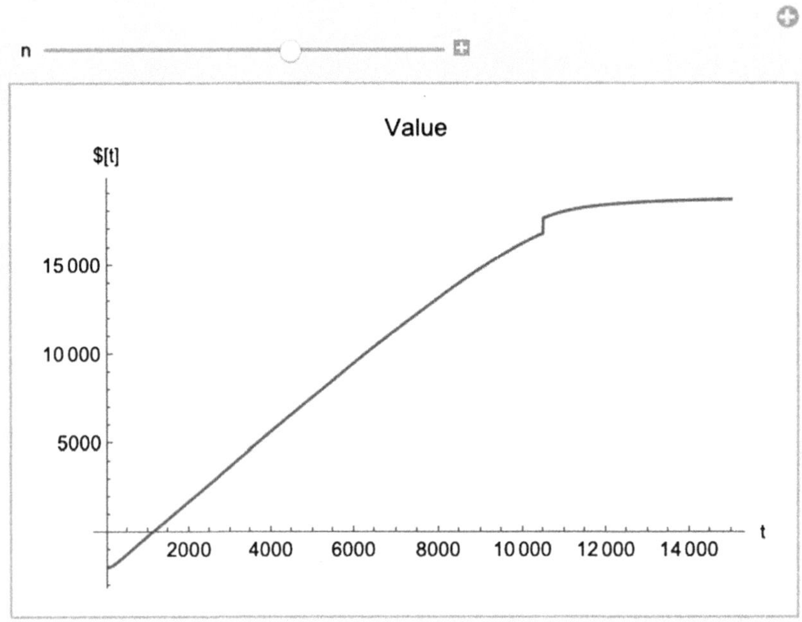

As with almost every topic that we have covered, there is more that can be said about the fed-batch of semi-batch reactors, but our coverage should provide sufficient grounding for the reader to consider other cases on their own.

Continuous flow reactors

The two most useful idealizations of chemical reactors are the continuously stirred tank reactor (CSTR) and the plug flow reactor (PFR). Both are idealizations and they represent two different and quite distinct extremes of idealized mixing. Real reactors are more complex, but often they can be analyzed approximately in terms of these idealizations. Furthermore, when starting from scratch to consider the design of a new reactor system, these simplified models are used to estimate the size of the system that will be required and, in some cases, which mixing regime will lead to better results. Finally, the ideal reactors allow us to do analyses when the specific kinetics are complex that will give us real insight into how actual reactors operate, which factors are most important and how to control them for better performance. So although the CSTR and PFR are idealizations, they are quite powerful models for chemically reacting systems. In the laboratory, in many cases experimental reactors can be constructed that behave as nearly ideal. In developing kinetics, we have much to gain from studies using them. We begin with the perfectly mixed system first because it is a powerful tool for the kineticist.

Continuous flow stirred tank reactor

The name continuous flow stirred tank reactor is nicely descriptive of the type of reactor that is used frequently for production and for fundamental kinetic studies. But, this name, abbreviated as CSTR, doesn't quite get to the essence of the idealization that this reactor represents. The ideality arises from the assumption in the analysis that the reactor is perfectly mixed, that its measurable properties are *homogeneous*. A better name for this model might be the PMR for *perfectly mixed reactor*. Nonetheless, so long as we realize this and its mathematical consequences, and that the terminology refers to the mathematics of the idealized form of the reactor, then the more prevalent name CSTR is serviceable.

Because of the well-mixed assumption, it is natural to think of the CSTR as a liquid phase reactor with a mixer as shown in the schematic:

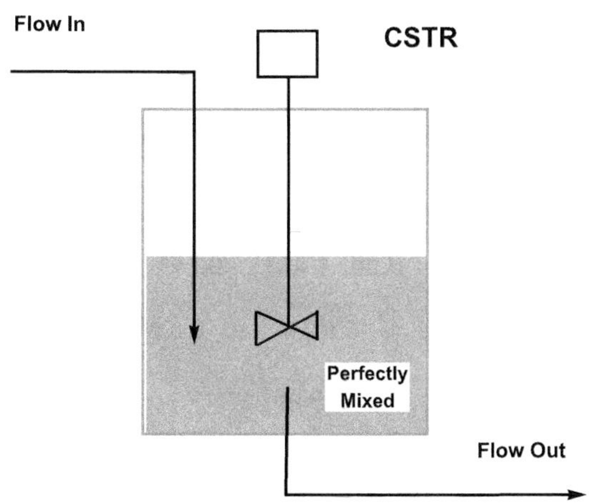

The chemistry in this case is the irreversible conversion of A to D and following simple, linear kinetics:

$$A \longrightarrow D$$

$$rA- = ka\ CA[t]$$

When we write the time-dependent mass balance for A in this system we have in words:

The rate of change of the mass of A is equal to the mass flow of A into the system, minus the mass flow of A out of the system minus the rate at which A is converted by chemical reaction into one or more other species, i.e. products.

The equation looks as follows with the product of concentration and volume flow rate being mass flow rates, per the usual. If the concentrations are in mass concentrations, then these are mass flow rates and they are mole flow rates, if the concentrations are molar. (Remember the molecular weight allows us to go back and forth between the two types of concentrations.) Here we assume molar concentrations, as is more typical when chemical reaction rates are involved:

$$\frac{d\text{Ca}[t]V[t]}{dt} = \text{Caf qaf - Ca}[t]\text{q - kad Ca}[t]\text{V}[t]$$

$$\frac{d\text{Cd}[t]V[t]}{dt} = -\text{Cd}[t]\text{q} + \text{kad Ca}[t]\text{Cb}[t]\text{V}[t]$$

$$\frac{d\rho[t]V[t]}{dt} = \rho\text{af qaf} - \rho[t]\text{q}$$

If the system is at steady state, then total mass flowing in must be balanced by the total mass out. Furthermore, if the densities of the feed and product are nearly the same, then we can take the flow rate in as equal to the flow rate out. The equations for the system at steady state become:

$$0 = (\text{Caf -Ca})\text{q - kad Ca V}$$

$$0 = -\text{Cdq} + \text{kad Ca V}$$

$$0 = \text{qaf - q}$$

We can divide both of the component balances by the product Caf V to find:

$$0 = \frac{(1 - \Phi\text{a})}{\theta} - \text{kad }\Phi\text{a}$$

$$0 = -\frac{\Phi\text{d}}{\theta} + \text{kad }\Phi\text{a}$$

where $\frac{V}{q} = \theta$, is the holding time for the CSTR. If we now solve for the dimensionless exit concentration, then we find:

Clear["Global`* "]

Solve[{

$$\{0 == \frac{(1-\Phi a)}{\theta} \text{ - kad } \Phi a, \{\Phi a\}]$$

$$\left\{\left\{\Phi a \to \frac{1}{1+\text{kad}\theta}\right\}\right\}$$

Solving both equations simultaneously to find Φd:

Clear["Global`* "]

Solve[{

$$\{0 == \frac{(1-\Phi a)}{\theta} \text{ - kad } \Phi a,$$

$$0 == -\frac{\Phi d}{\theta} + \text{kad } \Phi a\},$$

{Φa,Φd}]

$$\left\{\left\{\Phi a \to -\frac{1}{\theta(-\text{kad})-1}, \Phi d \to \frac{\theta \text{kad}}{\theta \text{kad}+1}\right\}\right\}$$

The concentration of **A** leaving the reactor is the product of the first order rate constant and the holding time over the sum of one plus the product of the first order rate constant and the holding time. The first order rate constant, we recall, has dimensions of reciprocal time, and the holding time is just time, so their product is dimensionless. In fact this product, kad θ, is actually the ratio of the holding time to the characteristic time required for the chemistry to occur. If the rate constant is taken to be of order unity, then let's see how the concentrations of **A** and **D** change with holding time.

$$\Phi a[\theta\_]:=\frac{1}{1+\text{kad}\theta}$$

$$\Phi d[\theta\_]:=\frac{\text{kad}\theta}{1+\text{kad}\theta}$$

kad = 1;

Plot[{Φa[θ], Φd[θ]}, {θ, 0, 50},

PlotStyle → {Hue[0], Hue[.6]},

AxesLabel → {"θ ", "Φi[t]"},

Epilog → {RGBColor[0, 1, 0], Line[{{0, 1}, {50, 1}}]},

PlotLabel → "St.St.CSTR 1st Ord. Irrev. Rxn."]

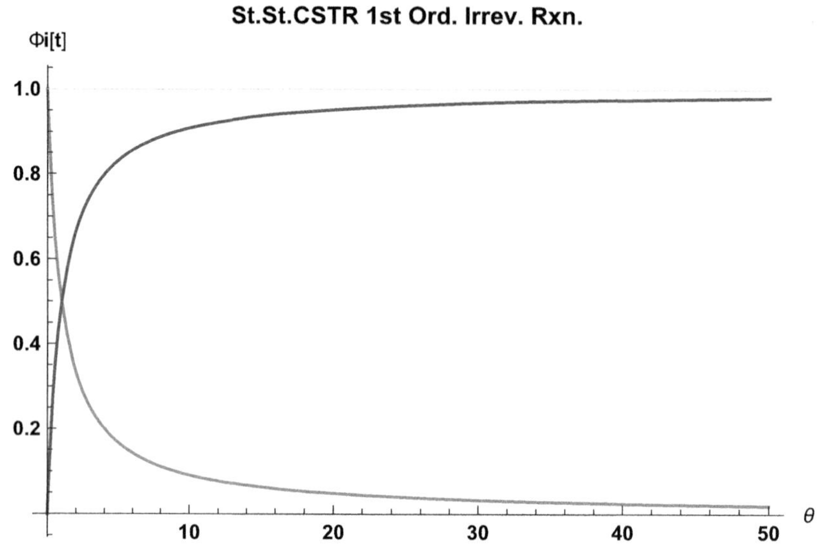

When the holding time has become a factor of ten larger than the characteristic time for the reaction chemistry, then we find that the concentration of **A** has dropped by ~90% of it feed value. Hence, the dimensionless concentration of product **D** tends toward unity asymptotically.

There is another important way to view these equations. If we go back to the dimensional form, it will be more evident. Typically, the CSTR is considered to be operated at steady state, which much simplifies the problem as we will see.

$$0 = (\text{Caf} - \text{Ca})\,\text{q} - \text{kad Ca V}$$

$$0 = -\text{Cd q} + \text{kad Ca V}$$

We can rearrange these as follows:

$$(\text{Caf} - \text{Ca})\,\frac{q}{V} = \text{kad Ca}$$

$$\frac{(\text{Caf} - \text{Ca})}{\theta} = \text{kad Ca}$$

$$\frac{(\text{Caf} - \text{Ca})}{\theta} = r_{\text{ad}}$$

This last equation shows why the CSTR at steady state is such a valuable tool to the experimentalist doing kinetics. The rate of reaction, *independent of the form of the kinetics*, is simply the change in concentration of A between the feed and exit streams, divided by the holding time. It has the beauty of sheer simplicity.

There is something important to learn from rearranging the second equation also:

$$0 = -\text{Cd q} + \text{kad Ca V}$$

$$\frac{\text{Cd}\, q}{V} = r_{\text{ad}}$$

This result, which is also very simple, can be used in several ways. Firstly, let's say we know the rate of a chemical reaction at process conditions, then knowing how much we want to produce in certain period of time, i.e. the target production rate is the product Cd q, and we can then compute the volume of the well-mixed reactor necessary to achieve this output.

$$\frac{\text{Cd}\, q}{r_{\text{ad}}} = \text{V}$$

Likewise, if we have a reactor available of a given volume and we know the reaction rate at process conditions, then we can compute the production rate that can be obtained:

$$\text{Cd}\, q = r_{\text{ad}}\, \text{V}$$

Lastly, maybe we do not know the rate of the chemical reaction, but we do know our competitors production rate and someone has managed to take a picture of their reactor, from which we can compute its size and volume, then we can get the rate from the original arrangement of the formula:

$$\frac{\text{Cd}\, q}{V} = r_{\text{ad}}$$

So, this becomes a useful equation to be employed in the earliest stages of the design for an economic viable reactor and it dropped out of the simplest analysis for a CSTR. It is rare to get something so simple and useful from such a straightforward analysis.

Steady state CSTR with higher order, reversible kinetics

The first order, irreversible chemical rate case is useful in terms of providing us with insight into what are the consequences of perfect mixing and with a sense of how the characteristic times for reaction and flow are related. On the other hand, it is limited in usefulness since it represents highly simplified chemistries and corresponding kinetics. Often the actual kinetics are far more complex. Let's consider the same chemistry as we had examined in the fed-batch reactor, namely that of **A** and **B** reacting to give **D**. The rate law will be second order overall, first order in each component, but this time let's assume that is it reversible and that the rate law for the reverse reaction will be second order in **D**:

$$\mathrm{A} + \mathrm{B} \Longleftrightarrow \mathrm{D}$$

$$\mathrm{rf} = \mathrm{kab\ Ca\ Cb}$$

$$\mathrm{rr} = \mathrm{kd\ Cd}^2$$

$$r_A = r_B = -r_D = -\mathrm{kab\ Ca\ Cb} + \mathrm{kd\ Cd}^2$$

and

$$\mathrm{Keq} = \frac{\mathrm{kd}}{\mathrm{kab}}$$

The transient balance equations for this system will be:

$$\partial_t \mathrm{Ca}[t]\ \mathrm{V}[\mathrm{t}] = \mathrm{Caf\ qf} - \mathrm{Ca}[\mathrm{t}]\ \mathrm{q} - \mathrm{kab\ Ca}[\mathrm{t}]\ \mathrm{Cb}[\mathrm{t}]\ \mathrm{V}[\mathrm{t}] + \mathrm{kd\ Cd}[\mathrm{t}]^2\ \mathrm{V}[\mathrm{t}]$$

$$\partial_t \mathrm{Cb}[t] V[t] = \mathrm{Cbf\ qf} - \mathrm{Cb}[\mathrm{t}]\ \mathrm{q} - \mathrm{kab\ Ca}[\mathrm{t}]\ \mathrm{Cb}[\mathrm{t}]\ \mathrm{V}[\mathrm{t}] + \mathrm{kd\ Cd}[\mathrm{t}]^2\ \mathrm{V}[\mathrm{t}]$$

$$\partial_t \mathrm{Cd}[t]\ \mathrm{V}[\mathrm{t}] = -\mathrm{Cd}[\mathrm{t}]\ \mathrm{q} + \mathrm{kab\ Ca}[\mathrm{t}]\ \mathrm{Cb}[\mathrm{t}]\ \mathrm{V}[\mathrm{t}] - \mathrm{kd\ Cd}[\mathrm{t}]^2$$

$$\partial_t \rho[t] V[t] = \rho \mathrm{f\ qbf} - \rho\ [\mathrm{t}]\ [\mathrm{t}]\ \mathrm{q}$$

At the steady state, the mass input must be the same as the mass output. Furthermore the net rate of change in concentration of each of the components is zero. This makes the differentials each zero in all four equations. The inlet densities of the liquid reactant streams are typically not too different from each other, or from the density of the outlet stream including products. The inlet flow rates and concentrations are also equal. Thus the equations reduce to:

$$0 = \mathrm{Caf\ qf} - \mathrm{Ca\ q} - \mathrm{kab\ Ca\ Cb\ V} + \mathrm{kd\ Cd}^2\ \mathrm{V}$$

$$0 = \mathrm{Cbf\ qf} - \mathrm{Cb\ q} - \mathrm{kab\ Ca\ Cb\ V} + \mathrm{kd\ Cd}^2\ \mathrm{V}$$

$$0 = - \text{Cd}\ \ q + \text{kab Ca Cb V} - \ \text{kd Cd}^2\text{V}$$

$$0 = \text{qf}\ - \text{q}$$

These are four equations that include a total of 9 variables and parameters, (qf = q), and only three of the four equations are independent, since three can be solved to find the fourth. We can solve the three component balances for the concentrations at the exit of the reactor using Solve.

```
Clear["Global`* "]
cstr1 = Simplify[
Solve[
{0 == (Caf − Ca)q − kabCaCbV + kdCd²V,
0 == (Cbf − Cb)q − kabCaCbV + kdCd²V,
== − Cdq + kabCaCbV − kdCd²V}, {Ca, Cb, Cd}]
    ]
```

$$\left\{\left\{\text{Ca} \rightarrow \frac{1}{2(\text{kab}-\text{kd})V}\right.\right.$$

$$(-q + \text{Caf kab } V - \text{Cbf kab } V - 2\text{Caf kd } V + \surd(-4\text{CafCbf}$$

$$\text{kab}(\text{kab}-\text{kd})V^2 + (q + (\text{Caf}+\text{Cbf})\text{kab } V)^2)),$$

$$\text{Cb} \rightarrow \frac{1}{2(\text{kab}-\text{kd})V}$$

$$(-q - \text{Caf kab } V + \text{Cbf kab } V - 2\text{Cbfkd}V + \surd(-4\text{Caf Cbf}$$

$$\text{kab}(\text{kab}-\text{kd})V^2 + (q + (\text{Caf}+\text{Cbf})\text{kab}V)^2)),$$

$$\text{Cd} \rightarrow \frac{1}{2(\text{kab}-\text{kd})V}$$

$$(q + \text{Caf kab } V + \text{Cbf kab } V - \surd(-4\text{CafCbfkab}$$

$$\left.(\text{kab}-\text{kd})V^2 + (q + (\text{Caf}+\text{Cbf})\text{kab } V)^2))\right\},$$

$$\left\{\text{Ca} \rightarrow -\frac{1}{2(\text{kab}-\text{kd})V}\right.$$

$$(q - \text{Caf kab } V + \text{Cbf kab } V + 2\text{Cafkd}V + \surd(-4\text{CafCbfkab}$$

$$(\text{kab}-\text{kd})V^2 + (q + (\text{Caf}+\text{Cbf})\text{kab } V)^2)),$$

$$\text{Cb} \to -\frac{1}{2(\text{kab}-\text{kd})V}$$
$$(q + \text{Caf kab } V - \text{Cbf kab } V + 2\text{Cbfkd}V + \sqrt{}(-4\text{Caf Cbf kab}$$
$$(\text{kab}-\text{kd})V^2 + (q + (\text{Caf}+\text{Cbf})\text{kab } V)^2)),$$

$$\text{Cd} \to \frac{1}{2(\text{kab}-\text{kd})V}$$
$$(q + \text{Caf kab } V + \text{Cbf kab } V + \sqrt{}(-4\text{CafCbfkab}$$
$$(\text{kab}-\text{kd})V^2 + (q + (\text{Caf}+\text{Cbf})\text{kab } V)^2))\}\}$$

The result is that we get two sets of symbolic solutions for the concentrations. The first set appears to be the appropriate one since the leading coefficient is positive, whereas for the second set the same term is negative, indicating that for real positive values of the parameters, it would return negative concentrations, which are unphysical. So we extract the first set of solutions with the bracketed number 1:

cstr1[[1]]

$$\Big\{\text{Ca} \to \frac{1}{2(\text{kab}-\text{kd})V}$$
$$(-q + \text{Caf kab } V - \text{Cbf kab } V - 2\text{Caf kd } V + \sqrt{}(-4\text{CafCbf}$$
$$\text{kab}(\text{kab}-\text{kd})V^2 + (q + (\text{Caf}+\text{Cbf})\text{kab}V)^2)),$$

$$\text{Cb} \to \frac{1}{2(\text{kab}-\text{kd})V}$$
$$(-q - \text{Caf kab } V + \text{Cbf kab } V - 2\text{Cbf kd } V + \sqrt{}(-4\text{Caf Cbf}$$
$$\text{kab}(\text{kab}-\text{kd})V^2 + (q + (\text{Caf}+\text{Cbf})\text{kab } V)^2)),$$

$$\text{Cd} \to \frac{1}{2(\text{kab}-\text{kd})V}$$
$$\Big(q + \text{Caf kab } V + \text{Cbf kab } V - \sqrt{}\Big(-4\text{Caf Cbf kab}(\text{kab}-\text{kd})V^2 +$$
$$(q + (\text{Caf}+\text{Cbf})\text{kab}V)^2\Big)\Big)\Big\}$$

We note that in order to solve for these concentrations, we would have to know the values of the two rate constants, the two inlet concentrations, one flow rate, and the reactor volume for a total of six known quantities plus three equations, which is nine for nine.

To extract the right-hand sides of the three solutions and to apply them as functions, we can use the sequence of bracketed numbers as follows for the concentration of **A**:

cstr1[[1, 1, 2]]

$$\frac{1}{2(\text{kab}-\text{kd})V}(-q+\text{Caf kab }V-\text{Cbf kab }V-2\text{Caf kd }V+$$
$$\sqrt{(-4\text{Caf Cbf kab }(\text{kab}-\text{kd})V^2+(q+(\text{Caf}+\text{Cbf})\text{kab }V)^2)})$$

Now, we can pause. Here we see why the CSTR operated at steady state is so different from the transient batch reactor and better for the experimentalist seeking kinetics. If the flow rates and feed concentrations are fixed, then, since the volume in the reactor is constant, by measuring the concentrations at the exit the kinetic parameters are defined. Or, in other words, if we need to evaluate kab and kd, we simply vary the flow rates and collect the corresponding concentrations in order to fit the data to these equations to obtain their magnitudes. If we want the activation energies, we do the same at different temperatures. To find the dependencies on concentration, the kinetics may be assumed, as we have done here and the concentration data under different experimental condition are fit to these solutions. We do not need to do any integration in order to obtain the result, because there we no derivatives in time or in space.

Significantly, we also do not need to have fast analysis of the exit concentrations, even if the kinetics are very fast, because these stay constant in time at constant conditions. We set up the reactor mass flows, fix the temperature, let the system come to steady state and then take as many measurements as we need of the steady state concentrations. Then, we set up a new set of flows and repeat the process. In fact most reactors of this kind as used in laboratories, are, and have been fully automated. We do this for as many points as necessary, in order to obtain a statistically valid set of rate parameters. Again, this is why the steady state flow reactor is considered to be the best experimental reactor type to be used for gathering chemical kinetics. That of course does not mean that is the most commonly used reactor in the laboratory.

Why is it that the flow rate should change the concentrations at the exit of the reactor? Perhaps, the answer is obvious to the reader by now. In case it is not, then we will non-dimensionalize our equations and seek the reason.

Firstly, we will divide each component balance by V and by Caf.

$$0=\frac{(\text{Caf}-\text{Ca})q-\text{kabCaCb}V+\text{kdCd}^2V}{\text{Caf}V}$$
$$=(1\text{ - }\Phi\text{a})\frac{q}{V}\text{ - kab }\Phi\text{a Cb}+\text{kd }\Phi\text{d Cd}^2$$

We can multiply the last two terms by $\frac{\text{Caf}}{\text{Caf}}$ in order to express each concentration in non-dimensional terms:

$$0=\ (1\text{ - }\Phi\text{a})\frac{1}{\theta}\text{ - kab }\Phi\text{a}\frac{\text{Caf}}{\text{Caf}}\text{ Cb}+\text{kd }\Phi\text{d}\frac{\text{Caf}}{\text{Caf}}\text{ Cd}$$

For each of the components we obtain by this procedure:

$$0 = (1 - \Phi a)\frac{1}{\theta} - \text{kab Caf } \Phi a\ \Phi b\ + \text{kd Caf } \Phi d^2$$

$$0 = (\frac{\text{Cbf}}{\text{Caf}} - \Phi b)\frac{1}{\theta} - \text{kab } \Phi a\ \Phi b\ + \text{kd } \Phi d^2$$

$$0 = -\ \Phi d\frac{1}{\theta} + \text{kab } \Phi a\ \Phi b\ -\ \text{kd } \Phi d^2$$

The flow rate through the reactor is q, and so the holding time is $\frac{V}{q}$, which is θ. Since the stoichiometry is 1:1 we can take Cbf = Caf:

$$0{=}(1 - \Phi a) - \text{kab Caf } \theta \Phi a \Phi b + \text{kd Caf } \theta \Phi d^2$$

$$0{=}(1 - \Phi b) - \text{kab Caf } \theta \Phi a \Phi b + \text{kd Caf } \theta \Phi d^2$$

$$0{=} - \Phi d + \text{kab Caf } \theta \Phi a \Phi b - \text{kd Caf } \theta \Phi d^2$$

And now we solve for the dimensionless concentrations:

Clear["Global`* "]

ndcstr1 = Simplify[

Solve[

{0== − kab CafθΦaΦb + kd Cafθ,

0== − kabCafθΦaΦb + kd Cafθ,

0== + kabCafθΦaΦb − kdCafθ},

{Φa, Φb, Φd}]]

{{Φa →

$-\left(\left(1+2\text{Caf kd }\theta-\sqrt{1+4\text{Caf kab }\theta+4\text{Caf }^2\text{kab kd }\theta^2}\right)\Big/ (2\text{Caf kab }\theta-2\text{Caf kd }\theta)\right),$

Φb →

$-\left(\left(1+2\text{Caf kd }\theta-\sqrt{1+4\text{Caf kab }\theta+4\text{Caf }^2\text{kab kd }\theta^2}\right)\Big/ (2\text{Caf kab }\theta-2\text{Caf kd }\theta)\right),$

Φd → $\left(1+2\text{Caf kab }\theta-\sqrt{1+4\text{Caf kab }\theta+4\text{Caf }^2\text{kab kd }\theta^2}\right)\Big/ (2\text{Caf kab }\theta-2\text{Caf kd }\theta)$},

{Φa →

$-\left(\left(1+2\text{Caf kd}\,\theta+\sqrt{1+4\text{Caf kab}\,\theta+4\text{Caf}^{2}\text{kab kd}\,\theta^{2}}\right)\Big/\ (2\text{Caf kab}\,\theta-2\text{Caf kd}\,\theta)\right)$,

$\Phi\text{b}\rightarrow$
$-\left(\left(1+2\text{Caf kd}\,\theta+\sqrt{1+4\text{Caf kab}\,\theta+4\text{Caf}^{2}\text{kab kd}\,\theta^{2}}\right)\Big/\ (2\text{Caf kab}\,\theta-2\text{Caf kd}\,\theta)\right)$,

$\Phi\text{d}\rightarrow\ \left(1+2\text{Caf kab}\,\theta+\sqrt{1+4\text{Caf kab}\,\theta+4\text{Caf}^{2}\text{kab kd}\,\theta^{2}}\right)\Big/\ (2\text{Caf kab}\,\theta-2\text{Caf kd}\,\theta)\}\}$

These are the same solutions as before, but in non-dimensionalized form they are more compact. Now, when we provide reasonable values of the parameters and look at these functions in terms of the holding time as the independent variable, we find that the longer the holding time, the higher the conversion of reactant, and the higher the formation of product. In the following code, we compute the equilibrium concentrations of reactant and product and mark these with dashed horizontal lines.

```
ϕa[θ_]:=ndcstr1[[1]][[1]][[2]]
ϕb[θ_]:=ndcstr1[[1]][[2]][[2]]
ϕd[θ_]:=ndcstr1[[1]][[3]][[2]]
Caf = .25;
kd = 1.5;
kab = 1.1;
tmin = 0.001;
tmax = 100;
```

$\alpha\text{eq}=\text{NSolve}\left[\frac{\text{kab}}{\text{kd}}==\frac{\alpha^{2}}{(1-\alpha)^{2}},\alpha\right]$

```
Plot[{ϕa[θ], ϕd[θ]},
{θ, tmin, tmax}, PlotRange → All,
PlotStyle → {Hue[0], Hue[.6]},
PlotLabel → "rd=ϕa; bl=ϕd; lines=eq",
AxesLabel → {"θ", "ϕi[θ]"},
Epilog → {
{RGBColor[0, 0, 1], Dashed,
```

```
Line[{{tmin, αeq[[2, 1, 2]]},
{tmax, αeq[[2, 1, 2]]}}]},
{RGBColor[1, 0, 0], Dashed,
Line[{{tmin, 1 − αeq[[2, 1, 2]]},
{tmax, 1 − αeq[[2, 1, 2]]}}]}}
]
```

$\{\{\alpha \to -5.96131\}, \{\alpha \to 0.461308\}\}$

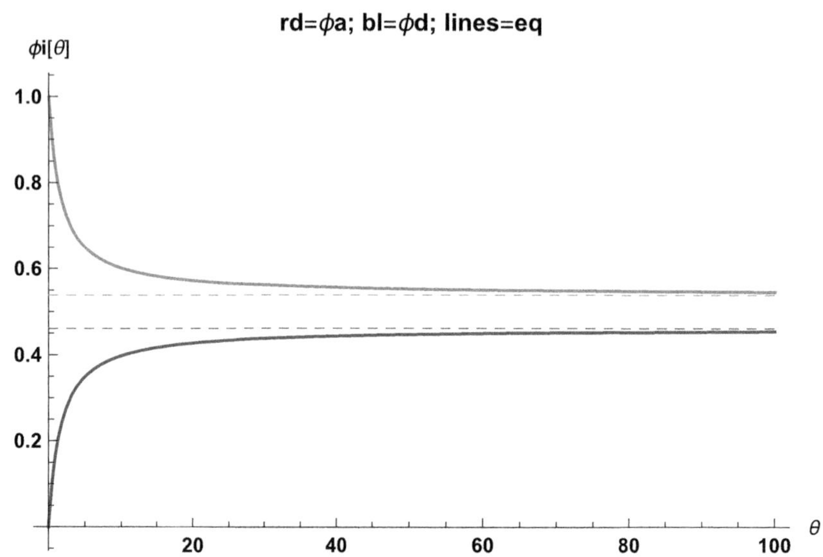

So, the dimensionless concentrations come to their equilibrium values asymptotically as the holding time is increased. In other words, for kinetics of this kind and with no other phenomena controlling the observed rates, a lower flow rate gives more time for the reaction to take place and leads to higher conversion. In fact, if the measured conversion increases with increasing the flow rate, rather than decreases, then it is possible that intrinsic chemical kinetics are being masked by transport phenomena, and most likely by mass transport phenomena if the system is operating isothermally.

Time dependence - the transient approach to steady state and saturation kinetics

Although the steady state CSTR is simple to operate, and to analyze, and even though it offers real advantages to the kineticist, it is also true that these systems must go through a start up. They do not start up and necessarily achieve steady state immediately. The time period in which the system moves toward a steady state condition is called the *transient*, meaning that the system is in transition from one which is time-dependent to one that is time-independent. If we do not know the rate constants, then we have no a-priori way of knowing how long it will take a given reaction, or set of reactions, to achieve a steady state in the CSTR. Thus, we either do an experiment, or solve the time dependent model equations with estimated rate constants, or we do both.

Consider the reaction of a molecule that takes place on a solid catalyst surface. This reaction simply involves converting one form of the molecule into another, in other words, it is an isomerization reaction. But, the reaction in question only takes place on the catalyst surface and nowhere else.

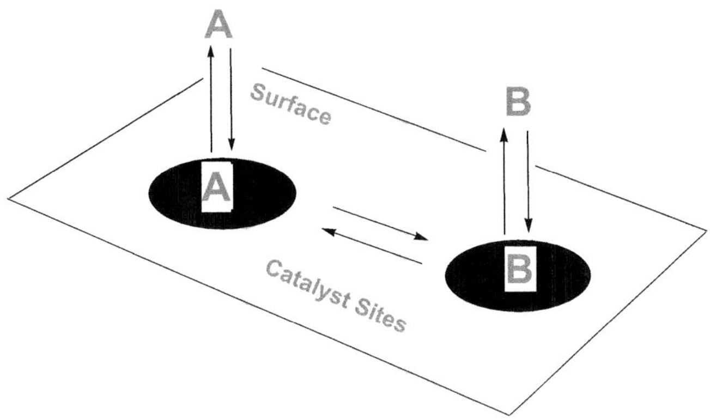

When we analyze a reaction of this kind, as we saw in Chapter 6, then we find that at least two steps are involved, adsorption and surface reaction. The adsorption equilibrium steps take place by the interaction of the molecule in the bulk phase with a what is called an adsorption site on the solid surface. The adsorption site is the locus of points on the surface that interact directly with the molecule, it is a group of atoms.

$$A + \text{site} \Longleftrightarrow A_{\text{surface}}$$

$$A_{\text{surface}} \longrightarrow B_{\text{surface}}$$

$$B_{\text{surface}} \Longleftrightarrow B + \text{Site}$$

Once **B** is formed, it too undergoes desorption and adsorption. The desorption carries **B** from the surface and into the bulk fluid phase. In this case, we will assume the reaction is irreversible and that the rate of this reaction is first order in surface concentration of **A**. It also is first order in the concentration of surface sites. Thus the kinetics follow a simple surface rate law:

$$r_A = k'_{\text{surface}} C_{A,\text{surface}} C_{\text{sites}}$$

The surface concentration is difficult to measure, so we need to re-express it in terms of the bulk phase concentration of species **A**. To do this we take advantage of the fact that the molecules often adsorb and desorb so quickly that they come to equilibrium rapidly with the surface sites. Therefore, subject to this assumption, we can express the surface concentration in terms of the equilibrium. The equilibrium gives rise to the following relationship for the surface concentration of **A** in terms of the bulk concentration of **A**:

$$C_{A,\text{surface}} = \frac{K_A C_A}{1 + K_A C_A}$$

We can substitute this expression into the rate expression for the reaction. This leads to this rate in terms of the bulk phase concentrations:

$$r_{A-} = k'_{\text{surface}} C_{\text{sites}} \frac{K_A C_A}{1 + K_A C_A}$$

The concentration of sites can be incorporated into the rate constant by rewriting the product of the surface site concentration and the surface rate constant simply as a rate constant.

$$\text{k} = k'_{\text{surface}} C_{\text{sites}}$$

We can do this because the surface site concentration is also a constant. Thus the overall rate for this catalytic reaction is:

$$r_{A-} = \frac{k K_A C_A}{1 + K_A C_A}$$

The time-dependent component balance equations for **A** and **B** in the CSTR are as follows:

$$\partial_t C_A \epsilon \text{ V} = (C_{\text{Af}} - C_A)\,\text{q} - (1\text{-}\epsilon)\, r_{A-}\text{V}$$

$$\partial_t C_{\text{Ab}} \epsilon V = -\, C_B\,\text{q} + (1\text{-}\epsilon)\, r_{A-}\text{V}$$

Recall that the solid catalyst occupies a fraction, 1- ϵ, of the reactor volume leaving a fraction ϵ for the fluid phase volume. We write the balances in terms of the fluid phase. The kinetics have been written also in terms of the fluid phase concentration, but they are written for a process that occurs within the second phase, which is the catalyst. This is called the *pseudo-homogeneous* approximation. In this case we take that phase as homogeneous and continuous, and occupying the (1- ϵ) of the reactor volume. We can substitute into these equations the kinetics we just derived:

$$\frac{dC_A\epsilon V}{dt} = (C_{Af} - C_A)\, q - (1 - \epsilon)\, \frac{kK_AC_A}{1 + K_AC_A} V$$

$$\frac{dC_B\epsilon V}{dt} = -C_B\, q + (1 - \epsilon)\frac{kK_AC_A}{1 + K_AC_A} V$$

The integration of these two equations in time will show us how long it will take the reactor to achieve a steady state conversion of **A** and production of **B**.

The first step is to set up a solution to these equations and a graphical display of the results. Using NDSolve, we can solve these time dependent equations to find the concentrations as functions of time. We make a new Module function, cstr4 to handle this.

```
cstr4[k_, K1_, q_, tmax_]:=
Module[
{Caf = 1, V = 1000, Cao = 0, Cbo = 0, ϵ = .4, solns,
Ca, Cb, CA, CB, t},
solns = NDSolve[{
ϵCa'[t]==(Caf − Ca[t]) q/V − (1 − ϵ) (kK1 Ca[t])/(1 + K1Ca[t]),
ϵCb'[t]== − Cb[t] q/V + (1 − ϵ) (kK1 Ca[t])/(1 + K1Ca[t]),
Ca[0]==Cao, Cb[0]==Cbo},
{Ca[t], Cb[t]}, {t, 0, tmax}];

CA[t] = Evaluate[Ca[t]/.solns];
CB[t] = Evaluate[Cb[t]/.solns];

Plot[{CA[t], CB[t]}, {t, 0, tmax},
Ticks → {{0, tmax/2, tmax}, {0, .5, 1}},
PlotStyle->{Hue[0], Hue[.6]},
PlotRange->{{0, tmax}, {0, Caf}},
```

```
PlotLabel → {k"=k", K1"=K1", q"=q"}]
]
```

We can examine the solution at a few extremes to understand how the parameter values affect the kinetic behavior. We can take k = 0 first, to see how the system responds to the flow of A. This gives us a sense of how long it takes the flow and mixing to come to steady state in the absence of reaction. Experimentally, we would do this with a non-catalytic solid present. Keeping all else the same, we vary the rate constant, k, from 0 to 100 in multiples of ten.

```
Show[GraphicsGrid[
{{cstr4[0., .01, 10, 1000], cstr4[1., .01, 10, 1000]},
{cstr4[10., .01, 10, 1000],
cstr4[100., .01, 10, 1000]}}]]
```

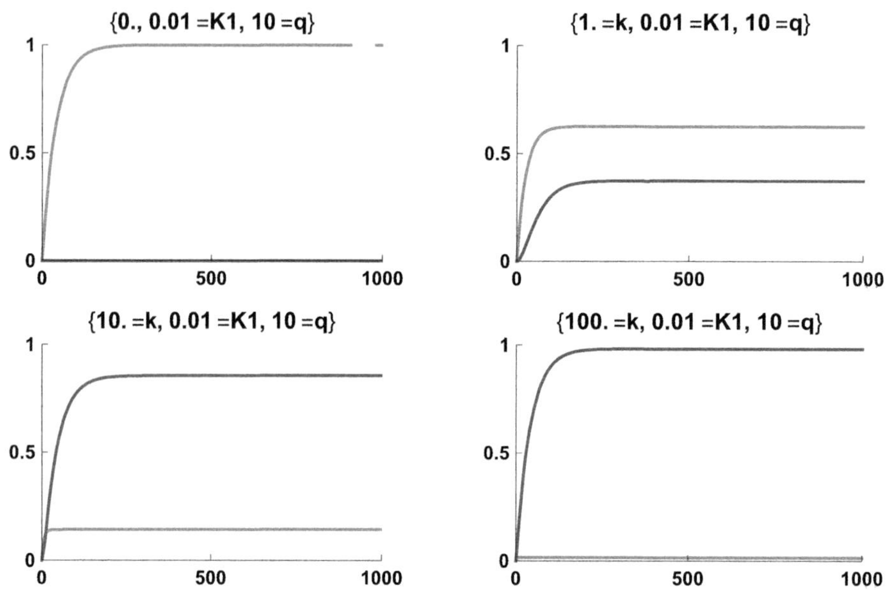

In the first plot, with k = 0, we note that it takes the system about 200 time units at this flow rate to reach a steady state. As we raise the rate constant from zero to unity to ten and then to 100, the steady state concentrations of **A** (red - mid gray in the print version) drop from 0.6 to less than 0.2 to nearly zero. We also see that the time to reach steady state for the product **B** (blue - dark gray in the print version) is about 200 time units in each case, whereas for **A** it is always less than that time and the time to steady state shortens as the rate of chemical reaction increases. This is because the concentration of **A** at steady state decreases as k increases, so the time required to reach the plateau is less.

Looking back at the rate expression we see:

$$r_{A-} = \frac{kK_A C_A}{1 + K_A C_A}$$

If $\boldsymbol{K_A C_A}$ is large compared to unity then the rate reduces to just k, that is a constant or "zeroth order" rate. Alternatively, when $\boldsymbol{K_A C_A}$ is small compared to unity, the rate becomes $\boldsymbol{kK_A C_A}$ or first order with respect to C_A. Letting k be unity for example, we can vary K_A from 10^{-2} to 10^3 over a range of C_A from zero to unity and then plot the results as an array to see the effect of the adsorption constant:

```
k = 1;
Show[
GraphicsGrid[
{Table[
Plot[k10^n Ca/(1 + 10^n Ca), {Ca, 0, 1}, AxesLabel → {Ca, r},
PlotLabel → 10^n "= Ka",
PlotRange → {{0, 1}, {0, 10.^n}},
Ticks → {{0, .5, 1.0}, {0, 5. 10^(n-1), 10.^n}}],
{n, -2., 0, 1}],
Table[Plot[k10^n Ca/(1 + 10^n Ca), {Ca, 0, 1},
AxesLabel → {Ca, r}, PlotLabel → 10^n "= Ka",
PlotRange → {{0, 1}, {0, 10.^n}},
Ticks → {{0, 0.5, 1.0}, {0, 5. 10^(n-1), 10.^n}}],
{n, 1., 3.}]}
        ]
  ]
```

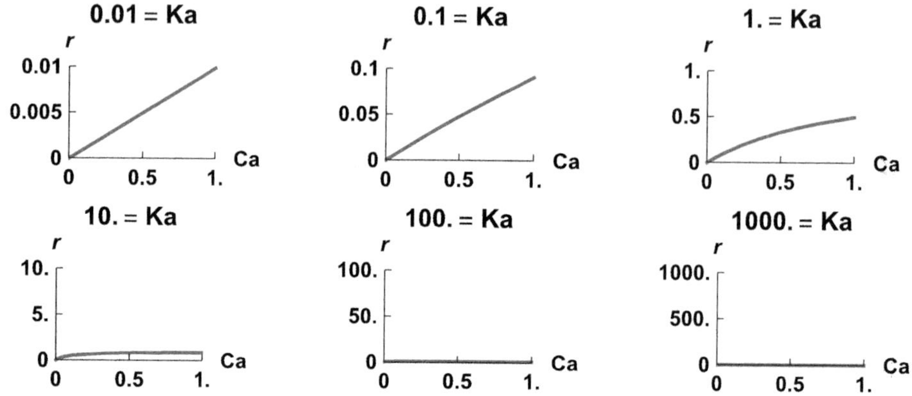

In the first two plots, $\boldsymbol{K_A C_A}$ is small compared to unity and we see that the rate is first order in concentration C_A over the whole range. The last two plots are cases in which $\boldsymbol{K_A C_A}$ is large compared to unity at most values of C_A, except for the very smallest ones. Hence the rate becomes constant at larger values of C_A. This is called saturation. It means that the rate cannot increase in magnitude even though the concentration of reactant has been increased and the rate is an apparently strong function of C_A. In fact when $\boldsymbol{K_A C_A}$ is large, we see that the rate is a very weak function of C_A. Hence Langmuir-Hinshelwood (and Michaelis-Menton) kinetics are often referred to as "saturation kinetics." The intermediate values of Ka lead to intermediate results and behavior in the rate.

How will the effect of saturation kinetics show up in the evolution to the steady state in a CSTR? We can find this out by letting K1 vary over this range of magnitudes, from 10^{-2} to 10^3, within the Module function "cstr4". We also have taken the rate constant down from 1 to 0.05 to make the differences more evident for the same values of q and V, i.e. the holding time. This does not change the effect of K1 since we are comparing its product with Ca to unity:

```
Show[
GraphicsGrid[
  Partition[
    Table [cstr4 [.05, 10^n, 10., 1000],
{n, -2., 3}],2]]]
```

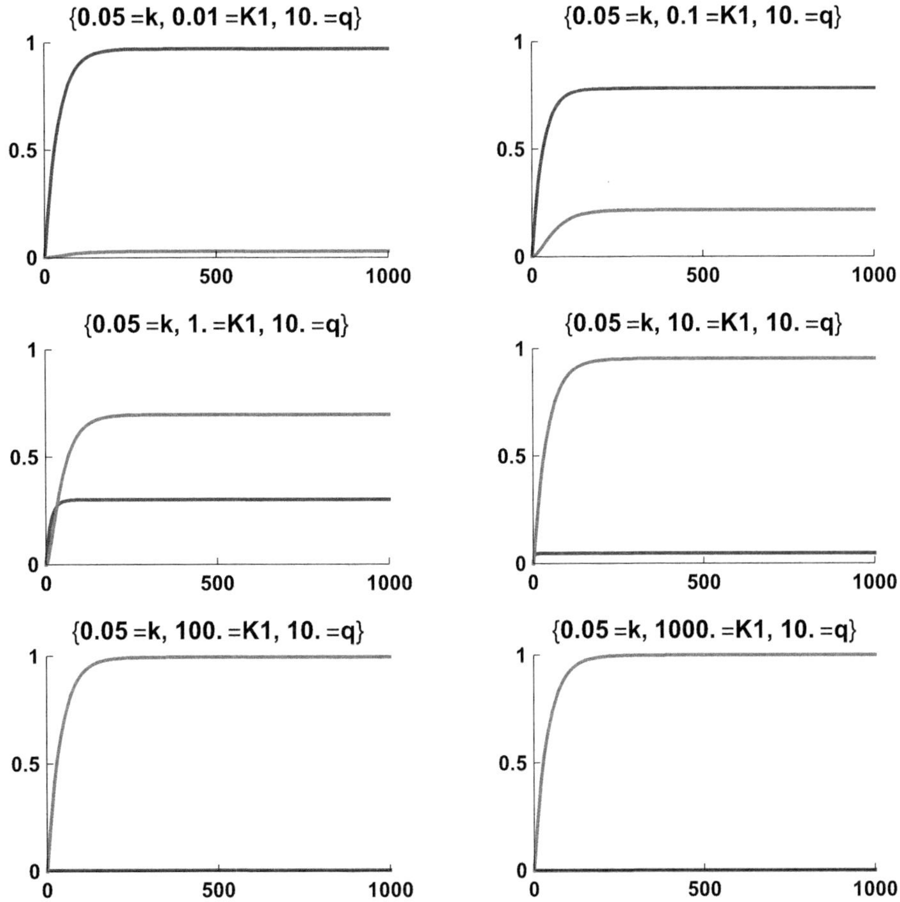

The results are not dramatically different than what we had seen before. The saturation kinetics at these flow rate, i.e. holding times, give rise to complete conversion as we see in the last two plots in which K1 has values of 10^2 and 10^3.

Finally, the last calculation prompts the question of holding time effect. If we vary the flow rate q at fixed V, keeping k and K1 constant we should see the conversion rise with longer holding times, that is lower flow rates:

```
Show[
GraphicsGrid[
  Partition[
    Table[cstr4[.05, 10^3, 10^n, 1000],
```

{n, −2., 3}],2]]]

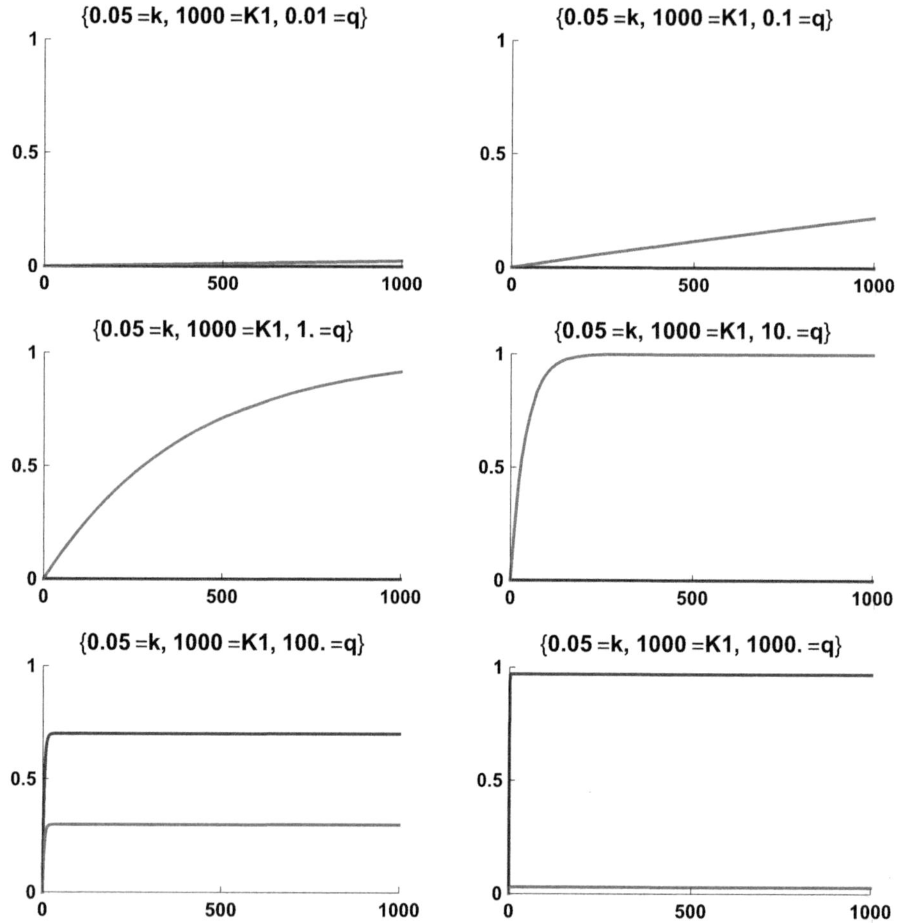

The effect is dramatic. We have taken K1 = 1000, which puts the kinetics in the zeroth order regime. We see in the upper-left plot that the conversion appears to be complete, but even after 1000 time units the system is still far from the steady state. At the lower right, the conversion is essentially zero, and the system comes to steady state nearly instantaneously. The other plots show self-consistent behaviors. Notice that with q = 10, the approach to steady state is fast and the conversion is essentially complete.

But what would happen if we took the adsorption constant K1 to be quite small, say on the order of 10^{-2}? Let's find out:

```
Show[
GraphicsGrid[
  Partition[
    Table[cstr4[.05, 1. 10^-2, 10^n, 1000],
{n, -2., 3}],2]]]
```

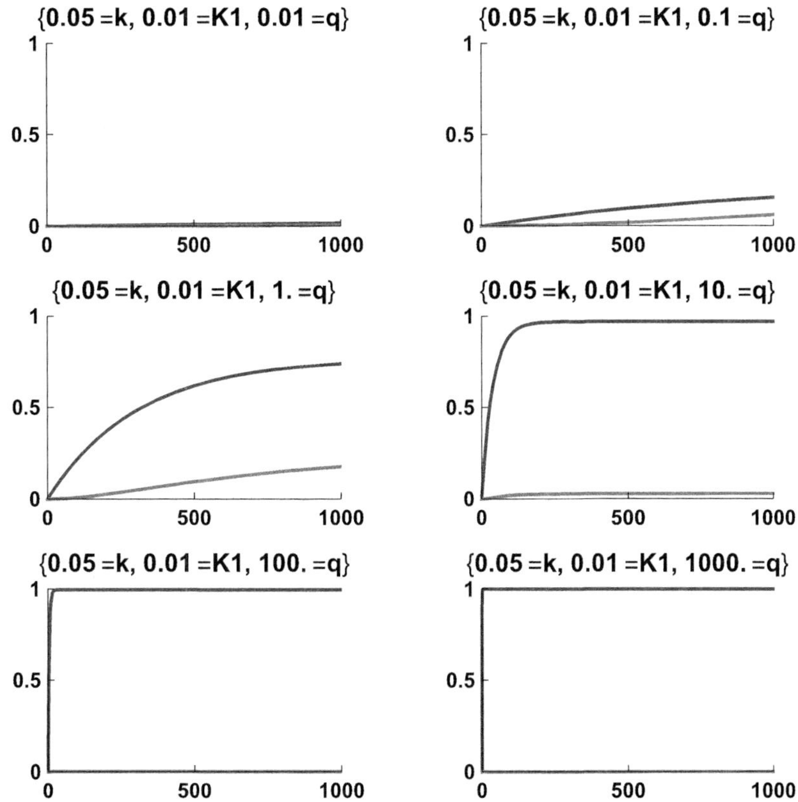

This is very interesting and instructive. Here we see that K1 is so small that at any of the holding times even the highest ones, the rate is so small that the conversion is effectively zero throughout the range. This means that the catalyst just lacks the adsorption forces necessary to make the reaction run fast. In other words for the reaction to take place efficiently, the reactant **A** must be adsorbed. If it is not adsorbed to an appreciable extent, then the rate is always going to be small unless the rate constant for the surface reaction is very high.

Before we go on to the next section we should do some "housekeeping." The command Names is shown below with its *Mathematica* explanation:

```
Symbol                                              $Failed

Names["string"] gives a list of the names
  of symbols that match the string.
Names[patt] gives a list of names matching
  the arbitrary string pattern patt.

$Failed
```

As we have worked through this session, or any other session, we have generated lots of new Names for functions. These show up in the Global context. If we ask for them we will get a list of those which we have created and used so far. (We show this below, but we have suppressed the output.)

Names["Global`* "];

To clean up the Global context, we can Remove everything we have created in the Global context as follows; asking for Names in this context once again returns an empty set:

Remove["Global`* "];

Names["Global`* "];

{}

Plug flow reactors

We turn now to the plug flow reactor. Here as we have said there is absolutely no mixing in the direction of flow, but perfect mixing perpendicular to it, i.e. between the center line and the walls. This special case of a tubular reactor can be simulated to be in a transient or steady state mode, but it is the latter mode that is most often considered for kinetics and design. Consider the following reactor in which A is converted to B irreversibly and with linear kinetics:

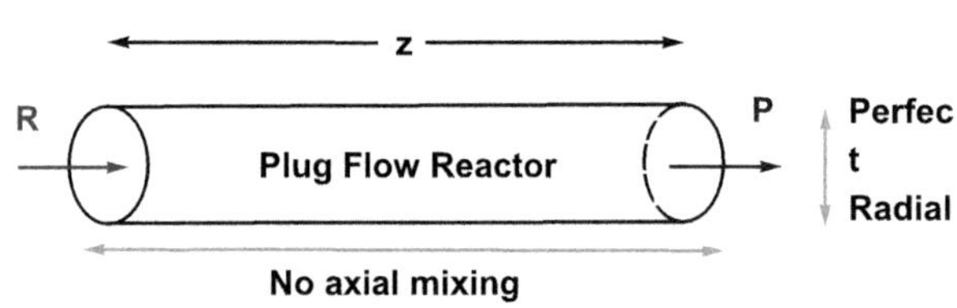

In contrast to the first two reactors, *concentrations within the plug flow reactor are characterized by their axial position dependencies, but they are independent of their radial positions.*

The term plug means that the fluid moves through the cylindrical tube as if it were a “plug” of material translating through the volume. We must connect the time-in-the-tube dependence to the axial position. Near the entrance of the tube, the fluid is nearly 100 % reactant, whereas at the exit, it is a mix of reactant and product, if the conversion is less than 100 % and at axial positions between the two ends the gas is a mix of products and reactants. Our goal will be to predict how the mix of concentration changes as a function of position and flow parameters. The holding time will figure in prominently once again.

Some very interesting consequences of complete mixing versus partial mixing can be defined in terms of reactor efficiencies. We will see that the fully-mixed CSTR reactor will require a larger volume than the partially mixed tubular system to achieve the same conversion and at the same holding time, assuming positive order kinetics. This is a very important result. At the same time although conversion may be higher, so too may selectivity be lower, if multiple reactions are involved. So, we learn much about the systems in which chemical reactions are conducted, by considering these two different kinds of ideal behavior.

For the first time we must, as a consequence of the plug flow, take into account spatial variation as well as time dependence. This means that the concentrations of A and B will have z and t dependence and the equations describing them will be made up of partial rather than ordinary differentials. We can derive the equation that describes the plug flow system by first visualizing a zone of reaction that corresponds to a differential control volume Acr dz.

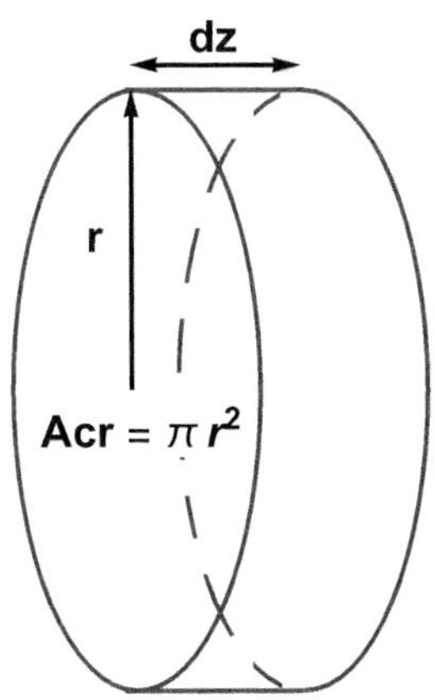

The total differential of the concentration is equal to the rate of chemical reaction in this zone over some differential time dt. In *Mathematica* the total differential of f [x, y] is given as Dt [f [x, y]]:

Dt[$f[x, y]$]

$\mathrm{Dt}[y]f^{(0,1)}[x, y] + \mathrm{Dt}[x]f^{(1,0)}[x, y]$

This output statement means:

$$\partial_y f[x, y]\mathrm{dy} + \partial_x f[x, y]$$

or

$$\frac{\partial f[x, y]}{\partial y}\mathrm{dy} + \frac{\partial f[x, y]}{\partial x}\mathrm{dx}$$

Taking the total differential, Dt, of the concentration and the rate we obtain:

$\mathbf{Dt[Ca[}z, t\mathbf{]] == -raDt[}t\mathbf{]}$

$\mathrm{Dt}[t]\mathrm{Ca}^{(0,1)}[z, t] + \mathrm{Dt}[z]\mathrm{Ca}^{(1,0)}[z, t] == -\mathrm{raDt}[t]$

Reading from the left to the right, this means the derivative of time multiplied by the derivative of the concentration of **A** with respect to time plus the derivative of time multiplied by the derivative of the concentration of **A** with respect to radial dimension, z, is equal to the rate of reaction of **A** times the derivative of time.

We can use the shortcut of dividing through by the derivative of time, Dt[t], to get the following:

$$\mathbf{Simplify}\left[\frac{\mathbf{Dt}[t]\mathbf{Ca}^{(0,1)}[z, t] + \mathbf{Dt}[z]\mathbf{Ca}^{(1,0)}[z, t]}{\mathbf{Dt}[t]}\right] ==$$

$$\frac{-\mathbf{raDt}[t]}{\mathbf{Dt}[t]}$$

$$\mathrm{Ca}^{(0,1)}[z, t] + \frac{\mathrm{Dt}[z]\mathrm{Ca}^{(1,0)}[z, t]}{\mathrm{Dt}[t]} == -\mathrm{ra}$$

In the second term on the left-hand side, we have the ratio of $\frac{\mathbf{Dt}[z]}{\mathbf{Dt}[t]}$, or more familiarly ($\frac{dz}{dt}$). This is the velocity of the plug of gas through the differential volume in the axial direction. We will use the symbol vz for the axial velocity. Making this substitution we have:

$\mathbf{Ca}^{(0,1)}[z, t] + \mathbf{vzCa}^{(1,0)}[z, t] == -\mathbf{ra}$

$\mathrm{Ca}^{(0,1)}[z,t] + \mathrm{vzCa}^{(1,0)}[z,t] == -\mathrm{ra}$

Written in traditional form this states:

$$\frac{\partial \mathrm{Ca}[z,t]}{\partial t} = - v_z \frac{\partial \mathrm{Ca}[z,t]}{\partial z} - r_{A-}$$

This mathematical "sentence" states that the partial derivative of the concentration of **A** with respect to time is equal to negative of the sum of the product of the axial velocity and the partial derivative of the concentration of **A** with respect to z and the rate of reaction of **A**.

For the product, B, we would have the following equation:

$$\frac{\partial \mathrm{Cb}[z,t]}{\partial t} = -v_z \frac{\partial \mathrm{Cb}[z,t]}{\partial z} + r_{A-}$$

There is also a more intuitive way to arrive at these equations that begins at the well-mixed approximation. Imagine that within the region Δz, the fluid phase is "well-mixed." The volume of this region is ΔV = Acr Δz. The flow rate across the volume is taken to be q, and the concentrations in the volume element are $C_{i,1}$ while those exiting are $C_{i,2}$. Writing the mass balance for **A** we have:

$$\frac{d\mathrm{Ca}\Delta \mathrm{V}}{dt} = \left(C_{a,1} - C_{a,2}\right) q - r_{A-} \Delta \mathrm{V}$$

$$\frac{d\mathrm{CaAcr}\Delta \mathrm{z}}{dt} = \left(C_{a,1} - C_{a,2}\right) q - r_{A-}\ \mathrm{Acr}\ \Delta \mathrm{z}$$

$$\frac{d\mathrm{Ca}}{dt} = \frac{\left(C_{a,1} - C_{a,2}\right) q}{\mathrm{Acr}\Delta \mathrm{z}} - r_{A-}$$

$$\frac{d\mathrm{Ca}}{dt} = v_z \frac{-\Delta C_a}{\Delta \mathrm{z}} - r_{A-}$$

We can write the corresponding differential difference equation for component **B**. When we take the limit as Δz→ 0, these two differential equations become partial differential equations:

$$\underset{\Delta \mathrm{z}\to 0}{\mathrm{Limit}} \left[\frac{d\mathrm{Ca}}{dt} = - v_z \frac{\Delta C_a}{\Delta \mathrm{z}} - r_{A-}\right] \Longrightarrow \frac{\partial \mathrm{Ca}}{\partial t} = - v_z \frac{\partial C_a}{\partial z} - r_{A-}$$

$$\underset{\Delta \mathrm{z}\to 0}{\mathrm{Limit}} \left[\frac{d\mathrm{Cb}}{dt} = - v_z \frac{\Delta C_b}{\Delta \mathrm{z}} + r_{A-}\right] \Longrightarrow \frac{\partial \mathrm{Cb}}{\partial t} = - v_z \frac{\partial C_b}{\partial z} + r_{A-}$$

Thinking physically, it is as if we have taken a slice out of a not well-mixed reactor, but across that slice the mixing is perfect. This is a "CSTR" of vanishing or differential height, dz, and cross sectional area, Acr.

Usually, we are interested in solving these equations for the concentrations at steady state. In this condition, the time differentials are as usual identically zero. Recall that v_z divided into a distance z would be the time, τ, required to translate across that distance. If we divide v_z into the differential distance dz, then we have the differential time, $d\tau$ required to translate across the infinitesimal distance. Thus $\frac{v_z}{dz}$ is just $\frac{1}{d\tau}$, and the equations transform at the steady state as follows:

$$\frac{q}{\text{Acr}}\frac{\partial C_a}{\partial z} = -r_{A-} \Longrightarrow v_z\frac{\partial C_a}{\partial z} = -r_{A-} \Longrightarrow \frac{\partial C_a}{\partial \tau} = -r_{A-}$$

$$\frac{q}{\text{Acr}}\frac{\partial C_b}{\partial z} = +r_{A-} \Longrightarrow v_z\frac{\partial C_b}{\partial z} = +r_{A-} \Longrightarrow \frac{\partial C_b}{\partial \tau} = +r_{A-}$$

Remarkably, the form of the steady state, PFR equations are similar to those for the fully transient well-mixed batch reactor with no volume change. The difference is that instead of the derivative with respect to real time, the PFR equations involve the derivative with respect to reduced time. This is a very significant result.

Solution of the steady state PFR

Firstly, we begin by solving for the concentrations with linear kinetics, and we so do in complete form. We will do this with DSolve as an exercise, even though the equations are trivial to solve:

```
Clear["Global`*"]
Simplify[
DSolve[
{vz∂zCa[z] == −kabCa[z],
vz∂zCb[z] == +kabCa[z],
Ca[0] == Caf,
Cb[0] == 0},
{Ca[z], Cb[z]},
z]]
```

$$\left\{\left\{\text{Ca}[z] \to \text{Caf}\, e^{-\frac{\text{kab}z}{\text{vz}}}, \text{Cb}[z] \to \text{Caf} - \text{Caf}\, e^{-\frac{\text{kab}z}{\text{vz}}}\right\}\right\}$$

Recall that kab is just $\frac{1}{\tau_{\mathbf{rxn}}}$, that is the reciprocal of the characteristic time for reaction and $\frac{z}{v_z}$ is the holding time τ for the PFR. Therefore these solutions become:

$$\mathrm{Ca}[z] = \mathrm{Caf}e^{-\frac{\mathrm{kab}z}{\mathrm{vz}}} \Longrightarrow \mathrm{Ca}[\tau] = \mathrm{Caf}e^{-\mathrm{kab}\tau}$$

$$\mathrm{Ca}[\tau] = \mathrm{Caf}e^{-\frac{\tau}{\tau_{\mathrm{rxn}}}}$$

$$\mathrm{Cb}[z] = \mathrm{Caf}\left(1 - e^{-\frac{\mathrm{kab}z}{\mathrm{vz}}}\right) \Longrightarrow \mathrm{Cb}[\tau] = \mathrm{Caf}\left(1 - e^{-\mathrm{kab}\tau}\right)$$

$$\mathrm{Cb}[\tau] = \mathrm{Caf}\left(1 - e^{-\frac{\tau}{\tau_{\mathrm{rxn}}}}\right)$$

When we solved the transient, well-mixed batch reactor with linear kinetics, we obtained the same solution functionally, but instead of kab τ, we had kab t as the argument of the exponential, that is in terms of real time instead of holding time.

We return now to the Langmuir-Hinshelwood kinetics from the CSTR section to see how the PFR will behave and to compare the CSTR and the PFR. As in the case of the steady state CSTR, we will write a steady state PFR Module function. Recall that the rate law was:

$$r_{A-} = \frac{kK_A C_A}{1 + K_A C_A}$$

Therefore, once again invoking the pseudo-homogeneous approximation, the equations that we must solve are:

$$v_z \frac{\partial C_a}{\partial z} = -\frac{(1-\epsilon)}{\epsilon} \frac{kK_A C_A}{1 + K_A C_A}$$

$$v_z \frac{\partial C_b}{\partial z} = +\frac{(1-\epsilon)}{\epsilon} \frac{kK_A C_A}{1 + K_A C_A}$$

The Module function for this PFR will be called "pfr1." The arguments in "pfr1" are the rate constant k, the adsorption equilibrium constant K1, the volume flow rate q, and the radius of the reactor cross-section, r. So that we can make comparisons to the CSTR, we have kept the total volume the same at 1000. Once we specify the radius, that fixes the circular cross-section. This divided into the volume gives the length of the reactor, zmax. Therefore instead of specifying the reactor length, we simply specify the reactor radius and at constant volume this fixes zmax. Everything else will be kept the same for comparison sake, especially ϵ and the holding time $\frac{V}{q}$.

```
pfr1[k_, K1_, q_, r_]:=
Module[
{Caf = 1, V = 10000, vz, Acr, Cao = 1, Cbo = 0,
ϵ = .4, Ca, Cb, CA, CB, zmax},
Acr = N[πr^2];
vz = q/Acr;
zmax = V/Acr;
pfr = NDSolve[{
vzCa'[z]== -((1 - ϵ)/ϵ) (kK1 Ca[z])/(1 + K1Ca[z]),
vzCb'[z]==((1 - ϵ)/ϵ) (kK1 Ca[z])/(1 + K1Ca[z]),
Ca[0]==Cao, Cb[0]==Cbo},
{Ca[z], Cb[z]}, {z, 0, zmax}];

CA[z] = Evaluate[Ca[z]/.pfr];
CB[z] = Evaluate[Cb[z]/.pfr];

Plot[{CA[z], CB[z]}, {z, 0, zmax},
Ticks → {{0, Round[zmax/2, 1], Round[zmax, 1]},
{0, .5, 1.}},
PlotStyle → {{Dashing[{0.15, 0.05}], Hue[0.6]},
{Dashing[{0.15, 0.05}], Hue[0]}},
PlotRange->{{0, Round[zmax, 1]}, {0, Caf}},
PlotLabel → {"PFR", k"=k", K1"=K1", q"=q", r"=r"}]
]
```

To see how the program works let's try it out for a radius of 10 units and a volume of 10,000. This makes zmax about 32 units.

Show[pfr1[0.5, .01, 10, 10]]

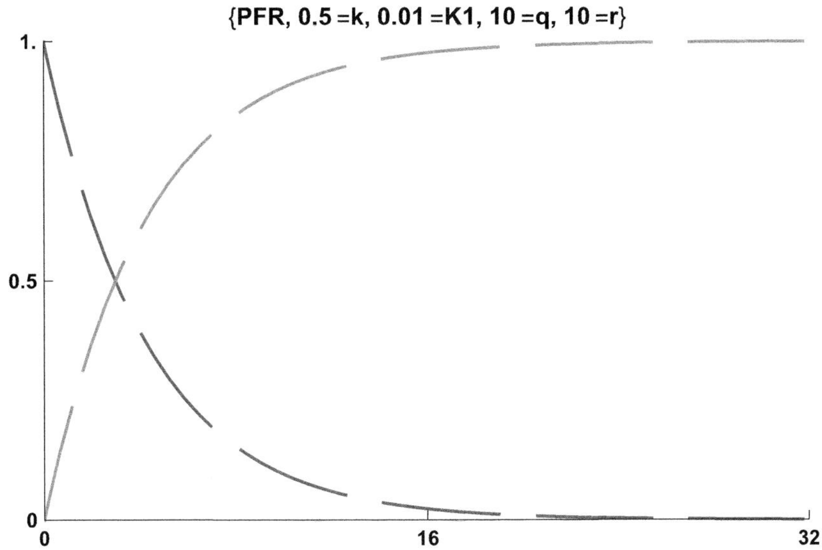

Now, for comparison to the CSTR, we can build a new Module function that allows us to create a plot of the exit concentration from a CSTR under the same conditions and also as a "function" of z. Of course there is no functional z-dependence for the CSTR, and the plots will be simply horizontal lines. But, they will be graphed over the same range as the axial distance through the PFR. In this way we can put the two on one graph for comparison. Here is the steady state CSTR Module:

cstrstst[k_, K1_, q_, r_]:=

Module[

{Caf = 1, Acr, V = 10000, ϵ = .4, ststsolns, Ca, Cb, t},

Acr $= N\left[\pi r^2\right]$;

zmax $= \frac{V}{\text{Acr}}$;

ststsolns = Solve[

$$\left\{0==(\text{Caf}-\text{Ca})\frac{q}{V}-(1-\epsilon)\frac{k\text{K1 Ca}}{1+\text{K1Ca}},\right.$$

$$\left.0==-\text{Cb}\frac{q}{V}+(1-\epsilon)\frac{k\text{K1 Ca}}{1+\text{K1Ca}}\right\},$$

```
{Ca, Cb}];
{Ca, Cb};
{Ca = ststsolns[[2, 1, 2]],
Cb = ststsolns[[2, 2, 2]]};
Plot[{Ca, Cb}, {z, 0, zmax},
Ticks → {{0, Round[zmax/2, 1], Round[zmax, 1]},
{0, .5, 1.}},
PlotStyle->{Hue[0], Hue[.6]},
PlotRange->{{0, Round[zmax, 1]}, {0, Caf}},
PlotLabel → {"CSTR", k"=k", K1"=K1", q"=q"}]
]
Show[cstrstst[0.5, .01, 10, 10]]
```

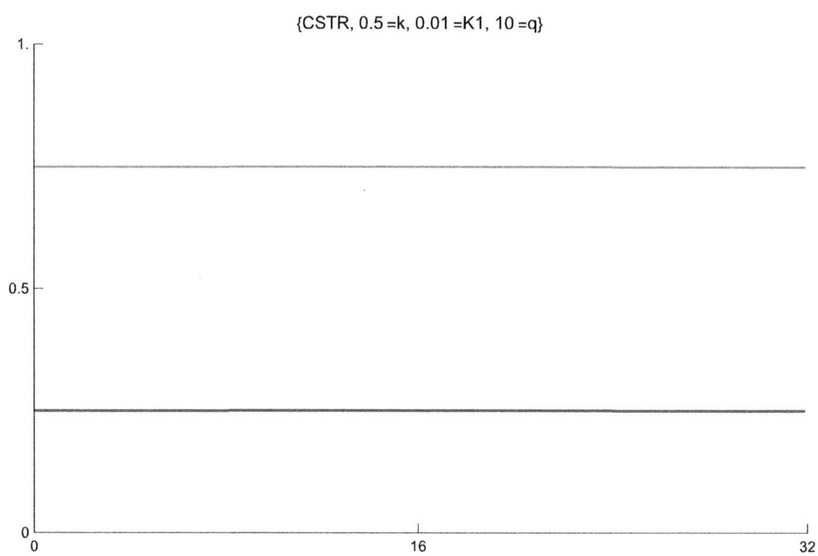

Putting the PFR and CSTR results on the same graph we find:

```
Show[{pfr1[0.5, .01, 10, 10],

cstrstst[0.5, .01, 10, 10]}]
```

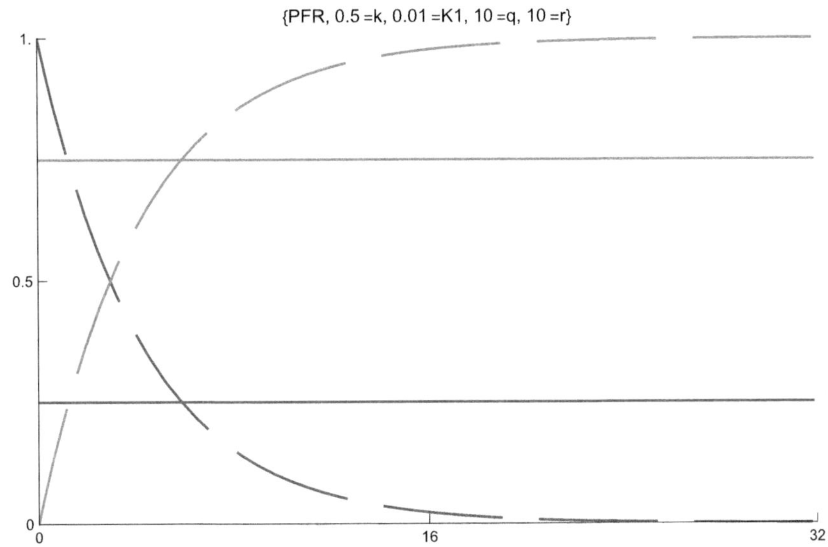

We see that conversion at the exit of the PFR is larger than that of the CSTR at the same condition. This is because there is no back mixing in the PFR.

We can vary the flow rate over a few orders of magnitude for both the PRF and the CSTR to see what will happen. This is done by combining the two functions into one Table and the plotting the results as a GraphicsArray.

```
Table[{pfr1 [0.5, .01, 1. 10^n, 10],

 cstrstst [0.5, .01, 1. 10^n, 10]}, {n, -1, 1, 1}];

Show[

GraphicsGrid[%]

]
```

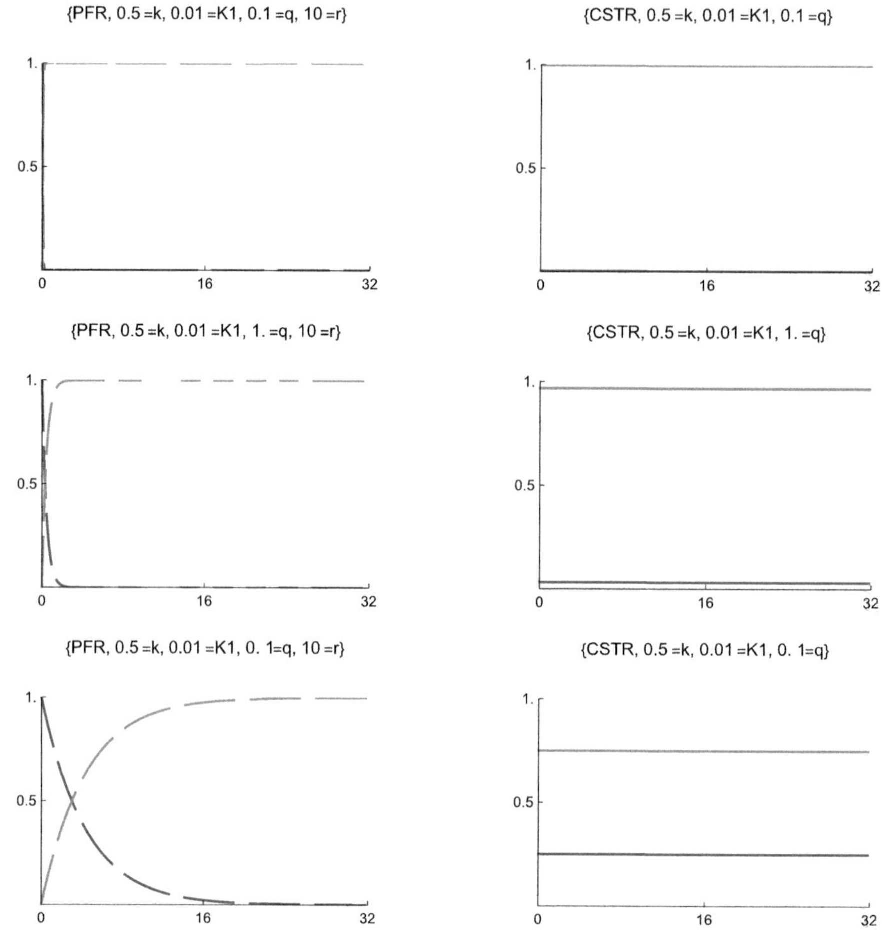

For the PFR at shorter q values the cross-over in concentrations moves to shorter holding times. For the CSTR the trend is the same, as q rises the conversion drops. In fact, we can find that even at q = 1, the CSTR has still not achieved full conversion of the feedstock. Remember the only difference between the two cases is that the CSTR is well-mixed throughout its volume, but the PFR is well-mixed only radially, and not at all axially. So this leads to the oft quoted rule-of-thumb that the PFR is more efficient than the CSTR. At the same volume of reactor, one achieves higher conversion, or to achieve equivalent conversion the PFR volume can be smaller than the CSTR volume.

Now in the previous calculations, we assumed that K1 was small. This forced the rate toward first order dependence. Therefore how does the comparison between PFR and CSTR work out if we vary K1 over several orders of magnitude to make the kinetics range from first order to zeroth order? To do this we will fix the flow rate at q =10 and vary K1 from 10^{-3} to 10^{0}.

comps2 =

```
Table[{pfr1[0.5, 1. 10^n, 10., 10],
 cstrstst[0.5, 1. 10^n, 10., 10]}, {n, -4, -1, 1}];
Show[GraphicsGrid[comps2]]
```

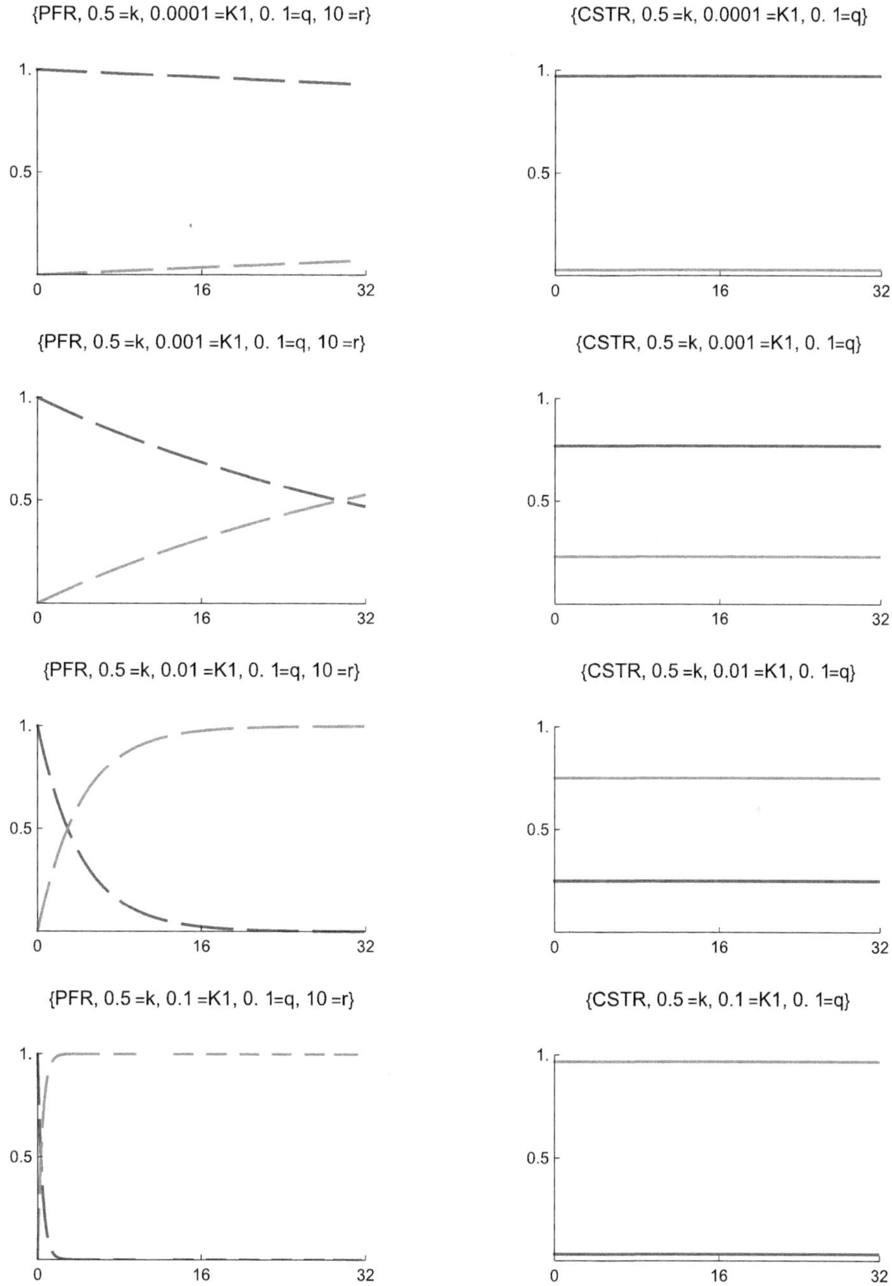

The results indicate that even if the adsorption equilibrium constant is increased, the PFR shows better results than the CSTR until the very last value at K1 = 0.1. The only time this is violated for an isothermal reaction system is if the kinetics are negative order overall, then the CSTR will actually be more efficient than the PFR, otherwise the PFR is better. Ceteris paribus, the PFR will give the same conversion but with a smaller volume, or higher conversion at the same volume.

Mixing effects on selectivities: series and series-parallel with CSTR and PFR

In the previous chapter, we examined series and series-parallel kinetics. The extent of mixing can have an effect on the selectivity to the various products in such reaction networks. The selectivity is the percentage of the products that are any one of the products. To compute the yield we take the product of the conversion and the selectivity. So the yield is a fraction of a fraction.

We can begin by computing the selectivities and yields for the series network in the CSTR versus the PFR first. Consider the simplest series reaction network:

$$\mathrm{A} \longrightarrow \mathrm{B} \longrightarrow \mathrm{D}$$

$$r_{\mathrm{ab}} = k_1 C_A;$$

$$r_{\mathrm{bd}} = k_2 C_B$$

The Module functions for the CSTR and PFR with these species and kinetics are written below as "cstrABD" and "pfrABD."

```
Clear["Global`* "]


cstrABD[k1_, k2_, q_, r_]:=

Module[

{cCaf = 1, V = 1000},


Acr = N[πr²];

zmax = V/Acr;

ststsolns = Solve[
```

$$\{0 == (\text{cCaf} - \text{cCa})\frac{q}{V} - \text{k1cCa},$$

$$0 == -\text{cCb}\frac{q}{V} + \text{k1cCa} - \text{k2cCb},$$

$$0 == -\text{cCd1}\ \frac{q}{V} + \text{k2cCb}\},$$

{cCa, cCb, cCd1}];

cCA = ststsolns[[1, 1, 2]];

cCB = ststsolns[[1, 2, 2]];

cCD1 = ststsolns[[1, 3, 2]];

{Print["CSTR Conversion of A = ",

$$\text{Round}\left[\left(1 - \left(\frac{\text{cCA}}{\text{cCaf}}\right)\right), .01\right] 100\text{"\%"}\Big];$$

Print["Selectivity to B =",

Round[(cCB/(cCA + cCB + cCD1)), 0.0001]100"%"];

Print["Selectivity to D1 =",

Round[(cCD1/(cCA + cCB + cCD1)), 0.01]100"%"];

Print["Yield of B =",

$$\text{Round}\left[\left(1 - \left(\frac{\text{cCA}}{\text{cCaf}}\right)\right)(\text{cCB}/(\text{cCA} + \text{cCB} + \text{cCD1})), 0.01\right]$$

100"%"];

Print["Yield of D1 =",

$$\text{Round}\left[\left(1 - \left(\frac{\text{cCA}}{\text{cCaf}}\right)\right)(\text{cCD1}/(\text{cCA} + \text{cCB} + \text{cCD1})), 0.01\right]$$

100"%"];

Plot[{cCA, cCB, cCD1}, {t, 0, zmax},

PlotStyle->{Hue[0.6], Hue[.0], Hue[0.8]},

```
AxesLabel → {z, "Cstr"},
PlotRange->{{0, zmax}, {0, cCaf}},
PlotLabel → {k1"=k1", k2"=k2", q"=q"}]}
]

Clear[CA, CB, CD1]

pfrABD[k1_, k2_, q_, r_]:=Module[
{Ao = 1, V = 1000},
Acr = N[πr^2];
vz = q/Acr;
zmax = V/Acr;

sln = NDSolve[{vz∂zA[z] == −k1A[z],
vz∂zB[z] == +k1A[z] − k2B[z],
vz∂zD1[z] == k2B[z],
A[0] == Ao, B[0] == 0, D1[0] == 0},
{A[z], B[z], D1[z]}, {z, 0, zmax}];

CA[z_]:=Evaluate[A[z]/.sln];
CB[z_]:=Evaluate[B[z]/.sln];
CD1[z_]:=Evaluate[D1[z]/.sln];

{Print["PFR Conversion of A =",
Round[1 − (CA[zmax]/Ao), .01]100"%"];

Print["Selectivity to B =",
Round[(CB[zmax]/(CB[zmax] + CD1[zmax])), .01]
```

```
100"%"];

Print["Selectivity to D1 =",
Round[CD1[zmax]/(CB[zmax] + CD1[zmax]), 0.01]
100"%"];

Print["Yield to B =",
Round[(1 − (CA[zmax]/Ao))
(CB[zmax]/(CB[zmax] + CD1[zmax])), .01]
100"%"];

Print["Selectivity to D1 =",
Round[CD1[zmax]/(CB[zmax] + CD1[zmax]), 0.01]
100"%"];

AA = Plot[{CA[z]}, {z, 0, zmax}, PlotRange → All,
PlotStyle → {Dashed, Hue[0.6]},
PlotLabel → "CA[z]"],

BB = Plot[{CB[z]}, {z, 0, zmax}, PlotRange → All,
PlotStyle → {Dashed, Hue[0]},
PlotLabel → "CB[z]"],

D1D1 = Plot[{CD1[z]}, {z, 0, zmax}, PlotRange → All,
PlotStyle → {Dashed, Hue[0.8]},
PlotLabel → "CD[z]"],
Show[{AA, BB, D1D1}, PlotLabel → " "]
}]

pfrABD[.4, .03, 10, 10]
```

PFR Conversion of A ={100.%}

Selectivity to B ={5.%}

Selectivity to D1 ={95.%}

Yield to B ={5.%}

Selectivity to D1 ={95.%}

{, , , }

cstrABD[.4, .03, 10, 10]

CSTR Conversion of A = 98.%

Selectivity to B =24.39%

Selectivity to D1 =73.%

Yield of B =24.%

Yield of D1 =71.%

{}

Show[

{pfrABD[.4, .03, 10, 10],

cstrABD[.4, .03, 10, 10]}

]

PFR Conversion of A ={100.%}

Selectivity to B ={5.%}

Selectivity to D1 ={95.%}

Yield to B ={5.%}

Selectivity to D1 ={95.%}

CSTR Conversion of A = 98.%

Selectivity to B =24.39%

Selectivity to D1 =73.%

Yield of B =24.%

Yield of D1 =71.%

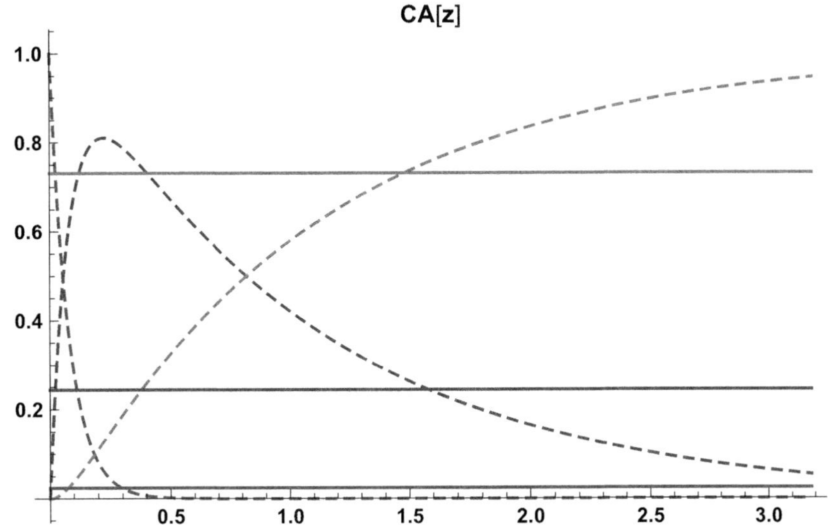

For both reactors the volumes are identical, and that is critical to this comparison. If B (red - dark gray in the print version) were the species that we wished to produce, then for the same volume the CSTR with its perfect mixing produces a higher yield than does the PFR at the same volume and flow conditions. This is a consequence of the higher efficiency of the PFR. Notice, however, that if we made the PFR smaller by cutting it off at z = 0.3 for volume of $\frac{1000}{10} = 100$, then it would be better for the production of B than the CSTR with a volume of 1000 for the production of B.

Next, we rewrite the two codes to handle another limiting case, that of parallel reactions of A to form either B or E with B reacting further to form E.

$$r_{ab} = k_1 C_A;\ r_{bd} = k_2 C_B;$$

$$\mathrm{A} \rightarrow \mathrm{B} \rightarrow \mathrm{D}$$

$$\downarrow r_{ae} = k_3 C_A$$

$$\mathrm{E}$$

We will take the magnitudes of the two rate constants to be the same as in the previous example, which makes the rate of formation of B a factor of two greater than the rate of formation of D. The codes are called "cstrABAD" and "pfrABAD."

Clear["Global`* "]

cstrABAD[k1_, k2_, k3_, q_, r_]:=

```
Module[
{Caf = 1, Acr, V = 1000},
Acr = N[π r^2];
zmax = V/Acr;
ststsolns = Solve[
{0==(Caf − Ca) q/V − k1 Ca − k3 Ca,
0== − Cb q/V + k1Ca − k2 Cb,
0== − Cd q/V + k2Cb, 0== − Ce q/V + k3 Ca},
{Ca, Cb, Cd, Ce}];
Ca = ststsolns[[1, 1, 2]];
Cb = ststsolns[[1, 2, 2]];
Cd = ststsolns[[1, 3, 2]];
Ce = ststsolns[[1, 4, 2]];

Plot[{Ca, Cb, Cd, Ce}, {t, 0, zmax},
PlotStyle->{Hue[.1], Hue[.3], Hue[0.5], Hue[.7]},
PlotRange->{{0, zmax}, {0, Caf}},
PlotLabel → {"CSTR", k1"=k1", k2"=k2", k3"=k3"}
]]

pfrABAD[k1_, k2_, k3_, q_, r_]:=
Module[
{Caf = 1, V = 1000, Cao = 1, Cbo = 0, Cdo = 0, Ceo = 0,
ϵ = .4},
Acr = N[π r^2];
vz = q/Acr;
```

```
zmax = V/Acr;
pfrsolns = NDSolve[{
vzCa'[z]== − k1Ca[z] − k3Ca[z],
vzCb'[z] == +k1Ca[z] − k2Cb[z],
vzCd'[z] == + k2Cb[z],
vzCe'[z] == + k3Ca[z],
Ca[0]==Cao, Cb[0]==Cbo, Cd[0] == Cdo,
Ce[0] == Ceo},
{Ca[z], Cb[z], Cd[z], Ce[z]}, {z, 0, zmax}];
CA[z] = Evaluate[Ca[z]/.pfrsolns];
CB[z] = Evaluate[Cb[z]/.pfrsolns];
CD[z] = Evaluate[Cd[z]/.pfrsolns];
CE[z] = Evaluate[Ce[z]/.pfrsolns];
Plot[{CA[z], CB[z], CD[z], CE[z]}, {z, 0, zmax},
PlotStyle → {
{Dashing[{0.15, 0.05}], Hue[0.1]},
{Dashing[{0.15, 0.05}], Hue[0.3]},
{Dashing[{0.15, 0.05}], Hue[.5]},
{Dashing[{0.15, 0.05}], Hue[.7]}},

PlotRange->{{0, zmax}, {0, Caf}},
PlotLabel → {"Pfr", k1"=k1", k2"=k2", k3"=k3"}]
]

a1 = pfrABAD[.05, .1, .025, 10, 10]
```

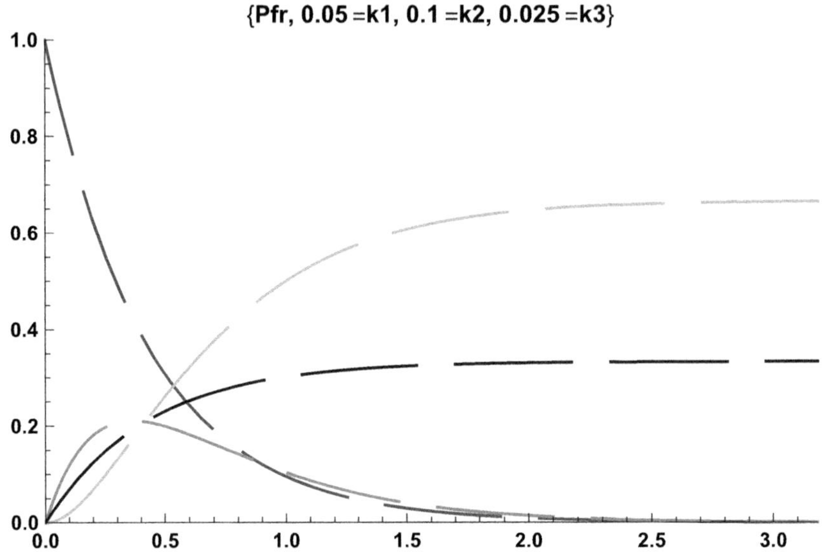

b1 = cstrABAD[.05, .1, .025, 10, 10]

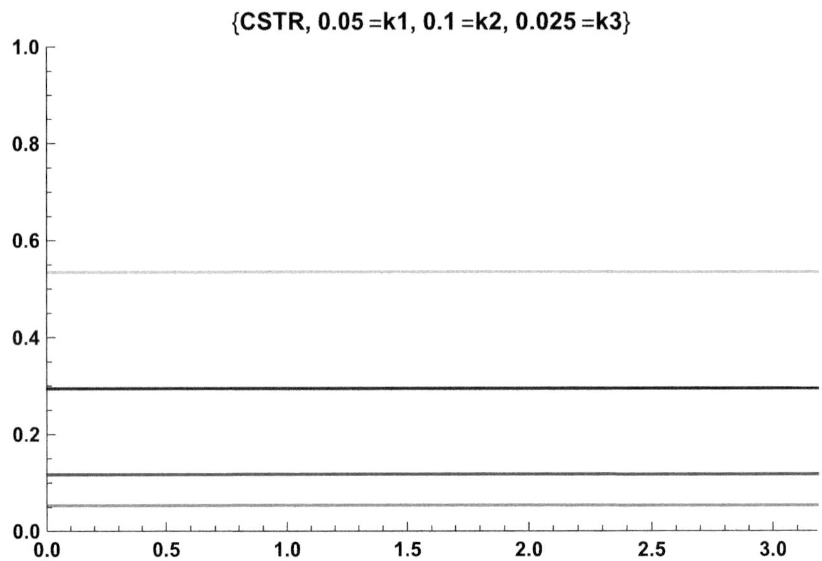

A = tan; **B** = green (mid gray in the print version); **D** = aqua (light gray in the print version); **E** = blue (dark gray in the print version).

Show[a1, b1]

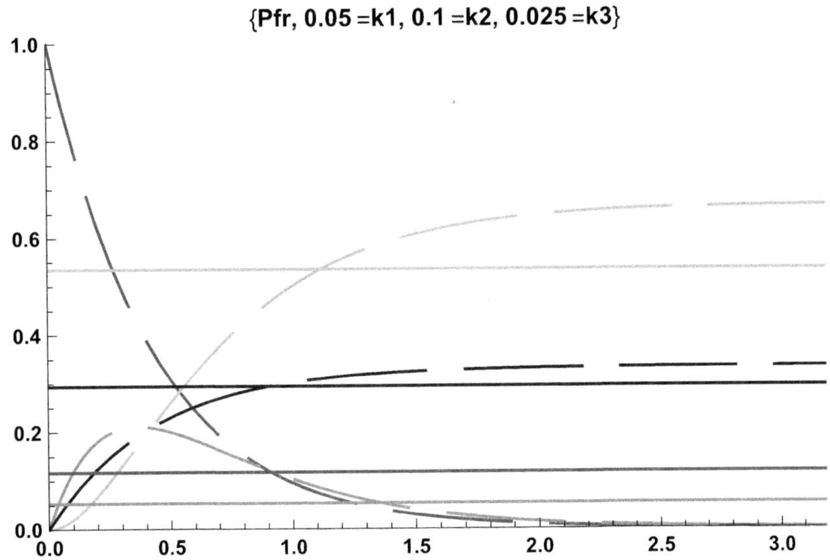

Where **A** is tan, **B** is green (mid gray in the print version), **D** is aqua (light gray in the print version), **E** is blue (dark gray in the print version), and the PFR is dashed. We see once again that the overall conversion is higher in the PFR than in the CSTR, the fractions of **D** and **E** in the products are therefore large than in the CSTR. The efficiency of the PFR with positive order kinetics versus the perfectly mixed CSTR raises an interesting question – if one were to divide the volume of one CSTR into two CSTRs in series of equal volume, would there be any change in efficiency? What if there were three or even more CSTRs in series with equal volumes that all summed up to that of the original one, would there be a significant difference? We address this in the next section.

PFR as a series of CSTRs

The main difference between the PFR and CSTR idealizations is the mathematical one of spatially distributed variables versus homogeneous ones, which leads to quite different equations. If one were to take the globally homogeneous CSTR and break it up into smaller homogeneous regions, then in the limit of an infinite number of these taken in series, they would become equivalent to one PFR of equal total volume. We can see this by comparing one, two, three, and more CSTRs in series with one PFR. As the number of CSTRs increases the results will approach the PFR. The simplest way to do this is to write a bit of code that allows us to specify the total volume and then to vary the number of contiguous homogeneous cells that are present. We can begin by taking a look at the case of $\mathbf{A} \longrightarrow \mathbf{B}$ with linear, irreversible kinetics, since this is simple, and we understand it well, and we can solve it exactly.

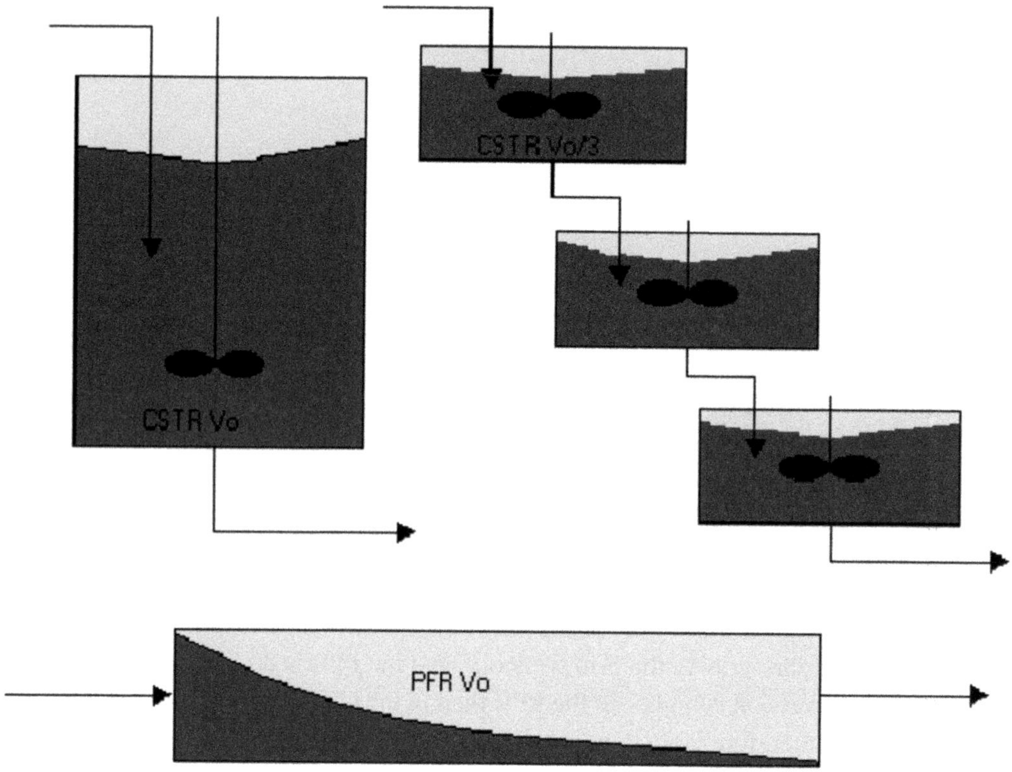

The steady state solutions for one CSTR are shown below:

Clear["Global`* "]

ststsolns = Solve[

$$\left\{0 == (\mathrm{Caf} - \mathrm{Ca})\frac{q}{V} - \mathrm{k1Ca},\right.$$

$$\left. 0 == -\mathrm{Cb}\frac{q}{V} + \mathrm{k1Ca}\right\},$$

{Ca, Cb}]

$$\left\{\left\{\mathrm{Ca} \rightarrow \frac{\mathrm{Caf}q}{q + \mathrm{k1}V}, \mathrm{Cb} \rightarrow \frac{\mathrm{Cafk1}V}{q + \mathrm{k1}V}\right\}\right\}$$

We can generalize this for any nth **CSTR** in a series as follows:

Clear[ca, cb, *q*, Caf, ntot, Vtot, *k*, *n*]

Solve[

$$\left\{0 == (\mathbf{Ca}[n-1]-\mathbf{Ca}[n])\frac{q}{V}-k\mathbf{Ca}[n],\right.$$

$$\left.0 == (\mathbf{Cb}[n-1]-\mathbf{Cb}[n])\frac{q}{V}+k\mathbf{Ca}[n]\right\},$$

{Ca[*n*], Cb[*n*]}

]

$$\left\{\left\{\mathrm{Ca}[n] \to \frac{q\mathrm{Ca}[-1+n]}{q+kV},\right.\right.$$

$$\left.\left.\mathrm{Cb}[n] \to -\frac{-kV\mathrm{Ca}[-1+n]-q\mathrm{Cb}[-1+n]-kV\mathrm{Cb}[-1+n]}{q+kV}\right\}\right\}$$

This is a recursion formula for the exact case. We would like to be able to apply this to any number, n, of CSTRs in series and find an analytical and then quantitative result for comparison to the exact PFR result. To do this we need recursive programming. There are three programming styles in *Mathematica*: Rule-Based, Functional, and Procedural. We will attack this problem in recursion with Rule-Based, Functional, and Procedural programming. We can begin by looking at the *rule-based* recursion codes for Ca and Cb in any n CSTRs.

The "seeds" for these rules are the solutions for the first CSTR, these exit concentrations become the inlet concentrations to the second CSTR, whose exit concentrations become the inlet concentrations for the third CSTR and so it goes on through to n CSTRs. We can have *Mathematica* assemble the equations and the variables that we will need for the Solve routine. This is illustrated with n = 3:

$n = 3$;

Clear[*q*, *V*, Vo, *k*, Caf, Cbf, Cdf, Ca]

$$\mathbf{Table}\left[\left\{(\mathbf{Ca}[i-1]-\mathbf{Ca}[i])\frac{q}{V}-k\mathbf{Ca}[i]{=}{=}0,\right.\right.$$

$$\left.(\mathbf{Cb}[i-1]-\mathbf{Cb}[i])\frac{q}{V}+k\mathbf{Ca}[i]{=}{=}0\right\},$$

{*i*, 1, *n*}]

vars = Table[{Ca[*i*], Cb[*i*]}, {*i*, 1, *n*}]

$$\left\{\left\{\frac{q(\mathrm{Ca}[0]-\mathrm{Ca}[1])}{V}-k\mathrm{Ca}[1]==0,\right.\right.$$

$$kCa[1] + \frac{q(Cb[0] - Cb[1])}{V} == 0\Big\},$$

$$\Big\{\frac{q(Ca[1] - Ca[2])}{V} - kCa[2] == 0,$$

$$kCa[2] + \frac{q(Cb[1] - Cb[2])}{V} == 0\Big\},$$

$$\Big\{\frac{q(Ca[2] - Ca[3])}{V} - kCa[3] == 0,$$

$$kCa[3] + \frac{q(Cb[2] - Cb[3])}{V} == 0\Big\}\Big\}$$

{{Ca[1], Cb[1]}, {Ca[2], Cb[2]}, {Ca[3], Cb[3]}}

In the way that we have written the equations in the code above, the volume is left unspecified. In fact, if we are planning to compare what the number of CSTRs in series does to concentrations of reactants and products as the number is increased, then we have to pay attention to this matter. For example if we want to compare the effect of going from one CSTR to five, then the volume of each of the five CSTRs should be one-fifth of the volume of the one CSTR. In following script, the volume of one CSTR is divided by n so that this is accounted for in the results:

```
Clear[q, V, Vo, k, Caf, Cbf, Cdf, n, Ca]

n = 4;

V = Vo/n;

Timing[

eqns = Table[{(Ca[i - 1] - Ca[i]) q/V - kCa[i]==0,

        (Cb[i - 1] - Cb[i]) q/V + kCa[i]==0},

  {i, 1, n}];

vars = Table[{Ca[i], Cb[i]}, {i, 1, n}];

Ca[0] = Caf;

Cb[0] = 0;

solns = Flatten[Solve[Flatten[eqns], Flatten[vars]]]]
```

$$\{0.016324, \{\mathrm{Ca}[1] \to \frac{4\mathrm{Caf}q}{4q + k\mathrm{Vo}},$$

$$\mathrm{Cb}[1] \to \frac{\mathrm{Caf}k\mathrm{Vo}}{4q + k\mathrm{Vo}},$$

$$\mathrm{Ca}[2] \to \frac{16\mathrm{Caf}q^2}{(4q + k\mathrm{Vo})^2},$$

$$\mathrm{Cb}[2] \to \frac{\mathrm{Caf}k\mathrm{Vo}(8q + k\mathrm{Vo})}{(4q + k\mathrm{Vo})^2},$$

$$\mathrm{Ca}[3] \to \frac{64\mathrm{Caf}q^3}{(4q + k\mathrm{Vo})^3},$$

$$\mathrm{Cb}[3] \to \frac{\mathrm{Caf}k\mathrm{Vo}\left(48q^2 + 12kq\mathrm{Vo} + k^2\mathrm{Vo}^2\right)}{(4q + k\mathrm{Vo})^3},$$

$$\mathrm{Ca}[4] \to \frac{256\mathrm{Caf}q^4}{(4q + k\mathrm{Vo})^4},$$

$$\mathrm{Cb}[4] \to \frac{\mathrm{Caf}k\mathrm{Vo}\left(256q^3 + 96kq^2\mathrm{Vo} + 16k^2q\mathrm{Vo}^2 + k^3\mathrm{Vo}^3\right)}{(4q + k\mathrm{Vo})^4}\}\}$$

We see that in this program the equations are first written explicitly from n = 1 to n, their output is suppressed, but then they are solved symbolically. We have enclosed the overall functions in "Timing" so as to see the CPU time required to conduct this. By supplying the necessary parameters and by changing Solve to NSolve, we can find the solution for the outlet of n-CSTRs. We make this into a Module function of n the number of CSTRs:

```
Clear[q, V, Vo, k, Ca, Cb, Caf, Cbf, Cdf, n]

cstr[n_]:=
Module[
{q = 10, Vo = 100, k = 0.1},
V = Vo/n;
eqns = Table[{(Ca[i - 1] - Ca[i]) q/V - k Ca[i]==0,
        (Cb[i - 1] - Cb[i]) q/V + k Ca[i]==0},
  {i, 1, n}];
```

```
vars = Table[{Ca[i], Cb[i]}, {i, 1, n}];
Ca[0] = 1;
Cb[0] = 0;
solns = NSolve[Flatten[eqns], Flatten[vars]][[1]];
{solns[[2n − 1]],
solns[[2n]]}]
```

We can ask for the result of placing 1000 CSTRs in series and we find that the concentrations of A and B are 0.37 and 0.63, respectively.

```
cstr[1000]//Timing
```

{2.2426, {Ca[1000] → 0.368063, Cb[1000] → 0.631937}}

When we have also taken just the solutions from the last CSTR by using {solns[[2n -1]], solns[[2n]]}. Now, we can use the "listability" of this function cstr[n] and Map it down a vector of values for n to see how the exit concentrations vary with the number of CSTRs. The *infix* form for Map is /@ and we use it as follows:

```
n = {1, 2, 3, 10, 20, 100, 1000};
cstr/@n//Timing
```

{2.17417, {{Ca[1] → 0.5, Cb[1] → 0.5}, {Ca[2] → 0.444444, Cb[2] → 0.555556}, {Ca[3] → 0.421875, Cb[3] → 0.578125}, {Ca[10] → 0.385543, Cb[10] → 0.614457}, {Ca[20] → 0.376889, Cb[20] → 0.623111}, {Ca[100] → 0.369711, Cb[100] → 0.630289}, {Ca[1000] → 0.368063, Cb[1000] → 0.631937}}}

We notice that there is a large change in Ca and Cb when we go from one CSTR to 2 or 3 or even to 10, but when we get beyond 10 to 20, 100, and even 1000, the increase in the number of CSTRs gives diminishing returns. To do this computation took ~2.2 seconds of CPU time. We added //Timing after the command line also in *postfix* form, in order to obtain this information so that as we try different approaches to the coding that we can see the effect on timing.

Before we examine the series of CSTRs further, we solve the equations for the PFR at the same conditions and with the same parameter values, especially that of the volume, Vo, so that we can make a comparison:

```
Clear["Global`*"]
Caf = 1;
Vo = 100;
q = 10;
k1 = .1;
zmax = 10;
A = Vo/zmax;
vz = q/A;
pfr = NDSolve[
{vzCa'[z]== - k1Ca[z],
vzCb'[z] == +k1Ca[z],
Ca[0]==Caf, Cb[0] == 0},
{Ca[z], Cb[z]},
{z, 0, zmax}];
CA[z_]:=Evaluate[Ca[z]/.pfr];
CB[z_]:=Evaluate[Cb[z]/.pfr];
{CA[zmax], CB[zmax]}//Timing
```

{0.000017, {{0.367879}, {0.632121}}}

Here we see that the limiting values of Ca and Cb are 0.37 and 0.63, respectively, exit concentrations which are nearly identical to those obtained with the large number of CSTRs (>100) and closely approached by even 10 CSTRs. The important point we learn from this is that even with two or three CSTRs in series, the math moves toward the PFR limit.

Let's return to the recursive programming part of this problem because it is a prototypical type problem that we can expect to encounter often in chemical engineering analysis and computations. We can now try *functional* programming to derive the solutions that we seek. We know the solution for Ca at the exit of the first CSTR and we have recursive relationship for the exit concentrations emerging from the next n-CSTRs. So we can put these two together into a functional program as follows:

Clear[ca, V, q, Caf, Vo, k]

ca[1] $= \frac{q\ \text{Caf}}{q+kV}$;

ca[n_]:=ca[n] $= \frac{q\ \text{ca}[n-1]}{q+kV}$

We can see how this works. We defined the seed for **ca [1]** first, then we created the function for any **ca [n]** as follows:

$$\text{ca[n\_]:=ca}[n] = \frac{q\text{ca}[n-1]}{q+kV}$$

Taking n = 4 we find:

ca[4]

$$\frac{\text{Caf}q^4}{(q+kV)^4}$$

Recall that in this recursion relation the symbol $V = \frac{\text{Vo}}{4}$, therefore we can redo the computation to derive:

V =;

ca[4]

Together[%]

$$\frac{\text{Caf}q^4}{\left(q+\frac{k\text{Vo}}{4}\right)^4}$$

$$\frac{256\text{Caf}q^4}{(4q+k\text{Vo})^4}$$

Referring back to the earlier solution that we derived for n = 4, we see the two agree perfectly. The function for ca[n] may be used in the Table function to derive the first four solutions symbolically:

n = 4;

Table[ca[x], {x, 1, n}]

Together[%]

$$\left\{\frac{\text{Caf}q}{q+\frac{k\text{Vo}}{4}},\frac{\text{Caf}q^2}{\left(q+\frac{k\text{Vo}}{4}\right)^2},\frac{\text{Caf}q^3}{\left(q+\frac{k\text{Vo}}{4}\right)^3},\frac{\text{Caf}q^4}{\left(q+\frac{k\text{Vo}}{4}\right)^4}\right\}$$

$$\left\{\frac{4\text{Caf}q}{4q+k\text{Vo}},\frac{16\text{Caf}q^2}{(4q+k\text{Vo})^2},\frac{64\text{Caf}q^3}{(4q+k\text{Vo})^3},\frac{256\text{Caf}q^4}{(4q+k\text{Vo})^4}\right\}$$

This is a simple and, yet, very powerful use of *Mathematica*'s set-delay utility : =. The right-hand side is the whole recursive equation with an equal sign, but by placing this after the set-delay, : =, it becomes a pattern that will not be evaluated until n is specified. When n is specified, then it begins with the seed and evaluates the function until it gets to the value of n. When we set n = 4, we get the concentration expression for the exit of the fourth CSTR. If we place the function in a Table, and set n = 4 we obtain the exit concentrations for all four of the CSTRs. We follow the same procedure for cb[n] as shown below:

Clear[cb, Caf, K, V, q]

$V =;$

$$\text{cb}[1]=\frac{\text{Caf}kV}{q+kV};$$

$$\text{cb[n\_]:=cb}[n]=\frac{kV\text{ca}[n-1]}{q+kV}+\text{cb}[n-1]$$

$s=4;$

cb[s]

Together[%]

$$\frac{\text{Caf k }q^3\text{Vo}}{4\left(q+\frac{k\text{Vo}}{4}\right)^4}+\frac{\text{Caf k }q^2\text{Vo}}{4\left(q+\frac{k\text{Vo}}{4}\right)^3}+\frac{\text{Caf k }q\text{Vo}}{4\left(q+\frac{k\text{Vo}}{4}\right)^2}+\frac{\text{Caf kVo}}{4\left(q+\frac{k\text{Vo}}{4}\right)}$$

$$\frac{\text{Caf kVo}\left(256q^3+96kq^2\text{Vo}+16k^2q\text{Vo}^2+k^3\text{Vo}^3\right)}{(4q+k\text{ Vo})^4}$$

Again, the solution for Cb at the exit of CSTR number 4 matches that previously derived. Note also that this solution for B was explicit in the parameters because we had already defined ca [n]. If we had not done this, then what follows is what we would have seen:

Clear["Global`*"]

$V =;$

$\mathbf{cb[1]} = \frac{\mathbf{Caf}kV}{q+kV};$

$\mathbf{cb[n\_]:=cb[}n\mathbf{]} = \frac{kV\mathbf{ca[}n-1\mathbf{]}}{q+kV} + \mathbf{cb[}n-1\mathbf{]}$

$s = 4$

cb[*s*]

Together[%]

4

$$\frac{\text{Caf k Vo}}{4\left(q+\frac{k\ \text{Vo}}{4}\right)}+\frac{k\ \text{Vo ca[1]}}{4\left(q+\frac{k\ \text{Vo}}{4}\right)}+\frac{k\ \text{Vo ca[2]}}{4\left(q+\frac{k\text{Vo}}{4}\right)}+\frac{k\ \text{Vo ca[3]}}{4\left(q+\frac{k\ \text{Vo}}{4}\right)}$$

$$\frac{\text{Caf k Vo} + k\ \text{Vo ca[1]} + k\ \text{V oca[2]} + k\ \text{Vo ca[3]}}{4q + k\ \text{Vo}}$$

To avoid any such problems, we simply place the two sets of functions together into one working cell as follows:

Clear["Global`*"]

$V = \frac{\mathbf{Vo}}{s};$

$\mathbf{ca[1]} = \frac{q\mathbf{Caf}}{q+kV};$

$\mathbf{ca[n\_]:=ca[}n\mathbf{]} = \frac{q\mathbf{ca[}n-1\mathbf{]}}{q+kV}$

$\mathbf{cb[1]} = \frac{\mathbf{Caf}kV}{q+kV};$

$\mathbf{cb[n\_]:=cb[}n\mathbf{]} = \frac{kV\mathbf{ca[}n-1\mathbf{]}}{q+k\ V} + \mathbf{cb[}n-1\mathbf{]}$

Testing the result we obtain:

$s = 4;$

{Together[ca[*s*]], Together[cb[*s*]]}

$$\left\{\frac{256\text{Caf}q^4}{(4q+k\text{Vo})^4}, \frac{\text{Caf}k\text{Vo}\left(256q^3+96kq^2\text{Vo}+16k^2q\text{Vo}^2+k^3\text{Vo}^3\right)}{(4q+k\text{Vo})^4}\right\}$$

We can apply values to the parameters and the values n. We use s to define the number of CSTRs rather than n per se in order to avoid a name clash that leads to an infinite loop. Before we try this, however, we shall set the $RecursionLimit to 1000. The default value of 256 is not sufficient for us to go out to numbers as large as n = 1000 CSTRs.

Clear["Global`*"]

$RecursionLimit = 2000;

$V = \frac{\text{Vo}}{s};$

$\text{ca}[1] = \frac{q\text{Caf}}{q+kV};$

$\text{ca}[\text{n\_}] := \text{ca}[n] = \frac{q\,\text{ca}[n-1]}{q+kV}$

$\text{cb}[1] = \frac{\text{Caf}kV}{q+kV};$

$\text{cb}[\text{n\_}] := \text{cb}[n] = \frac{kV\text{ca}[n-1]}{q+k\,V} + \text{cb}[n-1]$

$\text{Caf} = 1;$

$k = .1;$

$q = 10;$

$\text{Vo} = 100;$

$s = 1000$

{ca[*s*], cb[*s*]}//Timing

1000

{0.00866, {0.368063, 0.631937}}

We can see that this is huge speed up from the original code that we wrote in which first rewrote all the equations and then did the computations. The latter took 19.06 seconds of CPU time to reach 1000 CSTRs. This new program required only 2.91 seconds to do the same.

Solving this recursion problem can be done more traditionally using a *procedural* approach. The recursion in this case is within a "Do" loop, which is the classic structure in the procedural programming paradigm. The "Do" loop is placed within a Module function to keep all the variable names localized:

Clear["Global`*"]

ca1[n_]:=

Module[{m, ca},

$$V = \frac{\text{Vo}}{n};$$

$$\text{Do}\left[\text{ca}[1] = \frac{q\text{Caf}}{q + kV};\right.$$

$$\left.\text{ca}[m] = \frac{q\text{ca}[m-1]}{q + kV}, \{m, 0, n\}\right];$$

ca[n]

]

ca1[4]

Together[%]

$$\frac{\text{Caf}q^4}{\left(q + \frac{k\text{Vo}}{4}\right)^4}$$

$$\frac{256\text{Caf}q^4}{(4q + k\text{Vo})^4}$$

Within the Do loop, we have the same recursion that we had implemented by rules – we specify the seed as ca [1] and then the recursion relation as ca [m]. The set-delayed function argument of ca1 [n_] supplies n, as the limit of the iterative sequence. The last statement just outside the Do loop, writes the value of ca[n] and then, it stops. We implement the same approach for species **B**:

Clear[ca1]

cb1[n_]:=

Module[{m, cb},

$$V = \frac{\text{Vo}}{n};$$

$$\text{Do}\left[\text{cb}[1] = \frac{\text{Caf}kV}{q + kV};\right.$$

$$\left.\text{cb}[m] = \frac{\text{ca1}[m-1]kV}{q + kV} + \text{cb}[m-1], \{m, 0, n\}\right];$$

cb[n]

]

cb1[4]

Together[%]

$$\frac{\text{Caf}k\text{Vo}}{4\left(q + \frac{k\text{Vo}}{4}\right)} + \frac{k\text{Voca1}[1]}{4\left(q + \frac{k\text{Vo}}{4}\right)} + \frac{k\text{Voca1}[2]}{4\left(q + \frac{k\text{Vo}}{4}\right)} + \frac{k\text{Voca1}[3]}{4\left(q + \frac{k\text{Vo}}{4}\right)}$$

$$\frac{\text{Caf}k\text{Vo} + k\text{Voca1}[1] + k\text{Voca1}[2] + k\text{Voca1}[3]}{4q + k\text{Vo}}$$

To evaluate the expression fully, in terms of the parameters, we need the expression for ca1[n] to have been evaluated. We do the two in one cell to make this happens:

cacb[n_]:=

Module[{m, ca, cb},

$$V = \frac{\text{Vo}}{n};$$

$$\text{Do}\left[\left\{\text{ca}[0] = \text{Caf}, \text{ca}[1] = \frac{q\text{Caf}}{q + kV}, \text{cb}[0] = 0,\right.\right.$$

$$\left.\text{cb}[1] = \frac{\text{Caf}kV}{q + kV}\right\};$$

$$\left\{\text{ca}[m] = \frac{q\text{ca}[m-1]}{q + kV},\right.$$

$$\left.\text{cb}[m] = \frac{\text{ca}[m-1]kV}{q + kV} + \text{cb}[m-1]\right\},$$

{*m*, 1, *n*}];

{Together[ca[*n*]], Together[cb[*n*]]}

]

cacb[4]

$$\left\{\frac{256\mathrm{Caf}q^4}{(4q+k\mathrm{Vo})^4},\ \frac{\mathrm{Caf}k\mathrm{Vo}\left(256q^3+96kq^2\mathrm{Vo}+16k^2q\mathrm{Vo}^2+k^3\mathrm{Vo}^3\right)}{(4q+k\mathrm{Vo})^4}\right\}$$

This works very nicely indeed. Applying numerical values to these solutions, we solve for the concentration and also obtain the timing for a comparison to the rule-based case:

Caf = 1;

***k* = .1;**

***q* = 10;**

ntot = 10;

Vo = 100;

cacb[1000]//Timing

{0.011277, {0.368063, 0.631937}}

We see here that the solution is identical to the asymptotic solutions obtained earlier and the time to do n = 1000 is only 1.98 seconds of CPU, which is much faster than the first program and a little slower than the rule-based implementation. Once again the function cacb that we have created is *listable* so we can Map it down the vector of n values that we had used before in the rule-based computations and we can compare timings:

***n* = {1, 2, 3, 10, 20, 100, 1000};**

cacb/@*n*//Timing

{0.012809, {{0.5, 0.5}, {0.444444, 0.555556}, {0.421875, 0.578125}, {0.385543, 0.614457}, {0.376889, 0.623111}, {0.369711, 0.630289}, {0.368063, 0.631937}}}

When we did this via brute force using all the equations and no recursion, we utilized 17.54 seconds of CPU time, the procedural program requires just 1.26 seconds.

***n* = {1, 2, 3, 5, 10, 20, 30, 50, 60, 90, 100, 150};**

```
cstrconcs = cacb/@n
```

{{0.5, 0.5}, {0.444444, 0.555556}, {0.421875, 0.578125}, {0.401878, 0.598122}, {0.385543, 0.614457}, {0.376889, 0.623111}, {0.373927, 0.626073}, {0.371528, 0.628472}, {0.370924, 0.629076}, {0.369914, 0.630086}, {0.369711, 0.630289}, {0.369102, 0.630898}}

```
calist = Table[{n[[x]], cstrconcs[[x, 1]]},{x, 1, Length[n]}];

cblist = Table[{n[[x]], cstrconcs[[x, 2]]},{x, 1, Length[n]}];

ListPlot[
{calist, cblist},
PlotMarkers → Automatic,
Epilog → {
{RGBColor[0, 1, 1], Line[{{0, .63}, {150, 0.63}}]},
{RGBColor[1, 0, 0],
Line[{{0, 1 − .63}, {150, 1 − 0.63}}]}        },
AxesLabel → {"n-CSTRs", "Ca,Cb"},
PlotLabel → "nCSTRs (pts.) → PFR (lines)"]
```

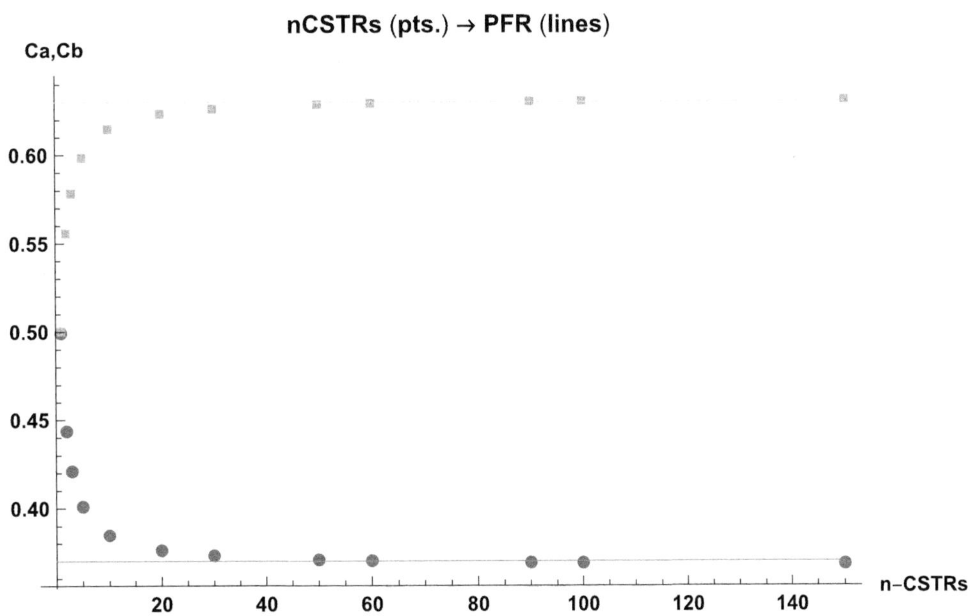

Residence time distribution

The problem of by-passing and less than perfect mixing was one that we first encountered in Chapter 4 on mixing in multicomponent systems. If we have two or more reactants that must mix in order for reaction to occur, then any deviations from a single valued residence time distribution will show us an apparent deviation from the predictions based upon perfect mixing. The spread in the residence time distribution leads to different extents of reaction for the different fluid elements with these different times.

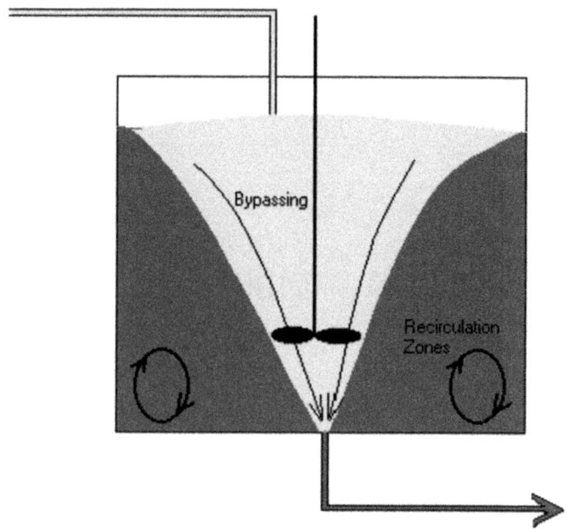

The lack of perfect mixing leads to this distribution. Recirculation zones may lead to longer than average residence times in some regions of the tank and by-passing to shorter than average time. Longer, or shorter times can translate into regions of higher or lower conversion. It is this type of problem we want to examine now.

We have developed the equations for a steady state CSTR in which the reversible reaction of **A** and **B** reacts to produce **D** and one mole of **D** reacts back to produce **A** and **B** but the kinetics are second order:

$$A + B \rightleftarrows D$$

$$0 = (\text{Caf} - \text{Ca})\frac{q}{V} - \text{kab Ca Cb} + \text{kd Cd}^2$$

$$0 = (\text{Cbf} - \text{Cb})\frac{q}{V} - \text{kab Ca Cb} + \text{kd Cd}^2$$

$$0 = -\text{Cd}\frac{q}{V} + \text{kab Ca Cb} - \text{kd Cd}^2$$

$$0 = \text{qbf} - \text{q}$$

$$0 = \frac{(\text{Caf} - \text{Ca})}{\theta} - \text{kab Ca Cb} + \text{kd Cd}^2$$

$$0 = \frac{(\text{Cbf} - \text{Cb})}{\theta} - \text{kab Ca Cb} + \text{kd Cd}^2$$

$$0 = -\frac{\text{Cd}}{\theta} + \text{kab Ca Cb} - \text{kd Cd}^2$$

Of course $\frac{q}{V} = \frac{1}{\theta}$ the holding time for the fluid in the reactor. When we derived these equations, we did so under the assumption of *perfect mixing*, or *complete back-mixing*, which is another term for this idealization. If there were perfect mixing, then there would be just one residence time and that would be the same as the holding time, θ. When the residence time distribution is a DiracDelta function at one time, then we have one holding time. But, what if this is not the case? What if instead the times spent by fluid elements are distributed about some mean value–then what? Well, then we would have to average these equations over the form of the distribution to get the average concentration emerging from the less than perfectly mixed reactor. We begin by solving the equation in terms of θ:

Clear["Global`*"]

cstr = Simplify[

Solve[

$$\left\{0 == \frac{(\text{Caf} - \text{Ca})}{\theta} - \text{kabCaCb} + \text{kdCd}^2,\right.$$

$$0 == \frac{(\text{Cbf} - \text{Cb})}{\theta} - \text{kabCaCb} + \text{kdCd}^2,$$

$$\left. == -\frac{\text{Cd}}{\theta} + \text{kabCaCb} - \text{kdCd}^2\right\},$$

{Ca, Cb, Cd}]];

TableForm[cstr, TableDirections->{Column, Column}]

$$\text{Ca} \to \frac{-1+\text{Caf kab }\theta-\text{Cbf kab }\theta-2\text{Caf kd }\theta+\sqrt{-4\text{CafCbf kab(kab}-\text{kd})\theta^2+(1+\text{Caf kab }\theta+\text{Cbf kab }\theta)^2}}{2(\text{kab}-\text{kd})\theta}$$

$$\text{Cb} \to \frac{-1-\text{Caf kab }\theta+\text{Cbf kab }\theta-2\text{Cbf kd }\theta+\sqrt{-4\text{Caf Cbf kab (kab}-\text{kd})\theta^2+(1+\text{Caf kab }\theta+\text{Cbf kab}\theta)^2}}{2(\text{kab}-\text{kd})\theta}$$

$$\text{Cd} \to \frac{1+\text{Caf kab }\theta+\text{Cbf kab }\theta-\sqrt{-4\text{Caf Cbf kab(kab}-\text{kd})\theta^2+(1+\text{Caf kab }\theta+\text{Cbf kab }\theta)^2}}{2(\text{kab}-\text{kd})\theta}$$

$$\text{Ca} \to -\frac{1-\text{Caf kab }\theta+\text{Cbf kab }\theta+2\text{Caf kd }\theta+\sqrt{-4\text{Caf Cbf kab (kab}-\text{kd})\theta^2+(1+\text{Caf kab }\theta+\text{Cbf kab }\theta)^2}}{2(\text{kab}-\text{kd})\theta}$$

$$\text{Cb} \to -\frac{1+\text{Caf kab }\theta-\text{Cbf kab }\theta+2\text{Cbf kd }\theta+\sqrt{-4\text{Caf Cbf kab (kab}-\text{kd})\theta^2+(1+\text{Caf kab }\theta+\text{Cbf kab }\theta)^2}}{2(\text{kab}-\text{kd})\theta}$$

$$\text{Cd} \to \frac{1+\text{Caf kab }\theta+\text{Cbf kab }\theta+\sqrt{-4\text{CafCbf kab (kab}-\text{kd})\theta^2+(1+\text{Caf kab }\theta+\text{Cbf kab }\theta)^2}}{2(\text{kab}-\text{kd})\theta}$$

The next step is to re-express these as function of θ. By examination, we can find that the second set of solutions is the one that leads to physically realistic values of the concentrations, thus we use these:

Clear[Ca, Cb, Cd]

Ca[θ_] = cstr[[2, 1, 2]];

Cb[θ_] = cstr[[2, 2, 2]];

Cd[θ_] = cstr[[2, 3, 2]];

Testing the solutions requires putting them back into the equations to see if they are valid:

{Simplify[

$$0 == \frac{(\text{Caf} - \text{Ca}[\theta])}{\theta} - \text{kabCa}[\theta]\text{Cb}[\theta] + \text{kdCd}[\theta]^2\Big],$$

Simplify[

$$0 == \frac{(\text{Cbf} - \text{Cb}[\theta])}{\theta} - \text{kabCa}[\theta]\text{Cb}[\theta] + \text{kdCd}[\theta]^2\Big],$$

$$\text{Simplify}\left[0 == -\frac{\text{Cd}[\theta]}{\theta} + \text{kabCa}[\theta]\text{Cb}[\theta] - \text{kdCd}[\theta]^2\right]\Big\}$$

{True, True, True}

This shows that the functions we are using do indeed satisfy the equations for the steady state CSTR.

We will use the NormalDistribution to make the representations of the residence time distribution. The Probability Density Function (PDF) is made up of the Normal Distribution and the variable θ. This can be integrated in closed form:

Clear[ndist, pdf, θ, δθ, θmin, θmax]

ndist = NormalDistribution[θm, δθ];

```
pdf = PDF[ndist, θ];

Integrate[pdf, {θ, θmin, θmax}]
```

$$\frac{1}{2}\left(\mathrm{Erf}\left[\frac{-\theta\mathrm{m}+\theta\mathrm{max}}{\sqrt{2}\delta\theta}\right]-\mathrm{Erf}\left[\frac{-\theta\mathrm{m}+\theta\mathrm{min}}{\sqrt{2}\delta\theta}\right]\right)$$

Given the Normal distribution PDF requires a mean value and a variance, we supply these and then Plot the result in order to visualize the distribution. We have purposefully chosen a very narrow distribution for the first case. Next, the parameter values are assigned and we Integrate the products of the concentration functions and the PDF in θ over the range of θ values. These are the residence time averaged values of the concentrations and the conversion of A and B:

```
θm = 10;

δθ = .2;

θmin = 0.001θm;

θmax =  50θm;

ndist = NormalDistribution[θm, δθ];

pdf = PDF[ndist, θ];

NIntegrate[pdf, {θ, θmin, θmax}]

NIntegrate[θpdf, {θ, θmin, θmax}]/

NIntegrate[pdf, {θ, θmin, θmax}]

Plot[pdf,

{θ, .0001θm , 2θm},

AxesLabel → {"θ", "PDF[θ]"},

PlotStyle → Hue[0.7],

PlotRange →

{{0, 2θm},

{0, Max[Table[N[pdf], {θ, θmin, θmax}]]}},


Epilog → {RGBColor[0, 1, 1],
```

Line[

{{θm, 0},

{θm, Max[Table[*N*[pdf], {θ, θmin, θmax}]]}}]},

PlotLabel → "θm"]

Caf = 1;

Cbf = 1;

kd = 1.1;

kab = 2.2;

$$\text{cave} = \frac{\text{NIntegrate[Ca}[\theta]\text{pdf}, \{\theta, \theta\text{min}, \theta\text{max}\}]}{\text{NIntegrate[pdf}, \{\theta, \theta\text{min}, \theta\text{max}\}]}$$

$$\text{cbave} = \frac{\text{NIntegrate[Cb}[\theta]\text{pdf}, \{\theta, \theta\text{min}, \theta\text{max}\}]}{\text{NIntegrate[pdf}, \{\theta, \theta\text{min}, \theta\text{max}\}]}$$

$$\text{cdave} = \frac{\text{NIntegrate[Cd}[\theta]\text{pdf}, \{\theta, \theta\text{min}, \theta\text{max}\}]}{\text{NIntegrate[pdf}, \{\theta, \theta\text{min}, \theta\text{max}\}]}$$

100(1 − cave)

Clear["Global`* "]

1.

10.

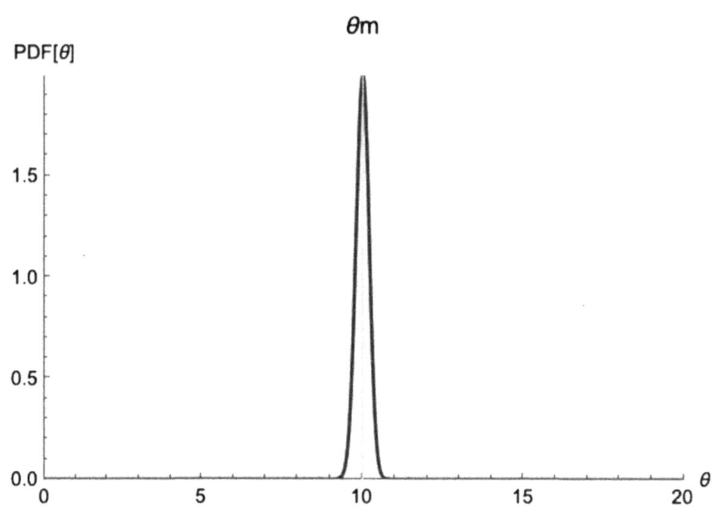

−2.52329

−2.52329

3.52329

352.329

Rather than copy this code cell multiple times and run it successively, it would be much better to create a Module function, but even better than a Module function would be a Package. A Package is a program that we can call at any time simply by loading it and then running it. This will allow us to run without variable name collisions and to use what we have developed outside of this notebook context. Normally, when we are running *Mathematica* we are operating in the Global context, which is why we have often started our codes cells with Clear["Global`*"]. This means clear all the variable names that we have made or used within the Global context. To find out which context we are in we input:

$Context

Global`

Contexts are very much like directories and you can find much written about this topic elsewhere. We will create a package called cstrresdist, which we will store in a folder called AddOns within the Applications folder. The inputs will be the forward and reverse rate constants, kab and kd, the mean residence time, θm, and the variance in the residence time, $\delta\theta$. The general form for a Package is:

```
Begin Package[ "Context`PackageName"]

PackageName::Usage="Narrative explanation of Package..."

Begin["`Private`"]

function name[ variables_]: = ...

End[ ]

EndPackage [ ]
```

We will use this format and the functions we have defined will be implemented as a Module Function. For the case at hand, instead of making the Context = AddOns`, it has been placed directly into the Global` context.

```
BeginPackage["Global`cstrrtd`"];
cstrrtd::Usage = "cstrrtd[kab,kd,θm,δθ] creates the r.t.d,
plots it and then computes the average exit concentrations of A and B and the
conversion of A";
Begin["`Private`"];
cstrrtd[kab_, kd_, θm_, δθ_]:=
Module[
{Caf = 1, Cbf = 1},
θmin = 0.001θm;
  θmax = 5θm;
```

$$ca[\theta\_] = \frac{1}{2(kab - kd)\theta}$$

```
(−1 − Cbfkabθ + Caf(kab − 2kd)θ+
√(−4CafCbfkab(kab − kd)θ^2+
(1 + Cafkabθ + Cbfkabθ)^2));
```

$$cd[\theta\_] = \frac{1}{2(kab - kd)\theta}$$

```
(1 + Cafkabθ + Cbfkabθ−
√(−4CafCbfkab(kab − kd)θ^2+
(1 + Cafkabθ + Cbfkabθ)^2));
distfunc[x_, y_, z_] = PDF[NormalDistribution[x, y], z];
{Plot[distfunc[θm, δθ, θ],
{θ, .000010θm , 2θm},
AxesLabel → {"θ", "PDF[θ]"},
PlotStyle → Hue[0.7],
PlotRange →
```

```
{{0, 2θm},
{0, Max[Table[N[distfunc[θm, δθ, θ]],
{θ, θmin, θmax}]]}},
Epilog → {RGBColor[0, 1, 1],
Line[{{θm, 0},
{θm, Max[Table[N[distfunc[θm, δθ, θ]],
{θ, θmin, θmax}]]}}]},
PlotLabel → "θm"],
a = NIntegrate[distfunc[θm, δθ, θ],
{θ, θmin, θmax}];
b = NIntegrate[θdistfunc[θm, δθ, θ],
{θ, θmin, θmax}]/
NIntegrate[distfunc[θm, δθ, θ],
{θ, θmin, θmax}];
caave = NIntegrate[ca[θ]distfunc[θm, δθ, θ],
{θ, θmin, θmax}]/
NIntegrate[distfunc[θm, δθ, θ],
{θ, θmin, θmax}];
cdave = NIntegrate[cd[θ]distfunc[θm, δθ, θ],
{θ, θmin, θmax}]/
NIntegrate[distfunc[θm, δθ, θ],
{θ, θmin, θmax}];
conversion = 100(1 − caave);
{a, b, caave, cdave, conversion}}End[]
EndPackage[]

Global`cstrrtd`Private`
```

To see how this works we can run a case as follows:

cstrrtd[2, 1, 100, 2]

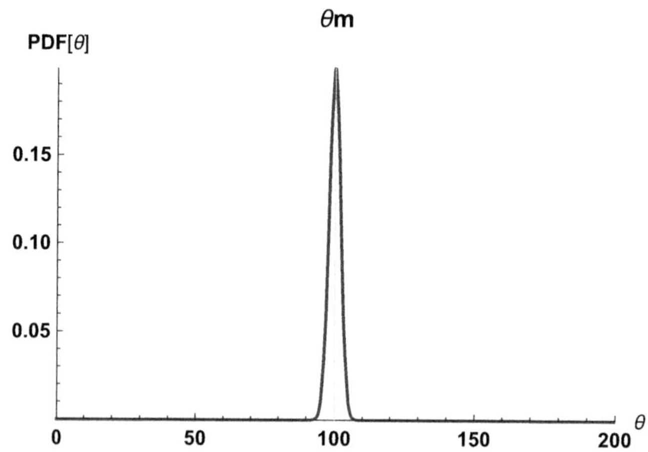

{1., 100., 0.416277, 0.583723, 58.3723}}

There are a few things in this code that are worthy of explanation. We let θmin be expressed as a fraction of the mean value 0.001θm. We did this to avoid having zero in the calculation. Secondly, we set the range of ordinate values with this expression in PlotRange:

PlotRange→ {{0,2θm},{0,Max[Table[N[distfunc[θm,δθ,θ]],{θ,θmin,θmax}]]}}

This computes the maximum value of the distribution function and that becomes the upper value of the range of the ordinate.

The graph is the distribution of residence times about the mean. The bottom line of output gives us the integral of the **PDF**, the mean normalized residence time, the concentration of **A** and **B** and, finally, the conversion of **A**. We see from this graph that with the mean residence time of 100 and variance of 2, this appears to be a sharp distribution. What would happen if the r.t.d. were broader–say 30?

cstrrtd[2, 1, 100, 30]

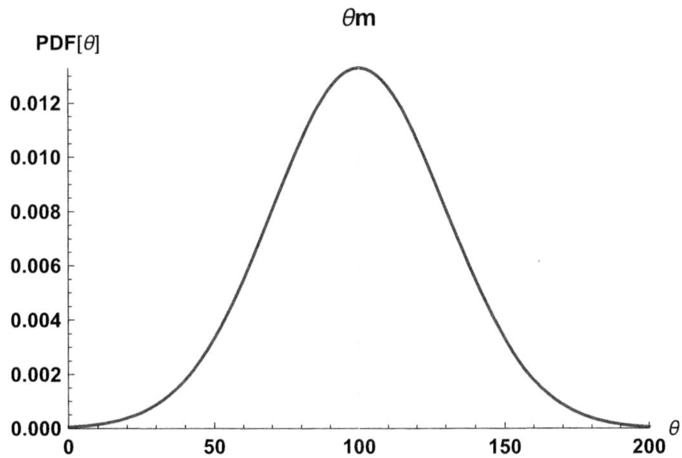

{0.999566, 100.047, 0.416577, 0.583423, 58.3423}

We see that even though the distribution has been broadened considerably, there is no evidence of any effect on the overall conversion. This makes sense, because there are just as many fluid elements below the average as there are above and so we get average behavior – just as we intuitively know we should. What, however, happens when the flow is so maldistributed that the residence time is actually bimodal?

In the following package, Global`rtdcstr`, two normal distributions are weighted (wf1 and wf2) and added together to give the overall residence time distribution. The second Gaussian distribution is taken to be centered at half the mean value of the first, but with the same variance. The function p, is their sum.

```
BeginPackage["Global`rtdcstr`"]
rtdcstr::usage = "testing 1,2,3..."
Begin["`Private`"]
rtdcstr[kab_, kd_, θm_, σθ_, wf1_, wf2_]:=Module[
{Caf = 1, Cbf = 1, x, xmin, xmax},
xmin = 0.001θm;
xmax = 5θm;
ndist1 = NormalDistribution[θm, σθ];
ndist2 = NormalDistribution[.5θm, σθ];
```

pdfunction1 = PDF[ndist1, x];

pdfunction2 = PDF[ndist2, x];

$$p = \frac{\text{wf1pdfunction1} + \text{wf2pdfunction2}}{\text{wf1} + \text{wf2}};$$

$$\text{ca[y\_]} = \frac{1}{2(\text{kab} - \text{kd})y}$$

$(-1 - \text{Cbfkab}y + \text{Caf}(\text{kab} - 2\text{kd})y+$

$\sqrt{}(-4\text{CafCbfkab}(\text{kab} - \text{kd})y^2+$

$(1 + \text{Cafkab}y + \text{Cbfkab}y)^2))$;

{Plot[p, {x, 0, xmax},

AxesLabel → {"θ", "PDF[θ]"},

PlotStyle → Hue[0.7],

PlotRange → {{xmin, 2θm},

{0, .01 + Max[Table[N[p], {x, xmin, xmax}]]}},

Epilog → {RGBColor[0, 1, 1],

Line[{{θm, 0},

{θm, Max[Table[N[p], {x, xmin, xmax}]]}}]},

PlotLabel → "θm,o"],

totp = NIntegrate[p, {x, 0, xmax}];

$$\text{thetam} = \frac{\text{NIntegrate}[xp, \{x, 0, \text{xmax}\}]}{\text{totp}};$$

$$\text{caave} = \frac{\text{NIntegrate}[\text{ca}[x]p, \{x, 0, \text{xmax}\}]}{\text{totp}};$$

conversion = 100(1 − caave);

{totp, thetam, caave, conversion}}

]

End[]

EndPackage[]

Global`rtdcstr`

testing 1,2,3...

Global`rtdcstr`Private`

Global`rtdcstr`Private`

"Global`rtdcstr`Private`"

Global`rtdcstr`Private`

rtdcstr[2, 1, 10, 1.5, 4, 2]

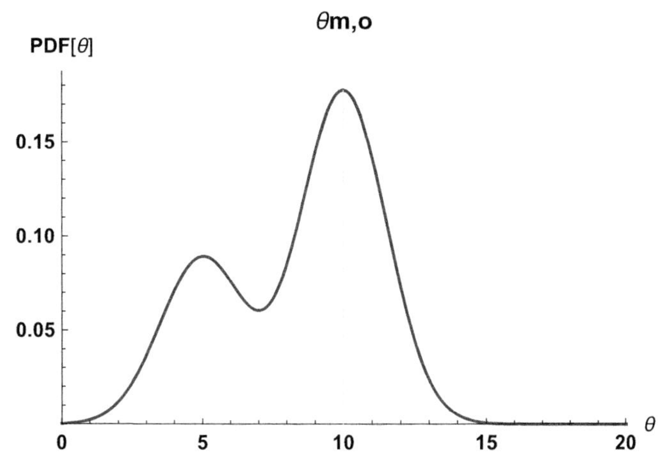

{0.999857, 8.33458, 0.441837, 55.8163}

This extreme case has some effect on the conversion, but not to any significant extent.

Let's turn now to the case of series reaction:

$$A \longrightarrow B \longrightarrow D$$

We can once again solve the equations for the concentrations as function so the residence time:

Clear["Global`*"]

cstrabd2 = Simplify[

Solve[

$$\left\{0 == \frac{(\text{Caf} - \text{Ca})}{\theta} - \text{kabCa}\,,\right.$$

$$0 == \frac{-\text{Cb}}{\theta} + \text{kabCa} - \text{kbdCb},$$

$$\left. == -\frac{\text{Cd}}{\theta} + \text{kbdCb}\right\},$$

{Ca, Cb, Cd}]];

TableForm[cstrabd2,

TableDirections->{Column, Column}]

$$\text{Ca} \to \frac{\text{Caf}}{1 + \text{kab}\theta}$$
$$\text{Cb} \to \frac{\text{Cafkab}\theta}{(1 + \text{kab}\theta)(1 + \text{kbd}\theta)}$$
$$\text{Cd} \to \frac{\text{Cafkabkbd}\theta^2}{(1 + \text{kab}\theta)(1 + \text{kbd}\theta)}$$

The package that we have just written for the bimodally distributed residence times can be adapted to give a new package we shall call rtdcstrabd2. The output of this Package will be the r.t.d. of the integrated PDF value, the average residence time, then the conversion followed by the selectivities to A and to B:

```
BeginPackage["Global`rtdcstrabd2`",
{"Statistics`ContinuousDistributions`",
    "Statistics`NormalDistribution`",
    "Statistics`Common`DistributionsCommon`",
    "Statistics`DescriptiveStatistics`"
 }]
rtdcstrabd2::usage = "testing 1,2,3..."
Begin["`Private`"]
```

rtdcstrabd2[kab_, kbd_, θm_, $\sigma\theta$_, wf1_, wf2_]:=

Module[

{Caf = 1, Cbf = 1, x, xmin, xmax},

xmin = 0.001θm;

xmax = 5θm;

ndist1 = NormalDistribution[θm, $\sigma\theta$];

ndist2 = NormalDistribution[.5θm, $\sigma\theta$];

pdfunction1 = PDF[ndist1, x];

pdfunction2 = PDF[ndist2, x];

$$p = \frac{\text{wf1pdfunction1} + \text{wf2pdfunction2}}{\text{wf1} + \text{wf2}};$$

$$\text{ca[y\_]} = \frac{\text{Caf}}{1 + \text{kab}y};$$

$$\text{cb[y\_]} = \frac{\text{Cafkab}y}{(1 + \text{kab}y)(1 + \text{kbd}y)};$$

$$\text{cd[y\_]} = \frac{\text{Cafkabkbd}y^2}{(1 + \text{kab}y)(1 + \text{kbd}y)};$$

{Plot[p, {x, 0, xmax},

AxesLabel → {"θ", "PDF[θ]"},

PlotStyle → Hue[0.7],

PlotRange → {{xmin, 2θm},

{0, .01 + Max[Table[N[p], {x, xmin, xmax}]]}},

Epilog → {RGBColor[0, 1, 1],

Line[{{θm, 0},

{θm, Max[Table[N[p], {x, xmin, xmax}]]}}]},

PlotLabel → "θm,o"],

totp = NIntegrate[p, {x, 0, xmax}];

$$\text{thetam} = \frac{\text{NIntegrate}[x p, \{x, 0, \text{xmax}\}]}{\text{totp}};$$

$$\text{caave} = \frac{\text{NIntegrate}[\text{ca}[x] p, \{x, 0, \text{xmax}\}]}{\text{totp}};$$

$$\text{cbave} = \frac{\text{NIntegrate}[\text{cb}[x] p, \{x, 0, \text{xmax}\}]}{\text{totp}};$$

$$\text{cdave} = \frac{\text{NIntegrate}[\text{cd}[x] p, \{x, 0, \text{xmax}\}]}{\text{totp}};$$

conversion = 100(1 − caave);

$$\text{selB} = \frac{\text{cbave}}{\text{caave} + \text{cbave} + \text{cdave}};$$

$$\text{selD} = \frac{\text{cdave}}{\text{caave} + \text{cbave} + \text{cdave}};$$

{totp, thetam, conversion, selB, selD}}]

End[]

EndPackage[]

Global`rtdcstrabd2`

Global`rtdcstrabd2`Private`

Global`rtdcstrabd2`Private`

rtdcstrabd2[2, 1, 10, 1.5, 1, .25]

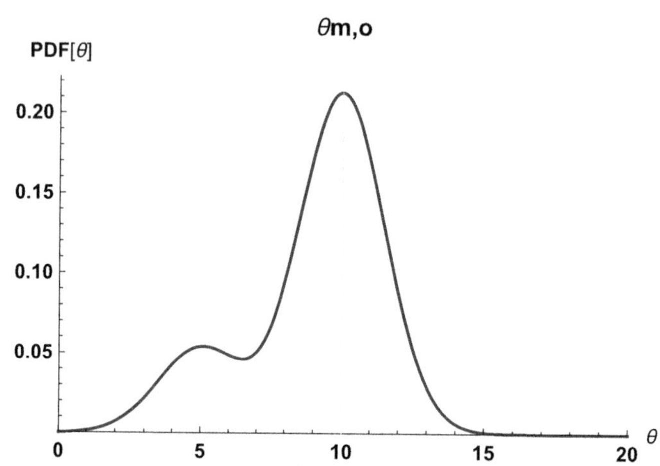

{0.999914, 9.00081, 94.1074, 0.102341, 0.838733}

{rtdcstrabd2[.2, .1, 10, 1.5, 1, 0],

rtdcstrabd2[.2, .1, 5, 1.5, 1, .25],

rtdcstrabd2[.2, .1, 10, 1.5, 1, 1],

rtdcstrabd2[.2, .1, 10, 1.5, 1, .25],}

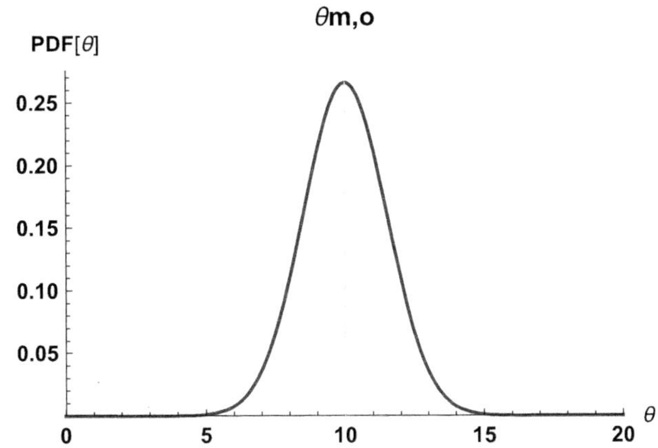

{1.,10.,66.3228,0.332179,0.331049},

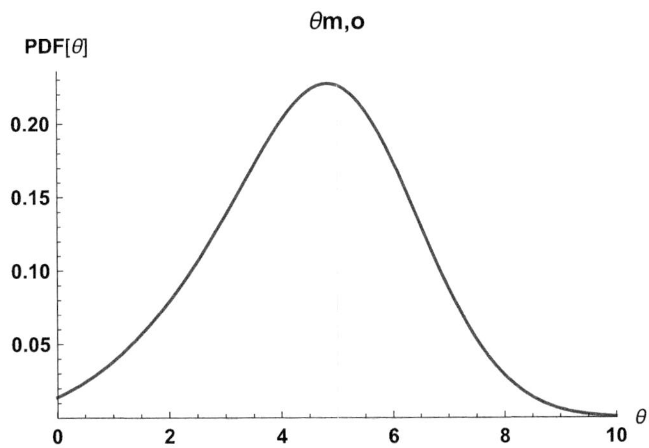

{0.990099,4.55114,45.6971,0.309069,0.147901}},

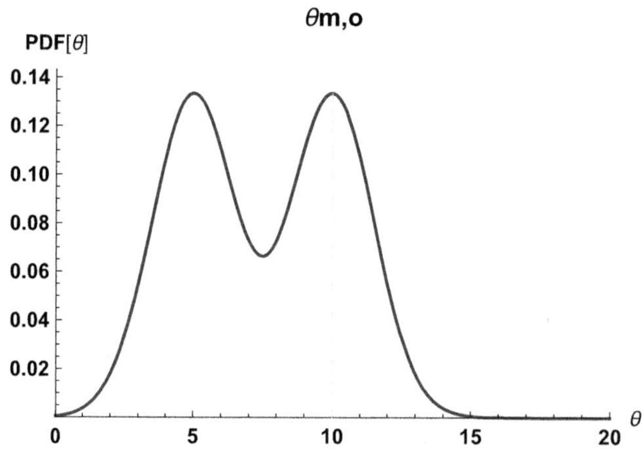

{0.999785,7.50169,57.5702,0.327614,0.248088}},

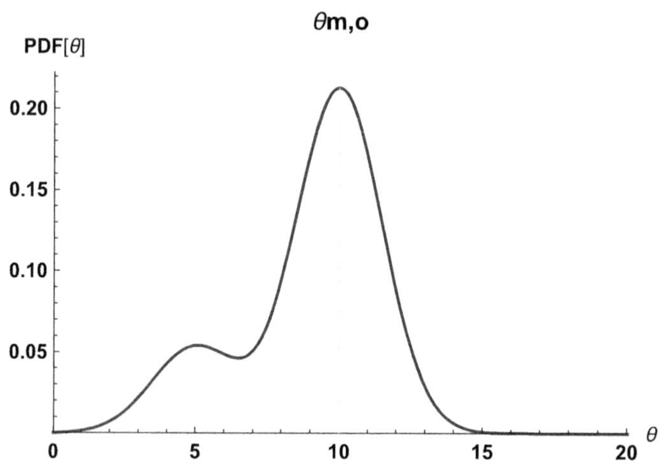

{0.999914,9.00081,62.8222,0.330353,0.297869}}

Here we can see a much stronger effect of the r.t.d. on the conversions and the selectivities. The cases we have chosen to examine are purposefully extreme. With the r.t.d. centered at θm = 10, the conversion of **A** is 66% and the selectivities to **A** and to **B** are virtually 1:1. When we push the r.t.d. to a shorter residence time and θm = 5 the whole picture changes to one of lower conversion and higher selectivity to **B** by almost 2:1 over **D** – as we would expect. Note, however that the yield to **B** is now under 16%, whereas in the first case it was closer to 22%. With a bimodal distribution having nearly equal probabilities at θm = 5 and 10, the picture changes once again, the conversion is lower than the narrowly distributed case centered at θm = 10, and so too is the selectivity to **D** rather than **B**. Finally, if the maldistribution is less severe, then too is the departure from the unimodal result.

For comparison sake, below, we compute the case of a very narrow distribution and the actual values for the base case of a single holding time, that is for DiracDelta Function of residence times.

rtdcstrabd2[.2, .1, 10, .2, 1, 0]

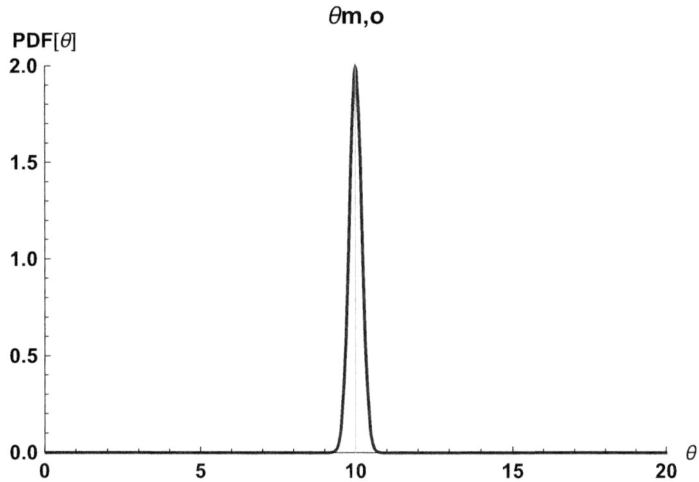

{1., 10., 66.6607, 0.333315, 0.333293}

kab = .2;

kbd = .1;

Caf = 1;

θ **= 10;**

$$\mathbf{Cd} \rightarrow N\left[\frac{\mathbf{Cafkabkbd}\theta^2}{(1+\mathbf{kab}\theta)(1+\mathbf{kbd}\theta)}\right]$$

$$\mathbf{Cb} \rightarrow N\left[\frac{\mathbf{Cafkab}\theta}{(1+\mathbf{kab}\theta)(1+\mathbf{kbd}\theta)}\right]$$

$$\mathbf{Ca} \rightarrow N\left[\frac{\mathbf{Caf}}{1+\mathbf{kab}\theta}\right]$$

{{{{Cd → 0.333333}, {Cb → 0.333333}, {Ca → 0.333333}}}}

The result is that the selectivities are equal for **A**, **B**, and **D** in this case.

Time dependent PFR - complete and numerical solutions

Introduction

The PFR equation that we derived had two partial derivatives–one in time and one in space. Recall this is the equation for component **A** being fed to the PFR:

$$\frac{\partial \text{Ca}}{\partial t} = \frac{-q}{\text{Acr}} \frac{\partial \text{Ca}}{\partial z} - r_{A-}$$

The time dependent derivative of concentration is the accumulation term for the differential volume, Acr dz. This essentially the same for all the reactor types we have studied. What makes the PFR different and, perhaps, more interesting is the spatial derivative.

Since it was not within our ability to solve the time dependent equation, we naturally solved the steady state problem so that the accumulation term went to zero leaving only the spatial derivative:

$$0 = \frac{-q}{A} \frac{\partial \text{Ca}}{\partial z} - r_{A-}$$

$$\frac{q}{A} \frac{\partial \text{Ca}}{\partial z} = - r_{A-}$$

This problem was more soluble because of the fact that it only involves this one derivative. In fact if we recall that $\frac{q}{A}$ is the velocity, v_z, in the PFR, then we simplify the equation further.

$$\frac{q}{A} \frac{\partial \text{Ca}}{\partial z} = - r_{A-}$$

$$v_z \frac{\partial \text{Ca}}{\partial z} = - r_{A-}$$

But $\frac{v_z}{\partial z}$ is the ratio of a constant velocity to a differential distance. This has units of reciprocal time. Formally, we can take the constant into the derivative and this gives us $\frac{1}{\partial \frac{z}{v_z}}$ which can be defined as the reciprocal of the differential holding time $\frac{1}{\partial \tau}$. The equation becomes:

$$\frac{\partial \text{Ca}}{\partial \tau} = - r_{A-}$$

This result is very nice because it shows us that at the steady state the PFR has the same form of governing equation as the transient batch reactor but instead of the real time the differential is given in terms of the differential holding time.

Transient PFR

The question arises as to how long it will take a reactor operated in the plug flow regime to reach a steady state for a specific set of reaction kinetics, volume, and flow rate. To solve this problem we need to solve both in time and in space. If the kinetics are simple, then we can solve the problem analytically, that is we can derive expression for the concentrations that are functions of time and position. But, often the kinetics are not straightforward and analytical solutions must be surrendered in favor of numerical solutions.

Equations, initial conditions, and boundary conditions

Consider the following reaction and its occurrence in a transient PFR.

$$\mathrm{A + B \longrightarrow D \longrightarrow E}$$

The first reaction takes place via second order kinetics, k1 Ca Cb, and the second is first order in D, k2 Cd. The concentrations of every species will be a function of both space and time:

$$Ci[z, t]$$

The differential equations for the concentrations of each species are of the form shown for species A:

$$\frac{\partial \mathrm{Ca}}{\partial t} = \frac{-q}{A}\frac{\partial \mathrm{Ca}}{\partial z} \text{ - } r_{A-}$$

We can write these as a set of equations for **A**, **B**, **D**, and **E** each of which are coupled through their concentrations. We also need four initial conditions for these concentrations. We can set the concentrations of **A** and **B** to their inlet values as if the tube were uniformly filled with them initially. For the two products, we can set each of them to zero at time zero at all z positions in the reactor. We do this as follows:

Creactant[z,0] == Creactant,o

Cproduct[z,0] == 0

For a partial differential equation, we also need to have a set of *boundary conditions*. The initial conditions are for the time differential and the boundary conditions arise for the spatial differential. The boundary conditions are analogous to initial conditions. The boundary conditions must be satisfied at some position or positions for all times. Boundary conditions for our problem can be fairly straightforward; we will let the concentrations of the reactants **A** and

B, be equal to a constant for all times. The other concentrations will be zero. Here is how we express that:

Ca[0,t]==Cao

Cb[0,t]==Cbo

Cd[0,t]==0...

The first part of this problem is to write the set of component equations, the initial conditions, and the boundary conditions, including the kinetics. Call this set **eqns**. Next write out the set of variables that will be solved for calling it these **vars**.

eqns = { component equations, initial conditions, boundary conditions},

vars = { Ci[z,t]....}

We make a vertical list of parameter names, and values, and then we use NDSolve with the set of equations and variables as follows:

solns = NDSolve[eqns, vars, {t,0,tmax},{z,0,zmax}]

We know that NDSolve does the numerical integration and then it automatically fits a set of polynomials to the numerical values of each variable at each grid point in time and position and the output is an interpolation function. Per the usual we will assign these interpolation functions to function names and patterns. We will solve numerically and then plot the concentrations for **A**, **D**, and **E** in z and t. Since this is unlike the other problems we have done to this point, we will present it in a highly interactive, step-by-step fashion. Putting these pieces together into a Module or package, only makes sense to do once the computation and the implemented code are understood. Here is the code we have just described. We will let the results for each step flow to output so that we can see how this works in detail:

```
Clear[ca, cb, cd, ce, q, A, k1, k2, cao, cbo, cdo,

ceo, tmax, zmax, cA, cB, cD, cE, vz]
eqns = {vz==q/A,

     D[ca[z, t], t]==

-vzD[ca[z, t], z] - k1ca[z, t]cb[z, t],

     D[cb[z, t], t]==

-vzD[cb[z, t], z] - k1cb[z, t]cb[z, t],
```

```
    D[cd[z, t], t]==
-vzD[cd[z, t], z] + k1ca[z, t]cb[z, t]-
k2cd[z, t],
    D[ce[z, t], t]==
-vzD[ce[z, t], z] + k2cd[z, t],
    ca[0, t]==cao,
    cb[0, t]==cbo,
    cd[0, t]==cdo,
    ce[0, t]==ceo,
    ca[z, 0] == cao,
    cb[z, 0] == cbo,
    cd[z, 0]==0.0,
    ce[z, 0]==0.0};
vars = {ca[z, t], cb[z, t], cd[z, t], ce[z, t]};
q = 7.5;
A = 10;
k1 = 0.15;
k2 = 0.04;
cao = 1;
cbo = 1;
cdo = 0;
ceo = 0;
tmax = 100;
zmax = 100;
solns = NDSolve[eqns, vars, {z, 0, zmax}, {t, 0, tmax}]

cA[z_, t_]:=Evaluate[ca[z, t]/.solns[[1]]]
```

```
cB[z_, t_]:=Evaluate[cb[z, t]/.solns[[1]]]

cD[z_, t_]:=Evaluate[cd[z, t]/.solns[[1]]]

cE[z_, t_]:=Evaluate[ce[z, t]/.solns[[1]]]

a = Plot3D[cA[z, t], {z, 0, zmax}, {t, 0, tmax},
AxesLabel->{"z ", " t", "cA "},
ViewPoint->{1, −2, 2},
PlotPoints->20,
Ticks → {Automatic, Automatic, {0, 0.5, 1}},
DisplayFunction → Identity];

b = Plot3D[cB[z, t], {z, 0, zmax}, {t, 0, tmax},
AxesLabel->{"z ", " t", "cB "},
ViewPoint->{1, −2, 2}, PlotPoints->20,
Ticks → {Automatic, Automatic, {0, 0.5, 1}},
DisplayFunction → Identity];

d = Plot3D[cD[z, t], {z, 0, zmax}, {t, 0, tmax},
AxesLabel->{"z ", " t", "cD "},
PlotPoints->20,
Ticks → {Automatic, Automatic, {0, 0.5, 1}},
DisplayFunction → Identity];

e = Plot3D[cE[z, t], {z, 0, zmax}, {t, 0, tmax},
AxesLabel->{"z ", " t", "cE "},
PlotPoints->20,
Ticks → {Automatic, Automatic, {0, 0.5, 1}},
DisplayFunction → Identity];
```

Show[GraphicsGrid[{{*a*, *b*}, {*d*, *e*}}]]

{{ca[*z*, *t*] → InterpolatingFunction [□] [*z*, *t*], /
cb[*z*, *t*] → InterpolatingFunction [□] [*z*, *t*],
cd[*z*, *t*] → InterpolatingFunction [□] [*z*, *t*],
ce[*z*, *t*] → InterpolatingFunction [□] [*z*, *t*]}}

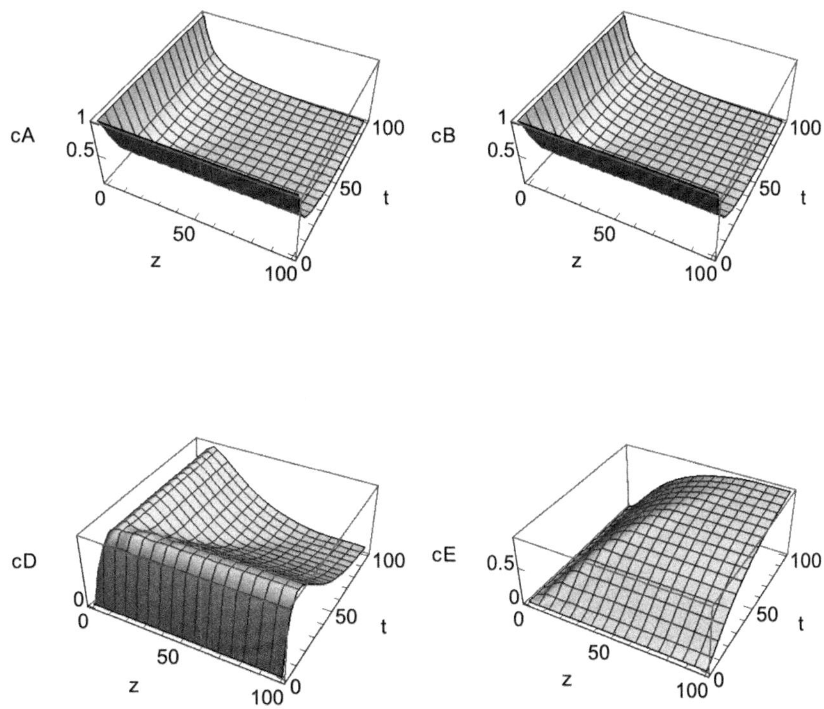

The solutions are in the form of the interpolation functions. The graphs of each of the species concentrations are plotted as function of position along the tube, z, and time, t. At the edges of the graphs for the concentrations of **A** and **B**, we see the boundary and initial conditions, all values are unit concentration as we had specified. As we move through time, we see the concentrations of both species drop monotonically at an position, furthermore, if we take any time slice we see that the concentrations of reactants drop exponentially with position–as we know they should. At the longer times the profiles of **A** and **B** through the reactor have reached their steady state values. Species **D** is the intermediate in the network. Its concentration rises sharply as a function of time and of position, maximizing across the reactor at short time and then falling to its steady state value. The only region of the reactor in which the concentration of **D** persists at relatively high levels is at a position of approximately 10 units. As time goes on the initially high **D** concentration beyond position $z \approx 10$ falls. Species **E** is the ultimate product of this network. At the lower left corner of its concentration graph, the concentration is zero. Through time and position this rises to a steady state level of nearly 0.8 in the upper right corner of the plot representing the reactor outlet at long time.

In the problem as we have solved it, the concentrations of **A** and **B** are uniform throughout the reactor and are equal to the inlet concentration:

```
ca[0,t]==cao,   ca[z,0]==cao,
cb[0,t]==cbo,   cb[z,0]==cbo,
cd[0,t]==cdo,   cd[z,0]==0.0,
ce[0,t]==ceo,   ce[z,0]==0.0
```

What if we wished to solve the problem of a reactor that initially does not contain either **A** or **B**. This would correspond to this set of conditions:

```
ca[0,t]== cao,   ca[z,0] == 0.0,
cb[0,t]== cbo,   cb[z,0] == 0.0,
cd[0,t]== cdo,   cd[z,0] == 0.0,
ce[0,t]== ceo,   ce[z,0] == 0.0
```

Let's just copy the code from the previous cell, change the boundary condition and ask *Mathematica* to solve this to see what happens:

Clear[ca, cb, cd, ce, *q*, *A*, k1, k2, cao, cbo, cdo,

ceo, tmax, zmax, cA, cB, cD, cE]

eqns = {

$$D[\mathrm{ca}[z,t],t] == -\frac{q}{A} D[\mathrm{ca}[z,t],z] - \mathrm{k1}\,\mathrm{ca}[z,t]\,\mathrm{cb}[z,t],$$

$$D[\mathrm{cb}[z,t],t] == -\frac{q}{A} D[\mathrm{cb}[z,t],z] - \mathrm{k1}\,\mathrm{cb}[z,t]\,\mathrm{cb}[z,t],$$

```
    D[cd[z, t], t]== -q/A D[cd[z, t], z] + k1ca[z, t]cb[z, t]- k2cd[z, t],
    D[ce[z, t], t]== -q/A D[ce[z, t], z] + k2cd[z, t],
    ca[0, t]==cao,
    cb[0, t]==cbo,
    cd[0, t]==cdo,
    ce[0, t]==ceo,
    ca[z, 0] == 0.0,
    cb[z, 0] == 0.0,
    cd[z, 0]==0.0,
    ce[z, 0]==0.0};

vars = {ca[z, t], cb[z, t], cd[z, t], ce[z, t]};

q = 7.5;
A = 10;
k1 = 0.15;
k2 = 0.04;
cao = 1;
cbo = 1;
cdo = 0;
ceo = 0;
tmax = 100;
zmax = 100;
solns = NDSolve[eqns, vars, {z, 0, zmax},
{t, 0, tmax}];
```

```
cA[z_, t_]:=Evaluate[ca[z, t]/.solns[[1]]]

cB[z_, t_]:=Evaluate[cb[z, t]/.solns[[1]]]

cD[z_, t_]:=Evaluate[cd[z, t]/.solns[[1]]]

cE[z_, t_]:=Evaluate[ce[z, t]/.solns[[1]]]

a = Plot3D[cA[z, t], {z, 0, zmax}, {t, 0, tmax},
AxesLabel->{"z ", " t", "cA "},
ViewPoint->{1, −2, 2},
PlotPoints->20,
Ticks → {Automatic, Automatic, {0, 0.5, 1}}];

b = Plot3D[cB[z, t], {z, 0, zmax}, {t, 0, tmax},
AxesLabel->{"z ", " t", "cB "},
ViewPoint->{1, −2, 2}, PlotPoints->20,
Ticks → {Automatic, Automatic, {0, 0.5, 1}}];

d = Plot3D[cD[z, t], {z, 0, zmax}, {t, 0, tmax},
AxesLabel->{"z ", " t", "cD "},
PlotPoints->20,
Ticks → {Automatic, Automatic, {0, 0.5, 1}}];

e = Plot3D[cE[z, t], {z, 0, zmax}, {t, 0, tmax},
AxesLabel->{"z ", " t", "cE "},
PlotPoints->20,
Ticks → {Automatic, Automatic, {0, 0.5, 1}}];

Show[GraphicsGrid[{{a, b}, {d, e}}]]
```

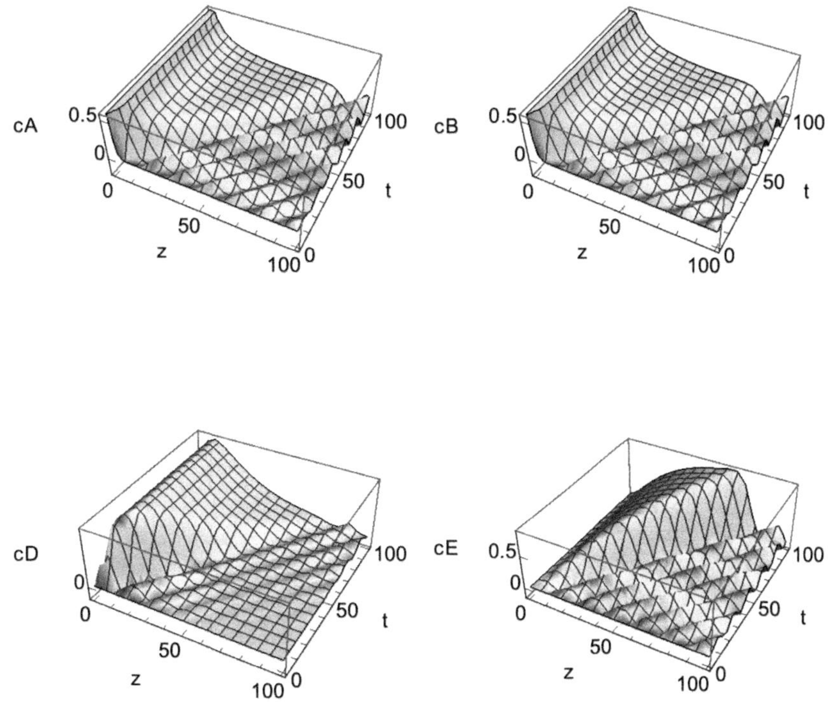

We notice that there are ripples in the concentration solutions that were not present when the initial conditions included having non-zero concentrations of **A** and **B**. These are the results of the shock wave at the inlet. If we repeat the calculations and go to longer times, these waves dampen out.

Clear[ca, cb, cd, ce, *q*, *A*, k1, k2, cao, cbo, cdo,

ceo, tmax, zmax, cA, cB, cD, cE]

eqns = {

$$D[\mathrm{ca}[z,t],t] == -\frac{q}{A}D[\mathrm{ca}[z,t],z] - \mathrm{k1ca}[z,t]\mathrm{cb}[z,t],$$

$$D[\mathrm{cb}[z,t],t] == -\frac{q}{A}D[\mathrm{cb}[z,t],z] - \mathrm{k1cb}[z,t]\mathrm{cb}[z,t],$$

$$D[\mathrm{cd}[z,t],t] == -\frac{q}{A}D[\mathrm{cd}[z,t],z] + \mathrm{k1ca}[z,t]\mathrm{cb}[z,t] - \mathrm{k2cd}[z,t],$$

$$D[\mathrm{ce}[z,t],t] == -\frac{q}{A}D[\mathrm{ce}[z,t],z] + \mathrm{k2cd}[z,t],$$

```
    ca[0, t]==cao,
    cb[0, t]==cbo,
    cd[0, t]==cdo,
    ce[0, t]==ceo,
    ca[z, 0] == 0.0,
    cb[z, 0] == 0.0,
    cd[z, 0]==0.0,
    ce[z, 0]==0.0};

vars = {ca[z, t], cb[z, t], cd[z, t], ce[z, t]};

q = 7.5;
A = 10;
k1 = 0.15;
k2 = 0.04;
cao = 1;
cbo = 1;
cdo = 0;
ceo = 0;
tmax = 1000;
zmax = 100;
solns = NDSolve[eqns, vars, {z, 0, zmax},
{t, 0, tmax}];

cA[z_, t_]:=Evaluate[ca[z, t]/.solns[[1]]]
cB[z_, t_]:=Evaluate[cb[z, t]/.solns[[1]]]
cD[z_, t_]:=Evaluate[cd[z, t]/.solns[[1]]]
cE[z_, t_]:=Evaluate[ce[z, t]/.solns[[1]]]
```

```
a = Plot3D[cA[z, t], {z, 0, zmax}, {t, 0, tmax},
AxesLabel->{"z ", " t", "cA "},
ViewPoint->{1, −2, 2},
PlotPoints->20, PlotRange → All,
Ticks → {Automatic, Automatic, {0, 0.5, 1}}];

b = Plot3D[cB[z, t], {z, 0, zmax}, {t, 0, tmax},
AxesLabel->{"z ", " t", "cB "},
ViewPoint->{1, −2, 2}, PlotPoints->20,
PlotRange → All,
Ticks → {Automatic, Automatic, {0, 0.5, 1}}];

d = Plot3D[cD[z, t], {z, 0, zmax}, {t, 0, tmax},
AxesLabel->{"z ", " t", "cD "},
PlotPoints->20, PlotRange → All,
Ticks → {Automatic, Automatic, {0, 0.5, 1}}];

e = Plot3D[cE[z, t], {z, 0, zmax}, {t, 0, tmax},
AxesLabel->{"z ", " t", "cE "},
PlotPoints->20, PlotRange → All,
Ticks → {Automatic, Automatic, {0, 0.5, 1}}];

Show[GraphicsGrid[{{a, b}, {d, e}}]]
```

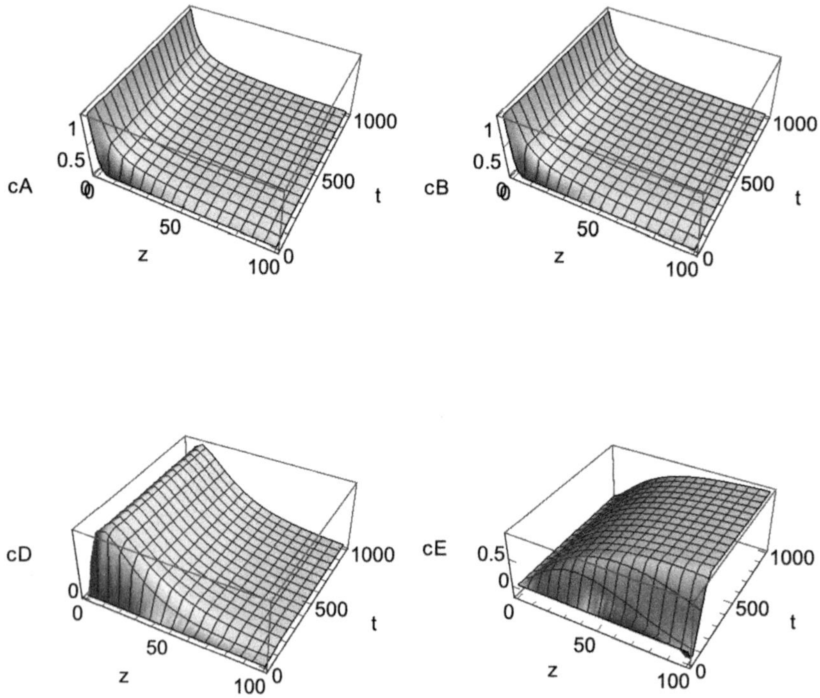

We can see that having a boundary condition with the concentrations of reactants everywhere zero at time zero within the control volume is problematic, since we have also said that the concentrations at the inlet are identically unit valued even at time zero. When we focus on the point z = 0 at time zero, we find that the concentration is unit valued at a point upstream of z by an infinitesimal amount, i.e. at z - δz, but at z + δz the values of the concentrations are zero! So in other words the boundary condition at the inlet at time zero looks like 1-UnitStep function:

```
Caf = 1;

Ca[z_, 0]:=Caf − UnitStep[z]

Plot[Ca[z, 0], {z, −10, 10},

AxesLabel → {z, "Ca[z,0]"},

PlotStyle → Hue[0]]
```

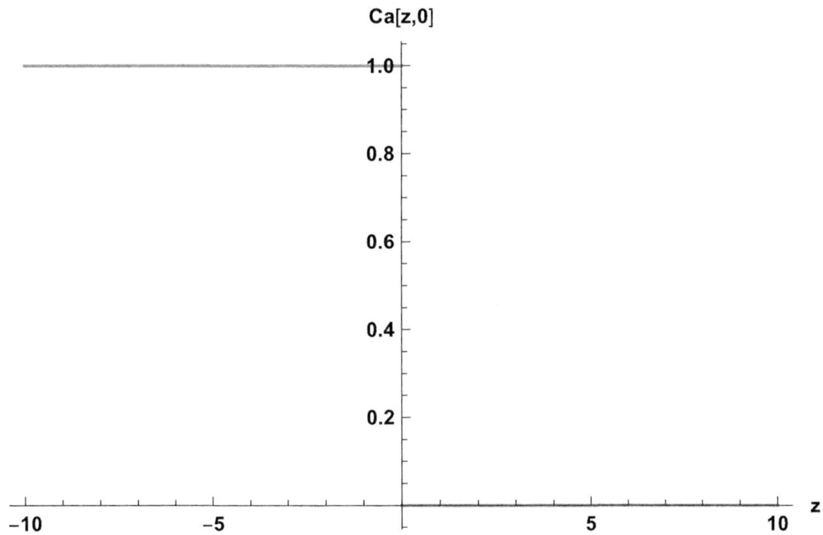

Now, we can try this function in the boundary conditions and then solve numerically once more:

Clear[ca, cb, cd, ce, q, A, k1, k2, cao, cbo, cdo,

ceo, tmax, zmax, cA, cB, cD, cE, eqns,

a, b, c, d, solns]

eqns = {

$$D[\text{ca}[z,t],t] == -\frac{q}{A}D[\text{ca}[z,t],z] - \text{k1ca}[z,t]\text{cb}[z,t],$$

$$D[\text{cb}[z,t],t] == -\frac{q}{A}D[\text{cb}[z,t],z] - \text{k1cb}[z,t]\text{cb}[z,t],$$

$$D[\text{cd}[z,t],t] == -\frac{q}{A}D[\text{cd}[z,t],z] + \text{k1ca}[z,t]\text{cb}[z,t] - \text{k2cd}[z,t],$$

$$D[\text{ce}[z,t],t] == -\frac{q}{A}D[\text{ce}[z,t],z] + \text{k2cd}[z,t],$$

ca[zmin, t]==cao,

cb[zmin, t]==cbo,

cd[zmin, t]==cdo,

ce[zmin, t]==ceo,

ca[z, 0] == 1 − UnitStep[z, 0],

```
    cb[z, 0] == 1 – UnitStep[z, 0],
    cd[z, 0]==0.0,
    ce[z, 0]==0.0};

vars = {ca[z, t], cb[z, t], cd[z, t], ce[z, t]};

q = 7.5;
A = 10;
k1 = 0.1;
k2 = 0.05;
cao = 1;
cbo = 1;
cdo = 0;
ceo = 0;
tmin = –1;
tmax = 500;
zmin = 0;
zmax = 100;

solns = NDSolve[eqns, vars, {z, zmin, zmax},
{t, tmin, tmax}, MaxSteps → 1000, AccuracyGoal → 5,
PrecisionGoal → 5, WorkingPrecision → 7]

cA[z_, t_]:=Evaluate[ca[z, t]/.solns[[1]]]
cB[z_, t_]:=Evaluate[cb[z, t]/.solns[[1]]]
cD[z_, t_]:=Evaluate[cd[z, t]/.solns[[1]]]
cE[z_, t_]:=Evaluate[ce[z, t]/.solns[[1]]]

a = Plot3D[cA[z, t], {z, 0, zmax}, {t, 0, tmax},
AxesLabel->{"z ", " t", "cA "},
```

```
ViewPoint->{1, −2, 2},
PlotRange → {{0, zmax}, {0, tmax}, {0, 1}},
PlotPoints->20, PlotRange → All,
Ticks → {Automatic, Automatic, {0, 0.5, 1}}];

b = Plot3D[cB[z, t], {z, 0, zmax}, {t, 0, tmax},
AxesLabel->{"z ", " t", "cB "},
ViewPoint->{1, −2, 2},
PlotRange → {{0, zmax}, {0, tmax}, {0, 1}},
PlotPoints->20, PlotRange → All,
Ticks → {Automatic, Automatic, {0, 0.5, 1}}];

d = Plot3D[cD[z, t], {z, 0, zmax}, {t, 0, tmax},
AxesLabel->{"z ", " t", "cD "},
PlotRange → {{0, zmax}, {0, tmax}, {0, 1}},
PlotPoints->20, PlotRange → All,
Ticks → {Automatic, Automatic, {0, .5, 1}}];

e = Plot3D[cE[z, t], {z, 0, zmax}, {t, 0, tmax},
AxesLabel->{"z ", " t", "cE "},
PlotRange → {{0, zmax}, {0, tmax}, {0, 1}},
PlotPoints->20, PlotRange → All,
Ticks → {Automatic, Automatic, {0, .5, 1}}];

Show[GraphicsGrid[{{a, b}, {d, e}}]]

{{ca[z, t] → InterpolatingFunction[□][z, t],
cb[z, t] → InterpolatingFunction[□][z, t],
cd[z, t] → InterpolatingFunction[□][z, t],
ce[z, t] → InterpolatingFunction[□][z, t]}}
```

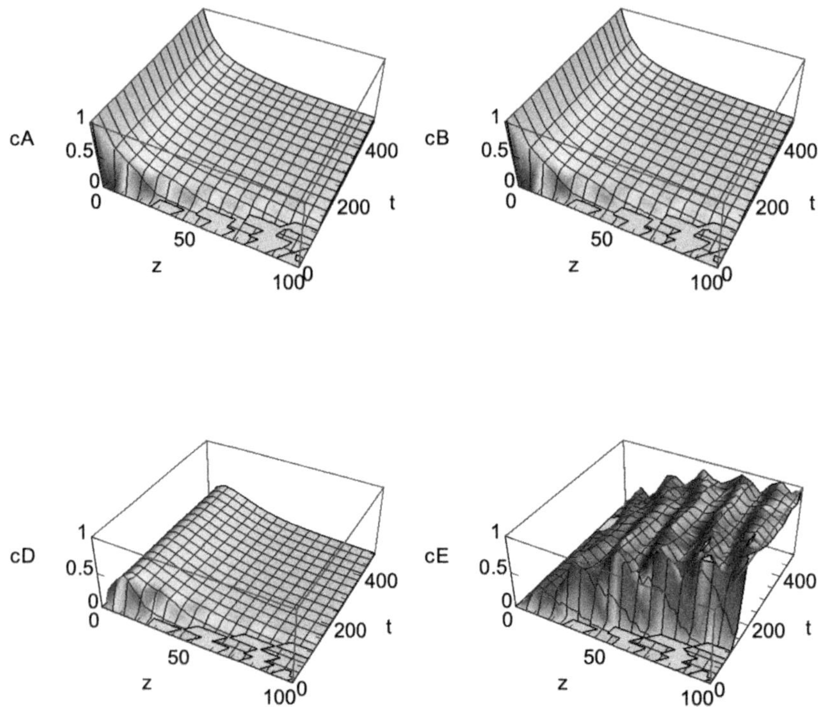

Interestingly, the results are about the same, except that the instabilities are evident at short times. The functions for concentrations of **A**, **B**, and **D** smooth out, but the function for **E** does not. We can see this even more clearly by creating two-dimensional plots of the concentration of **E** along the z-axis at different times. We do this by using the Table function to first create the functions for **E** at different times, and then again to create the separate and combined plots.

Table[cE[*z*, *t*], {*t*, 100, 500, 100}]

{InterpolatingFunction[□][z, 100],

InterpolatingFunction[□][z, 200],

InterpolatingFunction[□][z, 300],

InterpolatingFunction[□][z, 400],

InterpolatingFunction[□][z, 500]}

```
Table[Plot[cE[z, t], {z, 0, 100},
AxesLabel → {z, "CE(z)"}], {t, 100, 500, 100}]
Show[%]
```

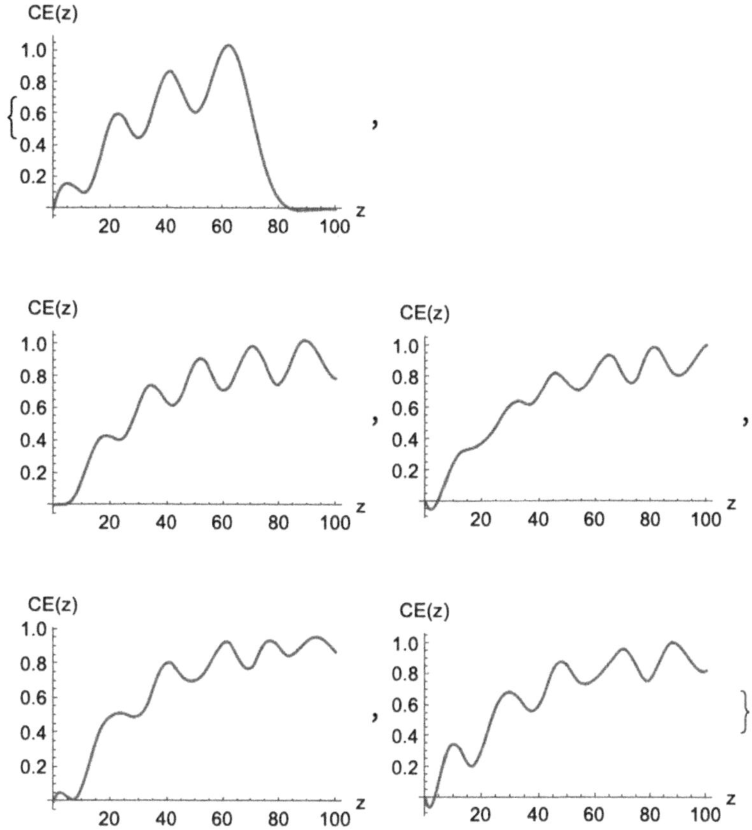

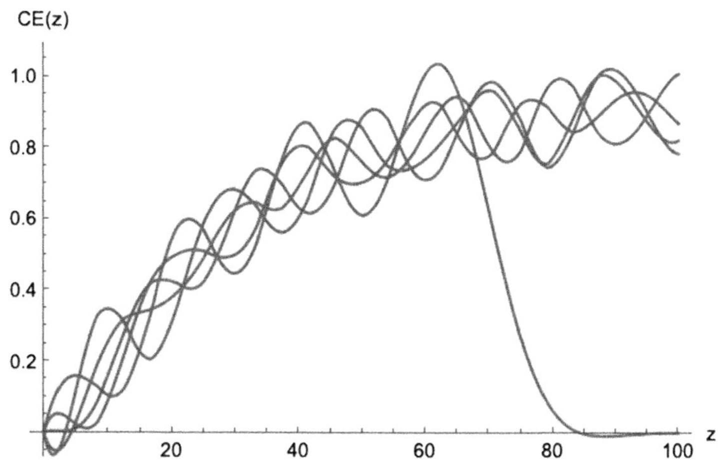

All-in-all the series network in a transient **PFR** appears to follow the trends that we would expect in evolving to the steady state condition. The time to reach the steady state is a function of the rate constants, the inlet concentrations, and the holding time, i.e. of the volume flow rate and reactor volume.

Conclusion

This chapter has been devoted to continuous reactors and their analyses. We have moved from the fed-batch system to the powerful idealizations of the CSTR and PFR. Pseudo-steady states and steady states have been covered as well as chemical equilibrium. We must remember that the steady state condition can be far from the equilibrium condition, and the two time independent situations must not be confused. The time dependent CSTR and PFR are interesting problems, but they are less often used than the steady state solutions.

We have seen how the kinetics fit into a reactor equation as a constitutive relationship. Flow and reaction come together in these systems to affect the rate of accumulation. Hence when one refers to the "rate" we must be careful to be specific about the reactor – if it is a batch reactor, then rate means the chemical rate. If the reactor is a transient CSTR or PFR, then the rate of change of the concentration at the exit of the reactor is not the chemical rate alone. Mixing effects are important and we have seen how to begin to account for the fluid mechanics in a reactor through the empirical measure of the r.t.d. The r.t.d. does affect the outcome from the reactor, but the sensitivity to the r.t.d. depends upon the kinetics and their functional form.

Finally, we have taken our *Mathematica* skills up another level by writing not just Module functions, but actual Packages. By writing Packages in the Global context we implement them immediately. We can, however, write and save packages to another context and then call them from our own library as we need them, either as stand alone computations, or as imbedded functions in another package.

CHAPTER 9

Additional examples

The following are exercises and examples that have been used with honors students and that seem to be both interesting and challenging. In many cases just the rudimentary coded solution is given. From this one can build much more sophisticated code and in many cases create Module functions that can be used repeatedly with different parameter values. Note: all of these separate problems were run as if the Kernel had be restarted with each computation. Running sequentially without adding statements to clear variables will result in incorrect output.

Another look at the level-controlled tank

The filling or draining of a tank is a relatively simple situation to model. If the tank has both input and output then it is a combination of the two previous cases and the differential equation for the rate of change of the level in the tank is as follows:

$$\partial_t h[t] = -\frac{(\text{qf} - q)}{A}$$

When the input flow rate is equal to the output flow rate, the system is at steady state and the level remains constant since $\frac{\text{dh}[t]}{\text{dt}} = 0$.

A tank could be operated in such a way that the inlet was always equal to the outlet when the two were at there designed magnitudes. So if q* were the designed flow rate in a CSTR for example, then the system could be maintained at some specific design level h* and design volume V*. As long as the inlet or outlet rates did not change, the system would remain at this condition. From our study of kinetics, we know that maintaining V* a constant would be desirable in order to keep the holding time, $\tau = \frac{V^*}{q}$, constant, and to remain at the design level of production.

Perturbation of the inlet flow rate & control

Suppose that the inlet flow rate was mostly a constant, but from time to time it suffered an upset. The upset would either increase the inlet flow, or decrease it. If this were to occur, then the level in the tank would either increase or decrease, unless there was some attempt made to

Introduction to Chemical Engineering Analysis Using *Mathematica*
https://doi.org/10.1016/B978-0-12-820051-3.00014-0

change the outlet flow rate. A simple control algorithm would be one in which the exit flow rate is adjusted automatically when there is an upset, either above or below, the design flow rate. To do this analysis we need to specify the upset and the system's response to it.

Let the upset be some additional flow rate over, or below the design value. Stated in simple mathematical terms:

$$\text{qf} = \text{q}^* + q_p$$

where q* is the designed magnitude and q_p is either a positive or negative perturbation. We will say more about the upset flow rate momentarily, but first we shall describe the controlled outlet flow rate. The outlet flow rate is typically at the setpoint, q*, until the upset q_p takes place. At that point the flow rate must be adjusted to compensate for the change in the inlet flow rate. We describe this by setting the adjusted flow rate equal to the set point plus a new flow with a rate that is proportional to the difference between the set point level, h*, and the actual level at that time h[t]:

$$\text{q} = \text{q}^* - \text{K} (\text{h}^* - \text{h}[\text{ t }])$$

Clearly, for this to be dimensionally consistent, the proportionality constant must have dimensions of area per time so that its product with h*- h(t) is in dimensions of volume per time. Putting all this together, during the upset the system responds transiently, that is it goes away from steady state. The governing equation is:

$$\partial_t h[t] = \frac{(q^* + q_p) - (q^* - K\,(h^* - h[t]))}{A}$$

$$\partial_t h[t] = \frac{(q_p) + K\,(h^* - h[t]))}{A}$$

This equation is the one that must be integrated and to do so we must know how the upset behaves in time. Before the upset from t=0 until some time just below t_1, q_p is zero and h* = h[t]:

$$0 < \text{t} < t_1 \quad \partial_t h[t] = \frac{(0) + K\,(h^* - h^*))}{A} = 0 \Longrightarrow \text{h}^* = \text{h}[\text{ t }]$$

The upset begins at time t_1, and it instantaneously rises to a value of q_p. The upset stays at the level q_p from t_1 until a time just less than t_2. At time t_2 it falls instantaneously back to zero. This kind of function is called a Unit Step. For the sake of making a graph of this type of disturbance, we plot the UnitStep function and 1 - UnitStep function using the parameters 1 and 6 in order to have a pulse of unit height that is five time-units wide:

```
a = Plot[UnitStep[x - 1], {x, -3, 6}]

b = Plot[1 - UnitStep[x - 6], {x, 1, 10}]
```

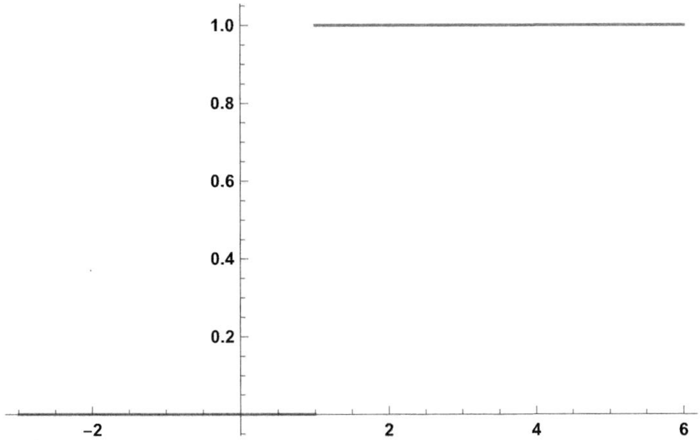

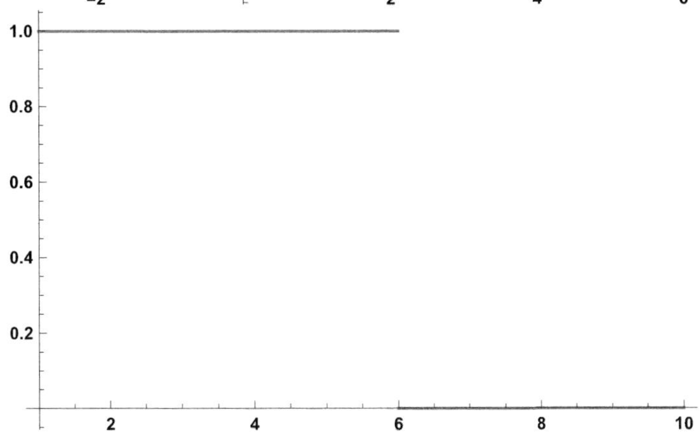

In general, the height of the disturbance is q_p and its width is t_2 - t_1. The question now becomes one of how to do the integrations during and after the disturbance.

Integration through and beyond the disturbance

We must now integrate the equation below from t_1 to t:

$$\partial_t h_2[t] = \frac{\left(q_p\right) + K\,(h^* - h_2[t])\big)}{A}$$

The initial condition is that $h_2\,[t_1]$= h* and we should integrate to t less than t2, so we integrate to t. When this is done we have an expression for $h_2[t]$ during the course of the disturbance. We must next integrate after the disturbance is complete, that is from t_2 out to any additional time t. Here we have the following equation to work with since q_p is zero:

$$\partial_t h_3[t] = \frac{K\,(h^* - h_3[t]))}{A}$$

The initial condition for this period is $h_3\,[t_2] = h_2\,[t_2]$. So we must evaluate the constant of integration very carefully.

Finally, we know that if the disturbance is positive, then the inlet flow increases and the tank level should rise. If there were no control, it would rise and stay at a new higher level. With control it should rise and then fall back to the control or design level. Tracking the change in level versus time we should see a sawtooth that looks like this:

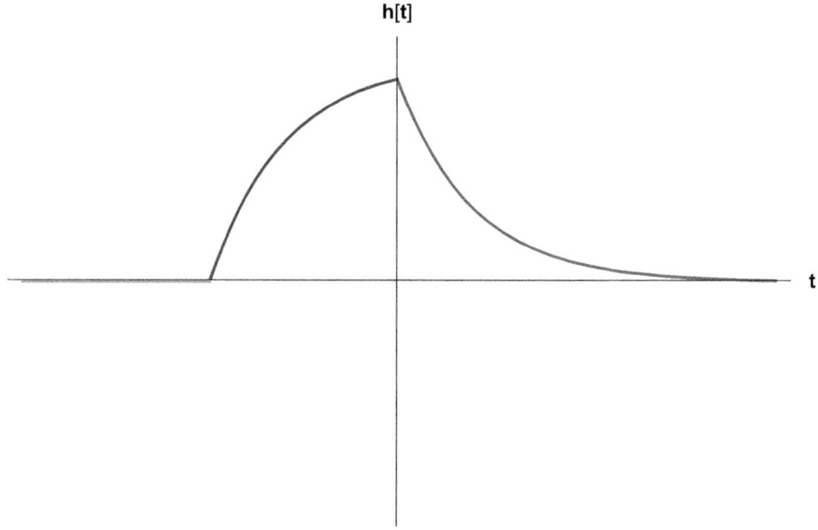

A) Find the expression for $h_2[t]$ by using DSolve. Evaluate the constant of integration using the initial condition that $h_2\,[t_1]$= h*.

$$\text{DSolve}\left[\partial_t h_2[t] == \frac{q_p + K\,(h^* - h_2[t])}{A}, h_2[t], t\right]$$

B) Find the expression for $h_3[t]$ using DSolve. Evaluate the constant of integration using the initial condition that $h_3\,[t_2] = h_2\,[t_2]$.

$$\text{DSolve}\left[\partial_t h_3[t] == \frac{K\,(h^* - h_3[t])}{A}, h_3[t], t\right]$$

C) Create functions for $h_2[t]$ and $h_3[t]$ and given the following parameter values Plot, $h_1[t]$, $h_2[t]$, and $h_3[t]$ consecutively in time.

hd = 50;

K = 5;

A = 10;

qp = 10;

t1 = 5;

t2 = 10;

tmax=30;

Find the general solutions for the levels in the tank in the three time domains of $0 < t < t_1$, $t_1 \leqslant t < t_2$, $t_2 \leqslant t$:

DSolve $\left[\left\{h_1'[t]==0, h_1[0]==h^*\right\}, h_1[t], t\right]$

$\{\{h_1[t] \rightarrow h^*\}\}$

Clear $\left[h_2, q_p, h^*, K, A, \mathrm{t1}, \mathrm{t2}\right]$

DSolve $\left[\left\{h_2'[t]==\frac{q_p + K(-h_2[t])}{A}\right\}, h_2[t], t\right]$

$$\left\{\left\{h_2[t] \rightarrow e^{-\frac{Kt}{A}} C[1] + \frac{q_p + Kh^*}{K}\right\}\right\}$$

Evaluate the constant of integration at the initial condition $h_2[t_1] = h^*$, using Solve:

Solve $\left[==e^{-\frac{Kt}{A}} C[1] + \frac{q_p + Kh^*}{K}, C[1]\right]$

$$\left\{\left\{C[1] \rightarrow -\frac{e^{\frac{Kt}{A}} q_p}{K}\right\}\right\}$$

Define the constant, C1, then replace it and simplify:

$\mathbf{C1} = -\frac{e^{\frac{Kt}{A}} q_p}{K}$;

Collect $\left[\mathbf{Simplify}\left[e^{-\frac{Kt}{A}} C[1] + \frac{q_p + Kh^*}{K}\right], q_p\right]$

$$e^{-\frac{Kt}{A}}C[1]+\frac{q_p}{K}+h^*$$

Now define a function for $h_2\,[t_1]$ to use in parts B and C:

$$\mathbf{h_2[t\_]:=h^*+\left(\frac{1}{K}-\frac{e^{\frac{K(-t+t1)}{A}}}{K}\right)q_p}$$

Find the general solution to the differential equation defining the rate of change in the level of the tank after t = t_2:

$$\mathbf{Clear[h^*, h_3]}$$

$$\mathbf{DSolve\left[\left\{\partial_t h_3[t]==\frac{K\,(h^*-h_3[t])}{A}\right\}, h_3[t], t\right]}$$

$$\left\{\left\{h_3[t]\to e^{-\frac{Kt}{A}}C[1]+h^*\right\}\right\}$$

The initial condition in this case is given as $h_3\,[t_2]=h_2\,[t_2]$. We find this by substituting $h_2\,[t_2]$ into the initial condition equation and evaluating C again:

$$\mathbf{h_2[t2]}$$

$$\left(\frac{1}{K}-\frac{e^{\frac{K(\mathrm{t1}-\mathrm{t2})}{A}}}{K}\right)q_p+h^*$$

$$\mathbf{Solve\left[h_2[t2]==h^*+e^{-\frac{Kt2}{A}}C[1], C[1]\right]}$$

$$\left\{\left\{C[1]\to-\frac{e^{\frac{K\mathrm{t2}}{A}}\left(-1+e^{\frac{K(\mathrm{t1}-\mathrm{t2})}{A}}\right)q_p}{K}\right\}\right\}$$

Replacing the expression for C into that derived for h3[t], we find the expression for h3[t]:

$$\textbf{Simplify}\left[h^* + e^{-\frac{Kt}{A}}\left(-\frac{e^{\frac{K\text{t2}}{A}}\left(-1+e^{\frac{K(\text{t1}-\text{t2})}{A}}\right)q_p}{K}\right)\right]$$

$$h^* - \frac{e^{\frac{K(-t+\text{t2})}{A}}\left(-1+e^{\frac{K(\text{t1}-\text{t2})}{A}}\right)q_p}{K}$$

Anticipating the need for this function, we define it:

$$\textbf{h3[t\_]:=}h^* - \frac{e^{\frac{K(-t+\text{t2})}{A}}\left(-1+e^{\frac{K(\text{t1}-\text{t2})}{A}}\right)q_p}{K}$$

The code that follows works with the three functions we have derived for the three intervals of time dependent behavior. After specifying the three functions, we list the parameter values. Each has a semi-colon at the end in order to prevent its value from being echoed back to the monitor. Note that t1 and t2 are constants that must be specified.

The first plot is made with the following command:

Plot[{ h_1[t],h_2[t], h_3[t] }, { t, 0, 50 }];

The syntax involves specifying the functions to be plotted first and the time interval over which they should be plotted next. The details show that the functions are implemented as a set included within the curly brackets { } as is the time range. The disadvantage of this approach is that each function is plotted over the whole time range, even though they each only apply to separate intervals of time. To overcome this problem, we use the code that follows. These have the following format:

pl1 = Plot[h_1[t], { t, 0, t_1 }, PlotStyle→Hue[.4]];

pl2 = Plot[h_2[t], { t, t_1, t_2 }, PlotStyle→Hue[.6]];

pl3 = Plot[h_3[t], { t, t_2, t_f }, PlotStyle→Hue[.9]];

The command structure is nearly the same as that used before with some differences. We include only one function in each command, in this case h_1[t]. Then we specify the interval we want to plot this function over, for h_1[t] it is from t = 0 to t = t1. For h_2[t] we set the interval to be from t_1 to t_2 and for h_3[t] from t_2 to t_f. We can distinguish between the three by setting PlotStyle to different Hue levels. Hue can have a value between 0 and 1. We use different values for the three different plots. Finally, we call each of the plots, pl1, pl2, and pl3, in the Show command. This makes one plot from the three separate plots and we have no overlapping of the functions.

h1[t_]:=h^*

$$\text{h2[t\_]:=} h^* + \left(\frac{1}{K} - \frac{e^{\frac{K(-t+\text{t1})}{A}}}{K} \right) q_p$$

$$\text{h3[t\_]:=} h^* - \frac{e^{\frac{K(-t+\text{t2})}{A}} \left(-1 + e^{\frac{K(\text{t1}-\text{t2})}{A}} \right) q_p}{K}$$

$h^* = 50;$

$K = 5;$

$A = 10;$

$q_p = 10;$

$\text{t1} = 5;$

$\text{t2} = 10;$

$\text{tmax} = 20;$

```
Plot[{h1[t], h2[t], h3[t]}, {t, 0, 50}]

pl1 = Plot[h1[t], {t, 0, t1}, PlotStyle → Hue[0.3]];

pl2 = Plot[h2[t], {t, t1, t2}, PlotStyle → Hue[0.6]];

pl3 = Plot[h3[t], {t, t2, tmax}, PlotStyle → Hue[0.9]];

Show[pl3, pl2, pl1, PlotRange->{{0, 20}, {48, 52}},
AxesLabel->{t, "h[t]"}]
```

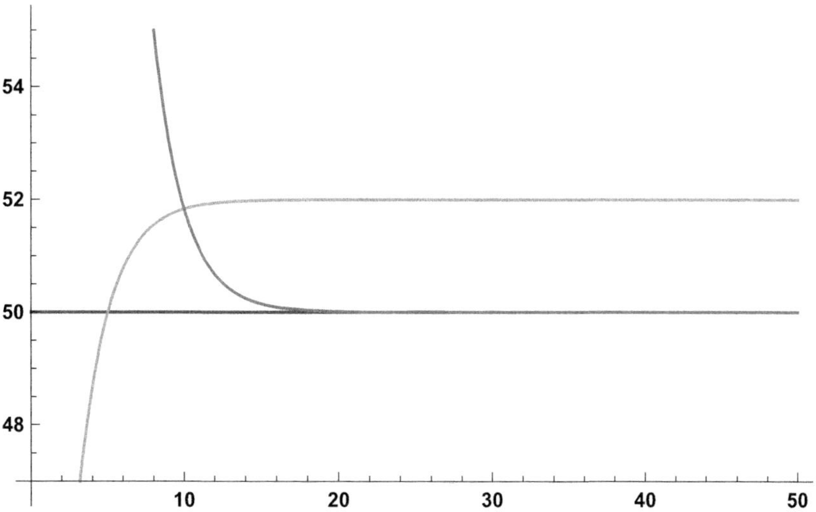

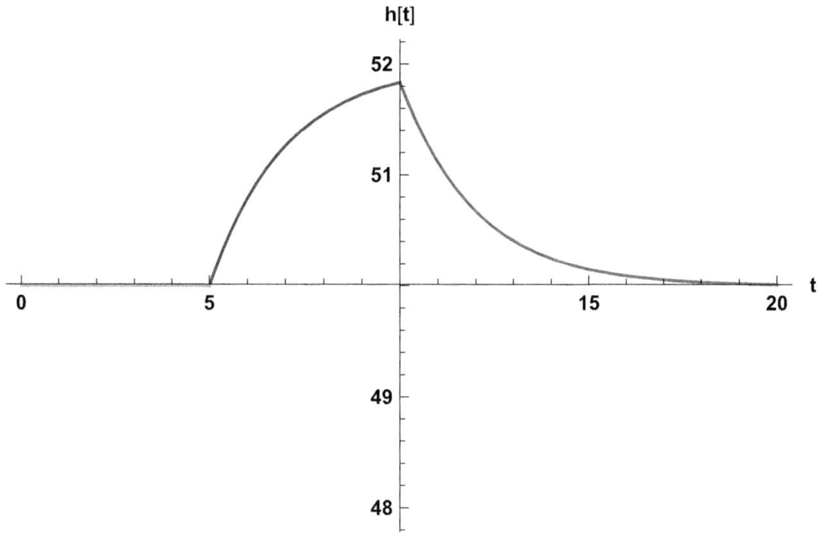

This gives us piece wise continuity before, during, and after the perturbation in the flow.

Proportional, integral, and differential control

Proportional, integral, and differential, or PID control can be combined to provide a system with even more stability. What we have done up until now has been proportional control, meaning that the control function was a linear or proportional to the difference between the actual level and the setpoint level:

$$K\left(h^* - h[t]\right)$$

but we can also include integral and differential control as follows:

$$K_P\left(h^* - h[t]\right) - K_I\int_0^t \left(h^* - h[t]\right) dt - K_D\partial_t\left(h^* - h[t]\right)$$

Note the use of the integral of the difference between the set point and the actual level and the derivative of the difference. We may wonder what this does when we use the whole of the expression in a differential equation expressing level. As in the case of purely proportional control, this PID function will be used to modify the computed exit flow, q_o, expressed as the product of some set point flow rate multiplied by the PID function.

$$\frac{q_o}{A}\left(\mathrm{Kp}\left(h^* - h[t]\right) - \mathrm{Ki}\int_0^t \left(h^* - h[t]\right) dt - \mathrm{Kd}\left(h^* - h[t]\right)\right)$$

Assume we have a system that has an initial level of 100 units. The inlet flow rate is fixed at 1 in arbitrary units. The goal is to bring the level down to 20 units and to hold it there. The valve controlling exit flow is set to provide 5 arbitrary units of exit flow and there is no provision for adjusting this setting. The following code shows us what happens:

```
qf = 1;
qo = 5;
A = 1;
Kp = .01;
Ki = 0.005;
Kd = 0.1;
h* = 20;
tmax = 30;

a = NDSolve[{∂t h[t] == (qf - qo)/A, h[0] == 100}, h[t],
{t, 0, tmax}]
Plot[Evaluate[h[t]/.a], {t, 0, tmax}, PlotRange → All,
Epilog → Line[{{0, h*}, {tmax, h*}}],
```

PlotLabel → "Uncontrolled",

AxesLabel → {*t*, "h[t]"}]

$\{\{h[t] \to \text{InterpolatingFunction}[\square][t]\}\}$

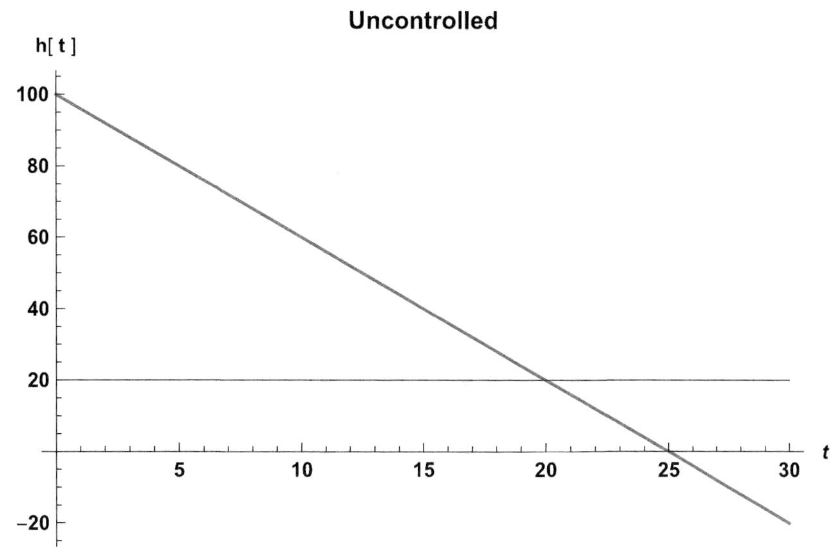

The plot tells the story. The inlet flow is so much smaller than the outlet that there is a linear decrease in the level in the unit until it goes to zero. As we knew it would, the tank simply empties. We can repeat this with inclusion of the PID control expression to see the results:

$\boldsymbol{q_f = 1;}$

$\boldsymbol{q_o = 5;}$

Kp = .01;

Ki = 0.005;

Kd = 0.1;

***h** = 20;**

***A* = 1;**

tmax = 50;

$a =$

NDSolve[

$\{\partial_t h[t] == \frac{q_f}{A} - \frac{q_o}{A}\left(\text{Kp}\,(h^* - h[t]) - \text{Ki}\int_0^t (h^* - h[t])\,dt - \right.$

$\left.\text{Kd}\,(h^* - h[t])\right), h[0] == 100\}, h[t],$

$\{t, 0, \text{tmax}\}]$

Plot[Evaluate[$h[t]$/.a], $\{t, 0, \text{tmax}\}$,

PlotRange → {{0, tmax}, {0, 100}},

Epilog → Line[{{0, h^*}, {tmax, h^*}}],

PlotLabel → "PID Controlled",

AxesLabel → {t, "h[t]"}]

{{$h[t]$ → InterpolatingFunction[□][t]}}

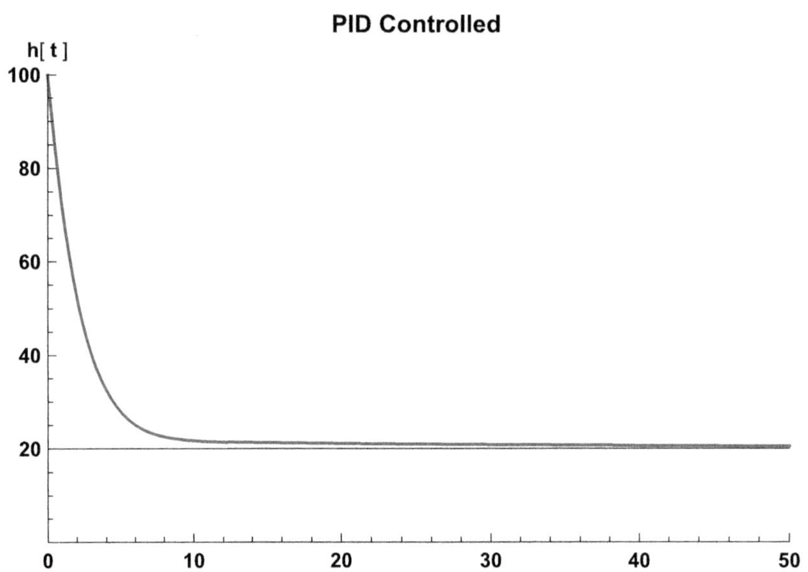

This result is quite different, and it displays the PID function's effect. It smoothly and rapidly adjusts the exit flow rate so that the level in the tank comes to the design level of 20 and remains there. Even though the initial condition has the level in the tank at the outset of the process as being larger than the set point level, this function brings the level down. How rapidly this system approaches the set point level and the steady state depends on the magnitude of the PID constants.

One more example of this control paradigm is to consider the input as fluxional with PID control of the output. A Cosine function can be used to express the input flow rate:

$$\frac{\mathrm{qo}(1+\alpha\mathrm{Cos}[t\beta])}{A}$$

The output flow rate can be given by Torricelli's law:

$$\frac{1}{A}\sqrt{2}\mathrm{Aoo}\, b\sqrt{g}\sqrt{\mathrm{hx}[t]}$$

and this can be modulated by the PID function, as if there were a controller on a valve that would vary, Aoo, the orifice area:

$$\frac{1}{A}\sqrt{2}\mathrm{Aoo}b\sqrt{g}\sqrt{\mathrm{hx}[t]}\left(1+\mathrm{Kp}(\mathrm{hx}[t]-\mathrm{hd})-\mathrm{Ki}\int_0^t(\mathrm{hd}-\mathrm{hx}[t])\,dt-\mathrm{Kdhx}'[t]\right)$$

First we compute the result with all the PID constants set to zero for the base case of no control:

Clear[α, β, g, q, qo, b, Ao, A, h, K, hd, Aoo, t]

Ao=.

$r = 2$;

hd = 3.5;

$A = N[\pi r^2]$;

Aoo = 0.01A;

$b = 1$;

$g = 9.80$;

qo = 1;

$\alpha = .25$;

$\beta = .05$;

Kp = .0;

Ki = 0.0;

Kd = 0.0;

tmax = 2000;

solns =

NDSolve[

$\left\{ hx'[t] == \frac{qo(1 + \alpha Cos[t\beta])}{A} - \right.$

$\frac{1}{A}\sqrt{2}Aoob\sqrt{g}\sqrt{hx[t]}$

$\left(1 + Kp(hx[t] - hd) - Ki \int_0^t (hd - hx[t])\, dt - \right.$

Kdhx′[t]), hx[0] == 0}, hx[t], {t, 0, tmax}];

fns[t_]:=Evaluate[hx[t]/.solns];

pluc = Plot[{fns[t]}, {t, 0, tmax}, PlotRange → All,

AxesLabel → {t, “h[t]”}]

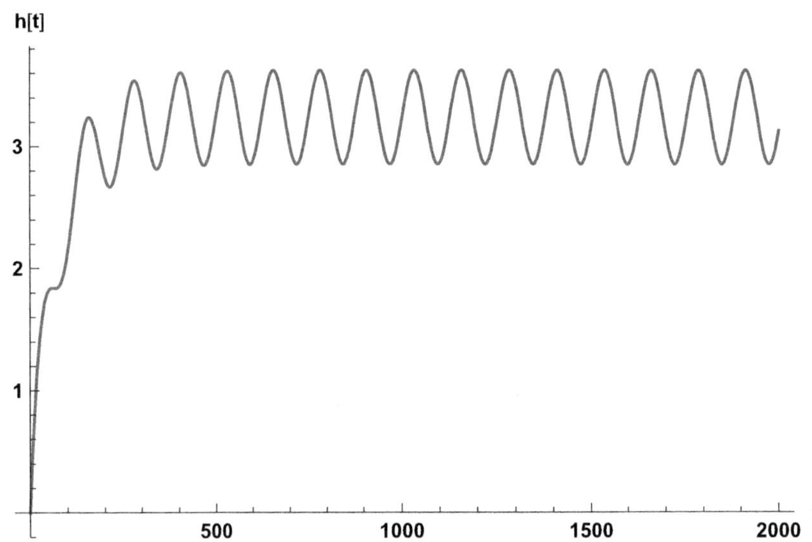

We repeat the calculation with the PID constants all set to 0.1:

```
Clear[α, β, g, q, qo, b, Ao, A, h, K, hd, Aoo, t]

Ao=.
r = 2;
hd = 3.5;
A = N[π r^2];
Aoo = 0.01 A;

b = 1;
g = 9.80;
qo = 1;
α = .25;
β = .05;
Kp = .1;
Ki = 0.1;
Kd = 0.1;
tmax = 2000;
solns =
NDSolve[
{hx'[t]== qo(1 + αCos[tβ])/A −
1/A √2 Aoo b √g √hx[t]
(1 + Kp(hx[t] − hd) − Ki ∫_0^t (hd − hx[t]) dt −
Kdhx'[t]), hx[0] == 0}, hx[t], {t, 0, tmax}];
fns2[t_]:=Evaluate[hx[t]/.solns]
plc = Plot[fns2[t], {t, 0, tmax}, PlotRange → All,
PlotStyle → Hue[0.1], AxesLabel → {t, "h[t]"}]
```

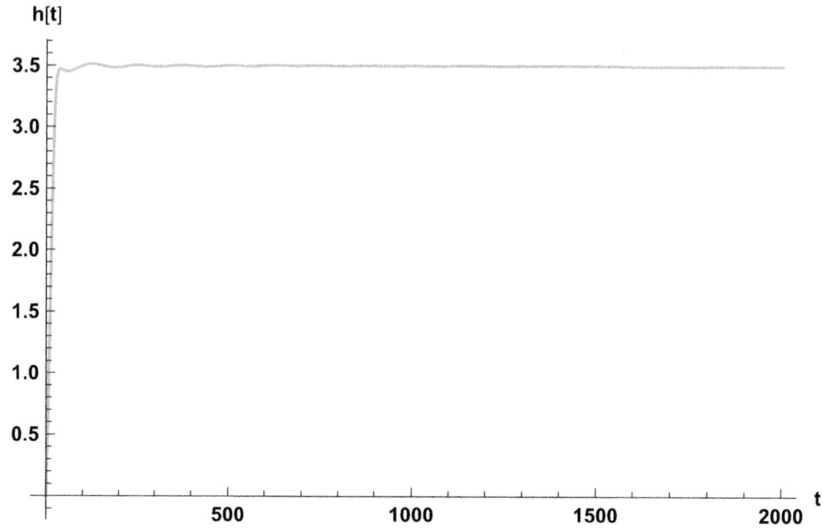

The difference in the degree of "ripple" in the level is quite evident. To make that very clear, we show the two results together:

Show[pluc, plc]

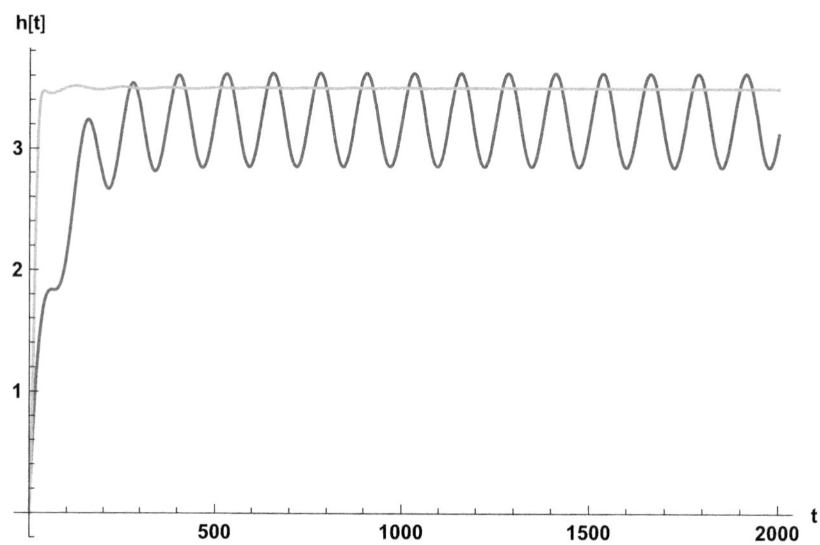

Finally, if the PID constants are increased by a factor of ten, the results become even more dramatically constant, or flat after a brief period of time.

```
Clear[α, β, g, q, qo, b, Ao, A, h, K, hd, Aoo, t]

Ao=.
r = 2;
hd = 3.5;
A = N[πr^2];
Aoo = 0.01A;

b = 1;
g = 9.80;
qo = 1;
α = .25;
β = .05;
Kp = 1;
Ki = 1;
Kd = 1;
tmax = 2000;
solns =
NDSolve[
```

$$\left\{\mathrm{hx}'[t]==\frac{\mathrm{qo}(1+\alpha\mathrm{Cos}[t\beta])}{A}-\right.$$

$$\frac{1}{A}\sqrt{2}\mathrm{Aoo}b\sqrt{g}\sqrt{\mathrm{hx}[t]}$$

$$\left(1+\mathrm{Kp}(\mathrm{hx}[t]-\mathrm{hd})-\mathrm{Ki}\int_0^t(\mathrm{hd}-\mathrm{hx}[t])\,dt-\right.$$

```
Kdhx'[t]), hx[0] == 0}, hx[t], {t, 0, tmax}];

fns2[t_]:=Evaluate[hx[t]/.solns]
```

```
plc2 = Plot[fns2[t], {t, 0, tmax}, PlotRange → All,
PlotStyle → Hue[0.1], AxesLabel → {t, "h[t]"}]
```

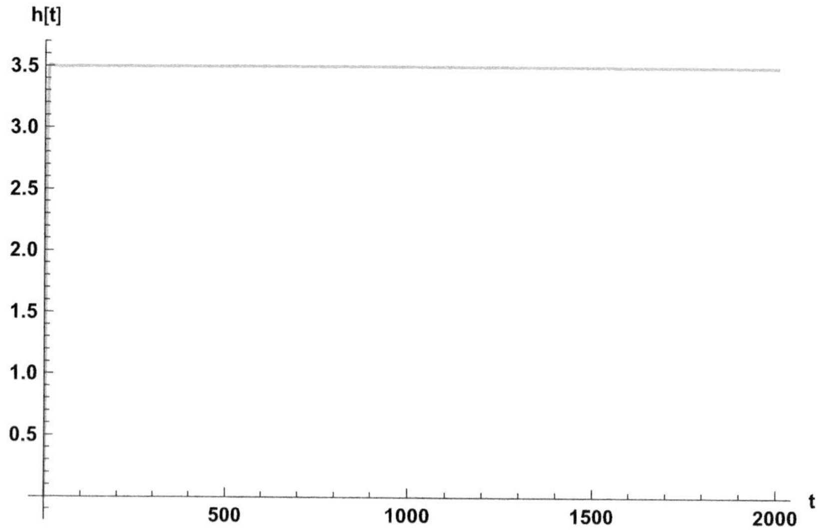

General model for titration of a strong acid with a strong base

There are a number of classic problems in ionic equilibria (see Ionic Equilibrium: Solubility and pH Calculations, James N. Butler and David R. Cogley) that are easily solved using *Mathematica.* One of these is the titration of a strong base with a strong acid to the equivalence point. The equivalence occurs when the number of moles of acid is equal to the number of moles of base. Consider the neutralization of NaOH by HCl:

$$HCl + NaOH \Longleftrightarrow HOH + NaCl$$

In this case a volume of acid, V_a, with concentration C_a is added to a volume of base, V_b, with a concentration of C_b, we want to know how the pH changes in solution as a function of the incremental volume of acid added to the volume of base.

The usual way to do this is to assemble the simultaneous equations for mass, the charge balances for the ions and the ionic equilibrium of water, which includes the concentration of hydrogen that we seek. Then in many cases simplifying assumptions are made in the regions of low and high pH that lead to approximate solutions. We will seek exact solutions across the whole of the range.

After all the acid has been added to the base, at the equivalence point the following will be the case:

Mass (Mole) Balance on Acid:

$$C_{Cl^-}\,(V_a + V_b) = C_a\,V_a$$

$$C_{Cl^-} = \frac{C_a V_a}{(V_a + V_b)}$$

Mass (Mole) Balance on Base:

$$C_{Na^+}\,(V_a + V_b) = C_b\,V_b$$

$$C_{Na^+} = \frac{C_b V_a}{(V_a + V_b)}$$

where C_a and C_b are the analytical concentrations of the acid and base.

The concentration equilibrium constant for water provides:

Water Equilibrium:

$$C_{H^+}\,C_{OH^-} = K_w = 10^{-14}$$

$$C_{OH^-} = \frac{10^{-14}}{C_{H^+}}$$

which is a good approximation if the ionic strength of the solution is not too high and if the temperature is near ambient.

At the equivalence point, and all along the process, there must be charge neutrality among the ions meaning the protons, sodium ions, hydroxyls and chlorides:

Charge Balance:

$$(C_{Na^+} + C_{H^+})\,(V_a + V_b) = (C_{Cl^-} + C_{OH^-})(V_a + V_b)$$

$$(C_{Na^+} + C_{H^+}) = (C_{Cl^-} + C_{OH^-})$$

We ideally want a solution for C_{H^+} that is expressed as the volume of acid added to solution. That means we can go back to the charge balance, make the substitutions for each ion concentration and then solve for the concentration of the protons:

$$(C_{Na^+} + C_{H^+}) = (C_{Cl^-} + C_{OH^-})$$

$$\frac{C_b V_b}{(V_a + V_b)} + C_{H^+} = \frac{C_a V_a}{(V_a + V_b)} + \frac{10^{-14}}{C_{H^+}}$$

Usually this equation is simplified to the low pH and high pH regions. At low pH the concentration of protons is large. Hence the equation can be simplified to:

Low pH:

$$C_{H^+} = \frac{C_a V_a}{(V_a + V_b)} - \frac{C_b V_b}{(V_a + V_b)}$$

And at High pH the concentration of protons is small, therefore we find:

High pH:

$$\frac{C_b V_b}{(V_a + V_b)} = \frac{C_a V_a}{(V_a + V_b)} + \frac{10^{-14}}{C_{H^+}}$$

$$\left(\frac{C_b V_b}{(V_a + V_b)} - \frac{C_a V_a}{(V_a + V_b)}\right) = \frac{10^{-14}}{C_{H^+}}$$

$$C_{H^+} = \frac{10^{-14}(V_a + V_b)}{C_b V_b - C_a V_a}$$

We can plot both of these equations to see how they behave and what they predict. Remember, in this computation, we are simulating the addition of base to acid, hence V_b is the independent variable.

CHL[Vb_]:= $C_a \frac{V_a}{V_a + V_b} - C_b \frac{V_b}{V_a + V_b}$

CHH[Vb_]:= $\frac{10^{-14}(V_a + V_b)}{(C_b\ V_b - C_a\ V_a)}$

$V_a = 50;$

$C_a = 0.1;$

$C_b = 0.2;$

a1 = Plot[–Log10[CHL[V_b]], {V_b, 0, 25},

PlotStyle → {Dashed, Hue[0.1]}]

b1 = Plot[–Log10[CHH[V_b]], {V_b, 25, 50},

PlotStyle → {Dashed, Hue[06]}]

Show[{a1, b1}, PlotRange → {{0, 50}, {0, 14}}]

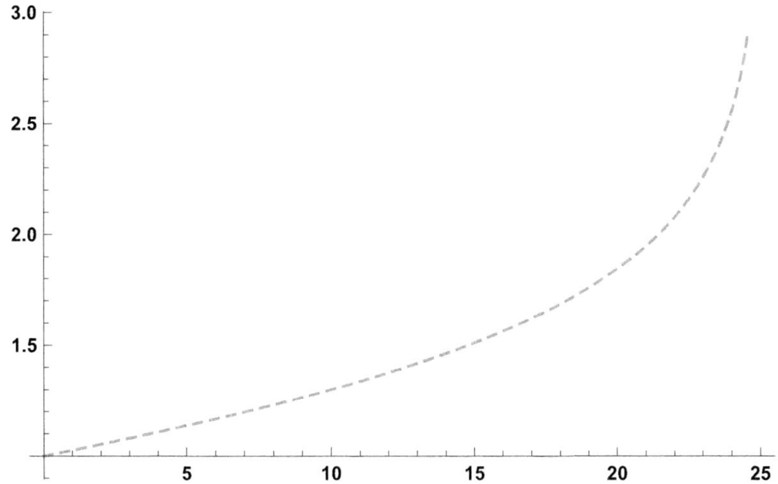
3.0
2.5
2.0
1.5
5
10
15
20
25

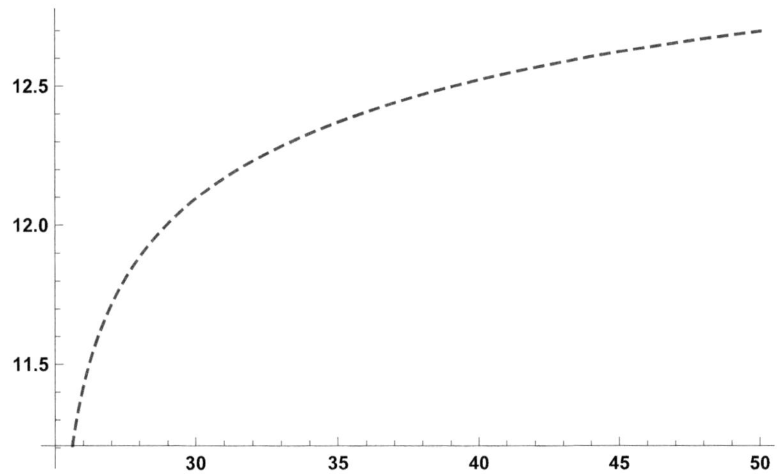
12.5
12.0
11.5
30
35
40
45
50

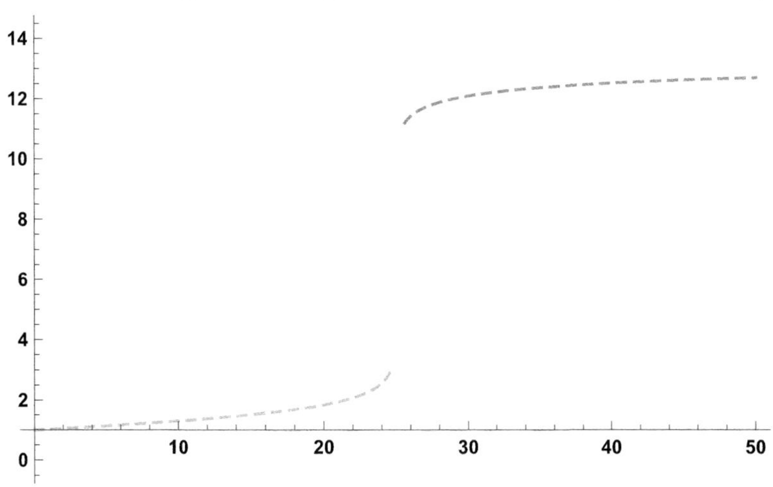
14
12
10
8
6
4
2
0
10
20
30
40
50

As we can see, the two equations that resulted from simplifications of the full equation provide reasonable results in both regions of pH. But, we have a very powerful tool in *Mathematica*, so we should solve the full equation, which is quadratic in the concentration of protons. Here is the code to do that. The subscripts used previously can cause some problems during computation, so we drop most of them here.

Clear[Va, Vb, Ca, Cb, Kw]

flsoln =

$$\text{FullSimplify}\left[\text{Solve}\left[C_{H^+}+\frac{\text{CbVb}}{\text{Va}+\text{Vb}}==\frac{\text{Kw}}{C_{H^+}}+\frac{\text{CaVa}}{\text{Va}+\text{Vb}},C_{H^+}\right]\right]$$

$$\left\{\left\{C_{H^+}\to-\frac{-\text{CaVa}+\text{CbVb}+\sqrt{4\text{Kw}(\text{Va}+\text{Vb})^2+(\text{CaVa}-\text{CbVb})^2}}{2(\text{Va}+\text{Vb})}\right\},\right.$$

$$\left.\left\{C_{H^+}\to\frac{\text{CaVa}-\text{CbVb}+\sqrt{4\text{Kw}(\text{Va}+\text{Vb})^2+(\text{CaVa}-\text{CbVb})^2}}{2(\text{Va}+\text{Vb})}\right\}\right\}$$

Since the equation is quadratic in C_{H^+}, we obtain two solutions for the proton concentration. The first solution includes a minus sign in front of all the terms, meaning that it predicts negative concentrations, which is unphysical. Therefore, the second solution is the physically realistic one. We can assign this to a function, and then plot the result, and compare it to the two approximate results:

C_{H^+}[Vb_]:=Evaluate [C_{H^+}/.flsoln[[2]]]

Va = 50;

Ca = 0.1;

Cb = 0.2;

Kw = 10^{-14};

c1 = Plot[−Log10[C_{H^+}[Vb]], {Vb, 0, 50},

AxesLabel → {"Vb", "pH"}]

Show[c1, b1, a1]

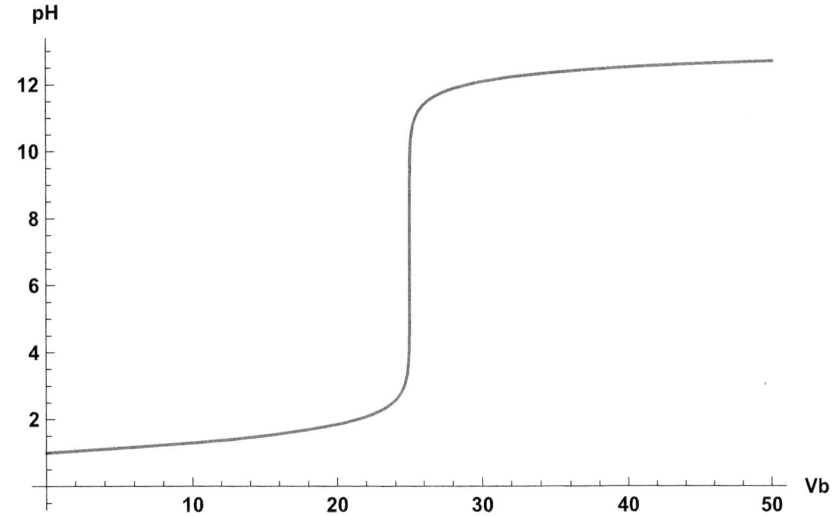

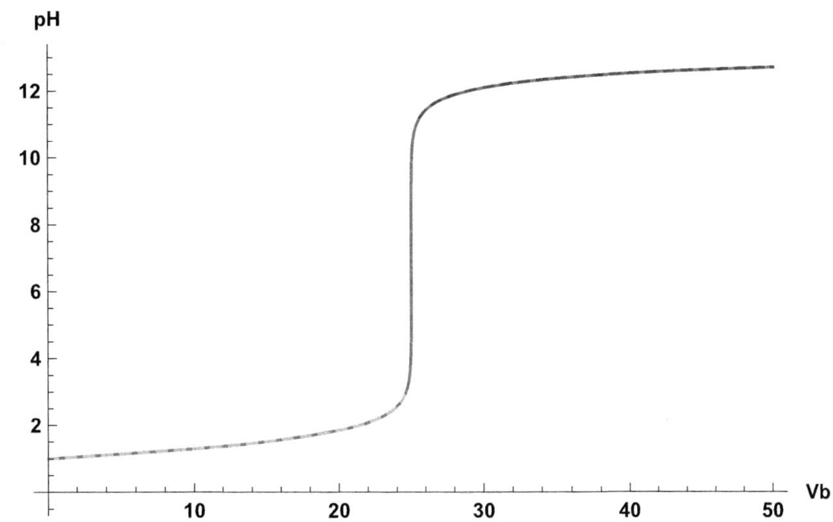

The dashed lines of the approximate solutions are superimposed on the exact solution. Over the range of their applicability, the agreement is good between the exact and approximate solutions. The exact solution obviously spans the entire range of base addition to the acid, including through and beyond the equivalence point.

At the equivalence point, the pH is exactly 7 and the derivative of the pH with respect to base volume, Vb, must get very large because the slope is so large in this region. Firstly, we find the derivative of the pH with respect to Vb, which is to say that we find the derivative of minus the Log10 of the function for the proton concentration:

Clear[Va, Vb, Ca, Cb, Kw]

$$D\left[-\text{Log10}\left[\frac{\text{CaVa}-\text{CbVb}+\sqrt{4\text{Kw}(\text{Va}+\text{Vb})^2+(\text{CaVa}-\text{CbVb})^2}}{2(\text{Va}+\text{Vb})}\right], \text{Vb}\right]\text{//FullSimplify}$$

$$\frac{(\text{Ca}+\text{Cb})\text{Va}}{(\text{Va}+\text{Vb})\sqrt{4\text{Kw}(\text{Va}+\text{Vb})^2+(\text{CaVa}-\text{CbVb})^2}\text{Log}[10]}$$

Here we see the derivative of the expression is fairly simple. When we apply the parameter values and plot the derivative, as expected at the equivalence point, the derivative is very large.

Va = 50;

Ca = 0.1;

Cb = 0.2;

Kw = 10^{-14};

$$\text{Plot}\left[\frac{(\text{Ca}+\text{Cb})\text{Va}}{(\text{Va}+\text{Vb})\sqrt{4\text{Kw}(\text{Va}+\text{Vb})^2+(\text{CaVa}-\text{CbVb})^2}\text{Log}[10]},\right.$$

{Vb, 0, 50}, PlotRange → {{0, 50}, {0, 120}},

AxesLabel → {"Vb", "∂_{Vb}pH"}]

$$\text{Plot}\left[\frac{(\text{Ca}+\text{Cb})\text{Va}}{(\text{Va}+\text{Vb})\sqrt{4\text{Kw}(\text{Va}+\text{Vb})^2+(\text{CaVa}-\text{CbVb})^2}\text{Log}[10]},\right.$$

{Vb, 24.95, 25.05}, PlotRange → All,

AxesLabel → {"Vb", "∂_{Vb}pH"}]

$$\text{Plot}\left[\frac{(\text{Ca}+\text{Cb})\text{Va}}{(\text{Va}+\text{Vb})\sqrt{4\text{Kw}(\text{Va}+\text{Vb})^2+(\text{CaVa}-\text{CbVb})^2}\text{Log}[10]},\right.$$

{Vb, 0, 50}, AxesLabel → {"Vb", "∂_{Vb}pH"}]

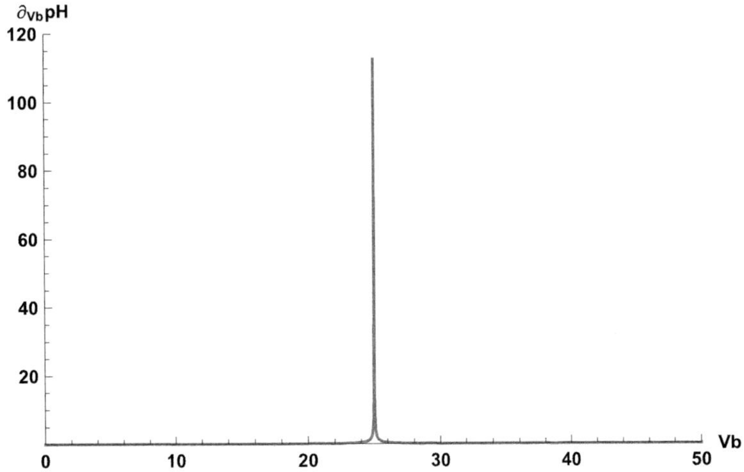
∂Vb pH
120
100
80
60
40
20
0
10
20
30
40
50
Vb

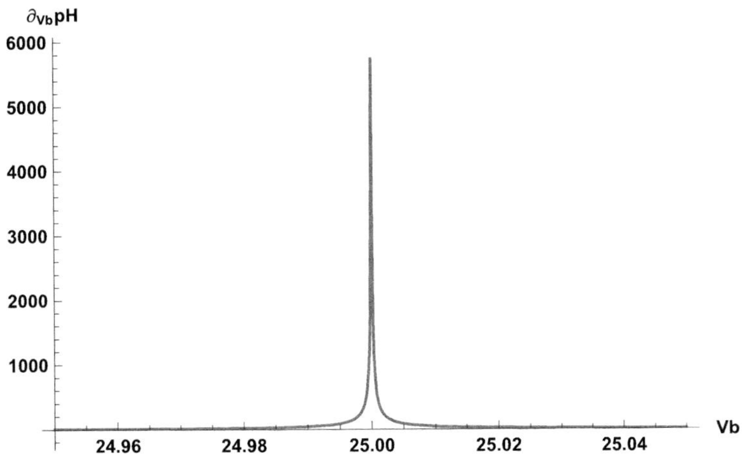
∂Vb pH
6000
5000
4000
3000
2000
1000
24.96
24.98
25.00
25.02
25.04
Vb

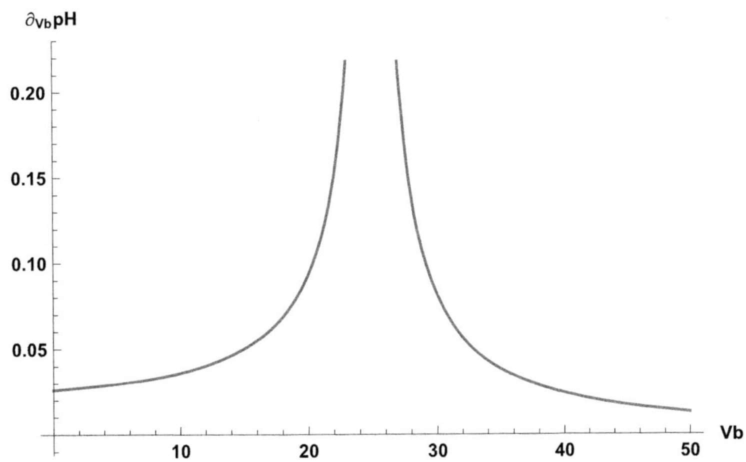
∂Vb pH
0.20
0.15
0.10
0.05
10
20
30
40
50
Vb

The first graph shows the narrowness and, the apparent, but not actual maximum of the derivative; it misses the true maximum. The second graph shows the true maximum of the derivative because we have narrowed the range of Vb to 0.05 below, and above, the equivalence point. The third plot shows that the derivative is not zero anywhere in the range of Vb. We can find the magnitude of the derivative at Vb = 25 and we see that it has a value of just under 5800.

Va = 50;

Ca = 0.1;

Cb = 0.2;

Kw = 10^{-14};

Table[

$$\left\{\mathbf{Vb}, \frac{(\mathbf{Ca}+\mathbf{Cb})\mathbf{Va}}{(\mathbf{Va}+\mathbf{Vb})\sqrt{4\mathbf{Kw}(\mathbf{Va}+\mathbf{Vb})^2+(\mathbf{CaVa}-\mathbf{CbVb})^2}\mathbf{Log[10]}}\right\},$$

{Vb, 24.95, 25.05, .01}]

{{24.95, 8.69167}, {24.96, 10.8631}, {24.97, 14.4822}, {24.98, 21.7204}, {24.99, 43.434}, {25., 5790.59}, {25.01, 43.4224}, {25.02, 21.7088}, {25.03, 14.4706}, {25.04, 10.8516}, {25.05, 8.68009}}

Going a step further, we can take the second derivative and evaluate it.

Clear[Va, Vb, Ca, Cb, Kw]

***D*[*D*[−Log10[**

$$\left.\frac{\mathbf{CaVa}-\mathbf{CbVb}+\sqrt{4\mathbf{Kw}(\mathbf{Va}+\mathbf{Vb})^2+(\mathbf{CaVa}-\mathbf{CbVb})^2}}{2(\mathbf{Va}+\mathbf{Vb})}\right], \mathbf{Vb}], \mathbf{Vb}]//\mathbf{FullSimplify}$$

$$-\left(\left((\text{Ca}+\text{Cb})\text{Va}\left(\text{Ca}^2\text{Va}^2+8\text{Kw}(\text{Va}+\text{Vb})^2+\text{Cb}^2\text{Vb}(\text{Va}+2\text{Vb})-\text{CaCbVa}(\text{Va}+3\text{Vb})\right)\right)/\left((\text{Va}+\text{Vb})^2\left(4\text{Kw}(\text{Va}+\text{Vb})^2+(\text{CaVa}-\text{CbVb})^2\right)^{3/2}\text{Log}[10]\right)\right)$$

We know that for an inflection point the second derivative should be equal to zero:

Va = 50;

```
Ca = 0.1;

Cb = 0.2;

Kw = 10^-14;

Table[

{Vb,

-(((Ca + Cb)Va

(Ca^2Va^2 + 8Kw(Va + Vb)^2 + Cb^2Vb(Va + 2Vb)-

CaCbVa(Va + 3Vb)))/

((Va + Vb)^2 (4Kw(Va + Vb)^2 + (CaVa - CbVb)^2)^3/2

Log[10]))}, {Vb, 24.95, 25.05, .005}]

ListPlot[%]
```

{{24.95, 173.717}, {24.955, 214.465}, {24.96, 271.433}, {24.965, 354.524}, {24.97, 482.545}, {24.975, 694.862}, {24.98, 1085.71}, {24.985, 1930.13}, {24.99, 4342.58}, {24.995, 17365.9}, {25., −154.417}, {25.005, −17365.9}, {25.01, −4342.58}, {25.015, −1930.13}, {25.02, −1085.71}, {25.025, −694.862}, {25.03, −482.545}, {25.035, −354.524}, {25.04, −271.433}, {25.045, −214.465}, {25.05, −173.717}}

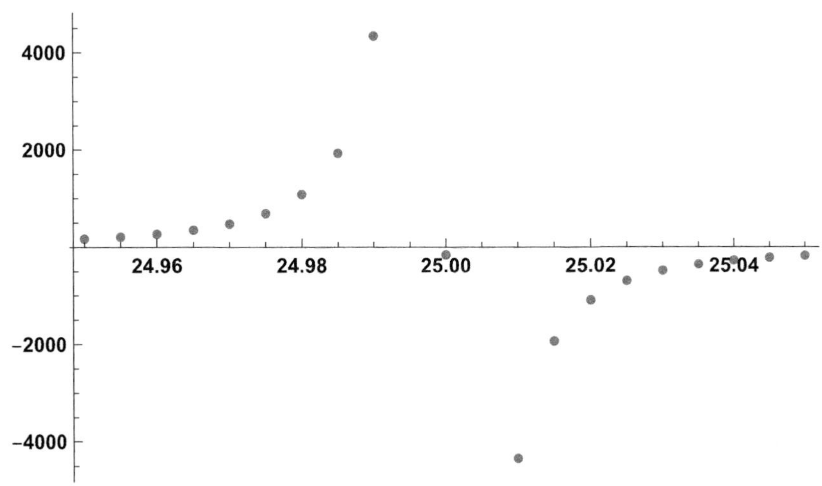

Interestingly, the value of the second derivative is off from zero as we see

```
Plot[
-(((Ca + Cb) Va
(Ca^2 Va^2 + 8Kw(Va + Vb)^2 + Cb^2 Vb(Va + 2Vb)-
CaCbVa(Va + 3Vb)))/
((Va + Vb)^2 (4Kw(Va + Vb)^2 + (CaVa - CbVb)^2)^3/2
Log[10])), {Vb, 24.95, 25.05}]
Show[%%, %]
Plot[
-(((Ca + Cb) Va
(Ca^2 Va^2 + 8Kw(Va + Vb)^2 + Cb^2 Vb(Va + 2Vb)-
CaCbVa(Va + 3Vb)))/
((Va + Vb)^2 (4Kw(Va + Vb)^2 + (CaVa - CbVb)^2)^3/2
Log[10])), {Vb, 24.99, 25.01}]
```

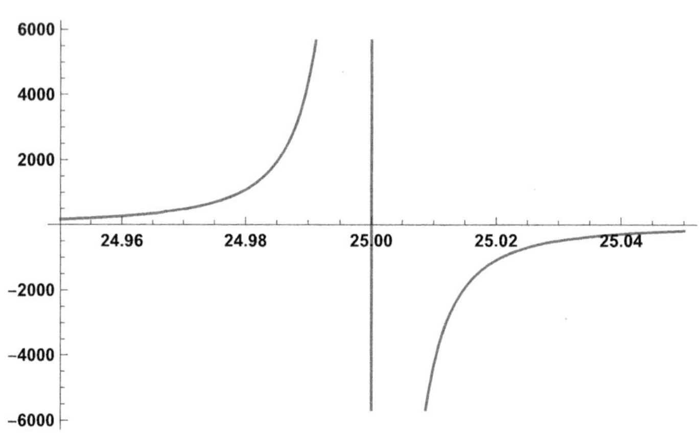

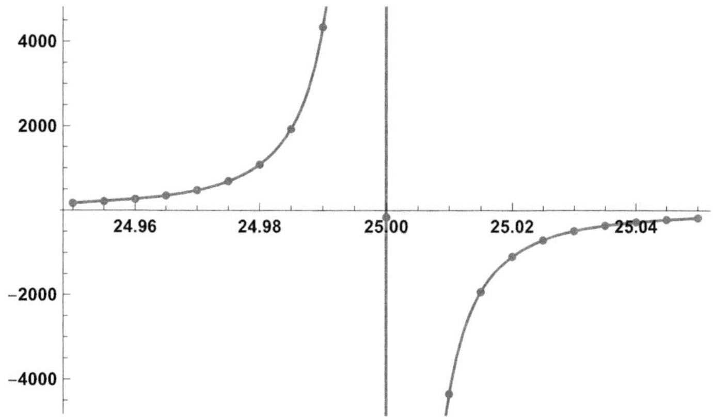

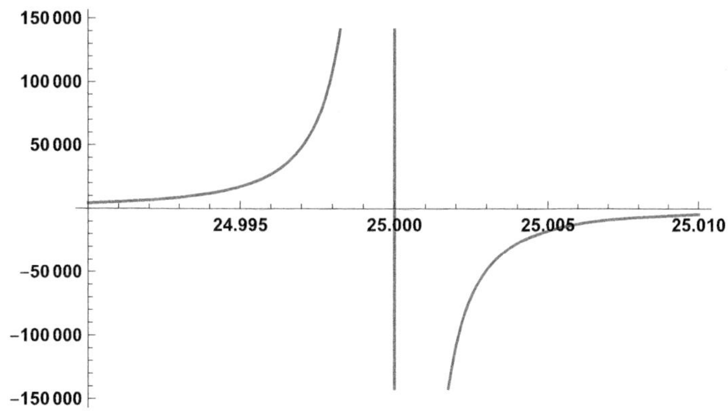

So, in the numerical evaluation of the second derivative, the value is not at zero, but when we use Plot, the value at 25 appears to be zero. We can see the problem by numerically evaluating Vb when the second derivative is set to zero.

NSolve[

$-(((\mathrm{Ca}+\mathrm{Cb})\mathrm{Va}$

$\left(\mathrm{Ca}^2\mathrm{Va}^2+8\mathrm{Kw}(\mathrm{Va}+\mathrm{Vb})^2+\mathrm{Cb}^2\mathrm{Vb}(\mathrm{Va}+2\mathrm{Vb})-\right.$

$\left.\mathrm{CaCbVa}(\mathrm{Va}+3\mathrm{Vb})\right))/$

$\left((\mathrm{Va}+\mathrm{Vb})^2\left(4\mathrm{Kw}(\mathrm{Va}+\mathrm{Vb})^2+(\mathrm{CaVa}-\mathrm{CbVb})^2\right)^{3/2}\right.$

Log[10])) == 0, Vb]

{{Vb → −12.5}}

This is not the expected value and it is not a physically realistic result, but it is mathematically valid. We can plot the second derivative over a wide range from negative to positive values of Vb:

Plot[

$$-(((\mathrm{Ca}+\mathrm{Cb})\mathrm{Va}$$

$$\left(\mathrm{Ca}^2\mathrm{Va}^2+8\mathrm{Kw}(\mathrm{Va}+\mathrm{Vb})^2+\mathrm{Cb}^2\mathrm{Vb}(\mathrm{Va}+2\mathrm{Vb})-\right.$$

$$\mathrm{CaCbVa}(\mathrm{Va}+3\mathrm{Vb})))/$$

$$\left((\mathrm{Va}+\mathrm{Vb})^2\left(4\mathrm{Kw}(\mathrm{Va}+\mathrm{Vb})^2+(\mathrm{CaVa}-\mathrm{CbVb})^2\right)^{3/2}\right.$$

$$\mathrm{Log}[10])),\{\mathrm{Vb},-50,50\}]$$

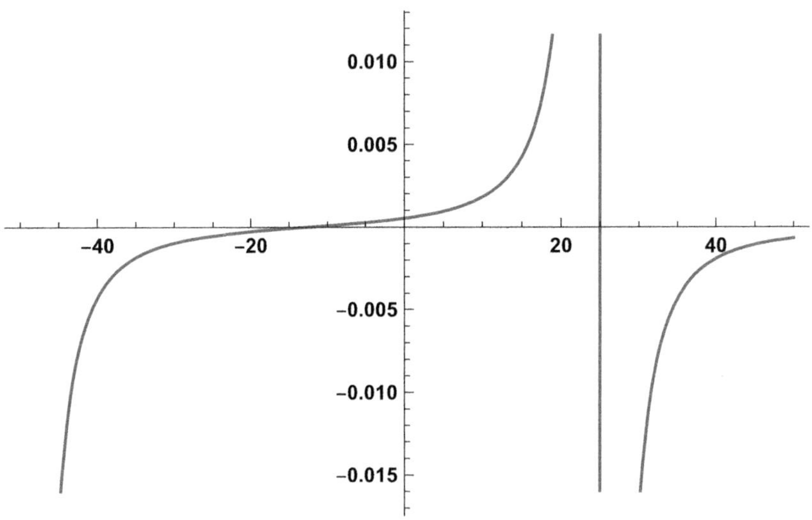

We can see the root at Vb = -12.5, but why did this root-finding calculation miss the value at Vb = 25? The reason is that we need to have higher working precision than the default for this problem. We can find the working precision by adding //Precision to the NSolve calculation.

NSolve[

$$-(((\mathrm{Ca}+\mathrm{Cb})\mathrm{Va}$$

$$\left(\mathrm{Ca}^2\mathrm{Va}^2+8\mathrm{Kw}(\mathrm{Va}+\mathrm{Vb})^2+\mathrm{Cb}^2\mathrm{Vb}(\mathrm{Va}+2\mathrm{Vb})-\right.$$

$$\mathrm{CaCbVa}(\mathrm{Va}+3\mathrm{Vb})))/$$

```
((Va + Vb)^2 (4Kw(Va + Vb)^2 + (CaVa - CbVb)^2))^(3/2)
Log[10])) == 0, Vb]//Precision
```

MachinePrecision

At machine precision, the inaccuracy of the result is too high and the second root is missed. $MachinePrecision is 53*Log[10,2]:

```
$MachinePrecision
```

15.9546

```
53 * Log[10, 2]//N
```

15.9546

We can set the working precision to a higher value to look for the positive root. Firstly, we set WorkingPrecision to the value of $MachinePrecision and we get the negative root back:

```
NSolve[
-(((Ca + Cb)Va
(Ca^2Va^2 + 8Kw(Va + Vb)^2 + Cb^2Vb(Va + 2Vb)-
CaCbVa(Va + 3Vb)))/
((Va + Vb)^2 (4Kw(Va + Vb)^2 + (CaVa - CbVb)^2))^(3/2)
Log[10])) == 0, Vb,
WorkingPrecision -> 15.954589770191001]
```

{{Vb → −12.5}}

But, when we set WorkingPrecision to 16, we get both roots:

```
NSolve[
-(((Ca + Cb)Va
(Ca^2Va^2 + 8Kw(Va + Vb)^2 + Cb^2Vb(Va + 2Vb)-
```

CaCbVa(Va + 3Vb)))/

$\left((Va + Vb)^2 (4Kw(Va + Vb)^2 + (CaVa - CbVb)^2\right)^{3/2}$

Log[10])) == 0, Vb, WorkingPrecision → 16]

{{Vb → 24.99999999985000}, {Vb → −12.49999999996250}}

This kind of problem happens with certain computations and it is something to be alert to. We might not have expected to see this issue in such a straightforward problem, but this is an incredibly good example.

We might wonder what the other solution to the equation provides. Let's go ahead and evaluate a few values of the pH based on that solution.

C_{H^+}Alt[Vb_]:=Evaluate [C_{H^+}/.flsoln[[1]]]

Va = 50;

Ca = 0.1;

Cb = 0.2;

Kw = 10^{-14};

Table [−Log10 [C_{H^+}Alt[Vb]], {Vb, 0, 50, 5}]

{13. − 1.36438*i*, 12.8617 − 1.36438*i*, 12.699 − 1.36438*i*, 12.4881 − 1.36438*i*, 12.1549 − 1.36438*i*, 7. − 1.36438*i*, 1.90309 − 1.36438*i*, 1.62839 − 1.36438*i*, 1.47712 − 1.36438*i*, 1.37566 − 1.36438*i*, 1.30103 − 1.36438*i*}

Obviously these are complex numbers that are not physical, however, we do notice that they each have a real part and the real part is decreasing as Vb increases. We can create a Table with many more values between 0 and 50, select only the real part and then ListPlot the results.

C_{H^+}Alt[Vb_]:=Evaluate [C_{H^+}/.flsoln[[1]]]

Va = 50;

Ca = 0.1;

Cb = 0.2;

Kw = 10^{-14};

Table[{Vb, Re[−Log10[$C_{H^+\text{Alt}}$[Vb]]]}, {Vb, 0, 50, .5}];

d1 = ListPlot[%, AxesLabel → {"Vb", "pH"}]

Show[c1, d1]

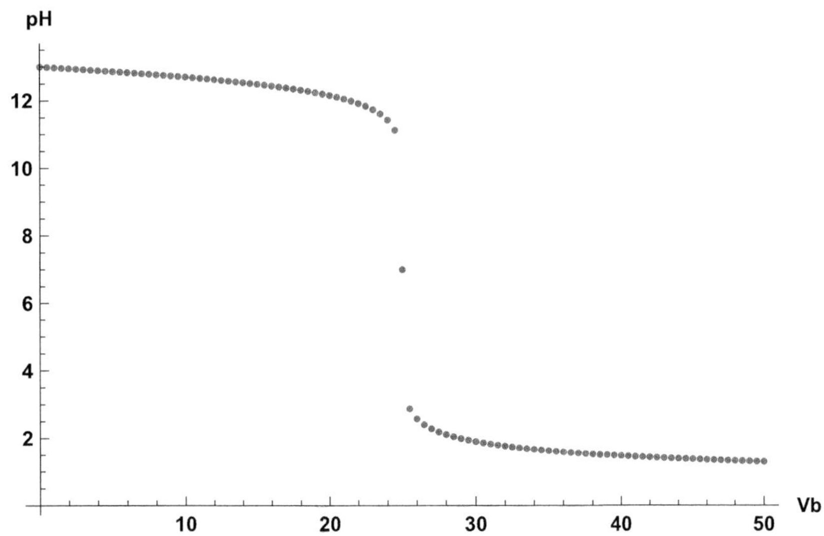

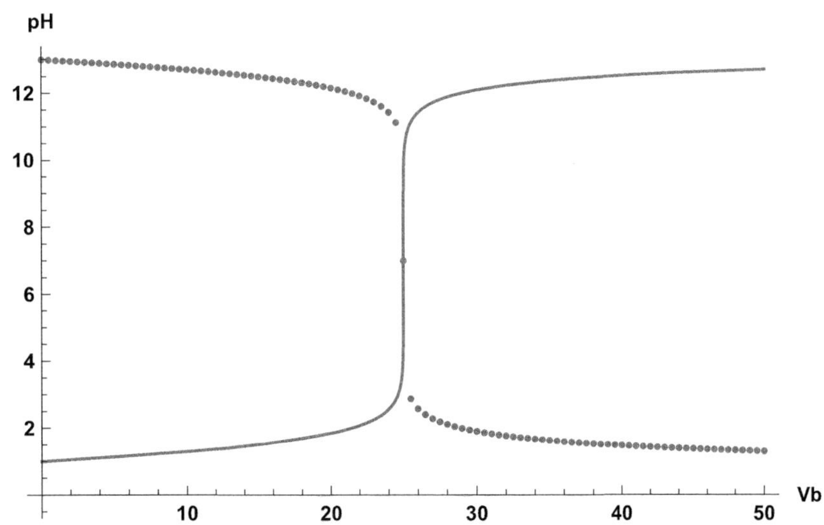

This is interesting. It represents the titration of a solution of base with acid. Hence, Vb is misnamed here, it should be Va.

The pH of a weak acid

How do we find the pH of the weak acid in solution? In the case of a strong acid that is monoprotic, it is safe to assume that the acid is fully dissociated and the concentrations of the anions are equal to the analytical concentration of the acid. We used that in the solution of the titration of the strong acid with a strong base. In the case of the weak acid, that is no longer a valid assumption. We know that some of the acid will be present in solution in its undissociated from and we must account for that using the pKa. We review this here.

The key equations that we use are the acid equilibrium, the water equilibrium, the mass balance, and the charge balance:

$$C_{H+}\, C_{A-} = K_a\, C_{\mathrm{HA}}$$

$$C_{H+}\, C_{\mathrm{OH}-} = K_w$$

$$C_{\mathrm{HAo}} = C_{A-} + C_{\mathrm{HA}}$$

$$C_{H+} = C_{A-} + C_{\mathrm{OH}-}$$

where C_{HAo} is the analytical concentration and C_{HA} is the concentration at equilibrium. We can do some preliminary rearranging and then seek solution that is analytical for C_{H^+}:

$$C_{H^+}C_{A^-} == K_a C_{\mathrm{HA}} \Longrightarrow C_{\mathrm{HA}} = \frac{C_{H^+}C_{A^-}}{K_a}$$

$$C_{\mathrm{HAo}} = C_{A-} + C_{\mathrm{HA}} \Longrightarrow C_{\mathrm{HAo}} = C_{A-} + \frac{C_{H^+}C_{A^-}}{K_a} = C_{A-}\,(1 + \frac{C_{H^+}}{K_a})$$

$$C_{H+} = C_{A-} + C_{\mathrm{OH}-} = \frac{C_{\mathrm{HAo}}}{\left(1 + \frac{C_{H^+}}{K_a}\right)} + \frac{K_w}{C_{H+}}$$

$$\mathbf{sln = Solve}\left[C_{H^+} == \frac{C_{\mathrm{HAo}}}{\left(1 + \frac{C_{H^+}}{K_a}\right)} + \frac{K_w}{C_{H^+}}, C_{H^+}\right] \mathbf{//FullSimplify}$$

$$\left\{\left\{C_{H^+} \to \tfrac{1}{6}\left(-2K_a + \left(2\, 2^{1/3}\left(3C_{\mathrm{HAo}}K_a + K_a^2 + 3K_w\right)\right)\right/\right.\right.$$

$$\left(-9C_{\mathrm{HAo}}K_a^2-2K_a^3+18K_aK_w+\sqrt{}\left(-4\left(3C_{\mathrm{HAo}}K_a+K_a^2+3K_w\right)^3+\right.\right.$$
$$\left.\left.\left(9C_{\mathrm{HAo}}K_a^2+2K_a^3-18K_aK_w\right)^2\right)\right)^{1/3}+2^{2/3}\left(-9C_{\mathrm{HAo}}K_a^2-2K_a^3+18K_aK_w+\right.$$
$$\left.\left.\sqrt{}\left(-4\left(3C_{\mathrm{HAo}}K_a+K_a^2+3K_w\right)^3+\left(9C_{\mathrm{HAo}}K_a^2+2K_a^3-18K_aK_w\right)^2\right)\right)^{1/3}\right)\Big\},$$

$$\Big\{C_{H^+}\to\tfrac{1}{12}\left(-4K_a-\left(4(-2)^{1/3}\left(3C_{\mathrm{HAo}}K_a+K_a^2+3K_w\right)\right)/\right.$$
$$\left(-9C_{\mathrm{HAo}}K_a^2-2K_a^3+18K_aK_w+\sqrt{}\left(-4\left(3C_{\mathrm{HAo}}K_a+K_a^2+3K_w\right)^3+\right.\right.$$
$$\left.\left.\left(9C_{\mathrm{HAo}}K_a^2+2K_a^3-18K_aK_w\right)^2\right)\right)^{1/3}+i2^{2/3}\left(i+\sqrt{3}\right)\left(-9C_{\mathrm{HAo}}K_a^2-2K_a^3+18K_aK_w+\right.$$
$$\left.\left.\sqrt{}\left(-4\left(3C_{\mathrm{HAo}}K_a+K_a^2+3K_w\right)^3+\left(9C_{\mathrm{HAo}}K_a^2+2K_a^3-18K_aK_w\right)^2\right)\right)^{1/3}\right)\Big\},$$

$$\Big\{C_{H^+}\to\tfrac{1}{12}\left(-4K_a+\left(4(-1)^{2/3}2^{1/3}\left(3C_{\mathrm{HAo}}K_a+K_a^2+3K_w\right)\right)/\right.$$
$$\left(-9C_{\mathrm{HAo}}K_a^2-2K_a^3+18K_aK_w+\sqrt{}\left(-4\left(3C_{\mathrm{HAo}}K_a+K_a^2+3K_w\right)^3+\right.\right.$$
$$\left.\left.\left(9C_{\mathrm{HAo}}K_a^2+2K_a^3-18K_aK_w\right)^2\right)\right)^{1/3}-2^{2/3}\left(1+i\sqrt{3}\right)$$
$$\left(-9C_{\mathrm{HAo}}K_a^2-2K_a^3+18K_aK_w+\sqrt{}\left(-4\left(3C_{\mathrm{HAo}}K_a+K_a^2+3K_w\right)^3+\right.\right.$$
$$\left.\left.\left.\left(9C_{\mathrm{HAo}}K_a^2+2K_a^3-18K_aK_w\right)^2\right)\right)^{1/3}\right)\Big\}\Big\}$$

We have three solutions because this was a cubic equation. The second and third solutions contain imaginary terms. The first solution, because it appears to be real, may be the proper solution. But, we should test all three:

$\mathbf{C_{HAo}=0.01;}$

$\boldsymbol{K_a=10^{-4.75};}$

$\boldsymbol{K_w=10^{-14};}$

$\boldsymbol{C_{H^+}=}$ **Evaluate** $\boldsymbol{[C_{H^+}}$**/.sln[[1]]]**

−Log10$\boldsymbol{[C_{H^+}]}$

$0.000412899-7.228014483236696*10^{-20}i$

$3.38416+7.6025*10^{-17}i$

$\boldsymbol{C_{H^+}=}$ **Evaluate** $\boldsymbol{[C_{H^+}}$**/.sln[[2]]]**

−Log10$\boldsymbol{[C_{H^+}]}$

$-0.000430682-7.2280*10^{-20}i$

$3.36584 + 1.36438i$

C_{H+} = Evaluate [C_{H+}/.sln[[3]]]

−Log10[C_{H+}]

$-1 * 10^{-12} + 1.4456 * 10^{-19} i$

$12. - 1.36438i$

The three concentrations contain a real part and an imaginary part. But, when we take - Log10 of the apparent concentration of the protons, we see that the imaginary part of the first solution is so small that we will ignore it and take the pH as 3.38. The other two solutions lead to pH values that have significant imaginary portions that cannot be neglected. And lastly, the CRC handbook in the table of approximate pH values, lists the pH of 0.01 Normal acetic acid as 3.4.

Following Butler (pg. 21-25) we can also ask what happens to the pH when we add in the salt of acetic acid, namely when we also add sodium acetate to the solution? Let C_{HAo} once again be the analytical concentration of the acid, and C_{NaAo} be the analytical concentration of the salt. Here is the analysis:

$$C_{H+}\, C_{A-} = K_a\, C_{\text{HA}}$$

$$C_{H+}\, C_{\text{OH}-} = K_w$$

$$C_{\text{HAo}} + C_{\text{NaAo}} = C_{A-} + C_{\text{HA}}$$

$$C_{\text{NaAo}} = C_{\text{Na}+}$$

$$C_{H+} + C_{\text{Na}+} = C_{A-} + C_{\text{OH}-}$$

We assume we know C_{HAo} and C_{NaAo} because we measured them analytically and we always know the K_a and K_w. We begin by clearing terms again:

$$C_{H+}C_{A-} == K_a C_{\text{HA}} \Longrightarrow C_{\text{HA}} = \frac{C_{H+}C_{A-}}{K_a}$$

$$C_{\text{HAo}} + C_{\text{NaAo}} = C_{A-} + C_{\text{HA}} \Longrightarrow C_{\text{HAo}} + C_{\text{NaAo}} = C_{A-} + \frac{C_{H+}C_{A-}}{K_a}$$

$$= C_{A-}\,(1 + \frac{C_{H+}}{K_a}) \Longrightarrow C_{A-} = \frac{C_{\text{HAo}} + C_{\text{NaAo}}}{\left(1 + \frac{C_{H+}}{K_a}\right)}$$

$$C_{\mathrm{OH-}} = \frac{K_w}{C_{H+}}$$

$$C_{H+} + C_{\mathrm{NaAo}} = \frac{C_{\mathrm{HAo}} + C_{\mathrm{NaAo}}}{\left(1 + \frac{C_{H+}}{K_a}\right)} + \frac{K_w}{C_{H^+}}$$

We can try to solve this analytically:

$C_{\mathbf{HAo}}$=.

$C_{\mathbf{NaAo}}$=.

K_a=.

K_w=.

C_{H+}=.

$$\mathbf{slns2} = \mathbf{Solve}\left[C_{H^+} + C_{\mathbf{NaAo}} == \frac{C_{\mathbf{HAo}} + C_{\mathbf{NaAo}}}{\left(1 + \frac{C_{H^+}}{K_a}\right)} + \frac{K_w}{C_{H^+}}, C_{H^+}\right] //$$

FullSimplify

$\Big\{\Big\{C_{H^+} \to \frac{1}{6}\big(-2\,(C_{\mathrm{NaAo}} + K_a) + \big(2\,2^{1/3}\,((C_{\mathrm{NaAo}} + K_a)^2 + 3\,(C_{\mathrm{HAo}}K_a + K_w))\big)\big)\,/$
$\big(-2C_{\mathrm{NaAo}}^3 - 6C_{\mathrm{NaAo}}^2K_a - 3C_{\mathrm{NaAo}}\,(3C_{\mathrm{HAo}}K_a + 2K_a^2 + 3K_w) +$
$K_a\,(-9C_{\mathrm{HAo}}K_a - 2K_a^2 + 18K_w) + \surd(((C_{\mathrm{NaAo}} + K_a)\,(9C_{\mathrm{HAo}}K_a + 2\,(C_{\mathrm{NaAo}} + K_a)^2) +$
$9\,(C_{\mathrm{NaAo}} - 2K_a)\,K_w)^2 - 4\,((C_{\mathrm{NaAo}} + K_a)^2 + 3\,(C_{\mathrm{HAo}}K_a + K_w))^3)\big)^{1/3} +$
$2^{2/3}\,\big(-2C_{\mathrm{NaAo}}^3 - 6C_{\mathrm{NaAo}}^2K_a - 3C_{\mathrm{NaAo}}\,(3C_{\mathrm{HAo}}K_a + 2K_a^2 + 3K_w) +$
$K_a\,(-9C_{\mathrm{HAo}}K_a - 2K_a^2 + 18K_w) + \surd(((C_{\mathrm{NaAo}} + K_a)\,(9C_{\mathrm{HAo}}K_a + 2\,(C_{\mathrm{NaAo}} + K_a)^2) +$
$9\,(C_{\mathrm{NaAo}} - 2K_a)\,K_w)^2 - 4\,((C_{\mathrm{NaAo}} + K_a)^2 + 3\,(C_{\mathrm{HAo}}K_a + K_w))^3)\big)^{1/3}\big)\Big\},$

$\Big\{C_{H^+} \to \frac{1}{12}\big(-4\,(C_{\mathrm{NaAo}} + K_a) - \big(4(-2)^{1/3}\,((C_{\mathrm{NaAo}} + K_a)^2 + 3\,(C_{\mathrm{HAo}}K_a + K_w))\big)\big)\,/$
$\big(-2C_{\mathrm{NaAo}}^3 - 6C_{\mathrm{NaAo}}^2K_a - 3C_{\mathrm{NaAo}}\,(3C_{\mathrm{HAo}}K_a + 2K_a^2 + 3K_w) +$
$K_a\,(-9C_{\mathrm{HAo}}K_a - 2K_a^2 + 18K_w) + \surd(((C_{\mathrm{NaAo}} + K_a)\,(9C_{\mathrm{HAo}}K_a + 2\,(C_{\mathrm{NaAo}} + K_a)^2) +$
$9\,(C_{\mathrm{NaAo}} - 2K_a)\,K_w)^2 - 4\,((C_{\mathrm{NaAo}} + K_a)^2 + 3\,(C_{\mathrm{HAo}}K_a + K_w))^3)\big)^{1/3} +$
$i2^{2/3}\left(i + \sqrt{3}\right)\big(-2C_{\mathrm{NaAo}}^3 - 6C_{\mathrm{NaAo}}^2K_a - 3C_{\mathrm{NaAo}}\,(3C_{\mathrm{HAo}}K_a + 2K_a^2 + 3K_w) +$

$K_a\left(-9C_{\text{HAo}}K_a-2K_a^2+18K_w\right)+\sqrt{}\left(\left(\left(\left(C_{\text{NaAo}}+K_a\right)\left(9C_{\text{HAo}}K_a+2\left(C_{\text{NaAo}}+K_a\right)^2\right)+\right.\right.\right.$
$\left.\left.\left.9\left(C_{\text{NaAo}}-2K_a\right)K_w\right)^2-4\left(\left(C_{\text{NaAo}}+K_a\right)^2+3\left(C_{\text{HAo}}K_a+K_w\right)\right)^3\right)\right)^{1/3}\Big)\Big\},$

$\Big\{C_{H^+}\to\frac{1}{12}\Big(-4\left(C_{\text{NaAo}}+K_a\right)+\Big(2i2^{1/3}\left(i+\sqrt{3}\right)\left(\left(C_{\text{NaAo}}+K_a\right)^2+3\left(C_{\text{HAo}}K_a+K_w\right)\right)\Big)/$
$\left(-2C_{\text{NaAo}}^3-6C_{\text{NaAo}}^2K_a-3C_{\text{NaAo}}\left(3C_{\text{HAo}}K_a+2K_a^2+3K_w\right)+\right.$
$K_a\left(-9C_{\text{HAo}}K_a-2K_a^2+18K_w\right)+\sqrt{}\left(\left(\left(\left(C_{\text{NaAo}}+K_a\right)\left(9C_{\text{HAo}}K_a+2\left(C_{\text{NaAo}}+K_a\right)^2\right)+\right.\right.\right.$
$\left.\left.\left.9\left(C_{\text{NaAo}}-2K_a\right)K_w\right)^2-4\left(\left(C_{\text{NaAo}}+K_a\right)^2+3\left(C_{\text{HAo}}K_a+K_w\right)\right)^3\right)\right)^{1/3}-$
$2^{2/3}\left(1+i\sqrt{3}\right)\left(-2C_{\text{NaAo}}^3-6C_{\text{NaAo}}^2K_a-3C_{\text{NaAo}}\left(3C_{\text{HAo}}K_a+2K_a^2+3K_w\right)+\right.$
$K_a\left(-9C_{\text{HAo}}K_a-2K_a^2+18K_w\right)+\sqrt{}\left(\left(\left(\left(C_{\text{NaAo}}+K_a\right)\left(9C_{\text{HAo}}K_a+2\left(C_{\text{NaAo}}+K_a\right)^2\right)+\right.\right.\right.$
$\left.\left.\left.9\left(C_{\text{NaAo}}-2K_a\right)K_w\right)^2-4\left(\left(C_{\text{NaAo}}+K_a\right)^2+3\left(C_{\text{HAo}}K_a+K_w\right)\right)^3\right)\right)^{1/3}\Big)\Big\}\Big\}$

Here too, we find three analytical solutions for the cubic equation and two have explicitly imaginary terms within them. We will evaluate all three to see what we can learn.

$\mathbf{C_{HAo} = 0.001;}$

$\mathbf{C_{NaAo} = 0.001;}$

$\mathbf{K_a = 10^{-4.75};}$

$\mathbf{K_w = 10^{-14};}$

C_{H^+} = Evaluate [C_{H^+}/.slns2[[1]]]

−Log10 [C_{H^+}]

0.000017182 + 0.*i*

4.76493 + 0.*i*

C_{H^+} = Evaluate [C_{H^+}/.slns2[[2]]]

−Log10 [C_{H^+}]

−0.00103496 + 3.614007241618348`*^-20*i*

2.98507 − 1.36438*i*

C_{H^+} = Evaluate [C_{H^+}/.slns2[[3]]]

−Log10[C_{H^+}]

−9.999988106098218`*^-12 + 0.i

11. − 1.36438i

The first solution is the valid one and gives a value of the pH of 4.765, which agrees with Butler's number exactly.

We may wonder how the pH will change if we increase the ionic strength and introduce activity coefficients. The activity-based equilibrium constant, K_a^o, is related to the concentration-based constant, K_a, as follows by definition:

$$K_a^o = \frac{C_{H^+} C_{A^-}}{C_{\mathrm{HA}}} \frac{\gamma_+ \gamma_-}{\gamma_o} = K_a \frac{\gamma_+ \gamma_-}{\gamma_o}$$

From which we obtain the result that:

$$\mathrm{p}K_a = \mathrm{p}K_a^o + \log \gamma_+ + \log \gamma_- - \log \gamma_o$$

Focusing on acetic acid with sodium acetate, Butler introduces the Davies equation as the constitutive equation for the single ion activity coefficients:

$$-\log \gamma_z = \mathrm{A}\, z^2 \left(\frac{\sqrt{I}}{1+\sqrt{I}} - 0.2\, \mathrm{I} \right)$$

Hence,

$$\mathrm{p}K_a = 4.7572 - 2 * 0.51 * \left(\frac{\sqrt{I}}{1+\sqrt{I}} - b' I \right) - \mathrm{b\, I}$$

where b is not known. Butler takes b = 0 and b' = 0.2 whereas Davies had b' = 0.2. If we let b' = 0.2, then b must be negative 0.1 for the fit to the experimental data to be adequate.

pKa[I_]:=4.757 − 2 ∗ 0.51 ∗ $\left(\frac{\sqrt{I}}{1+\sqrt{I}} - 0.2I\right) + .1I$

Plot[pKa[I], {I, 0, 4}, AxesLabel → {I, "pKa"}];

ListPlot[{{0, 4.749}, {0.092, 4.566}, {0.523, 4.509},

{1.015, 4.577}, {2.048, 4.806}, {3.014, 5.011},

{4.028, 5.247}}, AxesLabel → {I, "pKa"}];

Show[%, %%]

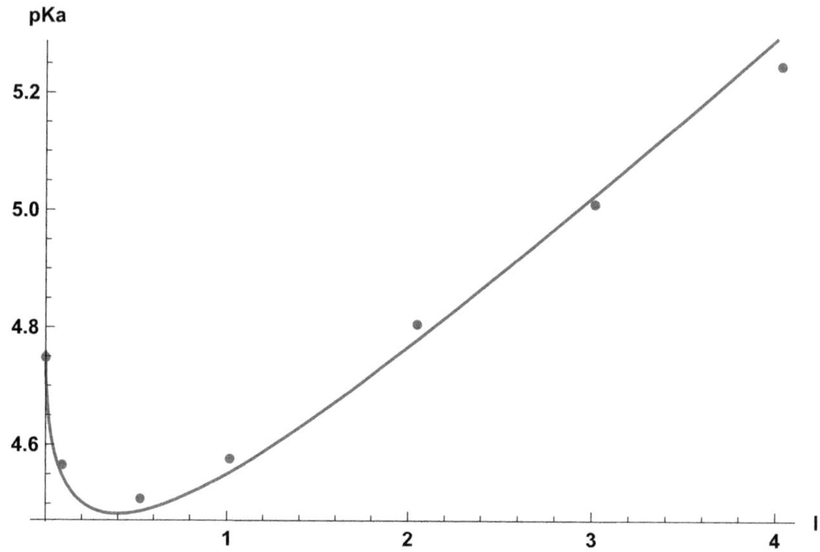

For the interested reader, much more can be learned about ionic equilibria from Butler and Cogley.

The partial molal volume of solution

From a more thermodynamic perspective, we have much that we can learn from the kinds of data and analyses that we have been using. The *partial molal volume* is one of many so-called molal quantities of solutions which form part of solution thermodynamics.

If a solution is made up of more than one component and if the sum of the moles of each component is unity then we have 1 mole of solution:

$$n_1 + n_2 + \ldots = 1$$

The molality of a solution is the quantity given by the number of moles of solute per 1000 g of water, which is equivalent to 55.51 moles of water. The advantage of a using the quantity molality is that, in contrast to concentration (mole of solute per liter of water), it is constant with varying temperature. The partial molal volume is defined by thermodynamics as the rate of change of the total volume of solution with the change in the amount of the solute component i:

$$\left(\frac{\partial V}{\partial n_i}\right)_T = \bar{v}_i$$

As with many thermodynamic quantities this is given as a partial rather than as an exact derivative. What this means physically is that at constant temperature, as we add substance i to solution, the volume of solution may either increase or decrease, that is the quantity $\left(\frac{\partial V}{\partial n_i}\right)_T$ may be either positive or negative, depending upon the nature of the intermolecular forces between it and the other species in solution.

The *apparent molal volume,* ${}^{\emptyset}V$, is related to the partial molal volume. If a solution contains $n_1$moles of solvent and $n_2$moles of a solute such as sodium chloride, then we can define the apparent molal volume as the difference between the solution volume and the equivalent volume of the same number of moles of pure solvent, $n_1 v_1^o$, (where v_1^o is the volume per mole of the pure solvent), divided by the number of moles of solute, n_2:

$$ {}^{\emptyset}V = \frac{V_{\text{soln}} - n_1 v_1^o}{n_2} $$

This is the change in the volume of solution produced by the corresponding moles of solute compared to the same number of moles of solvent. We can compute this from the data we have for NaCl. We can define each quantity as follows for 1cm^3 of salt solution:

$$ V_{\text{soln}} = 1\ \text{cm}^3 $$

$$ n_1 v_1^o = \left(\frac{(1-\text{wtfrsalt})\rho_{\text{soln}} V_{\text{soln}}}{\text{MW}_{\text{water}}}\right)\left(\frac{\text{MW}_{\text{water}}}{\rho_{\text{water}}}\right) = \left(\frac{(1-\text{wtfrsalt})\rho_{\text{soln}} V_{\text{soln}}}{\rho_{\text{water}}}\right) $$

$$ n_2 = \frac{wtfrsalt\,\rho_{\text{soln}} V_{\text{soln}}}{\text{MW}_{\text{salt}}} $$

$$ {}^{\emptyset}V = \frac{1 - \left(\frac{(1-\text{wtfrsalt})\rho_{\text{soln}} V_{\text{soln}}}{\rho_{\text{water}}}\right)}{\frac{\text{wtfrsalt}\,\rho_{\text{soln}} V_{\text{soln}}}{\text{MW}_{\text{salt}}}} = \frac{(\rho_{\text{water}} - (1-\text{wtfrsalt})\rho_{\text{soln}} V_{\text{soln}})\,\text{MW}_{\text{salt}}}{\text{wtfrsalt}\,\rho_{\text{soln}} V_{\text{soln}} \rho_{\text{water}}} $$

The molality of the solution at each weight fraction can be expressed as:

$$ \text{m} = \frac{\text{molessalt}}{1000g\text{water}} = \frac{\left(\frac{\text{wtfrsalt}\,\rho_{\text{soln}} V_{\text{soln}}}{\text{MW}_{\text{salt}}}\right)}{(1-\text{wtfrsalt})\rho_{\text{soln}} V_{\text{soln}}} 1000 $$

$$ = \frac{\text{wtfrsalt}}{(1-\text{wtfrsalt})\text{MW}_{\text{salt}}} 1000 $$

Given the data for sodium chloride, we can compute and plot the apparent molal volume of solution as a function of the molality at each weight fraction and density. Then we can Plot the first 10, 20, and 30 data points, and then the last 30 data points, and make a GraphicsGrid showing the four plots in a two by two format in order to compare the solution behavior in different regions of molality. To do this we can make up a function for partial molal volume that takes as its arguments the data from the set **rawdat**:

```
rawdat = {{.10, 1.0007}, {.20, 1.0014}, {0.3, 1.0021},
{0.4, 1.0028}, {0.5, 1.0036}, {.6, 1.0043},
{.7, 1.0050}, {0.8, 1.0057}, {0.9, 1.0064},
{1.00, 1.0071}, {1.10, 1.0078}, {1.2, 1.0085},
{1.30, 1.0093}, {1.40, 1.0100}, {1.50, 1.0107},
{1.60, 1.0114}, {1.70, 1.0121}, {1.8, 1.0128},
{1.90, 1.0135}, {2.0, 1.0143},
{2.1, 1.0150}, {2.20, 1.0157}, {2.3, 1.0164},
{2.4, 1.0171}, {2.5, 1.0178}, {2.60, 1.0185},
{2.70, 1.0193}, {2.80, 1.0200}, {2.9, 1.0207},
{3.00, 1.0214}, {3.1, 1.0221}, {3.2, 1.0229},
{3.3, 1.0236}, {3.4, 1.0243}, {3.5, 1.025},
{3.6, 1.0257}, {3.7, 1.0265}, {3.8, 1.0272},
{3.9, 1.0279}, {4.0, 1.0286}, {4.1, 1.0293},
{4.2, 1.0301}, {4.3, 1.0308}, {4.4, 1.0315},
{4.5, 1.0322},
{4.6, 1.0330}, {4.7, 1.0337}, {4.8, 1.0344},
{4.9, 1.0351}, {5.0, 1.0358}, {5.2, 1.0373},
{5.4, 1.0387}, {5.6, 1.0402}, {5.8, 1.0417},
{6.0, 1.0431}, {6.2, 1.0446}, {6.4, 1.0460},
{6.6, 1.0475}, {6.8, 1.0490}, {7.0, 1.0504},
{7.2, 1.0519}, {7.4, 1.0533}, {7.6, 1.0548},
{7.8, 1.0563}, {8.0, 1.0578}, {8.2, 1.0592},
```

{8.4, 1.0607}, {8.6, 1.0622}, {8.8, 1.0637},

{9.0, 1.0651}, {9.2, 1.0666}, {9.4, 1.0681},

{9.6, 1.0696}, {9.8, 1.0711}, {10.0, 1.0726},

{10.50, 1.0763}, {11.0, 1.0801}, {11.5, 1.0838},

{12.0, 1.0876}, {12.50, 1.0914}, {13.0, 1.0952},

{13.5, 1.0990}, {14.0, 1.1028}, {14.5, 1.1066},

{15.0, 1.1105}, {16.0, 1.1182}, {17.0, 1.1260},

{18.0, 1.1339}, {19.0, 1.1418}, {20.0, 1.1498},

{21.0, 1.1579}, {22.0, 1.1660}, {23.0, 1.1742},

{24.0, 1.1825}, {25.0, 1.1909}, {26.0, 1.1993}};

ListPlot[rawdat]

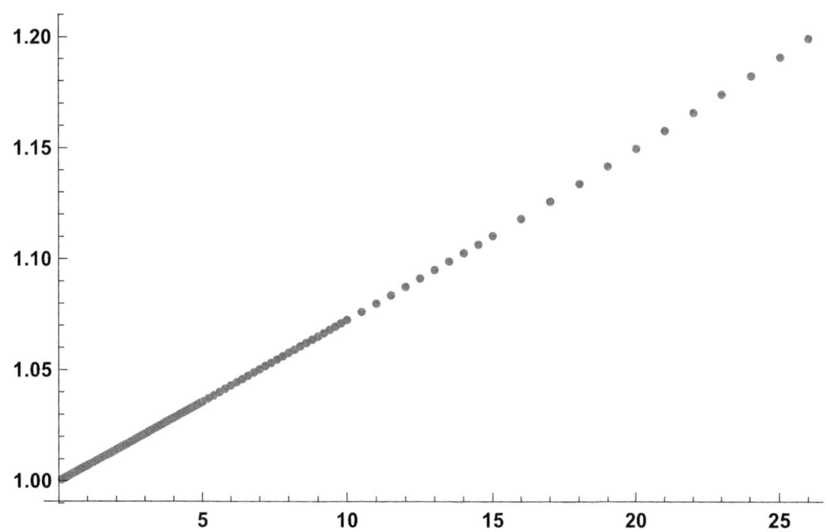

(* Partial Molal Volume *)

$$\phi\text{v[x\_\_]:=}\frac{(1-(1-.01\ x[[n,1]])x[[n,2]])58.44}{.01x[[n,1]]x[[n,2]]}$$

```
(* Molality *)

m[x__]:= .01x[[n, 1]] / ((1 - .01x[[n, 1]])58.44) 1000

(* First we compute the partial molal volume and each
corresponding molality *)

appmolvoldat = Table[{m[rawdat], φv[rawdat]},
{n, 1, Length[rawdat]}];

pl1 = ListPlot[appmolvoldat,
AxesLabel → {"m,mol/1000g", "φV:NaCl-H2O, cm3"},
PlotStyle → PointSize[0.01], Joined → True]
pl2 = ListPlot[appmolvoldat,
AxesLabel → {"m,mol/1000g", "φV:NaCl-H2O, cm3"},
PlotStyle → {Hue[0.1], PointSize[0.01]}]
Show[%, %%]
```

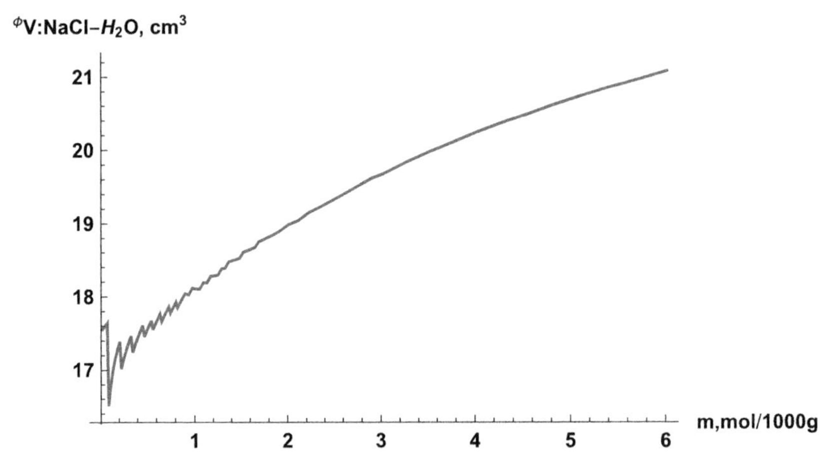

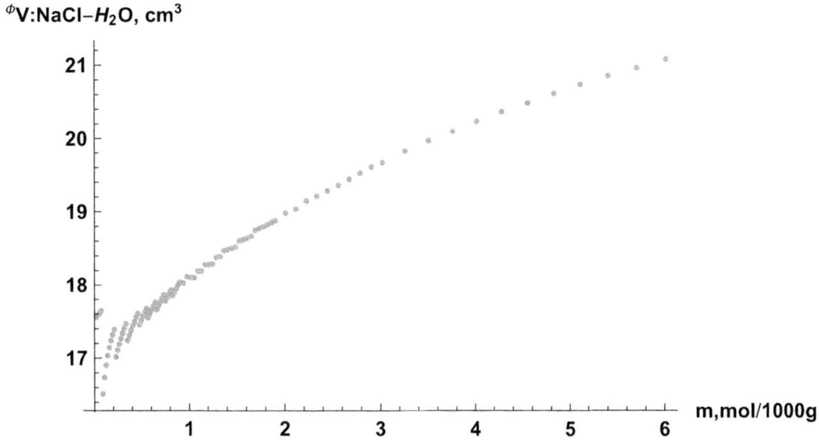

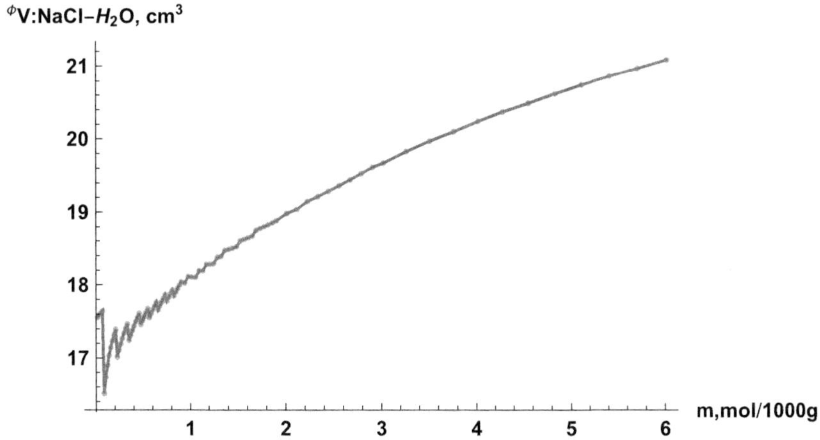

These data really are very interesting. They show that at low molality, the apparent molal volume of the NaCl solution appears to oscillate rather than to increase smoothly. To be sure that this is correct and that we have not made some mistake, we can check two points, say 9 and 10 in the data set. In what follows, we present first the numbers used in the plot for points 9 and 10, and then we have the numbers recomputed individually.

appmolvoldat[[9]](* point 9 from appmolvoldat *)

$$\left\{\left(\frac{.009}{1-.009}\right)\frac{1000}{58.44}, \frac{(1-(1-.009)1.0063)58.44}{(.009)1.0063}\right\}$$

appmolvoldat[[10]](* point 10 from appmolvoldat *)

$$\left\{\left(\frac{.01}{1-.01}\right)\frac{1000}{58.44}, \frac{(1-(1-.01)1.0071)58.44}{(.01)1.0071}\right\}$$

{15.5403, 17.1469}

{0.155403, 17.7881}

{17.2844, 17.2401}

{0.172844, 17.2401}

These numbers all check, so we can be confident that no mistake entered into our work to this point and the oscillations, whether physical or not are in the data. Next, we should plot regions of the data. We can do this by using the Take command to

```
pl2 = ListPlot[Take[appmolvoldat, 10],
AxesLabel → {"m,mol/1000g", "ϕV:NaCl-H2O, cm3"},
PlotStyle → {Hue[0.1], PointSize[0.02]}];

pl3 = ListPlot[Take[appmolvoldat, 50],
AxesLabel → {"m,mol/1000g", "ϕV:NaCl-H2O, cm3"},
PlotStyle → {Hue[0.3], PointSize[0.02]}];

pl4 = ListPlot[Drop[Take[appmolvoldat, 70], 40],
AxesLabel → {"m,mol/1000g", "ϕV:NaCl-H2O, cm3"},
PlotStyle → {Hue[0.6], PointSize[0.02]}];
pl5 =
ListPlot[Drop[appmolvoldat, Length[newdat] – 30],
AxesLabel → {"m,mol/1000g", "ϕV:NaCl-H2O, cm3"},
PlotStyle → {Hue[0.8], PointSize[0.02]}];

Show[
GraphicsGrid[{{pl2, pl3},
{pl4, pl5}}]
]
```

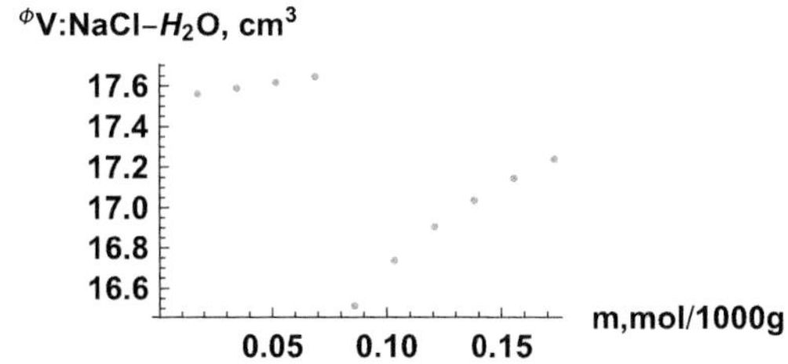

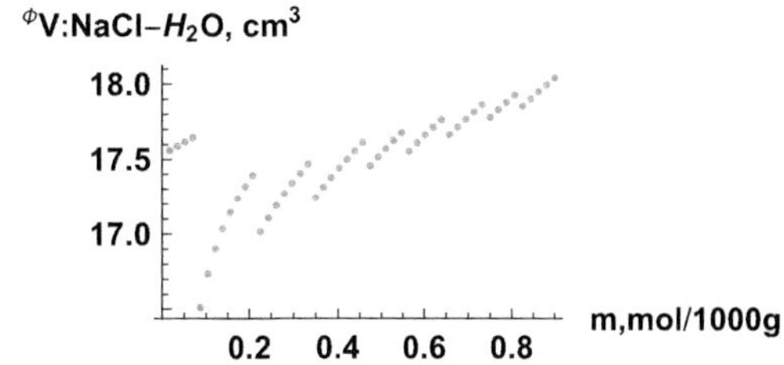

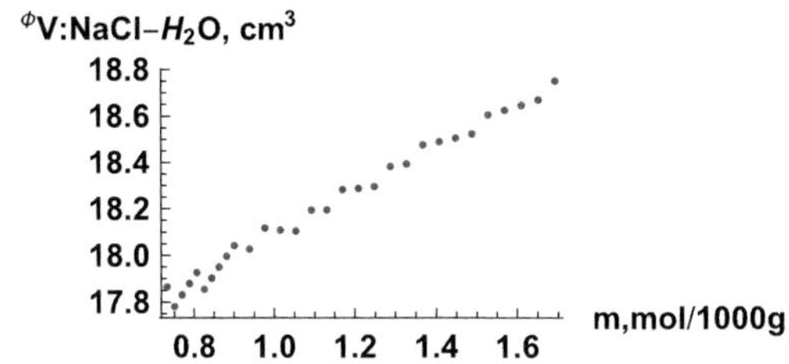

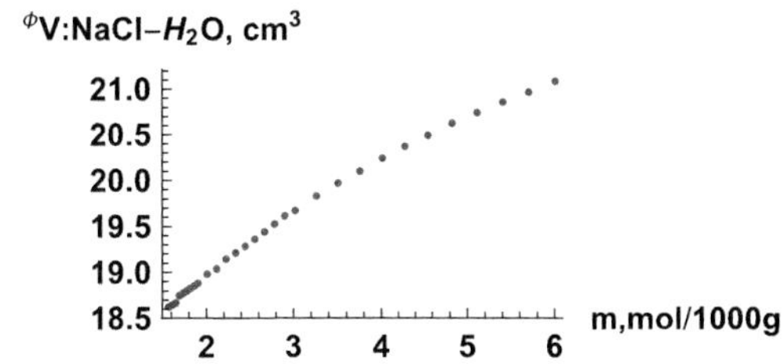

From these plots we see that the apparent molal volume data does oscillate at the lower molalities of sodium chloride. At the lowest levels it rises monotonically up to $\sim$ 17.8 cm^3 until 0.15 molal is reached, then it drops to $\sim$17.2 cm^3. The volume rises again and then falls back repeating this oscillation, but with smaller amplitudes of variation more than ten more times. At a molality of $\sim$2 the oscillations appear to dampen out in magnitude. Above 2 molal the apparent molal volume rises to a maximum of $\sim$21 cm^3. It is well known that water forms hydration spheres around the sodium cation, is it possible that this observation is related to that process? What about other alkali metal cation solutions - do they also do this?

We can go back to the CRC Handbook of Chemistry and Physics to find density versus weight percent solute data for LiCl, KCl, and CsCl and then create files for each of these. We begin with Lithium. We have to replace the MW of NaCl with that of LiCl in the apparent molal volume and molality calculations.

```
lidatset = {{.5, 1.0029}, {1.00, 1.0059}, {2, 1.0117},
{3, 1.0175}, {4, 1.0233}, {5, 1.0290}, {6, 1.0348},
{7, 1.0405}, {8, 1.0463}, {9, 1.0521}, {10, 1.0578},
{12, 1.0694}, {14, 1.0811}, {16, 1.0929},
{18, 1.1048}, {20, 1.1170}, {22, 1.1294},
{24, 1.1419}, {26, 1.1548}, {28, 1.1678},
```

```
{30, 1.1812}};
Length[lidatset];
lithiumappmolvoldat =
Table[{m[lidatset], φv[lidatset]}, {n, 1, Length[lidatset]}]

plLi = ListPlot[lithiumappmolvoldat,
AxesLabel → {"m,mol/1000g", "φV:LiCl-H2O, cm3"},
PlotStyle → {Hue[0.1], PointSize[0.02]}];

plLi2 = ListPlot[lithiumappmolvoldat,
AxesLabel → {"m,mol/1000g", "φV:LiCl-H2O, cm3"},
Joined → True];

Show[%%, %]
```

{{0.118517, 17.8791}, {0.238231, 17.5307}, {0.481325, 17.8829}, {0.72943, 18.0921}, {0.982704, 18.2644}, {1.24131, 18.5011}, {1.50542, 18.635}, {1.77521, 18.8234}, {2.05086, 18.9469}, {2.33257, 19.0706}, {2.62055, 19.2319}, {3.21612, 19.47}, {3.8394, 19.6808}, {4.49236, 19.8742}, {5.17717, 20.0555}, {5.89623, 20.1941}, {6.65215, 20.3184}, {7.44786, 20.4462}, {8.28659, 20.5397}, {9.17191, 20.6414}, {10.1078, 20.719}}

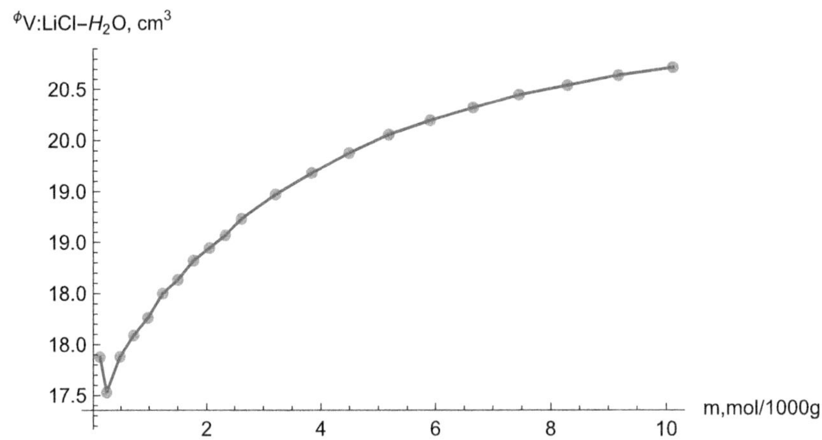

The lithium chloride data is far more sparse than that which we have for sodium chloride. We do notice that the apparent molal volume goes through a minimum and the rises monotonically. Since the data points are so sparse, we cannot determine if the apparent molal volume actually oscillates at the low end of concentration.

Now, we can analyze the data for potassium chloride in the same way, again using the proper molecular weight:

```
kdatset = {{.5, 1.0032}, {1, 1.0064}, {1.5, 1.0096},
{2, 1.0128}, {2.5, 1.0160}, {3, 1.0192},
{3.5, 1.0225}, {4, 1.0257}, {4.5, 1.0289},
{5, 1.0322}, {5.5, 1.0354}, {6, 1.0387},
{6.5, 1.0420}, {7, 1.0453}, {7.5, 1.0486},
{8, 1.0519}, {8.5, 1.0552}, {9, 1.0585},
{9.5, 1.0619}, {10, 1.0652}, {11, 1.0719},
{12, 1.0787}, {13, 1.0855}, {14, 1.0924},
{15, 1.0993}, {16, 1.1063}, {17, 1.1133},
{18, 1.1204}, {19, 1.1276}, {20, 1.1348},
{22, 1.1494}, {24, 1.1643}};
Length[kdatset];
potassiumappmolvoldat =
Table[{m[kdatset], ϕv[kdatset]}, {n, 1, Length[kdatset]}];

plK = ListPlot[potassiumappmolvoldat,
AxesLabel → {"m,mol/1000g", "ϕV:KCl-H2O, cm3"},
PlotStyle → {Hue[0.1], PointSize[0.02]}];

plK2 = ListPlot[potassiumappmolvoldat,
AxesLabel → {"m,mol/1000g", "ϕV:KCl-H2O, cm3"},
Joined → True];

Show[%,%%]
```

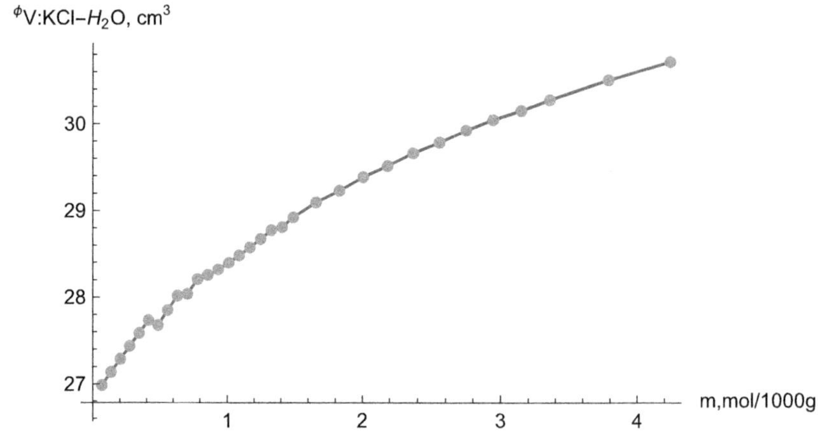

Here again, we have many fewer data at the low end of concentration, and so it is not possible to determine what solution structure there may be, but none is evident. The are three apparent breaks in the dependence of the apparent molal volume and they do appear to be real, but it is hard to be sure that they are.

Finally, we analyze the handbook data for cesium chloride. Here we have 52 data points to work with compared to 21 for Lithium and 32 for potassium and 96 for sodium. With many more points at the low end of concentration we do observe oscillations in the data that do appear to be quite real.

```
csdatset = {{.5, 1.0038}, {1, 1.0076}, {1.5, 1.0114},
{2.0, 1.0153}, {2.5, 1.0192}, {3.0, 1.0232},
{3.5, 1.0271}, {4.0, 1.0311}, {4.5, 1.0352},
{5, 1.0392}, {5.5, 1.0433}, {6.0, 1.0475},
{6.5, 1.0516}, {7, 1.0558}, {7.5, 1.0601},
{8.0, 1.0643}, {8.5, 1.0686}, {9, 1.0730},
{9.5, 1.0773}, {10, 1.0818}, {11, 1.0907},
{12, 1.0997}, {13, 1.1089}, {14, 1.1183},
{15, 1.1278}, {16, 1.1375}, {17, 1.1473},
{18, 1.1573}, {19, 1.1674}, {20, 1.1777},
{22, 1.1989}, {24, 1.2207}, {26, 1.2433},
```

```
{28, 1.2666}, {30, 1.2908}, {32, 1.3158},
{34, 1.3417}, {36, 1.3685}, {38, 1.3963},
{40, 1.4251}, {42, 1.4550}, {44, 1.4861},
{46, 1.5185}, {48, 1.5522}, {50, 1.5874},
{52, 1.6241}, {54, 1.6625}, {56, 1.7029},
{58, 1.7453}, {60, 1.7900}, {62, 1.8373},
{64, 1.8875}};
Length[csdatset];
cesiumappmolvoldat =
Table[{m[csdasett], ϕv[csdatset]}, {n, 1, Length[csdatset]}];

plCs = ListPlot[cesiumappmolvoldat,
AxesLabel → {"m,mol/1000g", "ϕV:CsCl-H2O, cm3"},
PlotStyle → {Hue[0.1], PointSize[0.01]}];

plCs2 = ListPlot[cesiumappmolvoldat,
AxesLabel → {"m,mol/1000g", "ϕV:CsCl-H2O, cm3"},
Joined → True];

Show[%, %%]
```

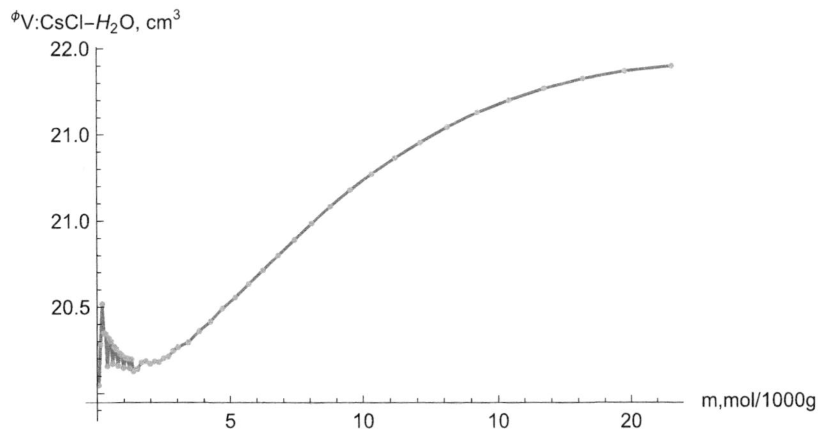

To see this behavior better, we can examine just those points up to and including a molality of 4. The points have been joined together for clarity, so that the variations and oscillations are made most evident.

plCs02 = ListPlot[Take[cesiumappmolvoldat, 30],

AxesLabel → {"m,mol/1000g", "$^{\phi}$V:CsCl-H_2O, cm^3"},

PlotStyle → {Hue[0.1], PointSize[0.02]}];

plCs22 = ListPlot[Take[cesiumappmolvoldat, 30],

AxesLabel → {"m,mol/1000g", "$^{\phi}$V:CsCl-H_2O, cm^3"},

PlotStyle → PointSize[0.02],

PlotJoined → True];

Show[%%, %]

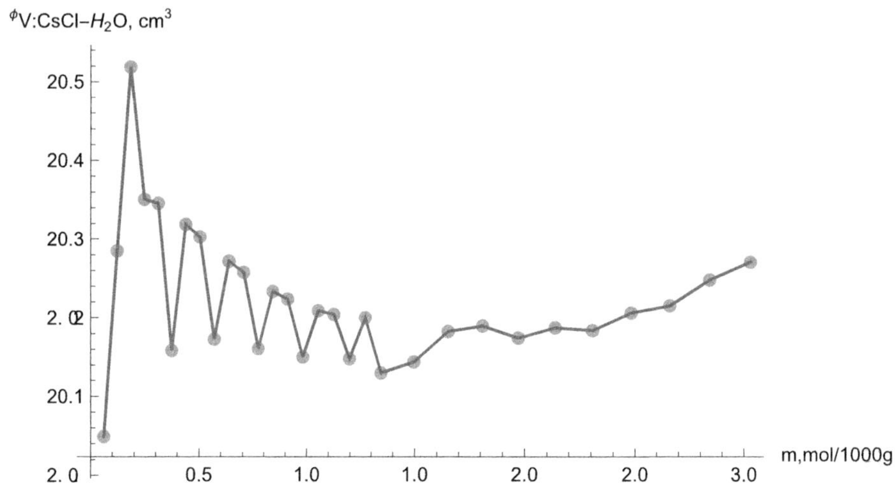

In summary, the partial molal volumes for sodium and cesium show some interesting apparent behavior at the lower end of measured molalities. If these oscillations in the data are real, they may suggest a structuring of water around the ions in solution. More work would need to be done to gain a deeper understanding of whether these are merely data artifacts that result from how the experiments were done, or if they are real.

The design of an optimal CSTR

Several questions arise with respect to the design of an optimal CSTR for a given chemistry. Chief among these are several equations that relate conversion, cost, and profitability to each other. The volume of the CSTR, as we can plainly see, controls the extent of conversion. So the magnitude of the volume goes up with the requirement of conversion. It would seem, then, that we simply might want the largest volume reactor that we can build since this will provide the highest conversion of reactant to product. But it is self evident that such logic is highly flawed since the cost of the reactor must scale with its size. Therefore how do we decide what level of conversion and reactor size is appropriate? The value of the product relative to that of the reactant must be a critical factor. We can do an analysis of the following reaction to see how this works:

$$\mathrm{A} \longrightarrow \mathrm{B}$$

$$r_{A-} = r_{B+} = \mathrm{k}\, C_A$$

The steady state CSTR equations for component A and B are as follows:

$$(C_{\mathrm{Af}} - C_A)\,\mathrm{q} - \mathrm{k}\, C_A \mathrm{V} = 0$$

$$-\, C_B\,\mathrm{q} + \mathrm{k}\, C_A \mathrm{V} = 0$$

The inlet concentration, $\boldsymbol{C_{\mathrm{Af}}}$, is a good reference point for reaction. We can normalize the equations by dividing through by $\boldsymbol{C_{\mathrm{Af}}}$:

$$\left(1 - \frac{C_A}{C_{\mathrm{Af}}}\right)\mathrm{q} - \mathrm{k}\, C_A \mathrm{V} = 0$$

$$-\frac{C_B}{C_{\mathrm{Af}}}\mathrm{q} + \mathrm{k}\,\frac{C_A}{C_{\mathrm{Af}}}\mathrm{V} = 0$$

The conversion of **A** can be written as $\left(\mathbf{1} - \dfrac{\boldsymbol{C_A}}{\boldsymbol{C_{\mathrm{Af}}}}\right)$ and $\boldsymbol{C_A} = \boldsymbol{C_{\mathrm{Af}}}\,(\mathbf{1} - \boldsymbol{X_A})$. Hence the equation for **A** can be rewritten in terms of $\boldsymbol{X_A}$:

$$X_A\,\mathrm{q} - \mathrm{k}\,\mathrm{V}\,(1 - X_A) = 0$$

$$X_A - \mathrm{k}\,\frac{V}{q}\,(1 - X_A) = 0$$

$$X_A - \mathrm{k}\,\theta\,(1 - X_A) = 0$$

Theta, is the holding time in the CSTR. We can see that then we can solve for it in terms of the conversion and the first order rate constant. The inverse of this rate constant has dimensions of time.

$$\theta = \frac{X_A}{k\,(1 - X_A)}$$

$$\mathrm{V} = \frac{X_A}{k\,(1 - X_A)}\mathrm{q}$$

We can immediately see the direct relationship between the volume of the reactor and the conversion of **A**. To make this absolutely clear, we can rearrange one more time to give:

$$\frac{1}{1 + \frac{q}{kV}} = X_A$$

When we consider the extremes we get a feeling for the behavior of this relationship:

$$\mathrm{V} \longrightarrow 0;\ \frac{q}{kV} \longrightarrow \infty\,;\ \frac{1}{1+\infty} \longrightarrow 0 \longrightarrow X_A$$

$$\mathrm{V} \longrightarrow \infty;\ \frac{q}{kV} \longrightarrow 0\,;\ \frac{1}{1+0} \longrightarrow 1 \longrightarrow X_A$$

On this basis we can see that all we really have learned is what we already knew intuitively–making the reactor as large as possible will provide the highest possible conversion. So we really need a better measure of our objective than simply the conversion.

When a chemical engineer balances an equation it first must be done from the perspective of stoichiometry, but, second on the basis of value and profit. So for a given reaction:

$$\mathrm{a\,A + b\,B \longrightarrow d\,D + e\,E}$$

As in chapter 8 we showed that the maximum profit potential is:

$$\text{max. profit potential} = [(\mathrm{d}\,\frac{\$}{\mathrm{mole}D} + \mathrm{e}\,\frac{\$}{\mathrm{mole}E}) - (\mathrm{a}\frac{\$}{\mathrm{mole}A} + \mathrm{b}\,\frac{\$}{\mathrm{mole}B})]$$

The conversion dictates how much product will be made and how much of the starting material will be left and so this dictates the value of the mixture. For the simpler case that we are considering we can write:

$$\text{max. profit potential} = [(\mathrm{b}\,\frac{\$}{\mathrm{mole}D}) - (\mathrm{a}\frac{\$}{\mathrm{mole}A})]$$

$$= [\,(C_b q \frac{\$}{\text{mole}B} t_{\text{prj}}) - (C_a q \frac{\$}{\text{mole}A} t_{\text{prj}})\,]$$

$$= [\,(C_b q \frac{\$}{\text{mole}B} t_{\text{prj}}) - (C_a q \frac{\$}{\text{mole}A} t_{\text{prj}})\,]$$

$$= [\,(C_{\text{af}} X_A q \frac{\$}{\text{mole}B} t_{\text{prj}}) - (C_{\text{af}}\, q\, (1 - X_A) \frac{\$}{\text{mole}A} t_{\text{prj}})\,]$$

$$= C_{\text{af}}\, q\, t_{\text{prj}} [\,(X_A \frac{\$}{\text{mole}B}) - (1 - X_A) \frac{\$}{\text{mole}A}]$$

$$= C_{\text{af}}\, q\, t_{\text{prj}} [\,X_A \frac{\$}{\text{mole}B} - \frac{\$}{\text{mole}A} + X_A \frac{\$}{\text{mole}A}\,]$$

$$= C_{\text{af}}\, q\, t_{\text{prj}} [\,X_A \frac{\$}{\text{mole}B} + X_A \frac{\$}{\text{mole}A} - \frac{\$}{\text{mole}A}\,]$$

$$= C_{\text{af}}\, q\, t_{\text{prj}} [\,X_A (\frac{\$}{\text{mole}B} + \frac{\$}{\text{mole}A}) - \frac{\$}{\text{mole}A}\,]$$

The maximum profit potential for a chemical reaction is only a crude, zeroth-order measure of value. To gain a better measure of the economics, we must have a fuller analysis of the process. The maximum profit potential can never be achieved since it costs money to invest in the process hardware and it costs money to operate the units. If we simply were to try and maximize the potential profit by maximizing the conversion X_A, then we would have to have a reactor of infinite volume:

$$X_A \longrightarrow 1$$

$$\Rightarrow \frac{q}{kV} \longrightarrow 0$$

$$\Rightarrow \text{V} \longrightarrow \infty$$

However, as V $\longrightarrow \infty$ the cost of the reactor also goes to infinity and the net profit must go to zero. Hence the net profit must be a better measure of value generation and this must consist of at least the cost of the reactor and its operation in addition to the cost of reactant and value of the product.

\$ Net Profit ={ \$ Value of Product - \$ Cost of Reactant - \$ Investment in Reactor -

\$ Cost of Operation }

This is a simplified accounting statement (there are many other costs to account for) that says in general, the net rate of accumulation is simply the difference between the rate input minus the rate output:

$$\text{Rate of Accumulation} = \{ \text{Rate of Input - Rate of Output} \}$$

The individual components of this equation for profit can be written in terms of a preset time over which we will evaluate the project, $\boldsymbol{t_p}$, and the relevant parameters and variables of the problems. Let's assume we are still considering the same simple reaction as before and the reactor is a steady state CSTR, then we have the following terms:

$$\$\text{ Value of Product} = C_B\, q \frac{\$}{\text{mol}B}\, t_p$$

$$\$\text{ Cost of Reactants} = C_{\text{Af}}\, q \frac{\$}{\text{mol}A}\, t_p$$

$$\$\text{ Investment in Reactor} = V_{\text{reactor}}\alpha$$

$$\$\text{ Cost of Reactor Operation} = t_p\, \beta$$

$$\$\text{ Net Profit} = C_B\, q \frac{\$}{\text{mol}B}\, t_p - C_{\text{Af}}\, q \frac{\$}{\text{mol}A}\, t_p - V_{\text{reactor}}\alpha - t_p\, \beta$$

$$C_B = C_{\text{Af}}\, X_A \qquad V_{\text{reactor}} = \frac{q X_A}{k(1 - X_A)}$$

$$\$\text{ Net Profit} = C_{\text{Af}} X_A\, q \frac{\$}{\text{mol}B}\, t_p - C_{\text{Af}}\, q \frac{\$}{\text{mol}A}\, t_p - \frac{q X_A}{k(1 - X_A)}\alpha - t_p\, \beta$$

$$\$\text{ Net Profit} = C_{\text{Af}}\, q\, t_p \left(X_A \frac{\$}{\text{mol}B} - \frac{\$}{\text{mol}A} - \frac{\beta}{C_{\text{Af}}q}\right) - \frac{\alpha q X_A}{k(1 - X_A)}$$

This equation now allows us to compute an optimum in profit as a function of the conversion, or in other words as a function of the reactor size or holding time.

```
netprofit[Caf _,q_,t_,$a_,$b_,k_,$v_,$t_,xa_]:=
```

$$\mathbf{Caf\,q\,t}\left(\left(\frac{\mathbf{xa}}{\mathbf{\$b}}\right) - \frac{\mathbf{1}}{\mathbf{\$a}}\right) - \frac{q(\mathbf{xa})}{k(1 - \mathbf{xa})\mathbf{\$v}} - \frac{t}{\mathbf{\$t}}$$

```
netprofit[Caf, q, t, $a, $b, k, $v, $t, x]

SetOptions[{Plot}, AxesStyle->{Thickness[0.01]},

PlotStyle->{Thickness[0.006]},

DefaultFont → {"Helvetica", 17}];
```

$$\text{Caf q t}\left(-\frac{1}{\$\text{a}}+\frac{x}{\$\text{b}}\right)-\frac{t}{\$\text{t}}-\frac{qx}{k(1-x)\$\text{v}}$$

Caf = 1;

$q = 10$;

$t = 100$;

\$a = 1;

\$b = .1;

$k = 0.01$;

\$v = 10;

\$t = 50;

```
Plot[netprofit[Caf, q, t, $a, $b, k, $v, $t, xa],
{xa, 0.01, .99}, AxesOrigin → {0, 0},
PlotStyle → Hue[0.5], AxesLabel → {t, "$[t]"}]
```

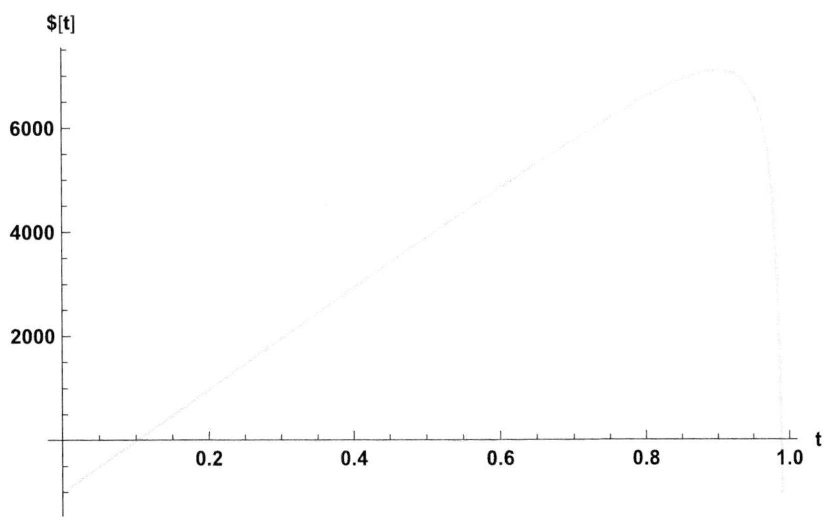

Caf = 1;

q=.

t = 100;

$a = 1;

$b = .1;

k=.

$v = 10;

$t = 50;

xa=.

b = Table[netprofit[Caf, 10^6, *t*, $a, $b, 10^−*n*, $v, $t, xa],

{*n*, 0, 6, 1}]; %;

Plot[{*b*[[1]], *b*[[2]], *b*[[3]], *b*[[4]]}, {xa, 0, .999},

AxesLabel → {t, "$[t]"}]

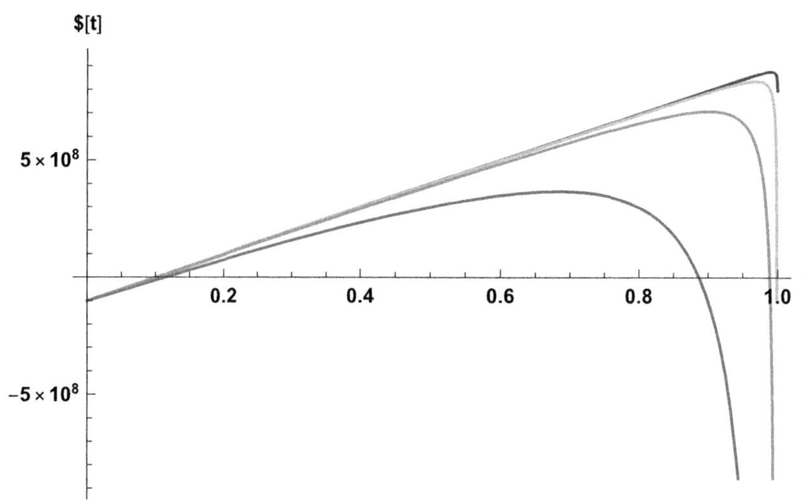

These curves show a clear optimum with respect to the holding time and parametric in the rate constant. This is a very simple model for an optimization, but it provides a basis for doing more thorough modeling.

A linear optimization

The Nittany Lion Pharmaceutical company is about to make two therapeutics, BBG1 and CCS3. The selling prices for BBG1 and CCS3 are \$ 300 per pound and \$ 650 per pound. Both BBG1 and CCS3 are made from the same four starting materials, toluene, chrysene, sulfur, and hydrazine, but they are used in different quantities. (This is clearly a made-up example.) It is known that 5 lbs of toluene, 3 lbs of chrysene, 4 lbs of sulfur, and 8 lbs of hydrazine are used to make 20 lbs of BBG1. It is also the case that 5 lbs of toluene, 5 lbs of chrysene, 3 lbs of sulfur, and 2 lbs of hydrazine are needed to make 15 lbs of CCS3. NLP Co. has in its inventory 50,000 lbs of toluene, 30,000 lbs of chrysene, 33,000 lbs of sulfur, and 45,000 lbs of hydrazine. These materials will be made in two batches and it is critical to maximize the gross profit on the combined batches. You have just been hired at NLP Co. with your degree from Lion Tech and the plant manager wants you to recommend how much of BBG1 and CCS3 to make, given that the reactants are all available now and have been paid for. How much of each should be made and can you justify the amounts you recommend making?

Let's collect what we really know:

$$\text{BBG1} = \$300 / \text{lb}$$

$$\text{CCS3} = \$650 / \text{lb}$$

How do we know that the following is true?

$$5 \text{ lb Tl} + 3 \text{ lb Chr} + 4 \text{ lb S} + 8 \text{ lb Hy} = 20 \text{ lb BBG1}$$

$$5 \text{ lb Tl} + 5 \text{ lb Chr} + 3 \text{ lb S} + 2 \text{ lb Hy} = 15 \text{ lb CCG3}$$

Conservation of mass.

$$\text{Tl}_{\text{max}} = 50{,}000 \text{ lb}$$

$$\text{Chr}_{\text{max}} = 30{,}000 \text{ lb}$$

$$S_{\text{max}} = 33{,}000 \text{ lb}$$

$$\text{Hy}_{\text{max}} = 45{,}000 \text{ lb}$$

Let x represent the number of pounds of BBG1 we will produce and y represents the number of pounds of CCS3 that will be produced. The gross profit, call it k, will be given as follows:

$$\text{Gross Profit} = k = x \text{ lbs BBG1 } \left(\frac{\$300}{\text{lbBBG1}}\right) + y \text{ lb CCS3 } \left(\frac{\$650}{\text{lbCCS3}}\right)$$

What else can we say? We can get the amounts of each reactant in pounds per reactant per pound of product from the previous statements for the masses of reactant to be used for each.

$$\frac{5\text{lbTl} + 3\text{lbChr} + 4\text{lb}S + 8\text{lbHy}}{20\text{lbBBG1}} = 1 \text{ lb BBG1}$$
$$= \frac{0.15\text{Chrlb}}{\text{BBG1lbs}} + \frac{0.4\text{Hylb}}{\text{BBG1lbs}} + \frac{0.2\text{lb}S}{\text{BBG1lbs}} + \frac{0.25\text{lbTl}}{\text{BBG1lbs}}$$

$$\frac{5\text{lbTl} + 5\text{lbChr} + 3\text{lb}S + 2\text{lbHy}}{15\text{lbCCG3}} = 1 \text{ lb CCG3}$$
$$= \frac{0.333333\text{Chrlb}}{\text{CCG3lbs}} + \frac{0.133333\text{Hylb}}{\text{CCG3lbs}} + \frac{0.2\text{lb}S}{\text{CCG3lbs}} + \frac{0.333333\text{lbTl}}{\text{CCG3lbs}}$$

We can also say that the total amount of each reactant that we ultimately use to make the two products must be less than the total amount of that component which we have in the inventory for example:

$$\text{Total Pounds of Chr Used in Making BBG1 and CCG3} \leq 100{,}000 \text{ lb}$$

But how can we express the total number of pounds of each reactant to be used based on the amounts of BBG1 and CCG3 we will produce, i.e. in terms of x and y? We can do so by realizing that we know how many pounds of each reactant are used to make each product and we know that this can never exceed the amount of reactant which we have:

$$\frac{0.1\text{Chrlb}}{\text{BBG1lbs}}\text{x} + \frac{0.33\text{Chrlb}}{\text{CCG3lbs}}\text{y} \leq 30{,}000 \text{ lb of Chrysene}$$
$$\frac{0.4\text{Hylb}}{\text{BBG1lbs}}\text{x} + \frac{0.13\text{Hylb}}{\text{CCG3lbs}}\text{y} \leq 45{,}000 \text{ lb of Hydrazine}$$
$$\frac{0.2\text{lb}S}{\text{BBG1lbs}}\text{x} + \frac{0.2\text{lb}S}{\text{CCG3lbs}}\text{y} \leq 33{,}000 \text{ lb of Sulfur}$$
$$\frac{0.25\text{lbTl}}{\text{BBG1lbs}}\text{x} + \frac{0.33\text{lbTl}}{\text{CCG3lbs}}\text{y} \leq 50{,}000 \text{ lb of Toluene}$$

These inequalities become constraints on the values of x and y we can produce and hence on our gross profit. So the equation we must solve is:

$$\text{Gross Profit} = \text{k} = \text{x lbs BBG1} \left(\frac{\$300}{\text{lbBBG1}}\right) + \text{y lb CCS3} \left(\frac{\$650}{\text{lbCCS3}}\right)$$

but subject to the constraints. The process of finding this maximum profit subject to the constraints is often called a Linear Program and *Mathematica* allows you to find this solution with a command called ConstrainedMax for constrained maximization (or minimization). Even if we did not use a program like *Mathematica* we can solve this problem graphically, since we have just two variables {x,y}. The solution is a two dimensional map.

To do this graphically we will plot each of the constraints as equations in x, y space to find the region of allowed solutions - that is allowed values of x and y, the pounds produced of BBG1 and CCG3. So in {x,y} space we need to plot the constraints as equations:

$$\frac{0.1\text{Chrlb}}{\text{BBG1lbs}}\text{x} + \frac{0.25\text{Chrlb}}{\text{CCG3lbs}}\text{y} = 30{,}000\text{ lb}$$
$$\frac{0.4\text{Hylb}}{\text{BBG1lbs}}\text{x} + \frac{0.13\text{Hylb}}{\text{CCG3lbs}}\text{y} = 45{,}000\text{ lb}$$
$$\frac{0.2\text{lb}S}{\text{BBG1lbs}}\text{x} + \frac{0.2\text{lb}S}{\text{CCG3lbs}}\text{y} = 33{,}000\text{ lb}$$
$$\frac{0.25\text{lbTl}}{\text{BBG1lbs}}\text{x} + \frac{0.33\text{lbTl}}{\text{CCG3lbs}}\text{y} = 50{,}000\text{ lb}$$

Plotting the constraint equations is more easily seen, if we rearrange the equations into y = m x + b form. We can use the Solve command to do this as follows:

Clear[*y*, *x*]

Solve[0.1 *x* + 0.25*y* == 30000, *y*]

Solve[0.4*x* + 0.13*y* == 45000, *y*]

Solve[0.2*x* + 0.2*y* == 33000, *y*]

Solve[0.25*x* + 0.33*y* == 50000, *y*]

Solve[300*x* + 650*y* == *k*, *y*]//*N*

$\{\{y \to 4.(30000. - 0.1x)\}\}$

$\{\{y \to 7.69231(45000. - 0.4x)\}\}$

$\{\{y \to 5.(33000. - 0.2x)\}\}$

$\{\{y \to 3.0303(50000. - 0.25x)\}\}$

{{y → 0.00153846(k − 300.x)}}

```
xm = 120000;
a = Plot["4."("30000." − "0.1"x), {x, 0, xm},
PlotStyle → Dashed, AxesLabel → {x, "y[x]"}];
b = Plot["7.69"("45000." − "0.4"x), {x, 0, xm}];
c = Plot["5."("33000." − "0.2"x), {x, 0, xm},
PlotStyle → Dashed];
d = Plot["3.03"("50000." − "0.25"x), {x, 0, xm}];

e1 = Plot["0.00154" (5 10^7 − "300."x), {x, 0, xm},
PlotStyle → Hue[0]];
e2 = Plot["0.00154" (7.5 10^7 − "300."x), {x, 0, xm},
PlotStyle → Hue[0]];
e3 = Plot["0.00154" (10^8 − "300."x), {x, 0, xm},
PlotStyle → Hue[0]];
e4 = Plot["0.00154" (9 10^7 − "300."x), {x, 0, xm},
PlotStyle → Hue[0]];
e5 = Plot["0.00154" (8 10^7 − "300."x), {x, 0, xm},
PlotStyle → Hue[0]];
e6 = Plot["0.00154" (8.1 10^7 − "300."x), {x, 0, xm},
PlotStyle → Hue[0.4]];
Show[a, c, e6, e5,
Graphics[{RGBColor[0, 0, 1],
Line[{{75000, 0}, {75000, 90000}}]}],
Graphics[{RGBColor[0, 0, 1],
Line[{{0, 90000}, {75000, 90000}}]}]]
```

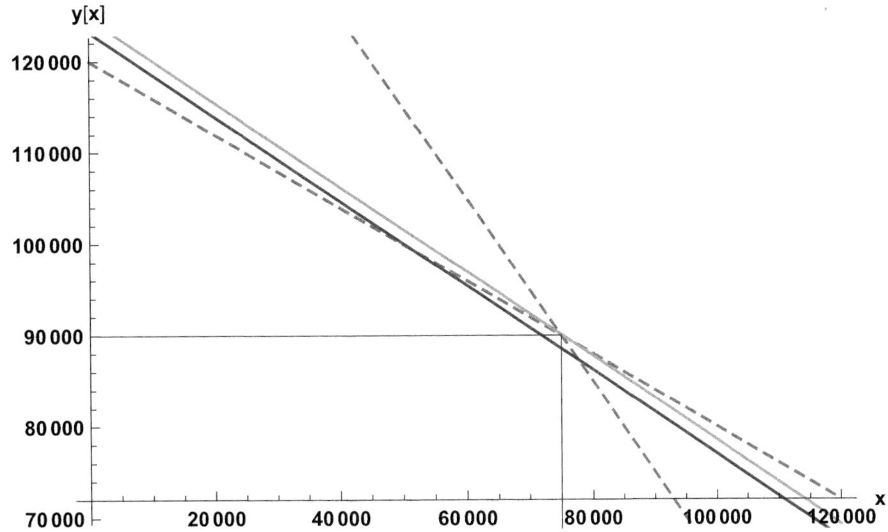

The solution point is the point that is at the vertex of intersection between the profit (or objective function) line, the constraint lines at the highest point along the upper boundary of the region of allowed solutions. In this case that point is found at x = 75000 and y = 90000 pounds.

We don't need to solve this problem graphically. Instead for these problems and those with more than two variables we can use the function called ConstrainedMax. This function calls for the objective function equation and then the constraints. It is rather simple to find this maximum directly:

ConstrainedMax[300x + 650y,

{0.1x + 0.25y < 30000, 0.4x + 0.13y < 45000,

0.2x + 0.2y < 33000, 0.25x + 0.33y < 50000},

{x, y}]

$\{8.1 \times 10^7, \{x \to 75000., y \to 90000.\}\}$

We can check these results by setting them equal to x and y and then evaluating the constraints and the objective function as follows:

x = 75000;

y = 90000;

0.1x + 0.25y

$\mathbf{0.4x + 0.13y}$

$\mathbf{0.2x + 0.2y}$

$\mathbf{0.25x + 0.33y}$

$\mathbf{300x + 650y}$

30000.

41700.

33000.

48450.

81000000

Linear or non-linear?

In classical mechanics, linearization is introduced in the context of the equation of motion for the undamped pendulum. The equation is well-known and is covered in many different texts on mathematics as well as mechanics. The analysis of the angular momentum, which is conserved, leads to the equation of motion:

$$\partial_{t,t}\theta[t] = -\frac{g}{l}\text{Sin}[\theta[t]]$$

This is a simple equation with a long history. The solution of this equation is difficult because it is non-linear. If we try to solve it analytically with DSolve, it does not work.

$$\textbf{DSolve}\left[\left\{\partial_{t,t}\theta[t] == -\frac{g}{l}\textbf{Sin}[\theta[t]], \theta[0] == \theta_o\right\},\right.$$
$$\left.\theta[t], t\right]$$

$$\text{DSolve}\left[\left\{\theta''[t] == -\frac{g\text{Sin}[\theta[t]]}{l}, \theta[0] == \theta_o\right\}, \theta[t], t\right]$$

One way around this problem, is to linearize the equation. It is realized that if the angle of the motion is not too large, then to a first approximation, the Sine[θ] $\approx \theta$, and the equation becomes:

$$\partial_{t,t}\theta[t] == -\frac{g}{l}\,\theta[t]$$

This is a linear equation and is readily solved:

Clear[t, g, l, θ, θo]

DSolve$\left[\left\{\partial_{t,t}\theta[t] == -\frac{g}{l}\,\theta[t], \theta[0] == \theta\text{o}, \theta'[0] == 0\right\}\right.$,

$\theta[t], t]$

$$\left\{\left\{\theta[t] \to \theta\text{oCos}\left[\frac{\sqrt{g}t}{\sqrt{l}}\right]\right\}\right\}$$

and the solution is a simple and well-behaved function that we can see gives us oscillatory results:

θ**[t_]:=**θ**oCos**$\left[\frac{\sqrt{g}t}{\sqrt{l}}\right]$

$g = 9.8;$

$l = 1;$

$\theta\text{o} = 10;$

Plot[$\theta[t]$, {t, 0, 2}, AxesLabel → {t, "θ[t]"}]

dat = Table[$\theta[t]$, {t, 0, 2, .01}];

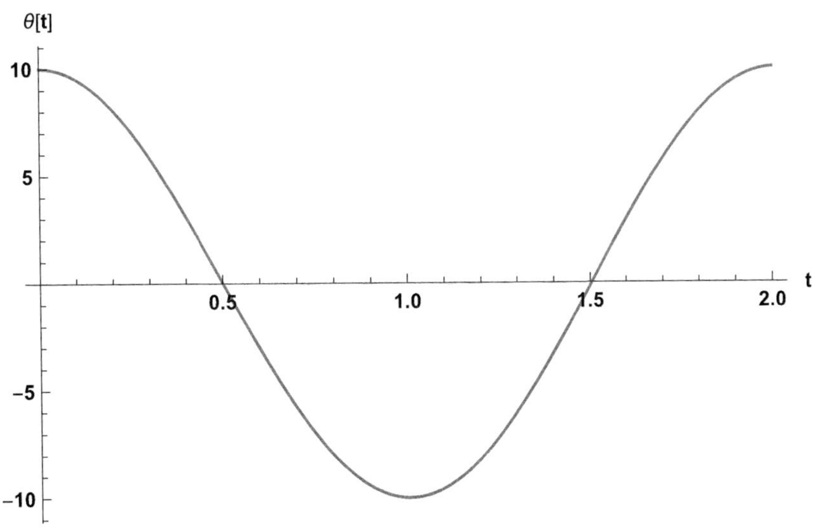

We can determine the x and y positions of the pendulum to plot the trajectory for an given set of constants. The analysis shows that this leads to elliptical trajectories.

```
Transpose[
Partition[
Join[lSin/@dat, lCos/@dat], Length[dat]]];
ListPlot[%, PlotRange → All]
```

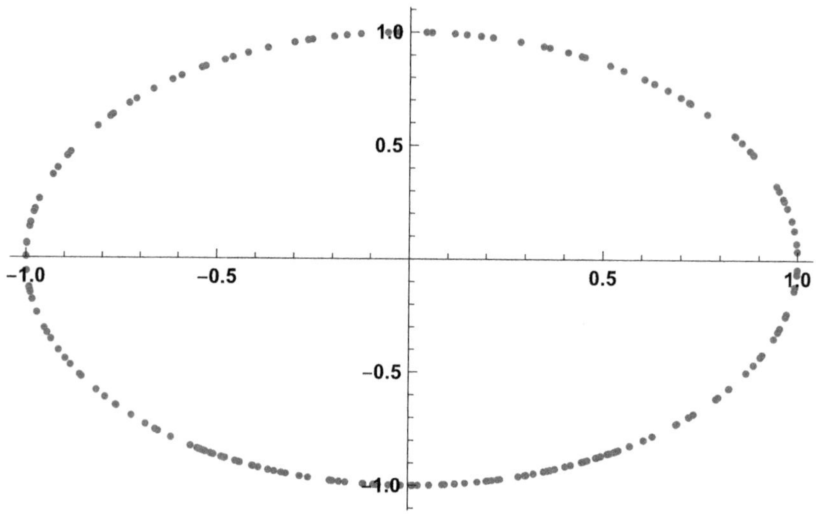

The linearized, undamped pendulum is a simplification of the real problem. We would rather not make the assumption or limit the results to small θ. Instead, while we cannot solve the equation analytically, we can solve it numerically and then compare the results to the linearized case:

```
Clear[t, g, l, θ]
g = 9.8;
l = 1;
θo = π/2;
θlin =
First[
θ/.NDSolve[{θ"[t] == -g/l θ[t], θ[0] == θo,
```

θ'[0] == .6}, θ, {t, 0, 10}]];

lin = Plot[θlin[t], {t, 0, 10}, PlotStyle → Hue[0]];

θnonlin =

First[

θ/.NDSolve$\left[\left\{\theta''[t] == -\frac{g}{l}\ \text{Sin}[\theta[t]], \theta[0] == \theta o,\right.\right.$

θ'[0] == .6}, θ, {t, 0, 10}]];

nonlin = Plot[θnonlin[t], {t, 0, 10},

AxesLabel → {t, "θ[t]"}];

Show[nonlin, lin]

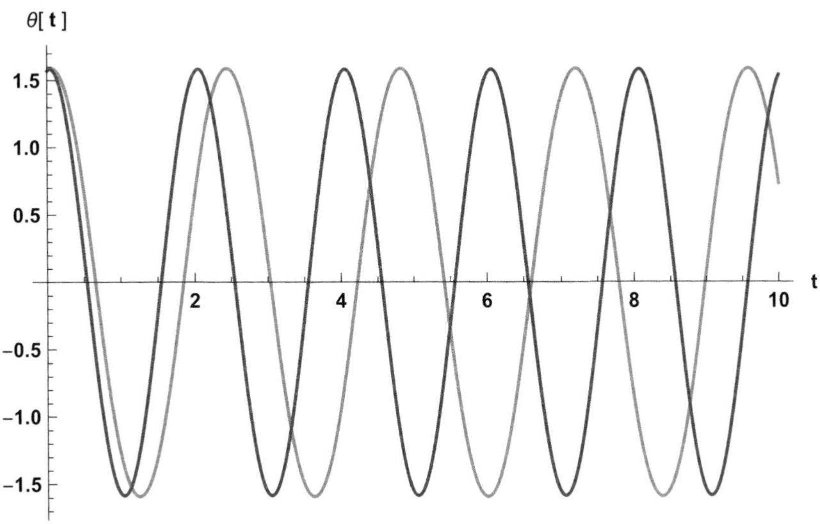

The red (mid gray in the print version) represents the linear pendulum, while the blue (dark gray in the print version) illustrates the effect of the nonlinear terms. It is typical to consider the effect of damping on the pendulum's behavior. Consideration of the angular momentum and the introductions of a damping force that is proportional to the angular velocity leads to these equations of motion:

$$\text{ml}^2\partial_{t,t}\theta[t] = -cl\partial_t\theta[t] - mgl\sin[\theta]$$

$$\partial_{t,t}\theta[t] + \frac{c}{ml}\partial_t\theta[t] + \frac{g}{l}\sin[\theta] = 0$$

```
Clear[t, g, l, θ]

g = 9.8;

l = 1;

m = 1;

c = 0.3;

θo = π/2;

tmax = 15;


θdnln =

First[

θ/.NDSolve[{θ''[t] + c/(m l) θ'[t] + g/l Sin[θ[t]] == 0,

θ[0] == θo, θ'[0] == 0}, θ, {t, 0, tmax}]]

dnln = Plot[θdnln[t], {t, 0, tmax},

AxesLabel → {t, "θ[ t ]"}]
```

InterpolatingFunction[□]

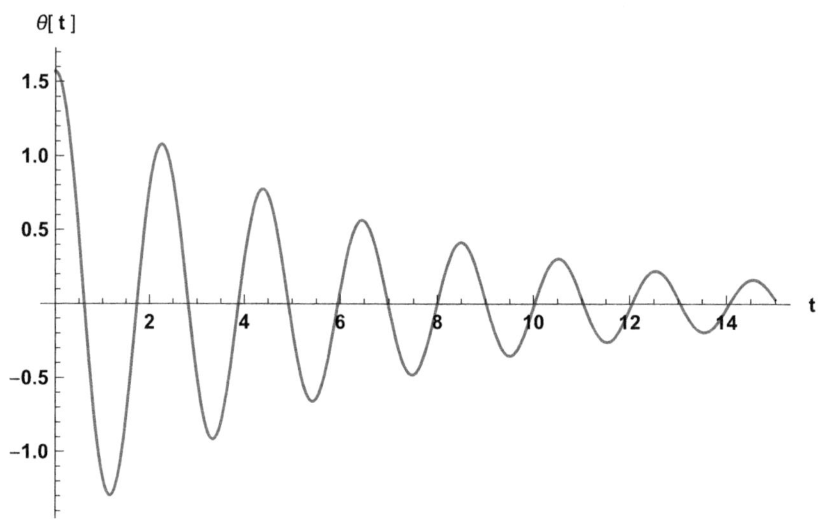

This looks very nice. The damping force diminishes the oscillations as a function of time. Just as importantly, NDSolve allowed us to get to the behavior of the more complex system directly.

Differential equations with impulse functions

We have already seen how impulse functions can be used to model systems undergoing a shock-like change. In their classic text (Elementary Differential Equations and Boundary Value Problems, 4th Edition, Wiley), Boyce and DiPrama devoted a whole section to the solution of impulse functions in differential equations. Not surprisingly, they use the Laplace transform to get the job done. What follows is an adaptation from their text. Take the following equation:

$\partial_{t,t}y[t] + 2\,\partial_t y[\,t] + 2\,\mathrm{y[t]} == \mathrm{DiracDelta[t - \pi]}$

The delta function will be everywhere zero except at t = π. The first step is to find the Laplace Transform. To do so we call the function LaplaceTransform and apply it to the equation to go from time to the transformed variable s:

LaplaceTransform[

y"[*t*] + 2y′[*t*] + 2y[*t*]==DiracDelta[*t* − π], *t*, *s*]

$$2\mathrm{LaplaceTransform}[y[t],t,s] + s^2\mathrm{LaplaceTransform}[y[t],t,s] + 2(s\mathrm{LaplaceTransform}[y[t],t,s] - y[0]) - sy[0] - y'[0] == e^{-\pi s}$$

Then, we solve the transformed equation for y [t]:

Solve[%, LaplaceTransform[*y*[*t*], *t*, *s*]]

$$\left\{\left\{\mathrm{LaplaceTransform}[y[t],t,s] \to \frac{e^{-\pi s}\left(1+2e^{\pi s}y[0]+e^{\pi s}sy[0]+e^{\pi s}y'[0]\right)}{2+2s+s^2}\right\}\right\}$$

Finally, we invert the result with InverseLaplaceTransform of the last result:

InverseLaplaceTransform[%, *s*, *t*]//FullSimplify

$$\{\{y[t] \to e^{-t}\left(\mathrm{Cos}[t]y[0] + \mathrm{Sin}[t]\left(-e^{\pi}\mathrm{HeavisideTheta}[-\pi+t] + y[0] + y'[0]\right)\right)\}\}$$

If the initial conditions are that y [0] and y’ [0] are both zero, then the solution is

$y[t] = e^{-t}\,(\mathrm{Sin}[t]\,(-e^{\pi}\,\mathrm{HeavisideTheta}[-\pi + t]))$

This is an elegant use of the Laplace transform. However, it proves unnecessary to do this as DSolve will do the whole process for us and do so directly:

fnc =

DSolve[

{y"[t] + 2y′[t] + 2y[t]==DiracDelta[$t - \pi$],

y[0] == 0, y′[0] == 0}, y[t], t]

y[t_]:=Evaluate[y[t]/.fnc]

$\{\{y[t] \to -e^{\pi - t}\text{HeavisideTheta}[-\pi + t]\text{Sin}[t]\}\}$

The plot of the solution is a perfect match to the one in the text:

Plot[y[t], {t, 0, 10}]

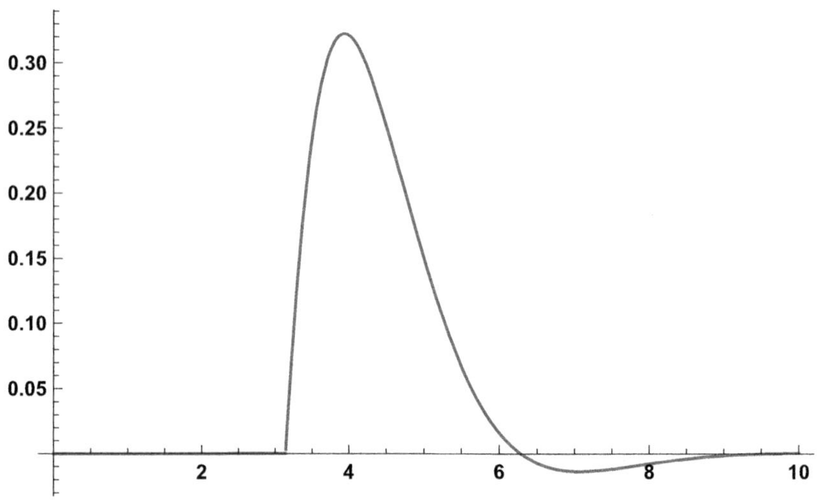

We can change the initial conditions and see the effect of that on the solution. In this case we have taken y [0] == 1:

Clear[fnc, y]

fnc =

DSolve[

{y"[t] + 2y′[t] + 2y[t]==DiracDelta[$t - \pi$],

y[0] == 1, y′[0] == 0}, y[t], t]

y[t_]:=Evaluate[y[t]/.fnc]

Plot[y[t], {t, 0, 10}]

$\{\{y[t] \to e^{-t}(\text{Cos}[t] + \text{Sin}[t] - e^{\pi}\,\text{HeavisideTheta}[-\pi + t]\text{Sin}[t])\}\}$

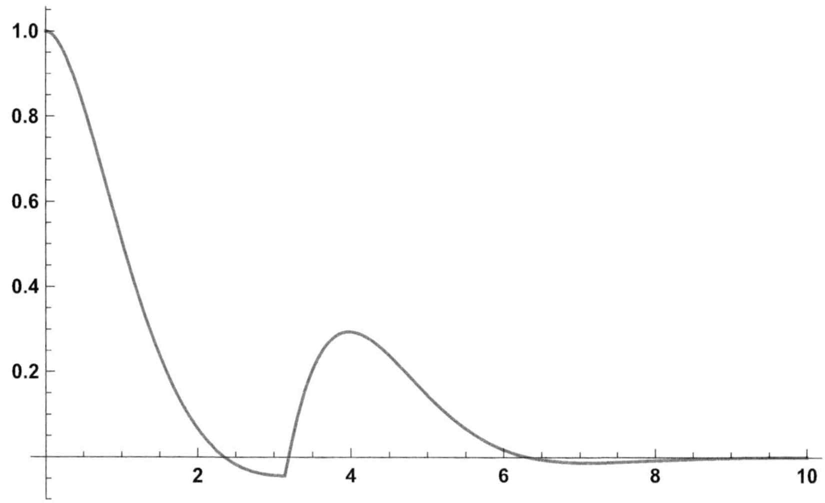

It is also possible to solve the equations using general constant coefficients:

```
Clear[fnc, y, a, b]
fnc =
DSolve[
{y"[t] + ay'[t] + by[t]==DiracDelta[t − π],
y[0] == 0, y'[0] == 0}, y[t], t]
y[t_]:=Evaluate[y[t]/.fnc]
a = 5;
b = 10;
Plot[y[t], {t, 0, 10}, PlotRange → All]
```

$$\left\{\left\{y[t] \to -\frac{1}{\sqrt{a^2-4b}} e^{-\frac{1}{2}\left(-a+\sqrt{a^2-4b}\right)\pi}\right.\right.$$

$$\left(-e^{\frac{1}{2}\left(-a+\sqrt{a^2-4b}\right)t}+\right.$$

$$\left.e^{\frac{1}{2}\left(-a+\sqrt{a^2-4b}\right)\pi+\frac{1}{2}\left(a+\sqrt{a^2-4b}\right)\pi+\frac{1}{2}\left(-a-\sqrt{a^2-4b}\right)t}\right) \times$$

HeavisideTheta[$-\pi+t$] }}

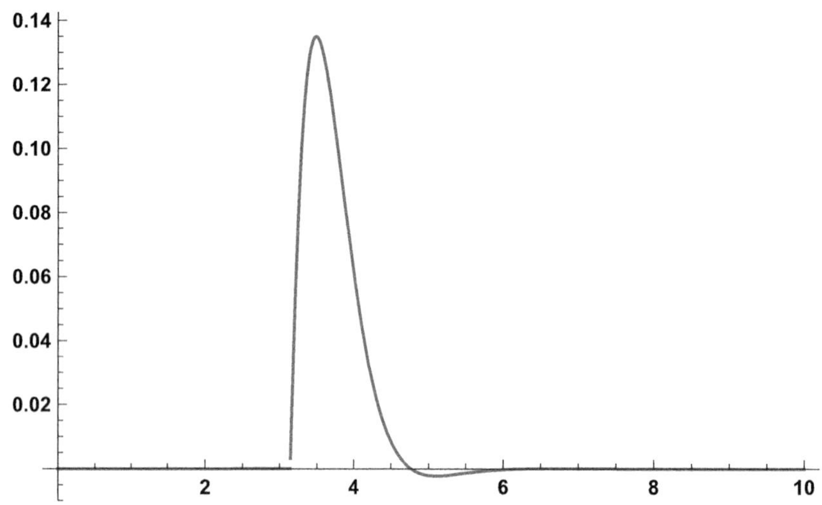

It is also possible to solve an equation that has both the impulse function and a sine function in it:

Clear[fnc, *y*, *a*, *b*]

fnc =

DSolve[

{y"[*t*] + 2*y*′[*t*] + 3*y*[*t*] ==

Sin[*t*] + DiracDelta[*t* − *π*], *y*[0] == 0, *y*′[0] == 1},

y[*t*], *t*]

y[t_]:=Evaluate[*y*[*t*]/.fnc]

Table[y[t]//N, {t, 1, 10, 1}]

Plot[y[t], {t, 0, 10}]

$$\left\{\left\{y[t]\to -\frac{1}{4}e^{-t}\left(-\mathrm{Cos}\left[\sqrt{2}t\right]+e^{t}\mathrm{Cos}[t]\mathrm{Cos}\left[\sqrt{2}t\right]^{2}+\right.\right.\right.$$

$$2\sqrt{2}e^{\pi}\mathrm{Cos}\left[\sqrt{2}t\right]\mathrm{HeavisideTheta}[-\pi+t]\mathrm{Sin}\left[\sqrt{2}\pi\right]-e^{t}\mathrm{Cos}\left[\sqrt{2}t\right]^{2}\mathrm{Sin}[t]-2\sqrt{2}\mathrm{Sin}\left[\sqrt{2}t\right]-$$

$$2\sqrt{2}e^{\pi}\mathrm{Cos}\left[\sqrt{2}\pi\right]\mathrm{HeavisideTheta}[-\pi+t]\mathrm{Sin}\left[\sqrt{2}t\right]+e^{t}\mathrm{Cos}[t]\mathrm{Sin}\left[\sqrt{2}t\right]^{2}-$$

$$\left.\left.\left.e^{t}\mathrm{Sin}[t]\mathrm{Sin}\left[\sqrt{2}t\right]^{2}\right)\right\}\right\}$$

{{0.346582}, {0.328654}, {0.245752}, {0.251148}, {−0.251931}, {−0.340639}, {−0.0357026}, {0.286559}, {0.332691}, {0.0735933}}

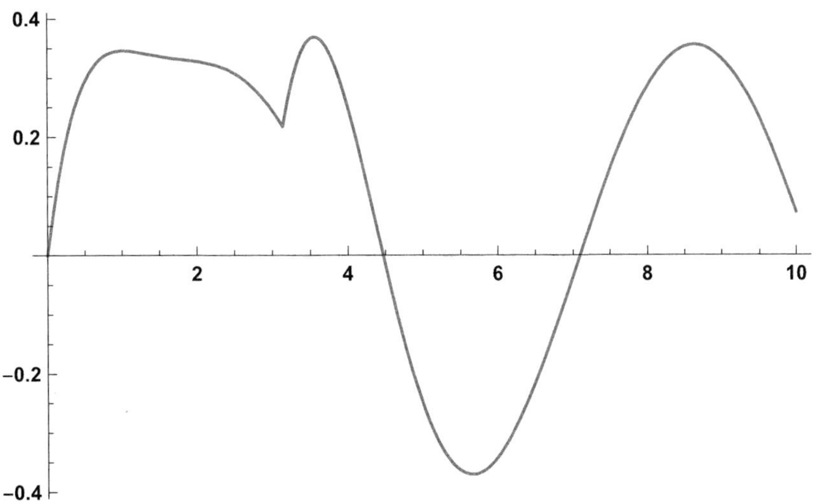

So, the versatility of DSolve is impressive and takes the emphasis off of solving the equations of this kind and places it instead on understanding their behavior and utility.

Perceptrons: primitive AI

This section is dedicated to the memory of Prasad Dhurjati a wonderful colleague and friend, who introduced me to artificial intelligence.

Artificial Intelligence, AI, or embedded neural networks have begun to make everything "smarter". After generating considerable interest decades ago, and then having an up and down research and development trajectory, artificial intelligence (AI) has now seen an explosion of research, development and application. While the topic of AI and *Mathematica* is one that is burgeoning, this section is not about recent ideas and work. Instead we are looking back to the beginning of this field and looking at the ancestor of all AI applications - and that is the *perceptron*.

The topic is interesting in and of itself and it lets us introduce other functions within *Mathematica*. In the 1950s, it was the introduction of the perceptron that led to electronic logic gates. While this is as an exercise in the archaeology of technology, I think it is also useful, as we can gain more experience in doing logic, graphics, and programming with *Mathematica* while learning about how these primitive AI functions worked.

The transfer function

Early in the development of AI, Frank Rosenblatt developed the perceptron. The perceptron was a clever device that could be used to create computational logic gates. The foundation for the perceptron is the sigmoidal transfer function, parametrically tweaked to produce a binary output - either a one or a zero. Let's see how this worked.

This is the sigmoidal transfer function, which is the basis for the perceptron or artificial neuron:

$$\mathrm{f}[\,\mathrm{x}\,] = \frac{1}{1+\mathrm{Exp}[-kx]}$$

We can evaluate and plot it with a fixed value for k, a constant that is also called the gain. If the input x is positive, then the outputs vary between 0.5 and 1.0. At x = 0, the output is exactly 0.5. With values of x less than zero, the outputs vary down from 0.5 toward zero. We see the full sigmoidal shape when we plot over a range of negative and positive values.

```
f[x_]:=1/(1+Exp[-kx])
(* Defines the Transfer Function *)

k=0.5;
```

Table[f[x], {x, 0, 10}]

Plot[f[x], {x, −10, 10}]

{0.5, 0.622459, 0.731059, 0.817574, 0.880797, 0.924142, 0.952574, 0.970688, 0.982014, 0.989013, 0.993307}

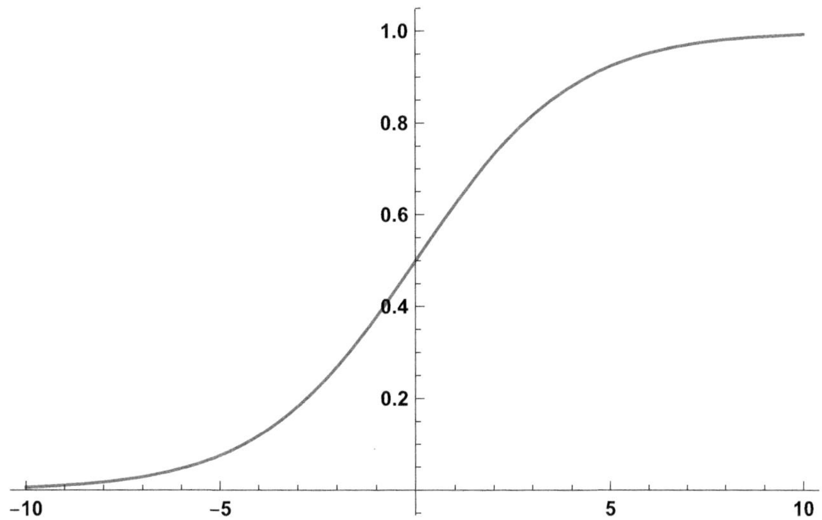

When we plot the whole function at this value of k, we see that the transfer function is a beautiful sweeping s-shaped or sigmoidal curve. This version of the curve has been used for many other purposes including as mathematical metaphors for such widely varying phenomena as learning and for the lifecycle of a business. But, that wonderful shape happens just for a range of values of k. It is very important to learn how a change in the value of k, changes the function. By increasing the value of k, that is by increasing the gain, the function becomes steeper, looking more like a step than a sigmoid. At a value of k equals zero, the function is single valued at y and becomes a straight line parallel to the x-axis.

Table[Plot[f[x], {x, −10, 10}, PlotLabel → {k"=k"}],

{k, 0, 15, 5}]

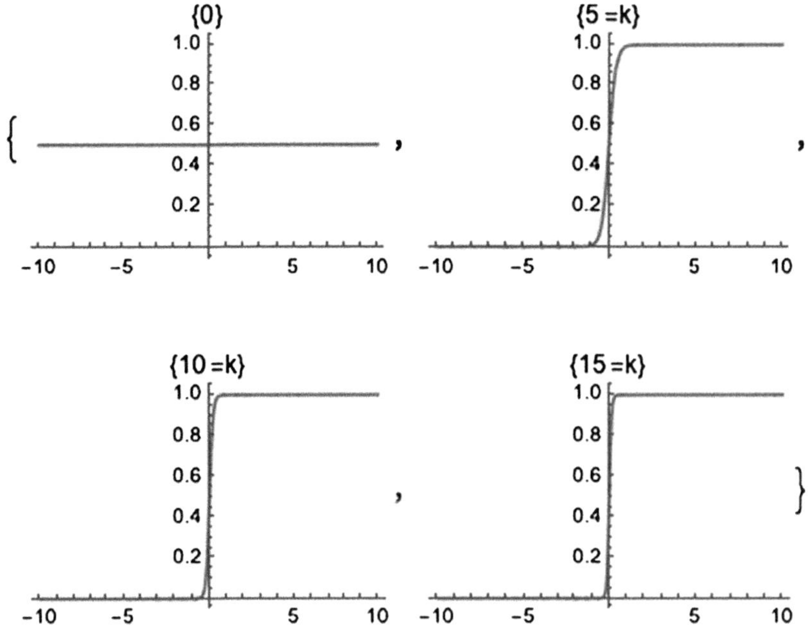

A more elegant way to look at the variation in the behavior of this transfer function is to use the slider function in *Mathematica.*

Our implementation will be simple and straightforward. We will look again at the Plot of the transfer function at different values of k, the gain. However, this variation in k will be done interactively by us from the mouse. Here is how it works:

k=. (* This clears k. *)

f[x_, k_]:=1/(1 + Exp[−kx])

(* Specifies the transfer function in terms of x and k. *)

(* Manipulate allows for the plot to be varied with smooth variation in k. *)

Manipulate[Plot[f[x, k], {x, −10, 10},

PlotLabel → {k"=k"}], {k, 0, 20}]

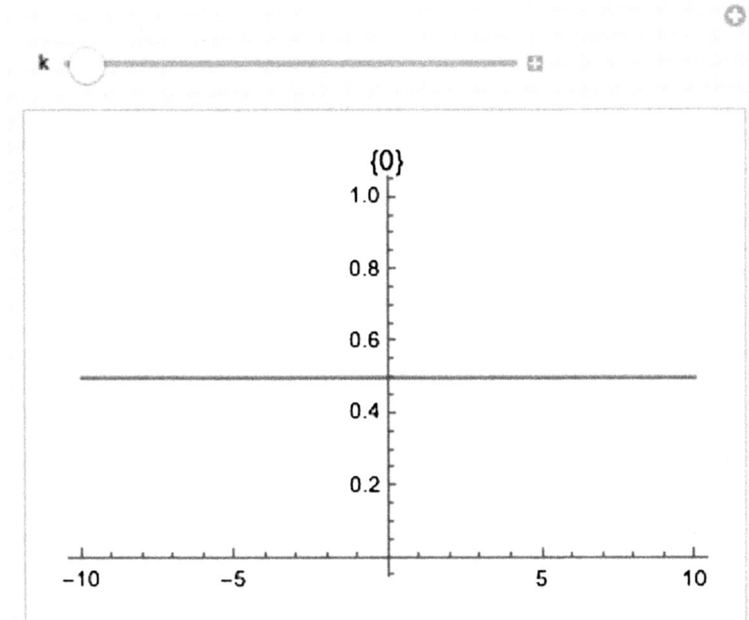
k
{0}
1.0
0.8
0.6
0.4
0.2
−10
−5
5
10

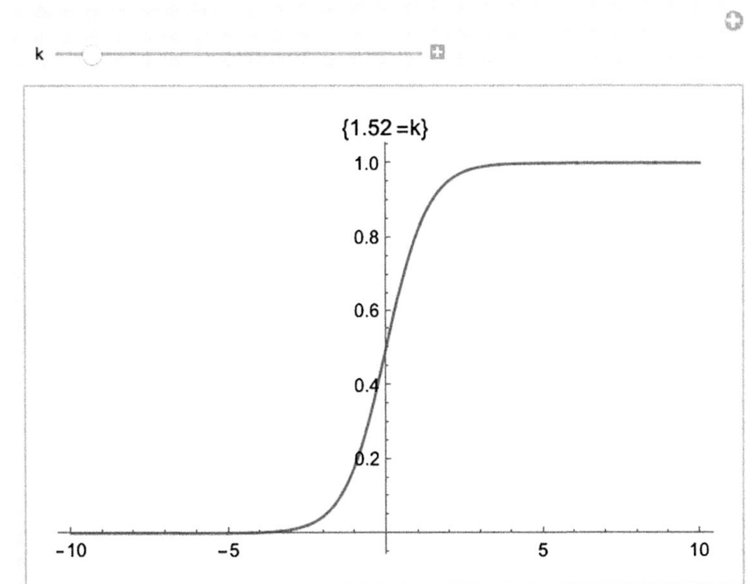
k
{1.52 =k}
1.0
0.8
0.6
0.4
0.2
-10
-5
5
10

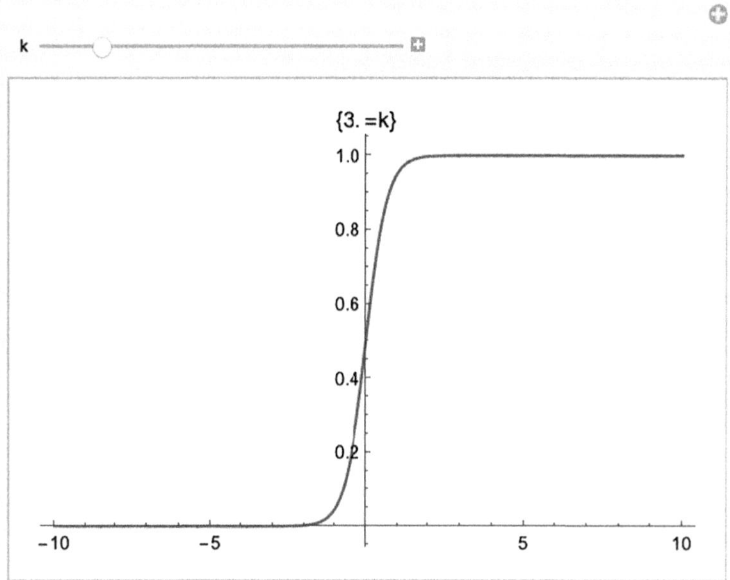

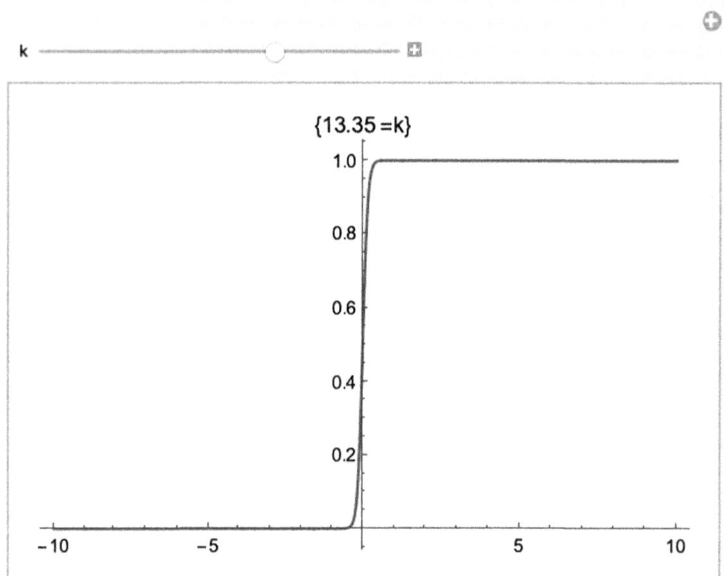

By placing the cursor over the button, holding the mouse button down and rolling the mouse to the right, we change the value of k, continuously, and see the variation in the behavior of the function. Thus, as the value of k becomes larger, and especially at values of say ten or more, the transfer function becomes less and less sigmoidal, or "soft", and more step-function like or "hard". What this means is that for any input above zero, the output is close to one and for any input close to, or less than zero, the output is zero. We can see that in the plot, but it is even easier to see if we list the values for a given large k, say k = 20:

$k = 20$;

Table[{x, $f[x]//N$}, {x, −10, 10}]

{{−10, 1.3838965267367376 ∗ 10^{-87}, {−9, 6.714184288211594 ∗ 10^{-79}},
{−8, 3.257488532207521 ∗ 10^{-70}}, {−7, 1.580420060273613 ∗ 10^{-61}},
{−6, 7.667648073722 ∗ 10^{-53}}, {−5, 3.7200759760208356 ∗ 10^{-44}},
{−4, 1.8048513878454153 ∗ 10^{-35}}, {−3, 8.75651076269652 ∗ 10^{-27}},
{−2, 4.248354255291589 ∗ 10^{-18}}, {−1, 2.0611536181902037 ∗ 10^{-9}},
{0, 0.5}, {1, 1.}, {2, 1.}, {3, 1.}, {4, 1.}, {5, 1.}, {6, 1.}, {7, 1.},
{8, 1.}, {9, 1.}, {10, 1.}}

For all values of x less than zero, the outputs are extraordinarily small and essentially zero. For all inputs above zero, the outputs are one, but at x = 0 point, the transfer function will return an output of 0.5 no matter the value of the gain. We don't want or need this value, we will make it the last value of x by rounding the value of the output at an input of 0.5 down to 0. We can repeat the calculation we just did, and nest { k "= k"} inside of the function **Round** and the output becomes quite tidy.

Table[{x, Round[$f[x]//N$]}, {x, −10, 10}]

{{−10, 0}, {−9, 0}, {−8, 0}, {−7, 0}, {−6, 0}, {−5, 0}, {−4, 0}, {−3, 0}, {−2, 0},
{−1, 0}, {0, 0}, {1, 1}, {2, 1}, {3, 1}, {4, 1}, {5, 1}, {6, 1}, {7, 1}, {8, 1}, {9, 1}, {10, 1}}

The artificial neuron (AN)

Now, we can build a mathematical representation of a neuron using this hard transfer function. Let's say that it will have two inputs, x_1 and x_2, and let's assume that these will be somehow summed together in a linear combination to create the one overall input to the neuron. We could sum them as they are, where in $x_{in} = x_1 + x_2$, or we could anticipate that the two separate inputs may eventually have different levels of impact on the outcome. In other words, if we leave x_1 and x_2 unweighted, they are actually equally weighted. In general, the inputs will need to have explicit weighing parameters, w_1 and w_2 adjusted for the neuron to behave properly, as we will see. Hence, the linear combination of the inputs will be $x_{in} = w_1 * x_1 + w_2 * x_2$.

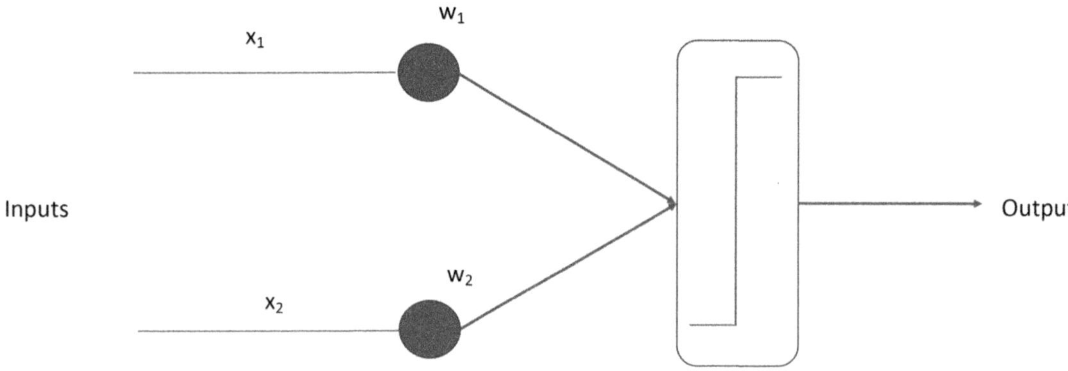

Next, let's assume that the inputs, x_1 and x_2, have values of either 0 or 1, only, which is to say that they are binary and consistent with "digital" operations. That means the possible input pairs, or dyads, will be completely specified by: { 1, 1 }, { 1, 0 }, { 0, 0 }, and { 0, 1 }. There are no other possibilities for two inputs with only these two values. Now, let's see what the transfer function, with k = 20, will produce from these inputs. First, we define a list (or vector) of dyads called **input**:

input = {{1, 1}, {1, 0}, {0, 0}, {0, 1}};

Because, the transfer function does not accept two values, since it is a function of only one variable, we have to convert the list of pairs into a list of single values. So we have to convert the pairs into single values as follows:

x1 * w1 + x2 * w2 - Threshold = Linear Combination = Transfer Function Input

Each pair will sum to a single value and those values will be the inputs to the transfer function. Since we have four pairs in this input list, the list length is four. With any list we can find its length with the Length function. We simply type **Length [listname]** and this comes back as a number.

Length[input]

4

What we want to do next is to sum the digits in each pair. When the inputs are equally weighted (w_1 and w_2= 1) and the threshold is zero, the first pair sums to 2, the second to

1, the third to 0 and the fourth to 1. We can do this in our heads. But, to do this computationally, we need to be able to access the numbers in each pair in the list. This can be done with the Part function. Each element of any list can be accessed from the list by writing **Part [Listname, part number]**, or, more simply, by placing a double-square-bracketed number after the list name as in **Listname [[n]]**. In this case the list input has four paired elements. We can access them as follows:

```
Part[input, 1](* This is the formal way to
use the Part function. *)

input[[1]](* This is the more elegant and
quicker method of pulling members out of a list. *)
input[[2]]
input[[3]]
input[[4]]
```

{1, 1}

{1, 1}

{1, 0}

{0, 0}

{0, 1}

Since we must the sum each digit within a pair, now we use two position numbers: **Listname [[n, m]]**. When we do this, we go first to the nth element of the list and then to the mth occupant of that element. Here is how that works for the first and fourth pairs:

```
(* Pick the first element from the list “input” and pick the first element from that pair. *)
input[[1, 1]]

(* Pick the first element from the list “input” and pick the second element from that pair *)
input[[1, 2]]
```

```
(* Pick the fourth element from the list "input" and the first element from that pair *)
input[[4, 1]]

(* Pick the fourth element from the list "input" and the second element of that pair *)
input[[4, 2]]
```

1

1

0

1

So we want to do the following, but we don't want to have to type each case individually as is done here to illustrate the underlying method:

```
{input[[1, 1]] + input[[1, 2]],
input[[2, 1]] + input[[2, 2]],
input[[3, 1]] + input[[3, 2]],
input[[4, 1]] + input[[4, 2]]}
```

{2, 1, 0, 1}

Instead, to do this automatically, we can use the **Table** function. We employ the same pattern as above, and the number of the pair becomes an iterator from 1 to the Length of the list that will end the computation:

```
Table[input[[n, 1]] + input[[n, 2]],
{n, 1, Length[input]}]
```

{2, 1, 0, 1}

This new list of single elements will be the input to the transfer function. Summing corresponds to a specific linear combination with the values of the weighting factors both w1 = w2 = 1 and the Threshold at 0, i.e. **x * 1 + y * 1 = 0**. While this linear combination was given as an arbitrary choice, it was actually chosen as the most straightforward to understand, among the many linear combinations of the input variables that could have been specified. Because **x * 1 + y * 1 = 0** is so easy to understand, we will use it to see what the result will be.

There are several ways to evaluate the transfer function for this set of inputs, but the easiest is to **Map** the transfer function onto the list of elements. When working interactively and line-by-line, the **%** means the last result, which in this case was { 2, 1, 0, 1 }. Remember this is our transfer function and we want it to be step-function like, so the gain is high at k = 20:

$k = 20$;

f[x_]:=1/(1 + Exp[$-kx$])

Then this f [x] is mapped over the results of the linear combination { 2, 1, 0, 1 }

Map[f, {2, 1, 0, 1}]

$$\left\{\frac{1}{1+\frac{1}{e^{40}}}, \frac{1}{1+\frac{1}{e^{20}}}, \frac{1}{2}, \frac{1}{1+\frac{1}{e^{20}}}\right\}$$

As we have seen, mapping over a list may also be done with a shorthand notation **/@**:

f/@{2, 1, 0, 1}

$$\left\{\frac{1}{1+\frac{1}{e^{40}}}, \frac{1}{1+\frac{1}{e^{20}}}, \frac{1}{2}, \frac{1}{1+\frac{1}{e^{20}}}\right\}$$

Interestingly, **Map** returns a new list of elements, but the elements are not numerically evaluated. When the results should be evaluated numerically, the function **N** is applied over the resultant list $N\left[\left\{\frac{1}{1+\frac{1}{e^{40}}}, \frac{1}{1+\frac{1}{e^{20}}}, \frac{1}{2}, \frac{1}{1+\frac{1}{e^{20}}}\right\}\right]$ or by appending it to the Map function as **f / @ { 2, 1, 0, 1 }//N**

$$N\left[\left\{\frac{1}{1+\frac{1}{e^{40}}}, \frac{1}{1+\frac{1}{e^{20}}}, \frac{1}{2}, \frac{1}{1+\frac{1}{e^{20}}}\right\}\right]$$

```
f/@{2, 1, 0, 1}//N
```

{1., 1., 0.5, 1.}

{1., 1., 0.5, 1.}

It is important to note that for the input values of { 2, 1, 0, 1 } the transfer function yields { 1, 1, 0.5, 1 }. This happens because when x = 0 the transfer function evaluates to 0.5. Since the input is binary, we also want the output to be binary; we want to know "did the neuron fire?", output = 1, or "did the neuron not fire?", output = 0. An output of 0.5 is an indeterminate outcome; we have to push it to zero, or to one to be determinant. If we push its value to one, then the neuron fires with any or no input. This is not of any value to us. So, we must push the output value of 0.5 to zero. This is easy to do since the function **Round** by definition, and by convention, pushes the value of 0.5 to zero. Hence by also appending Round to the mapping we get a valuable result, namely:

```
f/@{2, 1, 0, 1}//N//Round
```

{1, 1, 0, 1}

This is useful because the neuron discriminates among the four input values, it does not fire if both of the inputs are zero, but does fire if one, or both, are one.

We just executed these steps interactively, or line-by-line, to understand how the artificial neuron works. But, we would not want to do this for repetitive computations, especially, if we will have multiple, successive, or continuous inputs. Instead, we will bundle all those lines into a small program, or script that is annotated here for clarity:

```
(* This is the set of all possible inputs *)

input = {{1, 1}, {1, 0}, {0, 0}, {0, 1}};

(* The Table function applies the linear combination,
sums the pairs of binary inputs and creates a vector
of single inputs, xi, for the transfer function *)

xi = Table[input[[n, 1]] + input[[n, 2]],
{n, 1, Length[input]}]
```

```
(* f[x_]:=specifies the transfer function with a gain of 20 *)

f[x_]:=1/(1 + Exp[−20x])

(* Finally, we Map the transfer function onto
the vector of inputs x and then round down *)

f/@xi//N//Round

{2, 1, 0, 1}

{1, 1, 0, 1}
```

The output of the transfer function is { 1, 1, 0, 1 } after rounding. The original inputs were { { 1, 1 }, { 1, 0 }, { 0, 0 }, { 0, 1 } } which were transformed into { 2, 1, 0, 1} by the linear combination **x * 1 + y * 1 = 0** and those values were converted into { 1, 1, 0, 1} by the transfer function 1 /(1 + Exp[- 20 x]). What does this set of results tell us? Firstly, if the input is { 1, 1 }, then the output from the neuron is 1. We can think of the output of 1 as the firing of a biological neuron - it either fires, or it does not for a given set of inputs. We also note that if either one, or the other of the inputs is a 1, that is for either input pair { 1, 0 } or { 0, 1 }, then the neuron gives an output of one and also fires. In a sense, this means that the neuron has a threshold of one, where the threshold is the minimum value that must be reached on the input side in order for the neuron to fire.

$$x1 + x2 = 0 = \text{Threshold in this case}$$

And, when the inputs are both zero, the sum is zero, which is below the threshold of one and the neuron does not fire and gives an output of zero. Another way to think of this is that the neuron tells us if the two inputs are both equal to one, and, if they are, it tells us that for this input the answer to the question "are either or both of the values of a and b of any binary pair input equal to one?" is "**yes, at least one value of a or b is one**," or, more simply, the answer as output is **TRUE**.

The linear combination that we have used, **x * 1 + y * 1 = 0**, is not unique. There are many other linear combinations that can be specified that will give the same output from the transfer function. Which is to say that we can adjust the magnitudes of the weightings and get the same output of { 1, 1, 0, 1 } for the inputs of { { 1, 1 }, { 1, 0 }, { 0, 0 }, { 0, 1 } }, so long as we are careful in our choices. Let's repeat the computation that we just did, but let's combine the input values with the linear combination **x * 1.2 + y * 1.2 - .24 = 0** to see what the result will be. Here we annotated only the line of the script with the linear combination:

```
input = {{1, 1}, {1, 0}, {0, 0}, {0, 1}};

(* Next, the Table function applies the new
linear combination, x * 1.2 + y * 1.2 − .24 = 0,
sums the pairs of binary inputs and createsa
vector of single inputs, xi,
for the transfer function *)

xii = Table[1.2 * input[[n, 1]] + 1.2 * input[[n, 2]]−

0.24, {n, 1, Length[input]}];

f[x_]:=1/(1 + Exp[−20x])

f/@xii//N//Round

{1, 1, 0, 1}
```

Later, we will see why both of the linear combinations that we have chosen, work. By using a graphical interpretation of linear combination process, we can know if the one that we chose will provide the result that we want and from the transfer function. For now let's go on to other related matters, namely how to get different results from the same kind of process.

Consider a case in which the output from the transfer function is changed by the choice of weighting factors and threshold in the linear combination of the inputs. By adjusting w_1 and w_2, we can build a different neuron, one that will fire only when the sums of the inputs is two, that is only for the case of {1,1}, but that will not fire for the cases of {1,0}, {0,1} or for {0,0}, when the sum of the inputs are 1, 1, and 0. For this to be the outcome, we have to find values for **w1** and **w2** that will lead to a transfer function output of {1,0,0,0}. Let's try a threshold value of two with the same weightings:

x1 * w1 + x2 * w2 = 2 = Threshold

or that

x1 * w1 + x2 * w2 - 2 = 0

With the weightings equal to one and the threshold at 2, then for all the cases, the **LHS** will be zero or negative, to fire it must be positive:

$$\{1, 1\} \quad : \quad 1 * 1 + 1 * 1 - 2 = 0 \quad \Rightarrow \quad \text{no fire}$$

$$\{0, 1\} \quad : \quad 0 * 1 + 1 * 1 - 2 = -1 \quad \Rightarrow \quad \text{no fire}$$

$$\{0, 0\} \quad : \quad 0 * 1 + 0 * 1 - 2 = -2 \quad \Rightarrow \quad \text{no fire}$$

$$\{1, 0\} \quad : \quad 1 * 1 + 0 * 1 - 2 = -1 \quad \Rightarrow \quad \text{no fire}$$

In each case, the result, the output is zero or below zero/meaning that the neuron has not fired. So we need to change the weighting factors. We can test $w_1 = 1.5$ and $w_2 = 1.5$, $\mathbf{x1 * 1.5 + x2 * 1.5 - 2 = 0}$:

$$\{1, 1\} \quad : \quad 1 * 1.5 + 1 * 1.5 - 2 = 1 \quad \Rightarrow \quad \text{fire}$$

$$\{0, 1\} \quad : \quad 0 * 1.5 + 1 * 1.5 - 2 = -0.5 \quad \Rightarrow \quad \text{no fire}$$

$$\{0, 0\} \quad : \quad 0 * 1.5 + 0 * 1.5 - 2 = -2 \quad \Rightarrow \quad \text{no fire}$$

$$\{1, 0\} \quad : \quad 1 * 1.5 + 0 * 1.5 - 2 = -0.5 \quad \Rightarrow \quad \text{no fire}$$

What this tells us is that with this new linear combination, the new neuron will discriminate among the four pairs of inputs. It will only fire when two of the inputs are identically 1 and will not fire if one of the two is 1, or if both are zero. We can implement this in the script as follows:

```
input = {{1, 1}, {1, 0}, {0, 0}, {0, 1}};

xi = Table[input[[n, 1]] * 1.5 + input[[n, 2]] * 1.5 − 2,

{n, 1, Length[input]}];

f[x_]:=1/(1 + Exp[−20x])

f/@xi//N//Round
```

{1, 0, 0, 0}

There is something very interesting that has emerged here that we have to realize as important. In the first case with the weighting factors all set to unity, and the threshold at unity, the neuron fired if the inputs were both 1, or if either member of a pair had a value of one, but did not fire if they were both zero. In this case, the output is equivalent to telling us that for three of the inputs, it is the case that at least one of the inputs had a value of one, and this covers the case when both are one. Put another way, the proposition that at least one of the two inputs has a value of one is **TRUE** three of four times {1,1}, {0,1}, and {1,0}, but **FALSE** once for {0,0}. For the second neuron, the proposition that both input are one yields, **TRUE** but for all the other cases this **FALSE**. This is a perfect segue to thinking of these artificial neurons as logical operators or gates.

Truth Tables for logic gates

There is another way to think about what we have done, and that is to consider these two artificial neurons as logical operators. The first neuron can be used as if it knew that when x_1 **OR** x_2 is one, then it should fire. Put differently, the artificial neuron functions as logical gate **OR**, and asks the question, "Is it true that are either, or both, of the inputs that were just received, are equal to one?" If they are, then the reply, or output, is **True**; if not the reply is **False**. When we increased the weightings to 1.5 each and the threshold to 2.0, this new neuron fires only when x_1 **AND** x_2 are both one. As a logical gate, it asks the question, "Is it true that both of the inputs that were just received, are equal to one?" If they are, then the reply is **True**, and if not, the reply is **False**. In this way these two artificial neurons can be used to do the logical operations: **AND**, **OR**. This is why when we think of these as logic operators, it is usual to refer to them as logic gates, or just as gates.

We will see that there are four main logical gates: **AND**, **OR**, **NAND**, **NOR**. For each of these, we can design them to have two binary digital inputs of either 0 or 1 in pairs. The four possible pairs are {0,0}, {0,1}, {1,0}, and {1,1}. For each gate, we can assess the outcome for an binary pair input. These can be nicely arranged into tables that are referred to as Truth Tables:

AND

in1	in2	in1 AND in2
0	0	0
0	1	0
1	0	0
1	1	1

OR

in1	in2	in1 OR in2
0	0	0
0	1	1
1	0	1
1	1	1

NAND

in1	in2	in1 NAND in2
0	0	1
0	1	1
1	0	1
1	1	0

NOR

in1	in2	in1 NOR in2
0	0	1
0	1	0
1	0	0
1	1	0

87

There is also a **NOT** gate, but it is considered to be a simple inverter of any single input value: NOT 1 is 0 and NOT 0 is 1. We need this inverter function to be able to get to the NAND and NOR gates. We see that the output of the NAND gates is the same as inverting the output of the **AND** gate and in the same way, the output of the NOR gate is the inversion of the **OR** gate.

Maps and graphical interpretation of AND and OR

Now, let's look at these results and the linear combinations in terms of graphical maps. The graphs of the threshold equations take the form of lines:

$$\text{Case 1:} \quad x_1 * w_1 + x_2 * w_2 - 1 = 0$$

$$x_1 + x_2 - 1 = 0 \, ; \, w_1 = w_2 = 1$$

$$x_2 = - x_1 + 1$$

and

$$\text{Case 2:} \quad x1 * w1 + x2 * w2 - 2 = 0$$

$$x1 * 1.5 + x2 * 1.5 - 2 = 0 \, ; \, w_1 = w_2 = 1.5$$

$$x_2 = - x_1 + 1.33\overline{3}$$

When we plot the equations for the two different linear combinations, we see that in the first case, { 0,1}, {1,0}, and {1,1} are all above the line, on one side of the line, but {0,0} is below it. In the second case, the linear combination line intercepts the ordinate at {0,1.333} and the abscissa at { 1.333,0}. Now, only {1,1} is above line and the others are below the line. For both cases, points above the line are in the region of the graph in which the output is TRUE and vice versa. Hence, these maps explain graphically the logical functions of the two neurons.

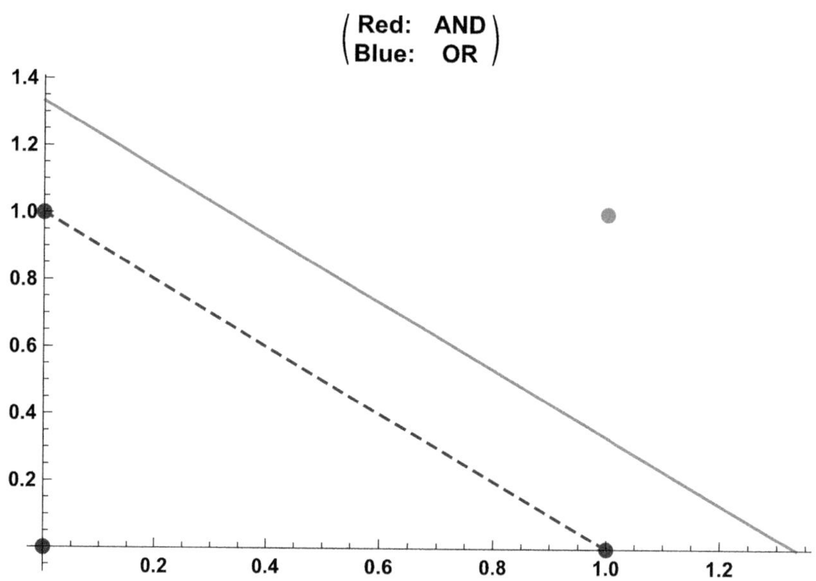

A graphical aside

Let's take minute to see how to make these plots. We will use Show, Plot, PlotStyle, PlotLabel, Graphics, PointSize, and Point in this script:

```
Show[Plot[(−x1 + 1.333), {x1, 0, 1.333},
PlotStyle → {Red},
PlotLabel → {{"Red:", "AND"}, {"Blue:", "OR"}}],
Plot[(−x1 + 1), {x1, 0, 1},
PlotStyle → {Blue, Dashed}],
Graphics[{PointSize[Large], Red, Point[{1, 1}]}],
Graphics[{PointSize[Large], Blue, Point[{1, 0}]}],
Graphics[{PointSize[Large], Blue, Point[{0, 0}]}],
Graphics[{PointSize[Large], Blue, Point[{0, 1}]}]
]
```

The script plots the two lines **(- x1 + 1.333)** and **(- x2 + 1)** with the first, the "**OR**" line, appearing blue (dark gray in the print version) and dashed, and the second, the "**AND**" line, appearing red (mid gray in the print version) and solid. This was accomplished by combining the two **Plot** functions into one **Show** command first. Then to add the labels and colors to the lines, we by added **PlotLabel→ { { "Red:", "AND" },{ "Blue:", "OR"} }** and **PlotStyle→ { Blue, Dashed }** commands. The latter command renders the **AND** line **{Red}** and the **OR** line **{Blue, Dashed}**. Also, added to the attributes of the first Plot is the **PlotLabel** given as { "AND", "OR"}. We also want to see the inputs. On this map, they are simply points at {1, 1}, {1, 0},{0, 0},{0, 1}. To present them with the two plots they are included as *graphics primitives* in the form of **Graphics[{ PointSize[Large], Pink, Point[{1, 1}] }]**, which specifies that the point {1, 1} be large and red (mid gray in the print version), and the others are specified to be large and blue (dark gray in the print version). The colors help to interpret what is included and what is excluded.

At this point, we have two gates that will be activated by the set {0, 1}, {1, 1}, and {1, 0}, or by {1, 1} alone, but we don't have a gate that will be activated when the two inputs are zero and only when both are zero. If we want a gate that will indicate when both inputs are zero, then it should be the complement of the **AND** gate. To do this we want the threshold to be just below zero in this case

$$x1 * w1 + x2 * w2 = -0.1$$

$$x1 * w1 + x2 * w2 + 0.1 = 0$$

The weightings have to be negative. Let's use w1 = -1, w2 = -1 and the threshold = - 0.1.

$$x1 * (-1) + x2 * (-1) + 0.1 = 0$$

{ 1, 1 }	:	**1 * (- 1) + 1 * (- 1) + 0.1 = - 1.9**	⇒	**no fire**
{ 0, 1 }	:	**0 * (- 1) + 1 * (- 1) + 0.1 = - 0.9**	⇒	**no fire**
{ 0, 0 }	:	**0 * (- 1) + 0 * (- 1) + 0.1 = + 0.1**	⇒	**fire**
{ 1, 0 }	:	**1 * (- 1) + 0 * (- 1) + 0.1 = - 1.9**	⇒	**no fire**

In this way, with inputs other than zero to the transfer function, its outputs will be negative, except in the case of {0, 0}, where it is positive. Hence, the output that we seek from the transfer function is {0, 0, 1, 0} with each value corresponding to the inputs of { {1, 1}, {1, 0}, {0, 0}, {0, 1} }, meaning the output would be equivalent to **{NO, NO, YES, NO}** or **{FALSE, FALSE, TRUE, FALSE}** in answer to the question posed "are the two inputs, a and b, within any input pair, { a, b }, equal to zero?". Here is a script for this:

```
input = {{1, 1}, {1, 0}, {0, 0}, {0, 1}}

(* Here we have adjusted the threshold to −
0.1 and the weighting factors, each to − 1 *)

xiii = Table[input[[n, 1]] ∗ −1 + input[[n, 2]] ∗ −1 + 0.1,

{n, 1, Length[input]}];

f[x_]:=1/(1 + Exp[−20x])

f/@xiii//N;

f/@xiii//N//Round
```

```
{{1, 1}, {1, 0}, {0, 0}, {0, 1}}

{0, 0, 1, 0}
```

This script does exactly what we expected, it can be interpreted as excluding, or rejecting all inputs other than {0, 0} by outputting zeroes or **FALSE** values. At the point {0, 0} this gate gives a **1** or **TRUE** rather than at the point {1, 1}. In another sense, this gate can be thought of as alerting that both inputs have gone to zero, or that they have shut off. It is another form of an AND gate that we will call AND2.

In terms of the graphical map that we constructed, we have a new line corresponding to x2 = - x1 + 0.1 with x and y intercepts at {0.1, 0} and {0, 0.1}. We can add to the original plot as a green line (light gray line in the print version) for AND2:

```
Show[
Plot[(−x1 + 1.333), {x1, 0, 1.333},
PlotStyle → {Red},
PlotLabel → {{Red, "AND"}, {Blue, "OR"},
{Green, "AND2"}}],

Plot[(−x1 + 1), {x1, 0, 1},
PlotStyle → {Blue, Dashed}],

Plot[(−x1 + 0.1), {x1, 0, 0.1},
PlotStyle → {Green, Dashed}],

Graphics[{PointSize[Large], Red, Point[{1, 1}]}],

Graphics[{PointSize[Large], Blue, Point[{1, 0}]}],

Graphics[{PointSize[Large], Black, Point[{0, 0}]}],

Graphics[{PointSize[Large], Blue, Point[{0, 1}]}],

Graphics[{PointSize[Large], Green,
Point[{0, 0.1}]}],

Graphics[{PointSize[Large], Green, Point[{0.1, 0}]}]
]
```

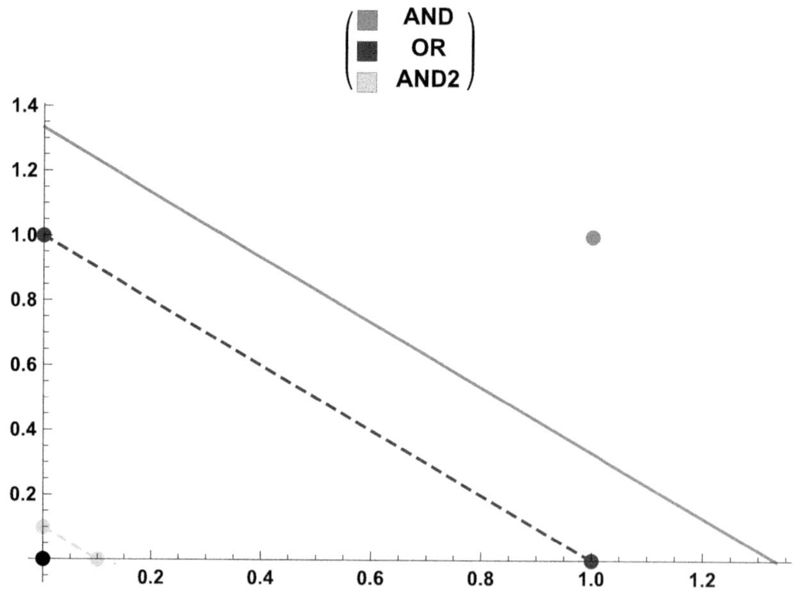

Before we go on, let's do one more thing and go back to the second **OR** gate built on the linear combination of x1 * 1.2 + x2 * 1.2 - .24 = 0, corresponding to x2 = - x1 + 0.2 and put its dividing line also on this graph (see the green line (light gray line in the print version) below). The key is that so long as the dividing line is above the origin, and at or below the line that includes { 0, 1 } and { 1, 0 }, then it will produce an **OR** gate. All four of the dividing lines, corresponding to AND, OR, AND2, and OR2 are shown in this graph:

```
Show[
Plot[(−x + 1.333), {x, 0, 1.333},
PlotStyle → {Red},

PlotLabel → {{"Red:", "AND"}, {"Blue:", "OR"},
{"Green:", "And2"}, {"Purple:", "OR2"}}],

Plot[(−x + 1), {x, 0, 1},
PlotStyle → {Blue, Dashed}],

Plot[(−1x + 0.2), {x, 0, 0.2},
PlotStyle → {Purple, Dashed}],
```

```
Plot[(-x + 0.1), {x, 0, 0.1},
PlotStyle → {Green, Dashed}],

Graphics[{PointSize[Large], Red, Point[{1, 1}]}],

Graphics[{PointSize[Large], Blue, Point[{1, 0}]}],

Graphics[{PointSize[Large], Black, Point[{0, 0}]}],

Graphics[{PointSize[Large], Blue, Point[{0, 1}]}],

Graphics[{PointSize[Large], Purple,
Point[{0, 0.2}]}],

Graphics[{PointSize[Large], Purple,
Point[{0.2, 0}]}],

Graphics[{PointSize[Large], Green,
Point[{0, 0.1}]}],

Graphics[{PointSize[Large], Green, Point[{0.1, 0}]}]]
```

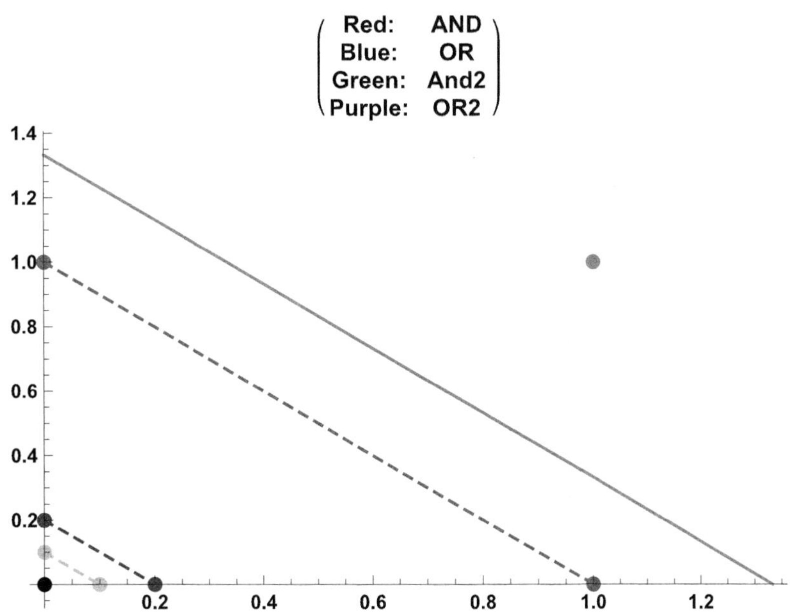

The NOT gate or inverter

Another logic function is the **NOT** gate. This is simply a negation function. If the input to a **NOT** gate is **1**, meaning **TRUE** or **YES**, its output is a **0**, for **FALSE** or **NO**. We will program this gate to take just one input at a time and it simply flips that digit from **1** to **0** or from **0** to **1**. Though this is simple in function, to make a script for this NOT gate, we need to implement some new approaches.

Recall from above, that when we had an output from the transfer function that was exactly **0.5**, we used **Round** as a short cut to round this down zero. The programming problem that we will face now with the **NOT** gate is that we will want to round **0.5** up to one. So, we cannot use **Round** to do this and, instead, we have to use a conditional statement. One of the interesting and powerful aspects of *Mathematica* is that we can do something called **functional programming**. For this problem that means we can create a function that has the conditional within it and then map that function onto a vector of inputs.

When we speak of a conditional, we typically mean a statement, or rule, of the type that evaluates an input by comparing it to a predetermined value, the condition. Depending on the value and the condition, the rule either changes the value, or leaves it the same. We do this all the time in our own thinking and activities. For example, "if the temperature is at 45F (7.2 C), then I will wear a coat". There is an "or else" in here that is unstated and that is "or else I will not." We can logically leave this unstated, if the only two options are to wear a coat, or to go without a coat. In the language of *Mathematica*, the statement would be **If [T ⩽ 45, "Wear a coat", "Don't wear a coat"]**. If we add an input for the T, the temperature, then we have a simple script that tells us what to do in the morning:

```
(* We put in a value for T *)

T = 35;

(* This compares the value that we input for
T to 45 (degrees F are implicit) and then
executes by stating what we should do. *)

If[T ⩽ 45, "Wear a coat", "Don't wear a coat"]
```

Wear a coat

```
T = 50;
```

If[$T \leqslant 45$, "Wear a coat", "Don't wear a coat"]

Don't wear a coat

Conditionals are extremely useful in programming. Note again how simple and elegant this is to do in *Mathematica*. We don't even have to use a print command and we don't have to specify "else," for the alternative output.

Now, let's go back to the problem we have with the **NOT** gate. To implement a function that will do what we want, we make the gain of the transfer function **- 10**, hence, the exponent will now be positive. For most values of **x** in **Exp [10 x]**, this function will be a large to a huge number, see the table function:

Table[Exp[10*x*]//*N*, {*x*, 1, 20, 5}]

$\{22026.5, 1.14201 \times 10^{26}, 5.92097 \times 10^{47}, 3.06985 \times 10^{69}\}$

That means that $\frac{1}{1 + \text{Exp}[10x]}$ will be very small, or effectively zero, as it goes as $\frac{1}{\text{Exp}[10x]}$:

Table[1/Exp[10*x*]//*N*, {*x*, 1, 20, 5}]

{0.0000453999, 8.75651076269652`*^-27, 1.6889118802245324`*^-48, 3.257488532207521`*^-70}

1/(1 + Exp[10 1])//*N*

0.0000453979

This is acceptable, because in keeping with our development to this point, we use only binary inputs, meaning that we only need to be concerned with two input values of **x**, 0 and 1. We also have written our scripts in such a way that the outputs are also binary, with values of 0 and 1. When x = 1 the transfer function produces a small number, 0.000045..., which we know how to handle with **Round**; it will be rounded down to zero. But, what about when the input value is zero? Now, the output of the transfer function will not be one, it once again will be 0.5. and we can't use **Round** to handle this.

Instead, we will use a conditional statement to make a function that can be mapped onto the inputs. The syntax is the same as the simple **If** statement script we wrote to decide whether or not to wear a coat. This will be **If [x == 0.5, 1, x]**, which means that if the input, **x**, is equal to 0.5 then it will be changed to 1, and if the value is anything else, it will remain the same. We can make this into a function **g [x _]: = If [x == 0.5, 1, x]**. Let's give this a try and work interactively, line-by-line, before we make a script.

First, we create the input **{ 0, 1 }** and we let this be equal to **xiv**. We once again define the transfer function, but now with the gain of - 10. Then we map it onto **{ 0, 1 }**:

```
input2 = xiv = {0, 1};
f[x_]:=1/(1 + Exp[10x])
a = f/@xiv//N
```

{0.5, 0.0000453979}

The outputs are as expected: an input of 0 gives an output of 0.5, and an input of 1 gives an output of ~0.000045. We want these outputs to be 1 and 0, respectively. Let's begin with the input of 1 and its output of 0.5. We can use this function **g[x_] := If [x == 0.5, 1, x]** to push 0.5 to 1:

```
g[x_]:=If[x == 0.5, 1, x]
g/@a
```

{1, 0.0000453979}

This conditional function has moved 0.5 up to exactly 1. What about the second value, the 0.000045... that we want to move down to exactly zero? We could use **Round** to do this. But, we could also use another conditional function. We create this function **h[x_]: = If [x < 0.5, 0, x]**; it is not unique and we could write this condition in many different ways, but this will work for our purpose. At this point, we need not be creative, we simply state that if the either number in the list is less than 0.5, push it to zero. Let's Map this new conditional function onto **a**

```
h[x_]:=If[x < 0.5, 0, x]
h/@a
```

{0.5, 0}

This does as is expected. It leaves the value of 0.5 intact, but it moves the value of ~0.000045 down to zero. But, how do we get both to be done? Again, there are many ways to do this. We could write a function that changes both values, but one of the features of *Mathematica* that is worth knowing and using, is that we can sequentially map functional programs. Which is to say, we can map **g[x_]** on **h[x_]** or, in this case, vice versa, to operate on **a**.

h/@g/@a

g/@h/@a

{1, 0}

{1, 0}

Perfect! When we input **{0, 1}** the combination of these conditionals will make a **NOT** gate that will return **{1, 0}** and vice versa.

Combinations of gates

It would be very useful now to implement these gates in combinations to see what kinds of behavior they will display when working together. To do that is tedious, if we choose to continue to write these scripts from scratch and interactively. Instead, we want to be able to call the gates to use over and over again. Therefore, we want to write a script for each gate once and then call them as functions for repeated, broader use. We will name these **andGate**, **orGate**, and the **notGate**. We used small letters at the beginning of these so as not to confuse our functions with those already in *Mathematica*. Let's create the function for the **andGate**. To do this we will use for the first time the Module command.

??Module

Variables defined within a Module, are local variables and are assigned values only for there use in that function.

The key to the **andGate** we created is that the sum of the inputs **{ x, y }** given by **x* 1.5 + y*1.5 - 2**. Thus the **andGate** is:

andGate[{x_, y_}]:=

Module[{*f*},

$f = 1/(1 + \mathrm{Exp}[-20(x * 1.5 + y * 1.5 - 2)])$//Round]

We can test this to be sure it is producing the correct results by mapping it over the **input** list with its dyadics in any order:

```
input = {{1, 1}, {1, 0}, {0, 0}, {0, 1}};
andGate/@input

input = {{1, 0}, {1, 1}, {0, 0}, {0, 1}};
andGate/@input

input = {{1, 0}, {0, 0}, {1, 1}, {0, 1}};
andGate/@input

input = {{1, 0}, {0, 0}, {0, 1}, {1, 1}};
andGate/@input
```

{1, 0, 0, 0}

{0, 1, 0, 0}

{0, 0, 1, 0}

{0, 0, 0, 1}

This output is exactly as it should be, **1**, the True or Yes response, happens only when both **x** and **y** are unity, for the other three cases the result is **0**, which is **False** or **No**. We also note that wherever the pair **{ 1, 1 }** appears in the list, the **andGate** works properly.

Now, we need to do the same for **orGate** and **orGate2** here they are:

```
orGate[{x_, y_}]:=
Module[{f},
f = 1/(1 + Exp[-20(x * 1 + y * 1)])//Round]
orGate2[{x_, y_}]:=
Module[{f},
f = 1/(1 + Exp[-20(x * 1.2 + y * 1.2 - .24)])//Round]

input = {{1, 1}, {1, 0}, {0, 0}, {0, 1}};
```

orGate/@input

orGate2/@input

{1, 1, 0, 1}

{1, 1, 0, 1}

This output, for both versions of the **orGate** is as we expect it to be. Next, we test **orGate** and **orGate2** on the input with the variation in the position of the pair { 0, 0 }

input = {{0, 0}, {1, 1}, {1, 0}, {0, 1}};

orGate/@input

orGate2/@input

input = {{1, 0}, {0, 0}, {1, 1}, {0, 1}};

orGate/@input

orGate2/@input

input = {{1, 0}, {1, 1}, {0, 0}, {0, 1}};

orGate/@input

orGate2/@input

input = {{1, 0}, {0, 1}, {1, 1}, {0, 0}};

orGate/@input

orGate2/@input

{0, 1, 1, 1}

{0, 1, 1, 1}

{1, 0, 1, 1}

{1, 0, 1, 1}

{1, 1, 0, 1}

{1, 1, 0, 1}

{1, 1, 1, 0}

{1, 1, 1, 0}

The outputs are as we expect them to be for both the **orGate** and **orGate2**. They are not dependent on the position of the pair **{ 0, 0 }**, wherever this is in the sequence, it always produces the output of 0 meaning **False** or **No**.

Now, we need to script as a **Module** the **notGate**. We did this earlier using the transfer function **f [x_] := 1 / (1 + Exp [10x])// N** and the two conditional functions **h [x_] := If [x < 0.5, 0, x]** and **g [x_] := If [x== 0.5, 1, x]**. We used two conditionals for understanding, but in actuality we needed only one **h [x_] := If [x < 0.5, 0, 1]**, which states that for any input value less than 0.5, output a 0, and when the value is equal to 0.5, or larger, output 1. Therefore, we need one function that will behave as the following functions do:

```
f=.(* Clears f of any previous assignment of f. *)
f[x_]:=1/(1+Exp[10x])//N
h[x_]:=If[x < 0.5, 0, 1]
input
f/@%
Partition[h/@Flatten[%], 2]
```

{{1, 0}, {0, 1}, {1, 1}, {0, 0}}

{{0.0000453979, 0.5}, {0.5, 0.0000453979}, {0.0000453979, 0.0000453979}, {0.5, 0.5}}

{{0, 1}, {1, 0}, {0, 0}, {1, 1}}

This sequential mapping of **f [x]** followed by **h [x]** on a vector of inputs simply "flips" their digits, this is why the **notGate** is also referred to as an **inverter**. The way to do this is to combine the transfer function:

y = 1 /(1 + Exp[10 x])

and the conditional into one Module function:

notGate[x_]:= Module[{ y, h }, y = 1 /(1 + Exp[10 x]) / / N; If [y < 0.5, 0, 1]]

This Module takes the input value **x**, and passes that through to the transfer function. The output of the transfer function is **y** and its value is passed to the conditional evaluation **If [y < 0.5, 0, 1]**, which then gives the final output as either 0 or 1. All this is done in sequence and the output is the alternative to the input. Remember, as we conceived of it, the **NOT** gate has just a single digit input. So to test it, we will flatten input and map it over the result. Each 1 from input should be a 0 and each 0 should become 1:

```
notGate[x_]:=
Module[{y, h}, y = 1/(1 + Exp[10x])//N;
If[y < 0.5, 0, 1]]
input
Flatten[input];
Partition[notGate/@%, 2]
```

{{1, 0}, {0, 1}, {1, 1}, {0, 0}}

{{0, 1}, {1, 0}, {0, 0}, {1, 1}}

The **notGate** has worked just as it should, it flipped the digits of **input**.

The NAND and NOR gates

We can also combine gates. For example, the **NOT** gate can be used with the **AND** gate and with the **OR** gate to produce **NAND** and **NOR** gates. That is to say if we place the **NOT** after the **AND** neuron, the combination becomes **NOT AND**, which is briefly written as **NAND**. Similarly, if the **NOT** is placed before **OR**, these two become **NOT**, **OR**, which we designate as **NOR**.

When we place a **NOT** gate on the output of the **AND** gate, we can consider the combination as one gate. Think about this in this way: imagine we did not build/program these gates, but instead they were implementable as "off the shelf" entities that are not transparent. We would test them and learn what they do by using them. The following diagram shows the **AND** gate before the **NOT** gate, thereby inverting its input. In other words

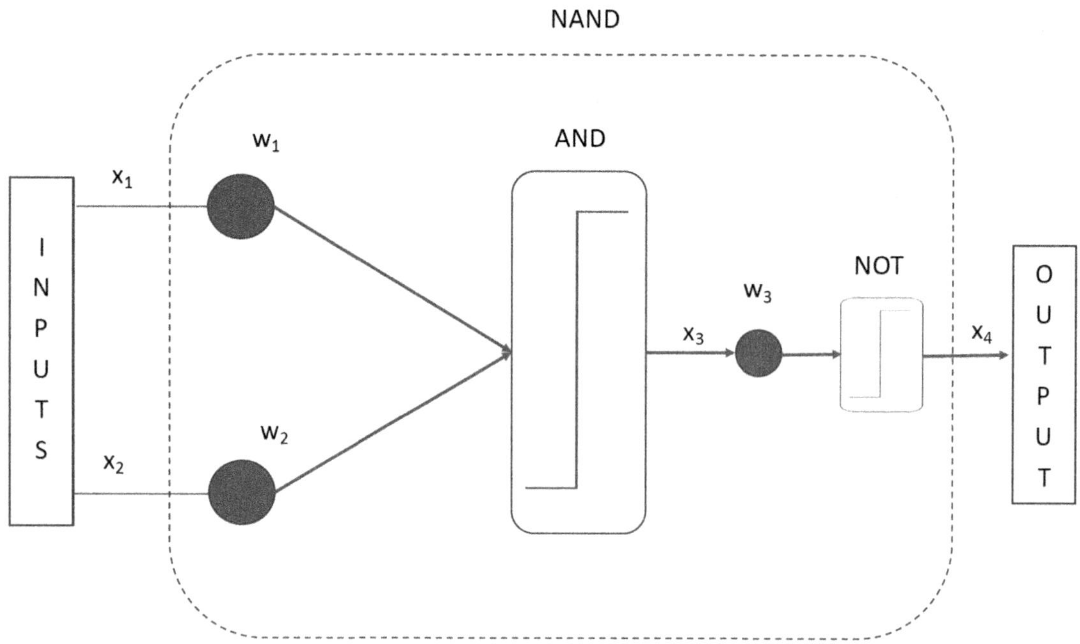

Let's put the pieces together interactively, line-by-line and then we will put the pieces together in one Module function. We use the following two separate sets of inputs, input1 and input2 corresponding to x_1 and x_2 in the diagram.

```
input1 = {0, 0, 1, 1}

input2 = {0, 1, 0, 1}
```

{0, 0, 1, 1}

{0, 1, 0, 1}

The **andGate** requires paired inputs. So, these two single lists need to be combined as follows:

```
Join[input1, input2];
Partition[%, 4];
Transpose[%]
```

{{0, 0}, {0, 1}, {1, 0}, {1, 1}}

Now, the **andGate** can be mapped onto this output and then the **notGate** can be mapped onto that output to give the overall **input1 NAND input2**:

andGate/@%

notGate/@%

{0, 0, 0, 1}

{1, 1, 1, 0}

We see that for the input **{ {0, 0}, {0, 1 },{1, 0},{1, 1} }** the **andGate** produces **{0, 0, 0, 1}**, but these are inverted after the **notGate** is applied. The **andGate** returns **1, True or Yes** when both inputs are **{ 1, 1 }, {True, True }** or **{ Yes, Yes }**, but the **nandGate** will return **{0, False }** for this same input and will return **{1, True}** or **Yes** for all the other inputs, **{0, 0}, {0, 1}, {1, 0}**.

A version of the same computation can be done to render the output from the combination of an **orGate** followed by the **notGate**. This can be done in one script:

input1;

input2;

z **= Join[%%, %];**

z1 = Partition[z, 4];

z2 = Transpose[z1]

notGate/@orGate/@z2

{{0, 0}, {0, 1}, {1, 0}, {1, 1}}

{1, 0, 0, 0}

This is the output that the **norGate** will provide. The only **1, True** or **Yes**, is returned when both inputs are **{0, False}** or **No**. The next step is to create **Module** functions for **NAND** and **NOR**.

We will use the **norGate** first. Recall that the **orGate2** looks as follows:

orGate2[x_, y_]:=

Module[{f},

$f = 1/(1 + \text{Exp}[-20(x * 1.2 + y * 1.2 - .12)])$//Round]

We will use this in the norGate in this way:

```
norGate[x_, y_]:=Module[{}, {x, y, z = Join[x, y];
z1 = Partition[z, 4]; z2 = Transpose[z1];
z3 = orGate2/@z2;
notGate/@z3}];
```

This takes the values of x and y and then creates the paired input needed as input for the **orGate** and then the results are inverted by the mapping of the **notGate**. In the way this function is written, the values of **x** and **y** will be displayed along with the **NOR** output. We can see how this works and then transpose the results and put them in **TableForm** to see the input values of **x**, **y** and the output **z**.

```
norGate[input1, input2];
TableForm[%//Transpose,
TableHeadings →
{{"NORTruth"}, {"in1", "inp2", "inp1NORinp2"}}]
```

	in1	inp2	inp1 NOR inp2
NOR Truth	0	0	1
	0	1	0
	1	0	0
	1	1	0

This is the first time that we have used the **TableForm** because it presents the results in the form of a **Truth Table**. We also can create the **Truth Table** for the **nandGate** and compare them. We use almost the same script for the **nandGate** as for the **norGate**, except that we operate on the internal variable, z2, with the **andGate** to produce **z3**, onto which we **Map** the **notGate**:

```
nandGate[x_, y_]:=Module[{}, {x, y, z = Join[x, y];
z1 = Partition[z, 4]; z2 = Transpose[z1];
z3 = andGate/@z2;
notGate/@z3}]
```

nandGate[input1, input2];

TableForm[%//Transpose,

TableHeadings →

{{"NANDTruth"}, {"in1", "inp2", "inp1NANDinp2"}}]

	in1	inp2	inp1 NAND inp2
NAND Truth	0	0	1
	0	1	1
	1	0	1
	1	1	0

andGate2[x_, y_]:=Module[{f}, {x, y, z = Join[x, y];

z1 = Partition[z, 4];

z2 = Transpose[z1];

$f = 1/(1 + \text{Exp}[-20(x * 1.5 + y * 1.5 - 2)])$//Round}]

andGate2[input1, input2];

TableForm[%//Transpose,

TableHeadings →

{{"ANDTruth"}, {"in1", "inp2", "inp1 AND inp2"}}]

	in1	inp2	inp1 AND inp2
AND Truth	0	0	0
	0	1	0
	1	0	0
	1	1	1

orGate2[x_, y_]:=Module[{f}, {x, y, z = Join[x, y];

z1 = Partition[z, 4];

z2 = Transpose[z1];

$f = 1/(1 + \text{Exp}[-20(x * 1.2 + y * 1.2 - .12)])$//

Round}]

orGate2[input1, input2];

TableForm[%//Transpose,

TableHeadings →

{{"ORTruth"}, {"in1", "inp2", "inp1ANDinp2"}}]

	in1	inp2	inp1 AND inp2
OR Truth	0	0	0
	0	1	1
	1	0	1
	1	1	1

Now that we have seen how the **OR** and the **AND** functions split the map that includes the points { {1, 1}, {1, 0}, {0, 0}, {0, 1} } with diagonal lines, it is logical to ask - why not split the map with a horizontal line or a vertical line. Taking the case of horizontal line first, we can use it to split the points { {0, 1}, {1, 1} } from the points { {0, 0}, {0, 1} }. To do this we let the linear combination be the line y = 0.5 or x * 0 +y * 1 - 0.5, while preserving the transfer function remains the same. We will call this **upGate**, since it will reply **True** when the upper two points are the input, meaning that when the value of the second input is 1, it will reply **True**.

upGate[x_, y_]:=Module[{f}, {$x, y, z =$ Join[x, y];

z1 = Partition[z, 4];

z2 = Transpose[z1];

$f = 1/(1 +$Exp$[-20(x * 0 + y * 1 - .5)]$)//Round}]

upGate[input1, input2];

TableForm[%//Transpose,

TableHeadings →

{{"UPTruth"}, {"in1", "inp2", "inp1ANDinp2"}}]

	in1	inp2	inp1 AND inp2
UP Truth	0	0	0
	0	1	1
	1	0	0
	1	1	1

The output is 1 when the input pair has a 1 at the second position. The map can be split vertically, by using a line at x = 0.5, or x * 1 +y * 0 -.5, and by keeping the transfer function the same. We will call this **rightGate** because it replies with True when the first value of each pair is equal to 1.

```
rightGate[x_, y_]:=Module[{f}, {x, y, z = Join[x, y];
z1 = Partition[z, 4];
z2 = Transpose[z1];
f = 1/(1 + Exp[−20(x * 1 + y * 0 − .5)])//Round}]
rightGate[input1, input2];
TableForm[%//Transpose,
TableHeadings →
{{"RIGHTTruth"}, {"in1", "inp2", "inp1ANDinp2"}}]
```

	in1	inp2	inp1 AND inp2
RIGHT **Truth**	0	0	0
	0	1	0
	1	0	1
	1	1	1

```
Show[

Plot[(−x + 1.333), {x, 0, 1.333}, PlotStyle → {Red},
PlotLabel → {{Red, "AND"}, {Blue, "OR"},
{Purple, "UP"}, {Green, "RIGHT"}}],

Plot[(−y + 1), {y, 0, 1},
PlotStyle → {Blue, Dashed}],

Graphics[
{PointSize[Large], Red, Point[{1, 1}]}
    ],
```

```
Graphics[
{PointSize[Large], Blue, Point[{1, 0}]}
      ],

Graphics[
{PointSize[Large], Black, Point[{0, 0}]}
      ],

Graphics[
{PointSize[Large], Blue, Point[{0, 1}]}
      ],
Graphics[
{GrayLevel[0.9], Rectangle[{−.01, −.01},
{.01, 1.01}]}
      ],

Graphics[
{GrayLevel[0.9], Rectangle[{.99, −.01},
{1.01, 1.01}]}
      ],

Graphics[
{{PointSize[Large], Red, Point[{1, 1}]},
   Text["{ 1, 1 }", {1.05, 1.05}]}
      ],

Graphics[
{{PointSize[Large], Blue, Point[{1, 0}]},
```

```
   Text["{ 1, 0 }", {1.07, 0.05}]}
      ],

Graphics[
{{PointSize[Large], Black, Point[{0, 0}]},
Text["{ 0, 0 }", {.07, 0.05}]}
      ],

Graphics[
{{PointSize[Large], Blue, Point[{0, 1}]},
   Text["{ 0, 1 }", {0.07, 1.05}]}
      ],

Graphics[
{{Thick, Dashed, Green,
Line[{{0.5, 0}, {0.5, 1}}]},
Text["x = 0.5 ", {0.52, 1.05}]}
      ],
Graphics[
{{Thick, Dashed, Purple,
Line[{{0, 0.5}, {1, 0.5}}]},
Text["y = 0.5 ", {0.32, 0.55}]}
      ]]
```

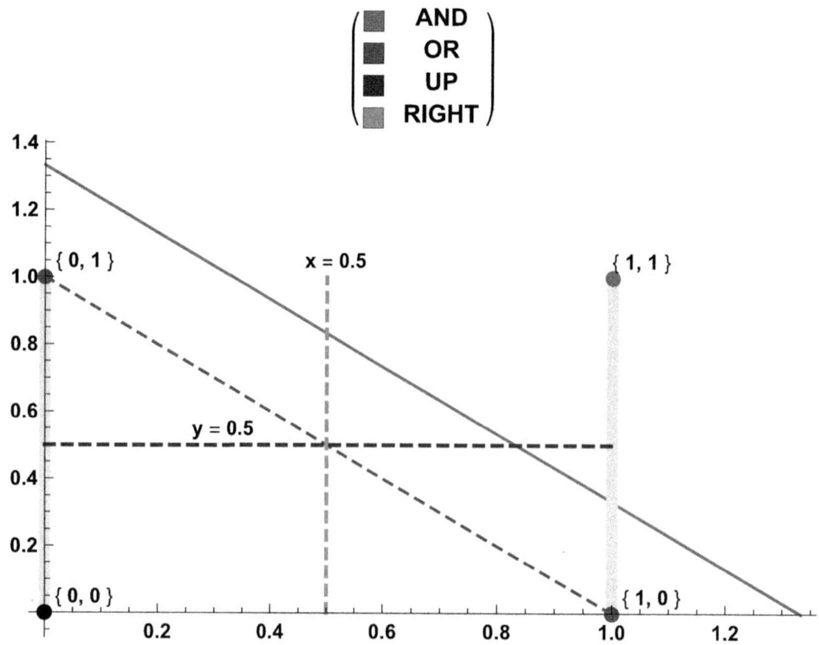

Summary of two input logic gates

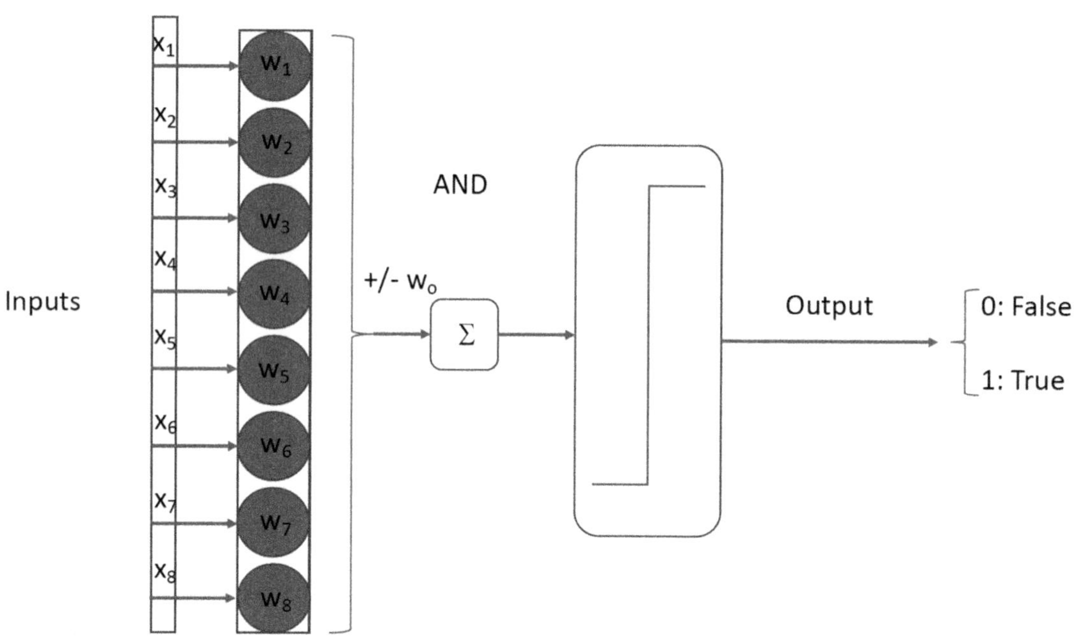

Consider the AND gate shown in the illustration above. In this case, there are eight inputs. Those eight raw inputs must be weighted and summed to one value that will be the single input to the transfer function. Remember, it is the combination of the values of the weighted sum and the gain of the transfer function that makes this an **AND** gate. Thus, even though there are eight inputs, if those inputs are conditioned as they were for the case of the two inputs, and if the gain is high, then the gate should function properly. To see how this works, we will once again work fully interactively, line-by-line and also use some new tools in *Mathematica*.

The vector of weights and the threshold will be an expanded version of that for the two input case. That means all the weights will be set to 1.5 but what should the threshold will be? The **AND** gate should return a **1** that is **TRUE**, only when all the input values are **1**. If even one value of the input is a zero, the AND gate must return a **0** or **FALSE**. When all seven of the input values are equal to **1**, the partial weighted sum will be 8 x 1.5 = 12. The transfer function is 1 / (1 + Exp[-20 (12 + (- threshold)]) and we want this to evaluate to one after rounding up from 0.5. For that to happen the argument of the exponential must be zero. Therefore the threshold has to be 12 or higher and it will go into the weighted sum as (-12). We can think of the weights and the negative of the threshold as a vector and the inputs as another vector.

wgts = {1.5, 1.5, 1.5, 1.5, 1.5, 1.5, 1.5, 1.5, −12};

Next, we create a vector of seven **Random** input values that are zeroes or ones and then we will append one value that will always be one corresponding to the threshold:

input8 = Append[Table[If[Random[] ≥ 0.5, 1, 0], {8}], 1];

Random[ ] actually generates random numbers between zero and one. The **If** function converts these into ones or zeroes. If the random number is equal to, or greater than 0.5, it becomes one, and if it is not, then it becomes zero. Appended to the vector of seven values is the value of one for the threshold. Every time we execute this command, it will give a new output vector that will be set equal to **input8**. But, once **input8** is set, it will remain set until the next time this command is executed. We see that this is the case here:

input8

{0, 1, 1, 1, 0, 0, 0, 0, 1}

The sum that we seek as the input to the transfer function, is the **Dot** product of these two vectors. To obtain the single transformed value from the inputs, we can simply **Dot** the weightings onto the inputs:

wgts.input8

−7.5

The value above is then input to the transfer function, which we couple with the **If** function to make the evaluation of the transfer function a zero or a one:

f[x_]:=1/(1+Exp[−20∗x])

If[f[wgts.input8] < 0.5, 0, 1]

0

The output is a zero because not all of the inputs were equal to one. In this case we can compute the value of the threshold in our heads. In more general cases, with many more inputs and different expectations of the gate and its purpose, the calculation of the threshold might not be this trivial. So, let's use this example and compute the minimum value of the threshold. To do this we first leave the threshold value out of the vector of weights; instead we put a place holder in as **th**:

wgts2 = {1.5, 1.5, 1.5, 1.5, 1.5, 1.5, 1.5, 1.5, th}

inputall1 = {1, 1, 1, 1, 1, 1, 1, 1, 1}

{1.5, 1.5, 1.5, 1.5, 1.5, 1.5, 1.5, 1.5, th}

{1, 1, 1, 1, 1, 1, 1, 1, 1}

We can use as inputs all ones and then apply the same methodology, except we will compute the value of **th** by putting the dot product into the transfer function and setting it equal to 0.5. The computation will find a value for **th** that satisfies this equation:

NSolve[f[wgts2.inputall1]==0.5, th]

{{th → −12.}}

We see that **NSolve** has back solved for the value if the threshold, in this case -12.

How could this AND gate be used? Many systems have numerous sensors. In chemical and petroleum plants, temperature sensors are ubiquitous on all units for many different operations. Thermal sensing is particularly important in controlling chemical reactors, and, especially, for reactors that contain solid catalyst pellets to promote the rate of exothermic reactions. Imagine an array of many thermocouples at different positions along or within a packed bed reactor. Catalysts are typically made up of materials such as alumina, Al_2O_3. Within this bed of particles, there may be tubes with a heat transfer fluid flowing within them to carry away the heat of the exothermic reaction. Monitoring and controlling for hot spots is critical within this bed is important. The reason is that if a hot spot develops, the rate of reaction in the region of the hot spot will increase temperature and the rate constant for the reaction. As the rate constant increases, so too does the rate of reaction and with that the rate or heat released. If rate of heat generated chemically is not matched by the rate of heat transfer away from that zone, the temperature will continue to rise. This rise in temperature will again raise the rate of reaction, and the rate of heat release and raising the temperature further. The rate of change in temperature will rise as the rate constant rises. If this is not check, it will lead to a condition referred to as thermal runaway. Depending on how long it is before this is detected and corrected, the reactor may be damaged, or, in the worse case scenario, the reactor could explode. Depending on the situation, the time for response may be minutes or seconds. Hence, the continuous monitoring of the temperature at different axial and radial positions in the catalyst bed is critical to safe operation.

Thermocouples are very simple and as a result they are very reliable. However, like all sensors, they do fail, or the electrical connections leading them to the monitoring system may cross, or break. In such a case, we can imagine that all of a sudden the system (or person) monitoring the reactor would notice the failure. To determine if the failure was at the sensor itself or in the connections after the sensor, the operator, or the control system, could switch away from the thermocouples and then send a signal - say a series of voltage spikes, a square wave, to all connections after the sensor. Every time the voltage spike was sensed by a field sensor, it should provide a 1 or a True to the inputs at the AND gate. If the inputs are all ones, or Trues, the AND gate would output 1 or True. This would indicate that the wiring between the thermocouples and the monitoring system was intact and the problem may lie in the sensor. If the AND gate returned a 0 or False, it would mean that somewhere in the circuitry after the thermocouples, a break or short had occurred.

Conclusion

This chapter shows how many different kinds of problems can be solved with *Mathematica.* It also, hopefully, illustrates how much can be learned by building models of this kind. In all cases, much of the lengthy but not necessarily interesting work that would have to be done, no longer needs to be done, the computer as an assistant does that part for us. The emphasis is the placed on translating the situation into mathematics and on how the mathematics predict the system will behave.

Index

9780128200513